MW01492796

THE BRITISH INTERNAL COMBUSTION ENGINE
RESEARCH ASSOCIATION
(B.I.C.E.R.A.)

A HANDBOOK ON
TORSIONAL VIBRATION

THE BRITISH INTERNAL COMBUSTION ENGINE
RESEARCH ASSOCIATION

MEMBER COMPANIES

Albion Motors Ltd   Leyland Motors Ltd
W. H. Allen Sons and Co. Ltd   J. and H. McLaren Ltd
Belliss and Morcom Ltd   Henry Meadows Ltd
The Bergius Company Ltd   Mirrlees, Bickerton and Day Ltd
Peter Brotherhood Ltd   The National Gas and Oil Engine Co. Ltd
Bryce Berger Ltd   F. Perkins Ltd
Crossley Brothers Ltd   Petters Ltd
Crossley-Premier Engines Ltd   Ruston and Hornsby Ltd
Davey, Paxman and Co. Ltd   Simms Motor Units Ltd
Wm. Doxford and Sons (Engineers) Ltd   Stuart Turner Ltd
The English Electric Co. Ltd   Tangyes Ltd
Gleniffer Engines Ltd   Vickers-Armstrongs (Engineers) Ltd
The Holset Engineering Co. Ltd   H. Widdop and Co. Ltd

ASSOCIATE MEMBERS

B. P. Trading Ltd   British Transport Commission
'Shell' Research Ltd

Grant-aid received from
Department of Scientific and Industrial Research

DIRECTOR OF RESEARCH
ENG. REAR-ADMIRAL D. J. HOARE, C.B., M.I.MECH.E.

MEMBERS OF THE TORSIONAL VIBRATION PANEL
DURING THE PREPARATION OF THE HANDBOOK

S. ARCHER	Lloyd's Register of Shipping
H. BAYES	Ruston and Hornsby Ltd
L. BERMAN	The English Electric Co. Ltd
C. H. BRADBURY (*Past Chairman*)	Simms Motor Units Ltd
D. F. BRAUND	Davey, Paxman and Co. Ltd
J. CALDERWOOD	J. and H. McLaren
W. CORNWALL	F. Perkins Ltd
F. CRESSWELL	Mirrlees, Bickerton and Day Ltd
P. J. F. CROSET	The Holset Engineering Co. Ltd
T. L. GARDENER	J. and H. McLaren Ltd
J. G. GUNN	Wm. Doxford and Sons (Engineers) Ltd
H. C. HILLS	Admiralty Engineering Laboratory
C. J. HIND	Ruston and Hornsby Ltd
P. JACKSON (*Past Chairman*)	Wm. Doxford and Sons (Engineers) Ltd
A. R. KENWORTHY	Crossley Brothers Ltd
L. C. LEIGH (*Chairman*)	W. H. Allen Sons and Co. Ltd
R. S. PUNT	H. Widdop and Co. Ltd
W. RILEY	Vickers-Armstrongs (Engineers) Ltd
F. SHEVLIN	The National Gas and Oil Engine Co. Ltd
J. SMITH	The National Gas and Oil Engine Co. Ltd
F. E. B. WEBB	W. H. Allen Sons and Co. Ltd
J. WRIGHT	Davey, Paxman and Co. Ltd

A HANDBOOK ON
TORSIONAL VIBRATION

COMPILED BY

E. J. NESTORIDES

OF THE

B.I.C.E.R.A. RESEARCH LABORATORY

CAMBRIDGE

AT THE UNIVERSITY PRESS

1958

CAMBRIDGE UNIVERSITY PRESS
Cambridge, New York, Melbourne, Madrid, Cape Town,
Singapore, São Paulo, Delhi, Tokyo, Mexico City

Cambridge University Press
The Edinburgh Building, Cambridge CB2 8RU, UK

Published in the United States of America by Cambridge University Press, New York

www.cambridge.org
Additional resources for this publication at www.cambridge.org/9780521203524

First published 1958
First paperback edition 2011

A catalogue record for this publication is available from the British Library

ISBN 978-0-521-04326-7 Hardback
ISBN 978-0-521-20352-4 Paperback

FOREWORD

By Lord Falmouth

The principal object of the British Internal Combustion Engine Research Association is to promote research and other scientific work in connexion with, or which will be of benefit to, the internal combustion engine trade or industry.

In authorizing the publication of this book, the Council of B.I.C.E.R.A. is seeking to contribute to the common stock of knowledge of a scientific subject which is of profound importance to all concerned with the design, manufacture and use not only of internal combustion engines but also of other machinery involving rotating parts. The Council believes, moreover, that the book provides an interesting example of what can be achieved by co-operative research.

The Association is in a particularly advantageous position for dealing with all internal combustion engine problems, since it is served by technical panels most of whose members are employed regularly in dealing with the subjects concerned. The practical and theoretical work carried out at the Association's well-equipped laboratory is supplemented by the contributions in kind of its members, who are able to conduct tests and make observations on a great variety of installations on test beds and in service.

General responsibility for the research activities of the Association is vested in the Research Committee. This committee appoints the specialist panels, among which is the Torsional Vibration Panel, the members of which have contributed material for the book and have taken part in the labour of its preparation.

Mr E. J. Nestorides, a Senior Research Engineer, who serves as Technical Officer of this Panel, has acted as compiler and much of the material is his own original work.

The Director of Research, Engineer Rear-Admiral D. J. Hoare, C.B., M.I.Mech.E., the Chief Research Engineer, W. P. Mansfield, Ph.D., B.Sc.(Hons.), A.M.I.Mech.E., and the Information Officer, J. White, M.I.Mech.E., have given valuable guidance and assistance in the production of the book.

The membership of the Torsional Vibration Panel is drawn from member companies, from the Admiralty Engineering Laboratory and from Lloyd's Register of Shipping. The long-standing co-operation of these concerns in all matters connected with the subject is warmly appreciated. In certain sections of the work it has been a privilege also to have the co-operation of the Research Association of British Rubber Manufacturers, the Electrical Research Association, the Electrical Engineering Department of City and Guilds College, London, and the Mechanical Engineering Laboratory of the University of Cambridge, as well as of a number of makers of mechanical and electrical components, accessories and machines, who in various ways, by contributing or checking information on special subjects or allowing tests to be carried out at their works, enabled valuable material to be added, and sincere appreciation of their help is expressed.

The Association is also indebted to the American Bureau of Shipping, Bureau Veritas, Germanischer Lloyd, Lloyd's Register of Shipping, Nippon Kaiji Kyokai, Det Norske Veritas, and Registro Italiano Navale, who have authorized the reproduction of extracts from their rules and recommendations on the subject of permissible torsional vibration stresses.

It is not claimed that the book provides a complete and final answer to all torsional vibration problems, but it is hoped and believed that the resources of the Association have enabled it to provide, in this book, a contribution to knowledge of the subject and some aid to its practical application.

FALMOUTH
President

The British Internal Combustion
Engine Research Association
111–112 Buckingham Avenue
Slough, Bucks

CONTENTS

[vii]

[x]

PART 2. EVALUATION AND PREDICTION OF TORSIONAL VIBRATION STRESSES

[xiii]

SCOPE OF HANDBOOK

This handbook on torsional vibration is written primarily to provide information in a convenient form for the staffs of design and development departments, although it is also intended to serve students of the subject.

For the engine designer, the recommended formulae, graphical procedures, numerical examples and tabulated values are given, and these are followed by remarks regarding the limitations of each method and its mathematical derivation.

For the development engineer, descriptions of test procedures, rigs and instrumentation are given in detail, together with remarks regarding precautions against possible sources of error, and the order of accuracy to be expected of the test results.

An endeavour has been made to present the required definitions and derivations with a reasonable minimum of mathematics, and illustrations have been added to facilitate the explanation wherever necessary.

Many subjects are treated more extensively than hitherto, and an indication is given of the direction in which further research is required. A fairly large amount of new information, based on accumulated experience obtained from member firms or at the B.I.C.E.R.A. Laboratory, is included in this handbook. The various sections also include discussions of problems associated with torsional vibration in its widest possible sense, and give at least an outline of the methods required in dealing with matters at present of secondary importance but which may become prominent in future.

All the derivations are given without reference to units. However, the numerical examples and formulae for practical applications are in all cases related to a definite system of units, indicated in the text, which determine the numerical constants. Throughout the text, both British and metric units have been included.

INTRODUCTORY NOTE ON DIMENSIONS AND UNITS

The dimensions used in the study of torsional vibrations are length $[L]$, time $[T]$, mass $[M]$ and others compounded from these three.

Examples with compounded dimensions are: force $[ML/T^2]$, pressure $[M/(LT^2)]$, mass density $[M/L^3]$, velocity $[L/T]$ and acceleration $[L/T^2]$.

Angular displacement being only a change of attitude, and not even that if revolution is complete, is dimensionless, but its rate of change involves time. Thus we have rotational velocity $[1/T]$ and rotational acceleration $[1/T^2]$.

In order to measure any quantity, a reference quantity of the same kind is required as a unit. Wherever it is necessary in this book to quote measurements of quantities, both British and metric units will generally be used.

The reference units of force are usually the accepted weights of standard masses. Unfortunately, the customary units of force have the same names as the corresponding units of mass. To distinguish between units of mass and force, Stroud's notation is used in this book. In this notation, Lb., Ton and Kg. are units of force, being the accepted weights of lb., ton and kg. masses.

The units of force and mass are related thus:

$$1\,\text{Lb.} = 1\,\text{lb.} \times g = 1\,\text{lb.} \times \frac{32 \cdot 2\,\text{ft.}}{\text{sec.}^2} = 32 \cdot 2\,\frac{\text{lb.ft.}}{\text{sec.}^2},$$

$$1\,\text{Kg.} = 1\,\text{kg.} \times g = 1\,\text{kg.} \times \frac{9 \cdot 81\,\text{m.}}{\text{sec.}^2} = 9 \cdot 81\,\frac{\text{kg.m.}}{\text{sec.}^2}.$$

Many practising engineers are more familiar with the measurement of force than with that of mass. This probably arises from the omnipresence of pressures and stresses and from the fact that mass is generally measured only by means or weight of force.

On this basis, the relation between force and mass would be expressed thus:†

$$1\,\text{lb.} = \frac{1\,\text{Lb.}}{g} = \frac{\text{Lb.sec.}^2}{32 \cdot 2\,\text{ft.}},$$

$$1\,\text{kg.} = \frac{1\,\text{Kg.}}{g} = \frac{\text{Kg.sec.}^2}{9 \cdot 81\,\text{m.}}.$$

† If a constant numerical value is used for gravity (for instance $g = 32 \cdot 2$ ft./sec.², or $g = 9 \cdot 81$ m./sec.²), the numerical error involved in determining mass from weight is generally less than 0·5 % at sea level at any latitude. This error can be eliminated by taking the 'local value' of g. However, engineering practice is to assign fixed average values to g, which are then used for all calculations.

The same process of thought leads to the common use of physically meaningless 'swinging moments' in place of moments of inertia. Moment of inertia, being the product of mass and radius of gyration squared, can only be expressed in such units as lb.ft.^{2} (or Lb.ft.sec.^{2}) and kg.m.^{2} (or Kg.m.sec.^{2}). The 'swinging moment' of an object is its weight times its radius of gyration squared or, according to habit, its weight times its diameter of gyration squared. The use of a 'swinging moment' thus requires, in order to obtain a moment of inertia, a division by g or by a multiple of g.

In order to make this book useful to as wide a range of readers as possible, common practices, as referred to above, are followed. The following are examples of the units in which the quantities commonly used are expressed:

Mass:

Lb.sec.^{2}/ft., Lb.sec.^{2}/in.; Kg.sec.^{2}/m., Kg.sec.^{2}/cm.

Specific weight:

Lb./ft.^{3}, Lb./in.^{3}; Kg./m.^{3}, Kg./cm.^{3}

Mass density:

(Lb.sec.^{2}/ft.) $\times$ (1/ft.^{3}) = Lb.sec.^{2}/ft.^{4}; Lb.sec.^{2}/in.^{4}; Kg.sec.^{2}/m.^{4}, Kg.sec.^{2}/cm.^{4}

Moment of inertia:

(Lb.sec.^{2}/ft.) $\times$ ft.^{2} = Lb.ft.sec.^{2}, Lb.in.sec.^{2}; Kg.m.sec.^{2}, Kg.cm.sec.^{2}

Swinging moment:

Lb.ft.^{2}, Lb.in.^{2}; Kg.m.^{2}, Kg.cm.^{2}

TABLES OF PHYSICAL CONSTANTS

Frequently used numerical values

$60/2\pi = 9{\cdot}55$	$\pi/2 = 1{\cdot}570796$	$2/\pi = 0{\cdot}63662$
$(60/2\pi)^2 = 91{\cdot}25$	$\pi/4 = 0{\cdot}785398$	$16/\pi = 5{\cdot}09296$
$4\pi^2 = 39{\cdot}478$	$\pi/8 = 0{\cdot}39270$	$32/\pi = 10{\cdot}18592$
$\pi^2 = 9{\cdot}8696$	$\pi/16 = 0{\cdot}19635$	
$\pi \cong 3{\cdot}141592$	$\pi/32 = 0{\cdot}098175$	
$2\pi/360 = 1° = 0{\cdot}17453\,\text{rad.}$	$\pi/64 = 0{\cdot}049088$	
$360°/2\pi = 57{\cdot}296° = 1\,\text{rad.}$		

For steel: $E/G = 2{\cdot}54$, $6E/G = 15{\cdot}2$; $(3\pi/4) \times G/E = 0{\cdot}93$;

where $E = 30 \times 10^6\,\text{Lb./in.}^2$ and $G = 11{\cdot}8 \times 10^6\,\text{Lb./in.}^2$.

Handy equivalents

1 h.p. (metric) $= 0{\cdot}9863$ h.p. (Brit. or U.S.) $= 735{\cdot}5$ watts
$= 32{,}550\,\text{ft.Lb./min.}$

1 h.p. (Brit. or U.S.) $= 1{\cdot}0139$ h.p. (metric) $= 746$ watts
$= 33{,}000\,\text{ft.Lb./min.}$

1 m.Kg. $= 7{\cdot}2330\,\text{ft.Lb.} = 86{\cdot}768\,\text{in.Lb.}$

1 ft.Lb. $= 0{\cdot}13826\,\text{m.Kg.}$

1 Kg. $= 2{\cdot}2046$ Lb. 1 Tonne (metric) $= 2205$ Lb.

1 Lb. $= 0{\cdot}45359$ Kg. 1 Ton $(= 2240\,\text{Lb.}) = 984{\cdot}2$ Kg.
2000 Lb. $= 907{\cdot}18$ Kg.

1 Kg./cm.^{2} $= 14{\cdot}2233\,\text{Lb./in.}^2$ $(= \text{psi})$.

1 Atm. $= 1{\cdot}033\,\text{Kg./cm.}^2 = 14{\cdot}7\,\text{Lb./in.}^2$

1 Lb./in.^{2} $(= 1\,\text{psi}) = 0{\cdot}0703\,\text{Kg./cm.}^2$

1 litre $= 1000\,\text{cm.}^3 = 61{\cdot}024\,\text{in.}^3$ 1 ft.^{3} $= 1728\,\text{in.}^3$

1 in.^{3} $= 0{\cdot}016387$ litre. 1 Lb./in.^{3} $= 27{\cdot}66\,\text{Grammes/cm.}^3$

1 cm.^{2} $= 3{\cdot}155\,\text{in.}^2$ 1 Lb.sec.^{2}/in.^{4} $= 10{\cdot}84\,\text{Grammes} \times \text{sec.}^2/\text{cm.}^4$

1 metre $= 1{\cdot}0936\,\text{yd.} = 3{\cdot}28084\,\text{ft.} = 39{\cdot}3701\,\text{in.}$

1 in. $= 2{\cdot}54$ cm. 1 in.^{5} $= 105{\cdot}8\,\text{cm.}^5$

1 poise $= 100$ centipoises $= 1\,\text{Gramme} \times \text{sec./cm.}^2$
$= 1{\cdot}447 \times 10^{-5}\,\text{Lb.sec./in.}^2$
$= 1{\cdot}447 \times 10^{-5}\,\text{reynolds.}$

1 reynold $= 1\,\text{Lb.sec./in.}^2 = 6{\cdot}9 \times 10^6\,\text{centipoises.}$

1 stoke $= 100$ centistokes $= 1\,\text{cm.}^2/\text{sec.} = 0{\cdot}155\,\text{in.}^2/\text{sec.}$

The following equivalents are for moments of inertia:

$$\frac{J}{\text{Lb.ft.sec.}^2} = 69\cdot565 \times \frac{W}{\text{Tons}} \times \frac{k^2}{\text{ft.}^2} = \frac{1}{12} \times \frac{J}{\text{Lb.in.sec.}^2} = 0\cdot03105 \times \frac{W}{\text{Lb.}} \times \frac{k^2}{\text{ft.}^2},$$

$$\frac{J}{\text{Lb.in.sec.}^2} = 834\cdot78 \times \frac{W}{\text{Tons}} \times \frac{k^2}{\text{ft.}^2} = 0\cdot3725 \times \frac{W}{\text{Lb.}} \times \frac{k^2}{\text{ft.}^2} = 0\cdot002588$$

$$\times \frac{W}{\text{Lb.}} \times \frac{k^2}{\text{in.}^2}.$$

(*Example.* If the Wk^2 of a flywheel is $2\,\text{Tons} \times \text{ft.}^2$, its moment of inertia is $69\cdot565 \times 2 = 139\cdot13\,\text{Lb.ft.sec.}^2$. This is also equivalent to $834\cdot78 \times 2 = 1669\cdot56\,\text{Lb.in.sec.}^2$.)

$$1\,\text{Kg.cm.sec.}^2 = 0\cdot86768\,\text{Lb.in.sec.}^2.$$

$$1\,\text{Lb.in.sec.}^2 = 1\cdot1525\,\text{Kg.cm.sec.}^2.$$

$$\frac{J}{\text{Kg.cm.sec.}^2} = 2\cdot55 \times \frac{W}{\text{Kg.}} \times \frac{D^2}{\text{m}^2}.$$

(*Example.* If the WD^2 of a flywheel† is $20\,\text{Kg.m.}^2$, its moment of inertia is

$$2\cdot55 \times 20 = 51\cdot0\,\text{Kg.cm.sec.}^2 \quad \text{or} \quad 0\cdot86768 \times 51\cdot0 = 44\cdot25\,\text{Lb.in.sec.}^2.)$$

Specific weights w and mass densities $\rho = w/g$

	w		ρ	
	Lb./ in.^{3}	Gramme/ cm.^{3}	Lb.sec.^{2}/ in.^{4}	Gramme $\times$ sec.^{2}/ cm.^{4}
Aluminium	0·097	2·68	2·51 $\times 10^{-4}$	2·72 $\times 10^{-3}$
Brass	0·300	8·28	7·76 $\times 10^{-4}$	8·41 $\times 10^{-3}$
Bronze	0·315	8·71	8·15 $\times 10^{-4}$	8·83 $\times 10^{-3}$
Cast iron	0·260	7·20	6·73 $\times 10^{-4}$	7·29 $\times 10^{-3}$
Copper	0·320	8·85	8·28 $\times 10^{-4}$	8·97 $\times 10^{-3}$
Duralumin	0·102	2·82	2·64 $\times 10^{-4}$	2·86 $\times 10^{-3}$
Gunmetal	0·315	8·71	8·15 $\times 10^{-4}$	8·83 $\times 10^{-3}$
Iron	0·260	7·20	6·73 $\times 10^{-4}$	7·29 $\times 10^{-3}$
Lead	0·412	11·40	10·67 $\times 10^{-4}$	11·53 $\times 10^{-3}$
Magnesium alloy	0·065	1·80	1·683 $\times 10^{-4}$	1·824 $\times 10^{-3}$
Micarta	0·049	1·35	1·268 $\times 10^{-4}$	1·374 $\times 10^{-3}$
Monel metal	0·323	8·93	8·36 $\times 10^{-4}$	9·05 $\times 10^{-3}$
Rubber	0·040	1·17	1·035 $\times 10^{-4}$	1·122 $\times 10^{-3}$
Silicone fluids	0·0355	0·982	0·917 $\times 10^{-4}$	0·993 $\times 10^{-3}$
Steel	0·283	7·85	7·32 $\times 10^{-4}$	7·93 $\times 10^{-3}$
Tungsten heavy alloy	0·600	16·60	15·54 $\times 10^{-4}$	16·83 $\times 10^{-3}$
Water	0·0361	1·000	0·932 $\times 10^{-4}$	1·019 $\times 10^{-3}$

Typical values for moduli of elasticity and rigidity

Note. In view of the considerable divergences in the values given by different sources, the values given below should be regarded only as approximate. Wherever possible, it is recommended to carry out tests to determine the moduli values, using specimens of adequate diameter (say 2 in.), or even complete crankthrows. For cast irons, the tests should

† WD^2 is often referred to as GD^2 in continental practice.

preferably be carried out with loads giving deflexions and stresses of the same order as those occurring under normal operating conditions. In the following, G = modulus of rigidity, and E = modulus of elasticity:

Steels

	G		E	
	Lb./in.^{2}	Kg./cm.^{2}	Lb./in.^{2}	Kg./cm.^{2}
Medium carbon steel (34–36 tsi, U.T.S.)	$11\cdot8\times10^6$	$8\cdot3\times10^5$	$30\cdot0\times10^6$	$21\cdot09\times10^5$
0·4 carbon steel, oil-hardened to 45 tsi, U.T.S.	$11\cdot5\times10^6$	$8\cdot09\times10^5$	$29\cdot5\times10^6$	$20\cdot74\times10^5$
Nickel chrome steel, hardened by nitriding	$11\cdot3\times10^6$	$7\cdot95\times10^5$	$29\cdot0\times10^6$	$20\cdot38\times10^5$
Alloy steel	$11\cdot5\times10^6$	$8\cdot09\times10^5$	$29\cdot3\times10^6$	$20\cdot60\times10^5$
National 1 D steel	$11\cdot4\times10^6$	$8\cdot02\times10^5$	$29\cdot0\times10^6$	$20\cdot38\times10^5$
Tungsten heavy alloy	—	—	$32\cdot6\times10^6$	$22\cdot92\times10^5$

Cast irons (applicable to low rates of stress only)

	G		E	
	Lb./in.^{2}	Kg./cm.^{2}	Lb./in.^{2}	Kg./cm.^{2}
Acicular iron	$7\cdot3\times10^6$	$5\cdot13\times10^5$	$18\cdot3\times10^6$	$12\cdot87\times10^5$
Cr-Mo iron	$8\cdot6\times10^6$	$6\cdot05\times10^5$	$20\cdot1\times10^6$	$14\cdot2\times10^5$
Cu-Cr iron	$11\cdot4\times10^6$	$8\cdot0\times10^5$	$26\cdot8\times10^6$	$18\cdot84\times10^5$
CWC Proferall iron	$9\cdot4\times10^6$	$6\cdot61\times10^5$	—	—
Grey flake cast irons	$6\cdot32\times10^6$	$4\cdot44\times10^5$	$14\cdot9\times10^6$	$10\cdot47\times10^5$
H.T.I. cast iron	$8\cdot6\times10^6$	$6\cdot04\times10^5$	$21\cdot35\times10^6$	$15\cdot0\times10^5$
Inoculated (S.G.)†	$8\cdot6\times10^6$	$6\cdot04\times10^5$	$21\cdot7\times10^6$	$15\cdot25\times10^5$
G.M. Meehanite	$8\cdot4$–$8\cdot9\times10^6$	$5\cdot9$–$6\cdot3\times10^5$	$22\cdot1\times10^6$	$15\cdot54\times10^5$
National 31·5 tsi cast iron	$6\cdot68\times10^6$	$4\cdot70\times10^5$	—	—
Ni-Cu iron	$7\cdot0\times10^6$	$4\cdot92\times10^5$	$18\cdot7\times10^6$	$13\cdot15\times10^5$
Nodular iron (S.G.)†	$8\cdot38\times10^6$	$5\cdot89\times10^5$	$22\cdot1\times10^6$	$15\cdot54\times10^5$

† S.G. = spheroidal graphite cast irons.

Other materials

	G		E	
	Lb./in.^{2}	Kg./cm.^{2}	Lb./in.^{2}	Kg./cm.^{2}
Aluminium	$3\cdot8\times10^6$	$2\cdot67\times10^5$	10×10^6	$7\cdot03\times10^5$
Brass	$5\cdot0\times10^6$	$3\cdot52\times10^5$	14×10^6	$9\cdot83\times10^5$
Bronze	$6\cdot0\times10^6$	$4\cdot22\times10^5$	15×10^6	$10\cdot55\times10^5$
Duralumin	$3\cdot8\times10^6$	$2\cdot67\times10^5$	$10\cdot5\times10^6$	$7\cdot38\times10^5$
Gunmetal	$5\cdot0\times10^6$	$3\cdot52\times10^5$	$14\cdot0\times10^6$	$9\cdot83\times10^5$
Micarta	$0\cdot4\times10^6$	$0\cdot28\times10^5$	$1\cdot4\times10^6$	$0\cdot98\times10^5$
Monel metal	$9\cdot0\times10^6$	$6\cdot33\times10^5$	$25\cdot0\times10^6$	$17\cdot57\times10^5$
Wood	$0\cdot08\times10^6$	$0\cdot06\times10^5$	$1\cdot5\times10^6$	$1\cdot055\times10^5$
Wood, compressed	$0\cdot32\times10^6$	$0\cdot23\times10^5$	$3\cdot9\times10^6$	$2\cdot74\times10^5$
Wrought iron	$11\cdot0\times10^6$	$7\cdot73\times10^5$	$28\cdot0\times10^6$	$19\cdot68\times10^5$

Natural rubber† (relation between Shore hardness S, rigidity modulus G, and elasticity in compression E_c. Average values, based on curves of J. F. D. Smith, published in *J. Appl. Mech.* March 1938, p. A-13, and Dec. 1939, p. A-159).

S	22	27	30	40	50	60	67	70
G [Lb./in.^{2}]	—	40	45	64	89	126	154	170
E_c [Lb./in.^{2}]	100	123	137	187	265	367	455	—
G [Kg./cm.^{2}]	—	2·81	3·16	4·5	6·25	8·85	10·83	11·95
G_c [Kg./cm.^{2}]	7·03	8·65	9·63	13·15	18·64	25·8	31·98	—

† Very few rubbers used for couplings are without fillers (see section 1·2). The E_c- and G-values thus obtained may be 2 or 3 times higher than for natural rubber, and reference should be made to coupling manufacturers on this subject.

PART 1

PRELIMINARY CALCULATIONS
AND MEASUREMENTS

An engine with its driven machine and auxiliaries is a complicated system. The shafting varies in shape and cross-section and may include, besides crankthrows, keyed or shrunk-on portions, tapered and splined sections, flanged couplings, etc. Some of the masses are 'distributed masses', such as the shafting and some driven machines, whereas others, such as flywheels and gearwheels, are more similar to 'concentrated masses'.

Therefore, in order to investigate the vibration conditions of an engine system, it is first necessary to make preliminary calculations which serve to determine a simplified 'equivalent dynamic system'. This equivalent system consists of a number of concentrated masses† or disks, situated at various points along an equivalent straight shaft of circular cross-section.

In the following, section 1·1 deals with the determination of the polar mass moments of inertia of the various masses usually included in engine systems. The determination of the equivalent shaft systems is dealt with in section 1·2.

---

† There are various methods by means of which distributed masses can be taken directly into account. These methods generally involve additional calculations and are not standard practice, but may be required in certain applications (see section 1·33).

1·1 MOMENTS OF INERTIA

NOTATION

Symbol	Brief definition	Typical units	Symbol	Brief definition	Typical units
A	area	in.², cm.²	S	length × (diameter)⁴	in.⁵, cm.⁵
B	width or breadth	in., cm.	t	time	sec., min.
D	diameter	in., cm.	t_m	mean thickness	in., cm.
e	eccentricity	in., cm.	T	torque	Lb.in., Kg.m.
F	frequency of vibration	vib./min.	U	energy	in.Lb., m.Kg.
g	acceleration due to gravity	in./sec.², cm./sec.²	v	velocity	in./sec., cm./sec.
h	height	in., cm.	$\dot{v}$	acceleration	in./sec., cm./sec.²
J	mass moment of inertia	Lb.in.sec.², Kg.cm.sec.²,	V	volume	in.³, cm.³
k	radius of gyration	in., cm.	w	specific weight	Lb./in.³, Kg./cm.³
L	length or thickness	in., cm.	W	weight	Lb., Kg.
m	mass	Lb.sec.²/in., Kg.sec.²/cm.	x, y, z	distances parallel to x-, y- and z-axes, respectively	in., cm.
M	bending moment	Lb.in., Kg.m.	Z	sectional modulus	in.³, cm.³
n	number of revolutions	—	α	angle	deg., rad.
$N_{cyl.}$	number of cylinders	—	θ	angle of twist	deg., rad.
p	pressure; stress	Lb./in.², Kg./cm.²	$\dot{\theta}$	angular velocity	rad./sec.
P	force	Lb., Kg.	$\ddot{\theta}$	angular acceleration	rad./sec.²
r	radius	in., cm.	ρ	mass-density	Lb.sec.²/in.⁴, Kg.sec.²/cm.⁴
r	ratio	—	τ	period of time	sec., min.
R	radius	in., cm.	ϕ	angle	deg., rad.
s	length	in., cm.	ω	phase velocity	rad./sec.

Subscripts are used in the text to define more precisely, e.g. $D_{cl.}$ = crankpin inner diameter. Their meaning is fully explained in their context in each instance.

1·11 Moment of inertia of cylinder masses

The standard procedure is to concentrate all the moments of inertia of each crankthrow, including connecting rod and piston, at the corresponding cylinder centre position on the equivalent shaft representing the crankshaft. The result is conventionally termed the moment of inertia per cylinder line.

The moment of inertia per cylinder line $J_{cyl.}$ [Lb.in.sec.²] is obtained by evaluating the individual terms of the following expression:

$$J_{cyl.} = J_{journal} + J_{crankpin} + J_{2\ webs} + J_{balance\ weights}$$
$$+ J_{conn.\ rod\ (rotating\ part)} + J_{conn.\ rod\ (reciprocating\ part)}$$
$$+ J_{piston\ and\ gudgeon\ pin.}$$

The reference axis is the centre-line of the journal in all cases. Methods for the evaluation of the mass moment of inertia of each part are described in the following, with numerical or graphical examples where necessary.†

Note. For some parts it is sometimes more convenient to calculate the 'swinging moment' Wk^2 [Lb.in.²], where W = weight of part (Lb.) and k = corresponding radius of gyration [in.]. The mass moment of inertia J is then obtained by dividing the result by $g = 386 \cdot 4$ [in./sec.²], and this is also dimensionally correct since [Lb.in.²]/[in./sec.²] = Lb.in.sec.².

1·111 *Moment of inertia of journal*

This is given by

$$J_{\text{journal}} = \frac{\pi}{32} \frac{w}{g} (D_{jo}^4 - D_{ji}^4) L_j \quad \text{[Lb.in.sec.}^2\text{]},$$

where w = specific weight [Lb./in.³], $g = 386 \cdot 4$ [in./sec.²], D_{jo} = journal outer diameter [in.], D_{ji} = journal inner diameter. [in.], L_j = length of journal [in.] from web face to web face.

1·112 *Moment of inertia of crankpin*

Let L_c = length of crankpin [in.] from web face to web face, D_{co} = crankpin outer diameter [in.], D_{ci} = crankpin inner diameter [in.], and e = eccentricity [in.] of crankpin bore (see accompanying figure). Further, let

$$R_0 = \text{crank radius [in.].}$$

Then

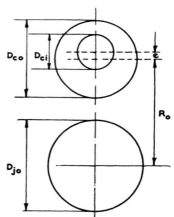

$$J_{\text{crankpin}} = \frac{\pi}{4} \frac{w}{g} L_c$$

$$\times \left[\frac{D_{co}^4 - D_{ci}^4}{8} + R_0^2 D_{co}^2 - (R_0 + e)^2 D_{ci}^2 \right]$$

$$\text{[Lb.in.sec.}^2\text{].}$$

Metric units. The formulae in sections 1·111 and 1·112 can be used directly to obtain J-values in Kg.cm.sec.². For this, all lengths should be in cm., specific weights in Kg./cm.³ (i.e. $\frac{1}{1000}$th of value in Gramme/cm.³), and $g = 981 \cdot 0$ cm./sec.². Thus, for steel, the specific weight is

$$w = 0 \cdot 00785 \, \text{Kg./cm.}^3.$$

1·113 *Moment of inertia of crankweb*

The most accurate and reliable method is the following semi-graphical method, which has been adopted as a standard by B.I.C.E.R.A. members. The procedure is as follows:

(1) Scale drawings are made of the web face and the web section in

† For *Vee engines*, the moment of inertia per line will also be denoted by '$J_{\text{cyl.}}$', although it includes the inertias of two connecting rods, two pistons and two gudgeon pins.

[4]

the plane passing through the centre lines of the crankpin and journal (see figure).

(2) Taking the journal centre as centre, a number of circular arcs are described on the drawing of the web face, with increasing radii, so as to cover the entire drawing. The web face is thus divided up into a number of circular ring portions: A, B, C, D, etc.

MOMENT OF INERTIA OF A CRANKWEB: METHOD USING GRAPH AND TABULATION

Evaluation based on formula: $J = \dfrac{Wk^2}{g} = \dfrac{w}{g} Vk^2.$

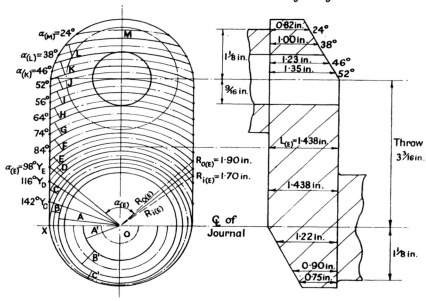

(The above drawings show all the numerical values used in line 'E' of the table overleaf.)

(3) Circular arcs corresponding to the mean radius of each ring portion are drawn. These are shown dotted in the accompanying figure. The value of the angles $Y_C OX$, $Y_D OX$, $Y_E OX$, etc., is determined by means of a protractor, and written on the drawing beside the corresponding point Y_C, Y_D, Y_E, etc., which is the intersection of the arc of corresponding radius and the edge of the web face.

(4) The radius of gyration k, the volume V, the weight W and hence the Wk^2, are determined for each ring portion. The radius of gyration is obtained from
$$k^2 = \tfrac{1}{2}(R_0^2 + R_i^2).$$

The volume is calculated by means of the relation
$$V = \frac{\pi}{4}(D_0^2 - D_i^2) L \frac{\alpha^\circ}{360^\circ},$$

[5]

where L = web thickness at the corresponding mean radius

$$R_m = \tfrac{1}{2}(R_0 + R_i),$$

as determined from the drawing of the web section, and $\alpha°$ = angle in degrees occupied by the ring portion under consideration

$$(\alpha = 180° - 2 \times Y_C OX, \text{ for instance}),$$

while $D_0 = 2R_0$ and $D_i = 2R_i$.

(5) The deductions for hollow crankpin and journal are evaluated separately and subtracted from the Wk^2 of the solid web.

An example of the tabulation, illustrating the manner in which this method is applied, is shown in the following:

Tabulation

[*Note.* $D_0 = 2R_0$; $D_i = 2R_i$.]

Segment	$k^2 = \tfrac{1}{2}(R_0^2 + R_i^2)$ [in.²]	$V = \dfrac{\pi}{4}(D_0^2 - D_i^2) \times \dfrac{\alpha°}{360°} \times L$ [in.³]	Weight [Lb.] $W = 0.283V$	Wk^2
A	$\tfrac{1}{2} \times 1.25^2$ $= 0.781$	$4.913 \times (180/360) \times 1.438 = 14.130$	3.999	3.123
B	$\tfrac{1}{2} \times (1.35^2 + 1.25^2)$ $= 1.693$	$0.817 \times (180/360) \times 1.438 = 2.350$	0.665	1.126
C	$\tfrac{1}{2} \times (1.52^2 + 1.35^2)$ $= 2.067$	$1.534 \times (142/360) \times 1.438 = 0.869$	0.246	0.508
D	$\tfrac{1}{2} \times (1.70^2 + 1.52^2)$ $= 2.600$	$1.822 \times (116/360) \times 1.438 = 0.844$	0.239	0.621
E	$\tfrac{1}{2} \times (1.90^2 + 1.70^2)$ $= 3.250$	$2.264 \times (98/360) \times 1.438 = 0.886$	0.251	0.816
F	$\tfrac{1}{2} \times (2.15^2 + 1.90^2)$ $= 4.116$	$3.183 \times (84/360) \times 1.438 = 1.066$	0.302	1.243
G	$\tfrac{1}{2} \times (2.45^2 + 2.15^2)$ $= 5.313$	$4.339 \times (74/360) \times 1.438 = 1.283$	0.363	1.929
H	$\tfrac{1}{2} \times (2.75^2 + 2.45^2)$ $= 6.783$	$4.905 \times (64/360) \times 1.438 = 1.254$	0.355	2.408
I	$\tfrac{1}{2} \times (3.05^2 + 2.75^2)$ $= 8.433$	$5.471 \times (56/360) \times 1.438 = 1.224$	0.346	2.918
J	$\tfrac{1}{2} \times (3.35^2 + 3.05^2)$ $= 10.263$	$6.040 \times (52/360) \times 1.350 = 1.177$	0.333	3.418
K	$\tfrac{1}{2} \times (3.65^2 + 3.35^2)$ $= 12.272$	$6.602 \times (46/360) \times 1.230 = 1.038$	0.294	3.608
L	$\tfrac{1}{2} \times (3.95^2 + 3.65^2)$ $= 14.463$	$7.168 \times (38/360) \times 1.000 = 0.757$	0.214	3.095
M	$\tfrac{1}{2} \times (4.27^2 + 3.95^2)$ $= 16.918$	$8.270 \times (24/360) \times 0.820 = 0.452$	0.128	2.166
A'	$\tfrac{1}{2} \times 0.375^2$ $= 0.070$	$0.443 \times (180/360) \times 1.220 = 0.270$	0.076	0.005
B'	$\tfrac{1}{2} \times (0.919^2 + 0.375^2)$ $= 0.493$	$2.213 \times (180/360) \times 0.900 = 0.996$	0.282	0.139
C'	$\tfrac{1}{2} \times (1.313^2 + 0.919^2)$ $= 1.285$	$2.765 \times (180/360) \times 0.750 = 1.037$	0.293	0.377

(Solid web) Total Wk^2 = 27.500 Lb.in.²

Crankpin bore: $3.188^2 + \tfrac{1}{2} \times 0.563^2 = 10.322$; $0.995 \times 1.438 = 1.431$; 0.405; -4.180

For one crankweb: Wk^2 = 23.320 Lb.in.²

For two crankwebs: $\begin{cases} Wk^2 = 46.640 \text{ Lb.in.}^2 . \\ J_{2\text{ webs}} = 0.121 \text{ Lb.in.sec.}^2 . \end{cases}$

For *solid rectangular crankwebs*, the moment of inertia can be calculated directly by means of the relation

$$J_{2\text{ webs}} = 2\frac{w}{g} hB \left[\frac{B^2 + h^2}{12} + \left(\frac{h}{2} - h_0\right)^2 \right] L_W$$

[Lb.in.sec.²],

where w = specific weight [Lb./in.³], $g = 386.4$ [in./sec.²], h = total height of web, B = web width, h_0 = distance from journal centre to bottom edge of web (see figure), and L_W = web thickness, all linear dimensions being in inches.

[6]

1·1131 *Alternative methods for calculating crankweb inertia*

Various other methods are used for the evaluation of the Wk^2 or moment of inertia of crankwebs, and these are briefly as follows:

Equivalent disk method. The crankweb is replaced by an equivalent disk, and its moment of inertia is determined by graphical constructions, using 'derived figures'. Full details of this method are given under 'Moment of inertia of marine propellers', section 1·13, page 24.

Planimetry method. This method is similar to the standard method for crankwebs given at the beginning of this section. The web is divided up into a number of circular segments, for various values of the radius R from the journal centre.

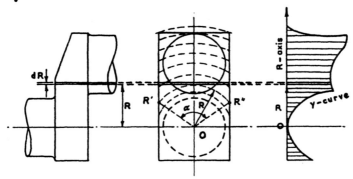

If dR = width of segment, $L_W(R)$ = crankweb thickness at the radius R, and $\alpha°$ = angle $R'OR''$ in the accompanying figure, the polar mass moment of inertia of the annular segment may be written as

$$dJ_{(R)} = \frac{\alpha°}{360°} \frac{w}{g} \times R^2 \times 2\pi R\, dR \times L_W.$$

The procedure is to calculate for each radius R the value of

$$y(R) = \alpha° R^3 L_W \quad [\text{in.}^4],$$

and to plot these successive values in a diagram of $y(R)$ against R, so as to obtain a y-curve.

The area under this curve $y(R)$ is then determined by means of a planimeter, giving a value

$$Y = \int dR\, y(R) \quad [\text{in.}^5].$$

The moment of inertia of one web is then obtained from

$$J_{1\,\text{web}} = \frac{2\pi}{360°} \times \frac{w}{g} \times Y \quad [\text{Lb.in.sec.}^2].$$

Metric units. The formulae and procedures given in section 1·113 can be used directly to obtain J-values in Kg.cm.sec.^{2}. For this, all lengths should be in cm., specific weights in Kg./cm.^{3} (i.e. $\frac{1}{1000}$th of the usual tabulated values, which are in Gramme/cm.^{3}), and $g = 981\cdot0$ cm./sec.^{2}. Thus, for steel, the specific weight is $w = 0\cdot00785$ Kg./cm.^{3}.

1·114 *Moment of inertia of balance weights*

The moment of inertia of balance weights can be determined by the tabulation method, previously described in its application to crankwebs. Alternatively, a purely graphical method may be used for weights of uniform thickness.†

A typical balance weight is shown in the accompanying figure. To determine its polar mass moment of inertia about the point O, corresponding to the centre of the crank journal, the procedure is as follows:

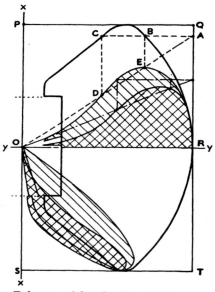

The two principal axes, yy and xx, are drawn, both passing through the point O. On one half of the balance weight, describe a rectangle $OPQR$. Draw any line OA on to the line RQ. Erect a perpendicular at A to cut the edge of the weight at B and C. Draw lines from C and B parallel to Ox, so as to cut OA at D and E.

Following the same procedure, draw a number of lines similar to OA. Then, by joining up the points R, E, D, etc., what is known as the 'first derived figure' is obtained (this is shown by the shading in the upper part of the figure).

In the next step, the derived figure is treated in exactly the same

Balance weight of uniform thickness.

manner as the curve of the balance-weight contour, in order to obtain a 'second derived figure' shown double-shaded. Denoting by A_0, A_1 and A_2 the areas of the original, the first derived, and the second derived figures, respectively, the radius of gyration k_{xx} of the balance weight about the xx-axis is obtained from

$$k_{xx}^2 = \frac{A_2}{A_0}(PQ)^2 \quad [\text{in.}^2].$$

† See, *Torsional vibration in diesel engines,* by C. H. Bradbury (C. Griffin and Co. Ltd. London, 1938), pp. 46–8.

[8]

If the balance weight has a thickness h and a specific weight w [Lb./in.³] its weight is given by

$$W = 2wA_0h \quad \text{[Lb.]}.$$

Therefore its moment of inertia about the xx-axis is

$$J_{xx} = Wk^2_{xx}/g \quad \text{[Lb.in.sec.}^2\text{]}.$$

For cast iron, $w = 0.260$ [Lb./in.³] and the value of g in the units chosen is 386.4 [in./sec.²]. (In metric units: $g = 981.0$ cm./sec.², $w_{(\text{cast iron})} = 0.0072$ Kg./cm.³, k is in cm., and J is in Kg.cm.sec.².)

The moment of inertia J_{yy} about the yy-axis now has to be determined. The method is the same as previously, using, however, ST as a base-line (instead of RQ) for lines to be drawn from O. The first and second derived figures are again shown shaded in the above figure. Denoting the areas of the first and second derived figures by A'_1 and A'_2, respectively, we obtain

$$k^2_{yy} = \frac{A'_2}{A_0}(OS)^2 \quad \text{[in.}^2\text{]},$$

and

$$J_{yy} = Wk^2_{yy}/g \quad \text{[Lb.in.sec.}^2\text{]}.$$

The polar moment of inertia J_0 of the balance weight, about the O-axis passing through the crank journal centre, is, therefore,

$$J_0 = J_{xx} + J_{yy} = \frac{W}{gA_0}[A_2(PQ)^2 + A'_2(OS)^2] \quad \text{[Lb.in.sec.}^2\text{]}.$$

The derivation of this method is based on considerations similar to those given in a further section for the determination of the moment of inertia of a marine propeller. (See section 1·13.)

1·115 *Moment of inertia of connecting rod, piston and gudgeon pin*

As shown in the accompanying Fig. 1, the crank end of a connecting rod describes a circle (C_1, C_2, C_3, C_4), whereas the piston end has a reciprocating motion (P_1, P_2, P_3, P_4). To avoid unnecessary calculations, for practical purposes the moment of inertia of the connecting rod is obtained with sufficient accuracy by the following method:

The connecting rod is split up into two concentrated weights, viz. a rotating weight $W_{\text{rot.}}$ at the crank end, and a reciprocating weight $W_{\text{recip.}}$ at the piston end, in such a way that $W_{\text{conn. rod.}} = W_{\text{rot.}} + W_{\text{recip.}}$.

For this purpose, the connecting rod is placed on two scale pans and supported at the centre-line positions of both of its end bearings by knife-edges, which are also in line with the centre lines of the scales.

The scale supporting the small end must be raised on blocks until the connecting-rod shaft is in a horizontal position. Then the large-end scale will indicate the rotating weight $W_{rot.}$ and the small-end scale the reciprocating weight $W_{recip.}$.†

If a single weighing machine is used, the ends of the connecting rod should again be supported on knife-edges, one of which is placed on the platform and the other on blocks adjusted until the connecting rod is horizontal, this being checked by means of a spirit-level. A further check of the results is that they should give: $W_{rot.} + W_{recip.} = W_{conn.\,rod}$.

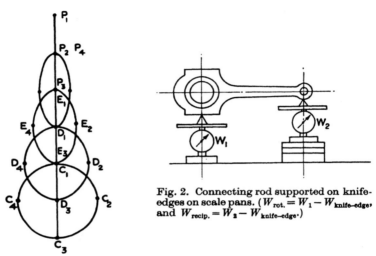

Fig. 1. Elliptical paths described by various points of a connecting rod.

Fig. 2. Connecting rod supported on knife-edges on scale pans. ($W_{rot.} = W_1 - W_{knife-edge}$, and $W_{recip.} = W_2 - W_{knife-edge}$.)

Another method consists in suspending the connecting rod from spring balances attached to wires passing through the centre lines of the connecting-rod ends.

The moment of inertia of the rotating part of the connecting rod is

$$J_{conn.\,rod(rot.)} = R_0^2 \frac{W_{rot.}}{g} \quad [\text{Lb.in.sec.}^2],$$

where $g = 386 \cdot 4$ (in./sec.^{2}) and R_0 = crank radius (in.).

The weight of the reciprocating portion of the connecting rod is added to the reciprocating weights of the piston W_P and gudgeon pin W_{GP},

---

† In design-stage calculations, it is necessary to evaluate the weight of the various portions into which the connecting rod is subdivided. This is done by a semi-graphical method. The centre of gravity is then determined graphically, the connecting-rod length L being thus divided into two portions of length $L_{rot.}$ and $L_{recip.}$. If W_{CR} = total weight of connecting rod, the weights $W_{rot.}$ and $W_{recip.}$ of its rotating and reciprocating portions are then obtained as:

$$W_{rot.} = \frac{L_{recip.}}{L} \times W_{CR} \quad \text{and} \quad W_{recip.} = \frac{L_{rot.}}{L} \times W_{CR}.$$

and the effective moment of inertia of these three masses is determined from

$$J_{\text{recip. parts}} = \frac{1}{2g} (W_{CR\,\text{recip.}} + W_P + W_{GP}) R_0^2 \quad [\text{Lb.in.sec.}^2].$$

The reason for taking one-half of the full values of the reciprocating weights is as follows. We assume for simplicity that the velocity $v_{\text{recip.}}$ of a reciprocating part $W_{\text{recip.}}$ varies sinusoidally with time, so that

$$v_{\text{recip.}} = \omega R_0 \sin \omega t \quad [\text{in./sec.}],$$

where $\omega = (2\pi/60) \times \text{rev./min.}$ This simplifying assumption disregards the effect of higher harmonics of $v_{\text{recip.}}$, which occur owing to the obliquity of the connecting rod. The kinetic energy at any instant is

$$U_{k(\text{recip.})}(t) = \tfrac{1}{2}(W_{\text{recip.}}/g) \times (\omega R_0 \sin \omega t)^2 \quad [\text{in.Lb.}],$$

and its average value during one revolution is (with $\tau = 2\pi/\omega$):

$$\overline{U}_{k\,(\text{recip.})} = \frac{\omega}{2\pi} \int_0^{2\pi/\omega} dt\, U_{k\,(\text{recip.})}(t) = \frac{W_{\text{recip.}}}{4g} \omega^2 R_0^2.$$

The kinetic energy of a rotating weight W_e with a constant peripheral velocity ωR_0 is

$$U_{k(\text{rot.})} = \frac{1}{2}\frac{W_e}{g}\omega^2 R_0^2.$$

If W_e is to be equivalent to $W_{\text{recip.}}$, the kinetic energies of both masses must have the same average value, so that

$$U_{k(\text{rot.})} = \overline{U}_{k(\text{recip.})} \quad \text{and hence} \quad W_e = \tfrac{1}{2}W_{\text{recip.}}.$$

Note on the moment of inertia per cylinder line

According to Haug,† the expression for the moment of inertia per cylinder line $J_{\text{cyl.}}$ (see p. 3) may give slightly high values. Haug suggests that it is preferable to use Grammel's relation‡

$$J_{\text{cyl.}} = \frac{1}{\dfrac{1}{\pi}\displaystyle\int_0^\pi \dfrac{d\phi}{J(\phi)}},$$

where $J(\phi)$ = value of moment of inertia per cylinder line at a crank-angle position ϕ. The integral in the denominator can be evaluated by graphical or numerical methods.

In the numerical example given by Haug, the value obtained for $J_{\text{cyl.}}$ with the above expression is about 1 % smaller than that calculated by the usual formula.

† K. Haug, *A.T.Z.* 10 August 1942, pp. 407–14.
‡ R. Grammel, 'The vibrations of I.C. engines', *Ingen.-Arch.* vol. VI (1935), pp. 59–68.

1·12 Moment of inertia of a flywheel

A flywheel of fairly simple shape can be calculated by dividing it up into a number of rings of rectangular section and evaluating the Wk^2 of each individual part. In many cases, the contribution of the hub is so small that it can be neglected in the calculation.

Thus, for instance, for the part denoted by '1' in the figure, the calculation is as follows:

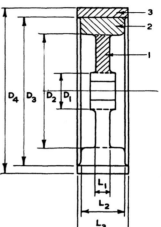

weight:

$$W_1 = \frac{\pi}{4}(D_2^2 - D_1^2)L_1 w \quad [\text{Lb.}],$$

radius of gyration squared:

$$k_1^2 = \tfrac{1}{8}(D_2^2 - D_1^2) \quad [\text{in.}^2],$$

where $w = 0\cdot260$ Lb./in.^{3} for cast iron, and $w = 0\cdot283$ Lb./in.^{3} for steel.

Proceeding in a similar manner for parts '2' and '3', the moment of inertia of the flywheel is finally obtained as

$$J_F = \frac{1}{g}(W_1 k_1^2 + W_2 k_2^2 + W_3 k_3^2) \quad [\text{Lb.in.sec.}^2],$$

where $g = 386\cdot4$ in./sec.^{2}.

For flywheels with a large number of sections, each of which has an outer diameter D_o, an inner diameter D_i, and a breadth L, it is convenient to tabulate the terms $L(D_o^4 - D_i^4)$, so as to obtain their sum

$$S = \Sigma L(D_o^4 - D_i^4) \quad [\text{in.}^5].$$

The moment of inertia of the flywheel is then

$$J_F = \frac{\pi}{32} \times \frac{w}{g} \times S \quad [\text{Lb.in.sec.}^2].$$

Substituting the values for w given above, we obtain

$$J_F = 6\cdot605 \times 10^{-5} \times S \quad [\text{Lb.in.sec.}^2] \quad \text{for cast iron,}$$

and $\quad J_F = 7\cdot186 \times 10^{-5} \times S \quad [\text{Lb.in.sec.}^2] \quad \text{for steel.}$

Metric units. Expressing linear dimensions in cm., specific weights in Kg./cm.^{3}, and using $g = 981\cdot0$ cm./sec.^{2}, we obtain (with S in cm.^{5}):

$$J_F = 7\cdot155 \times 10^{-7} \times S \quad [\text{Kg.cm.sec.}^2] \quad \text{for cast iron,}$$

and $\quad J_F = 7\cdot784 \times 10^{-7} \times S \quad [\text{Kg.cm.sec.}^2] \quad \text{for steel.}$

1·121 *Moment of inertia of grooves and fillets*

The moment of inertia of grooves, fillets, etc., of comparatively small cross-section can be estimated with sufficient accuracy by multiplying

[12]

the mass of material removed in forming the groove by the square of the radius R to the centre of gravity of the groove cross-section, that is,

$$J = 2\pi R^3 A \frac{w}{g} \quad [\text{Lb.in.sec.}^2],$$

where A = area of groove cross-section [in.^{2}], w = specific weight [Lb./in.^{3}], and $g = 386\cdot4$ [in./sec.^{2}].

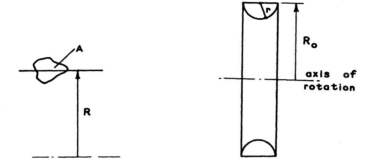

The moment of inertia of *peripheral grooves of semicircular cross-section* can be calculated by means of the formula

$$J = \frac{\pi w}{g} r R_0^4 \left[\pi \frac{r}{R_0} - 4 \left(\frac{r}{R_0} \right)^2 + \frac{3\pi}{4} \left(\frac{r}{R_0} \right)^3 - \frac{8}{15} \left(\frac{r}{R_0} \right)^4 \right],$$

where R_0 = outer rim radius and r = groove radius. This expression is obtained by evaluating the integral

$$J = 2 \int_0^r \frac{w}{g} \times \frac{\pi}{2} [R_0 - \sqrt{(r^2 - x^2)}]^4 \, dx,$$

and subtracting the result from $\pi \dfrac{w}{g} r R_0^4$.

For flywheels, pulleys and cover-plates of more complicated shape a more detailed evaluation may be necessary, using tabulation or graphical methods with derived figures as in the treatment of crankwebs, balance-weights and propellers.

The best way to obtain accurate figures is, however, to carry out swinging tests, with the flywheel mounted in a trifilar suspension or on knife-edges. Somewhat less accurate figures are obtained by means of the 'running-down' test or by the 'falling-weight' method. The various test methods are described in the following section.

1·122   *Moment of inertia of various parts of a flywheel*

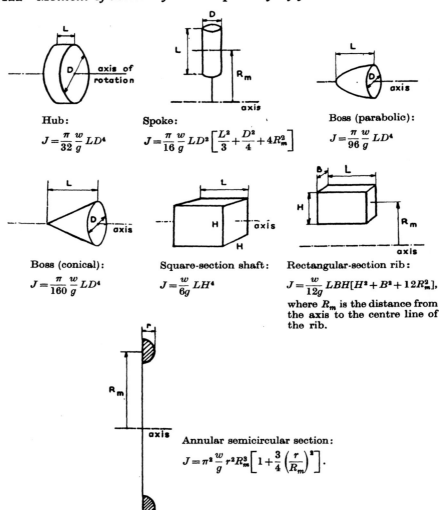

Hub:

$$J = \frac{\pi}{32}\frac{w}{g} LD^4$$

Spoke:

$$J = \frac{\pi}{16}\frac{w}{g} LD^2 \left[\frac{L^2}{3} + \frac{D^2}{4} + 4R_m^2 \right]$$

Boss (parabolic):

$$J = \frac{\pi}{96}\frac{w}{g} LD^4$$

Boss (conical):

$$J = \frac{\pi}{160}\frac{w}{g} LD^4$$

Square-section shaft:

$$J = \frac{w}{6g} LH^4$$

Rectangular-section rib:

$$J = \frac{w}{12g} LBH[H^2 + B^2 + 12R_m^2],$$

where R_m is the distance from the axis to the centre line of the rib.

Annular semicircular section:

$$J = \pi^2 \frac{w}{g} r^2 R_m^3 \left[1 + \frac{3}{4}\left(\frac{r}{R_m} \right)^2 \right].$$

1·123   *Vertical pendulum method for determining moments of inertia*

Application to rotors

The rotor is suspended horizontally on knife-edges as shown in the figure. The system is moved out of its equilibrium position through an angle θ of less than 5° and allowed to swing freely.

If τ = duration of one complete oscillation† [sec.], W = weight of rotor and shaft [Lb.], and L = length of suspension [in.], i.e. the distance from

† Obtained as an average value from 20 to 30 oscillations.

knife-edge to centre-line of shaft, then the moment of inertia J_o [Lb.in.sec.²] of the rotor system about its shaft axis OO is given by

$$J_o = WL \times \left[\frac{\tau^2}{39\cdot48} - \frac{L}{386\cdot4}\right] \quad \text{[Lb.in.sec.²]}.$$

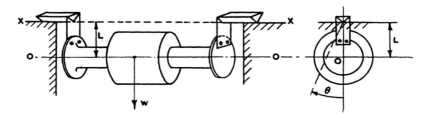

Derivation of formula. The equation of motion of the system is

$$J_x\ddot{\theta} = -WL\sin\theta \cong -WL\theta,$$

where $J_x =$ moment of inertia of system about the XX-axis. Substituting $\theta = \theta_0 \cos \omega t$, where $\omega = (2\pi/60)\, F$, F being the vibration frequency [vib./min.], we obtain

$$\theta_0(-\omega^2 J_x + WL)\cos \omega t = 0,$$

for all values of t. Also $\tau = 2\pi/\omega$, hence

$$\omega^2 = \frac{WL}{J_x} = \frac{4\pi^2}{\tau^2} = \frac{39\cdot48}{\tau^2},$$

so that $J_x = \dfrac{WL}{39\cdot48}\tau^2$ and $J_0 = J_x - \dfrac{L^2 W}{g} = WL\left(\dfrac{\tau^2}{39\cdot48} - \dfrac{L}{g}\right)$,

which is the above formula.

Applications to flywheels

The above formula can also be used to determine the moment of inertia of a flywheel when its period of vibration is measured in a swinging test. Three arrangements suitable for such tests are shown in the following figures.

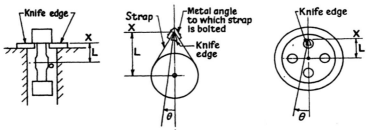

1·124 *Multifilar suspension method for determining moments of inertia*

Moments of inertia can be determined by torsional swinging tests using bifilar, trifilar or quadrifilar suspensions.

To obtain good results, the following precautions should be observed:

The length of the wires should be at least six times the radial distance from the centre of the object to its attachment points. The attachment points should all be at the same horizontal level. The time τ of a complete oscillation should be determined as the mean value of several series of 20–40 counts of vibration cycles, with oscillations not exceeding 5° in amplitude.

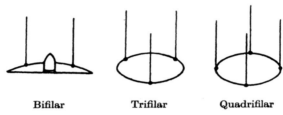

Bifilar   Trifilar   Quadrifilar

Knowing the weight W, the length of the wires L, and the radius R, as well as the period of vibration τ, the moment of inertia J of the object tested is then obtained from

$$J = \frac{W R^2 \tau^2}{39 \cdot 48 L} \quad [\text{Lb.in.sec.}^2],$$

where W is in Lb., R and L are in inches, and τ is in seconds.

This expression is obtained simply enough as follows:

During the oscillation the disk moves through an angle θ and the wire rotates through an angle α. As $s = R\theta \simeq L\alpha$, we have (approximately, since θ and α are small)

$$\alpha = \frac{R}{L}\theta.$$

The length L of the wire remaining constant during the motion, the disk is raised to a height h, which is given by

$$h = L(1 - \cos\alpha) \simeq \tfrac{1}{2}L\alpha^2 \simeq \frac{R^2}{2L}\theta^2.$$

The maximum potential energy is

$$U_{\text{pot.}} = Wh = \frac{W R^2}{2L}\theta^2,$$

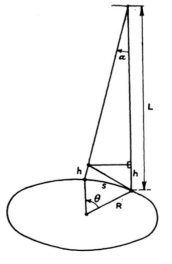

and the maximum kinetic energy is

$$U_{\text{kin.}} = \tfrac{1}{2}J\omega^2\theta^2, \quad \text{where} \quad \omega = \frac{2\pi}{\tau}.$$

Equating these two expressions we obtain

$$J = \frac{WR^2}{2L \times \dfrac{1}{2}\left(\dfrac{2\pi}{\tau}\right)^2} = \frac{WR^2\tau^2}{39\cdot48L}.$$

This expression is independent of the number of wires used. It assumes that the wires are of negligible stiffness, i.e. that their strain energy is zero.

Inertia-measuring apparatus

A simple arrangement, such as that shown in the figure, can be set up as a permanent apparatus.

After setting up, the moment of inertia of the lower disk is determined experimentally and should agree closely with the calculated value.

Any other mass of which the moment of inertia is required can then be placed on the lower disk and the total inertia is determined by a swinging test. The required inertia J_{mass} is then obtained as $J_{\text{total}} - J_{\text{disk}}$.

The unknown mass should be carefully centred on the lower disk, which may be provided with a number of concentric circular grooves to make sure of accurate centring.

The torsional swinging test should be repeated a number of times in order to obtain an accurate mean value for the vibration period τ.

1·125 *Multifilar suspension with non-parallel wires*

If the suspension wires are not parallel, the moment of inertia of the object is determined by means of the following formula:

$$J = \frac{WR_2^2 L_0}{39\cdot48L^2}\tau^2 \quad [\text{Lb.in.sec.}^2],$$

where $\tau =$ period of vibration [sec.], $W =$ weight of object [Lb.], $R_2 =$ radius at which wires are attached to object [in.], $L =$ length of a suspension wire [in.], and $L_0 =$ vertical distance [in.] between object and the fixed upper portion of the apparatus.

Metric units. The above formula can be used directly, with the same numerical constant, to obtain J-values in Kg.cm.sec.^{2} if all linear dimensions are in cm. and the weight W is in Kg.

Derivation. For small values of θ (less than 5°), we have

$$s = R_2\theta \cong L\alpha;$$

moreover,
$$h' = L(1 - \cos\alpha) \simeq L\frac{\alpha^2}{2} \simeq R_2^2\frac{\theta^2}{2L}.$$

The vertical displacement is
$$h = h'\cos\phi = h'\frac{L_0}{L},$$

hence the potential energy of the object is
$$U_{\text{pot.}} = WR_2^2\frac{\theta^2 L_0}{2L^2}.$$

The kinetic energy is
$$U_{\text{kin.}} = \tfrac{1}{2}J\omega^2\theta^2,$$

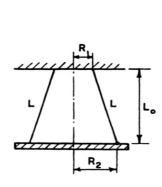
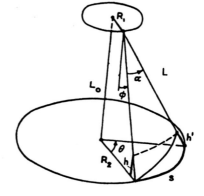

where $\omega = 2\pi/\tau$, so that by equating these two energy expressions we obtain
$$WR_2^2\frac{L_0}{L^2} = J\omega^2 = J \times \frac{4\pi^2}{\tau^2},$$

which gives the above formula. Since $L^2 = L_0^2 + (R_2 - R_1)^2$, this can also be written as
$$J = \frac{WR_2^2\tau^2}{39\cdot48 \times \left[L_0 + \dfrac{(R_2 - R_1)^2}{L_0}\right]}.$$

1·126 *Moment of inertia determined by rolling method*

The moment of inertia of a mass can also be determined by measuring the time t_B which it takes to roll down a slope inclined at an angle α to the horizontal.

Let $s = \overline{AB} = $ length of ramp [in.],

$h = \overline{OA} = $ height of ramp [in.],

$R = $ radius of shaft [in.], and

$W = $ weight of mass [Lb.],

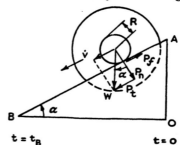

[18]

then the moment of inertia J [Lb.in.sec.^{2}] of the mass can be obtained from

$$J = WR^2\left[\frac{h}{2s^2}t_B^2 - \frac{1}{386\cdot4}\right] \quad \text{[Lb.in.sec.}^2\text{]}.$$

For J in Kg.cm.sec.^{2}, the constant $386\cdot4$ in./sec.^{2} is replaced by $981\cdot0$ cm./sec.^{2}, linear dimensions are in cm., and W is in Kg.

Derivation. The notation used is:

$P_t = W\sin\alpha = $ tangential force on shaft,

$P_n = W\cos\alpha = $ normal force on ramp,

$P_f = -\mu P_n = -\mu W\cos\alpha = $ frictional force on shaft
($\mu = $ coefficient for rolling friction),

and $\quad\dot{v} = R\dot\omega = $ linear acceleration of shaft.

We have

for motion parallel to slope: $\quad \dot{v} = (P_t - P_f)g/W,$ (1)

for rotation about shaft axis: $\quad \dot\omega = \dot{v}/R = RP_f/J.$ (2)

Hence $\qquad\qquad\qquad P_f = \dot{v}J/R^2.$ (3)

Substituting (3) in (1), we obtain

$$\dot{v} = \frac{g}{W}\left[P_t - \frac{\dot{v}J}{R^2}\right],$$

or $\qquad\qquad \dot{v}\left[\frac{W}{g} + \frac{J}{R^2}\right] = P_t = W\sin\alpha = W\frac{h}{s}.$ (4)

Starting from rest, with the acceleration $\dot{v}$, the distance travelled in the time t_B is $s = \frac{1}{2}\dot{v}t_B^2$. Hence $\dot{v} = 2s/t_B^2$, and eq. (4) gives

$$\frac{2s}{t_B^2}\left[\frac{W}{g} + \frac{J}{R^2}\right] = W\frac{h}{s} \quad \text{or} \quad \frac{Wh}{2s^2}t_B^2 - \frac{W}{g} = \frac{J}{R^2};$$

hence J is obtained as in the above formula.

1·127 *Moment of inertia determined by vibration test on curved rails*

The flywheel and shaft assembly is placed on circular rails as indicated in the figure. If the weight of the flywheel is known, the moment of inertia of the flywheel can be determined from the period of the oscillations which the system performs when it is displaced a small distance from its equilibrium position.

If $R = $ radius of curvature of rails [in.], $r = $ radius of shaft [in.],

$W = $ weight of flywheel [Lb.],

and $\tau = $ period of oscillation [sec.],

then the moment of inertia J of the flywheel is obtained from

$$J = Wr^2 \left[\frac{\tau^2}{39 \cdot 48(R-r)} - \frac{1}{386 \cdot 4} \right] \quad [\text{Lb.in.sec.}^2].$$

Derivation. The maximum potential energy of the system (at B) is

$$U_p = Wh \simeq \tfrac{1}{2} W R_0 \theta^2,$$

since

$$h = R_0(1 - \cos\theta) \simeq \tfrac{1}{2} R_0 \theta^2.$$

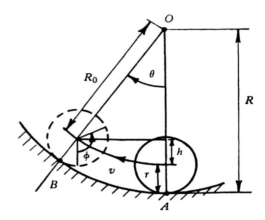

The maximum kinetic energy (at A) is the sum of the energies of rotation about O and about the shaft axis

$$U_k = U_{k1} + U_{k2} = \frac{W}{2g}(R_0 \omega\theta)^2 + \frac{J}{2}\left(\frac{R_0}{r}\omega\theta\right)^2,$$

where $\omega = 2\pi/\tau$. Equating these expressions, we find

$$WR_0 = \left[\frac{W}{g}R_0^2 + J\frac{R_0^2}{r^2}\right]\omega^2,$$

which gives

$$J = Wr^2 \left[\frac{1}{\omega^2(R-r)} - \frac{1}{g}\right],$$

since $R_0 = R - r$. As $4\pi^2/\tau^2 = \omega^2$, this gives the above formula.

Metric units. To obtain J-values in Kg.cm.sec.^{2}, using weights in Kg., and linear dimensions in cm., replace 386·4 by 981·0. The constant 39·48 remains unchanged.

1·128 *Moment of inertia determined by falling weight method*

This method can be applied to generators, etc. A cord carrying a weight W [Lb.] is wound round the shaft. The weight is allowed to fall

[20]

to the ground and in so doing rotates the machine which thereafter comes to rest.

Measurements are taken of the time t_1 required for the weight to reach the ground, and the time t_2 elapsed from this instant till the shaft stops.

The moment of inertia of the mass or masses on the shaft is then determined from

$$J = \frac{W\left[h - \dfrac{2h^2}{386 \cdot 4t_1^2}\right]}{\dfrac{2h^2}{R^2t_1^2} \times \left[1 + \dfrac{t_1}{t_2}\right]} \quad \text{[Lb.in.sec.}^2\text{]},$$

where time is measured in seconds. The mean value of several readings is taken. An alternative method requiring only one time measurement is given further on in this section.

Derivation. At the time t_1, the velocity of the mass as it touches the ground is

$$v_1 = 2h/t_1,$$

i.e. twice the mean velocity h/t_1. The shaft has performed n_1 revolutions

$$n_1 = h/2\pi R.$$

After the mass has touched the ground, the shaft performs a further n_2 revolutions, before coming to rest. At the time t_1, the kinetic energy of the system comprising the inertia mass and the falling weight is

$$\tfrac{1}{2}v_1^2\left(\frac{W}{g} + \frac{J}{R^2}\right) = Wh - n_1\Delta U_f,$$

where ΔU_f is the energy dissipated by friction in each revolution. After the weight has reached the ground, the kinetic energy of the inertia mass is dissipated by friction during the further number of revolutions n_2, so that

$$\tfrac{1}{2}v_1^2\frac{J}{R^2} = n_2\Delta U_f.$$

This assumes the same ΔU_f, i.e. the same friction, throughout the test, which is correct if W is small compared with the weight of the rotating mass. Therefore

$$n_1\Delta U_f = \frac{n_1}{n_2} \times \tfrac{1}{2}v_1^2\frac{J}{R^2},$$

and the substitution of this expression for $n_1\Delta U_f$ in the energy equation gives, after some rearranging,

$$\frac{v_1^2}{2R^2}\left[1 + \frac{n_1}{n_2}\right]J = W\left[h - \frac{v_1^2}{2g}\right].$$

Substituting $n_1/n_2 = t_1/t_2$ and $v_1 = 2h/t_1$, we obtain the above formula.

[21]

Note. Experiments indicate a practically linear fall in speed during, the second period t_2, so that the plot of $N = dn/dt$ (where n = number of revolutions) against time is approximately as shown in the accompanying figure. As a result, we have

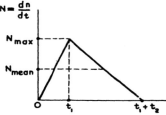

$$n_1 = N_{\text{mean}} \times t_1 \quad \text{and} \quad n_2 = N_{\text{mean}} \times t_2,$$

so that
$$\frac{n_1}{n_2} = \frac{t_1}{t_2}.$$

Alternative method. Instead of using a length of cord slightly longer than h, which detaches itself when the weight W reaches the ground, it is convenient to use a much longer length. If s is the unrolled length of cord corresponding to the number of revolutions n_2, the ratio n_1/n_2 is obtained directly by length measurements as

$$n_1/n_2 = h/s.$$

The moment of inertia is then obtained from

$$J = \frac{W\left[h - \dfrac{2h^2}{386 \cdot 4t_1^2}\right]}{\dfrac{2h^2}{R^2 t_1^2} \times \left[1 + \dfrac{h}{s}\right]} \quad [\text{Lb.in.sec.}^2].$$

This method does not use the ratio t_1/t_2 or its relation to n_1/n_2, and is therefore to be preferred.

1·129 *Moment of inertia determined from a running-down test*

The moment of inertia of an engine system consisting of cylinder inertias, flywheel and driven masses, can be determined by a running-down test.

The engine is run at no-load, preferably above the operating speed, and the fuel is then cut off suddenly. The decrease in engine speed per second is measured in the region of the operating speed. If the frictional torque of the engine at no-load is known, the total value of the moments of inertia in the system is then obtained as

$$J_{\text{total}} = T_f \Big/ \left(2\pi \frac{\Delta^2 n}{\Delta t^2}\right) \quad [\text{Lb.in.sec.}^2],$$

where T_f = frictional torque [Lb.in.], t = time [sec.], and n = number of revolutions.

The value of T_f can be determined from a curve of fuel consumption per b.h.p.-hr. plotted against brake mean effective pressure $p_{b(m)}$ and extrapolated to the negative side in order to obtain the frictional mean effective pressure $p_{f(m)}$ (see Fig. 1). Then

$$T_f = \frac{1}{\pi}\, p_{f(m)} A R_0 N_{\text{cyl.}} \quad \text{for} \quad 2 \text{ s.c. s.a engines,}$$

or
$$T_f = \frac{1}{2\pi}\, p_{f(m)} A R_0 N_{\text{cyl.}} \quad \text{for} \quad \text{4 s.c. s.a. engines,}$$

where T_f is in Lb.in., $N_{\text{cyl.}}$ = number of engine cylinders, A = piston area [in.^{2}], R_0 = crank radius [in.], and $p_{f(m)}$ = mean frictional effective pressure [Lb./in.^{2}]. (Further details on this subject are given in section 2·11.)

For accurate results, the value of $\Delta^2 n/\Delta t^2$ is best determined from well-spread-out photographic recordings obtained with a drum camera, using a tuning fork for a time signal on which a pulse is superimposed for each complete revolution of the shafting. The latter signal can be obtained by inserting a small projecting stud of steel on the periphery of the flywheel and placing close to the flywheel a stationary permanent-magnet

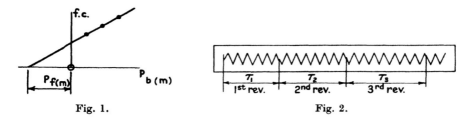

Fig. 1.   Fig. 2.

inductive type pick-up which generates a pulse each time the stud passes across its field.

The value of $\Delta^2 n/\Delta t^2$ is then determined from the photographic trace (Fig. 2) as
$$\frac{\Delta^2 n}{\Delta t^2} = \left[\left(\frac{1}{\tau_1}\right) - \left(\frac{1}{\tau_2}\right)\right] \Big/ \left[\frac{\tau_1 + \tau_2}{2}\right]$$

or
$$\frac{\Delta^2 n}{\Delta t^2} = \frac{2(\tau_2 - \tau_1)}{\tau_1 \tau_2 (\tau_1 + \tau_2)} \quad [\text{rev./sec.}^2],$$

where τ_1 and τ_2 are the times, in seconds, of the first and second engine revolutions, respectively.

If a drum camera is not available, the decrease in engine speed at the beginning of the running-down test can be determined from the number of revolutions in 1 sec. as indicated by a tachoscope reading taken at the free end of the engine, and the number of revolutions in 2 sec. by a tachoscope reading taken simultaneously at the driving end. Then the numerical value of $\Delta^2 n/\Delta t^2$ is obtained simply as

$$\frac{\Delta^2 n}{\Delta t^2} = 2n_1 - n_2 \quad [\text{rev./sec.}^2],$$

where n_1 and n_2 are the number of crankshaft revolutions occurring during the 1 sec. and the 2 sec. periods, respectively.

[23]

Derivation. The first tachoscope is stopped at t_1 sec., during which time it recorded n_1 revolutions, and the second instrument is stopped at t_2, recording n_2 revolutions from the beginning of the test. Then

$$\frac{\Delta^2 n}{\Delta t^2} = \left[\frac{n_1}{t_1} - \frac{n_2 - n_1}{t_2 - t_1}\right] \bigg/ \frac{t_2}{2} \quad [\text{rev./sec.}^2],$$

where $t_2/2$ is the mean time interval, i.e. $\frac{1}{2}(t_2 - t_1) + \frac{1}{2}t_1$.

If $t_2 = 2t_1$, the above equation reduces to

$$\frac{\Delta^2 n}{\Delta t^2} = \frac{2n_1 - n_2}{t_1^2},$$

and for $t_1 = 1$ sec., the result is, numerically, $\Delta^2 n / \Delta t^2 = 2n_1 - n_2$.

Metric units. To obtain J_{total} in Kg.cm.sec.^{2}, the above formulae are directly applicable, provided that R_0 is in cm., A in cm.^{2}, p in Kg./cm.^{2} and T_f in Kg.cm.

1·13   Moment of inertia of a marine propeller

The moment of inertia of a propeller is determined graphically. The first step is to reduce the propeller to an equivalent disk. For this the procedure is as follows (see Fig. 1):

(1) Take any radius r and determine the total cross-sectional area $A = \Sigma \Delta A$ of the blades at this radius r;

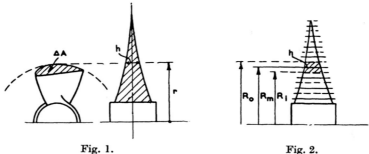

Fig. 1.   Fig. 2.

(2) Determine the equivalent disk thickness h at this radius from the relation

$$h = A/(2\pi r) \quad [\text{inches}].$$

(3) Repeat this process for different values of r to obtain the entire equivalent disk.

The moment of inertia of the equivalent disk can then be obtained either by tabulation or by a graphical method using derived figures.

1·131   *Tabulation method*

This is the same method that has been used for crankwebs. The equivalent disk is divided into a number of annular layers (see Fig. 2), each with a mean radius R_m, an outer radius R_o and an inner radius R_i.

[24]

Taking the value of h for any mean radius R_m, the value ΔJ of the moment of inertia of the corresponding annulus is determined as

$$\Delta J = \frac{\pi}{2}\frac{w}{g} h(R_o^4 - R_i^4) \quad \text{[Lb.in.sec.}^2\text{]},$$

where $g = 386\cdot4$ [in./sec.^{2}] and $w =$ specific weight of propeller material [for bronze: $w = 0\cdot315$ Lb./in.^{3}; and for cast iron: $w = 0\cdot260$ Lb./in.^{3}]. The total moment of inertia of the propeller about its axis of rotation is then
$$J_p = \Sigma\Delta J \quad \text{[Lb.in.sec.}^2\text{]},$$

where $\Sigma\Delta J$ is the sum of the tabulated values for the annuli at various values of the radius r. The boss is, of course, included in this tabulation.

1·132 *Graphical method with derived figures and equivalent disk*

This method has already been shown in its application to balance weights. The procedure for propellers is as follows:

(1) Draw a line zz parallel to xx, at any suitable radius, for instance, at $r_z = \overline{OS} + \tfrac{1}{2}\overline{ST}$ (see Fig. 3);

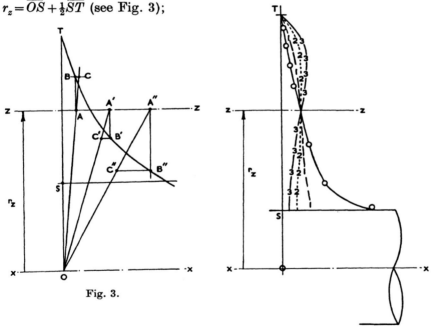

Fig. 3.

Fig. 4. The original figure is denoted by O-O, and the derived figures are indicated by their respective numbers.

(2) Draw from any point B on the disk contour a perpendicular on to line zz, which cuts zz at A;

(3) Draw the line $\overline{OA}$; its intersection C with a parallel to zz which passes through point B is a point of the 'first derived figure'.

[25]

(4) Take another point B' on the disk contour, and repeat procedures 2 and 3. This gives another point C'. The entire curve of the 'first derived figure' is thus obtained by determining the points C, C', C'', etc.

This 'first derived figure' is then treated in the same manner as the original figure, in order to obtain a 'second derived figure', which in turn is used to determine a 'third derived figure'. An example of the types of curves obtained is shown in Fig. 4.

Let A_0, A_1, A_2 and A_3 denote the areas enclosed by the corresponding curves in Fig. 4 from the disk tip T to the boss at S. Then, the weight of the blades is given by

$$W = 4\pi w r_z A_1 \quad [\text{Lb.}],$$

where w = specific weight of blade material [Lb./in.^{3}], r_z = distance from xx-axis to zz-axis [in.], and A_1 = area [in.^{2}] of the 'first derived figure' in Fig. 4. The latter can be obtained by planimetering the area within the limits T-S-1-1-1-T in the figure.

The moment of inertia J_{pb} of the propeller blades is obtained from

$$J_{pb} = 4\pi \frac{w}{g} r_z^3 A_3 \quad [\text{Lb.in.sec.}^2],$$

where w and r_z are as defined above, $g = 386\cdot4$ [in./sec.^{2}], and A_3 = area of the 'third derived figure'.

Derivation of method

We consider Fig. 5, which is similar to Fig. 3.

From similar triangles AOP and COQ,

$$\frac{x_C}{x_B} = \frac{r_B}{r_z}, \quad \text{hence} \quad x_C = x_B \frac{r_B}{r_z}.$$

The area A_0 under the disk contour curve is

$$A_0 = \int_S^T x_B \, dr.$$

The area under the first derived curve is

$$A_1 = \int_S^T x_C \, dr = \int_S^T \frac{x_B r_B}{r_z} \, dr = \frac{1}{r_z} \int_S^T x_B r_B \, dr.$$

Similarly, $\quad A_2 = \frac{1}{r_z^2} \int_S^T x_B r_B^2 \, dr \quad$ and $\quad A_3 = \frac{1}{r_z^3} \int_S^T x_B r_B^3 \, dr.$

Assuming the propeller equivalent disk to be symmetrical about a plane through O perpendicular to the x-axis, its weight W is

$$W = \int_S^T w \times 2x \times 2\pi r \, dr = 4\pi w \int_S^T xr \, dr = 4\pi w r_z A_1 \quad [\text{Lb.}].$$

[26]

The polar mass moment of inertia is obtained similarly

$$J_{pb} = \int_S^T r^2 \times \frac{w}{g} \times 2x \times 2\pi r\, dr$$

$$= \frac{4\pi w}{g} \int_S^T r^3 x\, dr = 4\pi \frac{w}{g} A_3 \quad \text{[Lb.in.sec.}^2\text{]},$$

where $g = 386{\cdot}4$ [in./sec.^{2}] and w = specific weight of the propeller material [Lb./in.^{3}].

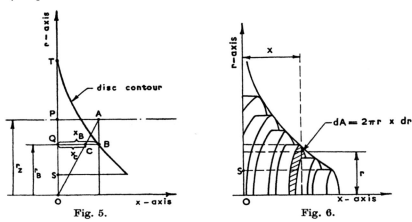

Fig. 5.   Fig. 6.

1·133   *Moment of inertia of a propeller boss*

The moment of inertia of the boss and the portion of the propeller shaft contained in it can be determined as

$$J_B \cong \frac{\pi}{32} D_B^4 L_B \frac{w}{g} \quad \text{[Lb.in.sec.}^2\text{]},$$

where D_B = boss diameter [in.], and L_B = boss length [in.], w and g being as previously defined.

1·134   *Effective moment of inertia of a marine propeller*

The moment of inertia of the propeller blades and boss is $J_P = J_{pb} + J_B$. This value is increased by the entrained water, and it is usual to add 25 % to the value of $J_{pb} + J_B$ in order to approximate the effective moment of the propeller under operating conditions. Therefore, for calculations of torsional vibration, the value to be used is

$$J_{P(\text{eff})} = (J_{pb} + J_B) \times 1{\cdot}25 \quad \text{[Lb.in.sec.}^2\text{]}.$$

1·135   *Estimation of moment of inertia of a marine propeller*

The following chart gives the moments of inertia of a number of marine propellers ranging from 18 to 210 in. diameter. The chart is based on

[27]

information received from Panel members. The measured values were determined from swinging tests using a bifilar or trifilar suspension.

This chart (Fig. 1) should be used for an approximate estimation of the moment of inertia of a propeller only when more exact data are not available. The indicated value must then be multiplied by 1·25 to include the entrained water allowance.

Note. The scatter is considerable, although not evident from the logarithmic scale. Hence the estimate can only be very approximate.

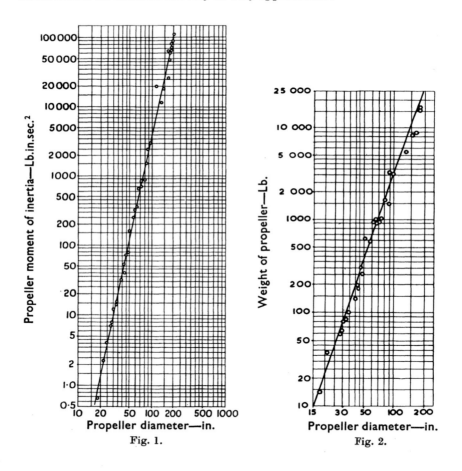

Fig. 1.

Fig. 2.

1·136 *Estimation of weight of a marine propeller*

The chart (Fig. 2) gives the measured weights of a number of marine propellers ranging in diameter from 18 to 189 in. This chart may be used to estimate the weight of a marine propeller when exact data are lacking.

[28]

1·137 Approximate formulae for assessment of the moment of inertia of a marine propeller

Accurate values for propeller inertias can only be obtained by detailed evaluation or from swinging tests. The following formulae, as well as the graph previously given, may, however, be used for an approximate assessment of the moment of inertia of a marine propeller, when more exact data are not available.

1·1371 Formula based on average values of k/D

If the weight W [Lb.] and the blade-tip diameter D [in.] of a propeller are known, the moment of inertia of the propeller may be estimated by means of the following relation:

$$J_P = \frac{W}{g} k^2 = \frac{W}{386\cdot 4} \times \left(\frac{k}{D}\right)^2 \times D^2 \quad \text{[Lb.in.sec.}^2\text{]},$$

where $k/D = 0\cdot 21$ for conventional propellers, and $k/D = 0\cdot 19$ for submarine propellers. The effective inertia with entrained water allowance is then $J_{P(\text{eff})} = 1\cdot 25 J_P$.

Metric units. Instead of 386·4 in./sec.², we take $g = 981\cdot 0$ cm./sec.²; the weight W being in Kg. and D in cm., we obtain J_P in Kg.cm.sec.².

1·1372 Whitaker's formula

For *bronze* propellers with diameters between 23 and 72 in., and blade-tip speeds of 5000–6000 ft./min., use may be made of the following formula, due to J. Whitaker:†

$$J_P = 1\cdot 04 \times 10^{-6} \times r^{\frac{3}{2}} \times D^5 \quad \text{[Lb.in.sec.}^2\text{]},$$

where D = blade-tip diameter [in.], and r = disk area ratio. The value should be multiplied by 1·25 to allow for increased inertia due to entrained water. If $A_0 = \pi D^2/4$, and A_s = propeller surface area, then $r = A_s/A_0$.

For propellers of *cast iron* or *other materials*, the formula is

$$J_P = \frac{w_{\text{C.I.}}}{w_{\text{Bronze}}} \times 1\cdot 04 \times 10^{-6} \times r^{\frac{3}{2}} \times D^5 \quad \text{[Lb.in.sec.}^2\text{]},$$

where $w_{\text{C.I.}}$ and w_{Bronze} are the specific weights of cast iron (or other material) and bronze, respectively.

Metric units. The range of blade-tip diameters is 58–183 cm., and the blade-tip speeds considered are between 1525 and 1830 m./min. Expressing D in cm., we obtain Whitaker's formula as

$$J_P = 1\cdot 13 \times 10^{-8} \times r^{\frac{3}{2}} \times D^5 \quad \text{[Kg.cm.sec.}^2\text{]}.$$

† See *The Shipbuilder and Marine Engine Builder*, March 1949, pp. 163–5.

Derivation. It is assumed that all propellers of the range considered operate under similar conditions as regards values of water pressure p_W. and root stress p_B, and, furthermore, that they have similar geometrical proportions as regards the ratios B/D and $\bar{x}/D$ (where B = blade-root breadth, D = blade-tip diameter, and $\bar{x}$ = distance from root to centre of pressure of blade).

Then, if t_m = overall mean thickness of propeller blades, and $r = A_s/A_o$ = disk area ratio ($A_o = \pi D^2/4$ and A_s = propeller surface area), we may write

$$J_P \propto D^4 t_m r,$$

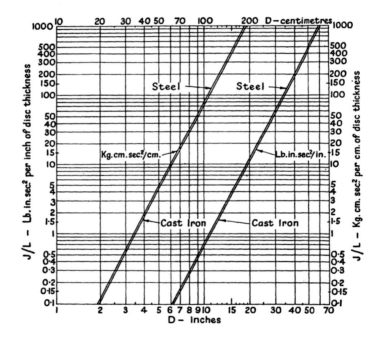

where the symbol $\propto$ means 'proportional to'. At the root section, the bending moment is

$$M_B = \bar{x} p_W A_s = p_B Z_B,$$

where Z_B = sectional modulus of the blade root. Since p_W and p_B are taken to be constant for the range of propellers considered, we have

$$\bar{x} A_s = \bar{x} r A_0 \propto D r A_0 \propto r D^3,$$

and

$$Z_B \propto t_m^2 D;$$

therefore

$$t_m^2 D \propto r D^3 \quad \text{or} \quad t_m \propto r^{\frac{1}{2}} D.$$

[30]

Substituting this in the above proportionality relation for J_P, we obtain

$$J_P = \text{const. } D^5 r^{\frac{3}{2}},$$

and the value of the constant is based on a number of known values of J_P for propellers of the class considered.

1·138 Chart for approximate determination of moments of inertia of cast-iron or steel disks

Lb.in.sec.² units:

The figure (p. 30) gives a chart of the relation

$$\frac{J}{L} = \frac{\pi}{32} \frac{w}{g} D^4 \quad \left[\frac{\text{Lb.in.sec.}^2}{\text{in.}} \right],$$

where J = polar mass moment of inertia [Lb.in.sec.²], L = disk thickness [in.], w = specific weight [Lb./in.³], $g = 386 \cdot 4$ [in./sec.²], and D = disk diameter [in.].

Kg.cm.sec.² units:

The figure (p. 30) gives a chart of the relation

$$\frac{J}{L} = \frac{\pi}{32} \frac{w}{g} D^4 \quad \left[\frac{\text{Kg.cm.sec.}^2}{\text{cm.}} \right],$$

where J = polar mass moment of inertia [Kg.cm.sec.²], L = disk thickness [cm.], w = specific weight [Kg./cm.³], $g = 981 \cdot 0$ [cm./sec.²], and D = disk diameter [cm.].

The chart is for first estimations only and more accurate values should be calculated in the usual manner.

Example of uses. A disk with $D = 15$ in., $L = 1$ in., will have a $J = 3 \cdot 4$ Lb.in.sec.² in cast iron, or $J = 3 \cdot 8$ Lb.in.sec.² in steel. J will, of course, be proportional to the thickness L.

Handy equivalents:
1 Lb.in.sec.²/in. $= 0 \cdot 45359$ Kg.cm.sec.²/cm. and 1 Lb.in.sec.² $= 1 \cdot 152$ Kg.cm.sec.².

1·14 Moments of inertia of generators

GRAPHS FOR ESTIMATING THE RADIUS OF GYRATION OF GENERATOR ARMATURES OR ROTORS

Reliable figures for the moment of inertia or Wk^2 of generator armatures are obtainable from the electrical manufacturers. These data should be based on swinging tests wherever possible.

In certain circumstances, however, for instance, in the case of second-hand machines, it may be difficult to obtain data. To assist in such cases, the following graphs are included for a rough estimation of the radius of gyration. The indications are only approximate, and, to emphasize this, it is pointed out that armatures for 110 and 440 V., otherwise generally similar, have been found to differ by 10 % in their respective Wk^2 values.

Note. As a rough average, $k \cong 0 \cdot 3D$ (k = radius of gyration, and D = outside diameter).

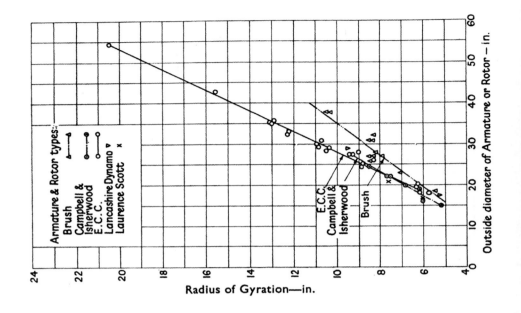

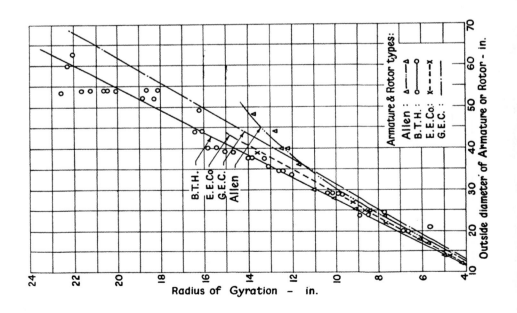

1·15   Moments of inertia of brakes

The following information refers to Heenan and Froude dynamometers (bucket-type water brakes) and the Mark VI eddy-current type 'Dynamatic' dynamometer also produced by them. For ease of reference, this section gives all the information grouped together, viz.

Brake power curves (Fig. 1),

Moments of inertia J_R of brake rotors (including allowance for entrained water), and

Shaft stiffnesses K for the shaft portion from the coupling flange to the brake rotor.

Brake data

The following figures refer to hydraulic dynamometers and the Mark VI eddy-current dynamometer produced by Heenan & Froude. They are given as indicative values for preliminary estimations.

As brake designs are available in various modifications, it is advisable in each case to calculate the stiffness of the brake shaft from dimensional drawings obtained from the brake manufacturer.

Brake type	Moment of inertia [Lb.in.sec.²] of		Shaft stiffness [Lb.in./rad.]
	Brake rotor†	Coupling flange	
DPX 2	0·10	—	0·815 × 10⁶
DPX 3	0·26	—	1·32 × 10⁶
DPX 4	0·88	—	2·45 × 10⁶
DPX 5 and DPY 5	3·50	—	10·05 × 10⁶
DPX 6 and DPY 6	11·20	—	20·13 × 10⁶
DPY 6 D	11·20	1·9	16·68 × 10⁶
DPX 645 and DPY 645	10·13	—	—
DPX 7 and DPY 7	20·0	—	33·04 × 10⁶
DPY 7 D	20·0	2·9	21·76 × 10⁶
DPX 8 D	55·18	7·25	41·5 × 10⁶
DPY R 7	47·4	—	16·5 × 10⁶
FA 5 and SA 5	138·7	—	74·9 × 10⁶
FA 6 and SA 6	329·1	—	92·7 × 10⁶
LS 4	69·2	—	18·8 × 10⁶
RSA 5	358·35	—	22·17 × 10⁶
RFA 6 and RSA 6	658·1	—	90·0 × 10⁶
Mark VI	624·0	14·2	66·7 × 10⁶

† Including allowance for entrained water.

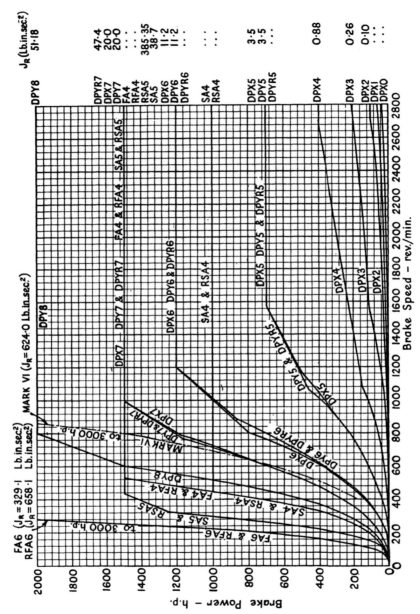

Fig. 1. Diagram showing power absorbed by Heenan and Froude dynamometers.

1·2 STIFFNESS

NOTATION

Symbol	Brief definition	Typical units	Symbol	Brief definition	Typical units
a	length	in., cm.	P	force	Lb., Kg.
a	numerical coefficient	—	P	perimeter	in., cm.
A	area	in.^{2}, cm.^{2}	q	shear stress	Lb./in.^{2}, Kg./cm.^{2}
b	length; width	in., cm.			
b	numerical coefficient	—	Q	parameter; 'Q-factor'	—
B	width	in., cm.	r	fillet radius	in., cm.
c	damping coefficient	Lb.in.sec./rad., Kg.cm.sec./rad.	r	ratio	—
			R	radius	in., cm.
C	relative overlap	—	s	displacement	in., cm.
C	constant or parameter	—	S	stroke	in., cm.
			t	temperature	°C., °F.
d	inner diameter	in., cm.	T	torque	Lb.in., Kg.cm.
D	diameter; cylinder bore	in., cm.	T	absolute temperature	°C.$_{abs.}$, °F.$_{abs.}$
e	eccentricity	in., cm.	$|T_n|$	nth order tangential-pressure component of gas torque	Lb./in.^{2}, Kg./cm.^{2}
E	modulus of elasticity	Lb./in.^{2}, Kg./cm.^{2}	U	strain energy	in.Lb., cm.Kg.
f	function of	—	w	ratio (couplings)	in.^{2}, cm.^{2}
F	frequency of vibration	vib./min.	W	weight	Lb., Kg.
g	acceleration due to gravity	in./sec.^{2}, cm./sec.^{2}	x	length	in., cm.
			X	parameter	—
G	modulus of rigidity	Lb./in.^{2}, Kg./cm.^{2}	y	distance; radius of curvature	in., cm.
h	height	in., cm.	y	parameter	in.^{4}, cm.^{4}
H	(total) height	in., cm.	Y	parameter (crankshafts)	in., cm.
I	second moment of area	in.^{4}, cm.^{4}	Y	diameter ratio	—
J	mass moment of inertia	Lb.in.sec.^{2}, Kg.cm.sec.^{2}	z	parameter	—
			Z	sectional modulus	in.^{3}, cm.^{3}
k	ratio; parameter	—	α	angle	rad., deg.
K	stiffness	Lb.in./rad., Kg.cm./rad.	β	belt coefficient	—
			γ	shear angle	rad., deg.
L	length; thickness	in., cm.	δ	deflexion, clearance	in., cm.
ΔL	increment of length	in., cm.	δ	loss angle	rad., deg.
m	diameter ratio	—	Δ	Holzer-table amplitude	rad.
M	bending moment	Lb.in., Kg.cm.			
n	order number of critical speed	—	θ	angle of twist	rad., deg.
N	engine speed	rev./min.	λ	parameter, ratio	—
N_s	number of springs, spokes, bushes, etc.	—	μ	friction coefficient	—
			ξ	diameter (at position x)	in., cm.
p	tensile stress	Lb./in.^{2}, Kg./cm.^{2}	σ	Poisson's ratio	—
			ϕ	angle	rad., deg.
			ψ	angle	rad., deg.
			ω	phase velocity	rad./sec.

Subscripts are used in the text to define more precisely, e.g. D_{ci} = crankpin inner diameter. Their meaning is fully explained in their context in each instance.

1·21   Stiffness of plain, tapered and stepped shafts

As an introduction to this subject, it will be useful to review the basic relations between shaft twist, stiffness and equivalent length.

Shaft twist and stiffness. Dealing first with the simplest case, consider a straight shaft of circular cross-section, rigidly clamped at one end and subject to torque at the other (Fig. 1).

Let L = shaft length [in.],

D = shaft diameter [in.],

G = modulus of rigidity of shaft material [Lb./in.²],

T = applied torque [Lb.in.], and

θ = shaft twist about its centre line [rad.].

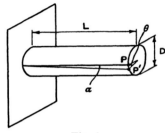

Fig. 1

Under the influence of the applied torque T, the point P in the accompanying figure rotates to a new position P', such that

$$\overline{PP'} = L\alpha = \frac{D}{2}\theta.$$

When the twist is equal to θ, the applied torque T is resisted by an equal and opposite shaft torque T_{sh}, which is given by

$$T_{sh} = \frac{GI_p}{L}\theta = Zq,$$

where $I_p = \frac{\pi}{32}D^4$ [in.⁴], $Z = \frac{\pi}{16}D^3$ [in.³], and q = torsional stress [Lb./in.²] on the shaft periphery at the clamped end. It is usual to define the expression

$$K = \frac{GI_p}{L} = \frac{T}{\theta} \quad \text{[Lb.in./rad.]}$$

as the 'torsional stiffness' of the shaft, so that the shaft twist at its 'free end' (the position at which the torque is applied) is

$$\theta = \frac{T}{K} = \frac{Zq}{K} \quad \text{[rad.]}.$$

Equivalent length. For a shaft of more complicated shape, for instance, a 'stepped shaft' with a change in diameter, it is also possible to determine the stiffness $K = T/\theta$. This can again be achieved experimentally, by clamping the shaft at one end, subjecting the other end to an applied torque T, and measuring the overall angle of twist θ (see Fig. 2).

Then the stiffness is again $K = T/\theta$.

[36]

Since there are two diameters to be considered, one, say D_2, may be taken as the 'reference diameter' D_e of an equivalent straight shaft, which has a second moment of area $I_p = \frac{\pi}{32} D_e^4$.

To be equivalent to the stepped shaft, the plain shaft must have the same value for stiffness, that is, its equivalent length must be such that

$$\frac{GI_p}{L_e} = K = \frac{T}{\theta} \quad \text{or} \quad L_e = \frac{GI_p}{K}.$$

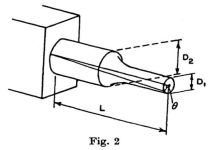

It is thus possible to take account of any shaft portion in calculations, either by means of its stiffness K or by indicating its equivalent length L_e with reference to a fixed value chosen as the equivalent diameter D_e.

In practice, it is generally found convenient to use equivalent lengths

Fig. 2

(which are additive, whereas shaft stiffnesses are not, see pp. 72–73) for evaluating complicated shaft portions, such as crankthrows, and the stiffness of the composite shaft between two masses is determined from the sum of the contributions of the individual lengths, i.e. from

$$K = GI_p \times \frac{1}{L_{e1} + L_{e2} + L_{e3} + \dots}.$$

It should be noted, however, that the torsional stress in the equivalent shaft is not the same as in the actual shaft at any point. In all cases, the true stress must be evaluated from the actual diameters and shapes considered.

Metric units. The above relations are, of course, independent of the system of units used, and the Lb.in. units given are indicated only as typical examples. In the metric system, the quantities defined in this section can be expressed in the following units:

L and D	cm.	G	Kg./cm.²
T	cm.Kg.	Z	cm.³
K	Kg.cm./rad.	I_p	cm.⁴
q	Kg./cm.²		

1·211 Equivalent length of straight cylindrical shafts

1·2111 Solid shaft

Let D, L, G = diameter, length and modulus of rigidity of shaft portion considered, and D_e, L_e, G_e = diameter, length and modulus of rigidity of equivalent shaft. [All linear dimensions in inches, G in Lb./in.²].

As the equivalent shaft must have the same torsional stiffness as the actual shaft

$$K_e = \frac{\pi}{32} G_e \frac{D_e^4}{L_e} = \frac{\pi}{32} G \frac{D^4}{L} \quad \text{[Lb.in./rad.]},$$

hence the equivalent length is

$$L_e = L \frac{G_e}{G} \left(\frac{D_e}{D}\right)^4 \quad [\text{in.}].$$

1·2112 Hollow shaft

Let D_o = outer diameter, and D_i = bore diameter of shaft; then, from stiffness considerations,

$$K_e = \frac{\pi}{32} G_e \frac{D_e^4}{L_e} = \frac{\pi}{32} G \frac{D_o^4 - D_i^4}{L} = K,$$

hence

$$L_e = L \frac{G_e}{G} \frac{D_e^4}{D_o^4 - D_i^4} \quad [\text{in.}].$$

Note. In this section, the shaft stiffnesses are expressed in Lb.in./rad. for convenience. The stiffnesses can readily be expressed in Lb.ft./rad., or in Ton.ft./rad., by multiplying by $\frac{1}{12}$ or $\frac{1}{12 \times 2240}$, respectively.

Engineers using the *metric system* express shaft stiffness in cm.Kg./rad. As G is then in Kg./cm.² and D and L are in cm., the formulae in this section can be used directly to obtain stiffness values in cm.Kg./rad. without requiring further numerical factors for this purpose.

1·2113 Straight shaft with eccentric bore

Let e = eccentricity, D_i = bore diameter, and D_o = outer diameter of shaft.

The equivalent length of a shaft with an eccentric bore is obtainable from

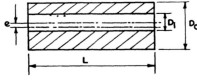

$$L_e = L \frac{G_e}{G} \frac{D_e^4}{(D_o^4 - D_i^4)} \times Q,$$

where, according to an investigation by T. S. Wilson,†

$$Q = 1 + \frac{16m^2\lambda^2}{(1-m^2)(1-m^4)} + \frac{384m^4\lambda^4}{(1-m^2)^2(1-m^4)^2} + \cdots,$$

with

$$m = D_i/D_o \quad \text{and} \quad \lambda = e/D_o.$$

Curves of Q as a function of e/D and d/D, calculated by Wilson, are given in the following graph, in which $D = D_o$ and $d = D_i$.

Note. The curves are for values of e/D between zero and two-fifths of the geometrical limit $(e/D)_{\max} = \frac{1}{2}(1-m)$.

† T. S. Wilson, 'The eccentric circular tube', *Aircr. Engng*, March 1952, pp. 76–9.

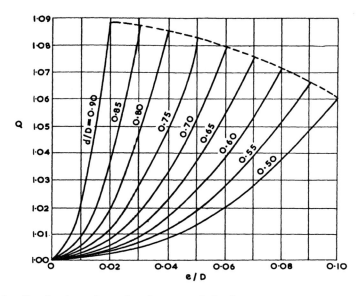

1·212 Equivalent length of tapered shafts

1·2121 Solid shaft with linear taper

The equivalent length of the solid tapered shaft shown in the accompanying figure is

$$L_e = L\frac{G_e}{G} \times \frac{D_e^4}{3(D_2-D_1)}\left[\frac{1}{D_1^3} - \frac{1}{D_2^3}\right],$$

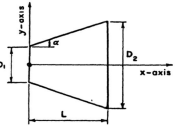

D_1 and D_2 being the diameters of the small and the large end of the shaft.

Derivation. This expression is obtained by noting that at any point x, along the shaft axis, the shaft diameter is

$$D(x) = D_1 + (D_2 - D_1)\frac{x}{L}.$$

The second moment of area of the shaft cross-section at x, about the shaft axis, is

$$I_p(x) = \frac{\pi}{32}D^4(x),$$

and therefore with

$$I_e = \frac{\pi}{32}D_e^4$$

$$\frac{1}{K_e} = \frac{L_e}{G_e I_e} = \frac{1}{G}\int_0^L \frac{dx}{I_p(x)} = \frac{1}{G}\int_0^L \frac{dx}{\frac{\pi}{32}\left[D_1 + (D_2 - D_1)\frac{x}{L}\right]^4}.$$

The evaluation of this integral gives the above expression.

Note. When the taper is less than a few degrees, or when the length of the tapered shaft does not greatly exceed the value of the smaller diameter D_1, it is sometimes considered sufficient to approximate the tapered shaft with a straight shaft having a diameter $D_m = \frac{1}{2}(D_1 + D_2)$.

In this case, the equivalent length is obtained as

$$L_e = L \frac{G_e}{G} \frac{D_e^4}{\left(\dfrac{D_1 + D_2}{2}\right)^4} = L \frac{G_e}{G} \frac{D_e^4}{D_m^4}.$$

1·2122 *Tapered shaft with tapered bore*

Let D_{o1}, D_{i1} = outer and inner diameters of small end, and D_{o2}, D_{i2} = outer and inner diameters of large end of the tapered shaft. The equivalent shaft length can be estimated from

$$L_e \simeq L \frac{G_e}{G} \frac{D_e^4}{3(A - B)},$$

where

$$A = D_{o1}^3 D_{o2}^3 / [D_{o1}^2 + D_{o1} D_{o2} + D_{o2}^2]$$

and

$$B = D_{i1}^3 D_{i2}^3 / [D_{i1}^2 + D_{i1} D_{i2} + D_{i2}^2].$$

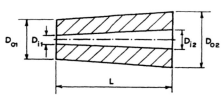

The above approximation is obtained by subtracting the stiffness of the bored shaft portion from that of the solid tapered shaft and expressing the result as an equivalent length.

1·2123 *Tapered shaft with straight bore*

Let D_i = bore diameter. The equivalent length of the shaft is obtained from

$$L_e = L \frac{G_e}{G} \times \frac{D_e^4}{4 D_i^3 (D_{o2} - D_{o1})} \left[\log_e Y - 2 \tan^{-1} \frac{D_{o2}}{D_i} + 2 \tan^{-1} \frac{D_{o1}}{D_i} \right],$$

where

$$\log_e Y = \log_e \frac{(D_{o2} - D_i)(D_{o1} + D_i)}{(D_{o2} + D_i)(D_{o1} - D_i)}.$$

(*Note.* $\log_e Y = 2 \cdot 303 \log_{10} Y$.)

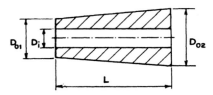 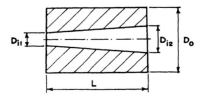

1·2124 *Cylindrical shaft with tapered bore*

Let D_o = outer diameter of shaft. The equivalent length is given by

$$L_e = L \frac{G_e}{G} \times \frac{D_e^4}{4 D_o^3 (D_{i2} - D_{i1})} \left[\log_e Y - 2 \tan^{-1} \frac{D_{i1}}{D_o} + 2 \tan^{-1} \frac{D_{i2}}{D_o} \right],$$

[40]

where
$$\log_e Y = \log_e \frac{(D_o - D_{i1})(D_o + D_{i2})}{(D_o + D_{i1})(D_o - D_{i2})}.$$

Derivation. To illustrate the method of obtaining the equivalent length of shafts of variable cross-section, details of the derivation are given below for the case of the tapered shaft with straight bore.

At any position x, the flexibility of a shaft element of length dx is given by
$$\frac{d\theta}{T} = \frac{dx}{G\frac{\pi}{32}(D^4 - D_i^4)},$$

where (see accompanying figure) $D = D_{o1} + mx$ and $m = (D_{o2} - D_{o1})/L$. Substituting $\xi = D_{o1} + mx$ and integrating partial fractions, we have
$$\int_0^\theta \frac{d\theta}{T} = \frac{1}{K} = \frac{32/\pi}{2D_i^3 Gm}\left[\frac{1}{2D_i}\log_e\left|\frac{\xi - D_i}{\xi + D_i}\right| - \frac{1}{D_i}\tan^{-1}\frac{\xi}{D_i}\right]_{D_{o1}}^{D_{o2}}.$$
Evaluating this for the limits D_{o1} and D_{o2} and using the relation
$$L_e = \frac{\pi}{32}G_e D_e^4 \times \frac{1}{K},$$

we obtain the above expression.

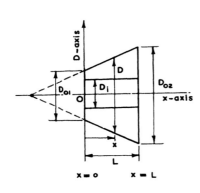

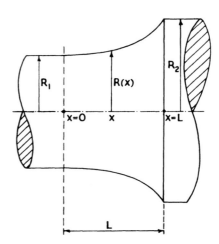

1·213   *Equivalent length with a gradual change in cross-section*

To determine the equivalent length of a shaft portion with a gradual change in cross-section, such as that shown in the accompanying figure, it is convenient to use a semi-graphical method. The procedure is as follows:

(1) Calculate the value of $R^4(x)$ for a number of positions x between $x = 0$ and $x = L$.

(2) Determine the reciprocal values $1/R^4(x) = y(x)$ for these positions.

(3) Plot these values $1/R^4(x)$ against x on graph paper and join the points together, so as to obtain a continuous curve $y(x) = 1/R^4(x)$.

(4) Determine the area under the curve (with a planimeter, or by counting squares). This area is

$$Y = \int_0^L dx\, y(x) = \int_0^L \frac{dx}{R^4(x)} \quad [\text{in.}^{-3}].$$

(5) The equivalent length L_e of this tapered portion, in terms of a reference shaft of equivalent diameter D_e and modulus of rigidity G_e, is then obtained from

$$L_e = \frac{D_e^4\, G_e}{16\, G} Y \quad [\text{in.}].$$

Derivation. The relation for stiffness is

$$\frac{1}{K} = \frac{1}{T}\int_0^{\theta_L} d\theta = \frac{1}{G}\int_0^L \frac{dx}{\frac{\pi}{2}R^4(x)} = \frac{2}{\pi}\frac{Y}{G},$$

and therefore

$$\frac{L_e}{G_e \frac{\pi}{32} D_e^4} = \frac{2}{\pi}\frac{Y}{G},$$

which gives the above expression for L_e.

Note. This method only applies to shafts with 'gradual' changes in cross-section. For 'abrupt' transitions, i.e. for fillets having a radius of curvature r smaller than $0 \cdot 5 R_1$ (where R_1 is the radius of the straight shaft portion at the small end), the evaluation of L_e should be made as explained in section 1·215.

EXAMPLE OF CALCULATION FOR SHAFT WITH GRADUAL CHANGE IN CROSS-SECTION. Consider a shaft with $R_1 = 1$ in., $R_2 = 2$ in., and a circular fillet of radius $r = 1$ in. The tabulation is as follows:

Shaft position ...	0	1	2	3	4	5	6	7	8
$R(x) =$	1·0	1·02	1·05	1·075	1·15	1·25	1·36	1·55	2·0
$R^4 =$	1·0	1·08	1·21	1·35	1·75	2·44	3·40	5·75	16·0
$y = 1/R^4 =$	1·0	0·93	0·826	0·74	0·572	0·41	0·294	0·174	0·063

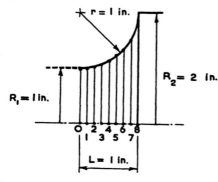
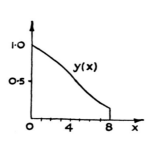

[42]

We then plot the values $1/R^4 = y(x)$ against x and join the points to obtain a continuous curve: The area under this curve is $Y = 0.55\,\text{in.}^2$ on the graph, or $Y = 0.55\,\text{in.}^{-3}$, with the units used for this calculation. Therefore, since $L = 1\,\text{in.}$, and assuming $G_e = G$ and $D_e = D_1$, the equivalent length of the transition portion is

$$L_e = \frac{D_e^4}{16} Y = \frac{2^4}{16} \times 0.55 = 0.55\,\text{in.}$$

1·214 Equivalent length of keyed and splined shafts

1·2141 Shaft with keyway   1·2142 Splined shaft

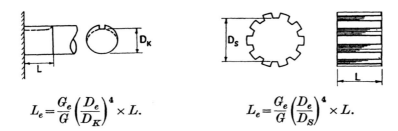

$$L_e = \frac{G_e}{G}\left(\frac{D_e}{D_K}\right)^4 \times L. \qquad\qquad L_e = \frac{G_e}{G}\left(\frac{D_e}{D_S}\right)^4 \times L.$$

Note. These are empirical formulae.

For the equivalent length of *keyed shafts, flanged couplings*, etc., see section 1·24 (Couplings).

1·2143 Shaft with flat side

$$L_e = C \times \left(\frac{D_e}{D_o}\right)^4 \times \frac{G_e}{G} \times L,$$

where the value of C, which depends on the angle α (see figure), is taken from the graph given below.

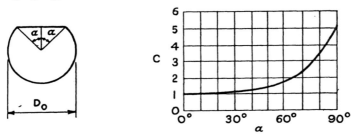

Reference: A. Weigand, 'The problem of torsion in prismatic members of circular segmental cross-section', *Tech. Memor. Nat. Adv. Comm. Aero., Wash.*, no. 1182 (1948).

1·215 Equivalent length of stepped shafts with abrupt change in cross-section

1·2151 Stepped shafts of circular cross-section

The following refers to stepped shafts of circular cross-section, with circular fillets having a radius r not greater than $0·5R_1$ (where R_1 is the shaft radius of the smaller, cylindrical shaft portion).

The equivalent length L_e of a stepped cylindrical shaft, such as that shown in the accompanying Fig. 1, is determined by the following method:†

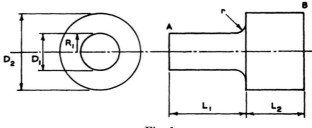

Fig. 1

(1) Calculate the ratios r/R_1 and D_2/D_1, where r = fillet radius [in.], $D_1 = 2R_1$ = diameter [in.] of the smaller cylindrical shaft portion, and D_2 = diameter of the larger cylindrical shaft portion.

(2) Refer to Fig. 2 and read the value of $\Delta L/D_1$ for the calculated ratio D_2/D_1 from the curve with the required fillet/shaft radius ratio r/R_1 (interpolating between the curves if necessary).

(3) The equivalent length L_e of the shafting from point A to point B, with reference to a cylindrical equivalent shaft of diameter D_e, is then

$$L_e = \left\{ \left[L_1 + \left(\frac{\Delta L}{D_1} \right) D_1 \right] \frac{D_e^4}{D_1^4} + L_2 \frac{D_e^4}{D_2^4} \right\} \frac{G_e}{G} \quad \text{[in.]}.$$

Note. The flexibility of the junction is thus expressed separately as an additional term in the equivalent length, viz.

$$\Delta L_e = \left\{ \left(\frac{\Delta L}{D_1} \right) \times D_1 \times \left(\frac{D_e}{D_1} \right)^4 \right\} \frac{G_e}{G} \quad \text{[in.]}.$$

For further clarity, the procedure is summarized as follows:

The value L_{e1} of the small-diameter shaft portion is calculated taking its full length up to the beginning of the larger diameter shaft. Then L_{e2} for the large-diameter shaft portion is calculated. Lastly, the value ΔL_e, an equivalent length representing the flexibility of the transition, assumed concentrated at the end of the small-shaft portion, is determined.

† All the torsion tests serving to determine the basic data used in this B.I.C.E.R.A. method were carried out at the B.I.C.E.R.A. Laboratory.

The total equivalent length is then

$$L_e = L_{e1} + L_{e2} + \Delta L_e.$$

Further considerations on the use of such 'equivalent lengths of junctions', and details of the experimental work done in this connexion, are given at the end of this section.

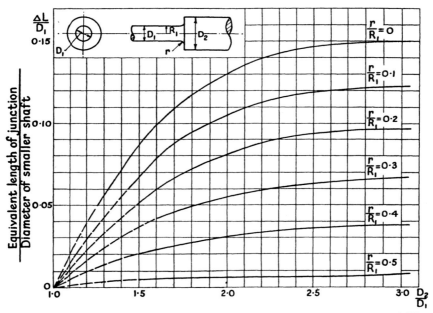

Fig. 2. Junction effect for cylindrical shafts with various diameter ratios and fillet radii.

NUMERICAL EXAMPLES

(1) Stepped shaft, with the following dimensions:

$$r = 0 \text{ in.}, \quad D_1 = 2 \text{ in.}, \quad L_1 = 10 \text{ in.}, \quad D_2 = 4 \text{ in.}, \quad L_2 = 10 \text{ in.}$$

Taking $D_e = D_1$, and using the curves of Fig. 2:

$$D_2/D_1 = 2\cdot0, \quad r/R_1 = 0, \quad \Delta L/D_1 = 0\cdot131.$$

$$\Delta L = 0\cdot131 \times D_1 = 0\cdot262 \text{ in.} = \Delta L_e.$$

$$L_{e1} = L_1 = 10\cdot0 \text{ in.}$$

$$L_{e2} = L_2 \times (D_1/D_2)^4 = 10 \times (\tfrac{1}{16}) = 0\cdot625 \text{ in.}$$

$$L_{e \text{ total}} = 10\cdot0 + 0\cdot262 + 0\cdot625 = 10\cdot887 \text{ in.}$$

(2) Shaft with coupling flange:

$$r = 0\cdot6 \text{ in.}, \quad D_1 = 6 \text{ in.}, \quad L_1 = 10 \text{ in.}, \quad D_2 = 12 \text{ in.}, \quad L_2 = 0\cdot75 \text{ in.}$$

Diameter of equivalent shaft: $D_e = 5\cdot5$ in.

It is convenient to use a tabulation, as follows:

Shaft portion	D	D_2/D_1	r	r/R_1	$\Delta L/D_1$	ΔL	$(D_e/D)^4$	L	L_e
1	6·0	—	—	—	—	—	0·705	10·00	7·050
Junction	—	2·0	0·6	0·2	0·081	0·486	0·705	0·486	0·343
2	12·0	—	—	—	—	—	0·044	0·75	0·033

$$L_{e\,\text{total}} = 7·426 \text{ in.}$$

Metric units. The curves, formulae and examples relating to junction effects can be used directly for evaluations in metric units. The curves and formulae are expressed in terms of non-dimensional ratios, and in the above examples the inch units can be replaced throughout by cm. without altering the numerical values.

1·2152 Stepped shafts with circular changing to square cross-section

The equivalent length ΔL_e of junctions between two concentric shaft portions, one having a circular and the other a square cross-section, is determined by means of the curves given in Fig. 4.†

The method of calculation is the same as for stepped shafts of circular cross-section, apart from the fact that the square-section shaft portion has to be evaluated as a length of an equivalent cylindrical shaft having the same torsional stiffness.

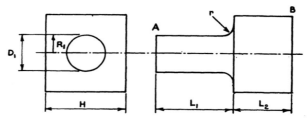

Fig. 3. Notation: r = fillet radius; H = side of square section; $D_1 = 2R_1$ = diameter of circular portion; D_e = diameter of equivalent shaft. [All linear dimensions in inches.]

The procedure for determining the equivalent length of a shaft such as that shown in Fig. 3 is as follows:

(1) After determining the values of r/R_1 and H/D_1, read the value of $\Delta L/D_1$ from Fig. 4 (interpolating between the curves if necessary) and calculate the equivalent length due to the flexibility of the junction:

$$\Delta L_e = \left\{ \left(\frac{\Delta L}{D_1} \right) \times D_1 \times \left(\frac{D_e}{D_1} \right)^4 \right\} \frac{G_e}{G} \quad \text{[in.]}.$$

(2) Calculate the equivalent length of the circular-section shaft portion for the entire length of L_1:

$$L_{e1} = L_1 \times \frac{G_e}{G} \times \left(\frac{D_e}{D_1} \right)^4 \quad \text{[in.]}.$$

† These curves are based on torsion tests carried out at the B.I.C.E.R.A. Laboratory.

(3) Calculate the equivalent length of the square-section shaft portion by means of the relation

$$L_{e2} = \frac{L_2}{1\cdot432} \times \frac{G_e}{G} \times \left(\frac{D_e}{H}\right)^4 \quad \text{[in.]}.$$

(4) The total equivalent L_e of the shafting from A to B is then

$$L_e = L_{e1} + \Delta L_e + L_{e2} \quad \text{[in.]}.$$

Note. The factor $1\cdot432$ represents the ratio of the stiffness of the square-sectioned shaft of side H to that of a circular shaft of diameter $D_2 = H$.

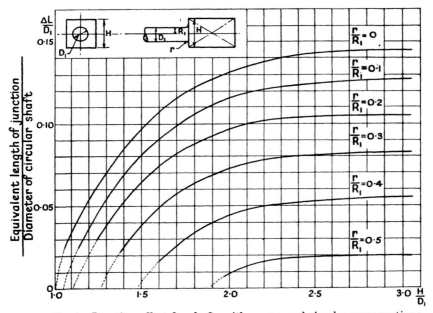

Fig. 4. Junction effect for shafts with square and circular cross-sections.

The theoretical value obtained by St Venant's theory of torsion is $1\cdot432 = \dfrac{32/\pi}{7\cdot11}$. This value has been verified experimentally at the B.I.C.E.R.A. Laboratory.

NUMERICAL EXAMPLES

(1) Stepped shaft, with a square-sectioned portion of side $H = 2\cdot5$ in., and a circular-sectioned portion of diameter $D_1 = 2\cdot0$ in. Other dimensions: fillet radius $r = 0$, $L_1 = 3$ in., $L_2 = 4$ in.

Taking $D_e = D_1$, and using the curves of Fig. 4,

$$H/D_1 = 1\cdot25, \quad r/R_1 = 0, \quad \Delta L/D_1 = 0\cdot063.$$

$$\Delta L = 0\cdot063 \times D_1 = 0\cdot126 \text{ in.} = \Delta L_e.$$

$$L_{e1} = L_1 = 3\cdot0 \text{ in.}$$

$$L_{e2} = L_2 \times \frac{1}{1\cdot432} \times \left(\frac{D_e}{H}\right)^4 = 4\cdot \times 0286 = 1\cdot144 \text{ in.}$$

$$L_{e \text{ total}} = 3\cdot00 + 0\cdot126 + 1\cdot144 = 4\cdot270 \text{ in.}$$

(2) Stepped shaft, with the following dimensions:

$$D_1 = 2\cdot0 \text{ in. (diameter of circular section),} \quad L_1 = 3\cdot0 \text{ in.}$$

$$H = 5\cdot0 \text{ in. (side of square section),} \quad L_2 = 4\cdot0 \text{ in.}$$

Fillet radius: $r = 0\cdot2$ in. Diameter of equivalent shaft: $D_e = 3\cdot0$ in. Using a tabulation, the results are as follows:

Shaft portion	D	H	H/D_1	r/R_1	$\Delta L/D_1$
1	2·0	—	—	—	—
Junction	—	—	2·5	0·2	0·1045
2	—	5·0	—	—	—

Shaft portion	ΔL	$(D_e/D)^4$	$(D_e/H)^4/1\cdot432$	L	L_e
1	—	5·063	—	3·00	15·189
Junction	0·209	5·063	—	0·209	1·059
2	—	—	0·0905	4·00	0·362

$$L_{e \text{ total}} = 16\cdot610 \text{ in.}$$

1·216 *Experimental determination of equivalent length of junctions*

The basic data used for the determination of the curves given in Figs. 2 and 4 of pp. 45 and 47 were obtained at the B.I.C.E.R.A. Laboratory by means of the test rigs and equipment described below.

Torsion machine

A photograph of the torsion machine is given in Fig. 5. The test shaft, flanged and spigoted at both ends, was bolted at one end to a bracket, and at the other end to a coupling and an extension shaft supported in pressure-lubricated plain bearings. This extension shaft was keyed on to the torque arm. To accommodate any axial movement possibly occurring under load conditions, the coupling was arranged as a sliding fit. For static tests the load was applied by means of screw jacks, the upward load being exerted by a screw jack mounted on the cast-iron bed plate, while the downward load was applied by a jack on an arch-shaped bracket over the other end of the torque lever arm. Load indication was by means of loop dynamometers placed between the jacks and the lever arm.

For dynamic tests, the load acting on the torque arm was generated by a mechanical vibrator with out-of-balance weights, driven by an electric motor through a driving shaft and flexible couplings.

Fig. 5. Torsion machine and optical measuring apparatus.

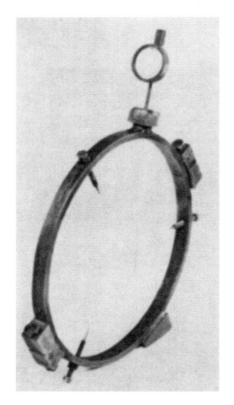

Type A for static tests.

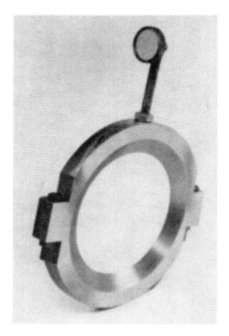

Type B for dynamic tests.

Fig. 7. Mirror supporting rings.

Test shafts. The shafts were machined from cylindrical mild steel bars to which flanges were welded at each end. A number of fillet radii were tested for each diameter ratio. Fine circumferential grooves, marked off on a lathe with the aid of slip gauges, were machined on the shaft at the various measuring positions, for the location of the mirror supporting rings.

Optical measuring equipment. Twist under a given torque was measured optically by means of autocollimators. The instruments, illustrated in Figs. 5 and 6, were designed and developed at the B.I.C.E.R.A. Laboratory to meet the special requirements of this investigation of shaft stiffness.

As shown in Fig. 6, the cross-line graticule *a* is illuminated by a beam of light coming from a bulb *b* and passing through condenser lenses c_1 and c_2, and a green filter *d*. The divergent rays from the graticule become

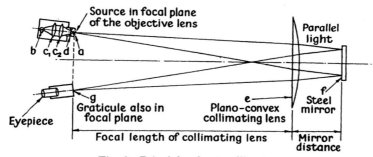

Fig. 6. Principle of autocollimator.

parallel after passing through the plano-convex objective lens *e*, since the graticule is situated in the focal plane of this lens. The rays are then reflected by the optically flat stainless steel mirror *f*. The reflected beam is refocused by means of the same lens *e* on to the graticule scale *g* of an eyepiece, the image then being magnified by the lenses in the eyepiece.

The advantage of the autocollimator is that the mirror need not be situated at a determined distance from the objective lens, since it receives and reflects parallel light. Thus, focusing is only necessary between the objective lens and the light unit. The image visible on the eyepiece graticule is reversed left to right and inverted, and is a full-size image of the source. For measurements with statically applied loads, a cross-line graticule was employed in the light unit, whereas for dynamic measurements with alternating loads a pin-hole light source was used.

The light source, graticule and eyepiece of each autocollimator were designed as a single unit. Eight such units were mounted on a common horizontal bar, at about 8 ft. (2·44 m.) from the test rig, as shown in Fig. 5. The lenses are carried on a rail secured to the bed plate of the torsion rig and are fully adjustable. The mirrors were adjustably mounted on supporting rings (see Fig. 7), ring *A* being used for static tests and ring B for dynamic tests. These supporting rings were positioned at the measuring points on the shaft, in planes at right angles to the shaft axis. The sensitivity of the instruments is such that angular displacements of a few seconds can be measured.

Test procedure. The test shafts were subjected to a considerable number of static tests for each value of the diameter ratio, or H/D ratio (H being the side of the square section), and fillet radius. Dynamic tests carried out as a check, at a frequency of 1500 load cycles per minute, showed close agreement with the results obtained with static loads. More than 600 values for equivalent lengths of junctions were determined as a basis for the curves of Figs. 2 and 4.

In the first method used, graticule readings at the various test positions were taken, first at no-load, then after applying a known load, and finally at zero load again. To obtain comparable readings, all the autocollimator scales were calibrated against the scale of autocollimator no. 1. This was done by measuring displacements when rotating the shaft through small angles, the shaft being in an unclamped condition.

The results were evaluated graphically, by plotting the angular deflexions $\alpha = s/(2L_f)$ (where s = screen reading and L_f = focal length) against distance along the shaft axis. The plot gave two straight lines, each with a slope depending on the shaft diameter. The distance between the point of intersection of these two lines and the junction of the two shaft portions gave the *interpenetration length x*. Although the method was not dependent on the applied load, the point of intersection was difficult to determine accurately, and the results showed considerable scatter.

Various other methods were tried which made use of the applied torque. These showed some improvement in the reduction of scatter, but were not considered satisfactory in view of the uncertain value of the applied torque owing to the frictional losses in the bearing.

Finally, the following method was derived, which was applied to all subsequent measurements on straight shafts and crankshafts:

The twist θ_0 of a straight portion L_0 of the small-diameter shaft was used as a measure of the value T/G. This avoided inaccuracies due to measurement of the torque T or assumed values for the modulus of rigidity G, and considerably reduced the scatter in the results.

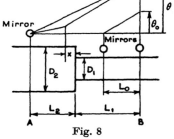

Fig. 8

Referring to Fig. 8, and using the notation θ = overall angle of twist [rad.] from point A to point B, D_e = diameter of equivalent shaft [in.], and L_e = length of equivalent shaft [in.], we may write

$$T = \theta \frac{\pi}{32} \frac{D_e^4}{L_e} G = \theta_0 \frac{\pi}{32} \frac{D_1^4}{L_0} G,$$

hence

$$\frac{L_e}{D_e^4} = \frac{L_0}{D_1^4} \frac{\theta}{\theta_0}.$$

As

$$\frac{G}{T} \times \frac{\pi}{32} \times \theta = \frac{L_e}{D_e^4} = \frac{L_1}{D_1^4} + \frac{L_2}{D_2^4} + \frac{\Delta L}{D_1^4},$$

where $\Delta L = $ *increment of length* due to junction flexibility, we have

$$\Delta L = L_0 \frac{\theta}{\theta_0} - L_1 - L_2 \left(\frac{D_2}{D_1}\right)^4 \quad [\text{in.}].$$

It was found convenient to use non-dimensional quantities, and the tests showed that it was justifiable to plot the 'relative equivalent length' of the junction, $\Delta L/D_1$, against the diameter ratio D_2/D_1 or the ratio H/D_1, in order to obtain curves for constant values of the fillet radius/ shaft radius ratio r/R_1.

Note on 'Interpenetration'. The 'interpenetration' x of the two shaft portions of different diameter, which was determined by the first method used in these tests, is useful for a visualization of the fact that there is partially 'idle' material at the beginning of the larger diameter shaft portion. It leads, however, to a further complication of calculations of equivalent length, and loses its visual advantages when attempts are made to apply it to more complicated shapes, such as crankshafts with chamfered webs. The increment in flexibility is, on the other hand, an equally proven fact which can be far more readily expressed as an 'equivalent length of the junction' for any given shape, applying the B.I.C.E.R.A. method outlined above.

The relation between an *interpenetration* x and the *corresponding equivalent length of the junction* ΔL is given by

$$x = \Delta L \frac{D_2^4}{D_2^4 - D_1^4},$$

using the notation of Fig. 8. This relation is derived as follows: Referring again to Fig. 8, it is seen that with an interpenetration x the smaller diameter shaft is apparently increased in length to a value $L_1 + x$, while the larger diameter shaft is shortened to a value $L_2 - x$. Therefore, in terms of an equivalent shaft of diameter D_e, we have

$$\frac{L_e}{D_e^4} = \frac{L_1 + x}{D_1^4} + \frac{L_2 - x}{D_2^4}$$

or

$$\frac{L_e}{D_e^4} = x\left[\frac{1}{D_1^4} - \frac{1}{D_2^4}\right] + \frac{L_1}{D_1^4} + \frac{L_2}{D_2^4}. \tag{1}$$

With the equivalent length of the junction, ΔL, the total equivalent length of the shafting is obtained as

$$\frac{L_e}{D_e^4} = \frac{L_1}{D_1^4} + \frac{L_2}{D_2^4} + \frac{\Delta L}{D_1^4}. \tag{2}$$

Subtracting eq. (1) from eq. (2) we obtain

$$\frac{\Delta L}{D_1^4} - x \left[\frac{1}{D_1^4} - \frac{1}{D_2^4} \right] = 0,$$

hence
$$\Delta L = x[1 - (D_1/D_2)^4]$$

and this gives the above relation.

1·217 *Remarks on the equivalent length of shafts with distributed masses*

The method generally employed is to concentrate the distributed mass at one point on the shaft axis. If the mass is extremely rigid, as, for instance, is the case for a solid flywheel, it is assumed to have only a moment of inertia, and no equivalent length (i.e. it is regarded as having 'infinite' stiffness). The shaft is then calculated up to the face (and not to the centre) of this mass. This also applies to the shaft on the other side of the mass, if it is a part of the system to be evaluated. If the mass is not extremely rigid, its stiffening effect on the shaft carrying it is taken into account by assuming that the shaft ends at a 'point of rigidity' located at a position within the mass on the shaft axis.

The following is a brief review of various cases occurring in practice. Detailed treatment will be found in the sections referred to in each instance.

(1) *Crankshafts.* The length of the crankthrow is evaluated from crankpin centre to crankpin centre (in most cases this is the same length as from journal centre to journal centre). The moment of inertia $J_{cyl.}$ is assumed to be concentrated at the crankpin centre (see section 1·1) and the crankthrow evaluation is by the methods of section 1·22.

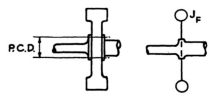

Actual system.   Simplified system used
for evaluation.

(2) *Flywheels.* The length of the shafts on either side of the flywheel is evaluated up to the corresponding face of the flywheel, flanged ends and bosses being included in the evaluation.

The flywheel is represented solely as a moment of inertia J_F (with zero equivalent length) situated between the flanged portions.

The flanges are treated as shaft lengths of a diameter equal to the pitch-circle diameter (P.C.D.) of the coupling bolts. For corrections for

[52]

fillet radii, see 'Stepped shafts' (section 1·21). For flywheel bosses and keyways, etc., see 'Stiffness of shafts of driven machines' (section 1·25).

(3) *Coupling disks and pinion disks.* (a) If the torque is transmitted through the disk by forces acting at the same radial distance, the disk is evaluated in the same way as a shaft of diameter D_1 ($=D_2$) and length L (equal to the disk thickness). Steel disks are usually extremely stiff so that their equivalent length can frequently be regarded as zero. Their moment of inertia in some cases is sufficiently large to require its being taken into account in the equivalent system. See also 'Disk-type couplings' in section 1·24.

(b) If the torque is transmitted through the disk by forces acting at different radial distances ($R_1 = \frac{1}{2}D_1$ and $R_2 = \frac{1}{2}D_2$), the disk itself will have a certain amount of flexibility in torsion. Its equivalent length can be calculated. (See 'Coupling disks' in section 1·24). If the disk inertia J_D is not small enough to be disregarded, the corresponding portion of the equivalent system will then consist of the following elements: (1) input shaft of diameter D_1, followed by (2) equivalent shaft length representing disk flexibility, (3) the moment of inertia J_D of the disk, and (4) the output shaft.

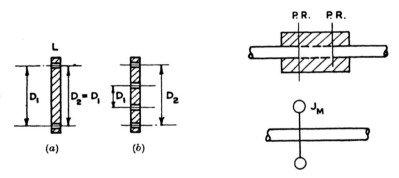

(a)   (b)

(4) *Shafts of driven machines.* Generally, a driven machine of the type shown in the above figure is represented solely as a moment of inertia J_M, which is assumed to be concentrated at a 'point of rigidity' (P.R.) on the shaft axis.

If the shafting on both sides of the driven machine has to be considered, the two points of rigidity are regarded as a single point at which J_M is situated. The intermediate length of shafting is assumed to be 'infinitely' stiff, i.e. to have zero equivalent length.

The assumption of a definite point of rigidity is used for convenience. For further details, see 'Shaft stiffness of driven machines' (section 1·25).

(5) *Force fits, shrink fits, keyed sections, etc.* The calculation of the equivalent length of shaft portions including force fits, shrink fits or keyed sections, is made with empirical formulae, usually based on the

results of tests of engine systems including the type of assembly considered. For details of evaluation, the reader should refer to section 1·24 (Couplings) and section 1·25 (Shaft stiffness of driven machines).

1·22   Stiffness of crankshafts.   Equivalent length and stiffness of a crankthrow

Note. As the cylinder masses are assumed to be concentrated at the crankpin centres, the length L of each crankshaft portion considered in torsional vibration calculations is the length between cylinder centres, as shown in Fig. 1. However, this length is the same as that of the crankthrow between journal centres, and the latter is always referred to in practice.

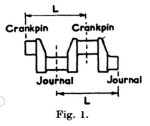

Fig. 1.

The length of shafting between the last cylinder and the flywheel is, therefore, a half-crankthrow plus the additional length of shaft to the flywheel. This also applies to the shafting at the forward end if it is fitted with an auxiliary mass.

The determination of the equivalent length of crankthrows is made either by means of formulae or on the basis of stiffness tests of crankthrows or crankthrow elements of a similar type.

1·221   *B.I.C.E.R.A. provisional formula*

This formula, which was developed on the basis of measurements of vibration frequencies and engine data supplied by B.I.C.E.R.A. members is as follows:

$$L_e = D_e^4 \left[\frac{L_j}{D_j^4 - d_j^4} + \frac{L_c}{D_c^4 - d_c^4} + \frac{0·07 L_c^3}{R_0^2 (D_c^4 - d_c^4)} + k \frac{R_0}{L_w B^3} \right] \quad \text{[inches or cm.]},†$$

where

$k = 4·559x + 0·439$ when all linear dimensions are in inches,

$k = 11·58x + 0·439$ when all linear dimensions are in centimetres,

with $x = L_w B^3 / R_0 D_j^4$ in both cases. The notation employed is [inches or cm.]

$L =$ length
$D =$ outer diameter
$d =$ inner diameter
$R_0 =$ crank radius
$B =$ web width
$L_w =$ web thickness

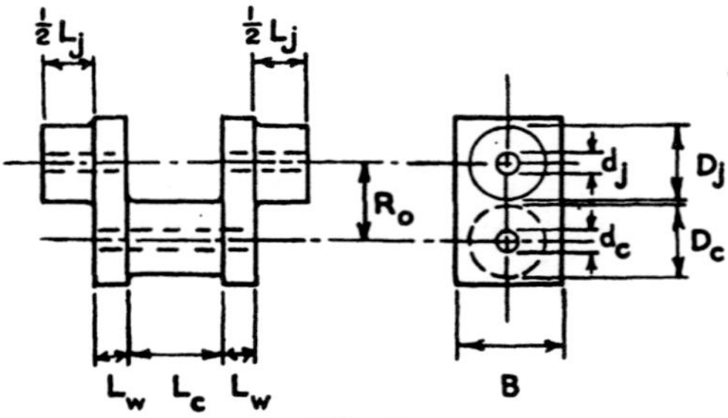

Fig. 2.

† All the formulae for crankshafts given in this section are equally valid for British and metric units, i.e. the equivalent length is obtained in inches if all the linear values used are in inches (or in centimetres if all values used are in centimetres).

In addition, the following subscripts are used:

$$e = \text{equivalent}, \quad c = \text{crankpin},$$

$$j = \text{journal}, \quad w = \text{web}.$$

The crankthrow stiffness is then determined from the relation

$$K = \frac{\pi}{32} \frac{D_e^4}{L_e} G \quad [\text{Lb.in./rad. or Kg.cm./rad.}],$$

where G = modulus of torsional rigidity of the material [Lb./in.^{2} or Kg./cm.^{2}].

This formula has been found to give very good results when applied to crankshafts of industrial engines. Many other crankshaft formulae are used, with good results for a certain range of applications. In a following section, twenty empirical or theoretical formulae for the equivalent length of crankshafts are given, together with references to literature on the subject.

Note on the evaluation of crankthrow stiffness

For all its accuracy in its application to industrial engine crankshafts, the B.I.C.E.R.A. provisional formula is not based on a knowledge of the individual crankshaft elements, but is a modification of simple theoretical expressions with the inclusion of a numerical factor k evaluated from the results of tests of a number of engines with crankshafts of various types. Like the other formulae previously mentioned, the provisional formula might be found deficient in the face of a radical change of crankshaft fashion.

Under the direction of the Torsional Vibration Panel of B.I.C.E.R.A., a programme of laboratory work was, therefore, undertaken in order to establish by direct measurement the effect of progressive changes in each of the design features of crankshafts.

The results of extensive tests were plotted in the form of basic curves, in the same way as those previously obtained for straight shafts of varying cross-section. A *method of evaluating crankshaft stiffness* was then developed, *using the B.I.C.E.R.A. basic curves*. This method is described below and followed by a numerical example.

It should be noted that the basic curves are true test results which stand irrespective of the method of stiffness evaluation. The evaluation method was also evolved at the laboratory, and has been found to give increased accuracy when applied to some thirty-two different types of engine crankshafts. However, whereas the curves can be regarded as complete in themselves, further increases in accuracy in their application to crankshafts of multi-cylinder engines may possibly be attained when more experimental information is available on subsidiary effects, such as bearing restraint for various crank-angle positions, etc., as well as on the effective value of the modulus of rigidity of crankshafts.

1·222 The B.I.C.E.R.A. method of crankthrow stiffness evaluation

In this method, the journals and crankpins are evaluated as plain cylindrical shafts right up to the web faces, and the additional flexibilities due to junction effects are included in the calculations for the webs.

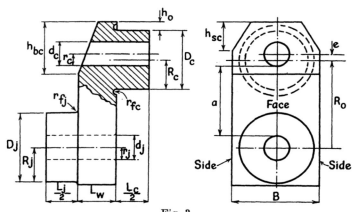

Fig. 3.

Symbols used

L_j = length of journal
L_{ej} = equivalent length of journal
$D_j = 2R_j$ = journal outer diameter
$d_j = 2r_j$ = journal inner diameter
r_{fj} = journal fillet radius
L_w = web thickness
L_{we} = equivalent web thickness
B = web width
B_e = equivalent web width
R_0 = crank radius (half stroke)
C = relative overlap $= (R_c + R_j - R_0)/R_0$
I = second moment of area

L_c = length of crankpin
L_{ec} = equivalent length of crankpin
$D_c = 2R_c$ = crankpin outer diameter
$d_c = 2r_c$ = crankpin inner diameter
r_{fc} = crankpin fillet radius
e = eccentricity of bore of crankpin (or journal)
a = thickness of metal between bores
h_0 = height of web above crankpin (or journal)
h_{bc} = height of back chamfer
h_{sc} = height of side chamfer

L_{eo} = equivalent length of one basic rectangular web
ΔL_{e1}, ΔL_{e2}, etc. = correction terms
$L_{e\,(web)}$ = equivalent length of one web
G = modulus of rigidity
K = stiffness
D_e = reference diameter (or 'equivalent shaft' diameter)

The larger of the two dimensions D_j (journal diameter) and D_c (crankpin diameter) is taken as reference diameter.

Equivalent length of crankweb

Many webs have non-parallel sides, and for such webs it is necessary to use an equivalent web width. Moreover, some webs are made with a tapering thickness, so that an equivalent value for thickness is required. Tests made at the laboratory have indicated that good agreement with test results can be obtained by determining the equivalent width and thickness by the following methods:

For circular, oval or biconcave webs, the equivalent web width B_e is determined from

$$\frac{1}{B_e^3} = \frac{1}{2}\left[\frac{1}{B_{max.}^3} + \frac{1}{B_{min.}^3}\right],$$

where $B_{max.}$ = maximum web width and $B_{min.}$ = minimum web width. These measurements are the maximum and minimum measurements which can be taken at right angles to a line joining the axes of crankpin and journal, and within the length of this line (see Fig. 4).

In cases where the thickness of the web between shaft centres is not uniform, the equivalent thickness L_{we} is determined from

$$L_{we} = 12I/B^3,$$

where I = flexural moment of inertia of transverse web section about the XX-axis. For the typical example shown in Fig. 5, the value of I is obtained as

$$I = \tfrac{1}{6}A_0(B/2)^3 + \tfrac{1}{12}A_1B^3 + \tfrac{1}{6}A_2(B/2)^3$$
$$= (B^3/12)\,[(A_0/4) + A_1 + (A_2/4)],$$

so that

$$L_{we} = \frac{12I}{B^3} = \frac{A_0}{4} + A_1 + \frac{A_2}{4}$$

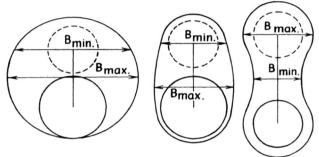

Fig. 4.

is the equivalent web thickness for this example. The difference between the maximum thickness and the equivalent thickness is regarded as a collar transition on each side of the web.

Having obtained the values for equivalent web width and thickness the equivalent length of the half crank-throw is then determined by taking account of the various design features and proceeding as indicated below:

(1) Referring to Fig. 7 (p. 61), calculate for one web

(a) the basic dimensions ratio

$$R_0 D_e^3/L_{we}B_e^3,$$

(b) the relative overlap

$$C = (R_c + R_j - R_0)/R_0,$$

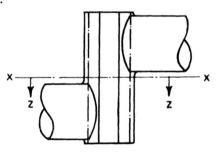

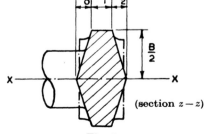

(section $z-z$)

Fig. 5.

and determine the value of the relative equivalent length L_{e0}/D_e for the basic rectangular web.

(2) Refer to Fig. 8 and calculate the value of B_e^3/D_e^3 for web-width correction, and read off the correction term $\Delta L_{e1}/D_e$.

(3) Referring to Fig. 9, for webs with crankpin and journal of unequal diameter, determine $\Delta L_{e2}/D_e$ for the corresponding values of D_e/D_c or D_e/D_j.

(4) From Fig. 10, determine the correction $\Delta L_{e3}/D_e$ for bored webs, using the ratio a/R_0, where $a = R_0 - (r_j + r_c - e)$, and $e =$ eccentricity of bore.

(5) Refer to Fig. 11. Determine the fillet ratio r_{fj}/R_j for the journal fillet and read off the correction term $\Delta L_{ef}/D_j$. Similarly, for crankpin fillets, determine r_{fc}/R_c and read off the value $\Delta L_{ef}/D_c$.
Evaluate the total correction for fillets:

$$\Delta L_{e4}/D_e = (\Delta L_{ef}/D_j)\,(D_e/D_j)^3 + (\Delta L_{ef}/D_c)\,(D_e/D_c)^3.$$

(6) Refer to Fig. 12. For back chamfer of webs at journal end, calculate $h_{bc}/(R_j + h_0)$ and L_{we}/D_j, and read off the corresponding value of $\Delta L_{ec}/D_j$.

Similarly, for back chamfer at the crankpin end of the web, determine $h_{bc}/(R_c + h_0)$ and L_{we}/D_c and obtain $\Delta L_{ec}/D_c$.
Evaluate the total back-chamfer correction:

$$\Delta L'_{e5}/D_e = (\Delta L_{ec}/D_j)\,(D_e/D_j)^3 + (\Delta L_{ec}/D_c)\,(D_e/D_c)^3.$$

Similar calculations for side-chamfer corrections should be made if the side chamfer goes beyond the centre-line of the corresponding crankpin or journal. In this case, refer to the curve for side chamfer on Fig. 14, and evaluate the side-chamfer correction:

$$\Delta L''_{e5}/D_e = (\Delta L_{ec}/D_j)\,(D_e/D_j)^3 + (\Delta L_{ec}/D_c)\,(D_e/D_c)^3.$$

Determine the total correction for chamfer:

$$\Delta L_{e5}/D_e = (\Delta L'_{e5}/D_e) + (\Delta L''_{e5}/D_e).$$

(7) Compute the equivalent length of the web:

$$L_{e(w)} = D_e\left[\frac{L_{e0}}{D_e} + \frac{\Delta L_{e1}}{D_e} + \frac{\Delta L_{e2}}{D_e} + \frac{\Delta L_{e3}}{D_e} + \frac{\Delta L_{e4}}{D_e} + \frac{\Delta L_{e5}}{D_e}\right].$$

Equivalent length of journal

$$L_{e(j)} = L_j\frac{D_e^4}{D_j^4 - d_j^4}.$$

Equivalent length of crankpin

$$L_{e(c)} = L_c\frac{D_e^4}{D_c^4 - d_c^4}.$$

A correction for crankpin bending is not required.

[58]

Resultant equivalent length of crankthrow

$$L_e = 2L_{e(w)} + L_{e(j)} + L_{e(c)}.$$

As previously stated, the equivalent shaft diameter is taken as the crankpin diameter if this is greater than the journal diameter. If it is desired to refer the equivalent length to the journal diameter, the conversion is achieved by multiplying L_e by $(D_j/D_c)^4$.

Stiffness of crankthrow

The stiffness of the crankthrow, i.e. the torque per radian of twist, is then determined as usual by

$$K = G \frac{\pi}{32} D_e^4 / L_e.$$

The accuracy of the result for K depends, therefore, on the accuracy with which the modulus of rigidity is known. It is desirable to have a correct figure for G available from torsion tests made on specimens of the crankshaft material to be used, and preferably not less than 2 in. diameter.

NUMERICAL EXAMPLE. In order to illustrate the application of this method, a detailed example of calculation is given in the following lines:

Engine type: National R4A8 (eight cylinders)

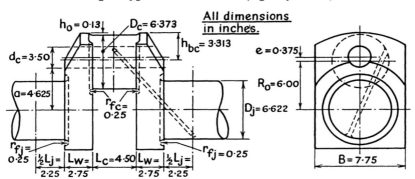

Fig. 6.

Reference diameter: $\qquad D_e = D_j = 6 \cdot 622$ in.

$(B_e/D_e)^3 = (7 \cdot 75/6 \cdot 622)^3 = 1 \cdot 606, \quad (D_e/D_c)^3 = (6 \cdot 622/6 \cdot 373)^3 = 1 \cdot 039^3 = 1 \cdot 122.$

Basic rectangular web:

$R_0 D_e^3/(B_e^3 L_{we}) = 6 \cdot 00/(1 \cdot 606 \times 2 \cdot 75) = 1 \cdot 363,$

$C = (R_j + R_c - R_0)/R_0 = (3 \cdot 311 + 3 \cdot 1865 - 6 \cdot 00)/6 \cdot 00 = +0 \cdot 083.$

From Fig. 7: $\qquad L_{e0}/D_e = \mathbf{0 \cdot 770}.$

Web-width correction:

$(B_e/D_e)^3 = 1 \cdot 606, \quad C = +0 \cdot 083.$

From Fig. 8: $\qquad \Delta L_{e1}/D_e = -0{\cdot}040.$

Web correction for unequal pin and journal diameters:
$$D_e/D_c = 1{\cdot}039, \quad C = +0{\cdot}083.$$
From Fig. 9: $\qquad \Delta L_{e2}/D_e = \mathbf{0{\cdot}020.}$

Correction for bored web:
$$a = 4{\cdot}625 \text{ in.,} \quad a/R_0 = 4{\cdot}625/6{\cdot}00 = 0{\cdot}771.$$
From Fig. 10: $\qquad \Delta L_{e3}/D_e = \mathbf{0{\cdot}060.}$

Corrections for fillets:
Journal fillet (undercut):
$$r_{fj}/R_j = 0{\cdot}25/3{\cdot}311 = 0{\cdot}076.$$
From Fig. 11: $\qquad \Delta L'_{e4}/D_j = 0{\cdot}057.$
Crankpin fillet (undercut):
$$r_{fc}/R_c = 0{\cdot}25/3{\cdot}186 = 0{\cdot}078.$$
From Fig. 11: $\qquad \Delta L''_{e4}/D_c = 0{\cdot}058.$
Total fillet correction:
$$\Delta L_{e4}/D_e = (\Delta L'_{e4}/D_j) + (\Delta L''_{e4}/D_c)\,(D_e/D_c)^3$$
$$= 0{\cdot}057 + 0{\cdot}058 \times 1{\cdot}122 = \mathbf{0{\cdot}122.}$$

Oil passage:
See Fig. 16: $\qquad$ no correction.

Correction for back chamfer:
Crankpin end: $\qquad h_{bc}/(R_c + h_0) = 1{\cdot}00.$
$$L_{we}/D_e = 2{\cdot}75/6{\cdot}373 = 0{\cdot}430.$$
From Fig. 12:
$$\Delta L_{e5}/D_e = (\Delta L_{ec}/D_c) \times (D_e/D_c)^3 = 0{\cdot}014 \times 1{\cdot}122 = \mathbf{0{\cdot}016.}$$
Journal end: $\qquad$ no correction.

Correction for side chamfer:
From Fig. 12: $\qquad$ no correction.

Equivalent length of one web:
$$L_{e(1\,\text{web})} = D_e[0{\cdot}770 - 0{\cdot}040 + 0{\cdot}020 + 0{\cdot}060 + 0{\cdot}122 + 0{\cdot}016]$$
$$= 6{\cdot}622 \times 0{\cdot}948 = \mathbf{6{\cdot}273 \text{ in.}}$$

Equivalent length of journal:
$$L_{e(j)} = L_j = \mathbf{4{\cdot}500 \text{ in.}}$$

Equivalent length of crankpin:
$$L_{e(c)} = L_c D_e^4/[D_c^4 - d_c^4] = 4{\cdot}50 \times 6{\cdot}622^4/[6{\cdot}373^4 - 3{\cdot}50^4] = \mathbf{5{\cdot}772 \text{ in.}}$$

Equivalent length of crankthrow:
$$L_{e\,\text{cyl.}} = 2L_{e(1\,\text{web})} + L_{e(j)} + L_{e(c)}$$
$$= 2 \times 6{\cdot}273 + 4{\cdot}500 + 5{\cdot}772 = \mathbf{22{\cdot}818 \text{ in.}}$$

Stiffness between cylinder centres:
$$K_{\text{cyl.}} = (\pi/32)\,GD_e^4/L_{e\,\text{cyl.}} = 0{\cdot}098175 \times 11{\cdot}8 \times 10^6 \times 6{\cdot}622^4/22{\cdot}818$$
$$= \mathbf{97{\cdot}60 \times 10^6 \text{ Lb.in./rad.}}$$

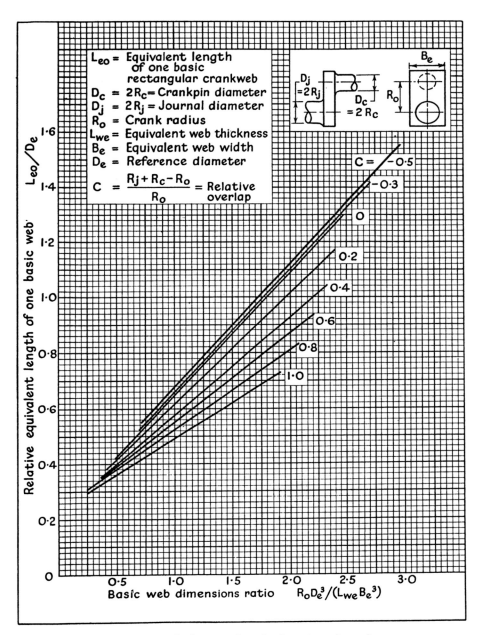

Fig. 7. Standard curves for a basic rectangular web.

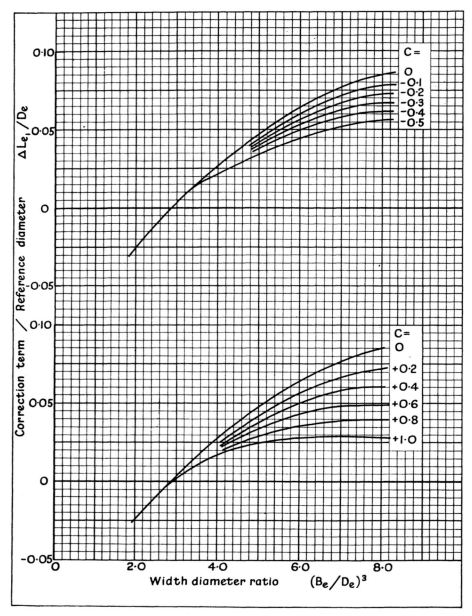

Fig. 8. Web width correction term.

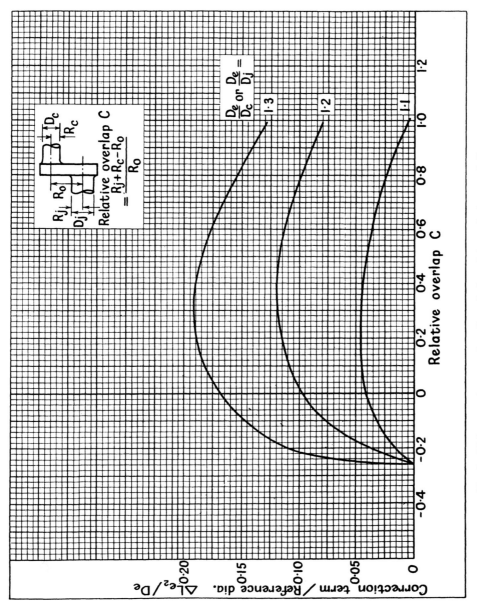

Fig. 9. Web correction term for effect of pin and journal of unequal diameter.

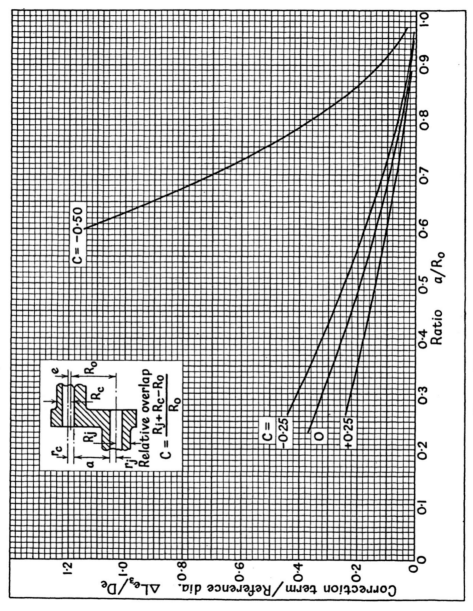

Fig. 10. Web correction term for bored web.

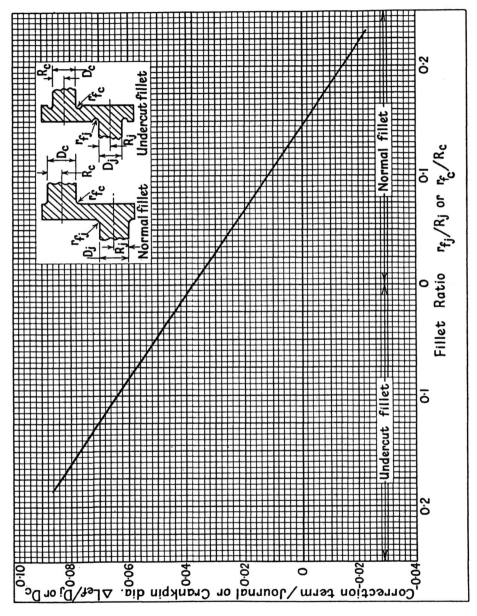

Fig. 11. Web correction term for one fillet.

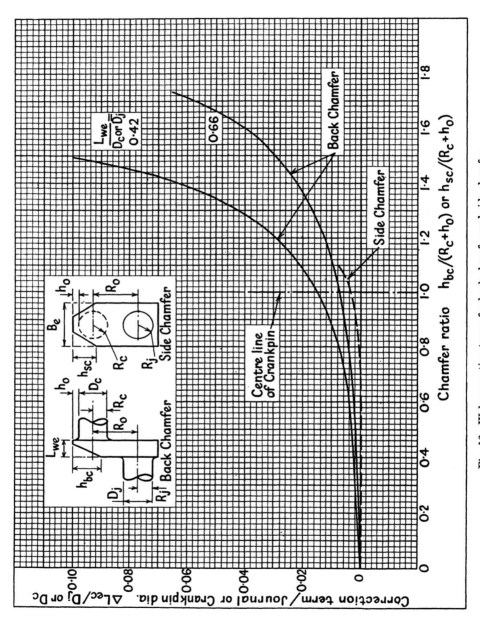

Fig. 12. Web correction terms for back chamfer and side chamfer.

Note: For chamfer on journal end of web R_c should be replaced by R_j in the above expressions.

Fig. 13. General view of test rig.

Fig. 14. Test rig with specimen and optical equipment.

[facing p. 67]

Stiffness of shafting between cylinder no. 8 and flywheel:
$$L_{ef} = 23 \cdot 115 \text{ in.},$$
$$K_f = (\pi/32)\, GD_e^4/L_{ef} = 97 \cdot 4 \times 10^6 \text{ Lb.in./rad.}$$

ONE-NODE NATURAL FREQUENCY

Calculated natural frequency based on above values for stiffness:
$$F_{calc.} = 4240 \text{ vib./min.}$$

Measured value: $\qquad F_{meas.} = 4320 \text{ vib./min.}$

Relative error: $\quad (F_{calc.} - F_{meas.}) \times 100/F_{meas.} = -1 \cdot 75 \%.$

Details of B.I.C.E.R.A. investigation of crankthrow stiffness

Test equipment

The rig used is illustrated in Figs. 13 and 14. The specimen to be tested was clamped at one end to a large cast-iron engine bearer bolted to a deep concrete foundation. The torque was applied to the specimen by means of a lever clamped to the other end. In order to avoid web bending in the axial direction, a light auxiliary lever and weight were arranged to place the centre of gravity of the lever system in the plane of the centreline of the web, perpendicular to the shaft axis. Angular deflexions at various points of the test specimen were measured optically by means of autocollimators of the type described in section 1·216, page 48.
The steel mirrors were mounted on knife-edge supporting rings at positions determined by means of gauge blocks.

Test pieces

The test pieces were machined from G.M. Meehanite cast in the form of half-crankthrows. Each test piece consisted basically of a rectangular web and two cylindrical portions of 3 in. diameter corresponding to a journal and crankpin. The 'crankpin' was not less than 6 in. long, which enabled accurate measurements to be made of the modulus of rigidity.

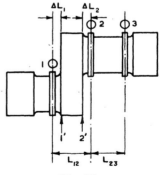

Fig. 15.

Test method

In the initial tests, a whole series of loads was applied. It was invariably found that an almost perfectly linear relationship was obtained between load and deflexion. In later tests, therefore, a single load was used, the value chosen depending on the crankweb dimensions. The corresponding nominal crankpin stress ranged from 1000 to 7000 Lb./in.[2].
The load was applied to the torque arm and the angular deflexions were measured at three points (1, 2 and 3 in Fig. 15). The load was then removed and the no-load readings were taken. This procedure was

repeated five times with the same load for each test to obtain good average values. The readings differed by not more than $\pm 1\%$.

From the difference in deflexion θ_{23}, between points 3 and 2, the twist per unit length of the crankpin and journal, θ_{23}/L_{23}, was determined. The deflexions $\Delta\theta$ between the measuring points and the sides of the web were determined from the relations

$$\Delta\theta_1 = (\theta_{23}/L_{23}) \times \Delta L_1 \quad \text{and} \quad \Delta\theta_2 = (\theta_{23}/L_{23}) \times \Delta L_2,$$

and the resultant deflexion across the web was obtained from

$$\theta_{1'2'} = (\theta_2 - \Delta\theta_2) - (\theta_1 + \Delta\theta_1).$$

To find the equivalent length of the web in terms of the journal diameter, it was only necessary to relate this web deflexion to the twist per unit length of the journal as follows:

$$L_{e\,(\text{web})\,1'2'} = \theta_{1'2'} \times (L_{23}/\theta_{23}).$$

In this evaluation method the effects of the junctions between main journal and web and crankpin and web are included in the equivalent length of the web.

For the main series of tests, a constant value of $\frac{7}{32}$ in. was used for the fillet radii, this being a good average value in relation to the size of shaft used. A separate study was made of the effect of varying the fillet radius.

Details of tests

The first series of tests was carried out to establish the variation of the equivalent length of webs with the variation of the principal dimensions (i.e. web width, thickness, and crank radius) for a range of values of overlap of crankpin and journal.

In addition, it was found necessary to establish by tests an equivalent web width for webs having special contours, such as circular, oval and biconcave webs, and the equivalent web thickness for webs tapering in thickness.

These were followed by tests determining the effects on the equivalent length of the web of the following: (1) chamfer on the back and side faces of the web, (2) boring of webs to obtain hollow crankpins and journals, (3) normal and undercut fillets of various radii, and (4) inequality of crankpin and journal diameters (see Fig. 16).

In all, some 200 specimens were tested, in order to cover adequately the variations in shape occurring in normal designs of diesel engine crankshafts.

Grammel, in 1933, put forward the theory that the stiffness of a crankshaft would depend to some extent on whether the torsion is the result of (*a*) torque applied at the journals, or (*b*) tangential forces applied at the crankpins. Tests made by Kimmel and Haug in 1938–40 showed, however, that the 'torque at the journals' assumption yields figures for stiffness in closer agreement with measured values than the theory of forces acting at the crankpin.†

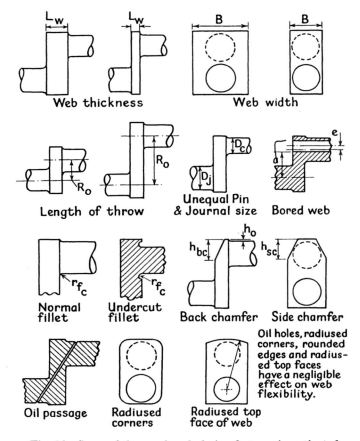

Fig. 16. Some of the crank web design features investigated.

The B.I.C.E.R.A. basic curves were obtained from tests of half-crankthrows with a torque acting on one of the cylindrical shaft portions of the throw. The load applied as a tangential force on a crankpin would still have been an idealization of the problem, since the maximum

† Bibliography: R. Grammel, *Ingen.-Arch.* vol. 4 (1933), p. 287, and vol. 11, no. 2 (1940), pp. 149–50; R. Grammel, K. Klotter and K. V. Sanden, *Ingen.-Arch.* vol. 7, no. 6 (1936), pp. 439–65; K. Haug, *Auto.-tech. Z.* vol. 43, no. 16 (1940), pp. 393–402, and vol. 45, no. 15 (1942), pp. 407–14; A. Kimmel, Thesis, 'On the torsional stiffness of crankshafts with unsupported intermediate webs', Stuttgart Engineering University (1935), also *Ingen.-Arch.* vol. 10 (1939), p. 196; J. Meyer, *Luftfahrtforsch.* vol. 17, no. 2 (1940), pp. 54–5; J. Meyer and Behrmann, *D.V.L. Research on Aero-engine Crankshafts* (1936).

[69]

reaction torque due to flywheel inertia in an engine system is effectively applied as a torque on the last journal, so that it appears that the torsion actually occurring in the crankshaft is due to to a combination of the two types of loading mentioned above.

The accuracy obtainable with the method of evaluation using the B.I.C.E.R.A. basic curves was investigated by calculating the crank-throw stiffnesses of thirty-two engines, covering a wide range of sizes

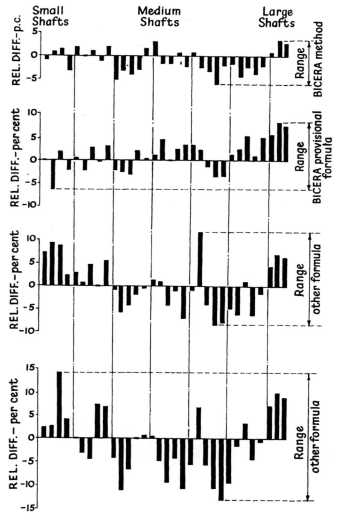

CRANKSHAFT STIFFNESS INVESTIGATION

Comparison of difference between measured natural frequencies of thirty-two engines and values calculated by different methods.

and shapes.† These engines were torsiographed by member firms of B.I.C.E.R.A., and in most cases the measured values of natural frequency were obtained with the engines running at zero load, without a driven machine, in order to have reliable figures for the value of the crankthrow stiffness. It was found that with the B.I.C.E.R.A. evaluation method, for the thirty-two engines tested, the average error and the range of error obtained were less than with the empirical or analytical formulae previously available.

Furthermore, an analysis was made of the trend of the calculated results as a function of the various parameters. It was found that the agreement with the values derived from engine measurements would generally be even better if, in Fig. 8 of this section, the negative portion of the $\Delta L_{e1}/D_e$-curve for positive C-values were given twice the slope actually obtained in the test curves. For the range of values $(B_e/D_e)^3 = 3\cdot0$ to $2\cdot0$, the $\Delta L_{e1}/D_e$-curve for $C > 0$ would then extend from zero to $-0\cdot05$ (instead of to $-0\cdot025$). This is possibly a stiffening effect due to bearing restraint. However, as all the basic curves are test curves, independent of the method of application, and of the method of frequency evaluation (concentrated masses and Holzer tables) it was decided not to alter the stiffness evaluation method until further experimental information on bearing restraint, etc. is available.

1·223 *Other formulae for the equivalent length of a crankthrow*

A considerable number of theoretical and empirical formulae have been evolved for the determination of the equivalent length of crankthrow. Some of these are widely used, and in some design offices it is usual to evaluate the equivalent length of a crankthrow by means of several formulae and then to take the average value. In the following, some twenty formulae are given, together with references to literature on the subject.

The notation has been standardized in all cases as follows:

Symbol	Brief description	Typical units
T	torque	Lb.in.
G	modulus of rigidity	Lb./in.^{2}
E	modulus of elasticity	Lb./in.^{2}
K	stiffness	Lb.in./rad.
I	polar second moment of area	in.^{4}
I^*	second moment of area for bending	in.^{4}
A	area of cross-section	in.^{2}
r	stiffening ratio	
θ	angle of twist	rad.

† See chart on p. 70. Since these first evaluations, the method has been applied to a large number of engines, with good results. In systems with driven machines, the stiffness of driven-machine shafting should also be evaluated in detail, taking account of junction effects, etc.

Symbol	Brief description	Typical units
L	length	in.
D	outer diameter	in.
d	inner diameter	in.
R_0	crank radius	in.
B	web width	in.
L_w	web thickness	in.
H	total height of web	in.
δ	diametral bearing clearance	in.

In addition, the following subscripts are used:

e equivalent c crankpin
j journal w web.

Metric units. If torque is expressed in Kg.cm., rigidity modulus in Kg./cm.², and the linear dimensions are in cm., the equivalent length is obtained in cm. and the stiffness in Kg.cm./rad.

1. Basic theoretical formula

$$L_e = D_e^4 \left[\frac{L_j}{D_j^4 - d_j^4} + \frac{L_c}{D_c^4 - d_c^4} + \frac{0\cdot 93 R_0}{L_w B^3} \right] \quad \text{[inches or cm.].}$$

Assumptions made: (i) no bearing constraint; (ii) no increment in flexibility due to junctions of webs with crankpin and journal; (iii) the bending of the webs is from journal centre to crankpin centre, with an effective length R_0; (iv) web stiffness is unaffected by chamfering and by overlap of crankpin and journal.

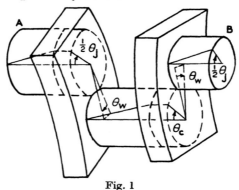

Fig. 1

Derivation. Considering Fig. 1, it is assumed that the crankthrow is clamped at point A and subjected to a torque T at point B. Then the total angle of twist is

$$\theta_{AB} = \theta_j + \theta_c + 2\theta_w.$$

Since $T/K = \theta$, we may write

$$\theta_{AB} = \frac{T}{K_e} = \frac{T}{K_j} + \frac{T}{K_c} + 2\frac{T}{K_w},$$

therefore
$$\frac{1}{K_e} = \frac{L_e}{G_e I_e} = \frac{L_j}{GI_j} + \frac{L_c}{GI_c} + \frac{2}{K_w}.$$

It should be noted that the applied couple T is a twisting action as concerns the crankpin and journal portions, but on the webs it is felt as a bending action M_b as shown in Fig. 1. Numerically, M_b has the same value as T.

The value of the web stiffness K_w is obtained by referring to Fig. 2, which shows a web subjected to a bending moment M_b. Let $y =$ radius of curvature of the strained web, and
$$I_w^* = L_w B^3 / 12.$$
Then, from

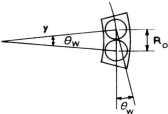

Fig. 2

$$\frac{M_b}{I_w^*} = \frac{E}{y} \quad \text{and} \quad y\theta_w = R_0 \quad \text{or} \quad \frac{E}{y} = \frac{E\theta_w}{R_0},$$

it follows that
$$\frac{M_b}{I_w^*} = \frac{E\theta_w}{R_0} \quad \text{and} \quad \frac{\theta_w}{M_b} = \frac{1}{K_w} = \frac{R_0}{EI_w^*}.$$

Substituting
$$I_e = \pi D_e^4 / 32, \quad I_c = \pi (D_c^4 - d_c^4)/32 \quad \text{and} \quad I_j = \pi (D_j^4 - d_j^4)/32,$$

we obtain (with $G_e = G$)
$$L_e = D_e^4 \left[\frac{L_j}{D_j^4 - d_j^4} + \frac{L_c}{D_c^4 - d_c^4} + \frac{2\pi}{\frac{32}{12} \times \frac{E}{G}} \times \frac{R_0}{L_w B^3} \right],$$

which gives the above equation when taking $E = 30 \times 10^6$ Lb./in.^{2} and $G = 11 \cdot 8 \times 10^6$ Lb./in.^{2}. The ratio E/G is about the same for steel and cast iron, so that the numerical coefficient also applies to cast-iron crankthrows.

B.I.C.E.R.A. provisional formula: see p. 54.

B.I.C.E.R.A. method: see p. 55.

2. Carter's formula

[B. C. Carter, 'An empirical formula for crankshaft stiffness in torsion', *Engineering*, 13 July 1928, p. 36.]

$$L_e = D_e^4 \left[\frac{L_j + 0 \cdot 8 L_w}{D_j^4 - d_j^4} + \frac{0 \cdot 75 L_c}{D_c^4 - d_c^4} + \frac{1 \cdot 5 R_0}{L_w B^3} \right] \quad \text{[inches or cm.].†}$$

3. Constant's 'stiffening ratio'

[H. Constant, 'On the stiffness of crankshafts', *Rep. Memor. Aero. Res. Comm., Lond.*, no. 1201 (E 29), October 1928.]

† This formula (and all those in this section) gives L_e in cm. if all the dimensions used are in cm.

Calculating the equivalent length $L_{e\,(\text{calc.})}$ by means of formula 1, as given above, the 'effective length' $L_{e\,(\text{eff.})}$ of the crankthrow taking account of bearing restraint is determined by means of a 'stiffening ratio' r, as follows:

$$L_{e\,(\text{eff.})} = \frac{1}{r} \times L_{e\,\text{calc.})},$$

where

$$r = a \times \frac{L^3 B_m}{D_c^4 - d_c^4} + b.$$

In the expression for the stiffening ratio, $L = L_j + L_c + 2L_w =$ total length of a crankthrow, $B_m = \frac{1}{2}(B_{\text{max.}} + B_{\text{min.}}) =$ mean web width. The graph originally published by Constant was evaluated at the B.I.C.E.R.A. Laboratory and gave the following values for the constants a and b:

Marine engines:
$$a = \tfrac{1}{328}, \quad b = 0.905.$$

Range: r between 1·0 and 1·14.

Motor-car and aircraft engines:
$$a = \tfrac{1}{101}, \quad b = 0.832.$$

Range: r between 1·0 and 1·40.

4. Geiger's formula

[J. Geiger, *Mechanical Vibrations and their Measurement* (published by Springer-Verlag, Berlin, 1927); 'Stressing of crankshafts', *Auto.-tech. Z.* 25 Feb. 1937, pp. 93–8.]

$$L_e = D_e^4 \left[\frac{L_j + 0.4 L_w}{D_j^4 - d_j^4} + \frac{L_c + 0.4 L_w}{D_c^4 - d_c^4} + \frac{0.912(R_0 - zD_j)}{L_w B^3} \right] \quad [\text{inches or cm.}],$$

where

$z = 0$ when $B/D_j = 1.60$ to 1.63 and $R_0/D_j = 1.20$ to 0.92;

$z = 0.4$ when $B/D_j = 1.49$ and $R_0/D_j = 0.84$

                                             (see Figs. 3, 4);

and $z = 0.3$ when $B/D_j = 1.33$ and $R_0/D_j = 1.07$

                                  (dotted curves in Figs. 3, 4).

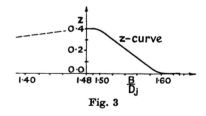

Fig. 3

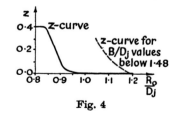

Fig. 4

5. Heldt's formula

[P. M. Heldt, 'High-speed combustion engines', *The Gasoline Motor*, 15th ed. (1951), pp. 316–17; 'Automotive engines', *The Gasoline Motor*, 8th ed. (1933), p. 266 (published by P. M. Heldt, Nyack, N.Y.).]

$$L_e = D_e^4 \left[\frac{L_j + 0.4 L_w}{D_j^4 - d_j^4} + \frac{1.096 L_c}{D_c^4 - d_c^4} + \frac{1.284 R_0}{L_w B^3} \right] \quad [\text{inches or cm.}].$$

6. Holzer-Föppl formula

[H. Holzer, *The Calculation of Torsional Vibrations* (*Die Berechnung der Drehschwingungen*) (Springer, Berlin, 1922), pp. 10–13.]

The symbols used are shown in Fig. 5. In addition,

$$\delta = \text{diametral bearing clearance [in.]},$$

$$G = \text{modulus of rigidity [Lb./in.}^2\text{], and}$$

$$T_0 = T_{\text{trans.}} + T_{\text{vib.}} \text{ [Lb.in.]},$$

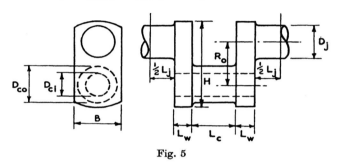

Fig. 5

where $T_{\text{trans.}} = $ transmitted torque, and $T_{\text{vib.}} = $ total vibratory torque $\left(\simeq \sum_1^k L\omega^2 \Delta \times \theta_1 \right)$ at the crankpin considered (kth crankpin), the latter being estimated from a preliminary frequency tabulation.

$$L_e = D_e^4 \left\{ \frac{L_j}{D_j^4} + 0.93 \frac{R_0}{L_w B^3} + 0.707 \times \frac{H^2 + B^2}{H^3 \times B^3} \times L_w + \frac{L_c}{D_{co}^4 - D_{ci}^4} - XY \right\}$$

[inches or cm.],

where

$$X = 4.724 \frac{R_0}{L_w B^3} + 0.354 \frac{H^2 + B^2}{H^3 \times B^3} \times L_w + \frac{L_c}{D_{co}^4 - D_{ci}^4} + 0.236 \times \frac{1}{L_w B R_0}$$

and

$$Y = \frac{R_0 \left[4.724 \frac{R_0}{L_w B^3} + 3.6 \frac{H^2 + B^2}{H^3 \times B^3} \times L_w + 10.186 \frac{L_c}{D_c^4} \right] - \frac{2G\delta}{T_0}}{3.150 \frac{R_0}{L_w B^3} + \frac{1.8}{B^3} \left[\frac{H^2 + B^2}{H^3} \times L_w R_0 + \frac{L_w^2 + B^2}{L_w^3} \times (L_c + L_w)^2 \right] + Z}.$$

with

$$Z = \frac{2.4}{B} \left[\frac{L_w}{H R_0} + \frac{1}{L_w} \right] + 10.186 \frac{L_c}{D_{co}^4} \left[R_0 + \frac{L_c^2}{15.24 R_0} + \frac{D_{co}^2}{6.745 R_0} \right]$$

These equations indicate that

(*a*) if Y is neglected, the calculated value of L_e will be higher than the effective value;

(*b*) if δ (diametral clearance) is fairly large, the value of Y will tend to be smaller, until $Y = 0$ is obtained (no bearing restraint);

[75]

(c) the larger the torque T_0 acting on the crankpin, the greater the effect of the restraint will be in stiffening the crankthrow; hence, with increasing values of T_0 towards the node of the system, the successive crankthrows become increasingly stiffer (and have smaller L_e-values).

Note on metric units. If metric units are used, G should be in Kg./cm.^{2}, T_0 in Kg.cm., and δ in cm.

7. Jackson's formula

[P. Jackson, 'The vibrations of oil engines', Paper no. S115 read before the Diesel Engine Users Association, 26 April 1933, pp. 19, 20.]

$$L_e = D_e^4 \left[\frac{L_w + 0 \cdot 27 D_j}{D_j^4 - d_j^4} + \frac{L_c + 0 \cdot 27 D_c}{D_c^4 - d_c^4} + \frac{0 \cdot 07 (L_c + 0 \cdot 27 D_c)^3}{R_0^2 (D_c^4 - d_c^4)} + \frac{0 \cdot 7 R_0}{L_w B^3} \right]$$

[inches or cm.].

8. Ker Wilson's formula

[W. Ker Wilson, *Practical Solution of Torsional Vibration Problems* (Chapman Hall, London, 1942), vol. 1, p. 192.]

$$L_e = D_e^4 \left[\frac{L_j + 0 \cdot 4 D_j}{D_j^4 - d_j^4} + \frac{L_c + 0 \cdot 4 D_c}{D_c^4 - d_c^4} \right.$$
$$\left. + \frac{R_0 - 0 \cdot 2 (D_j + D_c)}{L_w B^3} \right] \text{ [inches or cm.].}$$

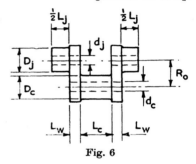

Fig. 6

9. Ker Wilson's formula for inclined centre crankweb

[W. Ker Wilson, *Practical Solution of Torsional Vibration Problems*, vol. 1, p. 194.]

The equivalent length L_e of the portion of length x (from crankpin centre to crankpin centre in the drawing of Fig. 7) is calculated from the relation

$$L_e = D_e^4 \left[\frac{0 \cdot 75 L_c}{D_c^4 - d_c^4} + \frac{0 \cdot 75 R_1}{L_{w_1} B_1^3} \right] \text{ [inches or cm.],}$$

where the subscript 1 refers to the centre crankweb.

10. Ker Wilson's formula for crankthrows of opposed-piston engines

[W. Ker Wilson, *Practical Solution of Torsional Vibration Problems*, vol. 1, p. 195.]

Using the notation indicated in Fig. 8, and the designations B = width of small webs (of thickness L_w), B_1 = width of large webs (of thickness L_{w_1}) taken at the centre section xx of the webs, the equivalent length of this triple crankthrow, from journal centre to journal centre, can be determined from the following formula:

$$L_e = D_e^4 \left[\frac{L_j + 0 \cdot 8 L_w}{D_j^4 - d_j^4} + \frac{1 \cdot 5 L_c}{D_c^4 - d_c^4} + \frac{0 \cdot 75 L_{c1}}{D_{c1}^4 - d_{c1}^4} + \frac{1 \cdot 5 R_0}{L_w B^3} + \frac{1 \cdot 5 R_1}{L_{w_1} B_1^3} \right]$$

[inches or cm.].

Note. This expression is based on Carter's formula suitably modified for this crankshaft type.

[76]

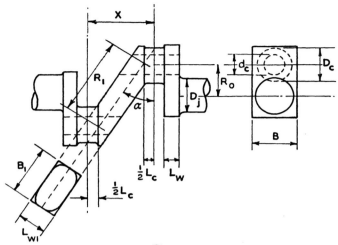

Fig. 7

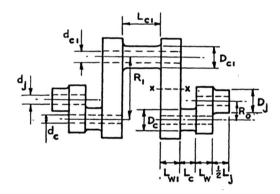

Fig. 8

11. Norman-Stinson formula

[C. A. Norman and K. W. Stinson, 'The angular distortion of crankshafts', *Bull. Engng Exp. Sta.* no. 43, Ohio State University (1928); and *S. A. E. Jl*, vol. 23 (July 1928), p. 83.]

$$L_e = D_e^4 \left[\frac{L_j}{D_j^4 - d_j^4} + \frac{L_c}{D_c^4 - d_c^4} + 0.49 \left(\frac{1}{D_j^3} + \frac{1}{D_c^3} \right) + \frac{0.475 R_0}{L_w B_m^3} \right] \quad \text{[inches or cm.]},$$

where $B_m^3 = \frac{1}{4}[B_{\max.}^3 + B_{\min.}^3 + B_{\max.} \times B_{\min.} \times (B_{\max.} + B_{\min.})]$ for webs of varying width.

For inclined centre crankwebs, replace R_0 by $R_0 \cos \alpha$, where $\alpha =$ angle between crank-radius and centre-web axis (see Fig. 7, above).

12. Seelmann's formula (simplified)

[—. Seelmann, *Z. Ver. dtsch. Ing.* 2 May 1925, p. 601. See also Strunz, *Die Drehschwingungen in Kolbenmaschinen (Torsional Vibration in Reciprocating Engines)*, p. 27.]

Using the notation given at the beginning of this section, Seelmann's formula can be written as follows:

$$L_e = D_e^4 \left[\frac{L_j}{D_j^4 - d_j^4} + \frac{0\cdot9 L_w}{D_c^4 - d_c^4} + k \left\{ \frac{0\cdot93 R_0}{L_w B^3} + \frac{L_c}{D_c^4 - d_c^4} + \frac{0\cdot9 L_w}{D_j^4 - d_j^4 + D_c^4 - d_c^4} \right\} \right]$$

[inches or cm.],

where k is a 'stiffening ratio' which can be determined from the graph of Fig. 9, as a function of the cylinder bore $D_{\text{cyl.}}$, for various values of the stroke-to-bore ratio $S/D_{\text{cyl.}}$ (where S = stroke).

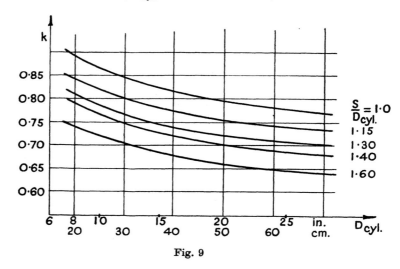

Fig. 9

13. Seelmann's formula

[Same references as above.]

Let $T_{\text{vib.}}$ = vibration torque [Lb.in.], H = total height of web [in.], δ = diametral bearing clearance [in.], G = modulus of rigidity [Lb./in.²]. Then Seelmann's theoretical formula taking account of bearing restraints can be written as follows:

$$L_e = D_e^4 \left\{ \left[\frac{L_c}{D_c^4 - d_c^4} + \frac{0\cdot9 L_w}{D_c^4 - d_c^4} + \frac{0\cdot93 R_0}{L_w B^3} \right] \right.$$
$$\left. - X \left[\frac{0\cdot452 R_0^2}{L_w B^3} + \frac{0\cdot354 L_w R_0}{B^2 H^2} \left(\frac{B}{H} + \frac{H}{B} \right) + \frac{R_0 L_c}{D_c^4 - d_c^4} \right] \right\} \quad \text{[inches or cm.]},$$

[78]

where

$$X = \cfrac{\left[\cfrac{0\cdot465R_0}{L_wB^3} + \cfrac{0\cdot9L_w}{D_c^4-d_c^4} + \cfrac{L_c}{D_j^4-d_j^4}\right]R_0 - \cfrac{\pi}{32} \times \cfrac{\delta G}{T_{\text{vib.}}}}{\cfrac{0\cdot31R_0^3}{L_wB^3} + 1\cdot8R_0^2\cfrac{B^2+H^2}{L_w^2H^3} + 1\cdot8R_0(L_c+L_w)^2 \times \cfrac{B^2+L_w^2}{B^3H^3} + 2\cdot4\cfrac{L_w}{BH} + Z}.$$

with
$$Z = \frac{L_j^2(L_j+R_0)}{D_c^4-d_c^4} + \frac{1\cdot186L_j}{\frac{1}{4}\pi(D_j^2-d_j^2)}$$

14. Shannon's formula for crankthrows of opposed-piston engines

[J. F. Shannon,† *J. R. Tech. Coll. Glasg.* vol. 2 (1932), pp. 638–56.]

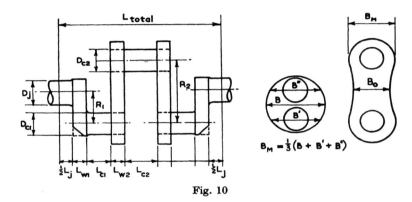

Fig. 10

Using the notation indicated in Fig. 10, the equivalent length of this triple crankthrow, from journal centre to journal centre, can be determined from

$$L_e = D_e^4\left[\frac{L_j^4}{D_j^4} + \frac{2L_{c1}}{D_{c1}^4} + \frac{L_{c2}^{'}}{D_{c2}^4} + \frac{0\cdot707L_{w1}}{B_m^3R_1^3}(B_m^2+R_1^2) + \frac{0\cdot707L_{w2}}{B_0^3R_2^3}(B_0^2+R_2^2)\right.$$
$$\left. + \frac{0\cdot942R_1}{L_{w1}B_m^3} + \frac{0\cdot942R_2}{L_{w2}B_0^3}\right] \quad \text{[inches or cm.]}.$$

For shrunk joints, add $0\cdot03L_{\text{total}}$; and for back chamfer of small webs add $0\cdot05L_{\text{total}}$. The resultant equivalent length is then

$$L_{e\,\text{res.}} = L_e + 0\cdot08L_{\text{total}},$$

where $L_{\text{total}} = $ overall length of triple crankthrow, from journal centre to journal centre.

---

† From author's abstract of the above paper: 'The results show that the tendency to lateral displacement of the journals due to the torque and the consequent web torque induced by bearing restraint are quite negligible. Torsion experiments carried out also show that it is quite accurate to apply the same straightforward deflexion rule for torsional rigidity as is employed for free shaft conditions.'

15. Southwell's formula

[R. V. Southwell, 'The effective rigidity of a crank', *Rep. Memor. Aero. Res. Comm., Lond.*, no. 1211 (E.30) (July 1927).]

This formula takes account of constraints due to rigid bearings which prevent any movement of the journals except pure torsion about the shaft axis. Southwell's equation can be written as follows:

$$
L_e = D_e^4 \left[\frac{L_j}{D_j^4 - d_j^4} + \frac{L_c}{D_c^4 - d_c^4} + \frac{0.93 R_0}{L_w B^3} \right.
$$
$$
\left. - \frac{R_0^2 \left(\dfrac{L_c}{D_c^4 - d_c^4} + \dfrac{0.465 R_0}{L_w B^3} \right)^2}{\dfrac{L_j(3L_c^2 + 3L_c L_j + L_j^2)}{30 \cdot 4(D_j^4 - d_j^4)} + \dfrac{0.588 R_0 L_c^2}{L_w B(L_w^2 + B^2)} + \dfrac{0.31 R_0^3}{L_w B^3} + \dfrac{L_c R_0^2}{D_c^4 - d_c^4} \left(1 + \dfrac{(L_c/R_0)^2}{15 \cdot 2} \right)} \right]
$$

[inches or cm.].

16. Southwell's formula (simplified)

[See H. Constant, *Rep. Memor. Aero. Res. Comm., Lond.*, no. 120 (E.29) (October 1928).]

It was suggested by Constant that the above formula can be simplified by retaining only the most important terms in the expression preceded by the minus sign, and disregarding the secondary stiffening effects due to journal, web and crankpin bending. The above formula is then reduced to the following relation:

$$
L_e = D_e^4 \left[\frac{L_j}{D_j^4 - d_j^4} + \frac{L_c}{D_c^4 - d_c^4} + \frac{0.93 R_0}{L_w B^3} - \frac{R_0 L_c}{(D_c^4 - d_c^4) \left[\dfrac{0.588 L_c(D_c^4 - d_c^4)}{L_w B(L_w^2 + B^2)} + R_0 \right]} \right]
$$

[inches or cm.].

17. Timoshenko's formula

[S. Timoshenko, *Vibration Problems in Engineering* (D. Van Nostrand and Co.), 2nd ed., 4th printing, p. 271.]

$$
L_e = D_e^4 \left[\frac{L_j + 0.9 L_w}{D_j^4 - d_j^4} + \frac{L_c + 0.9 L_w}{D_c^4 - d_c^4} + \frac{0.93 R_0}{L_w B^3} \right] \quad \text{[inches or cm.]}.
$$

This formula assumes that the bearing clearances are such that free displacements of the journal sections are possible during twist.

18. Timoshenko's formula for zero-clearance bearings

[S. Timoshenko, *Vibration Problems in Engineering*, pp. 271–2.]

For crankshafts in bearings giving complete lateral restraint (no bear-

[80]

ing clearance), the formula derived by Timoshenko can be written as follows:

$$L_e = D_e^4 \left[\frac{L_j + 0 \cdot 9 L_w}{D_j^4 - d_j^4} + \frac{L_c + 0 \cdot 9 L_w}{D_c^4 - d_c^4} \times \left(1 - \frac{R_0}{k} \right) + \frac{0 \cdot 93 R_0}{L_w B^3} \left(1 - \frac{R_0}{2k} \right) \right]$$

[inches or cm.],

where

$$k = \frac{\left\{ \dfrac{0 \cdot 177}{B^2} \left(\dfrac{L_c}{L_w} + 1 \right)^2 \times \left(\dfrac{L_w}{B} + \dfrac{B}{L_w} \right) + \dfrac{L_c R_0}{D_c^4 - d_c^4} \left[1 + \dfrac{(L_c/R_0)^2}{15 \cdot 2} \right] + \dfrac{0 \cdot 31 R_0}{L_w B^3} + 0 \cdot 118 \left[\dfrac{L_c}{R_0 A_c} + \dfrac{2}{A_w} \right] \right\}}{\dfrac{L_c}{D_c^4 - d_c^4} + \dfrac{0 \cdot 465 R_0}{L_w B^3}}.$$

In the expression for k, $A = \frac{1}{4}\pi(D_c^2 - d_c^2)$ = cross-sectional area of crank-pin, and A_w = cross-sectional area of web in plane perpendicular to the shaft axis (for rectangular-sectioned webs, $A_w = BH$, where H = total height of web).

19. Tuplin's formula

[W. A. Tuplin, 'The torsional rigidity of crankshafts', *Engineering*, 10 September 1937, pp. 275–7.]

$$L_e = D_e^4 \left[\frac{L_j + 0 \cdot 15 D_j}{D_j^4 \left[1 - \left(\dfrac{d_j}{D_j} \right)^4 \right]^2} + \frac{L_c + 0 \cdot 15 D_c}{D_c^4 \left[1 - \left(\dfrac{d_c}{D_c} \right)^4 \right]^2} + \frac{2 L_w - 0 \cdot 15 (D_j + D_c)}{B^4 - d_j^4} \right.$$

$$\left. + \frac{R_0}{L_w B^3} \left(0 \cdot 58 + \frac{0 \cdot 065 D_j}{L_w} \right) + \frac{0 \cdot 016}{B L_w^2} \right] \quad \text{[inches or cm.].}$$

20. Zimanenko's formula

[S. S. Zimanenko, *Engrs' Dig.* vol. 7, no. 11 (November 1946), pp. 337–40; and *Vestnik Ingenerov i Tehnikov.* no. 2 (February 1946), pp. 51–8.]

$$L_e = D_e^4 \left[\frac{L_j + 0 \cdot 6 \left(\dfrac{D_j}{L_j} \right) L_w}{D_j^4 - d_j^4} + \frac{0 \cdot 8 L_c + 0 \cdot 2 \left(\dfrac{B}{R_0} \right) D_j}{D_c^4 - d_c^4} + \sqrt{\left(\dfrac{R_0}{D_c} \right)} \times \frac{R_0}{L_w B^3} \right]$$

[inches or cm.].

As an alternative, the factor $\sqrt{(R_0/D_c)}$ can be replaced by the expression $(D_j + D_c)/\sqrt{(4 R_0 D_j)}$ in the last term of the equation.

1·23   Stiffness of gears, belts and chain drives

1·231   *Equivalent values of sub-systems rotating at different speeds*

(1) *Equivalent stiffnesses.* In systems comprising gearwheels, chains or belt drives, it is generally necessary to find the equivalent torsional stiffness of a shaft in relation to another part of the system running at a different speed.

Consider a system, such as that of Fig. 1, in which the wheel of effective radius R' rotates at a speed N' [rev./min.], while the wheel R'' rotates at a speed N''.

Both wheels act on each other with the same force P. The torques T' and T'', therefore, have different values:

$$T' = R'P \quad \text{and} \quad T'' = R''P \quad \text{[Lb.in.]},$$

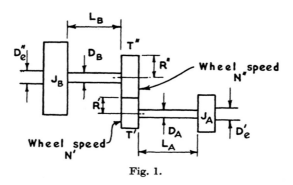

Fig. 1.

so that
$$\frac{T'}{T''} = \frac{R'}{R''} = \frac{N''}{N'}.$$

If the shaft L_A, which has an actual stiffness $K'_A = G_A \frac{\pi}{32} D^4_A / L_A$ [Lb.in./rad.], is vibrating with an amplitude $\pm \theta'_A$ [rad.], the vibratory shaft torque is
$$T' = K'_A \theta'_A.$$

Disregarding the flexibility of the wheel teeth, belt, chain, etc., the peripheral displacement of the two wheels is the same, so that we have
$$R'\theta'_A = R''\theta''_A,$$
therefore
$$T' = K'_A \theta'_A = K'_A \times \theta''_A \times \frac{R''}{R'}.$$

The torque transmitted through the wheel of effective radius R'' is, however,
$$T'' = T' \times \frac{R''}{R'}, \quad \text{hence} \quad T'' = K'_A \theta''_A \times \left(\frac{R''}{R'}\right)^2,$$

and the equivalent stiffness of K'_A, i.e. the torque T'' transmitted to wheel R'' while rotating it through an angle θ''_A, is

$$K''_A = \frac{T''}{\theta''_A} = K'_A \times \left(\frac{R''}{R'}\right)^2 = K'_A \times \left(\frac{N'}{N''}\right)^2.$$

(2) *Equivalent moments of inertia.* Assuming that the mass J_A acts

[82]

directly on the wheel R', the actual value of the inertia torque, when the mass J_A is vibrating with an amplitude $\pm\, \theta'_A$, is

$$T' = J_A \omega^2 \theta'_A \quad \text{[Lb.in.]},$$

where $\omega =$ phase velocity [rad./sec.] of the vibration. Using the relations $T' = T'' \times R'/R''$ and $\theta''_A = \theta'_A \times R'/R''$, we find that the torque transmitted to the wheel of effective radius R'' is

$$T'' = J_A \omega^2 \theta'_A \times R''/R' = J_A \omega^2 \theta''_A \times (R''/R')^2;$$

therefore J_A acts on the second wheel R'' as if it had an amplitude of vibration θ''_A and an equivalent moment of inertia determined by

$$J''_A \omega^2 \theta''_A = J_A \omega^2 \theta''_A \times (R''/R')^2,$$

or
$$J''_A = J_A \times \left(\frac{R''}{R'}\right)^2 = J_A \times \left(\frac{N'}{N''}\right)^2 \quad \text{[Lb.in.sec.}^2\text{]}.$$

Note on direction of rotation. If both wheels rotate in the same direction (belt or chain drive), the torques and angular deflexions have the same sign, whereas if the wheels rotate in opposite directions (gearwheels) the torques and deflexions have opposite signs:

Belt and chain drives:
$$T'' = +\frac{N'}{N''} \times T', \quad \theta'' = +\frac{N''}{N'} \times \theta'.$$

Gear systems:
$$T'' = -\frac{N'}{N''} \times T', \quad \theta'' = -\frac{N''}{N'} \times \theta'.$$

The stiffnesses and inertias do not change sign, since they are multiplied by the speed ratio squared. It should also be emphasized that θ' should be taken at the actual speed N' of the shaft section considered (this also applies to N'' for the value of θ'').

Equivalent values. Using the notation shown in Fig. 1, in terms of the equivalent shafts with the diameters D'_e, D''_e and the moduli of rigidity G'_e and G''_e, the following relations are valid:

$$K''_A = K'_A \times \left(\frac{R''}{R'}\right)^2 = K'_A \times \left(\frac{N'}{N''}\right)^2 \quad \text{[Lb.in./rad.]},$$

$$L''_{eA} = L_A \times \frac{G''_e}{G_A} \times \left(\frac{D''_e}{D_A}\right)^4 \times \left(\frac{R'}{R''}\right)^2 = L_A \times \frac{G''_e}{G_A} \times \left(\frac{D''_e}{D_A}\right)^4 \times \left(\frac{N''}{N'}\right)^2 \quad \text{[in.]},$$

$$J''_A = J_A \times \left(\frac{R''}{R'}\right)^2 = J_A \times \left(\frac{N'}{N''}\right)^2 \quad \text{[Lb.in.sec.}^2\text{]},$$

$$T'' = T' \times \left(\frac{\pm R''}{R'}\right) = T' \times \left(\frac{\pm N'}{N''}\right) \quad \text{[Lb.in.]},$$

$$\theta'' = \theta' \times \left(\frac{\pm R'}{R''}\right) = \theta' \times \left(\frac{\pm N''}{N'}\right) \quad \text{[rad.]}.$$

The + or − sign refers to the direction of rotation (see above Note) and the sign-value should always be included in subsequent calculations (Holzer tables, etc.) relating to the vibrations of the system.

The values of K'_B, L'_B, J'_B, T'_B and θ'_B are obtained directly from the above formulae by substituting B for A and $'$ for $''$.

Note. These relations can be used, whether the drive flexibility is taken into account or not, in natural frequency calculations. In a system with an equivalent length $L_{e(r)}$ representing the drive flexibility, the total equivalent length from point A to point B (Fig. 2) is

$$L_{e(AB)} = L_{e(A)} + L_{e(r)} + L_{e(B)}.$$

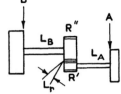

In systems with dissipative or non-linear elements in the driving arrangement, this relation is applicable only as an approximation. The treatment of systems with backlash, slip or non-viscous friction is dealt with, to the extent of the knowledge at present available, under section 1·3 (Calculation and measurement of natural frequencies).

Fig. 2.

Metric units. All the relations given in this section are directly applicable in metric units, after replacing Lb. by Kg., and in. by cm.

1·232 *Torsional stiffness of gearwheels*

The calculation can only give approximate values, since the point of contact, the direction of the applied force and the number of teeth in contact vary during the rotation.

For *spur and bevel gears*,† the stiffness calculation is based on the deflexion of teeth due to bending. Secondary effects, such as the depression of the tooth surface at the line of contact, torsional flexibility of pinion disks and the deformation in the part of the wheel body adjacent to the tooth, can be represented by a correction factor in the value of the constants used. The appreciable effect of backlash will be taken into account at a later stage, i.e. in the overall method of natural frequency calculation.

† *Worm wheels* can be regarded as absolutely rigid in torsional vibration calculations. For *spiral gear* the deflexion due to bending is practically zero. The flexibility due to the total depression δ_c at the line of contact is obtained from

$$\left(\frac{1}{K_s}\right)^3_{total} = \left(\frac{1}{K'_s} + \frac{1}{K''_s}\right)^3 = \left(\frac{\delta_c}{P}\right)^3 = \frac{C}{P\left[\dfrac{E_1}{1-\sigma_1^2} + \dfrac{E_2}{1-\sigma_2^2}\right]^2} \times \left(\frac{1}{r_1} + \frac{1}{r_2}\right) \quad [\text{in.}^3/\text{Lb.}^3],$$

where r_1 and r_2 = approximate radii of curvature at point of contact, $C = 9\cdot0$ for $r_1/r_2 = 1\cdot0$ and $C = 7\cdot0$ for $r_1/r_2 = 6\cdot0$.

For *helical gear*, the flexibility due to the total deflexion is given by

$$\frac{1}{K_{s\,total}} = \frac{\delta}{P_t} \simeq \frac{9}{L}\left[\frac{1}{E_1} + \frac{1}{E_2}\right] \times \frac{1}{\cos\psi} \quad [\text{in./Lb.}],$$

where ψ = helix angle at pitch line, and P_t = tangential force [Lb.].

[84]

The procedure of calculation is as follows:

(1) For each of the two gearwheels G' and G'' considered, the tooth outline is approximated by straight lines, as shown in Fig. 1.

(2) Using the notation

P = force [Lb.] acting on one tooth,

h = total height [in.] of triangle in Fig. 1,

B = chord at root of tooth [in.],

h_P = height from chord to pitch circle [in.],

$E = 2G(1 + \sigma)$ = modulus of elasticity [Lb./in.²],

δ = tooth deflexion [in.],

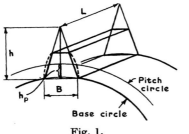

Fig. 1.

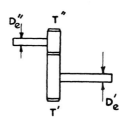

Fig. 2.

the linear flexibility (tooth deflexion per unit load) is calculated from

$$\frac{\delta}{P} = \frac{12}{EL} \left(\frac{h}{B}\right)^3 \times \left[2 \cdot 303 \log_{10} \left|\frac{h}{h - h_P}\right| - \frac{h_P}{h}\left(1 + \frac{h_P}{2h}\right) \right] + \frac{h_P}{GLB\left[1 - \dfrac{h_P}{2h}\right]}$$

[in./Lb.]

for each gearwheel, so as to obtain the values δ'/P and δ''/P.

(3) Owing to the additional deformations previously mentioned, it is necessary to multiply these values by a factor C. The linear flexibilities are then

$$\frac{1}{K'_s} = C\frac{\delta'}{P}, \quad \frac{1}{K''_s} = C\frac{\delta''}{P} \quad \text{and} \quad \frac{1}{K_{s\,(\text{total})}} = \frac{1}{K'_s} + \frac{1}{K''_s}.$$

The values of the constant factor are $C = 1 \cdot 3$ for plain spur gearing; $C = 1 \cdot 2 - 1 \cdot 3$ for bevel gearing (the higher value being for light loads); and $C = 1 \cdot 0$ for internal spur gearing.

(4) The torsional stiffness of the gear system depends on whether the torque $T' = 2R'P$ (acting with a force P on each of two teeth) of gearwheel G', or the torque $T'' = 2R''P$ of G'' is taken as a reference value (see Fig. 2). To estimate average loadings and stiffnesses, it is assumed that the load is equally shared between the two teeth, and that its average value is P, since obviously the load is fluctuating.

[85]

(a) With reference to the torque T', we have

$$K' = \frac{T'}{\theta} = \frac{2PR'}{C\left[\dfrac{\delta'}{R'} + \dfrac{\delta''}{R'}\right]}$$

or

$$K' = \frac{2PR'}{\dfrac{C}{R'}\left[\dfrac{\delta'}{P} + \dfrac{\delta''}{P}\right] \times P},$$

hence

$$K' = \frac{2R'^2}{\dfrac{1}{K_{s\,\text{total}}}} = \frac{2R'^2}{\dfrac{1}{K_s'} + \dfrac{1}{K_s''}} \quad \text{[Lb.in./rad.]},$$

and the equivalent length in terms of a plain shaft of diameter D_e' connected to the gearwheel T' is

$$L_e' = \frac{\pi}{32} G_e' D_e'^4 \times \frac{1}{K'} \quad \text{[in.]},$$

G_e' being the modulus of rigidity [Lb./in.²] of the equivalent shaft.

(b) With reference to the torque T'', we have the corresponding values

$$K'' = \frac{2R''^2}{\dfrac{1}{K_s'} + \dfrac{1}{K_s''}} = K' \times \left(\frac{R''}{R'}\right)^2 \quad \text{[Lb.in./rad.]}$$

and

$$L_e'' = \frac{\pi}{32} G_e'' D_e''^4 \times \frac{1}{K''} \quad \text{[in.]},$$

G_e'' being the modulus of rigidity of the shaft D_e''.

Metric units. The stiffnesses are obtained in Kg.cm./rad. if in the above relations all linear dimensions are in cm., torques in Kg.cm., and moduli of elasticity and rigidity are in Kg./cm.².

Derivation of the deflexion formula (Fig. 3)

The bending moment $M(x)$ at any point along the x-axis is

$$M(x) = P_n \times (x_P - x).$$

The flexural second moment of area is

$$I(x) = \tfrac{1}{12} \times L(2y)^3.$$

The strain energy in the tooth is, therefore,

$$U = \tfrac{1}{2}P_n \delta_n = \frac{1}{2} \int_0^{x_P} dx \, \frac{M^2(x)}{EI(x)}$$

or

$$U = \frac{12P_n^2}{2EL} \int_0^{x_P} dx \, \frac{(x_P - x)^2}{8y^3},$$

where $y = y(x) =$ half thickness of tooth at x and $I = 8Ly^3/12$.

[86]

Thus, the tooth deflexion δ_n, in the direction of the normal force P_n, is

$$\delta_n = \frac{\partial U}{\partial P_n} = \frac{12P_n}{EL} \int_0^{x_P} dx \, \frac{(x_P - x)^2}{(2y)^3} \, .$$

The integral for the true tooth contour can be evaluated, if necessary, by graphical methods. However, for most cases the tooth outline can be approximated by straight lines, as shown in Fig. 3. Then, from

$$\frac{y}{(B/2)} + \frac{x}{h} = 1,$$

we have

$$2y = \frac{B}{h}(h - x),$$

and substituting this expression in the relation for δ_n, we obtain the equation given at the beginning of this section.

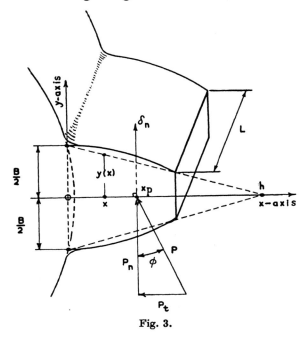

Fig. 3.

In most cases the pressure angle ϕ is small, so that $\cos\phi \cong 1$ and $P \cong P_n$. It is thus assumed that P_n has P as its average value during contact. Furthermore, we assume that the average position of the point of intersection of P with the centre-line of the tooth is on the pitch circle, so that $x_P = h_P$ (see Fig. 1).

Note. The above formula for tooth deflexion is only a first approximation. Extensive investigations of tooth deflexion have been carried out, and particular consideration should be given to recent results of research sponsored by the Department of Scientific and Industrial Research (Weber and others). References to these and other reports and information of interest are given in the following bibliography.

In addition, it may be noted that the formulae of Heywood and Lewis for stresses in gear-tooth fillets are included in section 2·32 of this handbook.

BIBLIOGRAPHY ON GEARS

(The numbers refer to the list given below.)

Books: 18, 19, 20, 21, 22, 61, 72
Backlash: 39, 56, 59, 73, 97
Bevel gears: 19, 95
Cyclic irregularity in gear drives: 84
Design: 6, 31, 51, 57, 61, 71, 72, 75, 85, 96
Durability: 1, 2, 3, 4, 35, 52
Elastically-mounted gearing: 15
Epicyclic gears: 12, 15, 16, 33, 82
Formulae: 9, 10, 16, 26, 27, 28, 42, 43, 45, 58, 71, 75, 81, 95, 96
Gearboxes: 50, 88
Gears, general: 8, 57, 68
Helical gears: 12, 18, 24, 25, 58, 74, 85, 86, 92, 93, 94, 98
Journal bearing displacement: 39, 97
Keyways for gear drives: 57
Load capacity and loading conditions: 44, 48, 49, 53, 76, 83, 96
Load distribution (number of teeth in contact): 74, 87, 96
Noise: 12, 23, 24, 29, 32, 46, 47, 63, 64, 80

Screw gears: 5
Shock loads: 40
Shrink fits on gears: 77
'Sira' gears: 81
Spiral bevel gears: 95
Spiral gears: 85
Spur gears: 14, 17, 18, 36, 38, 51, 69, 74, 79, 94
Stresses: 3, 4, 14, 26, 27, 28, 42, 43, 45, 48, 61, 79
Test apparatus: 9, 11, 60
Tooth deflexion and deformation: 13, 51, 54, 55, 58, 89, 91, 92, 96
Torque reversal: 41
Torsional vibration in gears: 15
Traction: 12
Transmissions: 7, 30, 31, 78
Turbine drives: 21
Vibration in gears: 87
Wear: 1, 12, 23, 31, 34, 35, 37, 52, 62, 65, 66, 67, 76, 90
Worm gears: 10, 20, 62, 70, 88, 91

1. Almen, J. O. and Boegehold, A. L. *Proc. Amer. Soc. Test. Mater.* vol. 35, pt. 11 (1935), pp. 99–135. (Wear; life; fatigue.)
2. Almen, J. O. and Straub, J. C. *Automot. Industr., N.Y.*, 25 Sept. and 9 Oct. 1937, pp. 426–32 and 488–93. (Durability.)
3. Almen, J. O. *S. A. E. Jl.* Feb. 1942, pp. 52–61. (Fatigue; stresses.)
4. Almen, J. O. *Machine Design*, July 1937. (Fatigue; stresses.)
5. Altmann, F. G. *Forschg IngWes.* Sept.–Oct. 1937, pp. 209–25. (Screw gears.)
6. Altmann, F. G. *Z. Ver. dtsch. Ing.* Feb. 1938, pp. 165–8. (Design of flanks.)
7. Altmann, F. G. *Z. Ver. dtsch. Ing.* 21 March 1951, pp. 205–8. (Transmissions.)
8. Altmann, F. G. *Z. Ver. dtsch. Ing.* 11 March 1954. (Gears, general.)
9. Altmann, F. G., Niemann, G., Rauh, K., Thomas, W. *et alii. Z. Ver. dtsch. Ing.* 21 Nov. 1950, pp. 897–958. (Tests and calculations.)
10. — *Automot. Industr., N.Y.*, 20 April 1929, p. 624. (Worm gears; formulae.)
11. — *Auto. Engr*, March 1941, pp. 97–9. (Tests.)
12. — *Z. Ver. dtsch. Ing.* 21 March 1951, pp. 205–52. (Noise; wear; fatigue strength; epicyclic gears for traction; helical gears.)
13. Banashek, K. D.S.I.R. Sponsored Research (Germany), Rep. no. 6 (1949). (Tooth deformation.)

14. Baud, R. V. and Hall, E. *Mech. Engng,* March 1931. (Stresses, spur gear teeth.)
15. Benz, W. *Luftfahrtforschung,* 27 Feb. 1943, pp. 46–7. (Torsional vibration of systems with elastically-mounted epicyclic gearing.)
16. Bernhard, J. M. *Glasers Ann. Gew.* 6 Feb. and 15 March 1941, pp. 35–43 and 112–17. (Calculation of epicyclic gears.)
17. Bradbury, G. V. *Amer. Mach., N.Y.,* 16 Aug. 1945, p. 117. (Spur gears.)
18. British Standards Institution. B.S. no. 436 (1940). (Helical and straight spur gears.)
19. British Standards Institution. B.S. no. 545 (1949). (Bevel gears.)
20. British Standards Institution. B.S. no. 721 (1937). (Worm gears.)
21. British Standards Institution. B.S. no. 1807, pt. 1 (1952). (Gears for turbine drives.)
22. Buckingham, E. *Analytical Mechanics of Gears,* 1st ed., McGraw-Hill, 1949.
23. Cady, E. L. *Mill Fact. Ill.* March 1948, pp. 100–3. (Wear; noise.)
24. Couling, S. A. *Proc. Instn Mech. Engrs, Lond.,* vol. 150, no. 5 (1943), pp. 172–94. (Helical gears; noise.)
25. Currie, E. M. *Mach. Design,* Jan. 1947, pp. 100–4. (Helical gears.)
26. Davey, H. T. *Engineering,* 1 May 1931, p. 580. (Stress formulae.)
27. Dolan, T. J. *J. Appl. Phys.* Aug. 1941, pp. 584–91. (Stress formulae.)
28. Dolan, T. J. A.G.M.A. meeting paper, Oct. 1941, 31 pp. (Stress formulae.)
29. Dorey, S. F. & Forsyth, G. H. *Trans. N.-E. Cst Instn Engrs Shipb.* vol. 63, pt. 6 (April 1947), pp. 267–330. (Noise.)
30. Douglas, L. M. *Engineering,* 17 and 31 Jan. 1941, pp. 57–60 and 98–100. (Transmissions.)
31. Findeisen, F. *Neuzeitliche Maschinenelemente.* Publ. by O. Elsner, Berlin, 1945. (Design; wear; transmissions.)
32. Firestone, F. A. and Abbott, E. J. *Iron Age,* 20 Dec. 1934, pp. 10–7. (Noise.)
33. Gaunitz, A. *Z. Ver. dtsch. Ing.* 1 April 1954, p. 334. (Epicyclic gears.)
34. Glaubitz, H. *Arch. tech. Mess.* Feb. 1948, T 43, 2 pp. (Wear.)
35. Glaubitz, H. D.S.I.R. Sponsored Research (Germany), Rep. nos. 1, 7 and 12. (Wear; fatigue.)
36. Glaubitz, H. D.S.I.R. Sponsored Research (Germany), Rep. no. 23 (1950). (Spur gears.)
37. Glaubitz, H. *Stahl u. Eisen. Düsseldorf,* 27 Sept. 1951, pp. 1041–4. (Wear.)
38. Guins, S. G. *Prod. Engng,* Jan. 1945, pp. 40–3. (Spur gears.)
39. Hain, K. *Z. Ver. dtsch. Ing.* 1954, pp. 250–4. (Backlash and journal bearing displacement.)
40. Halliday, W. M. *Mech. Handl.* Jan. 1941, pp. 8–9. (Shock loads.)
41. Hammer, A. *Prod. Engng,* April and May 1949, pp. 108–14 and 129–33. (Torque reversal.)
42. Heldt, P. M. *Automot. Industr., N.Y.,* 15 Oct. 1939, pp. 428–31. (Stress formulae.)
43. Heywood, R. B. *Proc. Instn Mech. Engrs, Lond.,* vol. 159 (1948). (Stress formulae.)
44. Hiersig. *Z. Ver. dtsch. Ing.* 11 March 1954. (Loading conditions; overload capacity.)
45. Hofer, H. *Werkstattstechnik,* 1 March 1931, pp. 128–31. (Stress formulae.)
46. Hofer, H. *Maschinenbau,* Aug. 1935, pp. 433–5. (Gear noise.)
47. Hofer, H. *Werkstattstechnik,* 1 March 1935, pp. 92–5. (Noise.)
48. Hofer, H. *B.I.O.S. Final Rep.* no. 1034, item 18. (Stresses; load capacity.)
49. Ide, H. *Luftfahrtforschung,* 20 May 1940, pp. 130–44. (Loading.)
50. Jennings, J. *Mech. World,* 4 March 1949, pp. 247–51. (Gearbox beam problems.)

51. Karas, F. *V.D.I.-Forschungsheft*, B, vol. 12 (Jan.–Feb. 1941), 17 pp. (Tooth deflexion of spur gears.)
52. Karas, F. *Z. Ver. dtsch. Ing.* 5 April 1941, pp. 341–4. (Wear; fatigue.)
53. Knibbe, K. *J. Aero. Sci.* Dec. 1939, pp. 68–71. (Loading.)
54. Lehnert, H. *D.S.I.R.* Sponsored Research (Germany), Rep. no. 22 (1950). (Tooth deformation.)
55. Lehnert, H. *Z. Ver. dtsch. Ing.* 11 March 1954. (Tooth deformation.)
56. Lofgren, K. E. *Prod. Engng*, July 1947, pp. 150–2. (Backlash.)
57. McArd, G. *Industr. Pwr, Lond.*, April 1946, pp. 148–52. (Gear drives; keyways; design.)
58. MacGregor, C. W. *Mech. Engng, N.Y.*, April 1935, p. 225. (Helical gear tooth, deflexion, formulae.)
59. Martin, L. D. *Prod. Engng*, July 1948, pp. 108–12. (Backlash.)
60. Meldahl, A. *Engineering*, 21 July 1939, pp. 63–6. (Test apparatus.)
61. Merritt, H. E. *Gears*, 2nd ed. (Pitman, 1946).
62. Merritt, H. E. *Proc. Instn Mech. Engrs, Lond.*, vol. 129 (1935), pp. 127–94. (Wear of worm gears.)
63. Mills, K. N. *Mach. Design*, Jan. 1949, pp. 88–92. (Gear noise.)
64. Moncrieff, D. *Instruments*, Oct. 1946, pp. 586–8. (Noise.)
65. Monk, I. *Pap. Amer. Soc. Mech. Engrs*, no. 48-A-50, 11 pp. (Wear.)
66. Niemann, G. *Werkst. u. Betr.* Feb. 1938, pp. 29–31. (Wear.)
67. Niemann, G. *Z. Ver. dtsch. Ing.* 21 Aug. 1943, pp. 521–3. (Wear.)
68. Niemann, G. *Z. Ver. dtsch. Ing.* 21 Nov. 1950. (Gears; general.)
69. Niemann, G. & Glaubitz, H. *Z. Ver. dtsch. Ing.* 21 March 1951, pp. 215–22. (Spur gears.)
70. Niemann, G. *Z. Ver. dtsch. Ing.* 21 Feb. 1953, pp. 147–57. (Worm gears.)
71. Nuebling, O. *Luftfahrtforschung*, 20 May 1940, pp. 145–53. (Design formulae.)
72. Peterson, R. E. *Mark's Mechanical Engineers' Handbook*, 4th ed. (McGraw-Hill, 1951), pp. 947–63.
73. Pohl, W. M. *Prod. Engng*, Jan. 1942, pp. 1–4. (Backlash.)
74. Poritsky, H., Sutton, A. D. and Pernick, A. *J. Appl. Mech.* (A.S.M.E. Trans.), June 1945, pp. A 78–A 86. (Helical gears; spur gears; load distribution.)
75. Rasmussen, A. C. *Prod. Engng*, July 1939, p. 301. (Design formulae.)
76. Rasmussen, A. C. *Machinist*, 12 and 26 Aug. 1939. (Loading; wear.)
77. Sawin, N. N. *Amer. Mach., N.Y.*, 19 June 1947, pp. 142–5. (Shrink fits on gears.)
78. Scharbach, K. *Z. Ver. dtsch. Ing.* 21 March 1951, pp. 223–39. (Transmissions and linkages.)
79. Schlesinger, G. *Engineering*, 23 Oct. and 20 Nov. 1936, pp. 457–9 and 567–8. (Spur gear stresses.)
80. Schmitter, W. P. *Mach. Design*, Oct. 1936, pp. 42–4. (Gear noise.)
81. "Sira" gears with trapezoidal teeth and parallel shafts. *Engineer*, 11 June 1954, p. 866.
82. Tank, G. *Z. Ver. dtsch. Ing.* 1 April 1954, pp. 305–8. (Epicyclic gears.)
83. Thum, A. *Schweiz. Arch.* Oct. 1952, pp. 309–20. (Loading conditions; overload capacity.)
84. Tuplin, W. A. *Engineer*, 18 May 1934, pp. 502–3 and 23 Aug. 1935, pp. 199–200. (Cyclic irregularity in gear drives.)
85. Tuplin, W. A. *Machinery*, 11 March, 24 June and 12 Aug. 1937, pp. 713–18, 393–4 and 612–24; *Engineering*, 23 Feb. 1940, pp. 187–90. (Helical and spiral gears, formulae.)
86. Tuplin, W. A. *Trans. Inst. Mar. Engrs*, vol. 60, no. 4 (May 1948), pp. 107–17. (Helical gears.)

87. Tymstra, S. R. *Bull. Wash. Univ. Exp. Sta.* no. 109, Sept. 1942, 10 pp. (Number of teeth in contact as vibration factor in gears.)

88. Walker, H. *Machinery*, 15 April 1937, pp. 71–3. (Worm gear boxes.)

89. Walker, H. *Engineer*, 14 and 18 Oct. 1938, pp. 409–12 and 434–6. (Tooth deformation.)

90. Walker, H. *Engineer*, 22 and 29 Dec. 1944, pp. 484–6 and 502–4. (Wear.)

91. Walker, H. *Auto. Engr*, June 1945, pp. 239–44. (Worm gear, deflexion.)

92. Walker, H. *Engineer*, 12 July 1946, pp. 46–8. (Tooth deformation, helical gears.)

93. Walker, H. *Engineer*, 26 July 1946, pp. 70–1. (Helical gears.)

94. Watts, F. G. *Machinery*, 16 Dec. 1948, pp. 94, 96, 98. (Helical spur gears.)

95. W. C., D. *Engineer*, 22 and 29 June 1923, pp. 649 and 677. (Spiral bevel gears, formulae.)

96. Weber, C. D.S.I.R. Sponsored Research (Germany), Rep. nos. 3, 4, 5, 6 and 15 (1949–1950). (Tooth deflexion formulae; loading conditions; overload capacity.)

97. Weber, C. D.S.I.R. Sponsored Research (Germany), Rep. no. 16 (1949). (Backlash and journal bearing displacement.)

98. Weber, C. D.S.I.R. Sponsored Research (Germany), Rep. nos. 5 and 15 (1949). (Helical gears.)

1·233 *Equivalent torsional stiffness of belts*

The arrangement considered is a belt drive with two pulleys.

The value of its torsional stiffness depends on whether the torque T' or the torque $T'' = (R''/R') \times T'$ is taken as reference value.

(*a*) *With reference to the torque T'*, the torsional stiffness of the belt is

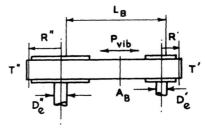

$$K' = \beta \times \frac{E_B A_B}{L_B} \times R'^2 \quad \text{[Lb.in./rad.]},$$

where $R' =$ pulley radius [in.], $L_B =$ effective length of belt [in.], $A_B =$ cross-sectional area of belt [in.^{2}], and $E_B =$ modulus of elasticity of belt [Lb./in.^{2}].

The value of the 'belt coefficient' β depends on the operating conditions:

(i) For belts with initial tension greater than the pull due to steady transmitted torque, and small vibratory forces: $\beta = 2·0$.

Examples. Vee-belts, belts with spring-loaded jockey pulleys, etc.

(ii) For belts with practically no initial tension, and small vibratory forces: $\beta = 1·0$.

Examples. Flat belts of non-elastic material, after considerable running.

[91]

(iii) For belts with large vibratory forces exceeding the initial tension $P_{\text{in.}}$ and the effective pull $P_{\text{eff.}}$ due to transmitted torque: $\beta \simeq 1\cdot5$.

Examples. All types of belts, when large vibratory forces occur which exceed the effective pull $P_{\text{eff.}}$.

| 1st half of cycle | 2nd half of cycle |

The equivalent length of the belt, in terms of a shaft of diameter D'_e and rigidity modulus G'_e connected to the pulley R', is

$$L'_e = \frac{\pi}{32} \times \frac{G'_e D'^4_e}{\beta E_B A_B R'^2} \times L'_B \quad \text{[in.]}.$$

(b) *With reference to the torque* T'' *and the shaft of diameter* D''_e *and rigidity modulus* G''_e, the corresponding values are

$$K'' = \beta \frac{E_B A_B}{L_B} \times R''^2 = K' \times \left(\frac{R''}{R'}\right)^2 \quad \text{[Lb.in./rad.]},$$

$$L''_e = \frac{\pi}{32} \times \frac{G''_e D''^4_e}{\beta E_B A_B R''^2} \times L_B \quad \text{[in.]},$$

$$L''_e = \frac{G''_e}{G'_e} \times \left(\frac{D'_e}{D''_e}\right)^4 \times \left(\frac{R'}{R''}\right)^2 \times L'_e \quad \text{[in.]}.$$

Metric units. All the relations given in this section are directly applicable in metric units, after replacing Lb. by Kg., and in. by cm.

Derivation. The initial tension $P_{\text{in.}}$ is usually determined from the effective pull $P_{\text{eff.}}$ [Lb.] corresponding to the steady transmitted torque by means of the formulae

$$P_{\text{in.}} = P_{\text{eff.}} \cdot \frac{e^{\mu\phi} + 1}{2(e^{\mu\phi} - 1)}, \quad \text{and} \quad P_{\text{eff.}} = 550H/v,$$

where H = transmitted power [h.p.], v = speed of belt [ft./sec.], μ = friction coefficient for pulley and belt materials ($0\cdot2$–$0\cdot5$) and ϕ = angle of contact of belt and pulley. It should be ascertained that the belt system actually has the required initial tension $P_{\text{in.}}$.

Case (i). High initial tension $P_{\text{in.}}$, effective pull $P_{\text{eff.}}$ due to steady transmitted torque, and small vibratory force $P_{\text{vib.}} \cos \omega t$, where ω = natural phase velocity of the vibration. As the vibratory force $P_{\text{vib.}}$ acts on both the 'taut' and 'slack' portions of the belt, in this case, the total torque acting on the pulley of radius R' is

$$T'_{\text{total}} = R'[P_{\text{eff.}} + 2P_{\text{vib.}}],$$

considering maximum values only, for convenience, and the vibratory torque is

$$T'_{\text{vib.}} = 2R'P_{\text{vib.}}.$$

The increase or decrease in belt elongation δ_B occurs both on the 'taut portion' and the 'slack portion' of the belt, and corresponds to an angular displacement θ', where

$$\theta' = \delta_B/R'.$$

Therefore, the equivalent torsional stiffness of the belt-and-pulley system is

$$K' = \frac{T'_{\text{vib.}}}{\theta'} = \frac{2R'P_{\text{vib.}}}{\delta_B/R'} = 2R'^2 P_{\text{vib.}}/\delta_B,$$

and since $\delta_B = P_{\text{vib.}} \times L_B/(E_B A_B)$, this gives the formula at the beginning of this section, with $\beta = 2$.

Case (ii). Practically no initial tension ($P_{\text{in.}} \cong 0$), and a vibratory force $P_{\text{vib.}} \cos \omega t$ always less than the pull $P_{\text{eff.}}$ due to steady transmitted torque.

In this case, the total torque is transmitted only by the 'taut portion':

$$T'_{\text{total}} = R'[P_{\text{eff.}} + P_{\text{vib.}}],$$

the vibratory torque is $T'_{\text{vib.}} = R'P_{\text{vib.}}$ and the stiffness is

$$K' = T'/\theta' = R'^2 P_{\text{vib.}}/\delta_B = R'^2 E_B A_B/L_B,$$

which is half the value obtained in the preceding case ($\beta = 1$).

Case (iii). Vibratory force $P_{\text{vib.}} \cos \omega t$, with a peak value exceeding $P_{\text{eff.}}$. (On the 'slack portion', the initial tension is reduced by the pull due to the steady transmitted torque.)

The vibratory torque during the first half of the cycle has an approximate value

$$T'_{\text{vib.}} \cong 2R'P_{\text{vib.}} \cos \omega t.$$

During the second half of the cycle, the vibratory torque is due to the 'taut portion', only and its approximate value is

$$T'_{\text{vib.}} \cong R'P_{\text{vib.}} \cos \omega t.$$

Therefore, for the entire cycle, we have $T'_{\text{vib.}} \cong 1 \cdot 5 R'P_{\text{vib.}} \cos \omega t$ and the average torsional stiffness is

$$K \cong 1 \cdot 5 E_B A_B R'^2/L_B.$$

The calculation of systems taking account of belt stiffness is described in detail in section 1·3 (natural frequencies).

Note. The effective length L_B is assumed to be the length between the points of contact as determined by the geometry of the stationary system. The dead weight does not appreciably affect this value in usual engine applications. The sagging length is a portion of a catenary curve.

The effect of the centrifugal forces is to increase the value of the effective length L_B, particularly at high speeds and light loads. The upward component of the centrifugal forces tends to raise the belt off the pulleys and the belt stiffness is decreased under these conditions.

If the pull P varies appreciably during the pulley rotation, owing to cyclic irregularity or torsional vibration of the engine system, this can cause variations in belt slack and centrifugal force, which may excite transverse belt vibrations (flapping) and result in abnormal wear. To reduce this possibility, it is advisable to take account of the possible modes of transverse vibration in the design calculations of the belt drive. The *flapping frequencies* can be calculated from

$$F_{\text{I, II, III}} = 30 \times C \times \sqrt{\frac{P/L_B}{W_B/g}} \quad [\text{vib./min.}],$$

where C has the values 1, 2 and 3 for F_{I}, F_{II} and F_{III}. In this equation† $P =$ pull on taut side of belt [Lb.], $L_B =$ effective belt length [in.], $W_B =$ weight [Lb.] of the effective belt length L_B, and $g = 386\cdot4$ in./sec.².

At running speeds which are sub-multiples of these frequencies, the possibility of belt resonance should be considered. Resonances in the running speed range may be avoided by altering the values of W_B or L_B. In tests, the transverse movement of the belt can be observed stroboscopically.

1·234 *Equivalent torsional stiffness of chain drives*

The equivalent length of a chain can be evaluated with the above formulae for belts, using, however, $\beta = 1\cdot0$. The value of A_C is determined as the average total cross-sectional area of the links.

The pinion tooth deflexion gives a further equivalent length $L_{e(P)}$ which should be added to the equivalent length $L_{e(C)}$ of the chain itself. The formulae for the stiffness of spur gearing, given in a previous section, are applicable to pinion wheels of chain drives, within certain limits. The force P per tooth may be regarded as equal to the chain pull divided by the number of teeth in one quadrant. The stiffness values of the formulae should be doubled (or the L_e-values halved), since the pinion tooth is not meshing with another tooth. In many instances the deflexion of the pinion, relative to the elongation of the chain, is small enough to be neglected.

Determination of the elasticity modulus

The value of E_B or E_C (for a belt or chain) can be determined statically, from a tensile test of the belt or chain, as the tangent of the stress/strain line, disregarding the curved portion at the bottom of the diagram, which represents initial permanent stretch.

A load/elongation diagram for the chain with oil can be determined to obtain the 'overall elasticity modulus' of the chain under normal working

† This is the frequency equation of a string with a uniformly distributed mass, clamped at its two ends. The derivation is given, for instance, in *Mechanical Vibrations*, by J. P. Den Hartog (McGraw-Hill), 2nd ed. p. 174.

conditions. Disregarding the curved portion of the diagram at very low loads, the slope of the straight-line portion is used to obtain

$$E = (\Delta P / \Delta \delta) \times (L/A).$$

The general indications given in the foregoing can be regarded as considerations to be borne in mind at the design stage. At present, experimental data on belt and chain drives are only beginning to be collected, and tests with stroboscopic observation are advisable for every new system developed.

1·24 Stiffness of couplings: Torsional characteristics of couplings and clutches

The problems arising from the use of couplings have received increasing attention as the accuracy of stiffness calculations and frequency measurements improved in regard to other elements of engine systems.

The difficulties occur because the action obtained with many coupling designs and the engineering properties of the materials employed in them are not fully determined in many instances. Moreover, it is not always a simple matter to predict the dynamic behaviour of couplings in vibrating systems, particularly for couplings having a stiffness which varies according to the operating conditions. Rubber, whether loaded in shear, tension or compression, has a stiffness which depends on strain and other factors. Steel spring elements also have a variable stiffness in many applications, mainly, however, because of the features incorporated in the supporting or clamping elements with which they are used.

1·241 *Formulae, test procedures and evaluation methods*

1·2411 *Keyed couplings on straight shaft*

(1) *Shaft and coupling of same material.* The keyed shaft is assumed to twist independently of the coupling for one-third of the total length of the coupling, as measured from the coupling boss end.

Thus the coupling in Fig. 1† is evaluated by subdividing it into three lengths L_1, L_2 and L_3, with corresponding diameters D_1, D_2 and D_3, as follows:

$$L_1 = \tfrac{1}{3}L, \quad L_{e1} = L_1 \times \left(\frac{D_e}{D_1}\right)^4,$$

D_e being the reference diameter. The remainder of the coupling twists with the shaft, so that

$$L_2 = L - (L_1 + L_3) = \tfrac{2}{3}L - L_3, \quad L_{e2} = [L_2 + \Delta L] \times \left(\frac{D_e}{D_2}\right)^4,$$

† Although Fig. 1 shows two keys at 180°, the above formulae apply equally well to shafts with one key or with two keys disposed at other angles.

where $\Delta L =$ additional length due to the flexibility of the junction (transition from D_2 to D_3), evaluated as shown under 'Stepped shafts' in section 1·21. Finally,

$$L_{e3} = L_3 \times \left(\frac{D_e}{D_3}\right)^4 \quad \text{and} \quad L_e = L_{e1} + L_{e2} + L_{e3},$$

D_3 being the pitch circle diameter, as shown in Fig. 1. The contribution of the spigot of length L_4 is assumed to be negligible, so that $L_{e4} = 0$.

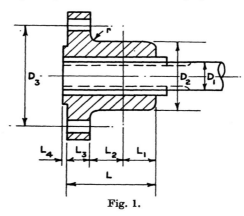

Fig. 1.

(2) *Shaft and coupling of different materials.* Let

$$G_s = \text{modulus of rigidity of shaft,}$$
$$G_c = \text{modulus of rigidity of coupling,}$$

and $\qquad G_e = \text{modulus of rigidity of reference shaft.}$

Subdividing into three sections, as previously, we have

$$L_{e1} = \frac{G_e}{G_s} \times \left(\frac{D_e}{D_1}\right)^4 \times L_1, \quad \text{where} \quad L = \tfrac{1}{3}L.$$

The shaft section $L_2 = \tfrac{2}{3}L - L_3$ consists of a shaft portion L_{2s} and a boss portion L_{2c}, so that

$$L_{2s} = \frac{G_e}{G_s} \times \left(\frac{D_e}{D_1}\right)^4 \times [L_2 + \Delta L] \quad \text{and} \quad L_{2c} = \frac{G_e}{G_c} \times \left(\frac{D_e^4}{D_1^4 - D_2^4}\right) \times [L_2 + \Delta L].$$

Therefore, since $\dfrac{1}{L_{e2}} = \dfrac{1}{L_{2s}} + \dfrac{1}{L_{2c}}$, the equivalent length of the L_2 section is obtained as

$$L_{e2} = \frac{\dfrac{G_e}{G_s} \times \left(\dfrac{D_e}{D_1}\right)^4 \times [L_2 + \Delta L]}{1 + \dfrac{G_c}{G_s}\left\{\left(\dfrac{D_2}{D_1}\right)^4 - 1\right\}}.$$

Similarly, for the third portion of the coupling

$$L_{e3} = \frac{\dfrac{G_e}{G_s} \times \left(\dfrac{D_e}{D_1}\right)^4 \times L_3}{1 + \dfrac{G_c}{G_s}\left\{\left(\dfrac{D_3}{D_1}\right)^4 - 1\right\}},$$

so that, for the entire coupling, $L_e = L_{e1} + L_{e2} + L_{e3}$.

Note. The additional length for the flexibility at the junction is determined from the corresponding graph in section 1·215 (Stepped shafts) for the diameter ratio D_3/D_2 and the fillet radius/shaft radius ratio r/R_2 considered.

1·2412 *Keyed coupling on tapered shaft*

Notation as indicated in Fig. 2. The subscripts s and c refer to the shaft and the coupling, respectively, and the subscript e refers to an 'equivalent shaft' used as a reference shaft in the calculation.

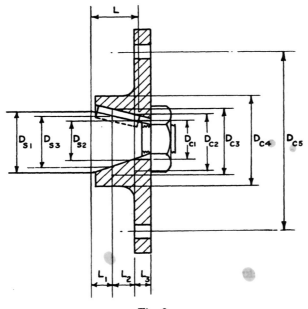

Fig. 2.

The procedure is similar to that used for a coupling keyed on a straight shaft. The evaluation is fairly detailed:

$$L_1 = \tfrac{1}{3}L,$$

where L = length from beginning of taper to end of key;

$$L_{e1} = L_1 \frac{G_e}{G_s} \times \frac{D_e^4}{3(D_{s1} - D_{s3})} \times \left[\frac{1}{D_{s3}^3} - \frac{1}{D_{s1}^3}\right];$$

L_2 = length of shaft from end of L_1 to face of flanged section.

Shaft portion:

$$L_{e2(s)} = L_2 \frac{G_e}{G_s} \times \frac{D_e^4}{3(D_{s3} - D_{s2})} \times \left[\frac{1}{D_{s2}^3} - \frac{1}{D_{s3}^3}\right].$$

Boss portion:

$$L_{e2(c)} = L_2 \times \frac{G_e}{G_c} \times \frac{D_e^4}{4D_{c4}^3(D_{c3} - D_{c2})} \times \left[\log_e Y - 2\tan^{-1}\frac{D_{c2}}{D_{c4}} + 2\tan^{-1}\frac{D_{c3}}{D_{c4}}\right],$$

where $\qquad Y = (D_{c4} - D_{c2})(D_{c4} + D_{c3})/[(D_{c4} + D_{c2})(D_{c4} - D_{c3})]$

and $\qquad\qquad\qquad \log_e Y = 2{\cdot}303 \log_{10} Y.$

Therefore $$L_{e2} = \frac{1}{\dfrac{1}{L_{e2(s)}} + \dfrac{1}{L_{e2(c)}}}.$$

Finally, $$L_{e3} \cong L_3 \times \frac{G_e}{G_c} \times \frac{D_e^4}{D_{c5}^4 - \left(\dfrac{D_{c2} + D_{c1}}{2}\right)^4},$$

and $$L_e = L_{e1} + L_{e2} + L_{e3}.$$

In some cases (thin coupling flange) it may be necessary to add a term L_{e4} for the flexibility of the coupling disk. For its calculation, see 'Coupling disks', in the following section.

1·2413 Torsional stiffness of a coupling disk: rubber annulus in concentric sleeves

The metal disk of Fig. 1, and the rubber annulus of Fig. 2, are both subjected to similar torsional conditions. For both cases the torsional stiffness can be determined as

$$K \cong \frac{4\pi LG}{\dfrac{1}{R_i^2} - \dfrac{1}{R_o^2}} \quad [\text{Lb.in./rad.}], \quad (1)$$

where L = disk thickness [in.], G = rigidity modulus of metal disk or rubber annulus [Lb./in.^{2}], R_i = inner radius [in.], and R_o = outer radius [in.].

In cases where the value of the expression $T/(2\pi LG)$ is greater than unity, a more accurate expression for stiffness is necessary, viz.

$$K \cong \frac{4\pi LG}{\left[\dfrac{1}{R_i^2} - \dfrac{1}{R_o^2}\right] + \dfrac{1}{9}\left[\dfrac{T}{2\pi LG}\right]^2 \times \left[\dfrac{1}{R_i^6} - \dfrac{1}{R_o^6}\right]} \quad [\text{Lb.in./rad.}]. \quad (2)$$

This expression shows that, for large twist deflexions, the torsional stiffness decreases with increasing torque T [Lb.in.].

In terms of a reference shaft of diameter D_e and rigidity modulus G_e, the equivalent length of the disk or rubber annulus is

$$L_e = \frac{32}{\pi} \times \frac{K}{G_e D_e^4} \quad \text{[in.]},$$

where K is the stiffness determined by either eq. (1) or eq. (2).

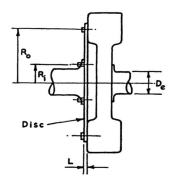

Fig. 1.

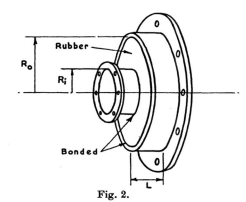

Fig. 2.

Metric units. In this section, the stiffness values can be obtained in Kg.cm./rad. when all linear dimensions are expressed in cm., torques in Kg.cm. and moduli of rigidity in Kg./cm.[2].

Derivation.† The torque, at any instant, has a constant value T throughout the disk, which is related to the shearing force P and the shear stress q, at any radius R, by the equations

$$T = RP = R \times 2\pi RLq.$$

The shear strain γ is given by

$$\gamma = q/G \quad \text{and} \quad \gamma = \tan^{-1}\left[R\frac{d\theta}{dR}\right],$$

the latter relation being obtained by consideration of Fig. 3. Therefore

$$\gamma = \frac{T}{2\pi LR^2} \times \frac{1}{G} = \tan^{-1}\left[R\frac{d\theta}{dR}\right],$$

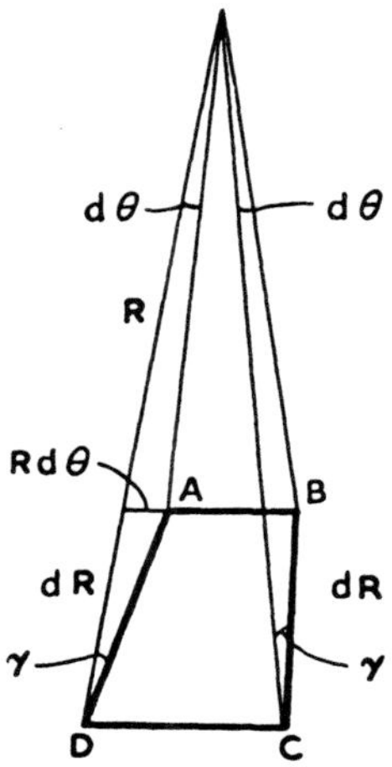

Fig. 3.

† These formulae have been derived by Carter and by J. F. D. Smith. See B. C. Carter: *Torsional Vibration in Aircraft Power Plants: Methods of Calculation*, Ministry of Supply, Aeronautical Research Council Technical Report, R. and M. no. 2739 (3519) (H.M.S.O. 1952); J. F. D. Smith, *India Rubb. World*, vol. 100 (August 1939), p. 48.

and the angular displacement contributed by the element $ABCD$ in Fig. 3 is

$$d\theta = \frac{dR}{R} \tan\left[\frac{T}{2\pi LG} \times \frac{1}{R^2}\right] = \frac{dR}{R} \tan\frac{w}{R^2},$$

where $w = T/[2\pi LG]$. Replacing $\tan(w/R^2)$ by its series expansion

$$\tan\frac{w}{R^2} = \frac{w}{R^2} + \frac{1}{3}\left(\frac{w}{R^2}\right)^3 + \frac{2}{15}\left(\frac{w}{R^2}\right)^5 + \cdots,$$

we have

$$d\theta = \left[\frac{w}{R^3} + \frac{w^3}{3R^7} + \frac{2}{15} \times \frac{w^5}{R^{11}} + \cdots\right]dR = f(R)\,dR,$$

so that, if $\theta = 0$ at $R = R_o$, and $\theta = \theta_0$ at $R = R_i$, the integration gives

$$\theta_0 = \int_{R_i}^{R_o} f(R)\,dR = \frac{w}{2}\left[\frac{1}{R_i^2} - \frac{1}{R_o^2}\right] + \frac{w^3}{18}\left[\frac{1}{R_i^6} - \frac{1}{R_o^6}\right] + \cdots.$$

Therefore, the stiffness is determined by

$$K = \frac{T}{\theta_0} = \frac{4\pi LG}{\left[\dfrac{1}{R_i^2} - \dfrac{1}{R_o^2}\right] + \dfrac{1}{9}\left[\dfrac{T}{2\pi LG}\right]^2 \times \left[\dfrac{1}{R_i^6} - \dfrac{1}{R_o^6}\right] + \cdots}.$$

Hyperbolic contours

If $R^2 L = R_o^2 L_o = R_i^2 L_i = $ const. (see Fig. 4), then the shear stress $q = q_0$ is constant throughout the disk. Denoting by γ_0 the constant strain $T/[2\pi R_o^2 L_o G] = \tan^{-1}(R\,d\theta/dR)$, one obtains the total angle of twist as

$$\theta_0 = \tan\gamma_0 \log_e(R_o/R_i) \cong \frac{T}{2\pi R_o^2 L_o G} \times \log_e(R_o/R_i)$$

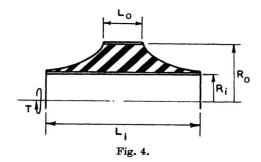

Fig. 4.

and the stiffness $\qquad K = 2\pi R_o^2 L_o G / \log_e(R_o/R_i);$

If $R^3 L = R_o^3 L_o = R_i^3 L_i$, then q increases with R, also

$$\gamma = TR/[2\pi L_o G R_o^3] = w^* R;$$

and, proceeding as in the previous derivations, the results are

$$\theta_0 \cong w^*(R_o - R_i) + \frac{w^{*3}}{9}(R_o^3 - R_i^3) + \dots$$

and $$K \cong 2\pi L_o GR_o^3 \Big/ \Big\{ [R_o - R_i] + \Big[\frac{T}{6\pi L_o GR_o^3}\Big]^2 \times [R_o^3 - R_i^3] + \dots \Big\}.$$

1·2414 *Hollow rubber sandwich coupling*

The coupling consists of a rubber annulus bonded to metal plates on its flat sides. The stiffness, depending on the strain angle γ_0 at the outer radius R_o, can be evaluated by means of the following expressions:

1. Shear angle

$$\gamma_0 \leqslant 5°$$

(corresponding to an angle of twist $\theta_0 \leqslant 0\cdot087L/R_0$ radians):

$$K = \frac{\pi G}{32L}[D_o^4 - D_i^4] \quad [\text{Lb.in./rad.}],$$

where $G =$ modulus of rigidity [Lb./in.^{2}] of rubber, $L =$ thickness [in.] of rubber, $D_o = 2R_o =$ outer diameter [in.], and $D_i = 2R_i =$ inner diameter [in.].

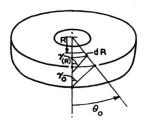

2. Shear angle $\qquad \gamma_0 \geqslant 5°$

(corresponding to an angle of twist $\theta_0 \geqslant 0\cdot087L/R_o$ radians):

$$K = \frac{\pi G}{32L}\Big[(D_o^4 - D_i^4) - \frac{\theta_0^2}{18L^2}(D_o^6 - D_i^6)\Big] \quad [\text{Lb.in./rad.}].$$

This latter formula takes account of the decrease in stiffness occurring at large strain angles.

Derivation. At any radius R, the strain in the rubber due to displacement of the upper plate through an angle θ_0 is $\gamma_{(R)} = \tan^{-1}[R\theta_0/L]$, hence the stress at R is

$$q_{(R)} = G\gamma_{(R)} = G\tan^{-1}[R\theta_0/L].$$

The annular element $2\pi R\,dR$ contributes to the resistance torque T by an amount $$dT = 2\pi R\,dR \times q_{(R)} \times R,$$

so that the total torque is determined by

$$T = 2\pi G \int_{R_i}^{R_o} R^2 \tan^{-1}\Big[\frac{R\theta_0}{L}\Big] dR$$

[101]

Substituting $\qquad \tan^{-1}[R\theta_0/L] \cong (R\theta_0/L) - \frac{1}{3}(R\theta_0/L)^3$

and integrating, we obtain

$$T \cong 2\pi G \left[\frac{R^4}{4L}\theta_0 - \left(\frac{\theta_0}{L}\right)^3 \frac{R^6}{18} \right]_{R_i}^{R_o}.$$

By taking the limits R_o and R_i, and using the relations $D_o = 2R_o$, $D_i = 2R_i$, and $K = T/\theta_0$, the above expression for stiffness is obtained.

It should be noted that both expressions are approximations for large angles. An exact expression valid for large strain angles (and necessary when $\gamma > 20°$) is given in the following lines.

3. Shear angle $\qquad\qquad \gamma \geqslant 20°$

(corresponding to an angle of twist $\theta_0 \geqslant 0\cdot 364 L/R_0$):

$$K = \frac{2\pi G}{3\theta_0}$$

$$\times \left[R_o^3 \tan^{-1}\frac{R_0\theta_0}{L} - R_i^3 \tan^{-1}\frac{R_i\theta_0}{L} - \frac{L}{2\theta_0}(R_o^2 - R_i^2) + \frac{L^3}{2\theta_0^3} \log_e \frac{\left(\frac{R_0\theta_0}{L}\right)^2 + 1}{\left(\frac{R_i\theta_0}{L}\right)^2 + 1} \right]$$

[Lb.in./rad.].

Derivation. Using $R\theta_0/L = x$, $\theta_0 dR/L = dx$, the integral already obtained becomes

$$T = 2\pi G \int_{x_i}^{x_o} \left(\frac{L}{\theta_0}\right)^3 x^2 \tan^{-1} x\, dx,$$

which can be readily integrated by parts to obtain the above result.

Thus this decrease in stiffness occurring with large θ_0 angles is characteristic of the geometrical conditions. There is also a further decrease due to the rubber properties, which give smaller G-values at large angles.

The exact expression, given above, reduces to the usual value

$$K = \frac{\pi}{2}\frac{G}{L}(R_o^4 - R_i^4)$$

for very small angles, as can be seen by substituting in it the approximations $\tan^{-1}(x) \cong x$ and $\log |1 + x^2| \cong x^2 - (x^4/2)$. The evaluation of the exact expression requires a desk-type computing machine and is only practical for fairly large angles.

EXAMPLE. Evaluation of a hollow sandwich coupling on the basis of the following data:

$$R_o = 3\cdot 5 \text{ in.}, \qquad L = 0\cdot 875 \text{ in.}, \qquad R_o/L = 4\cdot 0, \qquad G = 200 \text{ Lb./in.}^2$$
$$R_i = 1\cdot 75 \text{ in.}, \qquad\qquad\qquad\qquad R_i/L = 2\cdot 0,$$

Angle of twist	Stiffness	Evaluation by
$\theta_0 = 0\cdot020$ rad.	$K = 50\cdot52 \times 10^6$ Lb.in./rad.	Method 1
$\theta_0 = 0\cdot0375$ rad.	$K = 50\cdot25 \times 10^6$ Lb.in./rad.	Method 2
$\theta_0 = 0\cdot050$ rad.	$\begin{cases} K = 50\cdot06 \times 10^6 \text{ Lb.in./rad.} \\ K = 49\cdot99 \times 10^6 \text{ Lb.in./rad.} \end{cases}$	Method 2 Method 3
$\theta_0 = 0\cdot100$ rad.	$K = 48\cdot74 \times 10^6$ Lb.in./rad.	Method 3

Note. Generally, the moduli of rubber compounds are not constant but depend on the applied load. The modulus of rigidity G and the tensile modulus of elasticity E_t decrease with increasing loads (see section 1·244), whereas the compression modulus E_c increases with applied loads.

1·2415 *Peripheral spring couplings*

The equivalent length L_e of a peripheral-spring coupling of the type shown in Fig. 1 is given by

$$L_e = \frac{\frac{\pi}{32} G_e D_e^4}{N_s K_s R^2} \quad \text{[in.]},$$

where N_s = number of active springs, K_s = compression stiffness of each spring [Lb./in.], R = effective radius [in.], G_e = modulus of rigidity [Lb./in.²] of reference shaft, and D_e = diameter [in.] of reference shaft.

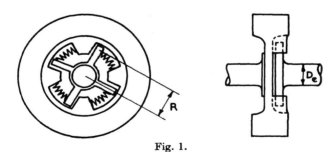

Fig. 1.

The derivation of this expression is based on the spring deflexion $s = R\theta$ in the peripheral direction, under an applied torque $T = N_s K_s . sR$. The coupling stiffness is therefore

$$K = T/\theta = N_s K_s R^2 \theta/\theta = N_s K_s R^2 \quad \text{[Lb.in./rad.]},$$

and the equivalent length of the coupling is obtained as usual in problems of this kind from $K = \frac{\pi}{32} G_e D_e^4/L_e$, which gives the above formula.

Couplings with springs on both sides of the torque arms

(*a*) If the springs are not clamped at their ends and not preloaded, the above formulae are valid, with

$$N_s = \text{number of torque arms,}$$

since the springs are only acted upon in compression.

(*b*) If the springs are preloaded on both sides, the above formulae are valid with
$$N_s = \text{number of springs,}$$

that is, the coupling is twice as stiff as in case (*a*).

The fact that the coupling is twice as stiff as in case (*a*) can be proved by energy methods, or by means of the following considerations:

If the coupling deflects a given amount from the neutral position, there is an increase in load on all the springs acting one way and an equal decrease on the other springs; the change in torque is thus proportional to the total number of springs, and this holds also for the stiffness, since $K = T/\theta$.

Note. Spring couplings are frequently incorporated in gear systems.

1·2416 *Couplings with non-linear characteristics: pin-type couplings*

Pin-type couplings with rubber bushes (see Fig. 1) have a variable stiffness, that is, a torque/deflexion characteristic which is not a straight line.

To determine the dynamic stiffness of such a coupling, it is necessary to consider (1) methods for obtaining the basic torque/deflexion characteristics, (2) the dynamic characteristics of couplings, and (3) the use of basic torque/deflexion characteristics in the evaluation of a system including a coupling of this type.

1. Methods for obtaining basic torque/deflexion characteristics

(*a*) '*Zero-frequency*' *tests of a single bush*

Note. There is no such thing as a 'static' value of stiffness for rubber. When subjected to a certain load, the rubber yields continuously, although at a gradually decreasing rate, and when the load is removed recovery is also gradual.

Thus, if a bush is tested by gradually increasing the applied load, and then decreasing it to zero load, the entire test may be carried out in, say, 10 min. The test frequency is then 0·1 vib./min. It is therefore suggested that such tests be called 'zero-frequency tests', implying that they are performed at a very low frequency, of the order of 0·1 vib./min.

The 'zero-frequency' test can be carried out as follows: Direct loads *P* [Lb.] are applied at both ends of the pin (see Fig. 2), while the bush is in its recess in the corresponding flange. The corresponding deflexions

δ [in.] are measured, for instance, with a dial gauge. Loads should be changed and readings taken at regular intervals (e.g. every 15 sec.). The test should be repeated several times in succession, and the results

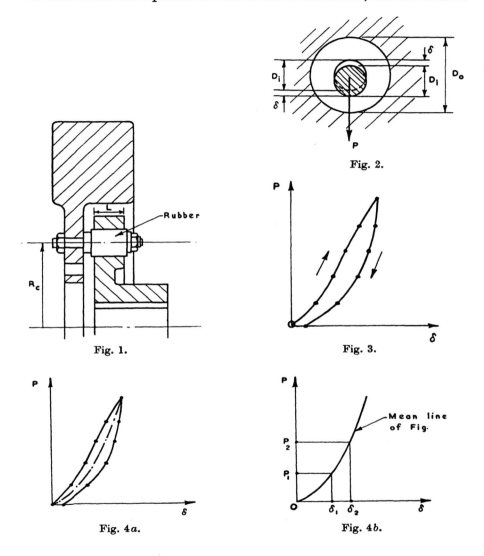

Fig. 2.

Fig. 1.

Fig. 3.

Fig. 4a.

Fig. 4b.

of the first loading cycle should be disregarded. The hysteresis loop finally obtained is of the type shown in Fig. 3.

A mean line is then drawn through the hysteresis loop (chain-dotted line in Fig. 4a) and this gives the 'zero-frequency' load/deflexion characteristic of the rubber bush (Fig. 4b).

The torque/deflexion characteristic of the coupling, at 'zero frequency' is obtained by means of the relations

$$T = N_B R_C P \quad \text{[Lb.in.]}$$

and
$$\phi = \delta/R_C \quad \text{[rad.]}$$

used for corresponding values (P_1 and δ_1, P_2 and δ_2, etc.) taken from Fig. 4b. In these equations

T = restoring torque [Lb.in.],

N_B = number of bushes,

R_C = effective radius (i.e. half the pitch circle diameter of the bush centres) [in.], as shown in Fig. 1,

P = applied load [Lb.] (from Fig. 4b),

ϕ = angular deflexion of coupling [rad.], and

δ = linear deflexion of one bush [in.] (from Fig. 4b).

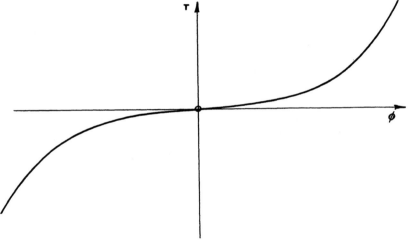

Fig. 5. Torque/deflexion characteristics of rubber-bush type coupling, obtained from 'zero-frequency' tests.

This evaluation gives the upper branch of the curve in Fig. 5. The lower branch is exactly the same curve, turned through 180° about the origin. It represents the restoring torques occurring when the loads and deflexions are in opposite direction to those of Fig. 4b. The use of this curve is explained at the end of this section.

(b) 'Zero-frequency' test of the entire coupling

This test is somewhat better than that of a single bush because it will indicate any stiffness variations due to backlash, etc., occurring in the coupling.

[106]

The coupling is mounted either vertically or horizontally on a rigid support. The horizontal arrangement may be such as that shown in Fig. 6a and b, which was used at the B.I.C.E.R.A. Laboratory. In this arrangement, the flywheel member is rigidly fixed to a heavy cast-iron table. The driving member, arranged uppermost, is keyed to a stub shaft bolted to a long steel plate serving as a torque arm. It is supported immediately below the stub shaft by a ball thrust bearing located on the table.

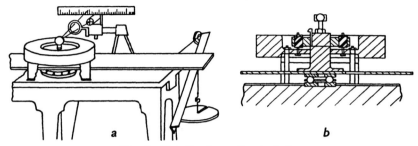

Fig. 6. Details of coupling test rig.

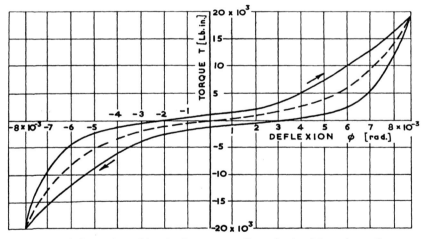

Fig. 7. Hysteresis loop of a rubber-bush type coupling, obtained from 'zero-frequency' tests. Its mean line is similar to that shown in Fig. 5.

'Zero frequency' tests are made by applying weights through cables and pulleys at both ends of the arm and noting the corresponding deflexions of the coupling. These are indicated on a graduated screen by an optical autocollimator system (description, see pp. 49 and 50). Loads are changed and readings taken at regular intervals, for instance, every 15 sec. The test is repeated several times in succession, and the results of the first loading cycle should be disregarded. In this manner, the hysteresis loop is obtained in terms of torque plotted against deflexion (see Fig. 7).

After drawing the mean line through the two branches, the curve obtained is again similar to that of Fig. 5.

However, the mean lines of the two branches may not correspond to absolutely identical values, owing to the effects of clearance, slightly eccentric positions of the pins in the recesses, etc. These mean lines should be retained as they are. The use of the 'zero-frequency' characteristic, derived from tests of either a single bush or the entire coupling, will now be explained in the following section.

2. The dynamic characteristics of couplings

The coupling with a torque arm, mounted on a rig as shown in Fig. 6, constitutes a one-mass system with a moment of inertia J (provided mainly by the torque arm) and a stiffness which is dependent on the torque/deflexion diagram of the coupling (see Fig. 8).

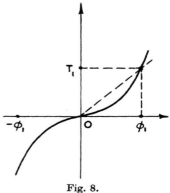

Under free vibration conditions, the restoring torque T_1 is in equilibrium with the inertia torque, and, since this is not linearly proportional to deflexion, the vibration is not sinusoidal and the phase velocity is not constant. However, for an approximate treatment, it is convenient to assume a sinusoidal variation of deflexion, $\phi = \phi_1 \sin \omega_1 t$, occurring at a mean phase velocity ω_1. This assumption will satisfy the equilibrium conditions for the restoring torque at three positions, viz. at the maximum-amplitude

Fig. 8.

positions (where $\pm T_1 = \pm \phi_1 \omega_1^2 J$) and at the zero position ($T_1 = 0$, $\phi = 0$).

Thus, if the maximum amplitude ϕ_1 of the deflexion is known, the corresponding maximum restoring torque T_1 can be determined by means of the graph of Fig. 8. The stiffness is then evaluated as $K_1 = T_1/\phi_1$, and the phase velocity is $\omega_1 = \sqrt{(K_1/J)}$.

Conversely, if ω_1 is the known value, the first step is to determine $K_1 = J\omega_1^2$. From the origin, a line is drawn with the slope $J\omega_1^2 = T_1/\phi_1$. Its intersection with the T-curve gives T_1 and ϕ_1. The above equations can, therefore, also be used to determine the relation between phase velocity ω_1 and the peak values ϕ_1 of deflexion.

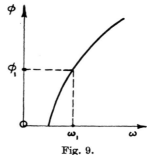

Fig. 9.

By plotting corresponding values of ϕ against ω, one obtains the curve of Fig. 9. Tests made at B.I.C.E.R.A. gave measured values in fairly good agreement with the figures determined by this method.

Under forced vibration conditions, the system is acted upon by an external torque T_* varying sinusoidally with a phase velocity ω_1. It may be assumed that the vibration of the inertia J is also at this same value ω_1. The equation of motion may thus be written

$$T_1 - J\omega_1^2\phi_1 = T_* \quad \text{or} \quad T_1 = J\omega_1^2\phi_1 + T_*.$$

To solve this equation graphically, plot $T_* = \overline{OP}$ along the $+T$-axis of the torque/deflexion diagram (Fig. 10). Assuming a value for ω_1^2, draw through P a line with the slope $\tan\alpha_1 = J\omega_1^2 \times s_x/s_y$, where s_x = scale conversion factor for horizontal axis [rad./in. of paper scale], and s_y = scale conversion factor for the vertical axis [Lb.in./in. of paper scale]. The intersection of this line with the T-curve gives ϕ_{1A} and its corresponding restoring torque T_{1A}.

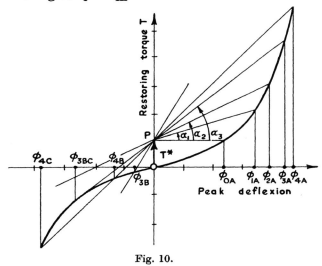

Fig. 10.

Proceeding in this manner, one finds for ω_2 the values ϕ_{2A} and T_{2A}. For ω_3 there are two points of intersection, ϕ_{3A} and ϕ_{3BC} (with their corresponding restoring torques T_{3A} and T_{3BC}). For ω_4, there are three values, ϕ_{4A}, ϕ_{4B} and ϕ_{4C}. By plotting coupling deflexions against phase velocity, one finally obtains the curves of Fig. 11 a.

Variations in coupling deflexion. Considering Fig. 11 b, it is seen that if the external torque T_* has a constant amplitude and its phase velocity is increased from zero to ω_1, ω_2, etc., the angular deflexion ϕ of the coupling will follow the points ϕ_{1A}, ϕ_{2A}, etc., on the ascending flank of the bent resonance curve and reach infinitely high values in the absence of damping. With damping, however, there will be a peak amplitude at some point such as ϕ_{5A}, which corresponds to the value ω_5 of the phase velocity.

As the phase velocity passes through ω_5, the coupling deflexion drops suddenly from ϕ_{5A} to ϕ_{5B}. It has come down the right-hand flank of the resonance curve without reaching intermediate values of deflexion, since these correspond to lower ω-values. For values greater than ω_5, the ϕ-values continue to decrease gradually.

The variation of ϕ with decreasing values of ω is somewhat different. Passing from ω_5 to ω_4, the angular deflexion increases from ϕ_{5B} to ϕ_{4B}. When passing through ω_3, the deflexion reaches ϕ_{3BC} and then jumps upwards to ϕ_{3A}. It has been found experimentally (see O. Mathiessen, *Phys. Z.* vol. 11 (1910), p. 448) that the portion of the curve between ϕ_{3BC} and ϕ_{5A} cannot be obtained in practice.

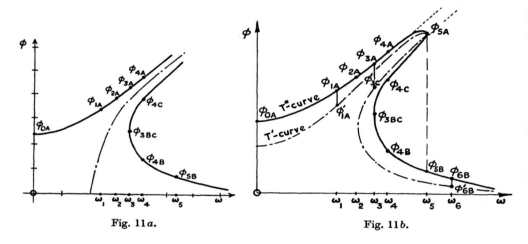

Fig. 11a.   Fig. 11b.

An outline of the reasons for this behaviour is obtained by considering, in Fig. 11b, the chain-dotted curve of deflexions, which corresponds to an amplitude of excitation torque T' somewhat lower than T_*. At ω_1, if the torque is increased from T' to T_*, the deflexion also increases from ϕ'_{1A} to ϕ_{1A}. A similar increase of deflexion with torque is obtained in the range of B-values, for instance at ω_6, where the deflexion increases from ϕ'_{6B} to ϕ_{6B}.

In the C-range, however, when the torque is increased from T' to T_*, the angular deflexion either decreases from ϕ'_{3C} to ϕ_{3BC} or increases to ϕ_{3A}. The value it takes depends on any small fluctuations in the value of ω. For decreasing ω-values, ϕ_{3A} is obtained. The range of deflexions on the right-hand flank of the curve, between ϕ_{3BC} and ϕ_{5A}, corresponds therefore to unstable conditions.

Peak shifts due to couplings. Thus, in the resonance region, the coupling can have two deflexions at the same frequency. These depend on whether the deflexion varies more or less in phase with the excitation torque

[110]

(case of ascending speeds) or in antiphase to it (decreasing speeds). The result is apparent as a shifting of the peak-amplitude positions of the vibration, or (if, ω_3 and ω_4 are very close to each other) as a resonance curve with a very steep flank on one side (see Fig. 12).

Peak shifts may occur when Holzer tabulations indicate that the resonance frequency of the system is hardly affected by changes in coupling stiffness (e.g. under one-node conditions with the node situated in the crankshaft near the main flywheel, or two-node conditions with the second node in the coupling). The amplitudes of vibration may also be modified by load changes of the engine.

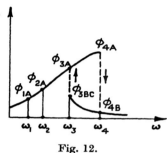

Fig. 12.

Detuning due to couplings. Under one-node frequency conditions with the node in the coupling, the 'resonance frequency' of the system is very sensitive to variations in coupling stiffness. In this case, a detuning effect occurs. In some cases, however, this one-node condition occurs with such low amplitudes that the one-node case may be disregarded.

Note. The above considerations apply to most types of couplings with variable stiffness, which can be treated by the methods set out in this section, on the basis of their 'zero-frequency' torque/deflexion characteristics.

This section describes the simplest methods for determining the dynamic characteristics of systems with non-linear couplings. Further analytical treatment of this subject will be found in the following references:

Den Hartog, J. P. *Mechanical Vibrations,* 2nd ed. (McGraw-Hill Book Company, Inc., 1940), pp. 480 ff.

Timoshenko, S. *Vibration Problems in Engineering,* 2nd ed. (D. Van Nostrand Company, Inc., 1944), pp. 119 ff.

McLachlan, N. W. *Ordinary Non-linear Equations in Engineering and Physical Sciences* (Oxford University Press, 1950), particularly pp. 48–62.

Stoker, J. J. *Non-linear Vibrations in Mechanical and Electrical Systems* (Interscience Publishers, Inc., 1950), particularly pp. 81–117.

3. Estimation of dynamic stiffness of non-linear couplings

The coupling is incorporated in an engine system such as that of Fig. 13, and it is required to determine the stiffness of the coupling from its torque/deflexion characteristic.

In Fig. 13, let $J_1 = J_2 = \ldots = J_6 = J_{\mathrm{cyl.}} =$ moment of inertia per line [Lb.in.sec.^{2}] of the engine; $J_F =$ inertia of main flywheel; J_{C1} and $J_{C2} =$ inertias of the driving and driven members of the coupling, respectively; and $J_M =$ inertia of driven machine.

The vibration amplitudes [rad.] at each of these masses are denoted by θ with corresponding subscripts. Thus, the deflexion of the coupling is

$$\phi = \theta_{C1} - \theta_{C2}.$$

No-Load Conditions

First approximation :

Assume a value $\phi_0 = 0.1\phi_{\max}$ (where $\phi_{\max}$ is the maximum permissible deflexion of the coupling) and, from the torque/deflexion characteristic (Fig. 14), evaluate the coupling stiffness as

$$K_{C(0)} = T_0/\phi_0 \quad [\text{Lb.in./rad.}].$$

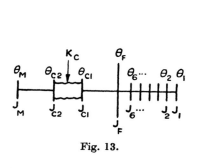

Fig. 13.

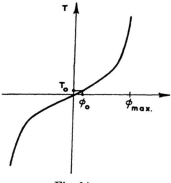

Fig. 14.

Second approximation (to be used only in cases where it is considered that the vibration frequency of the entire system is dependent to a considerable degree on coupling stiffness):

(1) Work out a Holzer tabulation for the entire system (see section 1·3), using the first approximate value $K_{C(0)}$ and starting with an ω^2 which is that of the engine-and-flywheel system alone (masses J_1 to J_F in Fig. 13). Thus, find the ω-value for the entire system.

(2) Assume a value θ_1 [rad.] for the vibration amplitude at mass no. 1 of the system.† Using the Holzer table values, calculate the inertia torque at J_{C1} (the driving member of the coupling) as

$$(J_1\omega^2\Delta_1 + J_2\omega^2\Delta_2 + \ldots + J_{C1}\omega^2\Delta_{C1}) \times \theta_1 = J_{\text{eff.}}\,\omega^2\theta_1 \quad [\text{Lb.in.}],$$

where $\Delta_1 = \theta_1/\theta_1$, $\Delta_2 = \theta_2/\theta_1$, etc., are the Holzer table amplitudes.

(3) This value $J_{\text{eff.}}\,\omega^2\theta_1$ should be of the same order of magnitude as

$$K_{C(0)} \times \phi_0 = T_0$$

within, say, $\pm 30\%$, since the value of θ_1 is also assumed.

---

† The value of θ_1 is assumed from previous experience or, for instance, estimated by means of the following empirical formula:

$$\pm\theta_1 = 0.05/n^* \quad \text{radians},$$

where $n^* =$ order number of critical speed for 4-stroke cycle engines,

$\quad\quad = \frac{1}{2} \times$ order number of critical for 2-stroke cycle engines.

(4) If (3) is not satisfied, take another value $\phi_1 = 0 \cdot 2\phi_{max.}$ (or $\phi_1 = 0 \cdot 05\phi_{max.}$) and repeat the evaluation procedures (1) to (4) with this new value.

The dynamic stiffness is thus obtained by successive approximations.

Full-Load Conditions

First approximation:

(1) Assume that the coupling deflexion is $\phi_A = \phi_{max.}$ (where $\phi_{max.}$ is the maximum permissible deflexion of the coupling) and, from the torque/deflexion characteristic, evaluate the coupling stiffness as

$$K_{A(0)} = 0 \cdot 5 T_{max.} / \phi_{max.} \quad \text{[Lb.in./rad.]}$$

(see Fig. 15). Using this first approximate value $K_{A(0)}$, determine, by means of Holzer tables, the natural phase velocity ω_n of the entire system.

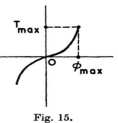

Fig. 15.

(2) Assume a value θ_1 [rad.] for the vibration amplitude at mass no. 1 of the system. Using the results of the Holzer table for ω_n, calculate the inertia torque at J_{C1} (the driving member of the coupling) as

$$(J_1\omega_n^2\Delta_1 + J_2\omega_n^2\Delta_2 + \ldots + J_{C1}\omega_n^2\Delta_{C1}) \times \theta_1 = J_{\text{eff.}}\omega_n^2\theta_1 \quad \text{[Lb.in]}.$$

This value $J_{\text{eff.}}\omega_n^2\theta_1$ should be of the same order of magnitude as

$$K_{A(0)} \times \phi_{max.} = 0 \cdot 5 T_{max.},$$

within, say, $\pm 30 \%$, since the value of θ_1 is also assumed.

Second approximation:

(1) Evaluate the steady transmitted torque T_P [Lb.in.] of the engine, for rated power and running speed.

(2) Evaluate the excitation torque (see section 2·1) as

$$T_E = |T_n| A R_0 \overrightarrow{\Sigma\Delta} \quad \text{[Lb.in.]},$$

where $|T_n| = n$th order tangential-pressure component of gas torque [Lb./in.²] (for the largest critical in the speed range of the engine; A = piston area [in.²]; R_0 = crank radius [in.]; and $\overrightarrow{\Sigma\Delta}$ = phase vector sum (see section 2·1) for the nth order critical, based on the Holzer table with the phase velocity ω_n already calculated.

(3) Plot the total external torque $T_* = T_P + T_E = \overline{OP}$ along the $+T$-axis of the torque/deflexion diagram of the coupling (see Fig. 16), so as to obtain the point P.

(4) Plot from P along the $+T$-axis the vector $\overrightarrow{PT_A} = \Sigma(J\omega_n^2\Delta_1) \times \theta_1$, thus obtaining the point T_A.

(5) Through T_A draw a line parallel to the horizontal axis of the diagram. The intersection of this line with the T-curve gives a point A, corresponding to a coupling deflexion ϕ_A.

(6) The second approximate value for the coupling stiffness is then

$$K_{A(1)} = (T_A - T_*)/\phi_A = \overline{PT}_A/\phi_A \quad \text{[Lb.in./rad.]}.$$

(7) Draw a line from point A to point P and extend this downwards beyond point P. If this line does not intersect the lower branch of the T-curve, there is only one possible coupling deflexion at the phase velocity ω_n. If there is an intersection at a point B, this gives a second possible coupling deflexion ϕ_B (see Fig. 17).

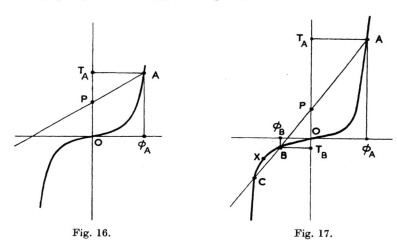

Fig. 16.   Fig. 17.

The stiffness is

$$K_{B(1)} = (-\overline{OP} - \overline{OT}_B)/ - \phi_B = K_{A(1)},$$

that is, it is the same as for ϕ_A.

(8) The deflexion ϕ_B corresponds to a vibration amplitude at mass no. 1 of the system

$$\theta_{1(B)} = \theta_1 \times \phi_B/\phi_A,$$

where θ_1 is the amplitude obtained with ϕ_A. Both values are possible at the same phase velocity ω_n.

Note. Disregard any further intersection C beyond the point X on the lower branch of the T-curve, X being the point at which a straight line from point A is tangent to the T-curve.

(9) With the new value $K_{A(1)}$ obtained as shown under (5) and (6), repeat the Holzer-table evaluations, which result now in a different value for the natural phase velocity, $\omega_{n(1)}$. If $\omega_{n(1)}$ is approximately equal

[114]

to ω_n (the figure from the first approximation), then $K_{A(1)}$ is the corresponding value for coupling stiffness.

(10) If condition (9) is not satisfied, calculate $\Sigma(J\omega_{n(1)}\Delta) \times \theta_1$ and repeat with this figure the procedure under (4), (5), and (6), which results in a third approximation $K_{A(2)}$.

1·242 Selection of couplings

The engine system is such that it will require a coupling capable of operating under conditions which may generally be specified as follows:

(1) maximum steady transmitted torque;

(2) maximum vibratory torque;

(3) maximum permissible deflexion across coupling (this includes deflexions due to steady torque, vibratory resonant torque, and vibratory non-resonant torque caused by cyclic speed variation);

(4) maximum extent of parallel and angular misalinement.

In addition, the range of ambient temperatures in which the coupling is expected to operate should also be taken into account.

In assessing the suitability of a coupling for a practical application, it is necessary to consider whether the coupling has a constant or variable stiffness under the corresponding operating conditions.

At present, single values of stiffness are frequently quoted by manufacturers, for couplings of which the stiffness is constant only over a limited range of torques and deflexions. If the coupling has a variable stiffness, then the relation between restoring torque and deflexion should be established by tests, if necessary, on lines such as those indicated on pp. 104–108 of this section.

Further factors to be considered are: (1) whether the coupling in question requires protection to prevent contact with oil; (2) its ability, if necessary, to provide damping action, detuning or peak shift effects; (3) whether it is required to accommodate a certain amount of axial displacement.

1·243 Practical points on couplings

1. Couplings of the rubber-sandwich type

These couplings, in which the rubber is loaded in shear, are used in various sizes up to about 18 in. diameter. The Shore hardness is usually between 50 and 70. The bond strength of the rubber-to-metal bond is generally 1000–1500 Lb./in.^{2}. Typical values for shear stress at the outer periphery are 40–48 Lb./in.^{2}, with corresponding shear angles of about 33°. It is stated for one of these couplings that the maximum shear stress (see p. 116), including stresses due to steady transmitted torque, torsional vibration, cyclic speed variation and misalinements, should not exceed 60 Lb./in.^{2}.

The maximum shear stress $q_{max.}$ may be calculated as follows: Referring to Fig. 1 on p. 101, let

R_o = outer radius of rubber disk [in.],

R_i = inner radius [in.],

L = thickness of disk [in.],

$I_p = \pi(R_o^4 - R_i^4)/2$ [in.^{4}],

G = modulus of rigidity of rubber [Lb./in.^{2}], and

E = modulus of elasticity ($E \cong 3G$) [Lb./in.^{2}].

The shear stress due to the steady transmitted torque T_{st} [Lb.in.] and the vibratory torque $T_{vib.}$ is

$$q_{sv} = (T_{st} + T_{vib.}) R_0/I_p \quad [\text{Lb./in.}^2].$$

The shear stress due to cyclic speed variation of the engine, corresponding to an amplitude of vibration θ_{cv} at the free end of the crankshaft is

$$q_{cv} = GR_0\theta_{cv}/L.$$

(The determination of the value of θ_{cv} by measurement or calculation is described in section 2·6.)

The shear stress due to a parallel misalinement y [in.] is

$$q_{mis.} = Gy/L$$

and the tensile stress due to an angular misalinement α [rad.] is

$$p_{ang.} = ER_0\alpha/L.$$

After determining these values, the total maximum shear stress can be estimated as

$$q_{max.} \cong \sqrt{[(q_{sv} + q_{cv} + q_{mis.})^2 + (p_{ang.}/2)^2]} \quad [\text{Lb./in.}^2].$$

Couplings for operation under low-temperature conditions are made of natural rubber, since this is less affected than synthetic rubbers by low temperatures. For natural-rubber couplings, the working temperature range is approximately $+60$ to $-20°$ C.

For natural rubber, it has been found that the shear modulus of rigidity increases by 8 % at $-5°$ C. and 50 % at $-20°$ C. above its normal value at room temperature. At $-40°$ C., the rubber acts as a brittle material.

The temperature of a rubber coupling is raised by internal heating generated by the hysteresis energy dissipation of the rubber material. The working temperature of a coupling cannot be easily assessed, since it depends on the thermal conductivity of the rubber and other materials of the coupling, the design and the location of the coupling. Where necessary, it should be determined by measurement.

Rubber-sandwich type couplings have been employed to transmit up to 360 b.h.p. at 430 rev./min. They are used to take up possible misalinement, to increase the flexibility of a system at a particular position, and to provide additional damping. Stiffness values of up to 230 000 Lb.in./rad. have been quoted for single-layer type couplings. A three-layer type coupling, with intermediate metal plates alternately connected so as to give a parallel-spring arrangement (see Fig. 23) has also been used and gives a stiffness three times greater than that of a single element, while the maximum permissible deflexion remains the same as that of one element.

Couplings of the rubber-sandwich type have been used for driving winches and alternators in parallel operation, in marine applications, as well as as in diesel-electric locomotives and welding sets.

It may be noted that information based on experience with rubber-type engine mountings is not always directly applicable to transmission couplings, since the latter operate with much larger dynamic deflexions, for which the stiffness and other characteristics of the rubber may have different values (see section 1·244).

2. Pin-type couplings

The two coupling halves are usually of cast iron. The bushes used in couplings of this type are of rubber, dermatine (a rubber-like material) or leather. They may be either in one piece or built up of flat or tapered disks or washers. In the latter case, the tapered sections allow the disks to expand laterally under compression.

Certain types of rubber bushes are provided with outer and inner metal sleeves to prevent abrasion. The outer sleeve is then inserted as a press fit into the corresponding bore machined in the driven half of the coupling. In cases where an inner sleeve inside the bush is also provided, this sleeve is placed with a slide fit over the pin fastened to the driver or flywheel half of the coupling. The number of pins varies according to coupling size, and from 4 to 32 pins are used in various designs.

In couplings with bushes not provided with outer sleeves, the amount of clearance to be allowed between the bush and its corresponding bore in the driven half of the coupling is a question on which various views have been expressed. Some engine-makers provide for a tight fit, i.e. zero clearance, internally and externally, while others allow an external bush clearance of about $\frac{1}{16}$ in. In general, the amount of clearance depends on the inertias, stiffnesses and operating conditions of the entire engine system.

Some pin-type couplings are designed to accommodate additional 'tuning disks' on the driven side of the coupling. These disks are added in order to alter to a suitable value the moment of inertia of the coupling.

Dermatine rubber is fairly hard (about 70 Shore). The stiffness of rubber, dermatine rubber or leather bushes can easily be determined by compressing a sample to, say, an external pressure of 200 Lb./in.² (see section 1·2416, p. 105). It is reported that rubber-sheathed couplings are not always suitable for tropical climates.

The quoted values for bush loading vary from 40 Lb./in.² to a maximum of 600 Lb./in.², referred to the projected area of the inner surface of the bush. The value depends on the type of bush considered. Some engine-makers limit the maximum pressures to values between 40 and 150 Lb./in.². Furthermore, the cantilever bolts on pin-type couplings should be checked for bending stresses.

Pin-type couplings are used for a wide variety of applications, such as generator sets, marine propulsion systems, welding sets, etc. The horse-power ratings usually supplied are for pulsating drives, i.e. reciprocating engines. For non-pulsating drives, the rating is in some cases increased by an average of 25 % above the standard values. In some cases, momentary overloads of 100 % are allowed.

The dynamic stiffness of pin-type couplings is indicated by a corresponding load/deflexion curve by some coupling manufacturers, whereas others quote sets of figures. In some cases, errors of up to 100 % have been found in quoted stiffness values. Engine-makers can always verify the effective stiffness values (a) by load tests of bushes, and (b) by subsequent analysis of torsiograph tests of complete engine systems fitted with the type of coupling considered.

Pin-type couplings provide a certain amount of damping and are used to accommodate possible angular or parallel misalinement, as well as axial motion in some designs. Their torsional stiffness values, depending on the torque/deflexion characteristics, cover ranges from $0·5 \times 10^6$ to approximately 8×10^6 Lb.in./rad.

The wear of the rubber bushes depends on the design features, clearances, etc., of the coupling. Some engine-makers take considerable precautions to obtain minimum wear conditions, while others consider that wear after a reasonable period of running is an acceptable feature of these couplings and that the replacement of worn bushes is a fairly easy matter.

It is of interest to note that the stiffness of cylindrical rubber bushes follows approximately a square law as a function of deflexion, this being based on tests carried out at the B.I.C.E.R.A. Laboratory. Bushes with conically tapered disks, according to maker's data, have a stiffness which follows a square law, or in some cases a logarithmic law, when plotted against deflexion.

Various opinions have been expressed on the presence of detuning action or peak shifts, due to couplings having a variable stiffness, in the neighbourhood of running speeds. It seems, however, that, provided that low vibration amplitudes are maintained in the running range, there should be no objection to such detuning effects produced by couplings.

3. Couplings with rubber blocks in compression

Couplings of this type are used in main and auxiliary drives. Some designs are also incorporated in gearwheels.

The compression modulus of elasticity of rubber increases with compressive strain following approximately a square law. Therefore, rubber couplings operating under compression have a constant stiffness only for a small range of preloads (corresponding to steady transmitted torque) and vibratory loads. When operating outside this range, it is possible that a stiffness value appreciably higher than the nominal (zero preload) stiffness is obtained and, as a result, it may be found that there are two resonance peaks of the same order number, but corresponding to 1-node and 2-node vibration conditions, which occur in the speed range of an engine. Comparative torsiograph tests can be made to determine the type of rubber elements having the most suitable values of elasticity modulus for the installation considered.

4. Impact loading of couplings

Experience indicates that rubber couplings are not always suitable for impact loading. This is due to the fact that solid rubber does not reduce the very-high-frequency components of a shock, so as to transmit only the lower frequencies. Good reduction of shock loads is obtainable with foam rubber or rubber of ribbed construction with hollow recesses.

In cases where foam or ribbed rubber cannot be used, the shock can be reduced by using rubber-and-spring type elements. Where it is necessary to isolate a component, e.g. a gear-box, from impact loads from both the engine side and the driven-machine side, suitable arrangements for impact reduction should be provided on both sides of this component.

5. Detuner type couplings and clutches

In systems incorporating detuner type couplings (e.g. of the Bibby type), it is usual practice to ensure that all the shafting is very stiff, as compared with the coupling. This confirms the fact that detuners operate as described in section 1·2416. Bibby detuners are either grease-packed or oil-lubricated, depending on the size and application. For description of Bibby couplings, see section 1·245, p. 125.

Centrifugal clutch-type couplings can be arranged to allow a certain amount of slip under overload and thus lessen any impact due to shock loading. In clutches of the unidirectional type, the torque transmission capacity is reduced in the reverse direction, and this produces a detuning effect, which reduces large-amplitude vibrations.

1·244 *General remarks on the properties of rubber materials*

There is no such thing as a 'static' value of stiffness for rubber. The dependence of stiffness on vibration frequency is illustrated in Fig. 18 and is a typical property of rubbers. (This and the following figures

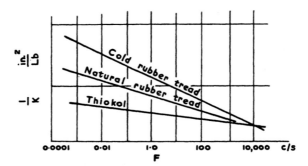

Fig. 18. Variation of $1/K$ as a function of frequency F. [Philippoff.] This relationship is valid for both shear and compression. K=stiffness.

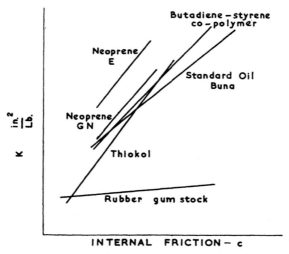

Fig. 19. Relation between stiffness and internal friction (K and c at constant frequency). [Stambaugh, Gehman, Woodford and Gemant.]

indicate the trend of the variations; exact values should be based on tests for each rubber compound considered.)

There is an interdependence between the stiffness K and the damping coefficient c, as shown in Fig. 19. Owing to the linear relationship between $1/K$ and $\log \omega$, indicated in Fig. 18, and the interdependence of K and c, there is also a linear relationship between $\log \omega$ and $\cot \delta$, where the 'loss angle' δ is defined by

$$\cot \delta = \frac{\text{restoring force}}{\text{damping force}} = \frac{K\theta}{c\omega\theta}$$

$$= K/c\omega = Q\text{-factor}$$

(with $\omega =$ phase velocity and $\theta =$ vibration amplitude).

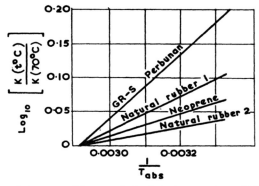

Fig. 20. Relation between stiffness and absolute temperature (°C abs.). [Waring.]

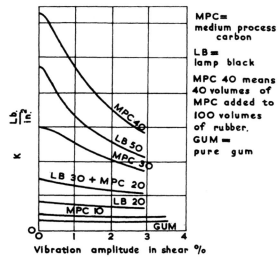

Fig. 21. Variation of stiffness in shear as a function of vibration amplitude. [Fletcher and Gent.]

At any temperature, the stiffness can be determined by means of a curve of the type shown in Fig. 20.

The curves of Figs. 18–20 are valid for pure, rubber-like materials. The addition of fillers (e.g. carbon black, zinc oxide, clay, etc.) results in a decrease in dynamic stiffness at increased vibration amplitudes. This is indicated in Fig. 21.

When a rubber has been strained to a large amplitude, and the 'structure' has been broken down to an equilibrium stiffness corresponding to the vibration, a considerable time (more than 3 hours) is required for the material to recover its initial value. Recovery is aided by elevated temperatures.

REFERENCES

Fletcher, W. P. and Gent, A. N. *I.R.I. Trans.* vol. 29, 1953, p. 266.
Gehman, S. D. *J. Appl. Phys.* vol. 13, no. 6 (1942), p. 402.
Gehman, S. D., Woodford, D. E. and Stambaugh, R. B. *Industr. Engng Chem.* vol. 33 (1941), p. 1032.
Gemant, A. *J. Appl. Phys.* vol. 12, no. 9 (1941), p. 680.
Philippoff, W. *J. Appl. Phys.* vol. 24 (1953), p. 685, fig. 2.
Stambaugh, R. B. *Industr. Engng Chem.* vol. 34 (Nov. 1942), p. 1358.
Waring, J. R. S. *Industr. Engng Chem.* vol. 43 (1951), p. 352.

1·245 *Examples of couplings and clutches of various types*

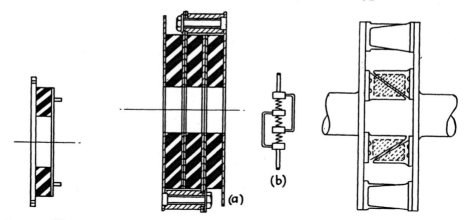

Fig. 22. Flanged rubber sandwich coupling. (Rubber Bonders.)

Fig. 23. Three-layer rubber sandwich coupling (*a*), and equivalent linear-spring system (*b*). (Dunlop.)

Fig. 24. Segmental coupling with rubber used in compression. (Dunlop.)

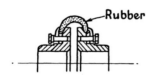

Fig. 25. Periflex 'tyre joint' type coupling. (Stromag; Stone-Wallwork.)

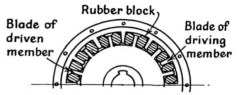

Fig. 26. Rubber block type coupling. (Holset.)

[122]

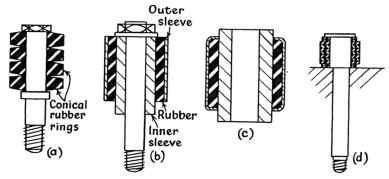

Fig. 27 a,b,c,d. Rubber bushes for pin-type couplings (see Fig. 1, p. 105, for a drawing of a coupling of this type). (a) David Brown; (b) and (c) Silentbloc; (d) Wigglesworth.

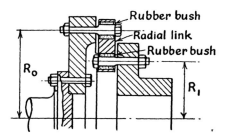

Fig. 28. Radial link type coupling with bonded-rubber bushes at the ends of each link. (Metalastik.)

For a deflexion ϕ of the coupling the required torque is

$$T = \tfrac{1}{4} \times N_L K_L \times \left(\frac{R_i R_o}{R_o - R_i}\right)^2 \times \phi^2 \quad \text{[Lb.in.]}$$

where $N_L =$ number of bushes, and $K_L =$ stiffness of one bush loaded in compression ($K_L \propto \phi^2$) [Lb./in.].

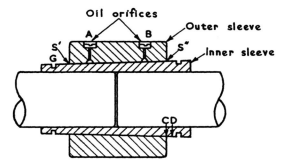

Fig. 29. Skefko coupling assembled by oil injection method.

The inner sleeve with a taper of 1 in 80 is pushed over the two plain shafts to its final position. The outer sleeve with a corresponding taper is pushed over it to a position C. Through two orifices A and B, oil is injected by means of hand pumps, until it flows out at the two ends S' and S''. This reduces friction; also the oil pressure compresses the inner sleeve and shaft, and expands the outer sleeve, so that the latter can be pushed further up the taper by means of a tool supported in groove G and pressing against the end face of the outer sleeve. This operation is continued until the outer sleeve reaches the required position D.

When the oil pressure is released, the oil film is forced out and the stiffness of the assembly is equivalent to that of a shrink fit. These couplings are used on shafts of up to 20 in. (500 mm.) diameter. The drive can be disconnected by applying oil pressure at A and B and reversing the process.

[123]

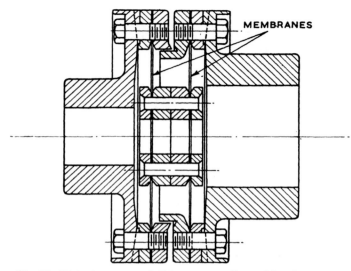

Fig. 30. Metastream metal-disk type coupling. (Metaducts Ltd.)

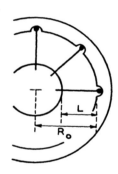

Fig. 31. Coupling with spokes pivoted at one end. Torsional stiffness:

$$K = \frac{3EI}{L^3} \times R_0^2 N_s.$$

Loading of spoke is similar to that of a cantilever beam.

NOTATION for Figs. 31 to 34

K = stiffness [Lb.in./rad.],
E = elasticity modulus [Lb./in.^{2}],
I = second moment of area in bending [in.^{4}],
R = pitch circle radius [in.],
N = number of rungs (N_r), spokes (N_s), clips (N_c) or plates (N_p),
L_o = total axial length [in.], and L = length [in.].

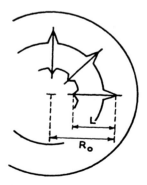

Fig. 32. Coupling with spokes fixed at both ends. Torsional stiffness:

$$K \cong \frac{4EI}{L^3} (3R_0^2 - 3R_o L + L^2) N_s.$$

Each spoke is treated as a cantilever with a direct load and a restraining couple at the outer rim.

[124]

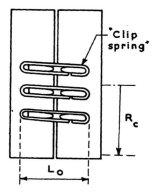

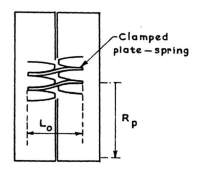

Fig. 33. Croft coupling. Formula for first approximation of torsional stiffness for maximum deflexion:

$$K \cong \frac{18EI}{L_0^3} R_c^2 N_c.$$

Fig. 34. Plate-spring coupling. Formula for first approximation of torsional stiffness for maximum deflexion:

$$K \cong \frac{12EI}{L_0^3} R_p^2 N_p.$$

Each half plate is treated as a cantilever and the two deflexions are added together.

Note. The formulae given for the couplings of Figs. 33, 34 and 35 are only approximate. The stiffness depends on a number of factors besides those indicated in the formulae, in particular on the profile of the curved teeth.

Bibby couplings and detuners

The Bibby coupling consists of driving and driven members, each provided with teeth and coupled together by means of continuous strip springs which are wound between the teeth so as to form flexible grids around the coupling periphery. The teeth are flared at their inner ends, so that, under no-load operating conditions, the stiffness of the coupling

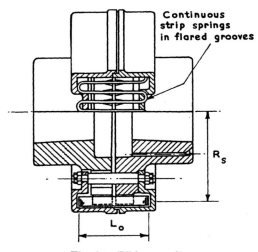

Fig. 35. Bibby coupling.

depends on the length of spring between the points of contact near the outer ends of the teeth. Under high-loading conditions, the torsional stiffness depends on the length of the springs between the inner ends of the flared teeth. Thus, under vibratory load conditions, the coupling

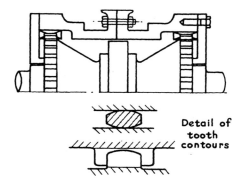

Fig. 36. Gear type coupling (Tacke). Curved tooth form allows sliding fit to be obtained in spite of some angular misalinement. Minimizes back-lash and avoids edge pressure.

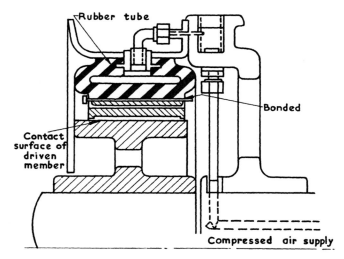

Fig. 37. Airflex clutch. The rubber tube is inflated with air at 70–100 Lb./in.^{2} until it presses against the inner hub (Croft; Kauermann).

has a detuning effect. By suitable design it is possible to detune frequencies and reduce the vibration amplitudes corresponding to two or sometimes three different modes of vibration.

One member of the coupling is sometimes also designed to incorporate a large flywheel mass. The coupling can thus be used either as a trans-

[126]

mission coupling or as a detuner at the free end of the engine. In such designs, the teeth on the flywheel member may have a much greater axial length than those of the driven member, to increase the coupling response to small-amplitude vibrations. Stops are sometimes provided at the ends of some of the spring loops in order to limit the axial motion of the springs. The coupling is also capable of absorbing impact loads.

The torsional stiffness characteristics can be altered by adjusting the gap between the two coupling members or altering the dimensions of the grid springs. Dog clutches are sometimes incorporated in Bibby couplings to disconnect the shaft. This arrangement may permit some axial motion between the two coupling members and hence modify the action of the coupling.

In practice it is found that a stiffness characteristic producing a change in frequency of about 15 % gives a reduction in front-end amplitude of about 70 %, and that a change in frequency of 30–40 % results in a vibration amplitude reduction of 85 %. To obtain 30 % frequency detuning it is necessary, however, to provide for a stiffness variation of 100 % under vibration conditions, by ensuring that the coupling stiffness varies through a range of values which are all lower than any other stiffnesses in the engine system, so that the node of the vibrations will be in the coupling. This is not always possible, but when the node is not situated in the coupling a frequency detuning of 10–15 % suffices to give an appreciable reduction of the vibration amplitudes.

In ship-propulsion systems fitted with a detuner at the free end, it is found that the reduced peak amplitude is at approximately the same frequency as when the detuner is locked. In stationary systems with a detuner used as a transmission coupling, the reduced peak frequency may be 10–15 % lower than the one-node frequency of the system with the coupling locked. In tests of marine systems incorporating a Bibby transmission type coupling, it has been found that, in addition to the one-node mode (with the node in the coupling), the two-node mode is also shown. The frequency of the two-node mode is similar to that obtained when it is assumed that the engine system is severed after the driving member of the coupling.

A first estimation of the coupling stiffness can be obtained by means of the formula applying to a beam clamped at both ends, viz.

$$K = C \times \frac{12EI}{L_0^3} R_s^2 N_R,$$

where C is a constant between 0·5 and 1·0.

For other notes on practical experience with Bibby couplings, see section 1·243 (p. 119).

Fig. 38. Twiflex automatic centrifugal clutch coupling. Includes rubber bushes enclosed in inner and outer metal sleeves. (This coupling also has a detuning action.)

Fig. 39. Centrifugal powder coupling when the shaft rotates, centrifugal force causes the powder to pack tightly around the outer portion of the casing and rotor rim (Ranzi; Stone-Wallwork).

Hydroflex coupling

This clutch coupling consists of two parts:

(1) a primary with a pinion and crownwheel type hydraulic drive (somewhat similar to that of a Sandner gearwheel damper), and

(2) a secondary with an Airflex type friction ring, actuated by a fluid, which locks against the crownwheel. Resonant vibrations are reduced by detuning action of the slipping primary element. It is stated that normal slip is less than 1 % over the entire speed range (Lohmann & Stolterfoht, Witten/Ruhr).

Double-spring-rate coupling

(Modern Wheel Drive.) This coupling is of the peripheral-spring type (see section 1·241, p. 103), in which, however, two springs of different stiffness are provided on either side of the torque arms. The coupling is specially designed for marine or similar applications in which the steady transmitted torque increases as the square of the running speed (propeller law). When the engine is running at the lower end of the speed range, the preload on the coupling, due to the steady torque, is small, and the vibratory deflexions of the coupling are due to the soft spring elements, so that the coupling has a low stiffness. Above a certain speed, the steady preload torque is sufficient to produce complete compression of the soft springs, so that any vibratory deflexions of the coupling are sustained by the stiff spring elements, the coupling having a considerably increased stiffness under these conditions.

The coupling is used as follows: If, at low running speeds, the engine operates under vibration-free conditions, the engine system being such

[128]

that there is a critical speed higher up in the speed range, when the speed is increased the coupling becomes stiffer and raises the natural frequency of the system, so that vibration-free conditions are again obtained. The critical is then shifted to a value above the speed range. Conversely, any low-amplitude critical of higher order number at the bottom of the speed range would be unimportant because it would be shifted to values below the speed range with decreasing engine speed, since this also causes the coupling to operate at its lower stiffness value.

1·246   Notes on clutches
1. Friction clutches

Stiffness. Multi-disk-type clutches have a torsional stiffness which is determined by the equivalent length of the shaft portion transmitting torque in the engaged position and the elasticity of the disks. The disks can usually be regarded as being infinitely stiff (zero equivalent length). Alternatively, the torsional stiffness of each disk can be calculated as indicated under 'Coupling disks' (see section 1·2413), and their total stiffness is obtained as that of one disk multiplied by the number of disks. The effective stiffness, however, is reduced by slipping action, which occurs whenever the vibratory torque exceeds the rated torque of the coupling.

Slipping action. The three main types of friction clutches are

(1) the *axial type*, with pairs of disks pressed together by springs or centrifugal action,

(2) the *radial type*, with an outer annulus against which inner annular segments are pressed by springs or centrifugal action, and

(3) the *conical type*, which combines some of the features of the two previous types.

In some respects, the slipping action is comparable to that of a semi-dry friction damper. Tests indicate that these clutches slip under high vibratory torques but act as solid couplings for low vibratory torques. As a result, torsiograph recordings of systems with friction clutches frequently show minimum amplitudes where peaks were present without the clutch, and apparent peak amplitudes of reduced values where minimum flank amplitudes previously occurred (*detuning due to stiffness variation*).

A measure of the slip is the amount of heat dissipated. The breakaway value of the friction coefficient is generally greater than the value occurring during the slipping action. In many instances, the driven part of a system, after the clutch, is isolated from the engine vibration owing to clutch slip, but this possibility should be verified by torsiograph tests under full-load and part-load operating conditions in each case. Slipping clutches may thus be regarded as non-linear couplings with a curve of torque against deflexion first rising steeply for small torques and then proceeding parallel to the horizontal axis (slipping action) when the torque reaches a certain value.

If the mating surfaces are at all in contact with oil, it is advisable to ensure that only clean (i.e. filtered) oil, free from grit or deposits, comes into contact with these surfaces, in order to avoid changes in the slipping characteristics. If possible, tests should also be carried out with varying spring pressures and surfaces with different types of finish, in order to determine optimum conditions.

2. Dog clutches and tooth-type couplings

Clutches of this type may excite vibrations, under conditions of angular misalinement, at frequencies approximately equal to the running speed multiplied by the number of teeth. The magnitude of the effect is related to the backlash variation with misalinement. Torsiograph tests with the filter set at approximately n times the running speed (where $n =$ number of teeth) may serve to indicate any vibration of this kind.

1·247 *Hydraulic and electromagnetic couplings*

As regards torsional vibration, the behaviour of hydraulic and electro-magnetic couplings is, in some respects, similar to that of slipping clutches. In ordinary Holzer table evaluations without damping, it is usual to consider that the vibrating system ends at the driving half of an hydraulic coupling; electromagnetic couplings, however, are included in the tabulation as they have a definite stiffness.

Calculations with damping involve the use of an 'equivalent viscous-damping coefficient' for the torque acting between the two halves of the coupling. When this is known from previous tests or approximate assessments, a system incorporating a hydraulic coupling can be evaluated from Holzer tables taking account of damping. The treatment of electromagnetic couplings may follow similar lines, but also requires taking account of the curve of restoring torque as a function of vibration amplitude for a full assessment.

A detailed analysis of these types of couplings leads into considerations of hydrodynamics and electricity which are outside the scope of this handbook. Further basic information may be obtained from the references given below.

REFERENCES

Hydraulic couplings

Allison, N. L., Olson, R. G. and Nelden, R. M. Hydraulic couplings for engine applications. *Trans. Amer. Soc. Mech. Engrs*, 1941, pp. 81–90.

Beck, E. Nomograms for Föttinger couplings. *Z. Ver. dtsch Ing.* 11 March 1954, pp. 223–35.

Habicht, R. C. Variable-speed fluid couplings. *Prod. Engng*, Dec. 1951, pp. 125–8.

Martyrer, E. Characteristics of Föttinger type torque converters under starting conditions. *Z. Ver. dtsch Ing.* 11 Feb. 1952, pp. 127–34.

Schjolin, H. O. The V hydraulic transmission. *S. A. E. Quart. Trans.* Oct. 1949, pp. 649–55.

Sinclair, H. Recent developments in hydraulic couplings. *Proc. Instn Mech. Engrs, Lond.*, 1935, pp. 75–157. Some problems of transmission of power by fluid couplings. *Proc. Instn Mech. Engrs, Lond.*, 1938, pp. 83–157.

Spannhacke, E. W. Hydrodynamics of the torque converter. *S. A. E. Quart. Trans.* Oct. 1949, pp. 592–608.

Ziebart, E. Investigation of a Föttinger coupling. *Z. Ver. dtsch Ing.* 21 Oct. 1953, pp. 1027–36.

Electromagnetic couplings

—. Gearbox and electromagnetic couplings for twin-engined ship. (B.T.H. electromagnetic couplings.) *Engineering*, 5 Sept. 1952, pp. 299–301.

Falderbaum and Grebe. The A.E.G. magnetic powder coupling, type E.M.G. *A. E. G. Mitt.* vol. 42, nos. 3/4 (1952), pp. 104–9.

Kelley, O. K. The polyphase torque converter. *S. A. E. Quart. Trans.* April 1949, pp. 297–306.

Klamt, K. Electric slip couplings for marine propulsion. *Elektrotech. Z.* 21 July 1954, pp. 273–6.

Koffman, J. L. Slipping clutches for engine auxiliary drives. *Gas Oil Pwr*, Nov. 1952, pp. 255–9.

Pounder, C. C. The geared diesel engine. *Trans. Inst. Mar. Engrs*, vol. 59, Suppl., part no. 12A (Jan. 1948), pp. 266–74.

Stallings, H. B. The electromagnetic clutch: its operation, application and control. *A.S.M.E.* Paper 54-F-9, 7 pp.

Suhr, F. W. Reluctance-type magnetic couplings. *A.I.E.E. Proc.* Aug. 1952, pp. 581–4.

Watts, J. L. Slip coupling design. *Mech. World*, 23 Feb. 1951, pp. 185–6.

Weymouth, H. P. Some interesting examples of geared diesel engines. *Trans. Inst. Mar. Engrs*, Sept. 1949, pp. 163–8.

1·25   Stiffness of driven machines

This section provides information and methods for assessing the point of rigidity, equivalent length and stiffness of driven-machine shafts.

Numerical values of stiffness of various types of Heenan & Froude dynamometer shafts are given together with the corresponding values for moments of inertia in section 1·15 (pp. 33 and 34).

1·251   *Point of rigidity of armatures and other driven machines*

As previously stated on p. 52, the 'point of rigidity' of a shaft is defined as the point at which the moment of inertia of a distributed mass (such as that of an armature core) is concentrated.

Although the stiffening effect of a mass on the shaft upon which it is mounted is gradual along most of the shaft length, for simplicity of calculation it is assumed (a) that no stiffening effect is felt by the shaft supporting the mass up to a certain point on the shaft axis, viz. the point of rigidity, and (b) that the shaft ends at this point (and becomes 'infinitely rigid' beyond it).

Points of rigidity are conveniently assumed on the basis of experience with shafts of a similar type. Generally, the point of rigidity is determined from available results of vibration frequency tests, which are evaluated by various methods to assess the stiffness of shaft sections with complicated features. Therefore, it may be noted that the 'point of rigidity' also to some extent depends on the evaluation method used for its determination, so that different evaluation methods may give different points of rigidity.

The following information in regard to points of rigidity of shaft portions supporting an armature core is based on vibration tests of pairs of identical generators coupled back-to-back, which were carried out by B.I.C.E.R.A.

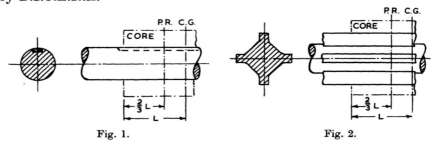

Fig. 1.   Fig. 2.

Plain keyed shafts (Fig. 1)

Point of rigidity is situated at two-thirds of the distance from the beginning of the armature core to the centre of gravity of the armature (including commutator).

Shafts with solid cruciform cross-section (Fig. 2)

Point of rigidity at two-thirds of the distance from the beginning of the armature core to the centre of gravity of the cross-section (including commutator).

Shaft with press fit or shrunk fit spiders (Fig. 3)

Point of rigidity at one-third of the distance from the beginning of the armature core to the centre of gravity of the armature (including commutator).

Shafts with welded-on arms (Fig. 4)

Methods of pp. 136 and 141 (webs in torsion): Point of rigidity at the beginning of the armature core. (The shaft portion carrying the armature core is regarded as infinitely rigid.) Griffith and Taylor method, p. 137, and method with webs in bending, p. 141: Point of rigidity at centre of gravity of armature core (including commutator).

If it is necessary to include in the vibrating system the shafting beyond the distributed mass, a second point of rigidity is assumed near the far end of the distributed mass. The shaft portion between these two points is taken as having an infinite stiffness, so that the equivalent system is as indicated in Fig. 5.

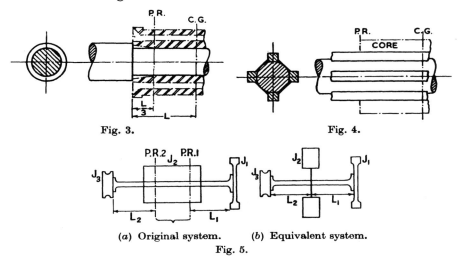

Fig. 3.   Fig. 4.

(a) Original system.   (b) Equivalent system.

Fig. 5.

Shaft with keyed and press fit portions (Fig. 6)

First point of rigidity half-way along the keyed portion effectively supporting the armature. For the shafting beyond the armature, the second point of rigidity is half-way along the effectively supporting press-fit portion.

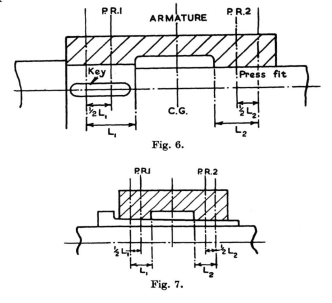

Fig. 6.

Fig. 7.

Long-hubbed pulley with bore relieved in centre (Fig. 7)

The points of rigidity are half-way along the effectively supporting portions of the hub.

1·252 Equivalent length of shafts of non-circular cross-section

1·2521 Rectangular and square cross-sections

From Saint-Venant's theory, the equivalent length of a shaft having a rectangular cross-section is obtained as

$$L_e = \frac{\pi}{32\alpha} \times \frac{G_e}{G} \times \frac{D_e^4}{ab^3} \times L \quad \text{[inches].}$$

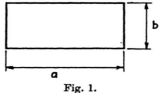

Fig. 1.

The notation used is as follows:

Equivalent shaft	Rectangular shaft
L_e = length [in.],	L = length [in.],
G_e = modulus of rigidity [Lb./in.²],	G = modulus of rigidity [Lb./in.²],
D_e = diameter [in.];	a = length of longer side of cross-section [in.],
	b = length of shorter side [in.],

and the value of the coefficient α for the corresponding a/b ratio is determined from the table below:

$a/b =$	1·0	1·2	1·4	1·6	1·8	2·0	2·5	3·0
$\alpha =$	0·1406	0·166	0·187	0·204	0·217	0·229	0·249	0·263
$a/b =$	4·0	5·0	6·0	8·0	10·0	20·0	∞	
$\alpha =$	0·281	0·291	0·298	0·306	0·312	0·322	0·333	

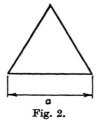

Fig. 2.

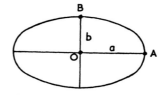

Fig. 3. Elliptical section with $\overline{OA} = a$ and $\overline{OB} = b$.

For *square-section* shafts, therefore, with $a = b = H$ and $\alpha = 0·1406$,

$$\pi/(32 \times 0·1406) = 1/1·431 \quad \text{and} \quad L_e = \frac{1}{1·431} \times \frac{G_e}{G} \times \frac{D_e^4}{H^4} \times L \quad \text{[inches].}$$

[134]

1·2522 Equilateral-triangle cross-sections

The equivalent length of a circular section shaft having the same stiffness is determined by

$$L_e = 4{\cdot}54 \times \frac{G_e}{G} \times \frac{D_e^4}{a^4} \times L \quad \text{[inches]}.$$

1·2523 Elliptical cross-sections

The equivalent length is

$$L_e = \frac{1}{32} \times \frac{G_e}{G} \times \frac{D_e^4}{a^2 b^2} \times \left(\frac{a}{b} + \frac{b}{a}\right) \times L \quad \text{[inches]}.$$

For a circular section, with $a = b = D/2$, this reduces to

$$L_e = \frac{G_e}{G} \times \left(\frac{D_e}{D}\right)^4 \times L.$$

Metric units. In this section, all equivalent lengths can be expressed in centimetres if G is in Kg./cm.² and linear dimensions are in cm.

1·2524 Thin-walled cylindrical cross-sections

By means of the Batho-Bredt formula for the polar second moment of area I_p of a thin-walled cylindrical section, viz.

$$I_p = 4A^2 / \Sigma\left(\frac{\Delta s}{h}\right) \quad \text{[in.}^4\text{]},$$

the equivalent length of a shaft of this type, in terms of an equivalent solid shaft of circular cross-section, can be determined as

$$L_e = \frac{G_e}{G} \times \frac{\pi}{128} \times \frac{D_e^4}{A^2} \times \Sigma\left(\frac{\Delta s}{h}\right) \times L \quad \text{[inches]},$$

the notation used being as follows:

Solid circular shaft	Thin-walled shaft

L_e = equivalent length [in.],

G_e = modulus of rigidity [Lb./in.²],

D_e = diameter [in.];

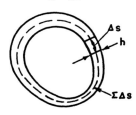

Note that the area A includes the entire area (wall and hollow portion) situated inside the broken line of the mean perimeter $\Sigma\Delta s$ of the cross-section.

L = length [in.],

G = modulus of rigidity [Lb./in.²],

A = area [in.²] enclosed within the perimeter $\Sigma\Delta s$ along the line of mean thickness (broken line in figure) of the cross-section,

h = thickness of wall at position of Δs [in.],

Δs = element (or any suitably short length) of *mean* perimeter of thin-walled section [in.].

The value of $\Sigma(\Delta s/h)$ is obtained by dividing the mean perimeter of the tubular cross-section into short lengths Δs, evaluating for each of these the ratio $\Delta s/h$, and adding together these contributions.

For the evaluation of the torsional stresses in shafts of this type, see section 2·3.

1·2525 Cross-sections having an entirely convex periphery

For shafts of symmetrical cross-section, without re-entrant angles or projections, the equivalent length can be determined by means of Saint-Venant's approximate formula, viz.

$$L_e = 1\cdot 25\pi \times \frac{G_e}{G} \times \frac{D_e^4 I_p}{A^4} \times L \quad \text{[inches],}$$

the notation used being as follows:

Equivalent shaft | Shaft with convex periphery

L_e = equivalent length [in.], $\qquad L$ = length [in.],

G_e = modulus of rigidity [Lb./in.^{2}], $\qquad G$ = modulus of rigidity [Lb./in.^{2}],

D_e = equivalent shaft diameter [in.]; $\qquad A$ = area of cross-section [in.^{2}],

$\qquad\qquad\qquad\qquad\qquad\qquad\qquad I_p$ = polar second moment of area [in.^{4}] about a perpendicular axis through the centroid of the cross-section.

EXAMPLES. (1) Shaft of *circular* section, with $L = 1$ in.

Let $D_e = D$, $G_e = G$, then

$$L_e = 1\cdot 25\pi \times \frac{D^4}{\left(\frac{\pi}{4}D^2\right)^4} \times \frac{\pi}{32} D^4 \times 1 = \frac{10}{\pi^2} = 1\cdot 0132 \text{ in.,}$$

i.e. the error is about 1·3 %.

(2) Shaft of *rectangular* section, with $L = 1$ in., and sides of lengths a and b, where $a = 2b$. Let $D_e = a$, $G_e = G$, then

$$L_e = 1\cdot 25\pi \times \frac{a^4}{(ab)^4} \times \tfrac{1}{12}(ab^3 + ba^3) \times 1 = 3\cdot 27 \text{ in.}$$

The exact value, from the formula for rectangular sections (p. 134) is

$$L_e = \frac{\pi}{32 \times 0\cdot 229} \times \frac{a^4}{ab^3} \times 1 = 3\cdot 43 \text{ in.,}$$

the error of the approximation is therefore $100 \times (3\cdot 43 - 3\cdot 27)/3\cdot 43 = 4\cdot 7 \%$.

1·2526 Approximations for cruciform cross-sections

Shafts with cross-sections of the types shown in Figs. 1 and 2 are used for supporting armature windings.

The cruciform cross-section in Fig. 1 can be replaced by a square of which the sides h are equal to the diameter of the circular shaft. The equivalent length is then given by

$$L_e = \frac{1}{1 \cdot 431} \times \frac{G_e}{G} \times \left(\frac{D_e}{h}\right)^4 \times L \quad \text{[inches]}.$$

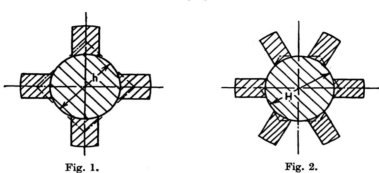

Fig. 1.   Fig. 2.

The cruciform cross-section in Fig. 2 (six-armed section) can be replaced by an hexagonal cross-section, of which the distance across the flats is equal to the diameter of the circular shaft. The equivalent length of the corresponding shaft portion is then given by

$$L_e = \frac{1}{1 \cdot 15} \times \frac{G_e}{G} \times \left(\frac{D_e}{H}\right)^4 \times L \quad \text{[inches]}.$$

Note. The above formulae are applicable to the normal case in which the torque is applied to the circular portion of the shaft.

1·253   *Equivalent length of shafts of irregular cross-section by the Griffith and Taylor method*

[References: A. A. Griffith and G. I. Taylor, *Rep. Memor. Adv. Comm. Aero., Lond.*, nos. 333, 334 and 392; G. W. Trayer and H. W. March, *N.A.C.A. Rep.* no. 334 (Appendix B).]

The torsional stiffness K of a shaft with any type of cross-section is equal to

$$T/\theta = GI_{p\,\text{eff.}}/L \quad \text{[Lb.in./rad.]},$$

where $T =$ applied torque [Lb.in.], $\theta =$ angle of twist [rad.], $G =$ modulus of rigidity [Lb./in.²], $I_{p\,\text{eff.}} =$ 'effective' polar second moment of area [in.⁴] of the cross-section, and $L =$ length of shaft [in.]. Its equivalent length in terms of a shaft of circular cross-section is therefore

$$L_e = \frac{G_e}{G} \times \frac{\pi}{32} \times \frac{D_e^4}{I_{p\,\text{eff.}}} \times L \quad \text{[in.]}.$$

To make use of these expressions, it is necessary to determine the 'effective' polar second moment of area of the cross-section, and this can readily be done by means of the Griffith and Taylor method, which is described below, using a four-armed cruciform section as an example (see Fig. 1).

Procedure for the determination of $I_{p\,\text{eff}}$.

(1) If the section includes long projections, split it up into a number of separate figures (five, in this example) by drawing dotted subdividing lines across these arms or projections, at a distance not greater than half the root thickness from the commencement of the projecting portion (see Figs. 1 and 2).

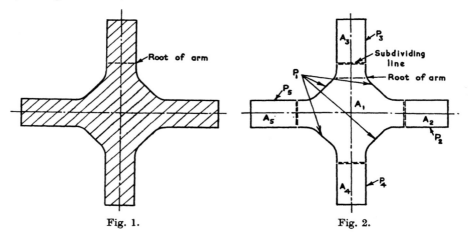

Fig. 1.   Fig. 2.

(2) For each of these portions, determine the area A and the 'outer perimeter' P. The latter should be taken to include only those lines which are parts of the boundary of the original cross-section, i.e. none of the subdividing lines should be included in P. (In Fig. 2, the perimeters P_1, P_2, etc., extend only over the heavy-line portions of the contour.)

(3) In each of these separate sections, inscribe a circle of the largest possible radius R, which will touch the original boundary of the cross-section at three (or more) points (see Fig. 3).

(4) Round off the *outer* corners with arcs of circles of radius r, using the r/R-values given in the following table as functions of the angle α (see Fig. 4) through which the tangent to the periphery at any point rotates as the point moves from one side of a corner to the adjacent side:

$\alpha =$	0°	18°	36°	54°	72°	90°	108°	126°	144°	162°	180°
$r/R =$	1·0	0·93	0·85	0·75	0·63	0·5	0·38	0·27	0·21	0·17	0·16

The resultant rounded-off corners related to the present example are shown in Fig. 5.

[138]

(5) Considering the central portion of the figure, determine the factor λ from the values given in the following table as functions of the ratio $RP/2A$ (where R = radius of largest inscribed circle, P and A = perimeter and area already obtained under (2) above):

$RP/2A =$	1·00	0·95	0·90	0·85	0·80	0·75
$\lambda =$	1·000	0·998	0·994	0·984	0·966	0·938

$RP/2A =$	0·70	0·65	0·60	0·55	0·50
$\lambda =$	0·897	0·848	0·793	0·732	0·667

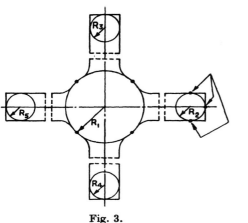

Fig. 3.

Fig. 4.

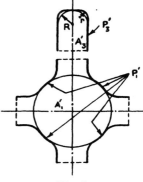

Fig. 5.

In this example
$$A_2' = A_3' = A_4' = A_5'$$
and $P_2' = P_3' = P_4' = P_5'.$

After determining the perimeter P_1' and the area A_1' of the rounded-off central figure (it so happens that in this example $P_1' = P_1$ and $A_1' = A_1$), evaluate its 'torsion constant'†

$$I_{p1} = \tfrac{1}{2}A_1\lambda\left(\frac{2A_1'}{P_1'}\right)^2 \quad [\text{in.}^4].$$

(6) Considering the outer figures, determine their corresponding values of λ, A' and P', and for each projection evaluate its 'torsion constant':

$$I_{p2} = \tfrac{1}{2}A_2\lambda\left(\frac{2A_2'}{P_2'}\right)^2 \quad [\text{in.}^4],$$

$$I_{p3} = \tfrac{1}{2}A_3\lambda\left(\frac{2A_3'}{P_3'}\right)^2, \quad \text{etc.}$$

(7) Evaluate the effective polar second moment of area of the cross-section as
$$I_{p\,\text{eff.}} = I_{p1} + I_{p2} + I_{p3} + \dots,$$

† This name is used because the quantities concerned are not the polar moments of inertia of the part sections although their total is the polar moment of inertia of the complete section.

[139]

and determine the equivalent length of the irregular-sectioned shaft by the relation

$$L_e = \frac{G_e}{G} \times \frac{\pi}{32} \times \frac{D_e^4}{I_{p\,\text{eff.}}} \times L \quad [\text{in.}].$$

NUMERICAL EXAMPLE. Square-sectioned shaft with sides $a = 2$ in., and length $L = 10$ in. It is required to determine the equivalent length of a circular shaft of diameter $D_e = 2$ in.

The radius of the inscribed circle (Fig. 6) is

$$R = 1 \text{ in.};$$
$$\alpha = 90°, \text{ hence } r/R = 0.5 \text{ and } r = 0.5 \text{ in.};$$
$$A_1 = A = 4 \text{ in.}^2;$$
$$P_1 = P = 8 \text{ in.}$$

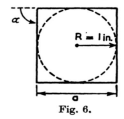

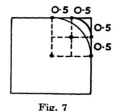

Fig. 6.   Fig. 7

The rounded-off contour with $r = 0.5$ in. at the corners is shown in Fig. 7.

$$\tfrac{1}{4}P_1' = 2 \times 0.5 + \frac{\pi}{2} \times 0.5 = 1.7855 \text{ in.}$$
$$P_1' = 7.142 \text{ in.}$$
$$\tfrac{1}{4}A_1' = 3 \times (0.5)^2 + \frac{\pi}{4} \times (0.5)^2 = 0.946 \text{ in.}^2$$
$$A_1' = 3.784 \text{ in.}^2$$
$$RP_1/2A_1 = 1 \times 8/2 \times 4 = 1; \quad \lambda = 1.00.$$

Thus $\quad I_{p\,\text{eff.}} = \tfrac{1}{4}A_1'\lambda \left(\frac{2A_1'}{P_1'}\right)^2 = \frac{4}{2} \times 1 \times \left(\frac{2 \times 3.784}{7.142}\right)^2 = 2.25 \text{ in.}^4.$

The equivalent length is

$$L_e = \frac{\pi}{32} \times \frac{D_e^4}{I_{p\,\text{eff.}}} \times L = \pi \times \frac{16}{32} \times \frac{1}{2.250} \times 10 = \frac{10}{1.431} = 6.988 \text{ in.}$$

This is the same value as with the exact formula, see p. 134.

Remarks on the Griffith and Taylor method

(1) The method permits a separate assessment to be made of the contributions to stiffness due to the projecting portions of a cross-section.

(2) If the projections are long and narrow, the rounding-off of the corners at their outer ends may be omitted without appreciable alteration of the results.

(3) The stresses at the re-entrant angles of the contour can be assessed by the Griffith and Taylor stress formulae, or by other formulae, which are given in section 2·3 (Stress evaluation).

[140]

1·254 *Further methods for determining the equivalent length of cruciform cross-sections*

In addition to the two methods already described (equivalent-square or equivalent-hexagon approximation, p. 136, and Griffith and Taylor method, p. 137), two other methods can be employed in assessing the equivalent length of cruciform-sectioned shafts. These are described below:

Shaft with arms effective in bending

In this method, it is assumed that the ribs or arms supporting the armature are effective in bending only.

The polar second moment of area of the ribs is estimated, on the basis of tests carried out on a number of machines with cruciform shafts, as $I_R = N_R \times R_m h b^2$ [in.^{4}],

where N_R = number of ribs, R_m = mean radius of rib [in.], h = actual rib height [in.], and b = rib width [in.].

The central circular portion of the shaft of diameter D [in.] has a second moment of area $I_S = \pi D^4/32$ [in.^{4}],

and the total effective I of the cross-section is obtained as $I = I_R + I_S$.

For an armature shaft of length L [in.], the equivalent length L_e (in terms of an equivalent circular-section shaft of diameter D_e and rigidity modulus G_e) is therefore estimated as

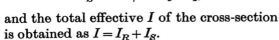

$$L_e = \frac{G_e}{G} \times \frac{\pi}{32} \times \frac{D_e^4}{I} \times L = \frac{G_e}{G} \times \frac{D_e^4}{\left[\dfrac{32}{\pi} N_R R_m h b^2 + D^4\right]} \times L \quad \text{[in.].}$$

Shaft with arms effective in torsion

If it is assumed that the ribs are effective in torsion, then with the above notation the polar second moment of area of the arms is expressed as

$$I_R = N_R \times bh \left[\frac{b^2 + h^2}{12} + \left(\frac{D}{2} + \frac{h}{2}\right)^2\right] \quad \text{[in.}^4\text{].}$$

The equivalent length of the armature or rotor shaft is therefore (with $I = I_R + I_S$)

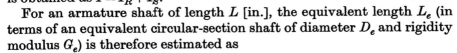

$$L_e = \frac{G_e}{G} \times \frac{D_e^4}{\left[\dfrac{8}{\pi} N_R bh \left\{\dfrac{b^2 + h^2}{3} + (D+h)^2\right\} + D^4\right]} \times L \quad \text{[in.].}$$

Example of calculation of armature shaft stiffness and comparison with test results

Two armature shafts of the type considered were coupled back-to-back so as to form a two-mass system, as shown schematically in Fig. 1. The pick-up was placed close to a projection (e.g. nut and bolt) tangentially to the periphery of the rotor, near the antinode of the vibration.

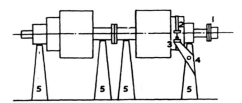

Fig. 1. Arrangement for back-to-back tests of generator shafts. 1, out-of-balance type mechanical vibration exciter; 2, strap round commutator; 3, permanent-magnet pick-up; 4, pick-up support; 5, bearing brackets.

Frequencies caused by bending vibrations were detected by moving additional pick-ups along the rotor towards the junction of the two armatures coupled back-to-back. Frequencies due to the pick-up support were eliminated by repeating the tests with an additional mass clamped, near the pick-up, on its support, so as to alter the bracket frequency. It was generally found that the indicated natural frequency of the armature system being tested remained unchanged.

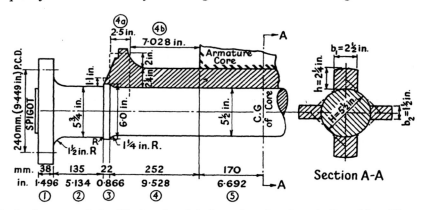

Fig. 2. Armature shaft. The lengths of shaft sections 1 to 5 are indicated in millimetres and inches. For evaluations by methods (*b*), (*c*) and (*d*), shaft section 4 is subdivided into sections 4*a* and 4*b*.

The type of armature shaft considered in this example is shown in Fig. 2. The calculated moment of inertia of the armature was 262·5 Lb.in.sec.² and the experimental value, determined in a pendulum test, was 243·0 Lb.in.sec.². The exciter was run at speeds up to 10,000 rev./min., and the test showed the natural frequency, with a large-amplitude torsional vibration, at 5750 vib./min.

Since the system comprises identical *J*-values and identical *K*-values, the relations applying are

$$\omega^2 = \frac{K}{2} \times \frac{J_1 + J_2}{J_1 \times J_2} = \frac{K}{2} \times \frac{2}{J} = \frac{K}{J},$$

Preliminary calculation: Lengths of shaft sections taking account of additional flexibilities at junctions

Shaft sections	1	2	3	4	5	Remarks
L	1·496	5·314	0·866	9·528	6·692	Section 5 is used only in methods (b) and (d)
D	9·449	5·75	6·0	5·5	5·5	
H	—	—	—	—	—	

Junctions	1–2	2–3	3–4	Remarks
Diameter ratios	$D_1/D_2 = 1·641$	$D_3/D_2 = 1·042$	$D_3/D_4 = 1·090$	$\Delta L/D$-values taken from p. 147 (for round-to-square transitions). $R = D/2$ for the corresponding shaft section
Fillet ratios	$r_{12}/R = 0·522$	$r_{23}/R_2 \simeq 0$	$r_{34}/R_4 = 0·091$	
$\Delta L/D$	$\Delta L/D_2 = 0$	$\Delta L/D_2 = 0·022$	$\Delta L/D_4 = 0·014$	
ΔL	$\Delta L_2 = 0$	$\Delta L_2 = 0·022D_2$ $= 0·127$	$\Delta L_4 = 0·014D_4$ $= 0·077$	

Shaft sections	1	2	3	4	5
Effective lengths $L + \Delta L$	$1·496 + 0·0$ $= 1·496$	$5·314 + 0·127$ $= 5·441$	$0·866 + 0$ $= 0·866$	$9·528 + 0·077$ $= 9·605$	$6·692 + 0$ $= 6·692$

Method (a): The four-armed sections are replaced by equivalent square sections as stated on p. 137. The evaluation is tabulated as follows [with $D_e = 5.5$ in., and the point of rigidity taken at the end of section 4 (see p. 131)]:

Shaft sections	1	2	3	4	Remarks
D_e/D	0·5821	0·9565	—	—	—
$C_1 = (D_e/D)^4$	0·1148	0·8371	0·9167	—	See p. 134
D_e/H	—	—	—	1·0	—
$C_2 = (D_e/H^4) \times \dfrac{1}{1·431}$	—	—	0·4931	0·6983	
$L_e = C_1(L + \Delta L)$	0·172	4·556	0·427	—	See p. 47
$L_e = C_2(L + \Delta L)$	—	—	—	6·707	

Total equivalent length: $L_e = 0·172 + 4·556 + 0·427 + 6·707 = 11·862$ in.

Stiffness (with rigidity modulus $G_s = 11·8 \times 10^6$ Lb./in.³):

$$K = (\pi/32) \times G_s D_e^4/L_e = 0·098175 \times 11·8 \times 10^6 \times (5·5)^4/11·862 = 8·937 \times 10^7 \text{ Lb.in./rad.}$$

[143]

where ω^2 = natural phase velocity squared, K = stiffness of one shaft, $J_1 = J_2 = J$ = moment of inertia of one armature. Thus, the stiffness can be determined experimentally from a vibration test. In this example, the resultant value is

$$K = J\omega^2 = 243 \cdot 0 \times \left(\frac{5750}{60/2\pi}\right)^2 = 8 \cdot 81 \times 10^7 \text{ Lb.in./rad.}$$

The stiffness of the armature shaft will now be calculated by the four different methods given in the previous pages of this section, viz.
(a) approximation by means of an equivalent square,
(b) webs considered effective in bending only,
(c) webs considered effective in torsion, and
(d) Griffith and Taylor method.

Method (b): Webs effective in bending only

Shaft sections 1 and 2 have no webs, and their equivalent lengths are therefore the same as for method (a), that is, $L_{e1} = 0 \cdot 172$ in., $L_{e2} = 4 \cdot 556$ in.

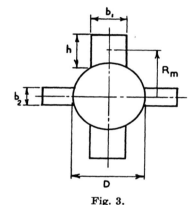

Fig. 3.

Shaft section 3. As the webs are not all of the same width, the formula for the effective length of cruciform sections has to be modified as follows:

$$L_e = \frac{D_e^4}{\frac{32}{\pi}\left(\frac{N_R}{2}\right) \times R_m h(b_1^2 + b_2^2) + D^4} \times (L + \Delta L),$$

with the notation indicated in Fig. 3. For shaft section 3, the above formula becomes

$$L_{e3} = \frac{(5 \cdot 5)^4}{10 \cdot 186 \times 2 \times 3 \cdot 55 \times 1 \cdot 1 \times (2 \cdot 25^2 + 1 \cdot 25^2) + 6^4} \times 0 \cdot 866$$

$$= 0 \cdot 4873 \times 0 \cdot 866 = 0 \cdot 417 \text{ in.}$$

Shaft section 4a. As shown in Fig. 2, it is necessary to split up section 4 into two sections 4a and 4b when using this method (or the following ones), since

[144]

different web heights have to be considered. The formula for equivalent length given above is applicable to shaft section 4a, and gives

$$L_{e4a} = \frac{915 \cdot 063}{10 \cdot 186 \times 2 \times 4 \cdot 875 \times 4 \cdot 25 \times (5 \cdot 750 + 1 \cdot 563) + 915 \cdot 063} \times (2 \cdot 5 + 0 \cdot 077)$$

$$= 0 \cdot 229 \times 2 \cdot 577 = 0 \cdot 590 \text{ in.}$$

Shaft sections 4b and 5. These can be considered together, since their web heights are the same. As stated on p. 132, the point of rigidity for webs effective in bending is taken at the centre of gravity of the armature core (including commutator). Thus, $L_{4b} = 7 \cdot 025$ in. and $L_5 = 6 \cdot 692$ in.

The equivalent length of these two shaft sections is, therefore,

$$L_{e4b} + L_{e5} = \frac{915 \cdot 063}{10 \cdot 186 \times 2 \times 3 \cdot 875 \times 2 \cdot 25 \times (5 \cdot 750 + 1 \cdot 563) + 915 \cdot 063} \times (7 \cdot 025 + 6 \cdot 692)$$

$$= 0 \cdot 413 \times 13 \cdot 717 = 5 \cdot 665 \text{ in.}$$

Total equivalent length:

$$L_e = 0 \cdot 172 + 4 \cdot 556 + 0 \cdot 417 + 0 \cdot 590 + 5 \cdot 665 = 11 \cdot 400 \text{ in.}$$

Stiffness:

$$K = (\pi/32) \times G_e D_e^4 / L_e = 0 \cdot 098175 \times 11 \cdot 8 \times 10^6 \times 915 \cdot 063 / 11 \cdot 40$$

$$= 9 \cdot 299 \times 10^7 \text{ Lb.in./rad.}$$

Method (c): Webs effective in torsion

Shaft sections 1 and 2 (without webs) have the same equivalent lengths as for the previous method, that is, $L_{e1} = 0 \cdot 172$ in. and $L_{e2} = 4 \cdot 556$ in.

Shaft section 3. For webs not all of the same width, the formula for the effective length becomes

$$L_e = \frac{D_e^4}{\dfrac{8}{\pi}\left(\dfrac{N_R}{2}\right) \times \left[\dfrac{b_1^2 + h^2}{3} + \dfrac{b_2^2 + h^2}{3} + (D + h)^2\right] + D^4} \times (L + \Delta L),$$

which can be rewritten as

$$L_e = \frac{D_e^4}{\dfrac{16}{\pi}\left[\dfrac{b_1^2 + b_2^2}{3} + \tfrac{2}{3} h^2 + (D + h)^2\right] + D^4} \times (L + \Delta L),$$

with $N_R = 4$ and using the notation indicated in Fig. 3. For shaft section 3, the formula gives

$$L_{e3} = \frac{(5 \cdot 5)^4}{5 \cdot 093\left[\dfrac{5 \cdot 750 + 1 \cdot 563}{3} + \tfrac{2}{3}(1 \cdot 1)^2 + (6 \cdot 0 + 1 \cdot 1)^2\right] + 6^4} \times 0 \cdot 866$$

$$= 0 \cdot 5831 \times 0 \cdot 866 = 0 \cdot 505 \text{ in.}$$

Shaft section 4a. With $h = 4 \cdot 25$ in., the evaluation gives

$$L_{e4a} = \frac{915 \cdot 063}{5 \cdot 093[2 \cdot 438 + \tfrac{2}{3}(4 \cdot 25)^2 + (5 \cdot 5 + 4 \cdot 25)^2] + 5 \cdot 5^4} \times 2 \cdot 577$$

$$= 0 \cdot 6212 \times 2 \cdot 577 = 1 \cdot 601 \text{ in.}$$

Shaft sections 4b and 5. As stated on p. 132, when applying this method with the webs effective in torsion, the point of rigidity is taken at the beginning of the armature core. Thus $L_5 = 0$, and only shaft section $4b$ requires evaluation. With $h = 2\cdot25$ in. and $D = 5\cdot5$ in., the equivalent length is determined as

$$L_{e4b} = \frac{915\cdot063}{5\cdot093[2\cdot438 + \frac{3}{4}(2\cdot25)^2 + (5\cdot5 + 2\cdot25)^2] + 915\cdot063} \times 7\cdot028$$

$$= 0\cdot7317 \times 7\cdot028 = 5\cdot142 \text{ in.}$$

Total equivalent length:

$$L_e = 0\cdot172 + 4\cdot556 + 0\cdot505 + 1\cdot601 + 5\cdot142 = 11\cdot976 \text{ in.}$$

Stiffness:
$$K = (\pi/32) \times G_e D_e^4 / L_e = 0\cdot098175 \times 11\cdot8 \times 10^6 \times 915\cdot063/11\cdot976$$

$$= 8\cdot852 \times 10^7 \text{ Lb.in./rad.}$$

Method (d): Griffith and Taylor method

Shaft sections 1 and 2 (without webs) have the equivalent lengths previously evaluated, viz. $L_{e1} = 0\cdot172$ in. and $L_{e2} = 4\cdot556$ in.

Shaft section 3. Referring to Fig. 4, which gives a cross-section of shaft portion 3, it is seen that the central figure of the cross-section contains the only circle that can be inscribed.

For the large webs, the subdividing line (at half the root width, see pp. 137–138) is at $b_1/2 = 1\cdot25$ in. from the circular centre-section. The web height, however, is only $1\cdot1$ in., so that the 'boundary' is outside the figure. This means (1) that there are no outer figures for the large webs and (2) that their areas and perimeters are integral parts of the central figure.

For the narrow webs, the subdividing line is at $b_2/2 = 0\cdot75$ in., which is within the length of these webs. This means that the areas and perimeters of these webs up to a distance of $0\cdot75$ in. along the webs belong to the central figure. Now, to have outer figures for the narrow web ends, it would be necessary to inscribe in these end-sections circles touching them at three points. The circle to be inscribed, however, has a radius $b_2/2 = 0\cdot75$ in. Measuring from the outer edge of the web inwards, it is found that the centre of this circle is on the other side of the sub-dividing line, i.e. within the web-root portion which belongs to the central figure. Thus, (1) there are no outer figures for the narrow webs, and (2) their areas and perimeters, up to a distance of $0\cdot75$ in. along the webs, belong to the central figure.

The central figure has an area

$$A = \pi D^2/4 + 2b_1 h + 2b_2 \times b_2/2 = 28\cdot274 + 5\cdot5 + 2\cdot25 = 36\cdot024 \text{ in.}$$

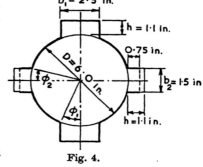

Fig. 4.

Its perimeter (including only the parts which are also boundaries of the original cross-section, i.e. the heavy lines in Fig. 4) can be evaluated as follows:

$$\sin\phi_1 = \tfrac{1}{2}b_1/(\tfrac{1}{2}D) = 1\cdot25/3\cdot00 = 0\cdot4167; \quad \phi_1 = 24° 37';$$

$$\sin\phi_2 = \tfrac{1}{2}b_2/(\tfrac{1}{2}D) = 0\cdot75/3\cdot00 = 0\cdot250; \quad \phi_2 = 14° 29';$$

$$P \cong \pi D - \frac{4}{360°}(\phi_1 + \phi_2)\pi D + (2b_1 + 4h) + (4b_2/2)$$

$$= 18\cdot850[1 - (39\cdot05°/90°)] + 9\cdot4 + 3\cdot0 = 18\cdot850[1 - 0\cdot434] + 12\cdot4 = 23\cdot069 \text{ in.}$$

Therefore $\qquad RP/2A = 3 \times 23{\cdot}069/(2 \times 36{\cdot}024) = 0{\cdot}961,$

and from p. 139, $\qquad \lambda = 0{\cdot}998.$

Since this central figure cannot be rounded off, $P' = P$ and $A' = A$. The polar second moment of area of the cross-section is therefore obtained as

$$I_p = \frac{\lambda}{2} A \times \left(\frac{2A}{P}\right)^2 = 0{\cdot}499 \times 36{\cdot}024 \times \left(\frac{72{\cdot}048}{23{\cdot}069}\right)^2 = 175{\cdot}35 \text{ in.}^4.$$

Accordingly, the equivalent length of shaft section 3 is

$$L_{e3} = \frac{\dfrac{\pi}{32} D_e^4}{I_p} (L + \Delta L) = \frac{(5{\cdot}5)^4 \times 0{\cdot}866}{10{\cdot}186 \times 175{\cdot}35} = 0{\cdot}444 \text{ in.}$$

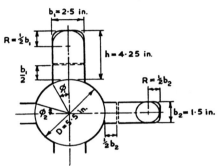

Note: $L_{4a} + \Delta L_{4a} = 2{\cdot}577$ in.

Fig. 5.

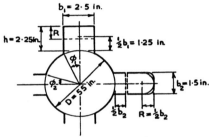

Note: $L_{4b} + L_5 = 13{\cdot}717$ in.

Fig. 6.

Shaft section 4 a (see Fig. 5)

In this case, the contributions of the outer arms have to be considered.

Central figure.

$$\begin{aligned}
\text{Area } A &= \pi D^2/4 + 2(b_1/2)\, b_1 + b_2^2 \\
&= (\pi/4) \times (5{\cdot}5)^4 + 6{\cdot}25 + 2{\cdot}25 \\
&= 32{\cdot}258 \text{ in.}^2.
\end{aligned}$$

Outer perimeter:

$$\sin \phi_1 = 1{\cdot}25/2{\cdot}75 = 0{\cdot}4545,$$
$$\sin \phi_2 = 0{\cdot}75/2{\cdot}75 = 0{\cdot}2727,$$
$$\phi_1 + \phi_2 = 27^\circ\, 02' + 15^\circ\, 50' = 42{\cdot}86^\circ,$$

$$P \cong \pi D \left[1 - \frac{4}{360^\circ}(\phi_1 + \phi_2)\right]$$
$$\qquad\qquad + 4(b_1/2) + 4(b_2/2)$$
$$= 9{\cdot}054 + 5{\cdot}0 + 3{\cdot}0 = 17{\cdot}054 \text{ in.},$$
$$RP/2A = 2{\cdot}75 \times 17{\cdot}054/64{\cdot}516 = 0{\cdot}727,$$
$$\lambda = 0{\cdot}919 \quad \text{(from p. 139)}.$$

Shaft sections 4 b and 5 (see Fig. 6)

These can be considered together, since their cross-sections are identical.

The central figure includes the P- and A-contributions of the large webs, since the latter have a circle centre at $R = b_1/2 = 1{\cdot}25$ in. from their outer edges, i.e. in the web-root area which is a part of the central portion of the figure.

Central figure.

$$\begin{aligned}
\text{Area } A &= \pi D^2/4 + 2hb_1 + 2(b_2/2)\, b_2 \\
&= 23{\cdot}758 + 11{\cdot}25 + 2{\cdot}25 \\
&= 37{\cdot}258 \text{ in.}^2.
\end{aligned}$$

Outer perimeter:

$$\phi_1 + \phi_2 = 42{\cdot}86^\circ,$$
$$P \cong \pi D \left[1 - \frac{4}{360^\circ}(\phi_1 + \phi_2)\right]$$
$$\qquad\qquad + 4h + 2b_1 + 4(b_2/2)$$
$$= 9{\cdot}054 + 9{\cdot}0 + 5{\cdot}0 + 3{\cdot}0 = 26{\cdot}054 \text{ in.},$$
$$RP/2A = 2{\cdot}75 \times 26{\cdot}054/74{\cdot}516 = 0{\cdot}962,$$
$$\lambda = 0{\cdot}998 \quad \text{(from p. 139)}.$$

Shaft section 4a (*cont.*)

Torsion constant of the central figure (with $P' = P$ and $A' = A$):

$$I_p = \frac{\lambda}{2} A \left(\frac{2A}{P}\right)^2 = 168\cdot04 \text{ in.}^4.$$

Large outer arms. For simplicity, the rounding-off is omitted. For one arm:

$$A = [h - (b_1/2)] b_1$$
$$= (4\cdot25 - 1\cdot25)\, 2\cdot50 = 7\cdot5 \text{ in.}^2,$$
$$P = 2[h - (b_1/2)] + b_1$$
$$= 2 \times 3\cdot0 + 2\cdot5 \qquad = 8\cdot5 \text{ in.,}$$
$$RP/2A = (b_1/2)\, P/2A$$
$$= 1\cdot25 \times 8\cdot5/15\cdot0 \quad = 0\cdot708,$$
$$\lambda = 0\cdot904.$$

Torsion constant for two arms:

$$2I_p = 0\cdot904 \times 7\cdot5 \times (15\cdot0/8\cdot5)^2$$
$$= 21\cdot12 \text{ in.}^4.$$

Small outer arms. For one arm:

$$A = [h - (b_2/2)] b_2$$
$$= (4\cdot25 - 0\cdot75) \times 1\cdot5 = 5\cdot25 \text{ in.}^2,$$
$$P = 2[h - (b_2/2)] + b_2$$
$$= 2 \times 3\cdot5 + 1\cdot5 \qquad = 8\cdot5 \text{ in.,}$$
$$RP/2A = (b_2/2)\, P/2A$$
$$= 0\cdot607,$$
$$\lambda = 0\cdot800.$$

Torsion constant for two arms:

$$2I_p = 0\cdot8 \times 5\cdot25(10\cdot5/8\cdot5)^2 = 6\cdot40 \text{ in.}^4.$$

Total:

$$I_p = 168\cdot04 + 21\cdot12 + 6\cdot40 = 195\cdot56 \text{ in.}^4.$$

$$L_{e4a} = \frac{\dfrac{\pi}{32} \times (5\cdot5)^4}{195\cdot56} \times 2\cdot577 = 1\cdot184 \text{ in.}$$

Shaft sections 4b and 5 (*cont.*)

Torsion constant of the central figure (with $P' = P$ and $A' = A$):

$$I_p = \frac{\lambda}{2} A \left(\frac{2A}{P}\right)^2 = 159\cdot08 \text{ in.}^4.$$

Small outer arms. For one arm:

$$A = [h - (b_2/2)] b_2$$
$$= (2\cdot25 - 0\cdot75) \times 1\cdot5 \quad = 2\cdot25 \text{ in.}^2,$$
$$P = 2[h - (b_2/2)] + b_2$$
$$= 2 \times 1\cdot5 + 1\cdot5 \qquad = 4\cdot5 \text{ in.,}$$
$$RP/2A = (b_2/2)\, P/2A$$
$$= 0\cdot75 \times 4\cdot5/(2 \times 2\cdot25) = 0\cdot75,$$
$$\lambda = 0\cdot938.$$

Torsion constant for two arms:

$$2I_p = 0\cdot938 \times 2\cdot25 \times (4\cdot5/4\cdot5)^2 = 2\cdot11 \text{ in.}^4$$

Total inertia of cross-section:

$$I_p = 159\cdot08 + 2\cdot11 = 161\cdot19 \text{ in.}^4.$$

Equivalent length of shaft sections 4b and 5:

$$L_{e4b} + L_{e5} = \frac{\dfrac{\pi}{32} \times (5\cdot5)^4}{161\cdot19} \times 13\cdot717$$
$$= 7\cdot645 \text{ in.}$$

Total equivalent length of armature shaft:

$$L_e = 0\cdot172 + 4\cdot556 + 0\cdot444 + 1\cdot184 + 7\cdot645 = 14\cdot00 \text{ in.}$$

Stiffness:

$$K = (\pi/32) \times G_e D_e^4/L_e = 0\cdot98175 \times 11\cdot8 \times 10^6 \times 915\cdot063/14\cdot00$$
$$= 7\cdot57 \times 10^7 \text{ Lb.in./rad.}$$

1·255 *Information on welded generator shafts*

The following information regarding welded generator shafts has been received from the Electrical Research Association (E.R.A.) and member companies.

The shaft forging should be rough turned before welding on arms. Webs must be flat and welding edges straight. It is good practice to chamfer the corners of the longitudinal bars or plates (e.g. to 60°) on each side. The surfaces are descaled so that clean metal is obtained at the joints. The parts to be welded are preheated and temperatures are checked by a thermometer. No welding is done unless the temperature at the weld position is 200° F. or more. After welding, the stresses are relieved by heat treatment of the shaft and spider at temperatures of 600–800° C., depending on the type of welding.

No transverse welding of webs to shaft should be made at the ends of the bars, since welding at these positions may start fatigue cracks by a notch effect. The welded sections should preferably stop short half an inch before and after a change of section. Where a welded construction is used, it should be ensured that the bending stresses of the installed system reach only low values.

Although not employed by some makers of electrical machines, welded constructions are used in certain applications, particularly in designs where any other form of construction would restrict the air passages for ventilation, or where for some reason the shaft size normally associated with the hub or sleeve has to be increased and the radial space between shaft and core punching is, consequently, restricted.

Welding has been successfully used for the spiders of d.c. and induction motors. In many cases, however, the bars, etc., are not welded directly on to the shaft but to steel hubs or rings so as to obtain a spider. Direct welding of the bars to the shaft is used with small armatures. Particular care is required when the generator is to run at a high speed or to be driven with a pulsating torque, e.g. by an internal-combustion engine.

Note. For shrink fits and press fits, see p. 132 and section 2·3.

1·256 *Information on water brakes*

For convenience, the data on water brakes (moments of inertia, shaft stiffness, and curves of power absorption) are grouped together and given at the end of section 1·1 (Moments of inertia).

1·3 NATURAL FREQUENCY CALCULATION

NOTATION

Symbol	Brief definition	Typical units	Symbol	Brief definition	Typical units
A	piston area	in.², cm.²	$\dot{q}$	current	amp.
b	number of propeller blades	—	$\ddot{q}$	rate of change of current	amp./sec.
c	damping coefficient	Lb.in.sec./rad., Kg.cm.sec./rad.	r	speed ratio	—
			R	radius	in., cm.
			R	resistance	ohms
C	cyclic speed variation	—	s	displacement	in., cm.
			t	time	sec.
C	capacitance	farad	T	torque	Lb.in., Kg.cm.
d	inner diameter	in., cm.	δT	increment in torque	Lb.in., Kg.cm.
D	diameter	in., cm.	$|T_m|$	mth order component of tangential pressure due to gas torque	Lb./in.², Kg./cm.²
f	natural frequency	cycles/sec.			
F	natural frequency	vib./min.			
g	acceleration due to gravity	in./sec.², cm./sec.²	u	parameter	—
G	modulus of rigidity	Lb./in.², Kg./cm.²	v	parameter	—
			V	voltage	volts
G_1	transfer function	(rad./sec.)/volt	w	specific weight	Lb./in.³, Kg./cm.³
G_2	transfer function	rad./volt			
H	power	h.p., watts	W	weight	Lb., Kg.
I_v	second moment of area	in.⁴, cm.⁴	x	distance	in., cm.
			x	$=\omega^2 =$ phase velocity squared	rad.²/sec.²
j	$=\sqrt{-1}$	—			
J	mass moment of inertia	Lb.in.sec.², Kg.cm.sec.²	x	$=\omega \sqrt{(J/K)}$	—
			Y	velocity admittance	(rad./sec.)/Lb.in.
k	radius of gyration	in., cm.			
k	parameter	in.⁻¹, cm.⁻¹	y	displacement admittance	rad./Lb.in.
k_e	proportionality factor	volt/(rad./sec.)	Z	number of blower deliveries per revolution	—
k_t	proportionality factor	Lb.in./amp.			
K	stiffness	Lb.in./rad., Kg.cm./rad.	Z	velocity impedance	Lb.in./(rad./sec.)
L	length	in., cm.	z	displacement impedance	Lb.in./rad.
L	inductance	henries			
m	mass	Lb.sec.²/in., Kg.sec.²/cm.	α	vibration amplitude	deg., rad.
			β	vibration amplitude	deg., rad.
m	order number	—	β	$=\omega \sqrt{(w/Gg)}$	in.⁻¹, cm.⁻¹
n	number of masses	—	γ	vibration amplitude	deg., rad.
N	number of masses	—	γ	phase angle	deg., rad.
N	speed	rev./min.	δ	$=\omega^2/\omega_0^2 =$ phase velocity ratio squared	—
p	$=j\omega = j \times$ phase velocity	rad./sec.			
P	force	Lb., Kg.	Δ	Holzer-table amplitude	rad.
q	shear stress	Lb./in.², Kg./cm.²	$\overrightarrow{\Sigma\Delta}$	phase vector sum of relative amplitudes	rad.
q	charge (electrical)	coul.			

[150]

Symbol	Brief definition	Typical units	Symbol	Brief definition	Typical units
θ	vibration amplitude	deg., rad.	ρ	mass density	Lb.sec.^{2}/in.^{4}, Kg.sec.^{2}/in.^{4}
$\dot{\theta}$	angular velocity	rad./sec.			
$\ddot{\theta}$	angular acceleration	rad./sec.^{2}	ϕ	phase angle	rad.
λ	$= J\omega^2/K$	—	Φ	'equivalence function'	—
λ	$= K_{cyl.}/K_1 =$ stiffness parameter	—	Φ	flux	voltsec.
μ	$= J_1/J_{cyl.} =$ inertia parameter	—	ω	phase velocity	rad./sec.
			Ω	phase velocity	rad./sec.

Subscripts are used in the text to define more precisely, e.g. $D_{ci} =$ crankpin inner diameter. Their meaning is fully explained in the context in each instance.

In all cases dealt with in this section, the systems are considered to be vibrating freely (without external torque, except where otherwise stated) and without damping. The moments of inertia J_1, J_2, ..., are expressed in Lb.in.sec.^{2} and the shaft stiffnesses K_{12}, K_{23}, ..., in Lb.in./rad., or in Lb.ft.sec.^{2} and Lb.ft./rad., respectively. The relations given are, however, equally valid with other consistent units, for instance, Ton.ft.sec.^{2} (or Kg.cm.sec.^{2}) for inertias, and Ton.ft./rad. (or Kg.cm./rad.) for shaft stiffnesses.

1·31   Preliminary estimation of natural frequencies

1·311   *Formulae for systems with 1, 2, 3 or 4 masses*

One-mass system

The natural frequency is determined by the relation

$$F = 9{\cdot}55\sqrt{(K/J)} \quad \text{[vib./min.]}.$$

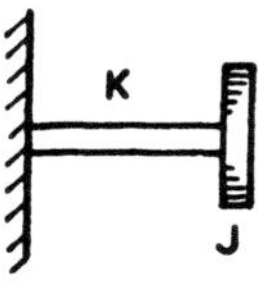

Derivation. The equation of motion is

$$J\ddot{\theta} + K\theta = 0.$$

Let $\theta = \theta_0 \cos \omega t$. Then

$$-\omega^2 J\theta_0 + K\theta_0 = 0,$$

Fig. 1.

and $\omega^2 = K/J$, which gives the above relation, using $F = (60/2\pi) \times \omega$ to obtain F from the natural phase velocity ω.

Two-mass system

The natural frequency is obtained as

$$F = 9{\cdot}55\sqrt{\left(K\frac{J_1 + J_2}{J_1 J_2}\right)} \quad \text{[vib./min.]}.$$

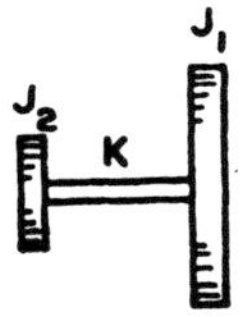

Derivation. For this system (see Fig. 2), the equations of motion are

$$J_1 \ddot{\theta}_1 + K(\theta_1 - \theta_2) = 0,$$
$$J_2 \ddot{\theta}_2 + K(\theta_2 - \theta_1) = 0.$$

Fig. 2.

Let $\theta_1 = \theta_{01} e^{j\omega t}$ and $\theta_2 = \theta_{02} e^{j\omega t}$; then

$$(K - \omega^2 J_1)\,\theta_{01} - K\theta_{02} = 0,$$
$$-K\theta_{01} + (K - \omega^2 J_2)\,\theta_{02} = 0,$$

are the corresponding characteristic equations. The determinant Det_s of the coefficients of θ_{01} and θ_{02}, known as the 'system determinant', is then

$$\mathrm{Det}_s = \begin{vmatrix} K - \omega^2 J_1 & -K \\ -K & K - \omega^2 J_2 \end{vmatrix} = (K - \omega^2 J_1)(K - \omega^2 J_2) - K^2$$

$$= -\omega^2(J_1 + J_2)K + \omega^4 J_1 J_2.$$

Assuming the existence of an excitation torque, when Det_s is equal to zero the values of $\theta_{01} = \mathrm{Det}_1/\mathrm{Det}_s$ and $\theta_{02} = \mathrm{Det}_2/\mathrm{Det}_s$ reach infinite values, that is, resonance occurs. Therefore, $\mathrm{Det}_s = 0$ is the frequency equation, and this gives F for the two-mass system.

Note. Further details regarding the use of determinants in vibration problems are given on pp. 153 and 155, and in section 2·45.

Three-mass system

The one-node and two-node frequencies of the three-mass system shown in Fig. 3 are obtained from

$$F_{\mathrm{I}} = 9{\cdot}55 \sqrt{\{(b/2) - \sqrt{[(b/2)^2 - c]}\}}$$

and
$$F_{\mathrm{II}} = 9{\cdot}55 \sqrt{\{(b/2) + \sqrt{[(b/2)^2 - c]}\}},$$

where
$$b = (K_1/J_1) + (K_2/J_3) + \{(K_1 + K_2)/J_2\}$$

and
$$c = (K_1/J_1) \times (K_2/J_3) \times \left\{1 + \frac{J_1 + J_3}{J_2}\right\}.$$

Derivation. The equations of motion are

$$J_1 \ddot{\theta}_1 + K_1(\theta_1 - \theta_2) = 0,$$

$$J_2 \ddot{\theta}_2 + K_1(\theta_2 - \theta_1) + K_2(\theta_2 - \theta_3) = 0,$$

$$J_3 \ddot{\theta}_3 + K_2(\theta_3 - \theta_2) = 0.$$

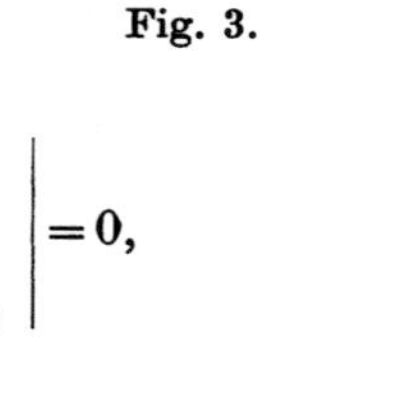

Using the substitutions $\theta_k = \theta_{0k}\, e^{j\omega t}$ ($k = 1, 2, 3$), the frequency equation is obtained from

Fig. 3.

$$\mathrm{Det}_s = \begin{vmatrix} K_1 - \omega^2 J_1 & -K_1 & 0 \\ -K_1 & K_1 + K_2 - \omega^2 J_2 & -K_2 \\ 0 & -K_2 & K_2 - \omega^2 J_3 \end{vmatrix} = 0,$$

which gives

$$\omega^4 - \omega^2 \left\{\frac{K_1}{J_1} + \frac{K_2}{J_3} + \frac{K_1 + K_2}{J_2}\right\} + \frac{K_1 K_2}{J_1 J_3} \times \left\{1 + \frac{J_1 + J_3}{J_2}\right\} = 0,$$

and this equation, with $F = (60/2\pi) \times \omega$, gives the above formulae.

[152]

Systems with a larger number of masses

For a four-mass system, the natural frequencies of the three modes of vibration are determined by a cubic equation, obtained in the same manner as the above expressions:

$$\omega^6 \times \frac{J_1 J_2 J_3 J_4}{K_1 K_2 K_3} - \omega^4 \times \left[\frac{J_1 J_2}{K_1 K_2}(J_3 + J_4) + \frac{J_1 J_4}{K_1 K_3}(J_2 + J_3) + \frac{J_3 J_4}{K_2 K_3}(J_1 + J_2) \right]$$

$$+ \omega^2 \times \left[\frac{J_1}{K_1}(J_2 + J_3 + J_4) + \frac{1}{K_2}(J_1 + J_2)(J_3 + J_4) + \frac{J_4}{K_3}(J_1 + J_2 + J_3) \right]$$

$$- (J_1 + J_2 + J_3 + J_4) = 0.$$

To find the three real roots of this equation, the computation work is already relatively extensive. In the general case, an n-mass system has a frequency equation of the $(n-1)$th order in ω^2. Therefore, for systems with more than three or four masses, it is necessary to use approximate methods of solution, such as Horner's method or Holzer tables. The latter are generally preferred, since they are also used to obtain the relative amplitudes and other characteristics of a multi-mass system.

Notes on the use of determinants

Some familiarity with the use of determinants is useful when dealing with simultaneous linear equations such as those occurring in vibration problems. A brief review of the properties of determinants is therefore included in the following.

Consider:
$$\text{(i)} \quad ax + by = E,$$
$$\text{(ii)} \quad cx + dy = F,$$

where $x, y =$ variables, and $a, b, c, d, E, F =$ constants.

The solution by gradual substitutions is as follows:

(i) $- \dfrac{b}{d}$ (ii) gives:

$$ax + by = E$$
$$- \frac{cb}{d}x - by = - \frac{b}{d}F$$
$$\overline{\left(a - \frac{bc}{d}\right)x \quad = E - \frac{b}{d}F} \quad \text{or} \quad x = \frac{Ed - bF}{ad - bc};$$

similarly (i) $- \dfrac{a}{c}$ (ii) gives:
$$y = \frac{aF - cE}{ad - bc}.$$

The denominator $ad - bc$ can be written as a 'determinant', i.e. a cross-product of the four factors

$$\begin{vmatrix} a & b \\ c & d \end{vmatrix} = ad - bc = \text{Det}_s, \quad \text{('system determinant')},$$

the upward product being multiplied with (-1). This can also be applied to the numerators:

$$dE - bF = \begin{vmatrix} E & b \\ F & d \end{vmatrix} = \text{Det}_x, \quad aF - cE = \begin{vmatrix} a & E \\ c & F \end{vmatrix} = \text{Det}_y,$$

[153]

so that the solutions of equations (i) and (ii) are

$$x = \frac{\text{Det}_x}{\text{Det}_s} \quad \text{and} \quad y = \frac{\text{Det}_y}{\text{Det}_s}.$$

Some of the properties of determinants will now be listed:

(1) If two adjacent lines (rows or columns) are interchanged, the value of the determinant becomes multiplied by (-1):

$$\begin{vmatrix} a & b \\ c & d \end{vmatrix} = (-1) \begin{vmatrix} c & d \\ a & b \end{vmatrix} = (-1) \begin{vmatrix} b & a \\ d & c \end{vmatrix}.$$

(2) If each term of a line (row or column) is multiplied by $\pm\lambda$ (where $\lambda = $ a constant) and added to another parallel line, term for term, the value of the determinant is unchanged:

$$\begin{vmatrix} a & b \\ c & d \end{vmatrix} = \begin{vmatrix} a+\lambda c & b+\lambda d \\ c & d \end{vmatrix} \begin{aligned} &= (a+\lambda c)d - c(b+\lambda d), \\ &= ad + \lambda cd - cb - c\lambda d = ad - bc. \end{aligned}$$

(3) Therefore, if two parallel lines (rows or columns) have the same elements, term for term, or proportional elements, the value of the determinant is zero:

$$\begin{vmatrix} a & b \\ \lambda a & \lambda b \end{vmatrix} = a\lambda b - \lambda ab = 0.$$

(4) A determinant which has a line (row or column) with two additive terms in each element can be split up into two determinants as follows:

$$\begin{vmatrix} a & c_1+c_2 \\ b & d_1+d_2 \end{vmatrix} = \begin{vmatrix} a & c_1 \\ b & d_1 \end{vmatrix} + \begin{vmatrix} a & c_2 \\ b & d_2 \end{vmatrix},$$

and this can be verified by direct evaluation of both sides of this equation.

(5) Multiplying a determinant by a number λ is the same as multiplying all the elements of one line (row or column) by this number:

$$\lambda \begin{vmatrix} a & b \\ c & d \end{vmatrix} = \begin{vmatrix} \lambda a & b \\ \lambda c & d \end{vmatrix} = \begin{vmatrix} \lambda a & \lambda b \\ c & d \end{vmatrix}.$$

This and all previous rules also apply to complex numbers. For instance, if $j = \sqrt{-1}$,

$$\begin{vmatrix} a & b_1+jb_2 \\ c & d_1+jd_2 \end{vmatrix} = \begin{vmatrix} a & b_1 \\ c & d_1 \end{vmatrix} + \begin{vmatrix} a & jb_2 \\ c & jd_2 \end{vmatrix} = \begin{vmatrix} a & b_1 \\ c & d_1 \end{vmatrix} + j \begin{vmatrix} a & b_2 \\ c & d_2 \end{vmatrix}.$$

For three equations with three unknowns x, y and z:

$$a_1 x + b_1 y + c_1 z = D_1,$$
$$a_2 x + b_2 y + c_2 z = D_2,$$
$$a_3 x + b_3 y + c_3 z = D_3,$$

the solutions are

$$\text{Det}_s = \begin{vmatrix} a_1 & b_1 & c_1 \\ a_2 & b_2 & c_2 \\ a_3 & b_3 & c_3 \end{vmatrix} = a_1 \begin{vmatrix} b_2 & c_2 \\ b_3 & c_3 \end{vmatrix} - a_2 \begin{vmatrix} b_1 & c_1 \\ b_3 & c_3 \end{vmatrix} + a_3 \begin{vmatrix} b_1 & c_1 \\ b_2 & c_2 \end{vmatrix},$$

$$\text{Det}_x = \begin{vmatrix} D_1 & b_1 & c_1 \\ D_2 & b_2 & c_2 \\ D_3 & b_3 & c_3 \end{vmatrix}, \quad \text{Det}_y = \begin{vmatrix} a_1 & D_1 & c_1 \\ a_2 & D_2 & c_2 \\ a_3 & D_3 & c_3 \end{vmatrix}, \quad \text{Det}_z = \begin{vmatrix} a_1 & b_1 & D_1 \\ a_2 & b_2 & D_2 \\ a_3 & b_3 & D_3 \end{vmatrix},$$

giving
$$x = \frac{\text{Det}_x}{\text{Det}_s}, \quad y = \frac{\text{Det}_y}{\text{Det}_s} \quad \text{and} \quad z = \frac{\text{Det}_z}{\text{Det}_s}.$$

The three determinants Det_x, Det_y and Det_z are evaluated by expanding them in the same manner as Det_s, shown above.

Note. If x, y and z are the vibration amplitudes θ_1, θ_2 and θ_3 of a three-mass system, these θ-values will reach infinity if the system determinant $\text{Det}_s = 0$, since then all the denominators are zero.

Therefore, $\text{Det}_s = 0$ is known as the 'frequency equation' which gives the natural frequencies at which resonance occurs.

1·312 *Methods for systems with any number of masses*

1·3121 *Bradbury's diagram for estimating one-node natural frequencies of Diesel engines*

A first approximation to the natural frequency of present-day diesel-engine systems can be obtained by means of the following diagram, due to C. H. Bradbury.†

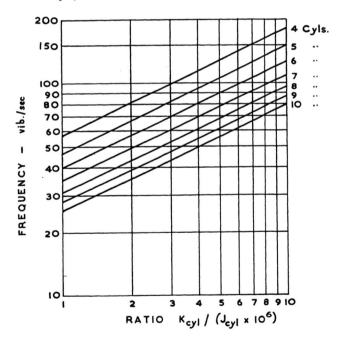

In this diagram, the frequency of the engine system, in cycles per second, is shown plotted against the ratio $K_{\text{cyl.}}/(J_{\text{cyl.}} \times 10^6)$, where $K_{\text{cyl.}}$ = stiffness per crankthrow, and $J_{\text{cyl.}}$ = moment of inertia per line.

† Bradbury, C. H. *Torsional Vibration in Diesel Engines* (C. Griffin and Co. Ltd, London, 1938), p. 29. This book sets forth the fundamentals and within its 118 pages contain much valuable information.

The diagram on the previous page gives the natural frequency of engines with 4–10 cylinders directly. It was calculated on the basis of the following assumptions:

(i) all moments of inertia per line are equal;

(ii) the ratio $J_F/J_{cyl.}$ is not less than 40 (where J_F = moment of inertia of flywheel);

(iii) all crankthrow stiffnesses are equal and the stiffness of the shaft portion K_F between the end cylinder centre and the flywheel is also equal to $K_{cyl.}$.

The values in the diagram above can also be calculated by means of the relation

$$f = \frac{u}{10^3} \sqrt{\frac{K_{cyl.}}{J_{cyl.}}} \quad [\text{cycles/sec.}],$$

where the value of u is determined from the following table:

No. of cylinders =	4	5	6	7	8	9	10
u =	56·9	46·9	40	34·9	31·05	27·9	25·45

Variations in the ratio $K_{cyl.}/K_F$ can also be taken into account by using a correction factor v in the above formula, which becomes:

$$f = uv \times 10^{-3} \sqrt{(K_{cyl.}/J_{cyl.})} \quad [\text{cycles/sec.}].$$

The value of v is determined from the following graph, also due to Bradbury:

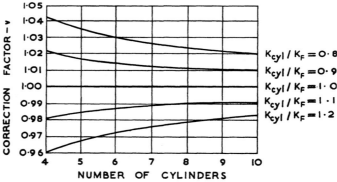

1·3122 One-node frequency estimation by B.I.C.E.R.A. formula

A quick and simple method of assessing the one-node natural frequency F of an engine-and-flywheel system with its node approximately at the flywheel position is by means of the following expression:

$$F \cong 9.55 \times \sqrt{\frac{K_{cyl.}}{J_{cyl.}}} \times \sqrt{\frac{A+B}{AB}} \quad [\text{vib./min.}],$$

where $K_{cyl.}$ = crankthrow stiffness [Lb.in./rad.], $J_{cyl.}$ = moment of inertia per cylinder [Lb.in.sec.^{2}], and $B = J_F/J_{cyl.}$ = ratio of flywheel inertia to moment of inertia per line.

The value of A, which depends on the number of cylinders N of the engine, is taken from the following table:

$N=$	1	2	3	4	5	6	7	8	9	10	11	12
$A=$	1	3	6	10	15	21	28	36	45	55	66	78

The value thus found can then be used as a first approximation in detailed Holzer tabulations to obtain an accurate value for the natural frequency.

Derivation. An engine comprising N cylinder inertias $J_{\text{cyl.}}$ and $N-1$ crankthrows of equal stiffness $K_{\text{cyl.}}$ (as shown in the accompanying figure) can be reduced to an equivalent inertia J_E situated at the position of the Nth mass, provided that the natural frequency $F = \omega \times 9\cdot55$, the relative amplitude Δ_N of the Nth mass and the total inertia torque at this Nth mass position remain unchanged. These conditions are expressed by the equation

$$J_E\,\omega^2\Delta_N = \sum_{K=1}^{N} J_{\text{cyl. }(K)}\omega^2\Delta_K = \omega^2 J_{\text{cyl.}} \sum_{K=1}^{N} \Delta_K,$$

where $\Delta_K = \Delta_1, \Delta_2, \ldots, \Delta_N$ are the relative vibration amplitudes of the individual masses. Therefore J_E is determined by

$$J_E = J_{\text{cyl.}} \frac{\sum\limits_{K=1}^{N} \Delta_K}{\Delta_N}.$$

The natural frequency $F_E = 9\cdot55\omega_E$ of the equivalent two-mass system is then obtainable from

$$\omega^2 = K_F\left[\frac{1}{J_E} + \frac{1}{J_F}\right].$$

If the node position is close to the flywheel, and $K_F \cong K_{\text{cyl.}}$, it may be assumed that the relative amplitudes Δ_K decrease linearly from $\Delta_1 = 1$ at mass no. 1 to $\Delta_{N+1} = \Delta_F = 0$ at the main flywheel. Then $\Delta_N = \Delta_1/N$ is the amplitude at the last cylinder, and

$$\sum_{K=1}^{N} \Delta_K = \Delta_1 + \Delta_1\left(1 - \frac{1}{N}\right) + \ldots + \Delta_1\left(1 - \frac{N-1}{N}\right)$$

$$= \Delta_1\left[N - \frac{1}{N}(1 + 2 + \ldots + (N-1))\right]$$

$$= \Delta_1(N+1)/2,$$

so that

$$J_E = J_{\text{cyl.}} \frac{\sum\limits_{K=1}^{N} \Delta_K}{\Delta_N} \cong J_{\text{cyl.}} \frac{(N+1)N}{2} = A J_{\text{cyl.}}.$$

The values of $A = N(N+1)/2$, for $N = 1$ to 12 cylinders, are given in the above table. Furthermore, if $K_F \simeq K_{\text{cyl.}}$, and $J_F = BJ_{\text{cyl.}}$, the frequency equation for the reduced two-mass system may be written as

$$\omega^2 \simeq (K_{\text{cyl.}}/J_{\text{cyl.}}) \left[\frac{1}{A} + \frac{1}{B} \right],$$

and since $F = (60/2\pi)\,\omega \simeq 9{\cdot}55\omega$, this expression is identical with that given at the beginning of this section.

Metric units. Natural frequencies in vib./min. are also obtained from the formulae in this section when the moments of inertia are expressed in Kg.cm.sec.^{2} and the shaft stiffnesses in Kg.cm./rad.

1·3123 *Lewis method for estimating the natural frequency when the approximate position of the node is known*

The following is an application of a method evolved by F. M. Lewis. Let

$$J_E = \Sigma J_{\text{cyl.}} \quad \text{and} \quad K_E = K_{\text{cyl.}} / [N_{\text{cyl.}} + \tfrac{1}{2}],$$

where $J_{\text{cyl.}} = $ moment of inertia per line and $K_{\text{cyl.}} = $ crankthrow stiffness (between cylinder centres). If the node of the vibration is situated close to the flywheel, as shown in the accompanying figure, the natural frequency of the entire system can be estimated from

$$\omega = \frac{\pi}{2} \sqrt{(K_E/J_E)} \quad \text{[rad./sec.]}.$$

This method can also be employed for the estimation of two- and three-node frequencies (see Example 3, below).

EXAMPLE 1. Nine-cylinder engine.

$$K_E = 64{\cdot}0/9{\cdot}5 = 6{\cdot}73 \times 10^6 \text{ Lb.ft./rad.},$$

$$J_E = 550 \text{ Lb.ft.sec.}^2,$$

$$\omega = 1{\cdot}5708 \times \sqrt{(6{\cdot}73 \times 10^6/5{\cdot}50 \times 10^2)} = 173{\cdot}7 \text{ rad./sec.}$$

From Holzer tabulation $\omega = 157{\cdot}0$. The node is situated between the flywheel and the driven machine, and the formula gives a slightly high value.

EXAMPLE 2. Six-cylinder engine.

$$K_E = 1{\cdot}56 \times 10^6/6{\cdot}5 = 0{\cdot}24 \times 10^6 \text{ Lb.ft./rad.},$$

$$J_E = 1{\cdot}116 \text{ Lb.ft.sec.}^2,$$

$$\omega = 1{\cdot}5708 \sqrt{(0{\cdot}24 \times 10^6/1{\cdot}116)} = 727{\cdot}5 \text{ rad./sec.}$$

From Holzer tabulation $\omega = 704{\cdot}9$. The node is situated between no. 6 cylinder and the flywheel.

These examples show that the approximation is fairly accurate, and the values for ω thus obtained can be used to obtain more accurate values by means of Holzer tables.

For systems in which the node is closer to the driven machine than to the flywheel, the approximate value of ω can be obtained by assuming a plausible value for Δ_F and using the relation

$$\omega = \cos^{-1} \Delta_F \times \sqrt{(K_E/J_E)} \quad [\text{rad./sec.}].$$

The arc of cosine Δ_F is expressed in radians.

EXAMPLE 3. Five-cylinder marine propulsion engine.

$$K_E = 519 \cdot 5 \times 10^6 / 5 \cdot 5 = 94 \cdot 4 \times 10^6 \text{ Lb.in./rad.},$$

$$J_E = 1432 \text{ Lb.in.sec.}^2.$$

We assume $\Delta_F = 0 \cdot 97$:

$$\sqrt{(K_E/J_E)} = 256 \cdot 7, \quad \cos^{-1} \Delta_F = \cos^{-1}\{0 \cdot 97\} = 14° \, 6' = 0 \cdot 24609 \text{ rad.},$$

$$\omega = 0 \cdot 24609 \times 256 \cdot 7 = 63 \cdot 14 \text{ rad./sec.}$$

Holzer table value $\omega = 72 \cdot 3$.

For the two-node vibration (with a node nearer to the main flywheel) we assume $\quad \Delta_F = 0 \cdot 36, \quad \cos^{-1}\{0 \cdot 36\} = 68° \, 54' = 1 \cdot 2025 \text{ rad.},$

$$\omega_{(2\text{-node})} = 1 \cdot 2025 \times 256 \cdot 7 = 308 \cdot 5 \text{ rad./sec.}$$

Holzer table value $\omega = 353$.

Derivation of formula

Consider an equivalent system, in which the engine inertias $\Sigma J_{\text{cyl.}}$ are assumed to be uniformly distributed over the entire crankshaft, from cylinder no. 1 to the flywheel position. This system can be regarded as a heavy shaft, with a modulus of rigidity G and a polar moment of inertia $I_p = \dfrac{\pi}{32} D_e^4$, where D_e is the equivalent shaft diameter. If L_E is the crankshaft length (in relation to D_e), each unit length of the system is capable of vibrating in accordance with the equation of motion

$$\frac{J_E}{L_E} \frac{\partial^2 \theta}{\partial t^2} = G I_p \frac{\partial^2 \theta}{\partial x^2},$$

where θ = vibration amplitude, and x = any position along the crankshaft, with $x = 0$ at cylinder no. 1. A suitable substitution for solving this equation is $\quad \theta = \Delta(x) \, \phi(t) = \Delta_1 \cos kx \, \cos(\omega t - \epsilon),$

where Δ_1 = relative amplitude at no. 1 cylinder, and k = a constant still to be determined. Inserting this expression in the equation of motion, one obtains $\quad -\dfrac{J_E}{L_E} \omega^2 = -G I_p k^2, \quad$ hence $\quad k = \omega \sqrt{\dfrac{J_E}{G I_p L_E}},$

or with
$$K_E = \frac{GI_p}{L_E}, \quad k = \omega \sqrt{\left(\frac{J_E}{K_E}\right)} \times \frac{1}{L_E}.$$

Therefore, at any position x along the crankshaft, the maximum amplitude is given by
$$\Delta(x) = \Delta_1 \cos\left\{\omega \sqrt{\left(\frac{J_E}{K_E}\right)} \frac{x}{L_E}\right\}.$$

If the node is at L_E, we must have $\Delta_{(x=L_E)} = 0$, and this requires that
$$\cos\left\{\omega \sqrt{\frac{J_E}{K_E}}\right\} = \cos\frac{\pi}{2} \quad \text{or} \quad \omega = \frac{\pi}{2}\sqrt{\frac{K_E}{J_E}},$$

which is the formula previously given.

Furthermore, the *inertia torque* of all the engine inertias ahead of the flywheel (i.e. the torque $(\Sigma J\omega^2\Delta)_{\text{max.}}$ of a Holzer table) is obtained as

$$\Sigma T_{\text{inertia}} = \int_0^{L_E} \frac{\omega^2 J_E}{L_E} \Delta(x)\,dx = \int_0^{L_E} \omega^2 J_E \frac{\Delta_1}{L_E} \cos\left\{\omega \sqrt{\left(\frac{J_E}{K_E}\right)} \frac{x}{L_E}\right\} dx$$

$$= \left[\omega \sqrt{(K_E J_E)} \Delta_1 \sin\left\{\omega \sqrt{\left(\frac{J_E}{K_E}\right)} \frac{x}{L_E}\right\}\right]_0^{L_E}$$

or
$$\Sigma T = \omega \sqrt{(K_E J_E)} \Delta_1 = \omega \sqrt{(K_E J_E)},$$

since $\Delta_1 = \text{unity}$. Using the above expression for ω, we obtain
$$\Sigma T = \frac{\pi}{2} K_E \cong \frac{\pi}{2} \frac{K_{\text{cyl.}}}{(N_{\text{cyl.}} + \frac{1}{2})} \cong (\Sigma J\omega^2\Delta)_{\text{max.}}.$$

Applications. Example 1 on p. 158 gives
$$\Sigma T_{\text{max.}} = 1\cdot5708 \times 6\cdot73 \times 10^6 = 10\cdot56 \times 10^6 \text{ Lb.ft.,}$$

compared with the Holzer table value of $9\cdot57 \times 10^6$; and Example 2 gives
$$\Sigma T_{\text{max.}} = 1\cdot5708 \times 0\cdot24 \times 10^6 = 0\cdot377 \times 10^6$$

as compared with a Holzer table value of $0\cdot376 \times 10^6$.

Metric units. Torques are obtained in Kg.cm. if J-values are expressed in Kg.cm.sec.^{2} and K-values in Kg.cm./rad.

1·32   Tabulation methods

1·321   *Holzer method*

Holzer's method is *the standard procedure* for the calculation of natural frequencies of multi-mass systems. It is useful not only for frequency determination but also for a number of other data, such as relative amplitudes, which are required in torsional vibration investigations. Holzer's method will, therefore, be treated in detail in this section.

The procedure is slightly different for straight systems and for branched systems. Both will be described by means of illustrative examples.

1·3211 *Straight systems*

As an example, we shall consider the engine system shown in Fig. 1, with the following numerical values:

Inertias†	Stiffnesses
$J_1 = J_2 = J_3 = 1{\cdot}58$ Lb.in.sec.^{2}	$K_1 = K_2 = 9{\cdot}19 \times 10^6$ Lb.in./rad.
$J_4 = 108{\cdot}6$ Lb.in.sec.^{2}	$K_3 = 7{\cdot}58 \times 10^6$ Lb.in./rad.
$J_5 = 6{\cdot}33$ Lb.in.sec.^{2}	$K_4 = 2{\cdot}30 \times 10^6$ Lb.in/rad.

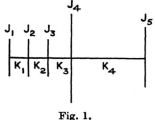

Fig. 1.

Determination of the one-node vibration frequency

An approximate value for the natural frequency F_I of the one-node vibration can be estimated by means of various methods previously described. For instance, noting that K_4 is the smallest stiffness value and that J_4 and J_5 are the largest masses, we may assume that under one-node conditions the value of F_I is approximately obtainable from

$$F_I = 9{\cdot}55 \sqrt{\bigg/\bigg(K_4 \times \frac{J_4 + J_5}{J_4 \times J_5}\bigg)}, \quad \text{which gives} \quad F_I \cong 5920 \text{ vib./min.}$$

The values of the one-node frequency F_I can now be calculated to the desired degree of accuracy with the aid of Holzer tables. The procedure is as follows:

(1) Using an approximate value for frequency, prepare a tabulation as shown in Table 1 a. In this table, Δ_1 (the *relative amplitude* of mass 1) is taken as 1 radian.‡ Thus, $J_1 \omega^2 \Delta_1$ and $\Sigma(J \omega^2 \Delta_1)$ at mass 1 have numerically the same value as $J_1 \omega^2$ and can be written directly.

(2) Calculate $\Sigma(J \omega^2 \Delta)/K = \Delta_{sh}$ and insert this in the corresponding column (7). In the numerical example, we have

$$0{\cdot}606 \times 10^6 / 9{\cdot}19 \times 10^6 = 0{\cdot}0659 \text{ rad.}$$

(3) Calculate the relative amplitude at mass 2 from

$$\Delta_2 = \Delta_1 - \Delta_{sh\,1} \quad [\text{rad.}].$$

---

† For brevity in explanations, polar mass moments of inertia are referred to as *inertias*, and torsional stiffnesses of shaft portions as *stiffnesses*.

‡ Alternative designations: Holzer-table amplitude, relative angular deflexion.

Table 1a. $F = 5920$ vib./min., $\omega = F/9{\cdot}55 = 619{\cdot}5$ rad./sec.,
$$\omega^2 = 0{\cdot}384 \times 10^6$$

(1)	(2)	(3)	(4)	(5)	(6)	(7)
		Deflexion				Shaft twist
		Inertia	at posi-	Inertia	Total	Δ_{sh}
		torque	tion of	torque of	torque at	Shaft
	Inertias	per rad.	mass m	mass m	mass m	stiffness	$= \dfrac{1}{K} \Sigma J_m \omega^2 \Delta_m$
Mass	J_m	$J_m \omega^2$	Δ_m	$J_m \omega^2 \Delta_m$	$\Sigma J_m \omega^2 \Delta_m$	K	
no.	[Lb.in.sec.^{2}]	[Lb.in./rad.]	[rad.]	[Lb.in.]	[Lb.in.]	[Lb.in./rad.]	[rad.]
1	1·58	$0{\cdot}606 \times 10^6$	1·0000	$0{\cdot}606 \times 10^6$	$0{\cdot}606 \times 10^6$	$9{\cdot}19 \times 10^6$	—
2	1·58	$0{\cdot}606 \times 10^6$	—	—	—	$9{\cdot}19 \times 10^6$	—
3	1·58	$0{\cdot}606 \times 10^6$	—	—	—	$7{\cdot}58 \times 10^6$	—
4	108·6	$41{\cdot}67 \ \times 10^6$	—	—	—	$2{\cdot}3 \ \times 10^6$	—
5	6·33	$2{\cdot}43 \ \times 10^6$	—	—	—	—	—

In the above example, this gives $\Delta_2 = 1{\cdot}0000 - 0{\cdot}0659 = 0{\cdot}9341$ rad. This is to be inserted in Table 1a in column (3), line 2.

(4) Calculate $J_2 \omega^2 \Delta_2$ and insert this value in column (4). We have

$$0{\cdot}606 \times 10^6 \times 0{\cdot}9341 = 0{\cdot}566 \times 10^6.$$

(5) In column (5), insert the value of the total inertia torque at mass 2, i.e.

$$\sum_1^2 (J \omega^2 \Delta) = J_1 \omega^2 \Delta_1 + J_2 \omega^2 \Delta_2.$$

This gives $\qquad (0{\cdot}606 + 0{\cdot}566) \times 10^6 = 1{\cdot}172 \times 10^6.$

(6) Divide this value by K_2, so as to obtain the shaft twist,

$$\Delta_{sh\,2} = \frac{1}{K_2} \times \sum_1^2 (J \omega^2 \Delta),$$

and write this in column (7). This gives

$$1{\cdot}172 \times 10^6 / 9{\cdot}19 \times 10^6 = 0{\cdot}1274 \quad \text{[rad.]}.$$

(7) Calculate $\Delta_3 = \Delta_2 - \Delta_{sh\,2}$ and insert the result in column (3). This gives

$$0{\cdot}9341 - 0{\cdot}1274 = 0{\cdot}8067 \quad \text{[rad.]}.$$

(8) Proceed in the same manner down to the last line of the table. The results are shown in Table 1b. Note the *change of sign* at the position of mass 5. This indicates that the node is between mass 4 and mass 5. The negative value for the inertia torque of mass 5 is slightly too great, so that the total torque at this last mass is not zero, but a negative value. This means that our value taken for ω^2 is too high, and we should take a slightly lower value.

Although the value of $-0{\cdot}03 \times 10^6$ Lb.in. for the residual torque is small, this is still about 5 % of the value of $J_1 \omega^2 \Delta_1 = 0{\cdot}606 \times 10^6$ Lb.in., and it is desirable to reduce it to about 1 % of the torque at mass no. 1.

	(1)	(2)	(3)	(4)	(5)	(6)	(7)
Mass no.	Inertias J_m [Lb.in. sec.^{2}]	Inertia torque per rad. $J_m\omega^2$ [Lb.in./rad.]	Deflexion at position of mass m Δ_m [rad.]	Inertia torque of mass m $J_m\omega^2\Delta_m$ [Lb.in.]	Total torque at mass m $\Sigma J_m\omega^2\Delta_m$ [Lb.in.]	Shaft stiffness K [Lb.in./ rad.]	Shaft twist Δ_{sh} $=\dfrac{1}{K}\Sigma J_m\omega^2\Delta_m$ [rad.]
1	1·58	$0\cdot606 \times 10^6$	1·0000	$0\cdot606 \times 10^6$	$0\cdot606 \times 10^6$	$9\cdot19 \times 10^6$	0·0659
2	1·58	$0\cdot606 \times 10^6$	0·9341	$0\cdot566 \times 10^6$	$1\cdot172 \times 10^6$	$9\cdot19 \times 10^6$	0·1274
3	1·58	$0\cdot606 \times 10^6$	0·8067	$0\cdot488 \times 10^6$	$1\cdot660 \times 10^6$	$7\cdot58 \times 10^6$	0·2191
4	108·6	$41\cdot67 \times 10^6$	0·5876	$24\cdot50 \times 10^6$	$26\cdot160 \times 10^6$	$2\cdot3 \times 10^6$	11·3700
5	6·33	$2\cdot43 \times 10^6$	$-10\cdot7824$	$-26\cdot19 \times 10^6$	$-0\cdot03 \times 10^6$	—	—

A negative value for the residual torque, for vibrations of the first mode, signifies that the value of ω^2 used in the calculation was too high (an explanation of this is given on p. 164). We therefore try a slightly lower value, for instance, $\omega^2 = 0\cdot95 \times 0\cdot384 \times 10^6 = 0\cdot3648 \times 10^6$, and make another tabulation (Table 1 c). To avoid unnecessary repetitions, the 10^6 factors are placed in the captions. It is seen that

$$T_{\text{res.}} = +1\cdot2868 \times 10^6 \text{ Lb.in.,}$$

which is a greater value than that of our first trial, but with a positive sign, so that we should take a slightly higher value for ω^2. The simplest way to obtain it is by linear interpolation using a graph of $T_{\text{res.}}$ plotted against ω^2, as shown below (Fig. 2). This gives an interpolated value $\omega^2 = 0\cdot3834 \times 10^6$, which is used in a final trial (Table 1 d).

Table 1 c. $F = 5764$ vib./min., $\omega = 603\cdot98$ rad./sec.,
$\omega^2 = 0\cdot3648 \times 10^6$

Mass no.	J [Lb.in.sec.^{2}]	$\dfrac{J\omega^2}{10^6}$ [Lb.in./rad.]	Δ [rad.]	$\dfrac{J\omega^2\Delta}{10^6}$ [Lb.in.]	$\dfrac{\Sigma}{10^6}$ [Lb.in.]	$\dfrac{K}{10^6}$ [Lb.in./rad.]	Δ_{sh} [rad.]
1	1·58	0·5764	1·0000	0·5764	0·5764	9·19	0·0627
2	1·58	0·5764	0·9373	0·5402	1·1166	9·19	0·1215
3	1·58	0·5764	0·8158	0·4702	1·5868	7·58	0·2093
4	108·6	39·6	0·6065	24·020	25·6068	2·30	11·1300
5	6·33	2·306	$-10\cdot5235$	$-24\cdot320$	$+1\cdot2868$	—	—

The residual torque is now practically zero, and we have determined the one-node frequency as $F_{\text{I}} = 5910$ vib./min.

Note on residual torque. When $\omega^2 = 0$, the system is not vibrating, and the residual torque is also zero. With low values of ω^2, all the $J\omega^2$-values are small, and $T_{\text{res.}} = \Sigma T$ has a positive value. At the resonant frequency, the deflexions must be such that $T_{\text{res.}} = \Sigma T = 0$. The residual-torque curve thus has a shape as shown in the diagram of Fig. 3. Therefore the trend of the calculation is indicated as follows:

(A) For one-node and three-node natural frequencies the value of the natural frequency is higher than the assumed value when $T_{\text{res.}}$ is positive, and lower when $T_{\text{res.}}$ is negative.

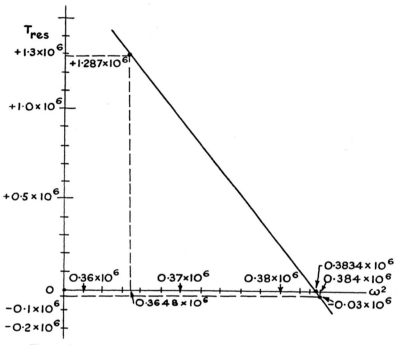

Fig. 2. Determination of ω^2 by means of residual-torque diagram.

Table 1 d. $F = 5910$ vib./min., $\omega = 619 \cdot 2$ rad./sec., $\omega^2 = 0 \cdot 3834 \times 10^6$

Mass no.	J [Lb.in.sec.²]	$\dfrac{J\omega^2}{10^6}$ [Lb.in./rad.]	Δ [rad.]	$\dfrac{J\omega^2\Delta}{10^6}$ [Lb.in.]	$\dfrac{\Sigma}{10^6}$ [Lb.in.]	$\dfrac{K}{10^6}$ [Lb.in./rad.]	Δ_{sh} [rad.]
1	1·58	0·6056	1·0000	0·6056	0·6056	9·19	0·0659
2	1·58	0·6056	0·9341	0·5653	1·1709	9·19	0·1273
3	1·58	0·6056	0·8068	0·4888	1·6697	7·58	0·2200
4	108·6	41·62	0·5868	24·4300	26·0997	2·30	11·3440
5	6·33	2·427	−10·7572	−26·0960	+0·0037	—	—

(B) For a two-node natural frequency the value of the natural frequency is higher than the assumed value when $T_{\text{res.}}$ is negative, and lower when $T_{\text{res.}}$ is positive.

[164]

In general (A) is valid for all modes having an odd number of nodes (1, 3, 5, etc.), and (B) for modes with an even number of nodes (2, 4, 6, etc.).

Determination of the two-node vibration frequency

Referring again to Fig. 3, it would be possible to determine the two-node frequency by using gradually increasing values of ω^2 in a number of Holzer tabulations. However, a saving in computation work can be obtained by estimating a fairly close value in a preliminary approximation.

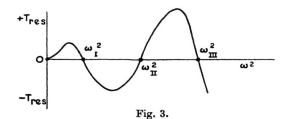

Fig. 3.

For this purpose it is useful to reduce the engine system to an equivalent three-mass system, as follows:

(1) Lump together all the cylinder inertias at the position of the middle cylinder. (For the system taken as an example (Figs. 1, 4), this gives $J_1 + J_2 + J_3 = \Sigma J_{\text{cyl.}} = 3 \times 1 \cdot 58 = 4 \cdot 74$ Lb.in.sec.^{2} at the position of J_2.)

(2) Determine the equivalent stiffness of the shafting between the middle cylinder and the flywheel. (We thus have:

$$\frac{1}{K_e} = \frac{1}{K_2} + \frac{1}{K_3} = \left[\frac{1}{9 \cdot 19} + \frac{1}{7 \cdot 58}\right] \times 10^{-6},$$

$$1/K_e = [0 \cdot 1088 + 0 \cdot 1319] \times 10^{-6} = 0 \cdot 2407 \times 10^{-6},$$

so that

$$K_e = 10^6/0 \cdot 2407 = 4 \cdot 15 \times 10^6 \text{ Lb.in./rad.})$$

(3) Determine the approximate value of ω_{II}^2 by means of one of the following relations:

$$(a) \quad \omega_{\text{II}}^2 = K_e \frac{J_F + \Sigma J_{\text{cyl.}}}{J_F \times \Sigma J_{\text{cyl.}}},$$

or $(b) \quad \omega_{\text{II}}^2 = \frac{1}{\omega_{\text{I}}^2} \times K_e K_M \times \frac{J_F + J_M + \Sigma J_{\text{cyl.}}}{J_F \times J_M \times \Sigma J_{\text{cyl.}}},$

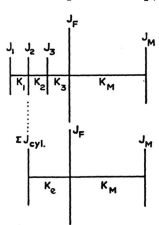

Fig. 4. Reduction of an engine system to an equivalent 3-mass system. J_F = flywheel inertia; J_M = driven-machine inertia.

where ω_{I}^2 is the value previously determined by Holzer tables for one-node frequency conditions.

[165]

For the example considered throughout this section, these relations give the values

$$(a) \quad \omega_{II}^2 = 4 \cdot 15 \times 10^6 \times \frac{108 \cdot 6 + 4 \cdot 74}{108 \cdot 6 \times 4 \cdot 74} = 0 \cdot 913 \times 10^6;$$

$$(b) \quad \omega_{II}^2 = \frac{1}{0 \cdot 3834 \times 10^6} \times 4 \cdot 15 \times 2 \cdot 3 \times 10^{12} \times \frac{108 \cdot 6 + 6 \cdot 3 + 4 \cdot 74}{108 \cdot 6 \times 6 \cdot 3 \times 4 \cdot 74}$$

$$= 0 \cdot 9125 \times 10^6.$$

It should be noted that (b) is based simply on the known relation that $\omega_I^2 \times \omega_{II}^2 =$ value of the constant term in the quadratic frequency equation for a three-mass system (which need not, therefore, be evaluated in full when the value of ω_I^2 is known).

Table 2a. $\quad F = 9120$ vib./min., $\omega^2 = 0 \cdot 913 \times 10^6$

Mass no.	J	$\dfrac{J\omega^2}{10^6}$	Δ	$\dfrac{J\omega^2\Delta}{10^6}$	$\dfrac{\Sigma}{10^6}$	$\dfrac{K}{10^6}$	Δ_{sh}
1	1·58	1·442	1·0000	1·442	1·442	9·19	0·1569
2	1·58	1·442	0·8431	1·216	2·658	9·19	0·2880
3	1·58	1·442	0·5551	0·7943	3·452	7·58	0·4550
4	108·6	99·15	0·1001	9·925	13·377	2·3	5·8120
5	6·33	6·042	−5·7119	−34·520	−21·143	—	—

Table 2b. $\quad F = 9550$ vib./min., $\omega^2 = 1 \cdot 00 \times 10^6$

Mass no.	J	$\dfrac{J\omega^2}{10^6}$	Δ	$\dfrac{J\omega^2\Delta}{10^6}$	$\dfrac{\Sigma}{10^6}$	$\dfrac{K}{10^6}$	Δ_{sh}
1	1·58	1·58	1·0000	1·58	1·580	9·19	0·1717
2	1·58	1·58	0·8283	1·308	2·888	9·19	0·3142
3	1·58	1·58	0·5141	0·812	3·700	7·58	0·4743
4	108·6	108·6	0·0398	4·320	8·020	2·3	3·4870
5	6·33	6·33	−3·4472	−21·820	−13·800	—	—

Table 2c. $\quad F = 10{,}015$ vib./min., $\omega^2 = 1 \cdot 10 \times 10^6$

Mass no.	J	$\dfrac{J\omega^2}{10^6}$	Δ	$\dfrac{J\omega^2\Delta}{10^6}$	$\dfrac{\Sigma}{10^6}$	$\dfrac{K}{10^6}$	Δ_{sh}
1	1·58	1·738	1·0000	1·738	1·738	9·19	0·1889
2	1·58	1·738	0·8111	1·411	3·149	9·19	0·3424
3	1·58	1·738	0·4687	0·814	3·963	7·58	0·5226
4	108·6	119·4	−0·0539	−6·431	−2·468	2·3	−1·0726
5	6·33	6·963	+1·0187	+7·091	+4·623	—	—

Having obtained the approximation for ω_{II}^2, the Holzer table calculation is carried out in the same manner as previously, taking account of all the inertias and stiffnesses of the system. The tabulations for the engine system used as an example are given in Tables 2a–e.

From the resultant curve for $T_{res.}$, constructed as in Fig. 5, we see that the two-node natural frequency corresponds to a value $\omega_{II}^2 = 1 \cdot 076 \times 10^6$, so that $\omega_{II} = 1037 \cdot 3$ rad./sec., and $F_{II} = 9900$ vib./min.

Table 2d. $F = 9940$ vib./min., $\omega^2 = 1\cdot084 \times 10^6$

Mass no.	J	$\dfrac{J\omega^2}{10^6}$	Δ	$\dfrac{J\omega^2\Delta}{10^6}$	$\dfrac{\Sigma}{10^6}$	$\dfrac{K}{10^6}$	Δ_{sh}
1	1·58	1·712	1·0000	1·712	1·712	9·19	0·1862
2	1·58	1·712	0·8138	1·393	3·101	9·19	0·3372
3	1·58	1·712	0·4766	0·815	3·916	7·58	0·5160
4	108·6	117·65	−0·0394	−4·633	−0·717	2·30	−0·3118
5	6·33	6·86	+0·2794	+1·867	+1·150	—	—

Table 2e. $F = 9915$ vib./min., $\omega^2 = 1\cdot078 \times 10^6$

Mass no.	J	$\dfrac{J\omega^2}{10^6}$	Δ	$\dfrac{J\omega^2\Delta}{10^6}$	$\dfrac{\Sigma}{10^6}$	$\dfrac{K}{10^6}$	Δ_{sh}
1	1·58	1·703	1·0000	1·703	1·703	9·19	0·1851
2	1·58	1·703	0·8149	1·390	3·093	9·19	0·3363
3	1·58	1·703	0·4796	0·816	3·909	7·58	0·5154
4	108·6	117·0	−0·0358	−4·188	−0·279	2·30	−0·1213
5	6·3	6·821	+0·0855	+0·583	−0·304	—	—

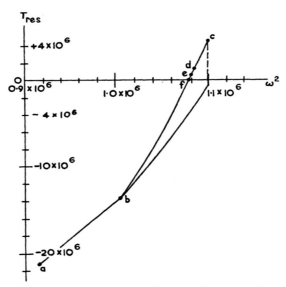

Fig. 5. Residual-torque curve with successive approximations.

Determination of the three-node vibration frequency

In many cases it is not necessary to determine the frequency of the third node of vibration. However, it is sometimes useful to have a fairly close approximation for its value, in order to verify that its significant major criticals do not occur in the running range and, in particular, do not coincide with those of the first two modes of vibration in the running range.

To obtain an approximation for ω^2_{III}, it is convenient to reduce the engine system to an equivalent four-mass system.

[167]

Assuming the presence of a node between the engine cylinder inertias, it is possible to simplify as follows (see Fig. 6):

(1) Draw an assumed curve for the relative amplitudes, with three nodal points x_1, x_2 and x_3.

(2) Lump together at the position of J_1 all the inertias ahead of the node x_1; concentrate at a position between x_1 and x_2 the inertias between these nodal points; leave J_F and J_M at their original positions.

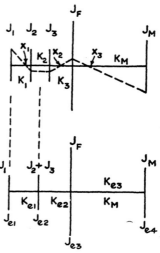

(3) Calculate the equivalent stiffnesses of the shaft portions between the positions of the equivalent inertias; determine the approximate value of ω^2_{III} from the relation

$$\omega^2_{III} = \frac{1}{\omega^2_I \omega^2_{II}} \times \frac{J_{e1} + J_{e2} + J_{e3} + J_{e4}}{J_{e1} \times J_{e2} \times J_{e3} \times J_{e4}} \times K_{e1} K_{e2} K_{e3},$$

where ω^2_I and ω^2_{II} are known values for the two first modes of vibration, already determined by Holzer tables.

Fig. 6. Reduction of an engine system to an equivalent 4-mass system.

EXAMPLE. Taking again the same engine system as previously, we may concentrate J_2 and J_3 at a position midway between these two masses. In this way, we have the four-mass sytem:

$$J_{e1} = J_1 = 1\cdot 58 \text{ Lb.in.sec.}^2,$$

$$J_{e2} = J_2 + J_3 = 3\cdot 16 \text{ Lb.in.sec.}^2,$$

$$J_{e3} = J_F = 108\cdot 6 \text{ Lb.in.sec.}^2,$$

$$J_{e4} = J_M = 6\cdot 33 \text{ Lb.in.sec.}^2,$$

$$1/K_{e1} = 1/K_1 + (1/2) \times (1/K_2) = 1\cdot 5/K_1,$$

$$\text{or} \quad K_{e1} = 9\cdot 19 \times 10^6/1\cdot 5 = 6\cdot 124 \times 10^6 \text{ Lb.in./rad.},$$

$$1/K_{e2} = (1/2) \times (1/K_2) + 1/K_3$$

$$= [0\cdot 5 \times 0\cdot 1088 + 0\cdot 1319] \times 10^{-6} = 0\cdot 1863 \times 10^{-6},$$

$$K_{e2} = 5\cdot 368 \times 10^6 \text{ Lb.in./rad.},$$

$$K_{e3} = K_M = 2\cdot 3 \times 10^6 \text{ Lb.in./rad.}$$

With these values, and $\omega^2_I = 0\cdot 3834 \times 10^6$ and $\omega^2_{II} = 1\cdot 076 \times 10^6$, we obtain

$$\omega^2_{III} = \frac{1}{0\cdot 4126 \times 10^2} \times \frac{120\cdot 07}{1\cdot 58 \times 3\cdot 56 \times 108\cdot 6 \times 6\cdot 33} \times 6\cdot 124 \times 5\cdot 368 \times 2\cdot 3 \times 10^{18}$$

$$= 5\cdot 684 \times 10^6 \text{ [rad./sec.]}^2.$$

It should be noted that the above formula is based on the relation $\omega^2_I \omega^2_{II} \omega^2_{III}$ = value of the constant term in the cubic frequency equation for a four-mass system (see p. 152).

From the numerical determination of the two-node frequency, we know that, for the system taken as an example, the first approximation for ω^2_{II} was about 10% on the low side, and we may expect a similar trend with the approximation of ω^2_{III}. To save computation work, we take for the first table the value

$$\omega^2_{III} = 1\cdot10 \times 5\cdot684 \times 10^6 = 6\cdot253 \times 10^6,$$

which corresponds to a frequency $F_{III} = 23{,}860$ vib./min.

Table 3. $F = 27{,}850$ vib./min., $\omega^2 = 8\cdot51 \times 10^6$

Mass no.	J	$\dfrac{J\omega^2}{10^6}$	Δ	$\dfrac{J\omega^2\Delta}{10^6}$	$\dfrac{\Sigma}{10^6}$	$\dfrac{K}{10^6}$	Δ_{sh}
1	1·58	13·45	1·0000	13·45	13·45	9·19	1·461
2	1·58	13·45	−0·461	−6·20	+7·25	9·19	0·789
3	1·58	13·45	−1·250	−16·20	−9·55	7·58	−1·260
4	108·6	924·0	+0·010	+9·24	−0·31	2·3	−0·136
5	6·33	53·90	+0·146	+7·86	+7·55	—	—

The tabulations give finally $\omega^2_{III} = 8\cdot511 \times 10^6$ and $F_{III} = 27{,}850$ vib./min. obtained by interpolation from the residual-torque diagram, based on five Holzer tables for this mode of vibration. The last calculation is reproduced in Table 3. An increase in ω^2 as small as to $8\cdot5126 \times 10^6$ makes the residual torque swing over to a negative value. With $\omega^2 = 8\cdot512 \times 10^6$, the first three lines of Table 3 are hardly changed, and the last two lines are as follows:

4	108·6	924·3	+0·012	+11·10	+ 1·54	2·30	+0·666
5	6·3	53·9	−0·654	−37·50	−35·96	—	—

Relative-amplitude diagram†

The relative amplitudes of the masses in a vibrating system depend on the mode of vibration considered. These relative amplitudes are the Δ-values obtained in the evaluation of Holzer tables. By plotting these

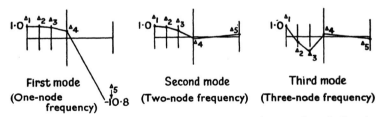

First mode
(One-node frequency)

Second mode
(Two-node frequency)

Third mode
(Three-node frequency)

Fig. 7. Relative-amplitude diagrams for the first three modes of vibration of the system calculated on pp. 161 to 168.

values in a schematic drawing of the equivalent system, at the positions of the corresponding masses, and joining together the points thus obtained, by means of a broken line, we have a 'diagram of the relative amplitudes'. This diagram has also been termed the 'swinging form' of the system, and the line of the relative amplitudes is also known as the 'mass-elastic line'. Diagrams for the first three modes of the above example are shown in Fig. 7.

† Alternative designations: Holzer-table amplitude diagrams, deflexion diagrams, diagrams giving normal elastic curves or swinging forms of system.

If the actual vibration amplitudes of the engine system, for a given mode, are denoted by θ_1, θ_2, θ_3, etc., the relations between these and the calculated relative amplitudes are

$$\frac{\theta_1}{\theta_1}=\Delta_1=1, \quad \frac{\theta_2}{\theta_1}=\Delta_2, \quad \ldots, \quad \frac{\theta_n}{\theta_1}=\Delta_n;$$

and conversely

$$\theta_1=\theta_1\times 1, \quad \theta_2=\theta_1\times\Delta_2, \quad \ldots, \quad \theta_n=\theta_1\times\Delta_n.$$

Thus, the amplitude ratio for the end masses is $\theta_n/\theta_1=\Delta_n/\Delta_1$. These relations can be verified in practice by amplitude measurements taken at both ends of an engine system. For values of Δ_n/Δ_1 between zero and 3 or 4, the agreement is generally fairly good. For larger values divergences occur, owing to the damping torque acting on the last mass. Holzer tabulations with damping or other methods (see section 2·4) are then required to determine Δ_n/Δ_1.

Basic theory of Holzer's method

(A) *Simple explanation, without any equations of motion*

For an assumed amplitude 1·0 at the first mass, the peak torque to accelerate it is $J_1\omega^2$, and thus the twist in the first shaft section is $J_1\omega^2/K_1$; the net peak torque required by J_2 is $J_2\omega^2\Delta_2$, so that the torque between J_2 and J_3 is $J_1\omega^2+J_2\omega^2\Delta_2$, since the torque $J_1\omega^2$ between J_1 and J_2 acts in the opposite direction, and so on to the last mass.

Finally, $\Sigma J\omega^2\Delta$ is the external torque required at the last mass to maintain the assumed vibration and this must be zero for a natural frequency.

(B) *Detailed mathematical derivation*

Under free-vibration conditions, i.e. without a forcing torque, the equations of motion of an n-mass system of the type shown in the figure are:

$$-J_1\omega^2\Delta_1+K_1\Delta_1-K_1\Delta_2=0, \tag{1}$$

$$-K_1\Delta_1+K_1\Delta_2-J_2\omega^2\Delta_2+K_2\Delta_2-K_2\Delta_3=0, \tag{2}$$

$$-K_2\Delta_2+K_2\Delta_3-J_3\omega^2\Delta_3+K_3\Delta_3-K_3\Delta_4=0, \tag{3}$$

$$\cdots\cdots\cdots\cdots\cdots\cdots\cdots\cdots\cdots\cdots\cdots\cdots\cdots$$

$$-K_{n-1}\Delta_{n-1}+K_{n-1}\Delta_n-J_n\omega^2\Delta_n=0, \tag{n}$$

where $K=$ shaft stiffness [Lb.in./rad.], $J=$ moment of inertia of mass [Lb.in.sec.^{2}], $\Delta=$ relative amplitude of vibration at a mass position [rad.], and $\omega=$ phase velocity of the vibration [rad./sec.].

By substituting $K_1\Delta_1 - K_1\Delta_2 = J_1\omega^2\Delta_1$ from eq. (1) in eq. (2), we have

$$K_2\Delta_2 - K_2\Delta_3 = J_1\omega^2\Delta_1 + J_2\omega^2\Delta_2 = \sum_1^2 (J\omega^2\Delta); \qquad (2')$$

introducing this in eq. (3) we obtain

$$K_3\Delta_3 - K_3\Delta_4 = J_1\omega^2\Delta_1 + J_2\omega^2\Delta_2 + J_3\omega^2\Delta_3 = \sum_1^3 (J\omega^2\Delta). \qquad (3')$$

Thus, all the terms with K-values are gradually eliminated. The result for eq. (n) is

$$\omega^2(J_1\Delta_1 + J_2\Delta_2 + \dots + J_n\Delta_n) = \omega^2 \sum_1^n (J\Delta) = 0. \qquad (n')$$

This sum represents the total inertia torque of the system, and eq. (n') shows that it must be zero under free-vibration conditions. Since $\omega^2 \neq 0$, the expression reduces to

$$J_1\Delta_1 + \dots + J_n\Delta_n = \sum_1^n (J\Delta) = 0.$$

However, this can only be zero for certain sets of values of the relative amplitudes Δ (*natural configurations* of the system). These values can be determined by means of Holzer tables, since, for any chosen value of ω^2,

$$\Delta_1 - \Delta_2 = \Delta_{sh\,1} = J_1\omega^2\Delta_1/K_1 \quad \text{from eq. (1),}$$

$\Delta_{sh\,1}$ being the twist of the shaft portion of stiffness K_1; furthermore,

$$\Delta_2 = \Delta_1 - \Delta_{sh\,1} = \Delta_1 - (J_1\omega^2\Delta_1/K_1),$$

a result which is entered in the second line of the Holzer table. From eq. $(2')$

$$\Delta_2 - \Delta_3 = \Delta_{sh\,2} = \frac{1}{K_2}\sum_1^2 (J\omega^2\Delta) = \frac{1}{K_2}(J_1\omega^2\Delta_1 + J_2\omega^2\Delta_2)$$

and

$$\Delta_3 = \Delta_2 - \Delta_{sh\,2}.$$

With this value, the third line of the Holzer table can be evaluated. Proceeding in this manner, if ω^2 is a value corresponding to a natural frequency of the system, the last line of the Holzer table should give

$$J_n\omega^2 \qquad \Delta_n \qquad J_n\omega^2\Delta_n \qquad J_n\omega^2\Delta_n + \sum_1^{n-1} (J\omega^2\Delta) = 0$$

If the resultant torque $\sum_1^n (J\omega^2\Delta)$ is not zero, the value assumed for ω^2 does not represent a natural frequency, and another tabulation is required, with a different value for ω^2.

It should be noted that, in a freely vibrating system, torque equilibrium can be verified at all mass positions. For instance, at any mass J_L,

$$\sum_1^{L-1} (J\omega^2\Delta) = -\sum_L^n (J\omega^2\Delta).$$

[171]

In dealing with straight systems, it is convenient to verify this condition at the last mass. For branched systems, which are considered in section 1·3215, the value of the resultant torque is determined for the common point of the various branches, using Holzer tables in which the first mass is that situated at the free end of each branch.

Curves for assessing the frequency of an engine coupled to various driven machines

A given engine may be coupled to any of a wide variety of driven machines, and it is frequently required to find the natural frequency of the various systems so formed. For this purpose it is very handy to have for rapid reference a set of constant-frequency curves, drawn in a diagram with K_M-values (driven-machine stiffness) on the horizontal axis and J_M-values (driven-machine inertia) on the vertical axis. For specified values of K_M and J_M, the natural frequency of the engine-and-driven machine system can then be read directly from these curves.

The procedure for obtaining these curves is as follows:

(1) Determine the natural frequency F_0 of the engine and flywheel, without driven machine, by means of a Holzer table. This gives the highest possible frequency.

(2) Assume a frequency value F_1 which is 50 vib./min. lower than F_0 and work out the corresponding Holzer table down to the flywheel inertia.

(3) Assume an arbitrary value for K_M and write $J_M \omega^2/10^6 = x$ in the table. Complete the table with this (unknown) value so as to obtain zero residual torque at the last mass J_M. This gives the relation

$$x\Delta_M + \sum_{1}^{M-1} (J\omega^2\Delta) \times 10^{-6} = 0,$$

which determines x and hence $J_M = x/\omega^2 \times 10^{-6}$.

(4) Repeat the calculation given in (3) for other values of K_M; these give corresponding values of J_M. Plot the resultant curve (with constant frequency as a parameter) in the J_M, K_M diagram.

(5) Assume other values of frequency F_2, F_3, etc., (each of these being, say, 50 vib./min. below the previous value), and repeat the calculations for various K_M-values as outlined under (3) and (4). For each of these sets, plot the corresponding curve $F = $ constant in the J_M, K_M diagram.

A typical example of the calculation (showing only the last two lines) of the tabulation) is given below:

$$F_1 = 13{,}800 \text{ vib./min.}, \quad \omega^2 = 2\!\cdot\!088 \times 10^6 \text{ rad.}^2/\text{sec.}^2$$

Mass no.	J [Lb.in. sec.²]	$\dfrac{J\omega^2}{10^6}$ [Lb.in./ rad.]	Δ [rad.]	$\dfrac{J\omega^2\Delta}{10^6}$ [Lb.in.]	$\dfrac{\Sigma(J\omega^2\Delta)}{10^6}$ [Lb.in.]	$\dfrac{K}{10^6}$ [Lb.in./ rad.]	Δ_{shaft} [rad.]
7	30·0	62·64	−0·044	−2·756	+2·464	2·0	1·232
8	J_M	x	−1·276	−1·276x	0	—	—

[172]

hence $\qquad -1 \cdot 276x + 2 \cdot 464 = 0, \quad x = 1 \cdot 931 = \omega^2 J_M/10^6,$

and $\qquad J_M = 1 \cdot 931/2 \cdot 088 = 0 \cdot 925 \, \text{Lb.in.sec.}^2.$

This gives one point of the curve for $F = 13,800$ vib./min. at

$$J_M = 0 \cdot 925 \, \text{Lb.in.sec.}^2 \quad \text{and} \quad K_M = 2 \cdot 0 \times 10^6 \, \text{Lb.in./rad.}$$

Other points are obtained by assuming, for instance, $K_M = 0 \cdot 2 \times 10^6$, $K_M = 10 \times 10^6$, etc. Further curves are determined for, say

$$F = 13,750, \ 13,700, \ \ldots, \ \text{vib./min.}$$

For systems including more than one mass in the driven-machine arrangement, use may be made of other semi-graphical methods for frequency determination, e.g. effective-inertia curves (section 1·3421).

Modes of vibration

A system of n masses and $n-1$ shaft elements has $n-1$ possible modes of vibration, i.e. $n-1$ different values of natural frequencies at which it can vibrate under resonance conditions. We denote these as follows:

$$F_{\text{I}}, \quad F_{\text{II}}, \quad F_{\text{III}}, \quad \ldots, \quad F_{n-1},$$

and the corresponding natural phase velocities are

$$\omega_{\text{I}}, \quad \omega_{\text{II}}, \quad \omega_{\text{III}}, \quad \ldots, \quad \omega_{n-1}.$$

Fortunately, in engine systems, the only frequencies usually of importance as regards torsional vibration amplitudes and stresses are the lowest two values, F_{I} and F_{II}. In some instances (systems with fairly rigid auxiliary drives) the three-node frequency F_{III} also has to be considered. In most cases the three-node criticals are too small to be apparent in torsiograph records and can be disregarded (although small amplitudes do not, of course, necessarily mean a small stress).

The relative amplitudes of a given engine system, determined, for instance, by means of Holzer tables, are

$$\Delta_1, \quad \Delta_2, \quad \ldots, \quad \Delta_F, \quad \Delta_n,$$

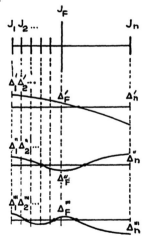

Fig. 1. Relative-amplitude curves for the first three modes of vibration.

where $\Delta_F =$ amplitude at flywheel, and $\Delta_n =$ amplitude at driven machine (see Fig. 1). It can be shown† that, if the amplitudes of the first mode are denoted by Δ^{I}, and those of the second mode by Δ^{II}, etc., the following relation is valid:

$$J_1 \Delta_1^{\text{I}} \Delta_1^{\text{II}} + J_2 \Delta_2^{\text{I}} \Delta_2^{\text{II}} + J_3 \Delta_3^{\text{I}} \Delta_3^{\text{II}} + \ldots + J_F \Delta_F^{\text{I}} \Delta_F^{\text{II}} + J_n \Delta_n^{\text{I}} \Delta_n^{\text{II}} = 0.$$

† See, for instance, T. von Kármán and M. A. Biot, *Mathematical Methods in Engineering* (McGraw-Hill, 1940), pp. 172–80.

This is known as the *orthogonality property* of the natural modes of vibration. Generally, we have

$$\Sigma J\Delta^I\Delta^{II} = 0, \quad \Sigma J\Delta^I\Delta^{III} = 0, \quad \Sigma J\Delta^{II}\Delta^{III} = 0, \quad \text{etc.}$$

This property can be used to determine higher modes of vibration when the Δ- and ω-values for lower modes are known. It can also be used to check the amplitude of the end-mass Δ_n.

For instance, if the value of Δ_n^{II} is small and does not vary appreciably when ω is slightly greater or smaller than ω_{II}, whereas variations of ω in the neighbourhood of ω_I cause large fluctuations in the approximate value of Δ_n^I (without appreciably affecting the other Δ^I-values), the correct value for Δ_n^I can be determined from

$$\sum_1^F J\Delta^I\Delta^{II} + J_n\Delta_n^I\Delta_n^{II} = 0,$$

which gives $\qquad \Delta_n^I = -\left(\sum_1^F J\Delta^I\Delta^{II}\right)\Big/(J_n\Delta_n^{II}).$

Extension of Holzer tables for relative amplitudes at intermediate points along the shafting

In marine installations, the tail shafting may consist of a large number of sections of different diameter, and the reduction of the shafting to equivalent lengths of crankshaft or any other diameter may become a very laborious task. In addition, if torsiograph records have to be taken at any position along the shafting, as is often necessary, the relative amplitude of vibration at that position may be difficult to determine from the usual Holzer table.

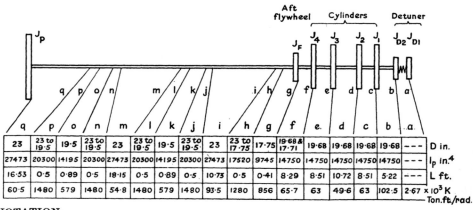

23	23 to 19·5	19·5	23 to 19·5	23	23 to 19·5	19·5	23 to 19·5	23	23 to 17·75	17·75	19·68 & 17·71	19·68	19·68	19·68	19·68	- - -	D in.
27473	20300	14195	20300	27473	20300	14195	20300	27473	17520	9745	14750	14750	14750	14750	14750	- - -	I_p in.⁴
16·53	0·5	0·89	0·5	18·15	0·5	0·89	0·5	10·73	0·5	0·41	8·29	8·51	10·72	8·51	5·22	- - -	L ft.
60·5	1480	579	1480	54·8	1480	579	1480	93·5	1280	856	65·7	63	49·6	63	102·5	2·67 × 10³ K	Ton.ft/rad.

NOTATION:

D = shaft diameter [in.], $\qquad I_p$ = second moment of area of shaft [in.⁴],
L = shaft length [ft.], $\qquad K$ = shaft stiffness [Ton.ft./rad.].

INERTIAS [Ton.ft.sec.²]:

$J_P = 4\cdot66, \qquad J_F = 0\cdot497, \qquad \begin{array}{l} J_1 = 2\cdot52, \\ J_2 = 2\cdot92, \\ J_3 = 2\cdot88, \\ J_4 = 2\cdot47, \end{array} \qquad \begin{array}{l} J_{D1} = 0\cdot311, \\ J_{D2} = 0\cdot124. \end{array}$

To facilitate the calculations and the interpretation of records, a simple extension of the Holzer table may be used, as shown in the following numerical example, which enables the relative amplitudes to be determined at each section of the shafting. The extension is self-explanatory, and assumes only that the intermediate shaft sections have their corresponding stiffnesses but no inertia masses at their ends.

NUMERICAL EXAMPLE. Natural frequency calculation of the marine propulsion system illustrated opposite, using an extended Holzer table giving the relative amplitudes at each of the eleven different sections of the shafting between the main flywheel and the propeller.

Note. The figures are determined with slide-rule accuracy.

(A) Standard tabulation, with an overall stiffness $K_M = 18\cdot9 \times 10^3$ Ton.ft./rad. for shafting between aft flywheel and propeller:

One-node frequency: $F = 550$ vib./min., $\omega = 57\cdot6$ rad./sec., $\omega^2 = 33,200$ rad.^{2}/sec.^{2}

Mass	J [Ton.ft. sec.^{2}]	$\dfrac{J\omega^2}{10^3}$ [Ton.ft./ rad.]	Δ [rad.]	$\dfrac{J\omega^2\Delta}{10^3}$ [Ton.ft.]	$\dfrac{\Sigma(J\omega^2\Delta)}{10^3}$ [Ton.ft.]	$\dfrac{K}{10^3}$ [Ton.ft./ rad.]	Δ_{sh} [rad.]
Detuner loose mass	0·311	1·03	1·625	1·675	1·675	2·67	0·625
Detuner fixed mass	0·124	0·41	1·0	0·41	2·085	102·5	0·0204
Cylinder 1	2·52	8·4	0·9796	8·2	10·285	63·0	0·163
Cylinder 2	2·92	9·7	0·8166	7·9	18·185	49·6	0·368
Cylinder 3	2·88	9·56	0·4486	4·28	22·465	63·0	0·357
Cylinder 4	2·47	8·2	0·0916	0·751	23·216	65·7	0·354
Flywheel	0·497	1·65	−0·2624	− 0·434	22·782	18·9	1·20
Propeller	4·66	15·4	−1·462	−22·51	0·272	—	—

(B) Extended tabulation, with stiffnesses for all shaft sections between aft flywheel and propeller (beginning at line-before-last of previous table):

Mass and/or shaft section	J	$\dfrac{J\omega^2}{10^3}$	Δ	$\dfrac{J\omega^2\Delta}{10^3}$	$\dfrac{\Sigma(J\omega^2\Delta)}{10^3}$	$\dfrac{K}{10^3}$	Δ_{sh}
Flywheel and section g	0·497	1·65	−0·2624	− 0·434	22·782	856·0	0·0266
section h	...	...	−0·289	...	...	1280·0	0·0177
section i	...	...	−0·3067	...	...	93·5	0·243
section j	...	...	−0·5497	...	...	1480·0	0·0153
section k	...	...	−0·565	...	...	579·0	0·0393
section l	...	...	−0·6063	...	...	1480·0	0·0153
section m	...	...	−0·6216	...	...	54·8	0·415
section n	...	...	−1·0366	...	...	1480·0	0·0153
section o	...	...	−1·0519	...	...	579·0	0·0393
section p	...	...	−1·0912	...	...	1480·0	0·0153
section q	...	...	−1·1065	...	...	60·5	0·36
Propeller	4·66	15·4	−1·466	−22·58	0·202	—	—

In this manner, the relative amplitudes are determined for all intermediate shaft sections. (For instance, the amplitude at the beginning of shaft section h is $\Delta_h = -0\cdot289$.) The corresponding stress in each shaft section can therefore be accurately evaluated without difficulty.

[175]

Continental method of frequency evaluation

Although basically similar, the methods used in some European countries, such as Germany, Switzerland, etc., for frequency evaluation differ somewhat in regard to definitions and details of computation from the methods usually employed in Britain and America. As it is sometimes necessary to study calculations of this type, an explanation of the definitions and general procedure is given below.

In the first place, the 'swinging moment' is defined as

$$WD^2 = 4Wk^2 \quad [\text{Kg.m.}^2],$$

where W = weight [Kg.], $D = 2k$ and k = radius of gyration [metres]. (The swinging moment is frequently denoted by GD^2, where G = weight.) The mass moment of inertia is then

$$J \quad = \quad Wk^2 \times 10^4 \times \frac{1}{g} \quad = \frac{10^4}{4 \times 981} \times WD^2 = \quad 2\cdot55 \quad \times \quad WD^2,$$

$$\text{Kg.cm.sec.}^2 \quad \text{Kg.m.}^2 \times \frac{\text{cm.}^2}{\text{m.}^2} \times \frac{\text{sec.}^2}{\text{cm.}} \qquad\qquad \frac{\text{cm.sec.}^2}{\text{m.}^2} \times \text{Kg.m.}^2$$

Furthermore, the practice is to reduce all the moments of inertia J to equivalent masses m at the crank radius R_0:

$$m \quad = J/R_0^2 = \quad 2\cdot55 \quad \times \quad WD^2 \quad / \quad R_0^2$$

$$\text{Kg.sec.}^2/\text{cm.} \qquad\qquad \frac{\text{cm.sec.}^2}{\text{m.}^2} \times \text{Kg.m.}^2 \times \frac{1}{\text{cm.}^2}$$

and to replace the inertia torque $T = J\omega^2\Delta$ by an inertia force $P = m\omega^2 s$ [Kg.], where $s = R_0\Delta$ = torsional displacement [cm.] of the mass m situated at the crank-throw radius.

Instead of the usual equation, for example,

Shaft torque between J_3 and J_4

$$= T_{3,4} = K_{3,4}(\Delta_4 - \Delta_3) = -\sum_1^3 (J\omega^2\Delta) = \text{total inertia torque at } J_3$$

one then uses the expressions

Torsional force on shaft between m_3 and m_4

$$= P_{3,4} = \frac{K^*}{L_{e_{3,4}}} (s_4 - s_3) = -\sum_1^3 (m\omega^2 s) = \text{total inertia force at } m_3,$$

where $\qquad\qquad L_{e_{3,4}} = $ equivalent length of shaft [cm.]

and $\quad K^* = \dfrac{GI_p}{R_0^2} = \dfrac{G}{R_0^2} \times \dfrac{\pi}{32} D_e^4 = $ stiffness per unit length [Kg./rad.],

[176]

G being the modulus of rigidity in torsion. [For steel $G = 8\cdot0 \times 10^5$ to $8\cdot28 \times 10^5$ Kg./(cm.^{2}rad.).] The relation between K^* [Kg./rad.] and our usual expression for stiffness $K_{3,4}$ [Kg.cm./rad.] is, therefore,

$$K_{3,4} = K^* \frac{R_0^2}{L_{e_{3,4}}} \text{ [Kg.cm./rad.]}.$$

The use of this method is illustrated by the following numerical examples.

EXAMPLE. Holzer-table determination of natural frequency using equivalent masses and inertia forces.
Eight-cylinder engine coupled to flywheel and generator.

Equivalent shaft diameter: $D_e = 10\cdot5$ cm., $I_p = \frac{\pi}{32} D_e^4 = 1194$ cm.^{4}.

Modulus of rigidity: $G = 8\cdot28 \times 10^5$ Kg./cm.^{2}. Crank radius: $R_0 = 11\cdot0$ cm.
Relative stiffness: $K^* = GI_p/R_0^2 = 8\cdot28 \times 10^5 \times 1194/121 = 81\cdot75 \times 10^5$ [Kg.].

$F = 4685$ vib./min., $\omega = 491\cdot0$ rad./sec., $\omega^2 = 2\cdot41 \times 10^5$ (rad./sec.)^2

Mass no.	m [Kg. sec.^{2}/cm.]	$m\omega^2$ [Kg./ cm.]	s [cm.]	$m\omega^2 s$ [Kg.]	$\Sigma m\omega^2 s$ [Kg.]	L_e [cm.]	$\frac{K^*}{L_e}$ [Kg./ cm.]	s_{shaft} $=\frac{\Sigma m\omega^2 s}{K^*/L_e}$ [cm.]
1	0·0362	8,720	1·0000	8,720	8,720	25·74	317,200	0·0275
2	0·0362	8,720	0·9725	8,470	17,190	25·74	317,200	0·0543
3	0·0528	12,710	0·9182	11,670	28,860	25·74	317,200	0·0910
4	0·0528	12,710	0·8272	10,520	39,380	25·74	317,200	0·1245
5	0·0528	12,710	0·7027	8,930	48,310	25·74	317,200	0·1520
6	0·0528	12,710	0·5507	6,995	55,305	25·74	317,200	0·1745
7	0·0362	8,720	0·3762	3,280	58,585	25·74	317,200	0·1845
8	0·0362	8,720	0·1915	1,670	60,255	27·65	295,400	0·2045
Flywheel	2·670	643,200	−0·0130	− 8,360	51,895	15·95	512,200	0·1015
Generator	1·790	431,200	−0·1145	−49,390	+2,505	—	—	—

Frequently, a tabulation is not used, and the calculation is further contracted as follows:

$$K' = \frac{K^*}{\omega^2} \text{ [Kg.sec.}^2], \quad P' = \frac{P}{\omega^2} = ms \quad \text{[Kg.sec.}^2],$$

$$s_1 = 100 \text{ cm.}, \quad P'_1 = m_1 \times 100, \quad s_{sh1} = \frac{P'_1}{K'_1/L_{e1,2}} = s_1 - s_2,$$

$$s_2 = 100 - P'_1 \times \frac{L_{e1,2}}{K'_{1,2}}, \quad \sum_1^2 P' = P'_1 + m_2 s_2,$$

$$s_3 = 100 - \left(\sum_1^2 P'\right) \times \frac{L_{e2,3}}{K'_{2,3}}, \quad \sum_1^3 P' = \sum_1^2 P' + m_3 s_3, \quad \text{etc.}$$

EXAMPLE. Thus, the tabulation in the above Example is represented in the following manner:

$s_1 = 100$ cm. $\quad P'_1 = 3\cdot62$ Kg.sec.^{2} $\quad K'_{1,2}/L_{e1,2} = 317,200/(2\cdot41 \times 10^5) = 1\cdot314.$

$s_2 = 100 \quad - \quad 2\cdot75 = 97\cdot25 \qquad \Sigma P'_2 = \quad 3\cdot62 + \quad 3\cdot52 = \quad 7\cdot14$

$s_3 = \quad 97\cdot25 - \quad 5\cdot43 = 91\cdot82 \qquad \Sigma P'_3 = \quad 7\cdot14 + \quad 4\cdot85 = 11\cdot99$

$s_4 = \quad 91\cdot82 - \quad 9\cdot10 = 82\cdot72 \qquad \Sigma P'_4 = 11\cdot99 + \quad 4\cdot37 = 16\cdot36$

$$s_5 = 82\cdot72 - 12\cdot45 = \quad 70\cdot27 \qquad \Sigma P_5' = 16\cdot36 + \; 3\cdot71 = 20\cdot07$$
$$s_6 = 70\cdot27 - 15\cdot2 \; = \quad 55\cdot07 \qquad \Sigma P_6' = 20\cdot07 + \; 2\cdot91 = 22\cdot98$$
$$s_7 = 55\cdot07 - 17\cdot45 = \quad 37\cdot6 \qquad \Sigma P_7' = 22\cdot98 + \; 1\cdot36 = 24\cdot34$$
$$s_8 = 37\cdot6 \; -18\cdot45 = \quad 19\cdot15 \qquad \Sigma P_8' = 24\cdot34 + \; 0\cdot69 = 25\cdot03$$
$$s_F = 19\cdot15 - 20\cdot45 = - \; 1\cdot3 \qquad \Sigma P_F' = 25\cdot03 - \; 3\cdot47 = 21\cdot56$$
$$s_G = -1\cdot3 \; -10\cdot15 = -11\cdot45 \qquad \Sigma P_G' = 21\cdot56 - 20\cdot56 = \; 1\cdot0.$$

Comparative check with (due allowance for slide-rule errors):

$$\frac{P_1'}{\Sigma P_G'} = \frac{3\cdot62}{1\cdot0} = 3\cdot62 \simeq \frac{8720}{2505} = \frac{m_1\omega^2 s_1}{\displaystyle\sum_1^G (m\omega^2 s)} = 3\cdot48.$$

Changing the natural frequencies of a system

It is sometimes necessary to alter the natural frequencies of an engine system, in order to obtain improved running conditions, by

(1) shifting a critical to a speed above or below the running range,
(2) shifting a node to a more suitable position,
(3) reducing vibration amplitudes at a certain position.

This can be achieved by altering either the stiffnesses or the moments of inertia, or both.

Changes in stiffness can be obtained by varying the coupling arrangements, or altering the lengths, diameters and materials used for certain shaft sections. The moments of inertia can be altered by employing different flywheels, balance weights and piston materials. Sometimes a worth-while change in moment of inertia can be obtained by chamfering crankwebs or hollow-boring crankpins (see section 1·2).

Alterations in frequency often affect the vibration amplitudes and hence the stress conditions of a system (see section 2·3). This should be borne in mind, both at the design stage and subsequently, when any modifications are to be made to the system.

Regarding the shifting of a node to a more favourable position, the following possibilities have also been considered for some applications. A large main flywheel (such as is required for flicker prevention in generator sets) may be replaced by two smaller flywheels, with the same total inertia, and an intermediate shaft, for instance, of cast iron, with a fairly large diameter, but relative 'soft' torsionally (i.e. of low stiffness). As regards cyclic irregularity and flicker, the modification retains the original features of the engine system, but its torsional behaviour may be improved (lower stress-per-degree values, possibly lower vibration amplitudes). The modification may in fact be designed to leave the one-node frequency unchanged, while altering the two-node vibration conditions.

1·3212 Straight-geared systems

The method is described in the following by means of a numerical example. The procedure is the same as for straight systems without reduction gears, up to the first gearwheel. After this mass, the effect of the speed ratio is taken into account.

NUMERICAL EXAMPLE. The engine system considered is shown in the accompanying figure. It is characterized by the following values:

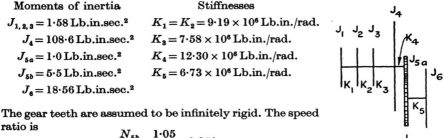

Moments of inertia

$J_{1,2,3} = 1·58$ Lb.in.sec.^{2}

$J_4 = 108·6$ Lb.in.sec.^{2}

$J_{5a} = 1·0$ Lb.in.sec.^{2}

$J_{5b} = 5·5$ Lb.in.sec.^{2}

$J_6 = 18·56$ Lb.in.sec.^{2}

Stiffnesses

$K_1 = K_2 = 9·19 \times 10^6$ Lb.in./rad.

$K_3 = 7·58 \times 10^6$ Lb.in./rad.

$K_4 = 12·30 \times 10^6$ Lb.in./rad.

$K_5 = 6·73 \times 10^6$ Lb.in./rad.

The gear teeth are assumed to be infinitely rigid. The speed ratio is

$$r = \frac{N_{5b}}{N_{5a}} = \frac{1·05}{3·0} = 0·350,$$

where N_{5b} = speed of driven wheel [rev./min.] and N_{5a} = speed of driving wheel [rev./min.].

The calculation is carried out as shown in the following table:

$$F = 5{,}920 \text{ vib./min.}, \quad \omega^2 = 0·3456 \times 10^6 \text{ (rad./sec.)}^2$$

Normal table

Mass no.	J [Lb.in./sec.^{2}]	$\dfrac{J\omega^2}{10^6}$ [Lb.in./rad.]	Δ [rad.]	$\dfrac{J\omega^2\Delta}{10^6}$ [Lb.in.]	$\dfrac{\Sigma}{10^6}$ [Lb.in.]	$\dfrac{K}{10^6}$ [Lb.in./rad.]	Δ_{shaft} [rad.]
1	1·58	0·546	1·0000	0·546	0·546	9·19	0·0594
2	1·58	0·546	0·9406	0·514	1·060	9·19	0·1153
3	1·58	0·546	0·8253	0·450	1·510	7·58	0·1991
4	108·6	37·532	0·6262	23·502	25·012	12·30	2·0325
5a	1·0	0·3456	− 1·4063	− 0·485	24·527	—	—

$$r = 0·350, \quad 24·527/r = 70·08$$

Transition portion

	J	$\dfrac{J\omega^2}{10^6}$	$r\Delta$	$\dfrac{J\omega^2 r\Delta}{10^6}$	$\dfrac{\Sigma}{10^6}$	$\dfrac{K}{10^6}$	Δ_{shaft}
5b	5·5	1·905	− 0·4922	− 0·936	69·144	6·73	10·274

Note. $- 0·4922 = − 1·4063 \times r$; $69·144 = 70·080 − 0·936$.

Normal table

	J	$\dfrac{J\omega^2}{10^6}$	Δ	$\dfrac{J\omega^2\Delta}{10^6}$	$\dfrac{\Sigma}{10^6}$	$\dfrac{K}{10^6}$	Δ_{shaft}
6	18·583	6·4223	− 10·7662	− 69·144	0	—	—

The speed ratio r is always taken as a positive value in straight-geared systems.

Derivation. The first part of the table is based on the same equations of motion as the tabulations for straight systems without gearwheels.

[179]

With reference to the figure below, instead of $\sum_1^5 (J\omega^2\theta) = K_5(\theta_5 - \theta_6)$, the equation for the two gearwheel inertias is

$$K_4(\theta_{5a} - \theta_4) + \left(J_{5a}\ddot{\theta}_{5a} + J_{5b}r^2\frac{\ddot{\theta}_{5b}}{r}\right) + K_5 r^2\left(\frac{\theta_{5b}}{r} - \frac{\theta_6}{r}\right) = 0$$

or $\quad K_4(\theta_{5a} - \theta_4) + (J_{5a}\ddot{\theta}_{5a} + J_{5b}r^2\ddot{\theta}_{5a}) + K_5 r^2\left(\theta_{5a} - \frac{\theta_6}{r}\right) = 0,$

since, with reference to the engine side of the system, the J- and K-values must be multiplied by r^2 and the θ-values after the gearwheels must be divided by r. (This is derived in section 1·2.) With the same substitutions as for simple straight systems (i.e. $\theta_1 = \theta_{01}\sin\omega t$, etc., and $\theta_{02}/\theta_{01} = \Delta_2$, $\Delta_1 = 1$, $\theta_{03}/\theta_{01} = \Delta_3$, etc.), the above equation gives

$$\sum_1^4 (J\omega^2\Delta) + J_{5a}\omega^2\Delta_{5a} + J_{5b}\omega^2 r^2\Delta_{5a} = K_5 r^2\left(\Delta_{5a} - \frac{\Delta_6}{r}\right)$$

or $\quad \dfrac{1}{r}\times\sum_1^{5a}(J\omega^2\Delta) + J_{5b}\omega^2 \times r\Delta_{5a} = K_5(r\Delta_{5a} - \Delta_6).$

We thus obtain

$$\Delta_{sh\,5} = r\Delta_{5a} - \Delta_6 = \frac{1}{K_5}\left[\frac{1}{r}\sum_1^{5a}(J\omega^2\Delta) + J_{5b}\omega^2 r\Delta_{5a}\right].$$

The equation for the last mass is

$$K_5 r^2\left(\frac{\theta_6}{r} - \theta_{5a}\right) + J_6 r^2(\ddot{\theta}_6/r) = 0$$

or $\quad K_5(\Delta_6 - r\Delta_{5a}) - J_6\omega^2\Delta_6 = 0,$

and inserting in this the above expression for the shaft twist Δ_{sh}, we have the equation for the resultant torque of the system:

$$\frac{1}{r}\sum_1^{5a}(J\omega^2\Delta) + \sum_{5b}^6 (J\omega^2\Delta) = 0,$$

and the Holzer tabulation is accordingly modified to comply with these relations.

Note. From the point of view of *stresses* and *twist amplitudes*, the effective value of the relative twist $\Delta_{sh\,5}$ of the shaft K_5 is

$$\Delta_{5b} - \Delta_6 = \Delta_{sh\,5} = r\Delta_{5a} - \Delta_6,$$

since $\Delta_{5b} = r\Delta_{5a} = (N_{5b}/N_{5a}) \times \Delta_{5a} = (R_{5a}/R_{5b}) \times \Delta_{5a}$, the effective gear radii being denoted by R_{5a} and R_{5b}, so that the value of $\Delta_{sh\,5}$ in the Holzer table gives the actual relative twist of shaft K_5.

[180]

1·3213 *Straight systems with belt or chain drive*

If the chain drive can be regarded as infinitely rigid, the Holzer tabulation to be used is the same as for straight-geared systems (see pp. 179 and 180). If the chain stiffness, determined as indicated in section 1·2, is not many times greater than that of a crankthrow, the evaluation procedure is the same as for engine systems with a belt drive.

NUMERICAL EXAMPLE. We consider the engine system shown in Fig. 1, which has the following J- and K-values:

Moments of inertia

$J_1 = J_2 = J_3 = 1 \cdot 58$ Lb.in.sec.^{2}

$J_4 = 108 \cdot 6$ Lb.in.sec.^{2}

J_{5a} (engine pulley) $= 1 \cdot 0$ Lb.in.sec.^{2}

J_{5b} (driven pulley) $= 5 \cdot 5$ Lb.in.sec.^{2}

$J_6 = 19 \cdot 5$ Lb.in.sec.^{2}

Stiffnesses

$K_1 = K_2 = 9 \cdot 19 \times 10^6$ Lb.in./rad.

$K_3 = 7 \cdot 58 \times 10^6$ Lb.in./rad.

$K_4 = 12 \cdot 30 \times 10^6$ Lb.in./rad.

Belt stiffness relative to driven system:

$K_B = 0 \cdot 90 \times 10^6$ Lb.in./rad.

$K_5 = 6 \cdot 5 \times 10^6$ Lb.in./rad.

The speed ratio (i.e. the ratio of the driven-machine speed N_b [rev./min.] to the engine speed N_a [rev./min.]) is

$$r = N_b / N_a = 0 \cdot 35.$$

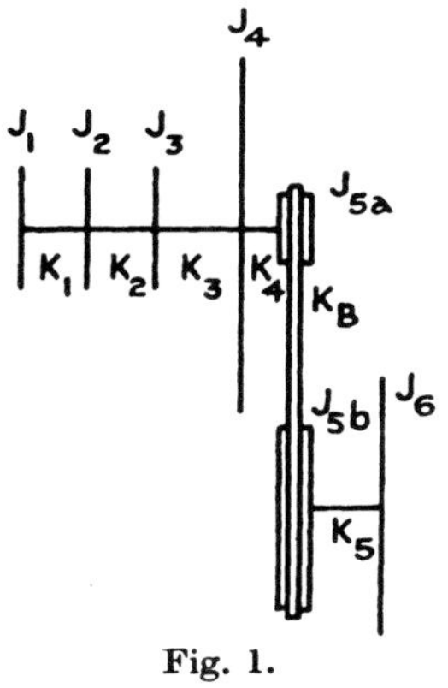

First approximation. In view of the low value of the belt stiffness, we may assume that (a) the first mode of vibration has a node in the belt, and (b) the second and third modes of vibration are approximately the modes of vibration of the engine, and the driven-machine system, each of these being considered separately.

To obtain the one-node frequency F_I, we lump together all the engine masses at the position of the engine pulley J_{5a}. The belt stiffness relative to the engine end is

$$K_{B(a)} = r^2 \times K_B = (0 \cdot 35)^2 \times 0 \cdot 9 \times 10^6 = 0 \cdot 112 \times 10^6 \text{ Lb.in./rad.}$$

The inertia of the driven machine is lumped with that of pulley J_{5b}, and their total inertia, relative to the engine system, is

$$r^2 \times (J_{5b} + J_6) = 0 \cdot 1225 \times 25 = 3 \cdot 06 \text{ Lb.in.sec.}^2.$$

These two formulae are derived in section 1·2. We thus have a two-mass system, with

$$J_{A1} = \overset{5a}{\underset{1}{\Sigma}} J = 112 \cdot 34 \text{ Lb.in.sec.}^2, \quad J_{B1} = 3 \cdot 06 \text{ Lb.in.sec.}^2$$

and

$$K_{B(a)} = 0 \cdot 112 \times 10^6 \text{ Lb.in./rad.}$$

Its natural frequency is given by

$$F_I = 9 \cdot 55 \sqrt{\left(K_{B(a)} \frac{J_A + J_B}{J_A \times J_B} \right)} = 9 \cdot 55 \times \sqrt{\left(0 \cdot 112 \times 10^6 \times \frac{115 \cdot 40}{343 \cdot 9} \right)} \cong 1850 \text{ vib./min.}$$

For an assessment of the two-node frequency (with one node in the engine crankshaft), we reduce the engine to an equivalent two-mass system. The cylinder

Fig. 1.

[181]

inertias are lumped together at the mid-point of shaft K_2 (see Fig. 1) and the engine pulley is added to the flywheel inertia. Thus we have

$$J_{A2} = \Sigma J_{cyl.} = 4 \cdot 74 \text{ Lb.in.sec.}^2 \quad \text{and} \quad J_{B2} = J_4 + J_{5a} = 109 \cdot 6 \text{ Lb.in.sec.}^2$$

The stiffness of the shafting from the flywheel J_4 to the centre of shaft K_2 is evaluated as

$$\frac{1}{K_E} = \frac{1}{K_3} + \frac{1}{2} \times \frac{1}{K_2} = \left[\frac{1}{7 \cdot 58} + \frac{1}{18 \cdot 38} \right] \times 10^{-6} = [0 \cdot 1319 + 0 \cdot 0544] \times 10^{-6}$$

or $\qquad 1/K_E = 0 \cdot 1863 \times 10^{-6} \quad \text{and} \quad K_E = 5 \cdot 364 \times 10^6 \text{ Lb.in./rad.}$

We thus obtain

$$F_{II} = 9 \cdot 55 \times \sqrt{\left(5 \cdot 364 \times 10^6 \times \frac{109 \cdot 6 + 4 \cdot 74}{109 \cdot 6 \times 4 \cdot 74} \right)} = 10,350 \text{ vib./min.}$$

If the equivalent shaft length is taken up to the position of J_2 (centre cylinder), this gives, $K_E = 4 \cdot 164 \times 10^6$ and $F_{II} = 9120 \text{ vib./min.}$ It is thus probable that F_{II} is at an intermediate value. The average value is

$$F_{II} = \tfrac{1}{2}[10,350 + 9120] = 9735 \text{ vib./min.}$$

The three-node frequency, with a node in the driven-machine system, is readily estimated without reduction to an equivalent system. We have

$$K_5 = 6 \cdot 5 \times 10^6 \text{ Lb.in./rad.}, \quad J_{5b} = 5 \cdot 5 \text{ Lb.in.sec.}^2 \quad \text{and} \quad J_6 = 19 \cdot 5 \text{ Lb.in.sec.}^2,$$

so that $\qquad F_{III} = 9 \cdot 55 \sqrt{\left(6 \cdot 5 \times 10^6 \times \frac{25 \cdot 0}{107 \cdot 25} \right)} = 11,750 \text{ vib./min.}$

These values are compared below with the exact values, determined by means of Holzer tabulations given in the following part of this section:

	One-node frequency	Two-node frequency	Three-node frequency
Approximate value [vib./min.]	1,850	9,735	11,750
Holzer-table value [vib./min.]	1,899	9,893	12,246

Note. In view of the somewhat uncertain value of the belt stiffness, the frequency values as calculated above are probably sufficiently good approximations. The Holzer-table calculations are necessary when it is required to determine the relative amplitudes of the system.

Systems such as belt drives, in which the stiffness of one element is considerably lower than the other stiffness values, require a certain amount of computation work in order to obtain correct relative amplitudes, since these are apt to vary considerably in the neighbourhood of the resonance frequency. The use of a calculating machine is recommended for such work.

Accurate values by Holzer-table method

The calculation is carried out as shown in the following table. The first part of the table ends at the engine pulley. The belt stiffness is referred to the driven-machine end of the system. The speed ratio must then be taken into account for the total torque at the engine pulley and the relative amplitude of this mass.

Derivation. The first part of the table is based on the same equations of motion as the tabulations for straight systems without a belt or chain drive. With reference to Fig. 1, we can write the equations of motion for mass 4 and the subsequent masses as follows:

$$K_3(\theta_4 - \theta_3) + J_4\ddot{\theta}_4 + K_4(\theta_4 - \theta_{5a}) = 0,$$

$$K_4(\theta_{5a} - \theta_4) + J_{5a}\ddot{\theta}_{5a} + K_{B(b)}r^2\left(\theta_{5a} - \frac{\theta_{5b}}{r}\right) = 0,$$

$$K_{B(b)}r^2\left(\frac{\theta_{5b}}{r} - \theta_{5a}\right) + J_{5b}r^2\frac{\ddot{\theta}_{5b}}{r} + K_5r^2\left(\frac{\theta_{5b} - \theta_6}{r}\right) = 0,$$

$$K_5r^2\left(\frac{\theta_6 - \theta_{5b}}{r}\right) + J_6r^2\frac{\ddot{\theta}_6}{r} = 0.$$

$$F_{\mathrm{I}} = 1899\cdot3 \text{ vib./min.}, \quad \omega_{\mathrm{I}} = 184\cdot1 \text{ rad./sec.}, \quad \omega_{\mathrm{I}}^2 = 0\cdot0339 \times 10^6$$

Normal table

Mass no.	J [Lb.in. sec.^{2}]	$\dfrac{J\omega^2}{10^6}$ [Lb.in./ rad.]	Δ [rad.]	$\dfrac{J\omega^2\Delta}{10^6}$ [Lb.in.]	$\dfrac{\Sigma}{10^6}$ [Lb.in.]	$\dfrac{K}{10^6}$ [Lb.in./ rad.]	Δ_{shaft} [rad.]
1	1·58	0·0536	1·0000	0·0536	0·0536	9·19	0·0058
2	1·58	0·0536	0·9942	0·0533	0·1069	9·19	0·0116
3	1·58	0·0536	0·9826	0·0527	0·1596	7·58	0·0211
4	108·6	3·6815	0·9615	3·5398	3·6994	12·30	0·3008
5a	1·0	0·0339	0·6607	0·0224	3·7218	—	—

$$r = 0\cdot350, \quad \Sigma/r = 10\cdot6337, \quad r\Delta_{5a} = 0\cdot2312$$

Transition portion

		$r\Delta$		$\dfrac{`\Sigma`}{r}$	$\dfrac{K}{10^6}$	Δ_{shaft}
—	—	0·2312	—	10·6337	0·90	11·8152

Normal table

	J	$\dfrac{J\omega^2}{10^6}$	Δ	$\dfrac{J\omega^2\Delta}{10^6}$	$\dfrac{\Sigma}{10^6}$	$\dfrac{K}{10^6}$	Δ_{shaft}
5b	5·5	0·1865	−11·5840	−2·1604	8·4733	6·50	1·3036
6	19·5	0·6611	−12·8876	−8·5200	−0·0467	—	—

Note. −11·5840 = 0·2312 − 11·8152; 8·4733 = 10·6337 − 2·1604.

The speed ratio is always taken as positive in straight systems.

With the same substitutions as for simple straight systems, these equations give the relations

$$-\sum_1^4 J\omega^2\Delta + K_4(\Delta_4 - \Delta_{5a}) = 0,$$

$$-\sum_1^4 J\omega^2\Delta - J_{5a}\omega^2\Delta_{5a} + K_{B(b)}r(r\Delta_{5a} - \Delta_{5b}) = 0,$$

$$K_{B(b)}r(\Delta_{5b} - r\Delta_{5a}) - J_{5b}r\omega^2\Delta_{5b} + K_5r(\Delta_{5b} - \Delta_6) = 0,$$

$$K_5r(\Delta_6 - \Delta_{5b}) - J_6r\omega^2\Delta_6 = 0,$$

hence $\quad K_{B(b)}r(r\Delta_{5a} - \Delta_{5b}) = \sum_1^{5a} J\omega^2\Delta \quad$ or $\quad K_{B(b)}(r\Delta_{5a} - \Delta_{5b}) = \dfrac{1}{r}\sum_1^{5a} J\omega^2\Delta,$

and therefore $\quad K_5(\Delta_{5b} - \Delta_6) = J_{5b}\omega^2\Delta_{5b} + \dfrac{1}{r}\sum_1^{5a} J\omega^2\Delta.$

From mass $5a$, the calculation is therefore as follows:

$$\Delta_{5a}, \quad J_{5a}\omega^2\Delta_{5a}, \quad \sum_1^{5a} J\omega^2\Delta, \quad \frac{1}{rK_{B(b)}}\sum = r\Delta_{5a} - \Delta_{5b} = \Delta_{sh\,(5a-5b)},$$

$$\Delta_{5b} = r\Delta_{5a} - \Delta_{sh\,(5a-5b)}, \quad J_{5b}\omega^2\Delta_{5b}, \quad J_{5b}\omega^2\Delta_{5b} + \frac{1}{r}\sum_1^{5a} = \sum',$$

$$\frac{1}{K_5}\sum' = \Delta_{5b} - \Delta_6 = \Delta_{sh\,(5b-6)},$$

$$\Delta_6 = \Delta_{5b} - \Delta_{sh\,(5b-6)}, \quad J_6\omega^2\Delta_6, \quad \sum' + J_6\omega^2\Delta_6 = 0.$$

This procedure is used step by step in the above table.

The tables for the two- and three-node frequencies are given below:

$$F_{II} = 9893\cdot5 \text{ vib./min.}, \quad \omega_{II} = 1\cdot03597 \text{ rad./sec.}, \quad \omega_{II}^2 = 1\cdot0732 \times 10^6$$

Mass no.	J [Lb.in.sec.^{2}]	$\dfrac{J\omega^2}{10^6}$ [Lb.in./rad.]	Δ [rad.]	$\dfrac{J\omega^2\Delta}{10^6}$ [Lb.in.]	$\dfrac{\Sigma}{10^6}$ [Lb.in.]	$\dfrac{K}{10^6}$ [Lb.in./rad.]	Δ_{sh} [rad.]
1	1·58	1·69566	1·00000	1·69566	1·69566	9·19	0·18451
2	1·58	1·69566	0·81549	1·38279	3·07845	9·19	0·33498
3	1·58	1·69566	0·48051	0·81478	3·89323	7·58	0·51362
4	108·6	116·550	−0·03311	−3·85897	+0·03426	12·30	+0·00279
5a	1·0	1·0732	−0·03590	−0·03853	−0·00427	—	—

$$r = 0\cdot350, \quad \Sigma/r = -0\cdot01220, \quad r\Delta_{5a} = -0\cdot01257$$

		$r\Delta$		$\dfrac{\text{`}\Sigma\text{'}}{r}$	$\dfrac{K}{10^6}$	Δ_{sh}
—	—	−0·01257	—	−0·01220	0·90	−0·01356

	J	$\dfrac{J\omega^2}{10^6}$	Δ	$\dfrac{J\omega^2\Delta}{10^6}$	$\dfrac{\Sigma}{10^6}$	$\dfrac{K}{10^6}$	Δ_{sh}
5b	5·5	5·9026	+0·00099	+0·00584	−0·00636	6·50	−0·01356
6	19·5	20·927	+0·00197	+0·04123	+0·03487	—	—

Note. $+0\cdot00099 = -0\cdot01257 + 0\cdot01356$, $\quad -0\cdot00636 = -0\cdot01220 + 0\cdot00584$.

$$F_{III} = 12{,}246 \text{ vib./min.}, \quad \omega_{III} = 1282\cdot3 \text{ rad./sec.}, \quad \omega_{III}^2 = 1\cdot6443 \times 10^6$$

Mass no.	J	$\dfrac{J\omega^2}{10^6}$	Δ	$\dfrac{J\omega^2\Delta}{10^6}$	$\dfrac{\Sigma}{10^6}$	$\dfrac{K}{10^6}$	Δ_{sh}
1	1·58	2·5980	1·0000	2·5980	2·5980	9·19	0·2827
2	1·58	2·5980	0·7173	1·8635	4·4615	9·19	0·4855
3	1·58	2·5980	0·2318	0·6022	5·0637	7·58	0·6880
4	108·6	178·571	−0·4362	−77·8920	−72·8290	12·30	−5·921
5a	1·0	1·6443	+5·4849	+9·0188	−63·8102	—	—

$$r = 0\cdot350, \quad \Sigma/r = -182\cdot314, \quad r\Delta_{5a} = +1\cdot9197$$

		$r\Delta$		$\dfrac{\text{`}\Sigma\text{'}}{r}$	$\dfrac{K}{10^6}$	Δ_{sh}
—	—	+1·9197	—	−182·314	0·90	−202·57

	J	$\dfrac{J\omega^2}{10^6}$	Δ	$\dfrac{J\omega^2\Delta}{10^6}$	$\dfrac{\Sigma}{10^6}$	$\dfrac{K}{10^6}$	Δ_{sh}
5b	5·5	9·0437	+204·49	+1849·35	+1667·0	6·50	+256·47
6	19·5	32·0639	−51·98	−1666·68	+0·32	—	—

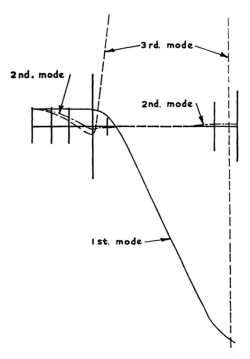

Fig. 2. Relative-amplitude curves of engine system with belt drive.
(First, second and third modes of vibration.)

1·3214 *Straight-geared systems including gear-teeth flexibility*

Note. Gear-teeth flexibility introduces additional modes of vibration in the frequency spectrum of an engine system, since in such a system the gearwheel inertias are not lumped together but are separated by an equivalent shaft representing gear stiffness. In some instances, the additional frequency or frequencies may be situated in the region between the one-node and two-node frequencies obtained for the system assuming fully rigid gearing. Such conditions may therefore require consideration. For the assessment of backlash effects, see section 1·35.

The stiffness of gear teeth can be assessed as indicated in section 1·23. If their stiffness value, referred to the driven-machine end of the system, is not many times greater than the crankthrow stiffness of the engine system considered, it is advisable to take account of gear teeth in the natural frequency calculations.

The Holzer-table evaluation for systems including gear-teeth stiffness is identical to the procedure used for systems with belt or chain drives. This is described in detail in section 1·3213. However, the first approximate values cannot be obtained in the manner indicated in that section, i.e. by assuming that the one-node frequency has its node between the gearwheel inertias. The tables can be evaluated with successive trial values for the assumed frequencies.

Alternatively, however, it is possible to use the graphical 'Effective-inertia method' for approximate values (see section 1·342). For this purpose, the system is split up into a 'basic system', with the engine gearwheel at one end (Fig. 1) and a 'subsidiary system' which ends with an equivalent shaft stiffness representing the gear-teeth stiffness referred to the driven end of the system. All the J- and K-values of the subsidiary system, should be multiplied by $r^2 = (N_S/N_E)^2$, where N_S = speed of subsidiary system, and N_E = speed of engine system [rev./min.].

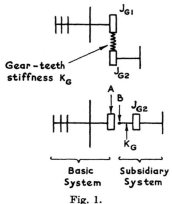

Fig. 1.

Apart from the inclusion of the shaft representing gear-teeth stiffness, the method of obtaining the tuning curves and the effective inertia curves is exactly the same as for rigid-geared systems. The value of J_p for the subsidiary system is calculated for the end 'B' (Fig. 1) of the driven system. The last lines of the tabulation are therefore as follows:

J	$\dfrac{J\omega^2}{10^6}$	Δ	$\dfrac{J\omega^2\Delta}{10^6}$	$\dfrac{\Sigma}{10^6}$	$\dfrac{K}{10^6}$	Δ_{sh}
...	...	Δ_{G2},	...	Σ,	K_G, $\dfrac{\Sigma}{K_G} = \Delta_{sh(G)}$,	
J_p,	$J_p\omega^2$,	$\Delta_p = \Delta_{G2} - \Delta_{sh(G)}$,		$\Sigma + J_p\omega^2\Delta_p = 0$,		

and J_p is determined by means of this last equation.

1·3215 *Gear-branched systems*

The natural frequencies of gear-branched systems can be determined by a number of methods, for instance by

(1) evaluation of the frequency equation for the simplified system (for a preliminary estimation),

(2) Holzer-table evaluations,

(3) effective-inertia curves,

(4) residual torque curves,

(5) mobility methods, etc.

In the following methods (1) and (2) will be used†. Examples of (3)

---

† It is advisable to use tabulations in conjunction with semi-graphical or approximate methods, the latter serving to verify that no important mode of vibration has been overlooked.

and (4) are given in sections 1·342 and 1·343, and for details of the application of mobility methods, references to literature are given in section 1·372.

Evaluation of frequency equation for the simplified system

This method is used in order to obtain the first approximate values for the natural frequencies.

The original system (Fig. 1 *a*) is first reduced to a simplified equivalent system (Fig. 1 *b*), by evaluating $J_{A2}^* = \sum_{i=2}^{n} J_{Ai}$ and locating this inertia at a position given by the 'centroid' of the various engine masses relative to mass J_{A2}, taking account of their relative values. Then K_A is determined from

$$\frac{1}{K_A} = \frac{1}{K_{A1}} + \frac{1}{K_{A2}} + \cdots$$

up to the position of the equivalent inertia J_{A2}^*. This procedure is repeated for the B and C branches, and we finally obtain a system such as that shown in Fig. 1 *b*.

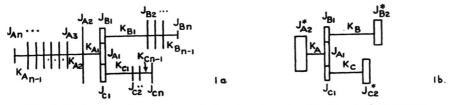

Fig. 1. Reduction of a gear-branched system to a simplified equivalent system.

The next step is to take account of the speed ratios, in order to relate all the K- and J-values to the vibration conditions of one branch, for instance, branch A. Let N_A, N_B, N_C be the running speeds [rev./min.] of branches A, B and C, respectively. Relative to branch A, the speed ratios are then

$$r_B = \frac{N_B}{N_A} \quad \text{and} \quad r_C = \frac{N_C}{N_A}.$$

We determine the equivalent values

$$J_{B1e} = J_{B1} r_B^2 \quad \text{and} \quad J_{C1e} = J_{C1} r_C^2,$$
$$J_{B2e} = J_{B2}^* r_B^2 \qquad J_{C2e} = J_{C2}^* r_C^2,$$
$$K_{Be} = K_B r_B^2 \qquad K_{Ce} = K_{C2} r_C^2,$$

and the entire system is thus related to the conditions in branch A.

The evaluation of the natural frequencies will now be illustrated by means of a numerical example, followed by the derivation of the required equations.

EXAMPLE. We consider the gear-branched system of Fig. 2, with the following values:†

$$J_{A2}=100 \qquad K_A=0{\cdot}1\times10^6$$
$$J_{A1}=340$$

$$J_{B2}=2{\cdot}22 \qquad K_B=0{\cdot}66\times10^6$$
$$J_{B1}=4$$

$$J_{C2}=8 \qquad K_C=1{\cdot}0\times10^6$$
$$J_{C1}=6$$

[Lb.in.sec.²] [Lb.in./rad.]

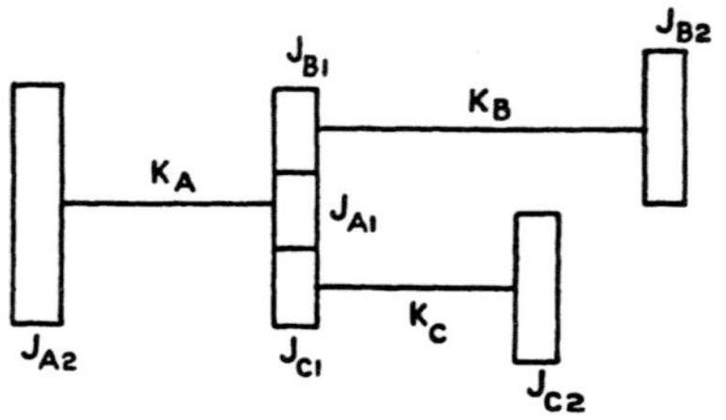

Fig. 2.

It is assumed that K_B runs at three times the speed of K_A, and shaft K_C at twice the speed of K_A. (Hence $r_B=3$ and $r_C=2$.) To obtain the dynamically equivalent system at the speed of shaft K_A, we have

$$J_{B1e}=J_{B1}r_B^2=4\times9=36, \qquad K_{Be}=K_Br_B^2=6{\cdot}0\times10^6,$$
$$J_{B2e}=J_{B2}r_B^2=2{\cdot}22\times9=20, \qquad K_{Ce}=K_Cr_C^2=4{\cdot}0\times10^6,$$
$$J_{C1e}=J_{C1}r_C^2=6\times4=24,$$
$$J_{C2e}=J_{C2}r_C^2=8\times4=32.$$

The equivalent mass moment of inertia of the combined gear masses is

$$J_G=J_{A1}+J_{B1e}+J_{C1e}=340+36+24=400\ \text{Lb.in.sec.}^2.$$

We determine the ω^2-values corresponding to the natural phase velocities of each branch, assuming each branch shaft fixed at the gearwheel J_{A1}:

$$\omega_A^2=K_A/J_{A2}=0{\cdot}1\times10^6/100=10^3\,\text{rad./sec.,}$$
$$\omega_B^2=K_{Be}/J_{B2e}=6{\cdot}0\times10^6/20=300\times10^3\,\text{rad./sec.,}$$
$$\omega_C^2=K_{Ce}/J_{C2e}=4{\cdot}0\times10^6/32=125\times10^3\,\text{rad./sec.,}$$

and then write the frequency equation, as follows:

$$J_G=\frac{K_A}{\omega^2-\omega_A^2}+\frac{K_{Be}}{\omega^2-\omega_B^2}+\frac{K_{Ce}}{\omega^2-\omega_C^2}\quad[\text{Lb.in.sec.}^2],$$

or

$$400=10^6\times\left\{\frac{0{\cdot}1}{\omega^2-10^3}+\frac{6{\cdot}0}{\omega^2-300\times10^3}+\frac{4{\cdot}0}{\omega^2-125\times10^3}\right\}.$$

Dividing both sides by 10^3, we have

$$\frac{400}{10^3}=\frac{0{\cdot}1}{\dfrac{\omega^2}{10^3}-1}+\frac{6{\cdot}0}{\dfrac{\omega^2}{10^3}-300}+\frac{4{\cdot}0}{\dfrac{\omega^2}{10^3}-125},$$

and substituting $x=\omega^2/10^3$, this gives

$$0{\cdot}4=\frac{0{\cdot}1}{x-1}+\frac{6{\cdot}0}{x-300}+\frac{4{\cdot}0}{x-125},$$

† *Metric units.* The above example is also valid, with the same numerical constants, when the J-values are taken to represent Kg.cm.sec² and the K-values are in Kg.cm/rad.

[188]

which after multiplying out becomes
$$x^3 - 452x^2 + 4291x - 50,200 = 0.$$

This is a cubic equation of the form
$$x^3 + ax^2 + bx + c = 0.$$

By means of the substitution
$$x = y - \frac{a}{3},$$

the quadratic term is eliminated and the equation is thus reduced to
$$y^3 + 3py + q = 0,$$

where
$$3p = -\frac{a^2}{3} + b \quad \text{and} \quad 2q = \frac{2a^3}{27} - \frac{ab}{3} + c.$$

The evaluation gives
$$x = y - (452/3),$$
$$3p = -[(-452)^2/3] + 42,900 = -25,100,$$
$$p = -8366,$$
$$2q = [2(-452)^3/27] - [(-452 \times 42,900)/3] - 50,200,$$
$$q = -200,100.$$

This is a case where $q^2 + p^2 \leqslant 0$. The solutions are
$$y_1 = 2\sqrt{(-p)}\cos\frac{\theta}{3}, \quad y_2 = -\sqrt{(-p)}\cos\left(\frac{\theta}{3} + 60^\circ\right), \quad y_3 = -2\sqrt{(-p)}\cos\left(\frac{\theta}{3} - 60^\circ\right),$$

where $\cos\theta = -\dfrac{q}{\sqrt{(-p^3)}}$. Therefore
$$\cos(\theta/3) = 200,100/(7{\cdot}61 \times 10^5) = 0{\cdot}2625,$$

so that $\theta = 74^\circ\, 48'$. We then obtain

$$\cos\frac{\theta}{3} = 0{\cdot}9068, \qquad\qquad \cos\left\{\frac{\theta}{3} + 60^\circ\right\} = 0{\cdot}0883,$$

$$y_1 = 182{\cdot}8 \times 0{\cdot}9068 = 165{\cdot}7, \qquad y_2 = -182{\cdot}8 \times 0{\cdot}0883 = -16{\cdot}1,$$

$$x_1 = 165{\cdot}7 + 151 = 316{\cdot}7, \qquad x_2 = -16{\cdot}1 + 151 = 134{\cdot}9,$$

$$\omega_{\mathrm{I}}^2 = 316,700, \qquad\qquad \omega_{\mathrm{II}}^2 = 134,900,$$

$$F_{\mathrm{I}} = 5370\ \text{vib./min.}, \qquad\qquad F_{\mathrm{II}} = 3500\ \text{vib./min.},$$

$$\cos\left\{\frac{\theta}{3} - 60^\circ\right\} = 0{\cdot}8185,$$

$$y_3 = -182{\cdot}8 \times 0{\cdot}8185 = -149{\cdot}4,$$

$$x_3 = -149{\cdot}4 + 151 = 1{\cdot}6,$$

$$\omega_{\mathrm{III}}^2 = 1600,$$

$$F_{\mathrm{III}} = 382\ \text{vib./min.}$$

A more correct figure for x_3 can be obtained from the two larger values, x_1 and x_2, and the constant term of the cubic. As $c = x_1 x_2 x_3$, we have

$$x_3 = 50{,}200/[316{\cdot}7 \times 134{\cdot}9] = 1{\cdot}172,$$

so that $\qquad \omega_{\mathrm{III}}^2 = 1{\cdot}172 \times 10^3 \quad$ and $\quad F_{\mathrm{III}} = 327 \text{ vib./min.}$

To obtain more accurate values, these approximations for F_{I}, F_{II} and F_{III} can be used in Holzer tables, as shown subsequently in this section.

Derivation of frequency equation. We consider the system of Fig. 3, with the speed ratios $r_B = N_B/N_A$ and $r_C = N_C/N_A$.

The equivalent stiffnesses are

$$K_{Ae} = K_A, \quad K_{Be} \dots r_B^2 K_{B1} \quad \text{and} \quad K_{Ce} = r_C^2 K_{C1},$$

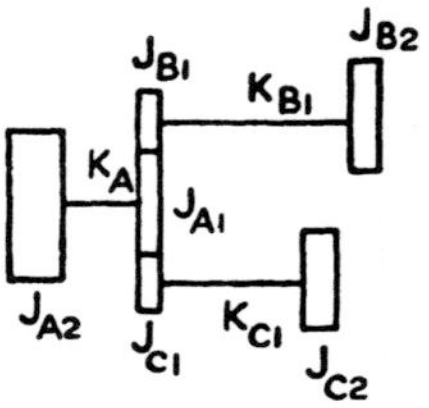

and the equivalent moments of inertia are

$$J_{A1e} = J_{A1}, \qquad J_{A2e} = J_{A2}, \qquad J_{B1e} = r_B^2 J_{B1},$$

$$J_{B2e} = r_B^2 J_{B2}, \quad J_{C1e} = r_C^2 J_{C1}, \quad J_{C2e} = r_C^2 J_{C2},$$

the total inertia of the gears being

$$J_G = J_{A1} + J_{B1e} + J_{C1e}.$$

Fig. 3.

The entire system vibrates with a natural phase velocity ω. In each branch of the system, the maximum inertia torque is equal to the maximum vibratory shaft torque, i.e.

$$J_{A2} \omega^2 \theta_{A2} = K_A(\theta_{A2} - \theta_{A1}) = T_A,$$

where $\theta_{A2} - \theta_{A1} = $ twist of shaft K_A. Dividing both sides of this equation by θ_{A1} gives

$$(K_A - J_{A2} \omega^2)\frac{\theta_{A2}}{\theta_{A1}} = K_A.$$

Now $K_A/J_{A2} = \omega_A^2$, where $\omega_A = $ natural phase velocity of branch A. Therefore

$$J_{A2}(\omega_A^2 - \omega^2)\frac{\theta_{A2}}{\theta_{A1}} = K_A$$

or $\qquad\qquad J_{A2}\dfrac{\theta_{A2}}{\theta_{A1}} = \dfrac{K_A}{\omega_A^2 - \omega^2}.$

Similarly, for the other branches

$$J_{B2e}\frac{\theta_{B2e}}{\theta_{B1e}} = J_{B2e}\frac{\theta_{B2e}}{\theta_{A1}} = \frac{K_{Be}}{\omega_B^2 - \omega^2} \quad \text{and} \quad J_{C2e}\frac{\theta_{C2e}}{\theta_{A1}} = \frac{K_{Ce}}{\omega_C^2 - \omega^2}.$$

As the system is in dynamic equilibrium, the sum of the vibratory torques at the junction J_{A1} must be zero.

$$T_A + T_G + T_B + T_C = 0,$$

which is equivalent to

$$J_{A2}\theta_{A2}\omega^2 + J_G\theta_G\omega^2 + J_{B2e}\theta_{B2e}\omega^2 + J_{C2e}\theta_{C2e}\omega^2 = 0,$$

[190]

where $\theta_G = \theta_{A1}$. Dividing by $\omega^2 \theta_G$, we obtain

$$J_{A2}\frac{\theta_{A2}}{\theta_G} + J_{B2e}\frac{\theta_{B2e}}{\theta_G} + J_{C2e}\frac{\theta_{C2e}}{\theta_G} + J_G = 0,$$

which, with the equations previously obtained, gives

$$J_G = \frac{K_A}{\omega_A^2 - \omega^2} + \frac{K_{Be}}{\omega_B^2 - \omega^2} + \frac{K_{Ce}}{\omega_C^2 - \omega^2}.$$

This is the 'frequency equation' for the three-branch system. For a system with a greater number of branches, this equation becomes

$$J_G = \frac{K_A}{\omega_A^2 - \omega^2} + \frac{K_{Be}}{\omega_B^2 - \omega^2} + \cdots + \frac{K_{ne}}{\omega_n^2 - \omega^2},$$

and the roots can be obtained by successive approximations.†

Holzer-table evaluation of gear-branched systems

The procedure will be illustrated by means of the engine system previously considered, with the values:

Branch A	Branch B	Branch C	Units
$J_{A2} = 100$	$J_{B2} = 2\cdot22$	$J_{C2} = 8\cdot0$	Lb.in.sec.²
$J_{A1} = 340$	$J_{B1} = 4\cdot0$	$J_{C1} = 6\cdot0$	Lb.in.sec.²
$K_A = 0\cdot1 \times 10^6$	$K_B = 0\cdot667 \times 10^6$	$K_C = 1\cdot0 \times 10^6$	Lb.in./rad.
—	$r_B = N_B/N_A = -3$	$r_C = N_C/N_A = -2$	—

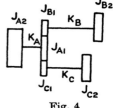

Fig. 4

In the Holzer tables, the K- and J-values are used without corrections for the different speed ratios, r_B and r_C. These are taken into account in the separate evaluations for each branch of the system.

In the following, the detailed tabulations are given for the one-node frequency of the system shown in Fig. 4. The correct value for this frequency is $F_I = 333$ vib./min.

$$F_I = 333 \text{ vib./min.} \qquad \omega^2 - 1\cdot22 \times 10^3 \text{ rad/sec}^3.$$

Main system (branch A)

Mass no.	J [Lb.in. sec.²]	$\dfrac{J\omega^2}{10^4}$ [Lb.in./ rad.]	Δ [rad.]	$\dfrac{J\omega^2\Delta}{10^4}$ [Lb.in.]	$\dfrac{\Sigma}{10^4}$ [Lb.in].	$\dfrac{K}{10^4}$ [Lb.in./ rad.]	Δ_{shaft} [rad.]
J_{A2}	100	12·20	1·000	12·20	12·20	10·0	1·220
J_{A1}	340	41·50	−0·220	−9·14	+3·06	—	—

Relative amplitude at J_{A1}: $\quad \Delta_{A1} = -0\cdot220$ rad.
Resultant specific torque: $\quad T_A = 30,600$ Lb.in.

† For various methods for determining the roots of polynomials, see page 230.

[191]

Branch B. Let β = assumed amplitude at J_{B2}.

Mass no.	J	$\dfrac{J\omega^2}{10^4}$	Δ	$\dfrac{J\omega^2\Delta}{10^4}$	$\dfrac{\Sigma}{10^4}$	$\dfrac{K}{10^4}$	Δ_{shaft}
J_{B2}	2·22	0·271	β	0·271β	0·271β	66·7	0·004β
J_{B1}	4·00	0·488	0·996β	—	—	—	—

But the amplitude at J_{B1} must be equal to $r_B \times \Delta_{A1} = -3 \times -0.220$, where the minus sign is prefixed to the speed-ratio value in order to take account of the opposite direction of rotation of the gears. Therefore

$$0.996\beta = -3 \times -0.220 = +0.660 \quad \text{and} \quad \beta = +0.660/0.996 = 0.663 \text{ rad.}$$

The tabulation is repeated with this numerical value for β:

J_{B2}	2·22	0·271	0·663	0·180	0·180	66·7	0·003
J_{B1}	4·00	0·488	0·660	0·322	0·502	—	—

Therefore the specific torque at J_{B1} due to J_{B2} is $T_B = -3 \times 5020 = -15,060$ Lb.in.

Branch C. Let γ = assumed amplitude at J_{C2}.

Mass no.	J	$\dfrac{J\omega^2}{10^4}$	Δ	$\dfrac{J\omega^2\Delta}{10^4}$	$\dfrac{\Sigma}{10^4}$	$\dfrac{K}{10^4}$	Δ_{shaft}
J_{C2}	8·00	0·976	γ	0·976γ	0·976γ	100·0	0·01γ
J_{C1}	6·00	0·731	0·990γ	—	—	—	—

The amplitude at J_{C1} must also be equal to $r_C \times \Delta_{A1} = -2 \times -0.220 = +0.440$. Therefore $\gamma = 0.440/0.990 = 0.444$ rad. Repeating the tabulation with this value, we obtain

J_{C2}	8·00	0·976	0·444	0·434	0·434	100·0	0·004
J_{C1}	6·00	0·731	0·440	0·322	0·756	—	—

The resultant specific torque at J_{C1} due to J_2 is $T_C = -2 \times 7560 = -15,120$ Lb.in.

Thus $\omega^2 = 1.22 \times 10^3$ is close to the correct value, since the sum of the specific torques is approximately zero: $T_A + T_B + T_C = (3.06 - 1.506 - 1.512) \times 10^4 \cong 0$.

The two- and three-node frequencies F_{II} and F_{III} can be determined accurately in the same manner.

TYPICAL EXAMPLES OF GEARED SYSTEMS

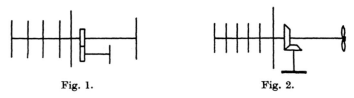

Fig. 1.   Fig. 2.

Engines with geared blower drives.

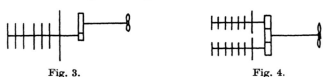

Fig. 3.   Fig. 4.

Marine propulsion systems.

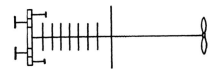

Fig. 5. Marine engine with pump
drives at free end.

Fig. 6. Motor-car engine
system.

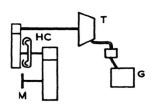

Fig. 7. Gas generator (G), turbine (T), hydraulic coupling (HC)
and driven machine (M).

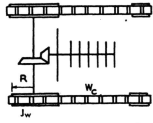

Fig. 8. Tracked vehicle (caterpillar) drive.

$J_W =$ moment of inertia of wheel.
$W_0 =$ weight of caterpillar.
$J_0 = R^2 W_0 / g$.
Total wheel inertia: $J = J_W + J_0$.

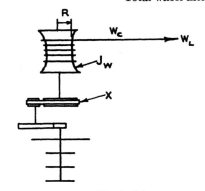

Fig. 9. Winch drive.

$W_0 =$ weight of cable. $W_L =$ weight of load.
$J_W =$ winch inertia.

Total winch inertia: $J = J_W + \dfrac{R^2}{g} [W_0 + W_L]$.

For a system of this kind, it is necessary to plot the natural frequency variation against
various values of the load W_L, unless there is sufficient belt slip or clutch slip at X to
isolate the engine as a vibrating system.

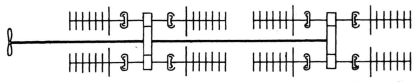

Fig. 10. Eight-engine system with hydraulic couplings, driving a single propeller.

1·3216 *Branched systems with flexible transmissions*

Branched systems with flexible transmissions are readily evaluated by the 'Effective-inertia method' (see section 1·3422), the subsidiary systems being computed up to and including the equivalent shafts of the transmissions. For this purpose the system is split up as shown in Fig. 1.

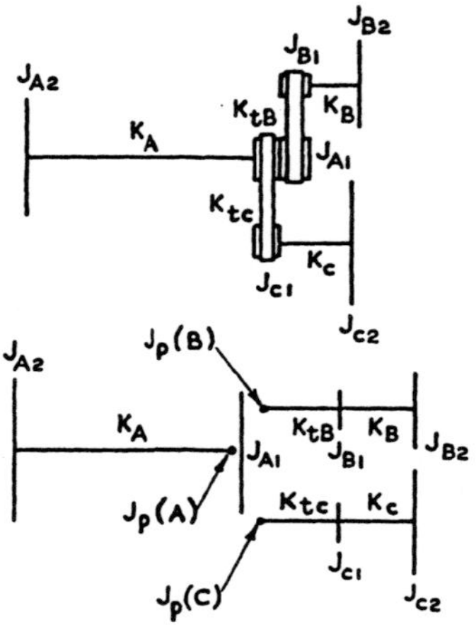

They can also be evaluated by means of Holzer tabulations. The computation procedure is outlined below.

Denoting by J the moments of inertia and by K the stiffnesses of the various branches, and using the speed ratios

$$r_B = N_B/N_A, \quad r_C = N_C/N_A,$$

we may write the equations of motion of the system shown in Fig. 1 as follows:

Fig. 1. Division of branched system with flexible transmissions into sub-systems.

Branch A:

$$J_{A2}\ddot{\theta}_{A2} + K_A(\theta_{A2} - \theta_{A1}) = 0,$$

$$K_A(\theta_{A1} - \theta_{A2}) + J_{A1}\ddot{\theta}_{A1} + K_{tB}r_B^2\left(\theta_{A1} - \frac{\theta_{B1}}{r_B}\right) + K_{tC}r_C^2\left(\theta_{A1} - \frac{\theta_{C1}}{r_C}\right) = 0.$$

†*Branch B:*
$$J_{B2}\ddot{\theta}_{B2}r_B + K_B r_B(\theta_{B2} - \theta_{B1}) = 0,$$

$$K_B r_B(\theta_{B1} - \theta_{B2}) + J_{B1}\ddot{\theta}_{B1}r_B + K_{tB}r_B^2\left(\frac{\theta_{B1}}{r_B} - \theta_{A1}\right) = 0.$$

†*Branch C:*
$$J_{C2}\ddot{\theta}_{C2}r_C + K_C r_C(\theta_{C2} - \theta_{C1}) = 0,$$

$$K_C r_C(\theta_{C1} - \theta_{C2}) + J_{C1}\ddot{\theta}_{C1}r_C + K_{tC}r_C^2\left(\frac{\theta_{C1}}{r_C} - \theta_{A1}\right) = 0,$$

where the transmission stiffnesses are denoted by K_{tB} and K_{tC}, and are referred to the speeds of the corresponding branches B and C. The substitutions $\ddot{\theta}_{A2} = -\omega^2\Delta_{A2}$, $\theta_{B2} = -\omega^2\Delta_{B2}$, etc., give, in conjunction with these equations, the torque equation

$$\sum_A J\omega^2\Delta + \sum_B J\omega^2\Delta + \sum_C J\omega^2\Delta = 0$$

for the common point J_{A1} of the system.

---

† In these equations the factors r_B and r_C are included to enable direct substitution into the second equation of branch A.

[194]

The Holzer tables can therefore be arranged as follows:

Branch A:

$$J_{A2}\omega^2, \quad \Delta_{A2}, \quad J_{A2}\omega^2\Delta_{A2}, \quad J_{A2}\omega^2\Delta_{A2}, \quad K_A, \quad \frac{1}{K_A}J_{A2}\omega^2\Delta_{A2} = \Delta_{sh\,A},$$

$$J_{A1}\omega^2, \quad \Delta_{A1} = \Delta_{A2} - \Delta_{sh\,A}, \quad J_{A1}\omega^2\Delta_{A1}, \quad J_{A2}\omega^2\Delta_{A2} + J_{A1}\omega^2\Delta_{A1}.$$

Branch B:

$$J_{B2}\omega^2, \quad \Delta_{B2} = \beta, \quad J_{B2}\omega^2\beta, \quad J_{B2}\omega^2\beta, \quad K_B, \quad \frac{1}{K_B}J_{B2}\omega^2\beta = \beta_{sh} = \lambda\beta,$$

$$J_{B1}\omega^2, \quad \Delta_{B1} = \beta(1-\lambda), \quad J_{B1}\omega^2\beta(1-\lambda),$$

$$\{J_{B2}\omega^2 + J_{B1}\omega^2(1-\lambda)\}\beta = \sum_B \times \beta, \quad K_{tB}, \quad \frac{\beta}{r_B K_{tB}}\sum_B = \Delta_{B1} - r_B\Delta_{A1},$$

hence

$$\frac{1}{r_B K_{tB}} \times \frac{1}{(1-\lambda)} \times \Delta_{B1} \times \sum_B = \Delta_{B1} - r_B\Delta_{A1}$$

and

$$\Delta_{B1} = \frac{r_B\Delta_{A1}}{1 - \dfrac{\sum_B}{r_B K_{tB}(1-\lambda)}};$$

also

$$\Delta_{B2} = \frac{\Delta_{B1}}{1-\lambda}.$$

Branch C: This is calculated in the same manner as branch *B*. The torque equation for the three sub-systems must be equal to zero if ω^2 corresponds to a natural frequency of the entire system.

If one of the transmissions is infinitely rigid, we have, for instance, $K_{tC} = \infty$, $\Delta_{C1} - r_C\Delta_{A1} = 0$. The other parts of the computation are unchanged.

1·322 *Gorfinkel's method*

Gorfinkel's tabulation method† is a simplified modification of Holzer's tables which shortens the computation work for engine systems comprising a large number of cylinder inertias $J_{\text{cyl.}}$ and crankthrow stiffnesses $K_{\text{cyl.}}$ with identical values.

The preparations for a Gorfinkel-table evaluation are as follows:

(1) Express all the inertias J_k of the engine system as non-dimensional ratios $J_k/J_{\text{cyl.}}$, where $J_{\text{cyl.}}$ = moment of inertia per line, so that

$$\mu_1 = J_1/J_{\text{cyl.}}, \quad \mu_2 = J_2/J_{\text{cyl.}}, \quad \text{etc.}$$

(2) Express all the stiffnesses K_k of the engine system as non-dimensional ratios $K_{\text{cyl.}}/K_k$, where $K_{\text{cyl.}}$ = stiffness of one crankthrow, so that

$$\lambda_1 = K_{\text{cyl.}}/K_1, \quad \lambda_2 = K_{\text{cyl.}}/K_2, \quad \text{etc.}$$

† Gorfinkel, A., 'Critical speeds of crankshafts', *Engineering*, 27 Dec. 1929, pp. 827–9.

(3) Assuming a value ω for the phase velocity of the engine system, evaluate

$$x^2 = \omega^2/(K_{cyl.}/J_{cyl.}).$$

(4) Lay out and evaluate a Gorfinkel tabulation as shown in the following example. If the total inertia torque (represented by $\Sigma\mu\Delta$) is approximately zero at the last mass, the value of ω used corresponds to a natural vibration frequency of the entire system, which is obtained from

$$F = 9.55x\sqrt{(K_{cyl.}/J_{cyl.})} \quad \text{[vib./min.]}.$$

EXAMPLE. Gorfinkel tabulation for an eight-cylinder engine with the following values:

$$J_1 = J_2 = \ldots = J_8 = J_{cyl.} = 5.62 \text{ Lb.ft.sec.}^2,$$

$$J_9 = 278 \text{ Lb.ft.sec.}^2,$$

$$J_{10} = 225.5 \text{ Lb.ft.sec.}^2,$$

$$K_1 = K_2 = \ldots = K_7 = K_{cyl.} = 12.6 \times 10^6 \text{ Lb.ft./rad.},$$

$$K_8 = 11.4 \times 10^6 \text{ Lb.ft./rad.},$$

$$K_9 = 29.8 \times 10^6 \text{ Lb.ft./rad.}$$

These values give: $\mu_1 = \mu_2 = \ldots = \mu_8 = 1,$

$$\mu_9 = 278/5.62 = 49.44, \quad \mu_{10} = 225.5/5.62 = 40.12, \quad \lambda_1 = \lambda_2 = \ldots = \lambda_7 = \ldots 1,$$

$$\lambda_8 = K_{cyl.}/K_8 = 12.6/11.4 = 1.105, \quad \lambda_9 = 12.6/29.8 = 0.423.$$

$$x^2 = 0.0348, \quad x = 0.1866$$

Mass no.	μ	Δ	$\Sigma\mu\Delta$	$x^2\lambda$	$\Delta_{sh} = x^2\lambda\Sigma\mu\Delta$
1	1·0	1·0000	1·0000	0·0348	0·0348
2	1·0	0·9652	1·9652	0·0348	0·0684
3	1·0	0·8968	2·8620	0·0348	0·0996
4	1·0	0·7972	3·6592	0·0348	0·1273
5	1·0	0·6699	4·3291	0·0348	0·1506
6	1·0	0·5193	4·8484	0·0348	0·1687
7	1·0	0·3506	5·1990	0·0348	0·1808
8	1·0	0·1698	5·3688	0·0348	0·2060
			[−1·7900]		
9	49·44	−0·0362	3·5788	0·0147	0·0525
			[−3·5700]		
10	40·12	−0·0888	+0·0088	—	—

$$x^2\lambda_8 = 0.0348 \times 1.105 = 0.0384, \quad x^2\lambda_9 = 0.0348 \times 0.423 = 0.0147,$$

$$K_{cyl.}/J_{cyl.} = \frac{12.6 \times 10^6}{5.62} = 2.242 \times 10^6, \quad \omega^2 = x^2 \times \frac{K_{cyl.}}{J_{cyl.}} = 0.078 \times 10^6,$$

$$\omega = 279.4 \text{ rad./sec.}, \quad F = 9.55 \times \omega = 2665 \text{ vib./min.}$$

The maximum torque at the node, for $\Delta_1 = 1$ rad. is,

$$T_{max.} = J_{cyl.} \times \omega^2 \times (\Sigma\mu\Delta)_{max.}$$

$$= 5.62 \times 7.8 \times 10^4 \times 5.3688 = 2.227 \times 10^6 \text{ Lb.ft.}$$

Derivation. In Holzer tabulations, the relative amplitude Δ_{n+1} of a mass J_{n+1} is given by the relation

$$\Delta_{n+1} = \Delta_n - \frac{1}{K_n} \sum_1^n (J\omega^2 \Delta),$$

where K_n = stiffness of shaft portion between mass J_n and J_{n+1}. The ω^2 can be placed in front of the summation sign, and, on introducing the substitutions given above under 1, 2 and 3, we obtain

$$\Delta_{n+1} = \Delta_n - x^2 \lambda_n \sum_1^n (\mu \Delta),$$

which forms the basis of Gorfinkel's tabulation method.

The following tables, for which we are indebted to C. H. Bradbury,[†] give data facilitating the construction of Gorfinkel tables for engines having five to eight cylinders with equal moments of inertia per cylinder line and stiffnesses equal between cylinders. Variation of flywheel moment of inertia and stiffness between the last cylinder and the flywheel is allowed for.

From these tables, the relative amplitude and the value of x can be read off for any orthodox system and the natural frequency calculated by inserting stiffness and moment of inertia values.

Gorfinkel calculation constants for frequency and relative amplitudes

$$[x = \omega / \sqrt{(K_{\text{cyl.}} / J_{\text{cyl.}})}, \quad \mu_k = J_k / J_{\text{cyl.}}, \quad \lambda_k = K_{\text{cyl.}} / K_k, \quad F = 9.55 x \times \sqrt{(J_{\text{cyl.}} / K_{\text{cyl.}})}]$$

Five-cylinder engine

	$\mu_6 = 40$				$\mu_6 = 60$			
x	0·305	0·300	0·294	0·290	0·300	0·295	0·289	0·285
λ_5	0·90	1·00	1·10	1·20	0·90	1·00	1·10	1·20
Δ_1	1·000	1·000	1·000	1·000	1·000	1·000	1·000	1·000
Δ_2	0·906	0·910	0·913	0·916	0·910	0·913	0·916	0·919
Δ_3	0·730	0·738	0·747	0·755	0·738	0·747	0·755	0·763
Δ_4	0·484	0·502	0·516	0·531	0·499	0·515	0·531	0·546
Δ_5	0·192	0·215	0·240	0·262	0·215	0·240	0·262	0·284
Δ_6	−0·087	−0·087	−0·087	−0·087	−0·057	−0·057	−0·057	−0·547

	$\mu_6 = 100$				$\mu_6 = 200$			
x	0·296	0·290	0·286	0·281	0·293	0·288	0·282	0·278
λ_5	0·90	1·00	1·10	1·20	0·90	1·00	1·10	1·20
Δ_1	1·000	1·000	1·000	1·000	1·000	1·000	1·000	1·000
Δ_2	0·912	0·915	0·918	0·921	0·914	0·917	0·920	0·922
Δ_3	0·744	0·753	0·761	0·769	0·750	0·758	0·766	0·773
Δ_4	0·512	0·527	0·542	0·556	0·522	0·536	0·552	0·565
Δ_5	0·235	0·257	0·280	0·301	0·245	0·270	0·292	0·312
Δ_6	−0·035	−0·035	−0·035	−0·035	−0·020	−0·020	−0·020	−0·020

† Bradbury, C. H., *Torsional Vibrations in Diesel Engines—Some Observations and Practical Aspects,* Diesel Engine Users Association (D.E.U.A.), Paper no. S 226 (March 1953).

Six-cylinder engine

	$\mu_7=40$				$\mu_7=60$			
x	0·2598	0·2563	0·2522	0·2490	0·2550	0·2510	0·2475	0·2440
λ_6	0·90	1·00	1·10	1·20	0·90	1·00	1·10	1·20
Δ_1	1·000	1·000	1·000	1·000	1·000	1·000	1·000	1·000
Δ_2	0·932	0·934	0·936	0·938	0·935	0·937	0·939	0·940
Δ_3	0·802	0·808	0·814	0·819	0·808	0·815	0·820	0·825
Δ_4	0·617	0·628	0·638	0·647	0·632	0·642	0·651	0·660
Δ_5	0·390	0·407	0·422	0·437	0·412	0·427	0·442	0·457
Δ_6	0·137	0·160	0·180	0·198	0·164	0·185	0·205	0·225
Δ_7	−0·100	−0·100	−0·100	−0·100	−0·070	−0·070	−0·070	−0·070

	$\mu_7=100$				$\mu_7=200$			
x	0·2507	0·2472	0·2437	0·2402	0·2483	0·2442	0·2407	0·2370
λ_6	0·90	1·00	1·10	1·20	0·90	1·00	1·10	1·20
Δ_1	1·000	1 000	1·000	1·000	1·000	1·000	1·000	1·000
Δ_2	0·937	0·939	0·940	0·942	0·939	0·940	0·942	0·944
Δ_3	0·815	0·820	0·825	0·830	0·819	0·824	0·830	0·835
Δ_4	0·642	0·652	0·661	0·671	0·650	0·660	0·670	0·689
Δ_5	0·427	0·442	0·458	0·472	0·442	0·457	0·470	0·480
Δ_6	0·186	0·207	0·227	0·245	0·202	0·222	0·242	0·262
Δ_7	−0·042	−0·042	−0·042	−0·042	−0·025	−0·025	−0·025	−0·025

Seven-cylinder engine

	$\mu_8=40$				$\mu_8=60$			
x	0·2270	0·2240	0·2212	0·2185	0·2222	0·2190	0·2165	0·2138
λ_7	0·90	1·00	1·10	1·20	0·90	1·00	1·10	1·20
Δ_1	1·000	1·000	1·000	1·000	1·000	1·000	1·000	1·000
Δ_2	0·948	0·950	0·951	0·952	0·951	0·952	0·953	0·954
Δ_3	0·848	0·852	0·856	0·859	0·854	0·858	0·862	0·865
Δ_4	0·704	0·712	0·718	0·725	0·716	0·723	0·730	0·736
Δ_5	0·524	0·534	0·546	0·556	0·542	0·553	0·563	0·574
Δ_6	0·318	0·332	0·346	0·360	0·342	0·357	0·372	0·386
Δ_7	0·094	0·113	0·132	0·148	0·125	0·144	0·162	0·179
Δ_8	−0·115	−0·115	−0·115	−0·115	−0·077	−0·077	−0·077	−0·077

	$\mu_8=100$				$\mu_8=200$			
x	0·2180	0·2152	0·2125	0·2100	0·2150	0·2122	0·2095	0·2070
λ_7	0·90	1·00	1·10	1·20	0·90	1·00	1·10	1·20
Δ_1	1·000	1·000	1·000	1·000	1·000	1·000	1·000	1·000
Δ_2	0·952	0·953	0·954	0·956	0·954	0·955	0·956	0·957
Δ_3	0·860	0·863	0·867	0·870	0·863	0·867	0·870	0·873
Δ_4	0·726	0·733	0·740	0·746	0·733	0·740	0·746	0·753
Δ_5	0·558	0·568	0·578	0·588	0·568	0·578	0·588	0·608
Δ_6	0·363	0·378	0·391	0·406	0·378	0·393	0·406	0·420
Δ_7	0·151	0·169	0·187	0·205	0·172	0·190	0·207	0·224
Δ_8	−0·048	−0·048	−0·048	−0·048	−0·022	−0·022	−0·022	−0·022

Eight-cylinder engine

	$\mu_9=40$				$\mu_9=60$			
x	0·2015	0·1990	0·1970	0·1950	0·1968	0·1947	0·1925	0·1902
λ_8	0·90	1·00	1·10	1·20	0·90	1·00	1·10	1·20
Δ_1	1·000	1·000	1·000	1·000	1·000	1·000	1·000	1·000
Δ_2	0·959	0·960	0·961	0·962	0·961	0·962	0·963	0·964
Δ_3	0·880	0·882	0·885	0·887	0·885	0·887	0·890	0·893
Δ_4	0·765	0·770	0·774	0·779	0·775	0·780	0·785	0·789
Δ_5	0·618	0·626	0·634	0·641	0·635	0·643	0·650	0·657
Δ_6	0·447	0·458	0·469	0·479	0·470	0·481	0·491	0·502
Δ_7	0·258	0·272	0·284	0·298	0·286	0·300	0·314	0·327
Δ_8	0·058	0·074	0·090	0·106	0·094	0·110	0·125	0·141
Δ_9	−0·128	−0·128	−0·128	−0·128	−0·088	−0·088	−0·088	−0·088

	$\mu_9 = 100$				$\mu_9 = 200$			
x	0·1930	0·1905	0·1885	0·1865	0·1900	0·1877	0·1855	0·1830
λ_8	0·90	1·00	1·10	1·20	0·90	1·00	1·10	1·20
Δ_1	1·000	1·000	1·000	1·000	1·000	1·000	1·000	1·000
Δ_2	0·963	0·964	0·964	0·965	0·964	0·965	0·966	0·966
Δ_3	0·890	0·892	0·895	0·897	0·893	0·895	0·897	0·900
Δ_4	0·783	0·788	0·790	0·797	0·790	0·795	0·799	0·803
Δ_5	0·648	0·656	0·663	0·670	0·658	0·666	0·673	0·680
Δ_6	0·489	0·500	0·510	0·520	0·503	0·514	0·524	0·534
Δ_7	0·311	0·325	0·339	0·351	0·329	0·344	0·356	0·369
Δ_8	0·122	0·138	0·155	0·170	0·144	0·162	0·178	0·192
Δ_9	−0·054	−0·054	−0·054	−0·054	−0·027	−0·027	−0·027	−0·027

1·323 Distributed-mass method

The usual Holzer tables assume that the shafting has rigidity but no mass; assumptions may be made by adding some of the mass of the shafting to the adjacent concentrated masses, but this is not always possible, for instance, in the case of the tail shafting of a marine propulsion system.

In the following method, the shafting has both mass and rigidity, and, when a mass is distributed along a section of the shafting (as is the case with a tandem armature of a main propulsion motor), the mass of the shafting is increased to include the mass of the windings, etc. A tabular calculation is then carried out to determine the natural frequencies of the system.

In a marine system where extra large shafting is fitted, torsional vibration up to the third mode may be experienced. The large line shafting may come at an anti-node for the three-node vibration and, when its distributed masses are taken into account, it may, therefore, lower the three-node frequency by up to 10 %, without appreciably affecting the one- and two-node frequencies. Frequencies of marine systems calculated by this distributed-mass method have been found to be very accurate.

Other distributed-mass methods are known,† but have not been used by B.I.C.E.R.A. members, so that a comparison is not possible of calculated frequencies and observed values.

Treatment of shafting of varying section, with a number of concentrated masses

The shafting is divided into lengths L_a, L_b, L_c, ..., each of which is of uniform cross-section, in such a way that any concentrated masses (of moments of inertia J_a, J_b, J_c, ...) are situated at the left-hand ends of the sections (see Fig. 1).

Notation:

w = specific weight [Lb./in.^{3}],

ρ = w/g = mass density [Lb.sec.^{2}/in.^{4}],

g = 386·4 [in./sec.^{2}],

† See, for instance, W. Ker Wilson, *Practical Solution of Torsional Vibration Problems*, vol. 1, pp. 685–6.

G = modulus of rigidity [Lb./in.²],

L = length of shaft [in.],

$I_p = (\pi/32) \times (D^4 - d^4)$ = polar second moment of area [in.⁴] of shaft,

D = shaft outer diameter [in.],

d = shaft inner diameter [in.],

Δ_1 = vibration amplitude [rad.], at left-hand end of shaft,

Δ_2 = vibration amplitude at right-hand end of shaft,

T_1 = torque at left-hand end of shaft [Lb.in.],

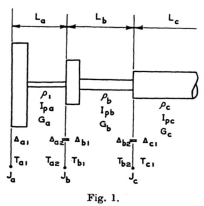

Fig. 1.

T_2 = torque at right-hand end of shaft,

J = moment of inertia of a concentrated mass [Lb.in.sec.²].

Thus the two ends of each shaft section are denoted by the subscripts 1 and 2, respectively, while the shaft sections themselves are denoted by the subscripts a, b, c, ..., etc.

Only those masses which are attached to a single flange are regarded as 'concentrated masses' and denoted by J_a, J_b, etc.

If a shaft section n, determined by the values L_n, ρ_n, I_{pn} and G_n, carries, for instance, an armature of mass moment of inertia J_n, this value J_n is 'distributed' over the length L_n and represented by an equivalent increase in mass density. The resultant density ρ'_n is calculated as

$$\rho'_n = \rho_n + \frac{J_n}{I_{pn}L_n} \quad \text{[Lb.sec.}^2/\text{in.}^4\text{]}.$$

Note. J_n is the moment of inertia outside the shaft diameter.

The same principle can be adopted for any mass whose attachment is distributed over a length of elastic shaft. Couplings, small wheels, or small local increases in diameter, may also be treated as giving rise to an equivalent density over their own length of shaft.

Cylinder masses are also treated as distributed over the shaft section corresponding to their crankthrow. Thus, if

$$J_{\text{cyl.}} = J_{\text{rot.}} + J_{\text{recip.}} = \text{moment of inertia per line} \quad \text{[Lb.in.sec.}^2\text{]},$$

when as usual $J_{\text{recip.}}$ includes only half the reciprocating masses, then the nth crankthrow, of equivalent length L_{en}, mass density ρ_n, and second moment of area I_{pn}, will have a resultant mass density

$$\rho'_n = \rho_n + \frac{J_{\text{cyl.}}}{I_{pn}L_{en}} \quad \text{[Lb.sec.}^2/\text{in.}^4\text{]}.$$

If all the $J_{\text{cyl.}}$ and L_{en} values are identical, the entire crankshaft can be replaced by an equivalent straight shaft section which has a mass density ρ'_n, obtained from the above equation.

Tabulation procedure

(1) Prepare a tabulation similar to that of Fig. 3, and fill in the lines J, G, I_p, w and L.

(2) Assume a value F for the natural frequency [vib./min.]; evaluate $\omega = (2\pi/60) \times F$, and for each length calculate β, $\beta L \times 180/\pi$, $\cos \beta L$, $\sin \beta L$ and $\beta G I_p$, where $\beta = \omega \sqrt{(w/Gg)}$.

(3) Assume that the left-hand end of the first shaft section (L_a) has a relative amplitude $\Delta_1 = 1$ rad. Write $T = 0$, since there is no torque to the left of this section. The inertia torque δT due to the concentrated mass is

$$\delta T = \omega^2 J \Delta_1 = \omega^2 J \times 1 \cdot 0.$$

Enter this value in the table, in its appropriate place in column 'a'. The torque on the right-hand side of L_a is, thus,

$$T_1 = T + \delta T = \omega^2 J.$$

Enter this value on line T_1 of column 'a'.

(4) Calculate the relative amplitude Δ_2 at the end of this first cross-section as

$$\Delta_2 = \Delta_1 \cos \beta L - \frac{T_1}{\beta G I_p} \sin \beta L = 1 \cdot 0 \times \cos \beta L - \frac{\delta T}{\beta G I_p} \sin \beta L.$$

(In the tabulated example of Fig. 3, $\Delta_2 = 0 \cdot 8707$.) This value is also the relative amplitude Δ_1 at the left-hand end of L_b, that is,

$$\Delta_{b1} = \Delta_{a2}.$$

Therefore, enter this value as Δ_1 under the column of L_b.†

(5) Calculate $\Delta_1 \cos \beta L$ and the following values in the table, down to the torque T_2 for L_a. This torque acts on the left-hand end of shaft section L_b. Rewrite its value, therefore, in column L_b, on line 'T'.

(6) Proceed with calculations for the column of shaft section L_b, in which Δ_1 and T have already been entered. Calculate Δ_2 and T_2, and enter these as Δ_1 and T, respectively, in the column for shaft section L_c.

(7) Continue the calculation for the remaining columns, up to the end of the last column. If the assumed value of F corresponds to a natural frequency, there should be $T_2 \cong 0$ in the last column (zero torque to the right of the last shaft section).

(8) If T_2 is not zero in the last column, repeat the entire tabulation with a new assumed value for F. The natural frequency is then obtained

† It should be remembered that Δ_1 and Δ_2 are the relative amplitudes at the left-hand end and the right-hand end, respectively, of the shaft section considered. In the table, Δ_1 is used to denote Δ_{a1}, Δ_{b1}, ..., etc., and Δ_2 to denote Δ_{a2}, Δ_{b3}, ..., etc.

[201]

graphically, by plotting the residual torque T_2 against F, and thus determining $T_2 = 0$.

(9) Having obtained F_I, F_{II} and F_{III}, for the modes of vibration investigated, make final tabulations for these values, in order to obtain their corresponding Δ_1 and T_1 curves (relative amplitude and torque-distribution curves), plotted on a base of shaft length.

NUMERICAL EXAMPLE (reproduced by courtesy of the Admiralty). Determination by means of tables of the one- and two-node frequencies of the marine system shown in Fig. 2.

Fig. 3. Table 1 (*distributed-mass method*). One-node frequency

$F_I = 762$ vib./min., $\omega = (2\pi/60) \times F_I = 79 \cdot 8$ rad./sec., $\omega^2 = 6 \cdot 368 \times 10^3$ rad.²/sec.²

Formula	Units	a	b	c	d	e
$J = Wk^2/g$	Lb.in.sec.²	$0 \cdot 991 \times 10^3$	—	—	—	$0 \cdot 285 \times 10^3$
G	Lb./in.²			$\longleftarrow \quad 1 \cdot 2 \times 10^7 \quad \longrightarrow$		
I_p	in.⁴	329·0	246·1	329·0	246·1	246·1
w	Lb./in.³	0·284	0·284	0·284	0·856	0·412
$\beta = \omega\sqrt{(w/Gg)}$	in.⁻¹	$0 \cdot 625 \times 10^{-3}$	$0 \cdot 625 \times 10^{-3}$	$0 \cdot 625 \times 10^{-3}$	$0 \cdot 082 \times 10^{-3}$	$0 \cdot 752 \times 10^{-3}$
L	in.	80	211	80	61·5	67·85
$\beta L \times 180/\pi$	deg.	2° 52′	7° 34′	2° 52′	3° 49′	2° 54′
$\cos \beta L$	—	0·9987	0·9913	0·9987	0·9978	0·9987
$\sin \beta L$	—	0·0500	0·1317	0·0500	0·0666	0·0506
βGI_p	Lb.in.	$2 \cdot 465 \times 10^6$	$1 \cdot 850 \times 10^6$	$2 \cdot 465 \times 10^6$	$3 \cdot 20 \times 10^6$	$2 \cdot 222 \times 10^6$
Δ_1	rad.	1·0000	0·8707	0·4075	0·2376	0·1350
T	Lb.in.	0	$6 \cdot 435 \times 10^6$	$6 \cdot 587 \times 10^6$	$6 \cdot 629 \times 10^6$	$6 \cdot 673 \times 10^6$
$\delta T = \omega^2 J\Delta_1$	Lb.in.	$6 \cdot 320 \times 10^6$	—	—	—	$0 \cdot 245 \times 10^6$
$T + \delta T = T_1$	Lb.in.	$6 \cdot 320 \times 10^6$	$6 \cdot 435 \times 10^6$	$6 \cdot 587 \times 10^6$	$6 \cdot 629 \times 10^6$	$6 \cdot 918 \times 10^6$
(1) $\Delta_1 \cos \beta L$	rad.	0·9987	0·8640	0·4070	0·2730	0·1348
(2) $(T_1/\beta GI_p) \sin \beta L$	rad.	0·1280	0·4575	0·1334	0·1380	0·1580
$\Delta_2 = \text{difference} = (1)-(2)$	rad.	0·8707	0·4075	0·2736	0·1350	−0·0232
(3) $T_1 \cos \beta L$	Lb.in.	$6 \cdot 312 \times 10^6$	$6 \cdot 375 \times 10^6$	$6 \cdot 579 \times 10^6$	$6 \cdot 615 \times 10^6$	$6 \cdot 909 \times 10^6$
(4) $\beta GI_p\Delta_1 \sin \beta L$	Lb.in.	$0 \cdot 123 \times 10^6$	$0 \cdot 212 \times 10^6$	$0 \cdot 050 \times 10^6$	$0 \cdot 058 \times 10^6$	$0 \cdot 015 \times 10^6$
$T_2 = \text{sum} = (3)+(4)$	Lb.in.	$6 \cdot 435 \times 10^6$	$6 \cdot 587 \times 10^6$	$6 \cdot 629 \times 10^6$	$6 \cdot 673 \times 10^6$	$6 \cdot 924 \times 10^6$

Fig. 4. Table 2 (*distributed-mass method*). Two-node frequency

$F_{II} = 1812$ vib./min., $\omega = (2\pi/60) \times F_{II} = 189 \cdot 5$ rad./sec., $\omega^2 = 3 \cdot 6 \times 10^4$ rad.²/sec.²

Formula	Units	a	b	c	d	e
$J = Wk^2/g$	Lb.in.sec.²	$0 \cdot 991 \times 10^3$	—	—	—	$0 \cdot 285 \times 10^3$
G	Lb./in.²			$1 \cdot 2 \times 10^7$		
I_p	in.⁴	329·0	246·1	329·0	246·1	246·1
w	Lb./in.³	0·284	0·284	0·284	0·856	0·412
$\beta = \omega\sqrt{(w/Gg)}$	in.⁻¹	$1 \cdot 484 \times 10^{-3}$	$1 \cdot 484 \times 10^{-3}$	$1 \cdot 484 \times 10^{-3}$	$2 \cdot 573 \times 10^{-3}$	$1 \cdot 788 \times 10^{-3}$
L	in.	80	211	80	61·5	67·25
$\beta L \times 180/\pi$	deg.	6° 48′	17° 56′	6° 48′	9° 4′	6° 53′
$\cos \beta L$	—	0·9930	0·9514	0·9930	0·9875	0·9930
$\sin \beta L$	—	0·1184	0·3079	0·1184	0·1576	0·1198
βGI_p	Lb.in.	$0 \cdot 585 \times 10^7$	$0 \cdot 4375 \times 10^7$	$0 \cdot 585 \times 10^7$	$0 \cdot 76 \times 10^7$	$0 \cdot 523 \times 10^7$
Δ_1	rad.	1·0000	0·2710	−2·2870	−2·975	−3·623
T	Lb.in.	0	$3 \cdot 6143 \times 10^7$	$3 \cdot 4765 \times 10^7$	$3 \cdot 2937 \times 10^7$	$2 \cdot 8987 \times 10^7$
$\delta T = \omega^2 J\Delta_1$	Lb.in.	$3 \cdot 5676 \times 10^7$	—	—	—	$-3 \cdot 7160 \times 10^7$
$T + \delta T = T_1$	Lb.in.	$3 \cdot 5676 \times 10^7$	$3 \cdot 6143 \times 10^7$	$3 \cdot 4765 \times 10^7$	$3 \cdot 2937 \times 10^7$	$-0 \cdot 8173 \times 10^7$
(1) $\Delta_1 \cos \beta L$	rad.	0·9930	0·2580	−2·271	−2·940	−3·6000
(2) $(T_1/\beta GI_p) \sin \beta L$	rad.	0·7220	2·5450	0·704	0·683	−0·1873
$\Delta_2 = \text{difference} = (1)-(2)$	rad.	0·2710	−2·2870	−2·975	−3·623	−3·4127
(3) $T_1 \cos \beta L$	Lb.in.	$3 \cdot 545 \times 10^7$	$3 \cdot 44 \times 10^7$	$3 \cdot 4521 \times 10^7$	$3 \cdot 255 \times 10^7$	$-0 \cdot 8116 \times 10^7$
(4) $\beta GI_p\Delta_1 \sin \beta L$	Lb.in.	$0 \cdot 0693 \times 10^7$	$0 \cdot 0365 \times 10^7$	$-0 \cdot 1584 \times 10^7$	$-0 \cdot 3563 \times 10^7$	$-0 \cdot 2270 \times 10^7$
$T_2 = \text{sum} = (3)+(4)$	Lb.in.	$3 \cdot 6143 \times 10^7$	$3 \cdot 4765 \times 10^7$	$3 \cdot 2937 \times 10^7$	$2 \cdot 8987 \times 10^7$	$-1 \cdot 0386 \times 10^7$

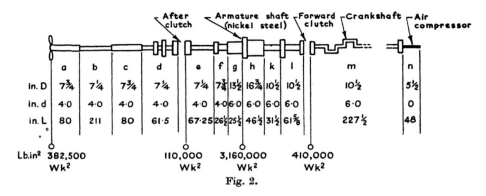

Fig. 2.

Fig. 3. Table 1 (*cont.*)

f	g	h	k	l	m	n
—	—	$8\cdot18\times10^3$	—	—	$1\cdot062\times10^3$	—
		$1\cdot12\times10^7$			← $1\cdot2\times10^7$ →	
329·0	3134·0	7600·0	1066·0	1066·0	1066·0	90·0
0·540	0·224	0·279	0·526	0·375	3·51	3·48
$0\cdot891\times10^{-3}$	$0\cdot574\times10^{-3}$	$0\cdot640\times10^{-3}$	$0\cdot880\times10^{-3}$	$0\cdot744\times10^{-3}$	$2\cdot195\times10^{-3}$	$2\cdot182\times10^{-3}$
26·5	25·5	46·5	31·5	61·625	227·5	48·0
1° 21′	0° 50′	1° 42′	1° 35′	2° 38′	28° 36′	6° 0′
0·9997	0·9999	0·9996	0·9996	0·9989	0·8821	0·9945
0·0235	0·0145	0·0297	0·0276	0·0459	0·4710	0·1045
$3\cdot280\times10^6$	$20\cdot18\times10^6$	$54\cdot5\times10^6$	$10\cdot5\times10^6$	$8\cdot875\times10^6$	$28\cdot10\times10^6$	$2\cdot36\times10^6$
−0·0232	−0·0728	−0·0778	−0·0793	−0·0864	−0·1004	−0·1215
$6\cdot924\times10^6$	$6\cdot920\times10^6$	$6\cdot898\times10^6$	$2\cdot719\times10^6$	$2\cdot695\times10^6$	$2\cdot657\times10^6$	$0\cdot413\times10^6$
—	—	$-4\cdot052\times10^6$	—	—	$-0\cdot680\times10^6$	—
$6\cdot924\times10^6$	$6\cdot920\times10^6$	$2\cdot846\times10^6$	$2\cdot719\times10^6$	$2\cdot695\times10^6$	$1\cdot977\times10^6$	$0\cdot413\times10^6$
−0·0232	−0·0728	−0·0778	−0·0793	−0·0864	−0·0885	−0·1215
0·0496	0·0050	0·0015	0·0071	0·0140	0·0330	0·0183
−0·0728	−0·0778	−0·0793	−0·0864	−0·1004	−0·1215	−0·1388
$6\cdot922\times10^6$	$6\cdot919\times10^6$	$2\cdot845\times10^6$	$2\cdot718\times10^6$	$2\cdot692\times10^6$	$1\cdot741\times10^6$	$0\cdot410\times10^6$
$-0\cdot002\times10^6$	$-0\cdot021\times10^6$	$-0\cdot126\times10^6$	$-0\cdot023\times10^6$	$-0\cdot035\times10^6$	$-1\cdot328\times10^6$	$-0\cdot030\times10^6$
$6\cdot920\times10^6$	$6\cdot898\times10^6$	$2\cdot719\times10^6$	$2\cdot695\times10^6$	$2\cdot657\times10^6$	$0\cdot413\times10^6$	$0\cdot380\times10^6$

Fig. 4. Table 2 (*cont.*)

f	g	h	k	l	m	n
—	—	$8\cdot18\times10^3$	—	—	$1\cdot062\times10^3$	—
		$1\cdot12\times10^7$			← $1\cdot2\times10^7$ →	
329·0	3134·0	7600·0	1066·0	1066·0	1066·0	90·0
0·540	0·224	0·279	0·526	0·375	3·51	3·48
$2\cdot115\times10^{-3}$	$1\cdot363\times10^{-3}$	$1\cdot521\times10^{-3}$	$2\cdot085\times10^{-3}$	$1\cdot704\times10^{-3}$	$5\cdot21\times10^{-3}$	$5\cdot19\times10^{-3}$
26·5	25·5	46·5	31·5	61·625	227·5	48
3° 13′	1° 59′	4° 3′	3° 46′	6° 1′	67° 54′	14° 16′
0·9984	0·9994	0·9975	0·9978	0·9945	0·3762	0·9692
0·0561	0·0346	0·0706	0·0657	0·1048	0·9265	0·2464
$0\cdot779\times10^7$	$4\cdot78\times10^7$	$12\cdot94\times10^7$	$2\cdot49\times10^7$	$2\cdot18\times10^7$	$6\cdot67\times10^7$	$0\cdot561\times10^7$
−3·4127	−3·3325	−3·3214	−2·77	−0·064	4·8663	13·409
$-1\cdot0386\times10^7$	$-1\cdot1859\times10^7$	$-1\cdot7362\times10^7$	$-102\cdot3\times10^7$	$-102\cdot55\times10^7$	$-102\cdot0\times10^7$	$-1\cdot34\times10^7$
—	—	$-97\cdot750\times10^7$	—	—	$18\cdot6\times10^7$	—
$-1\cdot0386\times10^7$	$-1\cdot1859\times10^7$	$-99\cdot50\times10^7$	$-102\cdot3\times10^7$	$-102\cdot55\times10^7$	$-83\cdot4\times10^7$	$-1\cdot34\times10^7$
−3·4073	−3·3300	−3·313	−2·764	−0·0637	1·829	12·990
−0·0748	−0·0086	−0·543	−2·700	−4·9300	−11·580	−0·589
−3·3325	−3·3214	−2·770	−0·064	−4·8663	13·409	13·579
$-1\cdot0369\times10^7$	$-1\cdot1852\times10^7$	$-99\cdot250\times10^7$	$-102\cdot10\times10^7$	$-102\cdot000\times10^7$	$-31\cdot40\times10^7$	$-1\cdot299\times10^7$
$-0\cdot1490\times10^7$	$-0\cdot5510\times10^7$	$-3\cdot035\times10^7$	$-0\cdot45\times10^7$	$-0\cdot015\times10^7$	$30\cdot06\times10^7$	$-1\cdot852\times10^7$
$-1\cdot1859\times10^7$	$-1\cdot7362\times10^7$	$-102\cdot285\times10^7$	$-102\cdot55\times10^7$	$-102\cdot015\times10^7$	$-1\cdot34\times10^7$	$+0\cdot553\times10^7$

[203]

Derivation. We consider a portion of shaft subjected to torsional vibration (see Fig. 5). If J = mass moment of inertia of shaft, and L = shaft length, its moment of inertia per unit length is J/L. The inertia torque dT of a shaft element of length dx is

$$dT = dJ \times \frac{\partial^2\theta}{\partial t^2} = \frac{J\,dx}{L}\frac{\partial^2\theta}{\partial t^2}, \quad \text{so that} \quad \frac{\partial T}{\partial x} = \frac{J}{L}\frac{\partial^2\theta}{\partial t^2}.$$

Assuming sinusoidal vibration, we have $\partial^2\theta/\partial t^2 = -\omega^2\theta$ at any instant (ω being the phase velocity of the vibration). The shaft element twists through an angle $d\theta$ over its length dx, when subject to a torque T. Therefore its flexibility is

$$\frac{d\theta}{T} = \frac{dx}{GI_p} \quad \text{or} \quad \frac{\partial\theta}{\partial x} = \frac{T}{GI_p},$$

where G = modulus of rigidity, and I_p = second moment of area of the shaft cross-section. From these expressions we obtain

$$\frac{\partial T}{\partial x} = GI_p\frac{\partial^2\theta}{\partial x^2} = -\omega^2\theta\frac{J}{L}.$$

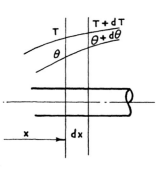

Fig. 5.

As $J = I_p L\rho$ (where ρ = mass density w/g of the shaft), this equation reduces to

$$\frac{\partial^2\theta}{\partial x^2} = -\omega^2\frac{\rho}{G}\theta = -\beta^2\theta,$$

where $\beta = \omega\sqrt{(\rho/G)}$. This equation has the solution $\theta = \theta_0\cos(\beta x + \gamma)$, where θ_0 and γ are constants depending on the end conditions. Thus the torque at x is

$$T = GI_p\frac{\partial\theta}{\partial x} = -\theta_0\beta GI_p\sin(\beta x + \gamma).$$

These relations apply at any section of a uniform portion of shaft, whether the vibration is 'natural' or 'forced'. Under 'natural' (i.e. free) vibration conditions, there are no end torques, so that $T = 0$ at each end.

First application. Uniform shaft of length L. Under free vibration conditions $T = 0$ at $x = 0$ and $x = L$. Hence

$$\text{at } x = 0: \quad T = \sin\gamma = 0 \quad \text{or} \quad \gamma = 0,\ \pi,\ 2\pi,\ \text{etc.}$$

$$\text{at } x = L: \quad T = \sin\beta L = 0 \quad \text{or} \quad \beta L = 0,\ \pi,\ 2\pi,\ \text{etc.}$$

[204]

For first mode (see Fig. 6):

$$\beta L = \pi = \omega \sqrt{(\rho/G)} \times L. \quad \text{Hence} \quad F_{\mathrm{I}} = \frac{\omega}{2\pi} = \frac{1}{2L}\sqrt{(G/\rho)} \quad \text{[cycles/sec.]}.$$

For second mode:

$$\beta L = 2\pi; \quad F_{\mathrm{II}} = \frac{1}{L}\sqrt{(G/\rho)} \quad \text{[cycles/sec.]}.$$

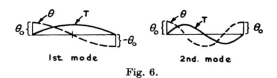

1st. mode   2nd. mode

Fig. 6.

Second application. Shafting of varying section, with a number of concentrated masses (see Fig. 7).

Having found that

$$\theta = \theta_0 \cos (\beta x + \gamma),$$

$$T = -\theta_0 \beta G I_p \sin (\beta x + \gamma),$$

we can establish formulae for θ_2 and T_2 at $x = L$, when θ_1 and T_1 at $x = 0$ have known values.

Disregarding, for convenience, the minus sign of torque, we write *for position* 1 $(x = 0)$:

$$\theta = \theta_1 = \theta_0 \cos \gamma, \quad T = T_1 = \theta_0 \beta G I_p \sin \gamma,$$

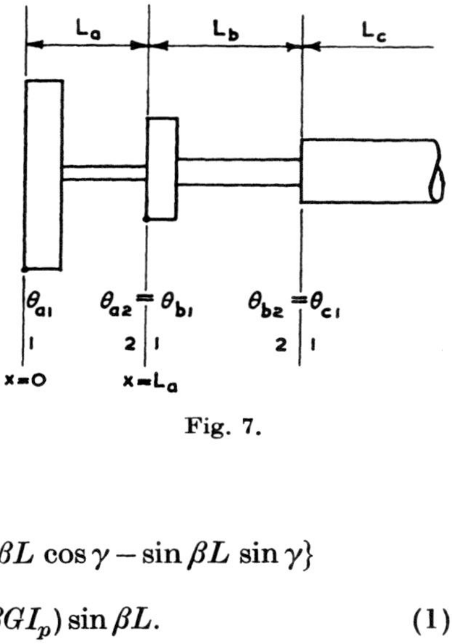

and *for position* 2 $(x = L)$:

$$\theta = \theta_2 = \theta_0 \cos (\beta L + \gamma) = \theta_0\{\cos \beta L \cos \gamma - \sin \beta L \sin \gamma\}$$

or
$$\theta_2 = \theta_1 \cos \beta L - (T_1/\beta G I_p) \sin \beta L. \qquad (1)$$

Furthermore
$$T_2 = \theta_0 \beta G I_p\{\sin \beta L \cos \gamma + \cos \beta L \sin \gamma\},$$

or
$$T_2 = \beta G I_p \theta_1 \sin \beta L + T_1 \cos \beta L. \qquad (2)$$

Equations (1) and (2) are the required formulae, which are used in the tabulations as indicated at the beginning of this section. The procedure consists in assuming a value for the frequency F, calculating β for each length, assuming $\theta_{1(a)} = \Delta_{1(a)} = 1$ rad. for the left-hand end of the first shaft section and making repeated use of the above formulae.

Fig. 7.

For position 1 of shaft L_a, the torque is $T_1 = 0 + \delta T$ (where δT is the inertia torque) since there is no torque from the left of this section.

For position 1 of shaft L_b, the torque is $T_{1(b)} = T_{2(a)} + \delta T_{(b)}$, etc. If F corresponds to a natural frequency of the system, the torque T_2 at the end of the last shaft section should be equal to zero.

1·33   Equivalent-mass or reduced-inertia method

1. Reduction of masses

A moment of inertia J_{1A} can be replaced by an equivalent moment of inertia J_{1B} at another position B (see Fig. 1), without altering the vibration frequency of the system. This may be seen from the following considerations.

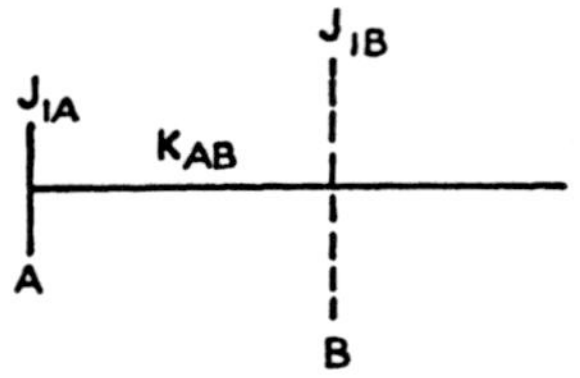

The inertia torque of J_{1A} at A is

$$T_A = J_{1A}\omega^2\Delta_A,$$

where $\omega =$ phase velocity of vibration, and $\Delta_A =$ amplitude of vibration of J_{1A} at position A. If there is no other inertia mass on the shafting between points A and B, the inertia torque at B must be the same as at A, so that $T_B = T_A$.

Fig. 1.

At B the vibration amplitude is Δ_B, hence the equivalent value of J_{1A} at B must be such that

$$J_{1B}\omega^2\Delta_B = J_{1A}\omega^2\Delta_A. \tag{1}$$

If $K_{AB} =$ stiffness of shafting between A and B, the maximum shaft torque must be equal to the maximum inertia torque, that is,

$$J_{1A}\omega^2\Delta_A = K_{AB}(\Delta_A - \Delta_B). \tag{2}$$

From these relations we obtain

$$\frac{J_{1B}}{J_{1A}} = \frac{\Delta_A}{\Delta_B} \quad \text{and} \quad J_{1A}\omega^2 = K_{AB}\left(1 - \frac{\Delta_B}{\Delta_A}\right),$$

so that

$$\frac{\Delta_B}{\Delta_A} = 1 - \frac{J_{1A}\omega^2}{K_{AB}},$$

from which

$$J_{1B} = J_{1A}\frac{\Delta_A}{\Delta_B} = \frac{J_{1A}}{1 - \dfrac{J_{1A}\omega^2}{K_{AB}}}. \tag{3}$$

If there is already a mass J_{2B} at B, the total inertia at B is

$$J_B' = J_{2B} + J_{1B} = J_{2B} + \frac{J_{1A}}{1 - \dfrac{J_{1A}\omega^2}{K_{AB}}}. \tag{4}$$

[206]

We have thus reduced the system of N masses to a system of $N-1$ masses, without altering its vibration conditions as regards relative vibration amplitudes and inertia torque.

Continuing this procedure, if J'_B is to be reduced to a position C further along the shaft (see Fig. 2), we have

$$J_C = \frac{J'_B}{1 - \dfrac{J'_B \omega^2}{K_{BC}}},$$

where K_{BC} is the shaft stiffness between B and C. If there is already an inertia J_{3C} at C, the total inertia at this position is

$$J'_C = J_{3C} + J_C = J_{3C} + \frac{J'_B}{1 - \dfrac{J'_B \omega^2}{K_{BC}}}. \qquad (5)$$

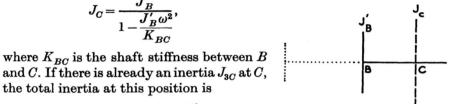

Fig. 2.

2. Natural frequency determination

The mass-reduction method can be used for determining natural frequencies, particularly when the engine system comprises a large number of identical inertia and stiffness values.

When $J_1 = J_2 = J_3 = J_{\text{cyl.}}$ and $K_{AB} = K_{BC} = K_{\text{cyl.}}$, the above equations (4) and (5) simplify as follows:

$$J'_B = J_{\text{cyl.}} \left[1 + \frac{1}{1 - \dfrac{\omega^2 J_{\text{cyl.}}}{K_{\text{cyl.}}}} \right] = J_{\text{cyl.}} \left[1 + \frac{1}{1 - \delta} \right],$$

where $\delta = \omega^2/\omega_0^2$ and $\omega_0^2 = K_{\text{cyl.}}/J_{\text{cyl.}}$, so that

$$J'_B = J_{\text{cyl.}} \times \left[\frac{2 - \delta}{1 - \delta} \right] = \frac{J_{\text{cyl.}}}{\Phi_2}, \qquad (6)$$

$\Phi_2 = (1 - \delta)/(2 - \delta)$ being termed the 'equivalence function'. Furthermore,

$$J'_C = J_{\text{cyl.}} \times \left[1 + \frac{\dfrac{2 - \delta}{1 - \delta}}{1 - \dfrac{J_{\text{cyl.}} \omega^2}{K_{\text{cyl.}}} \times \dfrac{2 - \delta}{1 - \delta}} \right],$$

$$J_{\text{cyl.}} \times \left[1 + \frac{\dfrac{2 - \delta}{1 - \delta}}{1 - \dfrac{\delta(2 - \delta)}{1 - \delta}} \right] = J'_C = J_{\text{cyl.}} \times \frac{3 - 4\delta + \delta^2}{1 - 3\delta + \delta^2} = \frac{J_{\text{cyl.}}}{\Phi_3}. \qquad (7)$$

Generally, for reducing N cylinders to the Nth (last cylinder) position,

[207]

the equivalence function to be used is obtained from the recurrence formula:

$$\Phi_N = \frac{\Phi_{N-1} - \delta + 1}{\Phi_{N-1} - \delta},$$

so that $J'_N = J_{\text{cyl.}}/\Phi_N$.

EXAMPLE. Reduction to a two-mass system of the engine shown in Fig. 3, which has the following values:

$$J_1 = J_2 = J_3 = 1 \cdot 0 \text{ Lb.in.sec.}^2, \quad J_4 = 50 \text{ Lb.in.sec.}^2,$$

$$K_1 = K_2 = 10 \times 10^6 \text{ Lb.in./rad.}, \quad K_3 = 5 \times 10^6 \text{ Lb.in./rad.}$$

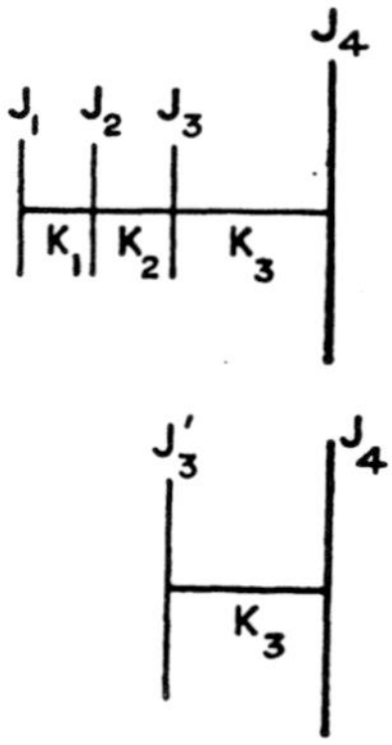

Using equation (7), and noting that

$$\delta = \frac{\omega^2}{K_{\text{cyl.}}/J_{\text{cyl.}}} = \frac{\omega^2}{10^7},$$

we obtain

$$J'_3 = J_{\text{cyl.}} \times \frac{3 - 4\delta + \delta^2}{1 - 3\delta + \delta^2} = 1 \times \frac{3 - 4\dfrac{\omega^2}{10^7} + \left(\dfrac{\omega^2}{10^7}\right)^2}{1 - 3\dfrac{\omega^2}{10^7} + \left(\dfrac{\omega^2}{10^7}\right)^2}.$$

The frequency equation for the reduced two-mass system is, therefore,

$$\omega^2 = \frac{K_3}{J'_3} + \frac{K_3}{J_4} = 5 \times 10^6 \left[\frac{1 - \dfrac{3\omega^2}{10^7} + \left(\dfrac{\omega^2}{10^7}\right)^2}{3 - \dfrac{4\omega^2}{10^7} + \left(\dfrac{\omega}{10^7}\right)^2} + \frac{1}{50} \right].$$

Fig. 3.

This equation can be solved by trial and error until the chosen value for ω^2 is equal to the resultant value of the right-hand side of this equation, using graphs for interpolation.

Note. Calculations of this kind can also be used in conjunction with 'effective-inertia diagrams' (see section 1·342). Apart from the bibliography given in section 1·36, reference can be made to the following:

Budinsky, F., *Méthode pratique pour déterminer la fréquence des vibrations propres des moteurs à cylindres en ligne.*

Ker Wilson, W., *Practical Solution of Torsional Vibration Problems* (Chapman and Hall, 1941), vol. 1, pp. 209–12.

Kleiner, A., *Sulzer Tech. Rev.* nos. 3–4 (1947), pp. 23–32.

Porter, F. P. *Practical Determination of Torsional Vibration in an Engine Installation which may be Simplified to a Two-mass System,* A.S.M.E. Paper AMP-51-22, 1929.

1·34 Graphical and semi-graphical methods

1·341 *Graphical equivalent of Holzer's method*

Graphical methods are sometimes used to estimate the natural frequencies of a system. The general procedure may be described as follows:

Use is made of three diagrams, viz. the 'deflexion diagram', the 'slope diagram' and the 'torque diagram'. The values considered in

[208]

the system are its moments of inertia J [Lb.in.sec.²] and equivalent shaft lengths L_e [in.].

(1) For the construction of the deflexion diagram, draw the engine system as shown in Fig. 1, using a convenient scale to represent the relative values of the equivalent lengths.

(2) With an assumed value for ω^2, calculate the base length

$$L_0 = \frac{G\frac{\pi}{32}D_e^4}{\omega^2 J_{\text{cyl.}}}\text{ in.,}$$

where G = modulus of rigidity [Lb./in.²], D_e = equivalent shaft diameter [in.], $J_{\text{cyl.}}$ = moment of inertia per line [Lb.in.sec.²] and $\omega = F/9{\cdot}55$, F being the assumed value for vibration frequency [vib./min.].

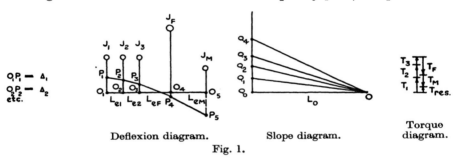

Deflexion diagram.　　Slope diagram.　　Torque diagram.

Fig. 1.

(3) Using the same scale for lengths as in the deflexion diagram, draw the base-line $L_0 = OQ_0$ for the slope diagram.

(4) Beginning at J_1, in the deflexion diagram, erect a vertical line $O_1P_1 = \Delta_1 = 1$ (of unit amplitude, using a convenient scale). Erect a line Q_0Q_1 of the same height Δ_1, at point Q_0 of the slope diagram. Join point Q_1 to O.

(5) In the deflexion diagram, draw through point P_1 a line having the same slope as the line Q_1O. This line P_1P_2 cuts the vertical at the position of J_2 at P_2.

(6) In the slope diagram, draw a vertical line of length $Q_1Q_2 = O_2P_2$, in the same direction as O_2P_2. This gives a further point Q_2. Join Q_2 to O.

(7) In the deflexion diagram, draw from point P_2 a line with the same slope as Q_2O. This intersects the vertical of J_3 at P_3.

(8) Repeat this procedure so as to obtain the parallel lines Q_3O and P_3P_4 in the two diagrams.

(9) If J_4 has a value differing from that of $J_{\text{cyl.}}$, calculate

$$\overline{O_4P_4} \times J_4/J_{\text{cyl.}} = \overline{Q_3Q_4},$$

and draw a vertical line of this length, upwards, in the slope diagram. Then, in the deflexion diagram, draw P_4P_5 parallel to Q_4O.

(10) Construct the torque diagram as follows: $T_1 = Q_0Q_1$, $T_2 = Q_1Q_2$, $T_3 = Q_2Q_3$. After T_3 the torque is negative, hence, from the tip of T_3 in the downward direction, draw $T_F = Q_3Q_4$. Then calculate

$$T_5 = T_M = \overline{O_5 P_5} \times J_M / J_{\text{cyl}}.$$

and plot this downwards again, from the tip of T_F. If the residual torque $T_{\text{res.}}$ (see Fig. 1) is zero, the value chosen for ω is that of a natural frequency of the system. If not, the entire procedure (items 2–10) should be repeated, with another value of ω.

Note. The $Q_n Q_{n+1}$ lines are plotted downwards when the total $\sum\limits_{1}^{n} T$ is negative, and upwards so long as it is positive. Provided this rule is observed, the method is applicable to two-node, three-node vibrations, etc.

Derivation. This method is based on the same general considerations as Holzer's tabulation method. The shaft torque between masses J_1 and J_2 is

$$T_{\text{shaft}} = \frac{\pi}{32} D_e^4 G \frac{\Delta_1 - \Delta_2}{L_{e1}} \quad [\text{Lb.in.}]$$

and is equal to the inertia torque of mass J_1, $T_{\text{inertia}} = J_1 \omega^2 \Delta_1 = J_{\text{cyl.}} \omega^2 \Delta_1$, so that

$$\frac{\Delta_1 - \Delta_2}{L_{e1}} = \frac{\Delta_1}{\dfrac{\pi}{32} GD_e^4/(\omega^2 J_{\text{cyl.}})} = \frac{\Delta_1}{L_0}.$$

This shows that, in the two corresponding diagrams of Fig. 1, the slope of the line $P_1 P_2$ must be the same as that of $Q_1 O$.

For the slope of the next line, $P_2 P_3$, we have

$$\frac{\Delta_2 - \Delta_3}{L_{e2}} = \frac{\Delta_2 + \Delta_1}{\dfrac{\pi}{32} GD_e^4/J_{\text{cyl.}} \, \omega^2} = \frac{\Delta_1 + \Delta_2}{L_0},$$

and the reasoning is the same as long as the inertias have the same value $J_{\text{cyl.}} = J_{1,2\,3}$. If the inertia considered is not equal to $J_{\text{cyl.}}$, as in the case of mass J_4 considered above, we have

$$T_{\text{shaft}} = \frac{\pi}{32} D_e^4 G \frac{\Delta_4 - \Delta_5}{L_{e4}} = J_4 \omega^2 \Delta_4 + J_{\text{cyl.}} \omega^2 (\Delta_1 + \Delta_2 + \Delta_3).$$

Therefore, the slope in this case is given by

$$\frac{\Delta_4 - \Delta_5}{L_{e4}} = \frac{J_4 \omega^2 \Delta_4}{\dfrac{\pi}{32} GD_e^4} + \frac{\Delta_1 + \Delta_2 + \Delta_3}{L_0} = \frac{J_4}{J_{\text{cyl.}}} \times \frac{\Delta_4}{\dfrac{\pi}{32} GD_e^4 \times \dfrac{1}{J_{\text{cyl.}} \omega^2}} + \frac{\Delta_1 + \Delta_2 + \Delta_3}{L_0}$$

$$= \frac{J_4}{J_{\text{cyl.}}} \times \frac{\Delta_4}{L_0} + \frac{\Delta_1 + \Delta_2 + \Delta_3}{L_0} = \frac{1}{L_0}\left[\Delta_1 + \Delta_2 + \Delta_3 + \frac{J_4}{J_{\text{cyl.}}} \times \Delta_4\right].$$

the effective inertia $J_{3p}(B)'$ required to obtain any frequency F_B would be correspondingly smaller, i.e.

$$J_{3p}(B)' = J_{3p}(B) - J_2.$$

On the other hand, if J_2 were included at p in the subsidiary system, the effective inertia for any frequency F_S would also be reduced to

$$J_{1p}(S)' = J_{1p}(S) - J_2 = -J_{3p}(S) - J_2.$$

In order to maintain torque equilibrium, $J_{1p}(S)'$ of the subsidiary system must correspond to a modified effective inertia $J_{3p}(S)'$ for the basic system, i.e.

$$J_{1p}(S)' = -J_{3p}(S)',$$

so that

$$J_{3p}(S)' = J_{3p}(S) + J_2.$$

Instead of subtracting J_2 from the basic system diagram, it is more convenient to add it to the curves of the subsidiary system. This is achieved by *displacing the abscissa axis of the subsidiary system diagram*

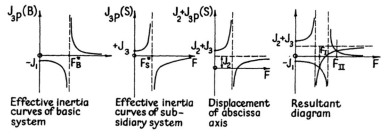

Fig. 6. Determination of natural frequency F_I and F_{II} of a three-mass system.

downwards, by an amount corresponding to J_2. After this adjustment, the latter diagram is superimposed on the diagram of the basic system, and the intersections of the curves in the resultant diagram give the natural frequencies for the two modes of vibration of the three-mass system (see Fig. 6).

Metric units. The 'effective-inertia method' is directly applicable with metric units, for instance, with J-values in Kg.cm.sec.^{2} and K-values in Kg.cm./rad.

Application to systems with a large number of masses

The n-mass system is divided into two sub-systems with a separation point p at the main flywheel (see Fig. 7), the engine stiffnesses and inertias forming the basic system, and the shafting and inertias at the driven end forming the subsidiary system.

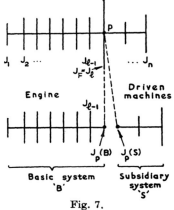

Fig. 7.

[215]

The calculation is based on the use of frequency tabulations. The effective inertia curves of $J_p(B)$ for the basic system are determined from Holzer tables for various non-resonant frequencies.

The tabulation, for any arbitrary value of frequency, is carried out in the usual manner up to the last line, which contains the unknown value J_p. This line gives the equation

$$\omega^2 J_p \Delta_p = -\sum_{k=1}^{l-1} J_k \omega^2 \Delta_k,$$

where the sum on the right-hand side is the total inertia torque due to the $l-1$ preceding masses. The basic system will vibrate at this natural

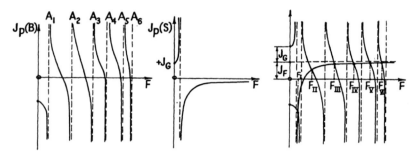

Fig. 8. Effective inertia curves of basic system.†

Fig. 9. Effective inertia curves of subsidiary system.

Fig. 10. Resultant diagram.

Figs. 8, 9 and 10. Effective-inertia curve method applied to the determination of the natural frequencies of a 6-cylinder engine with a main flywheel J_F coupled to a generator J_G.

frequency if the sum of the inertia torques is zero, so that $J_p(B)$ is obtained from

$$J_p(B) = -\left[\sum_{k=1}^{l-1} J_k \omega^2 \Delta_k\right] \Big/ (\omega^2 \Delta_p).$$

A number of tables are computed in this way for increasing values of ω. The values of $J_p(B)$ as a function of frequency are then plotted to obtain the effective inertia curves of the basic system. The number of asymptotes (for which $J_p(B) = \pm \infty$) will be equal to the number $(l-1)$ of masses in the basic system excluding the main flywheel (Fig. 8).

† In general, it is fairly easy to determine for the basic system (consisting of cylinder inertias $J_{cyl.}$ and stiffness K_{cyl} of constant value) the value of A_1, which is its one-node natural frequency. In regard to the asymptotes A_2, A_3 and A_4 for the multi-node natural frequencies, it is usually found that these can be expected to be situated approximately within the following ranges of values:

A_2	A_3	A_4
$2 \cdot 4 A_1$ to $3 \cdot 0 A_1$	$4 \cdot 0 A_1$ to $4 \cdot 5 A_1$	$5 \cdot 4 A_1$ to $6 \cdot 0 A_1$

Thus, for instance, the two-node frequency A_2 is about 2·4–3·0 times higher than the one-node frequency of the basic system. The higher values, such as 3·0, apply to engines having a large number of cylinders (e.g. 8- to 10-cylinder engines).

A similar set of calculations is made for

$$J_p(S) = -J_{1p}(S) = +\left[\sum_{k=n}^{l+1} J_k \omega^2 \Delta_k\right] \Big/ (\omega^2 \Delta_p),$$

where the Holzer tables are set out with the tail-end mass of the subsidiary system as mass no. 1. These give Fig. 9.

In plotting the effective inertia curves for the subsidiary system, the values obtained from the tables have been given opposite signs, in order to take account of equation (3), derived for two-mass systems but which is valid for any number of masses. The abscissa axis of this diagram is shifted downwards by a value corresponding to that of the main flywheel J_F and the figure is superimposed on Fig. 8. The intersections of the two sets of curves determine the natural frequencies $F_{\mathrm{I}}, F_{\mathrm{II}}, F_{\mathrm{III}}, \ldots, F_{n-1}$ of the n-mass system on the abscissa axis of the resultant diagram of Fig. 10.

Utilization of the effective inertia method

The method has the following advantages:

(1) The curves for a given type of engine need only be determined once for all applications.

(2) The resultant diagrams give the natural frequencies for all the modes of vibration included in the range of frequencies used for the determination of the curves, and these do not have to be computed for frequencies above the values normally requiring consideration for a given engine type.

(3) The effect on frequency of altering the moment of inertia of the main flywheel can be simply determined by displacing the abscissa axis of the subsidiary system diagram to the corresponding value.

The application of the method is extremely simple, as can be seen from the following numerical example.

NUMERICAL EXAMPLE. Three-cylinder engine with flywheel and brake (Fig. 11).

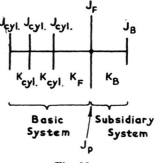

Fig. 11

Moments of inertia (Lb.ft.sec.²)	Stiffnesses (Lb.ft./rad.)†
$J_{\mathrm{cyl.}} = 0.132$	$K_{\mathrm{cyl.}} = 0.767 \times 10^6$
$J_F = 9.05$	$K_F = 0.632 \times 10^6$
$J_B = 0.528$	$K_B = 0.192 \times 10^6$

In view of the speed range of this engine only vibration frequencies up to 35,000 vib./min. have to be taken into account.

Calculation of effective inertia curves for basic system

The value of $J_p(B)$ for the basic system will be determined by means of frequency tabulations.

† *Metric units.* The above example can be regarded as being in metric units, by replacing the terms 'Lb.ft.' by 'Kg.m.', without altering the numerical values.

Table A. $F = 955 \cdot 0$ vib./min., $\omega^2 = 0 \cdot 01 \times 10^6$

Mass no.	J	$\dfrac{J\omega^2}{10^6}$	Δ	$\dfrac{J\omega^2\Delta}{10^6}$	$\dfrac{\Sigma}{10^6}$	$\dfrac{K}{10^6}$	Δ_{sh}
1	0·132	0·00132	1·0000	0·00132	0·00132	0·767	0·00172
2	0·132	0·00132	0·99828	0·001318	0·002638	0·767	0·00344
3	0·132	0·00132	0·99484	0·001314	0·003952	0·632	0·00625
J_p	J_p	$0·01 \times J_p$	0·98859	$0·009886 J_p$	0	—	—

$$0·009886 J_p + 0·003952 = 0, \quad J_p = -0·3998$$

Table B. $F = 7400 \cdot 0$ vib./min., $\omega^2 = 0 \cdot 6 \times 10^6$

Mass no.	J	$\dfrac{J\omega^2}{10^6}$	Δ	$\dfrac{J\omega^2\Delta}{10^6}$	$\dfrac{\Sigma}{10^6}$	$\dfrac{K}{10^6}$	Δ_{sh}
1	0·132	0·0792	1·0000	0·07920	0·07920	0·767	0·10326
2	0·132	0·0792	0·89674	0·07102	0·15022	0·767	0·19585
3	0·132	0·0792	0·70090	0·05551	0·20573	0·632	0·32552
J_p	J_p	$0·6 \times J_p$	0·37537	$0·22522 J_p$	0	—	—

$$0·22522 J_p + 0·20573 = 0, \quad J_p = -0·914$$

Table C. $F = 9550$ vib./min., $\omega^2 = 1 \cdot 0 \times 10^6$

Mass no.	J	$\dfrac{J\omega^2}{10^6}$	Δ	$\dfrac{J\omega^2\Delta}{10^6}$	$\dfrac{\Sigma}{10^6}$	$\dfrac{K}{10^6}$	Δ_{sh}
1	0·132	0·132	1·0000	0·13200	0·13200	0·767	0·1721
2	0·132	0·132	0·8279	0·10930	0·24130	0·767	0·3146
3	0·132	0·132	0·5133	0·06776	0·30906	0·632	0·48902
J_p	J_p	$1 \times J_p$	0·0243	$0·0243 J_p$	0	—	—

$$0·0243 J_p + 0·30906 = 0, \quad J_p = -12·7185$$

Table D. $F = 10{,}505$ vib./min., $\omega^2 = 1 \cdot 21 \times 10^6$

Mass no.	J	$\dfrac{J\omega^2}{10^6}$	Δ	$\dfrac{J\omega^2\Delta}{10^6}$	$\dfrac{\Sigma}{10^6}$	$\dfrac{K}{10^6}$	Δ_{sh}
1	0·132	0·15972	1·00000	0·15972	0·15972	0·767	0·20824
2	0·132	0·15972	0·79176	0·12646	0·28618	0·767	0·37312
3	0·132	0·15972	0·41864	0·06687	0·35305	0·632	0·558628
J_p	J_p	$1·21 J_p$	−0·13998	$−0·16938 J_p$	0	—	—

$$−0·16938 J_p + 0·35305 = 0, \quad J_p = +2·0844$$

Table E. $F = 11{,}460$ vib./min., $\omega^2 = 1 \cdot 44 \times 10^6$

Mass no.	J	$\dfrac{J\omega^2}{10^6}$	Δ	$\dfrac{J\omega^2\Delta}{10^6}$	$\dfrac{\Sigma}{10^6}$	$\dfrac{K}{10^6}$	Δ_{sh}
1	0·132	0·19008	1·0000	0·19008	0·19008	0·767	0·24782
2	0·132	0·19008	0·75218	0·14297	0·33305	0·767	0·43422
3	0·132	0·19008	0·31796	0·06044	0·39349	0·632	0·62261
J_p	J_p	$1·44 J_p$	−0·30465	$−0·4387 J_p$	0	—	—

$$−0·4387 J_p + 0·39349 = 0, \quad J_p = +0·8970$$

Table F. $F = 22{,}920$ vib./min., $\omega^2 = 5 \cdot 76 \times 10^6$

Mass no.	J	$\dfrac{J\omega^2}{10^6}$	Δ	$\dfrac{J\omega^2\Delta}{10^6}$	$\dfrac{\Sigma}{10^6}$	$\dfrac{K}{10^6}$	Δ_{sh}
1	0·132	0·76032	1·0000	0·76032	0·76032	0·767	0·99129
2	0·132	0·76032	0·00871	0·00662	0·76694	0·767	0·99992
3	0·132	0·76032	−0·99121	−0·75364	+0·01330	0·632	0·021044
J_p	J_p	$5·76 J_p$	−1·01265	$−5·833 J_p$	0	—	—

$$−5·833 J_p + 0·01330 = 0, \quad J_p = +0·00228$$

Table G. $F = 26{,}200$ vib./min., $\omega^2 = 7.5625 \times 10^6$

Mass no.	J	$\dfrac{J\omega^2}{10^6}$	Δ	$\dfrac{J\omega^2\Delta}{10^6}$	$\dfrac{\Sigma}{10^6}$	$\dfrac{K}{10^6}$	Δ_{sh}
1	0.132	0.99825	1.0000	0.99825	0.99825	0.767	1.30150
2	0.132	0.99825	−0.30150	−0.30097	+0.69728	0.767	0.90910
3	0.132	0.99825	−1.21060	−1.20848	−0.51120	0.632	−0.80886
J_p	J_p	$7.5625 J_p$	−0.40174	$-3.03816 J_p$	0	—	

$$-3.03816 J_p - 0.51120 = 0, \quad J_p = -0.1683$$

Table H. $F = 27{,}200$ vib./min., $\omega^2 = 8.1225 \times 10^6$

Mass no.	J	$\dfrac{J\omega^2}{10^6}$	Δ	$\dfrac{J\omega^2\Delta}{10^6}$	$\dfrac{\Sigma}{10^6}$	$\dfrac{K}{10^6}$	Δ_{sh}
1	0.132	1.07217	1.0000	1.07217	1.07217	0.767	1.39788
2	0.132	1.07217	−0.39788	−0.42660	+0.64951	0.767	0.84682
3	0.132	1.07217	−1.24470	−1.33453	−0.68502	0.632	−1.08389
J_p	J_p	$8.1225 J_p$	−0.16081	$-1.30618 J_p$	0	—	

$$-1.30618 J_p - 0.68502 = 0, \quad J_p = -0.5244$$

Table I. $F = 28{,}000$ vib./min., $\omega^2 = 8.6436 \times 10^6$

Mass no.	J	$\dfrac{J\omega^2}{10^6}$	Δ	$\dfrac{J\omega^2\Delta}{10^6}$	$\dfrac{\Sigma}{10^6}$	$\dfrac{K}{10^6}$	Δ_{sh}
1	0.132	1.14096	1.0000	1.14096	1.14096	0.767	1.48756
2	0.132	1.14096	−0.48756	−0.55629	+0.58467	0.767	0.76228
3	0.132	1.14096	−1.24984	−1.42602	−0.84135	0.632	−1.33125
J_p	J_p	$8.6436 J_p$	+0.08141	$+0.70368 J_p$	0	—	

$$0.70368 J_p - 0.84135 = 0, \quad J_p = +1.1956$$

Table J. $F = 29{,}600$ vib./min., $\omega^2 = 9.61 \times 10^6$

Mass no.	J	$\dfrac{J\omega^2}{10^6}$	Δ	$\dfrac{J\omega^2\Delta}{10^6}$	$\dfrac{\Sigma}{10^6}$	$\dfrac{K}{10^6}$	Δ_{sh}
1	0.132	1.26852	1.0000	1.26852	1.26852	0.767	1.65387
2	0.132	1.26852	−0.65387	−0.82945	+0.43912	0.767	0.57252
3	0.132	1.26852	−1.22639	−1.55570	−1.11658	0.632	−1.76674
J_p	J_p	$9.61 J_p$	+0.53835	$+5.17354 J_p$	0	—	

$$+5.17354 J_p - 1.11658 = 0, \quad J_p = +0.2158$$

The values of $J_p(B)$ as a function of frequency are plotted in Fig. 12. The calculation of $J_p(S)$ for the subsidiary system is easily carried out, and is given in the table on p. 220:

$$J_p(S) = \frac{J_{WB}}{1 - \dfrac{J_{WB}}{K_{WB}}\omega^2} = \frac{0.528}{1 - 2.75 \times 10^{-6} \times \omega^2},$$

since $J_{WB} = 0.528$ Lb.ft.sec.^{2}, $K_{WB} = 0.192 \times 10^6$ Lb.in./rad.

The change of sign of J_p from $+14.08$ to -5.28 in the table indicates that we have passed from the first to the second branch of the $J_p(S)$-curve. The value of the flywheel inertia J_F is added to each J_p-value and the results are plotted against frequency as shown in Fig. 12.

[219]

$\omega^2/10^6$ ...	0	0·1	0·2	0·3	0·35	0·4
ω ...	0	316	446	547	590	632
F [vib./min.] ...	0	3,015	4,260	5,210	5,620	6,020
$2·75 \times 10^{-6}\omega^2 = x$	0	0·275	0·550	0·825	0·963	1·100
$1-x$	1	+0·725	+ 0·450	+ 0·175	+ 0·0375	−0·100
J_p	0·53	+0·73	+ 1·17	+ 3·01	+14·08	−5·28
$J_p + J_F$	+9·58	+9·78	+10·22	+12·06	+33·13	+3·77

$\omega^2/10^6$ ...	0·45	0·50	0·70	1·25	3·0
ω ...	669	705	835	1,114	1,732
F [vib./min.] ...	6,380	6,720	7,950	10,600	16,500
$2·75 \times 10^{-6}\omega^2 = x$	1·238	1·375	1·925	4·297	24·75
$1-x$	−0·238	−0·375	− 0·925	− 3·297	−23·75
J_p	−2·22	−1·41	− 0·84	− 0·16	− 0·02
$J_p + J_F$	+6·83	+7·64	+ 8·21	+ 8·89	+ 9·03

$$(J_F = 9·05)$$

From the intersections of the two sets of curves in the resultant diagram of Fig. 12, the natural frequencies of the entire system are determined. The values for the first three modes of vibration are:

$$F_\mathrm{I} = 5700 \text{ vib./min.}, \quad F_\mathrm{II} = 9800 \text{ vib./min.} \quad \text{and} \quad F_\mathrm{III} = 28,500 \text{ vib./min.}$$

Increased accuracy could have been obtained by plotting the results to a larger scale. With the scale used in the figure, the reading accuracy is $2-2\frac{1}{2}$%.

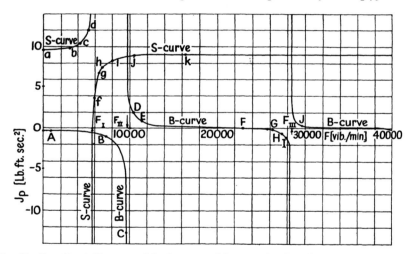

Fig. 12. Resultant diagram with the natural frequencies F_I, F_II and F_III corresponding to the first three modes of vibration.
The calculated points A to J for the basic curves, and a to k for the subsidiary curves are also indicated.

Significance of Asymptotes and Zeros in Effective-Inertia Curves.

Asymptotes occur at frequencies such that the effective inertia at the point of separation is infinite, so that for such frequencies the sub-system can be regarded as clamped at that point.

If the point of separation happens to be chosen so that, at a particular frequency, asymptotes occur in both the basic and subsidiary curves, this merely indicates that the point of separation is a node in the complete system as well as in each component system, and the corresponding frequency is a natural one as a whole and for each of the component systems considered clamped at the point of intersection.

If the point of separation happens to be chosen at one of the masses, for instance J_F in the example, then the foregoing indicates that, for that particular frequency, the flywheel could be removed or varied in inertia without in any way affecting the natural frequency of the whole system.

1·3422 *Gear-branched systems*

The method of effective inertias is particularly suitable for determining the natural frequencies of gear-branch systems, for which the usual Holzer table method involves much more labour.

The engine portion is taken as the basic system. The separation point p is taken at the end of the engine shaft.

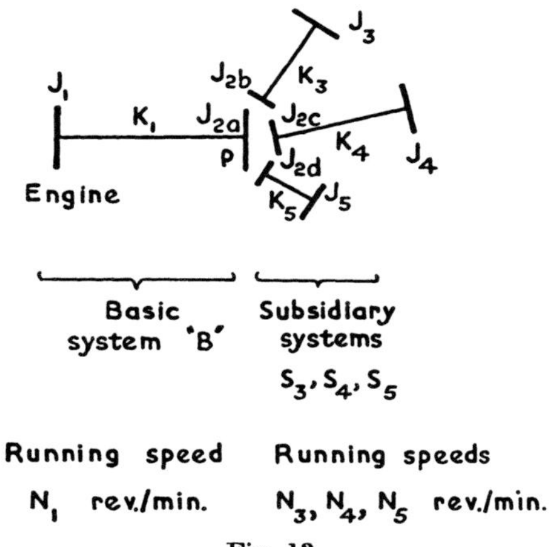

Fig. 13.

The stiffnesses and inertias are reduced to their equivalent values if the gear ratios are different from unity:

$$K_{e3} = K_3\left(\frac{N_3}{N_1}\right)^2, \quad J_{e3} = J_3\left(\frac{N_3}{N_1}\right)^2,$$

$$K_{e4} = K_4\left(\frac{N_4}{N_1}\right)^2, \quad J_{e4} = J_4\left(\frac{N_4}{N_1}\right)^2,$$

$$K_{e5} = K_5\left(\frac{N_5}{N_1}\right)^2, \quad J_{e5} = J_5\left(\frac{N_5}{N_1}\right)^2,$$

$$J_{e2} = J_{2a} + J_{2b}\left(\frac{N_3}{N_1}\right)^2 + J_{2c}\left(\frac{N_4}{N_1}\right)^2 + J_{2d}\left(\frac{N_5}{N_1}\right)^2.$$

The effective inertia diagram of the basic system is determined as usual, by means of frequency tabulations.

The effective inertia curves of each subsidiary system S_3, S_4, S_5 are determined separately, and their resultant effective inertia $J_{p\,\text{res.}}$ is obtained as the sum of the individual $J_p(S)$-values for each value of frequency

$$J_{p\,\text{res.}}(S) = J_p(S_3) + J_p(S_4) + J_p(S_5).$$

The abscissa axis of the resultant $J_{p\,\text{res.}}(S)$ curves is displaced downwards by the value corresponding to J_{e2}, and the $J_{p\,\text{res.}}$ diagram is superimposed on the effective inertia diagram of the basic system. Fig. 14 shows the graphs obtained for the four-branch system of Fig. 13, with the engine regarded as a single mass J_1 attached to the shaft K_1.

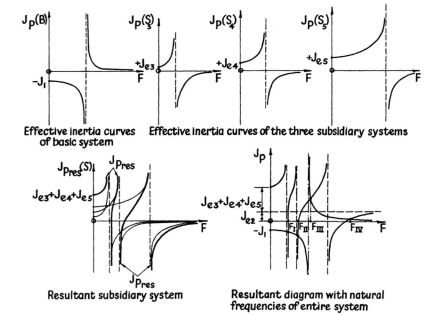

Fig. 14. Effective-inertia method applied to a four-branch system.

BIBLIOGRAPHY

The effective inertia method and other similar semi-graphical methods are described in the following references:

Behrens, H. Approximate calculation of natural frequencies of torsional vibration of multi-cylinder engines. *Werft, Reed., Hafen*, 7 Feb. 1930, pp. 50–9; 7 April 1930, pp. 141–2; 1930, p. 489; 1931, p. 94.

Grammel, R. *Ingen.-Arch.* 1931, p. 228; 1932, pp. 76, 277; 1934, p. 83.

Behrmann, W. *Werft, Reed., Hafen*, 1936, p. 41.

Kalichman, S. L. *Ann. Ass. Ing. Éc. Gand.* vol. 5 (1935), pp. 25, 57.

Geiger, J. *Mechanical Vibrations and their Measurement (Mechan. Schwingungen u. ihre Messungen)* (Springer, Berlin, 1927), p. 47.

Strunz, L. *Torsional Vibrations in Reciprocating Engines (Die Drehschwingungen in Kolbenmaschinen)* (1938), p. 76f.

Frank, B. Simplified torsional vibration calculations with the aid of equivalent masses and forces. *Ingen.-Arch.* vol. 10 (1939), pp. 371–94.

Ker Wilson, W. *Practical Solution of Torsional Vibration Problems,* vol. 1, pp. 671–707.

Manley, R. G. *Fundamentals of Vibration Study* (Chapman and Hall, 1952), pp. 50–62.

Manley, R. G. Dynamic stiffness and effective inertia methods in modern torsional vibration theory. *J. R. Aero. Soc.* vol. 47 (1943), pp. 5–26.

Note. The use of semi-graphical methods is also an advantage in conjunction with calculations by Holzer tables, since it serves as a check of the various modes of vibration and ensures that no important critical is missed. This is particularly so in the case of branched systems.

1·343   *Inertia-torque diagrams for branched systems*

If the cylinder inertias and crankthrow stiffnesses are such as to give a fairly high natural frequency of vibration for the one-node mode of the bare engine, it is possible to determine the natural frequencies of the entire system fairly rapidly by means of inertia-torque diagrams.

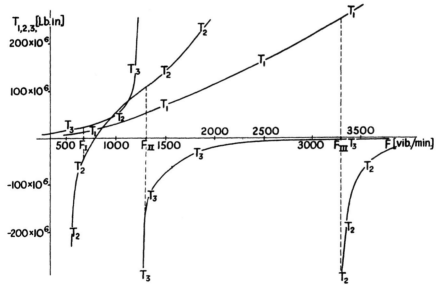

Natural frequencies of the first three modes of vibration, F_I, F_{II}, F_{III}, of a 3-branch system determined from inertia torque curves.

For this, it is first necessary to make Holzer tabulations for the engine only, at frequencies of say 500, 1000, 2000 and 3000 vib./min. These give a non-zero inertia torque $T_1 = \Sigma J \omega^2 \Delta$ as well as values of relative

[223]

amplitude Δ_E at the engine end on the last line of each table. By means of these results, the inertia-torque curve of the engine is plotted against frequency.

Similar tabulations are then made for each branch corresponding to a driven machine or auxiliary drive (see p. 191), and, by making use of a curve of Δ_E-values of the engine, plotted against frequency, the values of β, γ, etc., for the end-mass amplitudes of the driven-machine branches are obtained. This makes it possible to determine the inertia torques T_2, T_3, ..., for these branches at different frequencies.

A graph of the torque curves T_1, T_2, T_3, etc., is plotted against frequency. The curves have asymptotes with $+\infty$ and $-\infty$ values at the resonant frequencies of each sub-system. By inspection, the natural frequencies of the entire system are easily determined as being those values for which the sum of the inertia torques is zero. Complete Holzer tables need then only be calculated for these values. An example of this graphical method, which has been applied at the B.I.C.E.R.A. laboratory, is given in the figure on the previous page.

1·35 Additional frequencies due to excitation by driven machines

1·351 *Excitation by gearwheels*

The presence of backlash in gear systems may result in the appearance of additional resonance peaks at frequencies other than those determined by assuming that the gearwheels have an infinite stiffness.

The calculation of a system including backlash is not straightforward. At the design stage, a geared system such as that shown in the accompanying figure can be evaluated for three different conditions:

(*a*) as usual, assuming infinitely stiff gearwheels;

(*b*) assuming that backlash reduces K_B to zero, and that the main engine system vibrates independently of the K_B, J_B branch; and

(*c*) assuming a slight amount of coupling, and using, say, $K'_B = \tfrac{1}{4}K_B$ in place of K_B in frequency tabulations.

Computation is simplified by plotting the inertia torque T_1 for the engine branch and T_2 for the directly coupled driven-machine branch against frequency, and superimposing on this diagram the inertia torque curves T_3 of the gear-driven branch for the conditions (*a*) and (*c*). The positions at which $T_1 + T_2 + T_3 = 0$ indicate possible vibration frequencies of the system, and corresponding Holzer tabulations are then made to determine the relative amplitude β of the K_B, J_B branch at the gear position. Those frequencies which give large β-values are then regarded as important.

[224]

The above considerations also apply to systems with two sets of gear-wheels, one of which (G_{1B} and G_{2B}) is between two masses without intermediate shafting. In such a case, computation is based on stiffness variations occurring in this second set of gearwheels only, i.e. for the cases:

(a) $K_{(G_{1B}, G_{2B})} = \infty$,

(b) $K_{(G_{1B}, G_{2B})} = 0$,

(c) $K_{(G_{1B}, G_{2B})} = \frac{1}{4}K_A$.

An example of inertia torque curves and possible vibration frequencies obtainable with a system of this kind is shown in the following figure.

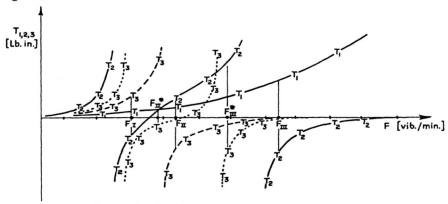

$T_1 \mathrel{\underline{}} T_1 =$ Engine torque.
$T_2 \mathrel{\underline{}} T_2 =$ Driven-machine torque.
$T_3 \mathrel{----} T_3 =$ Auxiliary-drive torque with $K_G = \infty$.
$T_3 \cdots T_3 =$ Auxiliary-drive torque with $K_G \cong \frac{1}{4}K_A$.

Figure showing inertia torque curves of three-branch system plotted against frequency. $F_{I, II, III}$ = natural frequencies with infinitely stiff gearwheels; $F^*_{II, III}$ = additional frequencies, with gearwheel backlash.

See also, W. A. Tuplin, 'Torsional vibrations with angular backlash', *Engineering*, 23 August 1935, p. 199.

1·352 *Excitation by resistance torque of auxiliaries*

1. Gear-driven blower

The mean value of the resistance torque offered by a gear-driven blower is given by

$$T_m = \frac{550 \times 12}{\frac{2\pi}{60}Nr} H_B = 6.303 \times 10^4 H_B/(Nr) \quad \text{[Lb.in.]},$$

where H_B = power absorbed by the blower [h.p.], N = engine speed [rev./min.] and r = ratio of blower speed divided by engine speed.

For Roots-type blowers (such as that shown in the accompanying figure), with a delivery pressure of about 4 Lb./in.^{2}, the maximum torque is about 1·2 times the mean torque, and the amplitude of the resistance torque variation is therefore

$$\Delta T = T_{\max.} - T_m = 0 \cdot 2 T_m.$$

If K is the stiffness of the quill shaft between the directly driven rotor and the bevel gear, the vibratory twist amplitude due to blower resistance-torque variation will be

$$\Delta\theta = \theta_B - \theta_G = \Delta T / K = 0 \cdot 2 T_m / K \quad [\text{rad.}].$$

This may be magnified by resonance. If r = number of blower revolutions per engine revolution and L = number of lobes of each rotor, the number of blower deliveries per engine revolution will be

$$Z = 2Lr$$

and the resistance-torque variation will have a frequency

$$F_B = 2rLN \quad [\text{vib./min.}].$$

Thus, resonance may be excited if the engine can run at a speed N, given by

$$N = F_{\text{I, II, III}} / 2rL \quad [\text{rev./min.}],$$

where $F_{\text{I,II,III}}$ is the one-, two- or three-node frequency of vibration of the entire engine system.

2. Camshaft drive

The torques from the suction, exhaust and fuel cams are not equal and are not equally spaced. If the fuel cam produces a torque considerably larger than the other two, as is sometimes the case, the predominant impulse will occur once per cycle for each cylinder. Thus it may be said that, in general, the resistance torque of a camshaft is very irregular and that these impulses may cause resonant vibration if the camshaft excitation frequency and the natural vibration frequency coincide.

It may also be noted that the camshaft torque will not have much influence on torsional vibration if the camshaft is driven from a point near the node, as is frequently the case.

1·353 Excitation by engine cyclic speed variation and gear wheel harmonics

(a) Frequencies of excitation

When the crankshaft and driven-machine shafting of an engine system are rotating with a cyclic speed variation due to the periodically varying engine torque, any auxiliary system coupled to the main shafting, by means of gearwheels, chains or belt drives, is subjected to a forced vibratory motion at its driven end.

The fundamental frequency of the excitation is

$$F_{(CV)} = Nm \quad \text{[vib./min.]},$$

where N = engine speed [rev./min.] and m = order number of the cyclic speed variation, so that Nm = frequency of the cyclic speed variation.

In addition, owing to cyclic pitch variations, angular misalinement, eccentricity and backlash effects, the driven gear of the auxiliary drive can be excited into vibration at frequencies which are multiples of $F_{(C.V.)}$. Thus, the frequencies of possible gear excitation will be

$$F_G = mN, \quad 2mN, \quad 3mN, \quad \dots \quad \text{[vib./min.]}.$$

Even in the absence of torsional vibration excitation from the engine, the auxiliary system is capable of vibrating at its own resonant frequency, which is determined as

$$F_A = \frac{60}{2\pi} \times \omega_A = 9 \cdot 55 \sqrt{(K_A/J_A)} \quad \text{[vib./min.]},$$

where ω_A = natural phase velocity [rad./sec.], K_A = stiffness [Lb.in./rad.], and J_A = moment of inertia [Lb.in.sec.²] of the *auxiliary system*.

Hence, resonant conditions in the auxiliary system may be excited when $F_G = F_A$, at the engine speeds

$$N = \frac{9 \cdot 55}{m} \sqrt{(K_A/J_A)}, \quad \frac{9 \cdot 55}{2m} \sqrt{(K_A/J_A)}, \quad \frac{9 \cdot 55}{3m} \sqrt{(K_A/J_A)}, \quad \dots \quad \text{[rev./min.]}.$$

(b) Amplitude of forced motion at driven gear

The amplitude of the induced vibration at the gears on the side of the auxiliary drive is

$$\pm \theta_G = rC/(2m) \quad \text{[rad.]},$$

where r = gear ratio (speed of auxiliary drive divided by engine speed), and C = cyclic speed variation of the engine at the speed considered.

The value of C can be obtained from a torsiograph test determining θ_1, the amplitude at the free end of the crankshaft (which is equal to θ_G/r if there is no torsional vibration at this speed). Alternatively, C may be determined by the excess energy method, or estimated by means of the equation†

$$C = 2AR_0 \Sigma\Delta_{(m)} |T_m| / [m\omega^2 \Sigma J],$$

where A = piston area [in.^{2}], R_0 = crank radius [in.], $\Sigma\Delta_{(m)}$ = the first mth order vector sum, with unit vectors, which does not give a zero resultant value, m = order of cyclic speed variation (obtained from vector sum considerations), $|T_m|$ = mth order tangential-pressure component of gas torque [Lb./in.^{2}], $\omega = N/9 \cdot 55$ [rad./sec.], where N = engine speed [rev./min.], and ΣJ = sum of all the moments of inertia in the engine system. Therefore

$$\theta_G = rAR_0 |T_m| \Sigma\Delta_{(m)} / [m^2 \omega^2 \Sigma J] \quad [\text{rad.}].$$

If the engine cyclic speed variation at any of the frequencies given by F_G above happens to coincide with one of the natural frequencies of the entire engine system, then the vibration amplitude θ_A at the end-mass J_A of the auxiliary system may be augmented by dynamic magnification.

(c) Shaft twist in auxiliary system

The forced rolling motion of amplitude $\pm \theta_G$ at the gear end causes a shaft twist $\Delta\theta_A$ in the auxiliary system. The value of $\Delta\theta_A$ is obtained from the relation‡

$$\Delta\theta_A = \frac{rAR_0 |T_m| \Sigma\Delta_{(m)}}{[\omega_A^2 - (\omega m)^2] \Sigma J} \quad [\text{rad.}].$$

† The derivation of this expression for C is given in section 2·6.

‡ *Derivation of formula for* $\Delta\theta_A$. The oscillating system consisting of J_A and K_A vibrates in accordance with the equation of motion

$$J_A \ddot\theta_A + K_A(\theta_A - \theta_G) = 0,$$

where $\theta_G = \theta_{0G} \sin m\omega t$, $\omega = N/9 \cdot 55$, N = engine speed [rev./min.], and m = order number of cyclic speed variation. θ_A will be of the form $\theta_A = \theta_{0A} \sin m\omega t$. Substituting we obtain

$$-m^2 \omega^2 J_A \theta_{0A} + K_A \theta_{0A} - K_A \theta_{0G} = 0,$$

so that

$$\theta_{0A}[K_A - m^2 \omega^2 J_A] = K_A \theta_{0G}$$

and

$$\theta_{0A} = \theta_{0G} \Big/ \Big[1 - \Big(\frac{m\omega}{\omega_A}\Big)^2 \Big],$$

with $\omega_A^2 = K_A / J_A$. The shaft twist is

$$\Delta\theta_A = \theta_{0A} - \theta_{0G} = \theta_{0G} \Big(\frac{m\omega}{\omega_A}\Big)^2 \Big/ \Big[1 - \Big(\frac{m\omega}{\omega_A}\Big)^2 \Big],$$

and by substituting in this the expression for θ_G already derived, we obtain the above relation for $\Delta\theta_A$.

For reference, it may be noted that if the auxiliary system has damping, represented by a dynamic magnifier M_A $(=\sqrt{(K_A J_A)}/c_A$, where c_A = damping coefficient), the corresponding expression for $\Delta\theta_A$ is

$$\Delta\theta_A = \theta_{0G} \times \frac{\sqrt{\Big\{ \Big(\frac{\omega m}{\omega_A}\Big)^2 + \frac{1}{M_A^2} \Big\}}}{\sqrt{\Big\{ \Big(\frac{\omega_A}{\omega m}\Big)^2 \times \Big(1 - \frac{\omega^2 m^2}{\omega_A^2} \Big)^2 + \frac{1}{M_A^2} \Big\}}},$$

which reduces to the previous expression when $M = \infty$.

(d) Resultant vibration amplitude at mass J_A

The resultant amplitude $\theta_{A\text{total}}$ of the mass J_A of the auxiliary system is due to

(i) a rolling motion θ_G at the gear end,
(ii) shaft twist $\Delta\theta_A$ caused by (i), and
(iii) shaft twist giving a vibration amplitude $\theta_{A(TV)}$ at mass J_A, as a result of torsional vibration of the entire engine system.

Thus, we may write

$$\theta_{A\text{total}} = \theta_G + \Delta\theta_A + \theta_{A(TV)} \quad [\text{rad.}].$$

The values of θ_G and $\Delta\theta_A$ are determined from the corresponding equations given under (b) and (c). The value of $\theta_{A(TV)}$ can be obtained by measurement of θ_1 at mass no. 1 of the engine and using the relative amplitude Δ_A at J_A obtained from a Holzer table (so that $\theta_{A(TV)} = \theta_1 \times \Delta_A$).

When the total vibration amplitude $\theta_{A\text{total}}$ is measured directly at J_A, it may be found that its value may differ somewhat from that obtained by the above equation. The divergence is to some extent due to the neglect of phase-angles in assuming zero damping (see also Graphical methods, section 2·43).

(e) Stresses

The amplitude of the resultant motion is

$$\theta_{A\text{total}} = \theta_G + \Delta\theta_A + \theta_{A(TV)} \quad [\text{degrees}].$$

The 'rolling' amplitude θ_G produces no stress and can be disregarded. The shaft stress due to $\Delta\theta_A$ is

$$q_{A(CV)} = K_A \Delta\theta_A / (57\cdot3 Z_A) \quad [\text{Lb./in.}^2],$$

where $Z_A = \pi D_A^3 / 16$, D_A being the minimum diameter [in.] of the shaft K_A.

The stress due to $\theta_{A(TV)}$, resulting from torsional vibration of the entire engine system, is

$$q_{A(TV)} = \theta_{A(TV)} \times q_A^* \quad [\text{Lb./in.}^2],$$

where $q_A^* = $ stress in auxiliary driving shaft per degree of vibration of the mass J_A in the auxiliary drive (see also section 2·3).

Owing to the difference in phase angles, it is not advisable to take the arithmetic sum of $q_{A(CV)}$ and $q_{A(TV)}$ to obtain a figure for the total stress. To avoid intricate expressions, it is then convenient to calculate the total stress either as

$$q_{\text{total}} = \theta_{A\text{total}} \times q_{A(CV)} / \Delta\theta_A \quad \text{or} \quad q_{\text{total}} = \theta_{A\text{total}} \times q_A^*,$$

using the value which corresponds to the type of resonance considered (i.e. sub-system resonance or resonance of the entire system).

1·354 Excitation by propeller

The blades of a propeller are subjected to variable loads during their rotation, owing to the non-uniform flow in a ship's wake, the variable depth of immersion and the proximity of the ship's brackets or supports which affect the hydrodynamic pressure of the water on a blade passing in their vicinity.

These load variations act on a propeller with a basic frequency

$$F_P = N_P \times b \quad [\text{vib./min.}],$$

where $N_P =$ propeller speed [rev./min.] and $b =$ number of blades of the propeller. Moreover, owing to the effects of the wake, the shape of the hull and its fittings, excitation can also occur at multiples of the basic frequency, so that we may write

$$F_P = N_P b, \quad 2N_P b, \quad 3N_P b, \quad \dots, \quad [\text{vib./min.}].$$

It has been suggested that by suitable phasing of the propeller blades and engine cranks it should be possible to reduce the torsional vibration due to a harmonic torque of the engine by an opposing torque of the propeller acting at the same frequency (e.g. a third-order engine harmonic by means of a three-bladed propeller, etc.).

Further information on this subject may be found in the following references:

Rabbeno, G. Torsional vibrations due to propellers. *Riv. Maritt.* vol. 82 (March 1950), pp. 551–8.

Panagopulos, E. Design-stage calculations of torsional, axial and lateral vibrations of marine shafting. *Pap. Soc. Nav. Archit.* no. 5, presented at annual meeting, 9 and 10 Nov. 1950.

Work, C. E. Singing propellers. *Trans. Soc. Nav. Archit., N.Y.*, vol. 63 (1951), p. 319.

1·36 Further methods of natural frequency determination

The natural frequencies of multi-mass systems may be determined from the corresponding frequency equation by any of the *methods developed for determining the roots of polynomials*, such as

Graeffe's method (for applications see Hansen, H. M. and Chenea, P. F., *Mechanics of Vibration* (J. Wiley and Sons and Chapman and Hall, 1952), pp. 234–46).

Horner's method (for applications, see Tuplin, W. A., *Torsional Vibration* (Chapman and Hall, 1934), pp. 71–7).

Lin's method (see Lin, Shi-Nge, 'Method of successive approximations for evaluating the real and complex roots of cubic and higher-order equations,' *J. Math. Phys.* vol. 20, no. 3 (August, 1941).

Various modifications of Holzer's tabulation method have also been evolved. Apart from Gorfinkel's method (see section 1·322), reference may be made to Boumard's method (Boumard, B., 'Détermination rapide des vitesses critiques dans les lignes d'arbres de moteurs', *Revue gén. Mécan.* October 1952, pp. 323–7).

The theory and practical application of the *reduced-inertia or mass-reduction method* (outlined in section 1·33) have been treated in great detail by Porter, and for these, reference should be made to the following publications:

Porter, F. P. Methods for calculating torsional vibration, Phase II of Evaluation of effects of torsional vibration. *S.A.E. War Engineering Board*, vol. 1, pp. 151–77. [This includes a method applicable to distributed-mass systems.]
Porter, F.P. Torsional vibration notes with solutions for an untuned viscous damper and a flexible coupling of non-linear elasticity. *A.S.M.E. Pap.* no. 52-OGP-7, 23–27 June 1952, 13 pp.

Vector methods, forced-frequency tables and Holzer tabulations with damping are mainly designed to determine vibration amplitudes and vibratory torques. They are included, therefore, in section 2·4.

Matrix methods, which are associated with evaluations of determinants, are finding increasing application in mechanical engineering. Bibliographic references relating to these methods are given in section 2·45.

Relaxation methods applied to engineering problems are extensively treated in the following textbooks:

Southwell, Sir R. V. *Relaxation Methods in Engineering Science* (Oxford University Press, 1946).
Allen, D. N. de G. *Relaxation Methods* (McGraw-Hill, 1954).
Shaw, F. S. *An Introduction to Relaxation Methods* (Dover Publications).

1·37   Electromechanical analogies †

1·371   *Analogous systems*

By means of electromechanical analogies it is possible to assess the behaviour of a mechanical system from the characteristics of a corresponding electrical system. These latter characteristics may either be calculated or determined experimentally as voltage differences, currents, etc.

Analogue computers are circuit analysers operating on this principle. They generally comprise a variable-frequency oscillator which is tuned to obtain resonance of the circuit. The measurements for the various stages give values of electrical quantities corresponding to the deflexions,

---

† Bibliography. The subject of electromechanical analogies has been extensively treated by G. D. McCann and his collaborators. Readers can refer to the following publications: *Mach. Design*, Dec. 1945, p. 137, and Feb. 1946, p. 129; *J. Appl. Mech., Trans. Amer. Soc. Mech. Engrs*, 1945, p. A-135; *Westingho. Eng.* March 1946, p. 49.
A detailed introduction to the subject is given in W. T. Thomson's *Mechanical Vibrations* (G. Allen and Unwin, Ltd. London, 1950).

torques, etc., occurring in the analogous mechanical system. In the following, a brief review of the theory is given, and this is illustrated with some examples of possible applications.

First analogy

Two systems are regarded as analogous when their behaviour is governed by similar equations of motion. Thus, the vibrations of a one-mass system subjected to a sinusoidal torque and the oscillations of an equivalent electrical circuit connected to a sinusoidal voltage may be represented by the following equations:

$$\text{Mechanical system:} \quad J\ddot{\theta} + c\dot{\theta} + K\theta = T_0 \sin \omega t.$$

$$\text{Electrical system:} \quad L\ddot{q} + R\dot{q} + \frac{1}{C}q = V_0 \sin \omega t.$$

These are used as a basis for the analogy given below:

Mechanical quantities	Symbol	Typical units
Torque	T	Lb.in.
Angular displacement	θ	radian
Stiffness	K	Lb.in./rad.
Angular velocity	$\dot{\theta}$	rad./sec.
Damping coefficient	c	Lb.in.sec./rad.
Angular acceleration	$\ddot{\theta}$	rad./sec.^{2}
Moment of inertia	J	Lb.in.sec.^{2}
Phase velocity	ω	rad./sec.

Electrical quantities	Symbol	Typical units
Voltage	V	volts
Charge	q	coul.
Inverse of capacitance	$1/C$	volt/coul. $= 1/$farad
Current	$\dot{q}$	coul./sec. $=$ amp.
Resistance	R	volt.sec./coul. $=$ ohm
Rate of change of current	$\ddot{q}$	coul./sec.$^2 =$ amp./sec.
Inductance	L	volt.sec.^{2}/coul. $=$ henries
Phase velocity	ω	rad./sec.

Some equivalent circuits are shown in the following figures:

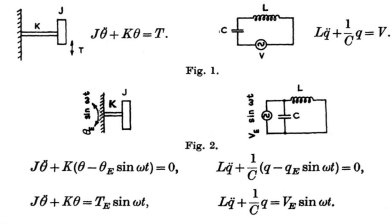

$$J\ddot{\theta} + K\theta = T. \qquad\qquad L\ddot{q} + \frac{1}{C}q = V.$$

Fig. 1.

Fig. 2.

$$J\ddot{\theta} + K(\theta - \theta_E \sin \omega t) = 0, \qquad L\ddot{q} + \frac{1}{C}(q - q_E \sin \omega t) = 0,$$

$$J\ddot{\theta} + K\theta = T_E \sin \omega t, \qquad L\ddot{q} + \frac{1}{C}q = V_E \sin \omega t.$$

Fig. 3.

$$J_1\ddot{\theta}_1 + K_1(\theta_1 - \theta_2) = T_1, \qquad\qquad L_1\ddot{q}_1 + \frac{1}{C_1}(q_1 - q_2) = V_1,$$

$$J_2\ddot{\theta}_2 + K_1(\theta_2 - \theta_1) + K_2(\theta_2 - \theta_3) = T_2, \quad L_2\ddot{q}_2 + \frac{1}{C_1}(q_2 - q_1) + \frac{1}{C_2}(q_2 - q_3) = V_2,$$

$$J_3\ddot{\theta}_3 + K_2(\theta_3 - \theta_2) + c_3\dot{\theta}_3 = 0, \qquad L_3\ddot{q}_3 + \frac{1}{C_2}(q_3 - q_2) + R_3\dot{q}_3 = 0.$$

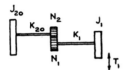 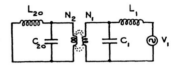

Fig. 4. Geared system.

$r = N_2/N_1 = $ speed ratio. $\qquad\qquad r = N_2/N_1 = $ turns ratio.

$$K_2 = K_{20} \times r^2, \quad \frac{1}{K} = \frac{1}{K_1} + \frac{1}{K_2}, \qquad \frac{1}{C_2} = \frac{1}{C_{20}} \times r^2, \quad C = C_1 + C_2,$$

$$J_2 = J_{20} \times r^2, \quad \theta_2 = \theta_{20}/r. \qquad L_2 = L_{20} \times r^2, \quad q_2 = q_{20}/r.$$

Equations of motion: $\qquad\qquad$ Equations of motion:

$$J_1\ddot{\theta}_1 + K(\theta_1 - \theta_2) = T_1, \qquad\qquad L_1\ddot{q}_1 + \frac{1}{C}(q_1 - q_2) = V_1,$$

$$J_2\ddot{\theta}_2 + K(\theta_2 - \theta_1) = 0. \qquad\qquad L_2\ddot{q}_2 + \frac{1}{C}(q_2 - q_1) = 0.$$

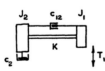 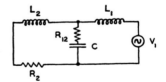

Fig. 5. System with shaft damping.

$$J_1\ddot{\theta}_1 + c_{12}(\dot{\theta}_1 - \dot{\theta}_2) + K(\theta_1 - \theta_2) = T_1, \qquad L_1\ddot{q}_1 + R_{12}(\dot{q}_1 - \dot{q}_2) + \frac{1}{C}(q_1 - q_2) = V_1,$$

$$J_2\ddot{\theta}_2 + c_2\dot{\theta}_2 + c_{12}(\dot{\theta}_2 - \dot{\theta}_1)$$
$$+ K(\theta_2 - \theta_1) = 0. \qquad L_2\ddot{q}_2 + R_2\dot{q}_2$$
$$+ R_{12}(\dot{q}_2 - \dot{q}_1) + \frac{1}{C}(q_2 - q_1) = 0.$$

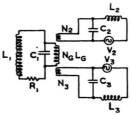

Fig. 6. Gear-branched system.

$$J_1\ddot{\theta}_1 + c_1\dot{\theta}_1 + K_1(\theta_1 - \theta_G) = 0,$$

$$\begin{aligned} J_G\ddot{\theta}_G &+ K_1(\theta_G - \theta_1) \\ &+ K_2(r^2\theta_G - r\theta_2) + K_3(r^2\theta_G - r\theta_3) = 0, \end{aligned}$$

$$J_2\ddot{\theta}_2 + K_2(\theta_2 - r\theta_G) = T_2,$$

$$J_3\ddot{\theta}_3 + K_3(\theta_3 - r\theta_G) = T_3.$$

$$\left[\text{Speed ratio:} \ r = \frac{-N_2}{N_G} = \frac{-N_3}{N_G}. \right]$$

Second analogy†

Instead of relating torque to voltage, another extensively used analogy relates torque to electric current. Thus, the sum of the currents in an electric circuit is analogous to the sum of the torques in a mechanical system (see Figs. 1*a* and 1*b*).

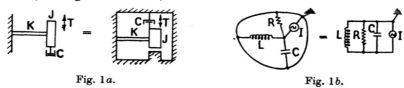

Fig. 1*a*.   Fig. 1*b*.

It is seen from Fig. 1*b* that the currents of the electrical circuit are in accordance with the equation

$$\frac{1}{L}\int V\,dt + \frac{V}{R} + C\frac{dV}{dt} = I = I_0\,e^{j\omega t}.$$

Introducing, for convenience, the symbol $\Phi = \int V\,dt$ which represents flux [volt sec.], this equation may be rewritten as

$$\frac{\Phi}{L} + \frac{\dot{\Phi}}{R} + C\ddot{\Phi} = I_0\,e^{j\omega t},$$

and by comparing it with the equation for the mechanical system of Fig. 1*a*

$$K\theta + c\dot{\theta} + J\ddot{\theta} = T_0\,e^{j\omega t},$$

† Bibliography: Firestone, F. A., *J. Acoust. Soc. Amer.* vol. 4 (1933), p. 249. Klotter, K., *Ingen-Arch.* vol. 18, no. 8 (1950), pp. 291–301.

[234]

the following analogy is obtained:

Mechanical quantities	Symbol	Typical units
Torque	T	Lb.in.
Angular displacement	θ	rad.
Shaft stiffness	K	Lb.in./rad.
Angular velocity	$\dot{\theta}$	rad./sec.
Damping coefficient	c	Lb.in.sec./rad.
Angular acceleration	$\ddot{\theta}$	rad./sec.^{2}
Moment of inertia	J	Lb.in.sec.^{2}
Phase velocity	ω	rad./sec.

Electrical quantities	Symbol	Typical units
Current	I	amp.
Flux	$\Phi = \int V\,dt$	volt.sec.
Inverse of inductance	$1/L$	coul./voltsec.^{2} = 1/henry
Voltage	$\dot{\Phi} = V$	volt
Inverse of resistance	$1/R$	1/ohm
Rate of change of voltage	$\ddot{\Phi} = \dot{V}$	volt/sec.
Capacitance	C	coul./volt = farad.
Phase velocity	ω	rad./sec.

The question, which analogy should be used, is a matter on which various views have been expressed. It should be borne in mind that each analogy has its limitations and advantages, which depend on the type of vibrating system considered and the variables requiring investigation.

EXAMPLES: *Note.* The symbol Φ (flux) is used as an abbreviation for $\Phi = \int V\,dt$ [volt sec.].

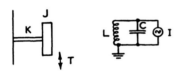

Fig. 1.

Fig. 2. Vibrometer with velocity excitation.

$$J\ddot{\theta} + K\theta = T. \quad C\ddot{\Phi} + \frac{1}{L}\Phi = I.$$

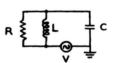

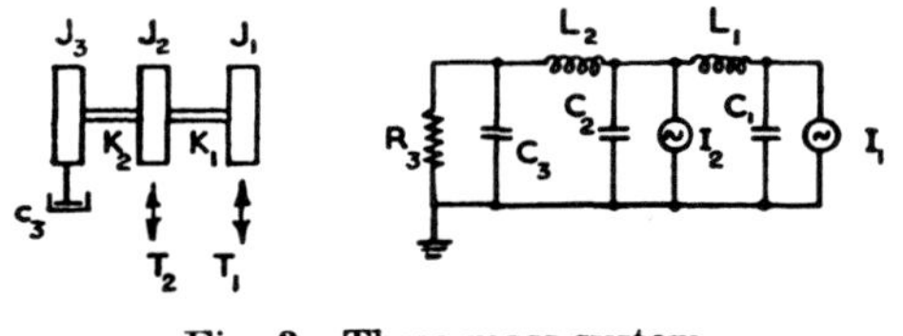

Fig. 3. Three-mass system.

$$C_1\ddot{\Phi}_1 + \frac{1}{L_1}(\Phi_1 - \Phi_2) = I_1,$$

$$C_2\ddot{\Phi}_2 + \frac{1}{L_1}(\Phi_2 - \Phi_1) + \frac{1}{L_2}(\Phi_2 - \Phi_3) = I_2,$$

$$C_3\ddot{\Phi}_3 + \frac{1}{L_2}(\Phi_3 - \Phi_2) + \frac{1}{R_3}\dot{\Phi}_3 = 0.$$

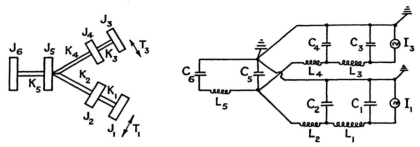

Fig. 4. Gear-branch system.

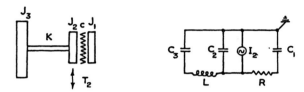

Fig. 5. System with viscous fluid damper.

1·372 *Impedance, admittance and mobility methods*

Electromechanical analogies not only make it possible to correlate measurements but also the methods of calculation dealing with mechanical and electrical systems. Considering the 'First analogy', for instance, one sees that the electrical circuits of Figs. 1, 2 and 4 could be evaluated by Holzer tables, and the circuits of Figs. 3, 5 and 6 by means of Holzer tables taking account of damping. Conversely, the mechanical systems of Figs. 1–6 can be evaluated by methods which originated in electrical calculations. As these are fairly frequently referred to in recent literature, they will be described in the following lines. It will be noted that they do not introduce new concepts but rather a different terminology, and the suitability of these methods depends to some extent on the type of problem considered.

Using the complex notation $\theta = \theta_0 e^{j\omega t}$, the equation of motion of the one-mass system shown in Fig. 1 can be expressed in the following manner:

$$\left.\begin{aligned} J\ddot{\theta} + c\dot{\theta} + K\theta &= T_0 e^{j\omega t}, \\ (-\omega^2 J + j\omega c + K)\,\theta_0 &= T_0. \end{aligned}\right\} \qquad (1)$$

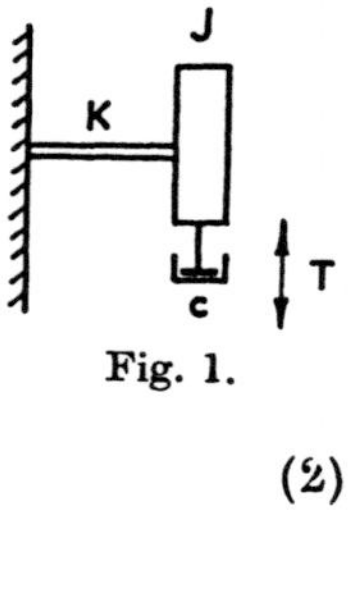

Alternatively, it can also be written as follows:

$$\left.\begin{aligned} J\ddot{\theta} + c\dot{\theta} + K\int\theta\,dt &= \int T_0 e^{j\omega t}\,dt, \\ \left(j\omega J + c + \frac{K}{j\omega}\right)\theta_0 &= \frac{T_0}{j\omega}. \end{aligned}\right\} \qquad (2)$$

Fig. 1.

[236]

The quantities in the parentheses are termed the 'mechanical impedances' of the system. From eqs. (1), we have the *displacement impedances*:

$$z_J = -\omega^2 J, \quad z_C = j\omega c, \quad z_K = K \quad \text{[Lb.in./rad.]},$$

so that the total impedance is

$$z = z_J + z_C + z_K = T_0/\theta_0.$$

From eqs. (2) we have the *velocity impedances*:

$$Z_J = j\omega J, \quad Z_C = c, \quad Z_K = -j\frac{K}{\omega} \quad \text{[Lb.in.sec.]},$$

and the total impedance is

$$Z = Z_J + Z_C + Z_K = T_0/\dot\theta_0.$$

Since $\dot\theta_0 = j\omega\theta_0$, we have

$$\theta_0 = \frac{T_0}{j\omega Z} \quad \text{[rad.]}.$$

The resonance frequency is obtained from $\partial\,|\,\theta\,|/\partial\omega = 0$, that is, from

$$\frac{\partial}{\partial\omega}\left\{\frac{T_0}{\sqrt{\{(K-\omega^2 J)^2 + (\omega c)^2\}}}\right\} = \frac{-T_0\{(K-\omega^2 J)(-2\omega J) + \omega c^2\}}{[(K-\omega^2 J)^2 + (\omega c)^2]^{\frac{3}{2}}} = 0,$$

which gives

$$-KJ + \omega^2 J^2 + \frac{c^2}{2} = 0 \quad \text{or} \quad \omega^2_{\text{res.}} = \frac{K}{J}\left(1 - \frac{c^2}{2KJ}\right) \simeq \frac{K}{J}.$$

Thus, for practical purposes, $\omega_{\text{res.}}$ is the value which is obtained by equating to zero the imaginary part of Z (or the real part of z).

The imaginary part X of

$$Z = c + j\left(\omega J - \frac{K}{\omega}\right) = c + jX$$

is termed the 'mechanical reactance' of the system. By plotting $Z = c + jX$ as a function of X, we obtain a straight line (see Fig. 2). The vector from O to any point on this line ('Z-locus') gives the impedance for the frequency corresponding to X, and the phase angle ($\tan\phi = X/c$) between the $\dot\theta$-vector ($\dot\theta = j\omega\theta$) and the applied torque vector T.

The calculation becomes more complicated for multi-mass systems. For the two-mass system of Fig. 3, including shaft damping, the impedance to rate of change of motion at the position of J_1 is

$$Z = \frac{T_1}{\dot\theta_1} = Z_{J1} + \frac{Z_{J2} \times (Z_C + Z_K)}{Z_{J2} + Z_C + Z_K},$$

using the corresponding rules for parallel impedances. Solving for θ_1 we obtain $\theta_{10} = \dot{\theta}_{10}/j\omega$ and

$$|\theta_{10}| = \frac{T_{10}}{\omega^2} \times \frac{\sqrt{\{(K - \omega^2 J_2)^2 + (\omega c)^2\}}}{\sqrt{\{[\omega^2 J_1 J_2 - K(J_1 + J_2)]^2 + (\omega c)^2 (J_1 + J_2)^2\}}}$$

and $\qquad \omega_{\text{res.}}^2 \cong K(J_1 + J_2)/(J_1 J_2) \quad$ from $\quad \dfrac{\partial |\theta_{10}|}{\partial \omega} = 0.$

Thus, for multi-mass systems, this method does not offer any advantages as regards rapidity or ease of computation when compared with the direct method using determinants obtained from the equations of motion.

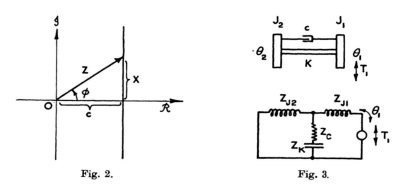

Fig. 2. Fig. 3.

Admittances are defined by expressions which are the reciprocals of corresponding elements used in impedances. We thus have the *displacement admittances* (also termed *receptances*) $y = \theta/T$ and the *velocity admittances* (also termed *mobilities*) $Y = \dot{\theta}/T$. The notation is as follows:

Displacement admittances or *receptances*:

$$y_J = \frac{1}{-\omega^2 J}, \quad y_C = \frac{1}{j\omega c}, \quad y_K = \frac{1}{K}.$$

Velocity admittances or *mobilities*:

$$Y_J = \frac{1}{j\omega J}, \quad Y_C = \frac{1}{c}, \quad Y_K = \frac{j\omega}{K}.$$

Mobilities lead to continued fractions of the type

$$Y_{\text{res.}} = \cfrac{1}{\cfrac{1}{Y_1} + \cfrac{1}{Y_2 + \cfrac{1}{\cfrac{1}{Y_3} + \cfrac{1}{Y_4}}}}$$

for multi-mass systems.

[238]

Note. Excellent examples of the use of admittances (receptances and mobilities) are given in the following references:

Duncan, W. J. Mechanical admittances and their applications to oscillation problems, Ministry of Supply. *Rep. Memor. Aero. Res. Council, Lond.,* no. 2000, (1946).

Freberg, C. R. and Kemler, E. N. *Elements of Mechanical Vibration* (J. Wiley and Sons, Inc., New York; Chapman and Hall, Ltd., London), 2nd ed.

Hansen, H. M. and Chenea, P. F. *Mechanics of Vibration* (J. Wiley and Sons, Inc., New York; Chapman and Hall, Ltd., London).

Bishop, R. E. D. Analysis and synthesis of vibrating systems. *J. R. Aero. Soc.* October, 1954, pp. 703–19.

Remarks on analogies

Some of the principles extensively employed for linear electrical network evaluations are directly applicable to mechanical systems. For instance, the *Principle of Superposition* can be used as follows:

Mechanical system	Electrical system

In an in-line multi-cylinder engine system, the deflexion θ_1 at any point can be obtained as the sum of the deflexions $\theta_{1(K)}$ at that point, determined for each cylinder torque T_K considered separately: In a linear network with a number of voltage generators V_K, the current I_1 at any point is the sum of the currents $I_{1(K)}$ flowing if each generator is considered separately (the other generators being replaced by their internal impedances Z_{V_K}):

$$\theta_1 = \theta_{1(1)} + \theta_{1(2)}.$$

$$I_1 = I_{1(1)} + I_{1(2)}.$$

The representation of a mechanical system by an analogous electrical network is useful for systems with up to about four masses. For more extensive systems, the network relations lose much of their simplicity.

Other electromechanical analogies. In some cases it is an advantage to have analogue models of a given system based on two or three different analogies, in order to determine various quantities by simple voltage or current measurements.

The table on p. 240 gives a comparison of the various analogies now used. The notation is the same as previously.

All these analogies are based on different interpretations of the equation of motion of a one-mass system and the equation for its equivalent electrical circuit. Analogies IV, V and VI are for an undamped system, and give only the natural frequency, i.e. ω_n. The damped

[239]

amplitude (at resonance only) could also be obtained if $c = R/\omega_n^2$ in analogy IV (with $\theta = \dot q$), and $c = 1/(R\omega_n^2)$ in V and VI (with $\theta = \Phi$ and $\dot V$, respectively), ω_n being used as a constant in the evaluation.

Mechanical quantities	Electrical analogies					
	Applicable for resonant and non-resonant conditions			Applicable for natural frequency only		
	I	II	III	IV	V	VI
T	V	I	$\dot V$	V	I	$\dot i$
θ	q	$\Phi=\int V\,dt$	I	$\dot q$	Φ	$\dot V$
K	$1/C$	$1/L$	$1/C$	L	C	C
$\dot\theta$	$\dot q$	$\dot\Phi=V$	$\dot I$	—	—	—
c	R	$1/R$	R	—	—	—
$\ddot\theta$	$\ddot q$	$\dot\Phi=\dot V$	$\dot I$	—	—	—
J	L	C	L	q, $1/C$	Φ, $1/L$	V, $1/L$
ω	ω	ω	ω	ω	ω	ω

Finally, it should be noted that networks consisting solely of resistances, or solely of inductances, are being developed as equivalent systems. The analogy is obtained 'topographically', that is, by placing elements in parallel (in phase or in opposition), in series or in other arrangements. These positional analogues appear superior to the formal analogies hitherto used, but their possibilities still require investigation.

Transformer analogues

An analogy which makes use only of voltage transformers for operations on complex-number quantities is described below.

Consider a system of four transformers connected together as shown. It can be arranged that a sinusoidal voltage V_p is supplied across the terminals a and c, and another voltage V_q, having the same frequency and peak value as V_p but with a 90° phase shift relative to V_p, is supplied across the terminals bd.

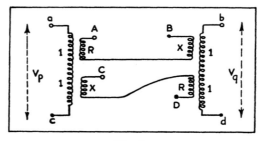

Fig. 1.

If the turns ratios of the transformer coils are $1:R$ and $1:X$, as shown in the arrangement of Fig. 1, the output voltage at the inner coil terminals AB is

$$E_{AB} \equiv E_p = RV_p - XV_q; \tag{1}$$

similarly, the output voltage at the coil terminals CD is

$$E_{CD} \equiv E_q = XV_p + RV_q. \tag{2}$$

Combining equations (1) and (2) and introducing the symbol $j = \sqrt{-1}$ to denote the 90° phase angle between in-phase and quadrature components, we have

$$(1) + j(2) = E_p + jE_q = (RV_p - XV_q) + j(XV_p + RV_q)$$
$$= (R + jX)(V_p + jV_q).$$

Hence this transformer analogue performs multiplications in accordance with the equation $\vec{E} = \vec{Z}\vec{I}$, where all three quantities are complex-number vectors. If the input voltages are applied across AB and CD, and output voltages are measured across ac and bd, the analogue performs divisions, giving $\vec{I} = \vec{E}/\vec{Z}$.

In effect, the impedance $\vec{Z}$ in these equations is represented by turns ratios, and the current $\vec{I}$ by a voltage vector $\vec{V}$.

A number of groups similar to that of Fig. 1, connected together in series (or in parallel), can be used to determine steady-state torsional vibration conditions of analogous straight (or branched) multi-mass systems excited by a sinusoidal torque.

Further information on this transformer analogy can be obtained from the following:

Humphrey-Davies, M. W. and Slemon, G. R. Transformer-analogue network analysers. *Proc. Inst. Elect. Engrs*, vol. 100, part II, no. 77 (October, 1953), pp. 469–86.

Cherry, C. E. The duality between intercoupled electrical and magnetic circuits and the formation of transformer equivalent circuits. *Proc. Phys. Soc.* vol. 62 (1949), p. 101.

1·373 *Transfer function and operator methods*

These methods are outlined in this section because of their close relation to mechanical impedance calculations. The use of transfer functions will be illustrated by means of the following simple example of an electro-mechanical system.

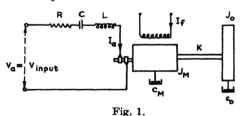

Fig. 1.

EXAMPLE. Two-mass system including a d.c. motor drive with variable input voltage and constant field current (see Fig. 1).

NOTATION:

Subscripts: M = motor, o = output, f = field, a = armature, c = counter (e.m.f.), t = torsional, m = mechanical, e = electrical.

Mechanical quantities: T = torque, J = moment of inertia, K = stiffness, c = damping constant, θ = angular amplitude.

Electrical quantities: R = ohmic resistance, L = inductance, C = capacitance, V = voltage, I = current.

Furthermore: ω = phase velocity of V_a (assumed sinusoidal); $p = j\omega$, where $j^2 = -1$, $\dot{\theta} = p\theta = \Omega$, $\ddot{\theta} = p^2\theta = \dot{\Omega}$.

The equations for the mechanical system are

$$J_M p^2 \theta_M + c_M p\theta_M + K\theta_M - K\theta_o = T_M, \quad (1)$$

$$J_o p^2 \theta_o + c_o p\theta_o + K\theta_o - K\theta_M = 0. \quad (2)$$

From (2)

$$\theta_o = \frac{(K/p)}{J_o p + c_o + \dfrac{K}{p}} \times \theta_M. \quad (3)$$

But $p\theta_M = \Omega_M$ = motor angular velocity, hence the relation between motor torque and motor speed is given by the 'mechanical impedance'

$$Z_{m(M)} = \frac{T_M}{\Omega_M} = J_M p + c_M + \frac{K}{p} - \frac{(K/p)^2}{\left(J_o p + c_o + \dfrac{K}{p}\right)}. \quad (4)$$

Equations for the electrical system

The motor torque T_M is proportional to the armature current I_a, so that, with a proportionality factor k_t, we can write

$$T_M = k_t I_a = \Omega_M Z_{m(M)} \quad (5)$$

(also using eq. (4)). The value of I_a depends on the electrical impedance

$$Z_e = R + pL + \frac{1}{pC}$$

of the armature circuit. If $V_a = V_i$ = input voltage, and V_c = counter-e.m.f. of armature circuit, we have the relation

$$V_a - V_c = Z_e I_a, \quad (6)$$

and therefore

$$I_a = \frac{V_a - V_c}{Z_e} = \frac{Z_{m(M)} \Omega_M}{k_t}. \quad (7)$$

Referring again to eqs. (1) and (3), we note that

$$T_M = p\left[\left(J_M p + c_M + \frac{K}{p}\right) \times \frac{(K/p)}{J_o p + c_o + \dfrac{K}{p}} - \frac{K}{p}\right] \times \theta_o; \quad (8)$$

thus the relation between motor torque and output velocity is given by the mechanical impedance

$$Z_{m(o)} = \frac{T_M}{\Omega_o} = \frac{T_M}{p\theta_o} = \frac{(K/p) \times \left(J_M p + c_M + \dfrac{K}{p}\right)}{J_o p + c_o + \dfrac{K}{p}} - \frac{K}{p}. \quad (9)$$

As the counter-e.m.f. V_c is also proportional to the motor velocity Ω_M, using a proportionality factor k_e we may write

$$V_c = k_e \Omega_M,$$

and therefore eq. (6) becomes, with eq. (7),

$$V_a - k_e \Omega_M = Z_e \times \frac{Z_{m(M)} \Omega_M}{k_t},$$

so that

$$V_a = \frac{k_e k_t + Z_e Z_{m(M)}}{k_t} \times \Omega_M. \tag{9}$$

As, from eq. (3),

$$\frac{\theta_o}{\theta_M} = \frac{p\theta_o}{p\theta_M} = \frac{\Omega_o}{\Omega_M} = \frac{(K/p)}{J_o p + c_o + \dfrac{K}{p}},$$

the value of Ω_o can be introduced in eq. (9), in order to obtain the 'overall transfer impedance' of the system

$$Z_T = \frac{V_a}{\Omega_o} = \frac{k_e k_t + Z_e Z_{m(M)}}{k_t} \times \frac{J_o p + c_o + \dfrac{K}{p}}{(K/p)}. \tag{10}$$

The 'overall transfer stiffness' is then

$$K_T = \frac{V_a}{\theta_o} = pZ_T.$$

The 'transfer function' as regards angular velocity is

$$G_1(p) = \frac{\text{output quantity}}{\text{input quantity}} = \frac{\Omega_o}{V_a} = \frac{1}{Z_T},$$

and the 'transfer function' as regards amplitude of vibration is

$$G_2(p) = \frac{\theta_0}{V_a} = \frac{1}{pZ_T} = \frac{1}{K_T}.$$

These functions are usually plotted in polar co-ordinates as functions of the phase angle ϕ between θ_o and V_a (or Ω_o and V_a), and on these G-curves the various values of ω (for the phase positions) are indicated as a parameter. Transient conditions (sudden input variation) can also be analysed by these methods, using transforms and conformal mapping to reduce calculations.†

† These methods originated in other fields and their use is also developing in mechanical vibration engineering. See, for instance, Kármán, T. v. and Biot, M. A. *Mathematical Methods in Engineering* (McGraw-Hill Book Co. 1940), pp. 423–39; Ahrendt, W. R. and Taplin, J. F., *Automatic Feedback Control* (McGraw-Hill, 1951); Kleiner, A., *Revue Sulzer*, no. 1 (1945), pp. 115–26 (French edition).

PART 2

EVALUATION AND PREDICTION OF TORSIONAL VIBRATION STRESSES

2·1 TANGENTIAL-PRESSURE COMPONENTS DUE TO GAS PRESSURE AND INERTIA

NOTATION

Symbol	Brief definition	Typical units	Symbol	Brief definition	Typical units
a	cosine component of Fourier series	—	r	radius	in., cm.
			R	radius	in., cm.
A	area	in.², cm.²	S	sine component of torque	Lb.in., Kg.cm.
b	sine component of Fourier series	—	S_n	sine term of tangential pressure component	Lb./in.², Kg./cm.²
C	conversion factor	—			
$|C|$	torque harmonic	Lb.in., Kg.cm.	t	time	sec., min.
C_F	fuel consumption	Lb./bhp.hr.	T	torque	Lb.in., Kg.cm.
C_n	cosine term of tangential pressure component	Lb./in.², Kg./cm.²	$|T|$	harmonic component of tangential pressure	Lb./in.², Kg./cm.²
F	natural frequency	vib./min.	U	work	in.Lb., Kg.cm.
g	acceleration due to gravity	in./sec.², cm./sec.²	V	volume	in.³, cm.³
H	power	h.p., watts	W	weight	Lb., Kg.
J	mass moment of inertia	Lb.in.sec.², Kg.cm.sec.²	y	displacement	in., cm.
			$\dot{y}$	velocity	in./sec., cm./sec.
k	ordinate number	—			
L	length; stroke	in., cm.	$\ddot{y}$	acceleration	in./sec.², cm./sec.²
m	order of harmonic	—			
n	number of revolutions	—	ϵ	phase angle	rad., deg.
n	critical speed order number	vib./rev.	η	efficiency	—
n^*	harmonic order number	vib./cycle	θ	vibration amplitude	rad., deg.
			θ	angular velocity	rad./sec.
N	crankshaft speed	rev./min.	τ	time	sec., min.
$N_{cyl.}$	number of cylinders	—	ϕ	crank angle	deg., rad.
N^*	number of power strokes per minute	min.⁻¹	ψ	angle	deg., rad.
			ω	$= (2\pi/60) \times N$	rad./sec.
p	pressure	Lb./in.², Kg./cm.²	ω	phase velocity	rad./sec.
P	force	Lb., Kg.	$\dot{\omega}$	angular acceleration	rad./sec.²

Subscripts are used in the text to define more precisely, e.g. $p_{i(m)}$ = mean indicated pressure. Their meaning is fully explained in the context in each instance.

2·11 Excitation torque, tangential force and tangential pressure†

In an engine, the *excitation torque* T [Lb.in.], which gives rise to crankshaft vibration, is due to the combined effects of variations of gas pressures, piston and connecting-rod inertia, etc.

† Alternative designations. The excitation torque is also referred to as 'harmonic torque', 'applied torque', 'shaking torque', 'input torque', and similar designations are used for the excitation force. The tangential pressure is also termed 'tangential effort'.

In general, it is useful to regard this torque as being produced by a *tangential force* P_t [Lb.] acting at the crankpin. Thus

$$\text{tangential force} = P_t = \frac{T}{R_0} \quad [\text{Lb.}],$$

R_0 being the crank radius [in.]. A further reduction is obtained by relating P_t to the piston area A [in.^{2}] of one cylinder. This is termed the

$$\text{tangential pressure} = p_t = \frac{P_t}{A} = \frac{T}{AR_0} \quad [\text{Lb./in.}^2].$$

It should be noted that, although this pressure p_t is obtained as a specific force per unit area of the piston, it does not act on the piston but on the crankpin, in a direction perpendicular to the crank radius and to the crankshaft axis.

We thus have to make a distinction between two types of pressures, viz.

$$p = \text{pressure acting on piston area } [\text{Lb./in.}^2],$$

and $\qquad p_t = $ tangential pressure acting on crankpin [Lb./in.^{2}].

The relations between p and p_t, as well as the expressions for torque, work per cycle and power, as functions of either p or p_t, are given in the following section, which deals with mean values only, i.e. mean pressures, mean transmitted torques, etc. The corresponding expressions for instantaneous values are set forth subsequently, in section 2·14.

Metric units. The formulae in this section can be employed with the following units:

torque (T)	Kg.cm.,	area (A)	cm.^{2},
force (P)	Kg.,	work (U)	cm.Kg.,
length (L)	cm.,	pressure (p)	Kg./cm.^{2} ('technical atmospheres').

The expressions relating to power (H), however, also require a different numerical constant. As 1 h.p. (British) = 550 ft.Lb./sec., whereas 1 h.p. (metric) = 7500 Kg.cm./sec., therefore in the following, eq. (1), for instance, becomes

$$H_{\text{cyl.}} = \frac{p_m ALN^*}{450{,}000} \quad [\text{h.p. (metric)}],$$

with all linear dimensions in centimetres. Thus in all the following expressions where 33,000 occurs for British units, it should be replaced by 450,000 for metric units. For direct conversions, it may be noted that 1 h.p. (British) = 746 watts, whereas 1 h.p. (metric) = 735·5 watts, hence

$$1 \text{ h.p. (British)} = 746/735 \cdot 5 = 1 \cdot 0143 \text{ h.p. (metric)}.$$

2·111 *Relations between mean pressure, power, and mean torque*

The work done by a piston of area A [in.²] when it is acted upon by a mean pressure p_m [Lb./in.²] while covering a distance L [ft.] equal to the stroke is
$$U_{\text{cyl.}} = p_m A L \quad [\text{ft.Lb.}].$$

If the piston of an engine cylinder performs N^* power strokes per minute, the *power developed per cylinder* is
$$H_{\text{cyl.}} = \frac{p_m A L N^*}{33,000} \quad [\text{h.p.}], \tag{1}$$

and if $N_{\text{cyl.}}$ = number of cylinders of the engine, the total power is
$$H = N_{\text{cyl.}} H_{\text{cyl.}} \quad [\text{h.p.}].$$

The work done by a mean tangential force P_t [Lb.] acting on a crank of radius R_0 [ft.] during one crankshaft revolution, while covering the peripheral distance $2\pi R_0$, is
$$U_{\text{cyl.}} = 2\pi R_0 P_t \quad [\text{ft.Lb.}],$$

and the power developed at N revolutions per minute is
$$H_{\text{cyl.}} = \frac{2\pi R_0 P_t N}{33,000} \quad [\text{h.p.}].$$

If a *mean equivalent tangential pressure*† p_t is introduced, by means of the relation
$$p_t = P_t / A,$$

the expression for power becomes
$$H_{\text{cyl.}} = \frac{2\pi R_0 A p_t N}{33,000} \quad [\text{h.p.}]. \tag{2}$$

By means of equations (1) and (2), it is therefore possible to relate the mean tangential crankpin pressure p_t to the mean piston pressure p_m as follows:
$$2\pi R_0 A p_t N = p_m A L N^*$$

and
$$p_t = p_m \times \frac{L N^*}{2\pi R_0 N}.$$

The formula for p_t depends, therefore, on the stroke/crank radius ratio L/R_0, and the number N^*/N of working cycles (of one cylinder) per crankshaft revolution. The following table gives the relation between p_t and p_m for various classes of engines:

Engine type	2 s.c. s.a.	2 s.c. d.a.	2 s.c.o.p.	4 s.c. s.a.	4 s.c. d.a.
$N^*/N =$	1	2	1	$\frac{1}{2}$	1
$L/R_0 =$	2	2	4	2	2
$p_t =$	p_m/π	$2p_m/\pi$	$2p_m/\pi$	$p_m/2\pi$	p_m/π

In the above table, the abbreviations used are:

2 s.c. = two-stroke cycle; 4 s.c. = four-stroke cycle;

s.a. = single acting; d.a. = double acting; o.p. = opposed piston.

† Also called 'mean tangential effort'.

[249]

The relation between the mean equivalent tangential pressure (on the crankpin) and the *mean engine torque* T_m is given by

$$T_m = p_t A R_0 N_{\text{cyl.}} \quad \text{[Lb.ft.],}$$

so that T_m can be determined from the mean piston pressure p_m by substituting the expression for p_t as a function of p_m given in the above table.

These relations are applicable to indicated, brake and frictional pressures, powers and torques, denoted by the subscripts i, b and f, respectively. For instance:

$$H_i = p_{i(m)} \times ALN * N_{\text{cyl.}}/33,000 \quad \text{[h.p.],}$$

$$H_b = p_{b(m)} \times ALN * N_{\text{cyl.}}/33,000 \quad \text{[h.p.],}$$

$$H_f = p_{f(m)} \times ALN * N_{\text{cyl.}}/33,000 \quad \text{[h.p.],}$$

values of mean pressure being used in these formulae. Moreover, as

$$H_b + H_f = H_i \quad \text{and} \quad p_{b(m)} + p_{f(m)} = p_{i(m)}$$

the *mechanical efficiency* η_m can be expressed as

$$\eta_m = H_b/H_i = p_b/p_i$$

$$= (H_i - H_f)/H_i = 1 - \frac{H_f}{H_i} = 1 - \frac{p_f}{p_i}.$$

Thus, the *frictional pressure* may be determined from

$$p_f = p_i(1 - \eta_m) = \frac{p_b}{\eta_m}(1 - \eta_m).$$

2·112 Formulae for brake power
Hydraulic brake and Prony brake

Brake power: $\quad H_b = \dfrac{\omega T}{550} = \dfrac{2\pi N R W}{33,000} \quad$ [b.h.p.],

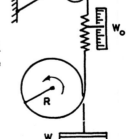

where $\omega = 2\pi N/60$, N being the running speed [rev./min.], R the torque arm [ft.], and W the applied weight [Lb.].

Rope brake

Brake power: $\quad H_b = \dfrac{2\pi N R(W - W_0)}{33,000} \quad$ [b.h.p.],

where $W_0 = $ reaction force of spring [Lb.] between rope and stationary end of system.

Hydraulic disk brake

$$H_b = \frac{N^3(R^5 - r^5)}{79 \times 10^6} \quad \text{[b.h.p.]},$$

$$T_b = \frac{N^2(R^5 - r^5)}{15,000} \quad \text{[Lb.ft.]},$$

where $R =$ outer radius of disk [ft.], and $r =$ radius of inner surface of rotating ring of water [ft.].

2·12 Determination of indicated and frictional mean effective pressures (i.m.e.p. and f.m.e.p.)

In torsional vibration investigations, it is necessary to know the value of the indicated mean effective pressure (i.m.e.p.) at which an engine is operating. From this value, it is then possible to assess the magnitude of the excitation torque, using data available for engines of a similar type (see sections 2·14, 2·17 and 2·18).

The various methods employed for the determination of i.m.e.p. are described below.

2·121 *Cylinder-pressure diagrams*

The pV diagram, from which the i.m.e.p. is evaluated as the ratio of the effective work of the cycle divided by the swept volume, is generally derived from a $p\phi$ diagram (Fig. 1), in which the cylinder pressure is plotted against crank angle degrees. This is the most accurate method for determining the indicated mean effective pressure p_{im}.

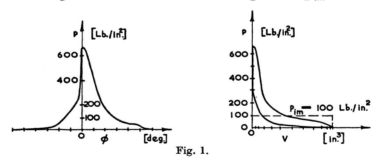

Fig. 1.

To obtain reliable values, the $p\phi$ diagram must be of a fairly large size (base length of 10 in. or more) and have a fully calibrated pressure scale. The equipment used to obtain the diagram must be free from phase-lag errors and have a sufficiently high sensitivity to indicate correctly the rapid variations in cylinder pressure.

Thus, small indicator cards, and photographs of complete diagrams on not absolutely flat oscilloscope screens, are generally of little value, and can only be used as approximate indications.

The standard methods recommended by the B.I.C.E.R.A. Torsional Vibration Panel are (i) the 'Sunbury point-by-point' method, and (ii) the direct-recording method using equipment consisting of a balanced disk valve indicating unit, a Farnboro' recording unit, and an electronic Standard-B.I.C.E.R.A. spark control unit.

It should be noted that the i.m.e.p. may vary from cylinder to cylinder, so that indicator diagrams should be taken on all cylinders if possible.

2·122 *Mechanical efficiency tests*

The following methods (methods 1–4) give results which are necessarily more approximate than those based on accurate indicator diagrams. In practice, it is advisable to use several methods and compare the results.

1. *Morse tests*

The engine coupled to a brake is run at a normal speed N [rev./min.] with a load giving a brake power H_b [b.h.p.]. The nozzle or the fuel supply to one cylinder is then cut out and the speed drops. It is raised immediately (to minimize errors due to temperature change) to its previous value by reducing the brake load to a value H_b'.

The difference between these two values gives the indicated horse-power of one cylinder:

$$H_i^{(1)} = H_b - H_b' \quad \text{[h.p.]}.$$

The reasoning on which this is based is as follows: With all the cylinders firing, we have in the first case

$$H_b = N_{\text{cyl.}} \times [H_i^{(1)} - H_f^{(1)}],$$

where $N_{\text{cyl.}}$ = number of cylinders, and $H_f^{(1)}$ = frictional horse-power of one cylinder. After cutting out one cylinder, the brake power is

$$H_b' = (N_{\text{cyl.}} - 1) \times H_i^{(1)} - N_{\text{cyl.}} \times H_f^{(1)}$$

(assuming the total frictional power is unchanged), and this leads to the above formula.

To obtain the total i.h.p. of the engine, the test should be repeated for each cylinder in turn, taking the sum of the actual $H_i^{(1)}$ values for the individual cylinders.

It is useful to make further checks at different engine speeds above and below the normal speed N, and a mean line can then be drawn through the test points (see Fig. 2).

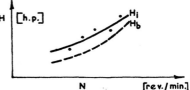

Fig. 2. Results of Morse tests with a constant load W [Lb.] on the brake.

The indicated mean effective pressure p_{im} is determined from

$$p_{im} = 33{,}000 \times \frac{H_i}{ALN^*N_{\text{cyl.}}} \quad [\text{Lb./in.}^2],\dagger$$

where H_i = total i.h.p. of the engine, A = piston area [in.^{2}], $L = 2R_0 =$ stroke [ft.], N^* = number of power strokes per minute, and $N_{\text{cyl.}} =$ number of cylinders.

Note. By carrying out a number of Morse tests at a constant speed but with brake loads varying from, for instance, half load to full load, it is possible to obtain a curve indicating the variation of $H_f = H_i - H_b$ (or $p_f = p_{im} - p_{bm}$) with different engine loads.

2. *Willans's fuel consumption diagram*

The engine coupled to a brake is run at a constant speed through the test. Beginning from normal load, the load is gradually reduced to zero b.h.p. while taking a series of fuel consumption and b.h.p. readings. The fuel consumption C_F per hour (or per cycle) is then plotted against H_b or p_b (brake power or brake mean pressure) and extrapolated to zero fuel consumption. The frictional power H_f or pressure p_f is thus obtained as shown in Fig. 3.

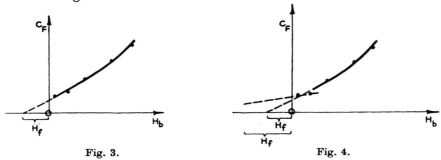

Fig. 3.   Fig. 4.

However, if the fuel consumption curve is not fairly straight near zero b.h.p., the extrapolation becomes uncertain (see Fig. 4).

Tests of this type have the advantage that all the engine cylinders are operating under normal conditions, but it is worth noting that, as the p_f and H_f values are extrapolated from test points for low engine loads, *strictly speaking they are valid only for low values of brake power and b.m.e.p.* This latter remark also applies to methods 3 and 4.

3. *Motoring method*

The engine is coupled to a swinging field electrical dynamometer. In the first part of the test, the dynamometer is used as an engine-driven generator and the normal b.h.p. of the engine is determined from the

† Metric units: see remarks on p. 248.

generator output. The engine is then stopped, the fuel supply is cut off and the connexions are reversed so that the still warm engine is driven by the generator at the same speed as previously. Making allowance for the dynamometer efficiency, the frictional power of the engine can thus be obtained. It should be noted that the motoring method can be used to confirm and extend Willans's line below the no-load value. [See also, Dicksee, C. B., *The high-speed compression-ignition engine*, pp. 303 and 323, Blackie and Son, Ltd, 1946 ed.]

4. *Running-down method*

The engine is run at its normal speed $\bar{N}$ [rev./min.] at no load (with no water in the brake) long enough for it to reach normal operating temperatures. The fuel is then cut off and the time taken by the engine to stop is measured. If $t =$ time from cutting off to standstill [sec.], the angular deceleration will be

$$\dot{\omega} \cong (2\pi/60) \times \bar{N}/t \quad [\text{rad./sec.}].$$

With $\Sigma J =$ sum of all the moments of inertia in the system [Lb.in.sec.²], we can then calculate the frictional torque as

$$T_f = \dot{\omega}\Sigma J \quad [\text{Lb.in.}],$$

The frictional horsepower is therefore:

$$H_f = (2\pi/60) \times \bar{N} \times T_f/(12 \times 550) \quad [\text{h.p.}],$$

and the frictional mean effective pressure is

$$p_{mf} = \frac{33,000}{ALN^*N_{\text{cyl.}}} \times H_f \quad [\text{Lb./in.}^2],$$

where $A =$ piston area [in.²], $L = 2R_0 =$ stroke [ft.], $N^* =$ number of power strokes per minute, and $N_{\text{cyl.}} =$ number of cylinders.

Further accuracy is obtained in the result if the deceleration process is recorded, together with a calibrated time signal, using drum camera equipment. In this case, $\dot{\omega}$ is evaluated from $(2\pi/60) \times \Delta N/\Delta t$.

2·123 *Frictional torque determined from a running-down test*

For this test, the total value of the moments of inertia of the engine system must be known with sufficient accuracy.

The engine is run at normal speed and zero brake load, and the fuel is then cut off suddenly. The decrease in engine speed per second, immediately after cutting off, is measured and the frictional torque T_f is then determined by means of the relation

$$T_f = J_{\text{total}} \times 2\pi \frac{\Delta^2 n}{\Delta t^2} \quad [\text{Lb.in.}],$$

with $\qquad J_{\text{total}} = N_{\text{cyl.}}J_{\text{cyl.}} + J_{\text{flywh.}} + J_{\text{brake}} \quad [\text{Lb.in.sec.}^2],$

where $N_{cyl.}$ = number of engine cylinders, n = number of crankshaft revolutions, and t = time [sec.].

For accurate results, the value of $\Delta^2 n/\Delta t^2$ is best determined from well spread out photographic recordings obtained with a drum camera, using a tuning fork for a time signal on which a pulse is superimposed for each complete revolution of the shafting. This latter signal can be obtained by inserting a small projecting stud of steel on the periphery of the flywheel and placing close to the flywheel a stationary permanent-magnet inductive type pick-up which generates a pulse each time the stud passes across its field.

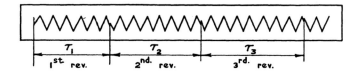

The value of $\Delta^2 n/\Delta t^2$ is then determined from the photographic trace as

$$\frac{\Delta^2 n}{\Delta t^2} = \left[\left(\frac{1}{\tau_1}\right) - \left(\frac{1}{\tau_2}\right)\right] \bigg/ \left[\frac{\tau_1 + \tau_2}{2}\right] = \frac{2(\tau_2 - \tau_1)}{\tau_1 \tau_2 (\tau_1 + \tau_2)} \quad [\text{rev./sec.}^2],$$

where τ_1 and τ_2 are the times, in seconds, of the first and second engine revolutions, respectively.

If a drum camera is not available, the decrease in engine speed at the beginning of the running-down test can be determined from the number of revolutions indicated by a 1 sec. tachoscope reading taken at the free end of the engine and a 2 sec. tachoscope reading taken simultaneously at the brake end. Then

$$\frac{\Delta^2 n}{\Delta t^2} = \left[\left(\frac{n_1}{t_1}\right) - \left(\frac{n_2 - n_1}{t_2 - t_1}\right)\right] \bigg/ \left(\frac{t_2}{2}\right) \quad [\text{rev./sec.}^2],$$

where n_1 = number of crankshaft revolutions during the time t_1 [sec.], immediately after cutting off, and n_2 = number of revolutions from cutting off up to the time t_2. The expression in the denominator is the mean time interval: $\frac{1}{2}(t_2 - t_1) + \frac{1}{2}t_1$.

2·13 Piston displacement, velocity and acceleration

Notation. R_0 = crank radius [in.], L = length of connecting rod [in.], ϕ = crank angle [deg.], N = crankshaft speed [rev./min.], $\omega = 2\pi N/60$ [rad./sec.], y_P, $\dot{y}_P$ and $\ddot{y}_P$ denoting piston displacement [in.], velocity [in./sec.] and acceleration [in./sec.²], respectively.

(1) Approximate expressions:

$$y_P \simeq R_0 \left[1 + \frac{R_0}{4L} - \cos\phi - \frac{R_0}{4L}\cos 2\phi \right] \quad \text{[in.]},$$

$$\dot{y}_P \simeq \omega R_0 \left[\sin\phi + \frac{R_0}{2L}\sin 2\phi \right] \quad \text{[in./sec.]},$$

$$\ddot{y}_P \simeq \omega^2 R_0 \left[\cos\phi + \frac{R_0}{L}\cos 2\phi \right] \quad \text{[in./sec.}^2\text{]}.$$

(2) Complete formulae:

$$y_P = R_0 + L - R_0\cos\phi - L\sqrt{\{1 - (R_0/L)^2\sin^2\phi\}} \quad \text{[in.]},$$

$$\dot{y}_P = \omega R_0 \sin\phi \left\{ 1 + \frac{R_0}{L} \times \frac{\cos\phi}{\sqrt{\{1 - (R_0/L)^2\sin^2\phi\}}} \right\} \quad \text{[in./sec.]},$$

$$\ddot{y}_P = \omega^2 R_0 \left\{ \cos\phi + \frac{(R_0/L)}{[1 - (R_0/L)^2\sin^2\phi]^{\frac{3}{2}}} \right.$$
$$\left. \times [\cos 2\phi + (R_0/L)^2\sin^4\phi] \right\} \quad \text{[in./sec.}^2\text{]}.$$

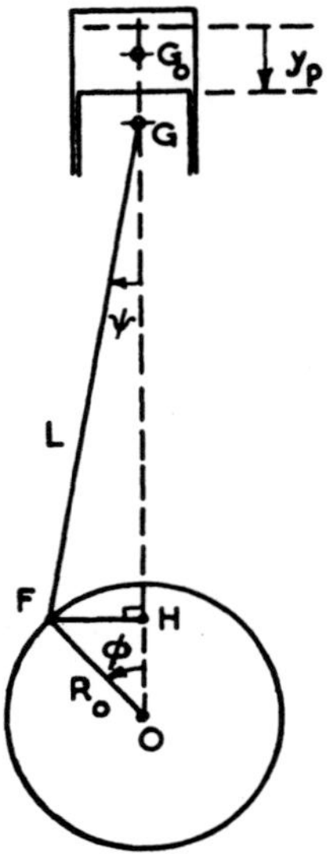

Fig. 1.

The derivation of these equations is simple: y_P is determined from Fig. 1 as

$$y_P = R_0 + L - \overline{OG} = R_0 + L - (R_0\cos\phi + L\cos\psi)$$

and noting that $\overline{FH} = R_0\sin\phi = L\sin\psi$, the exact expression is obtained readily.

The approximation contains the first terms of the power series

$$\sqrt{\{1 - (R_0/L)^2\sin^2\phi\}} = 1 - (R_0^2/2L^2)\sin^2\phi + - \dots.$$

As $\phi = \omega t$, the derivatives can be obtained as

$$\dot{y}_P = \omega\,\partial y_P/\partial\phi \quad \text{and} \quad \ddot{y}_P = \omega\,\partial\dot{y}_P/\partial\phi.$$

Note. As piston velocities are usually in ft./sec., we may replace ωR_0 [in./sec.] by $0.00873 \times N R_0 = 2\pi N R_0/(12 \times 60)$, where necessary; the numerical coefficient for acceleration then being

$$0.000914 N^2 R_0 = (2\pi/60)^2 \times N^2 R_0/12,$$

for acceleration in ft./sec.^{2}.

Metric units. The formulae in this section are also valid in metric units, for instance with y, R_0 and L expressed in cm., $\dot{y}$ in cm./sec., and $\ddot{y}$ in cm./sec.^{2}.

[256]

Table 1. *Values of relative piston displacement y_P/R_0*

For a given value of L/R_0 (connecting rod/crank radius ratio), the piston displacement y_P is obtained by multiplying the corresponding figures in this table by the value for the crank radius R_0.

$L/R_0 =$	3·0	3·2	3·4	3·6	3·8	4·0	4·2	4·4	4·6	4·8	5·0
Crank angle $\phi =$					Values of y_P/R_0						
0°	0·000	0·000	0·000	0·000	0·000	0·000	0·000	0·000	0·000	0·000	0·000
5°	0·005	0·005	0·005	0·005	0·005	0·005	0·005	0·005	0·005	0·005	0·005
10°	0·020	0·020	0·020	0·019	0·019	0·019	0·019	0·019	0·019	0·018	0·018
15°	0·045	0·045	0·044	0·043	0·043	0·043	0·042	0·042	0·041	0·041	0·041
20°	0·080	0·079	0·078	0·077	0·076	0·075	0·074	0·074	0·073	0·073	0·072
25°	0·124	0·122	0·120	0·118	0·117	0·116	0·115	0·114	0·113	0·112	0·112
30°	0·176	0·173	0·171	0·169	0·167	0·165	0·163	0·162	0·161	0·160	0·159
35°	0·236	0·233	0·230	0·227	0·224	0·222	0·220	0·218	0·216	0·215	0·214
40°	0·304	0·299	0·295	0·292	0·289	0·286	0·283	0·281	0·279	0·277	0·276
45°	0·377	0·372	0·367	0·363	0·359	0·356	0·353	0·350	0·347	0·345	0·343
50°	0·457	0·450	0·445	0·440	0·435	0·431	0·427	0·424	0·421	0·418	0·416
55°	0·540	0·533	0·527	0·521	0·516	0·511	0·507	0·503	0·500	0·497	0·494
60°	0·628	0·619	0·612	0·606	0·600	0·595	0·590	0·586	0·582	0·579	0·576
65°	0·718	0·708	0·700	0·693	0·687	0·681	0·676	0·672	0·668	0·664	0·660
70°	0·809	0·799	0·790	0·783	0·776	0·770	0·765	0·760	0·755	0·751	0·747
75°	0·901	0·890	0·881	0·873	0·866	0·860	0·854	0·849	0·844	0·839	0·835
80°	0·993	0·982	0·972	0·964	0·956	0·949	0·943	0·938	0·933	0·928	0·924
85°	1·083	1·072	1·062	1·053	1·046	1·039	1·033	1·027	1·022	1·017	1·013
90°	1·172	1·160	1·150	1·142	1·134	1·127	1·121	1·115	1·110	1·105	1·101
95°	1·257	1·246	1·236	1·228	1·220	1·213	1·207	1·201	1·196	1·192	1·187
100°	1·340	1·329	1·319	1·311	1·304	1·297	1·291	1·285	1·280	1·276	1·272
105°	1·419	1·408	1·399	1·391	1·384	1·377	1·371	1·366	1·361	1·357	1·353
110°	1·493	1·483	1·475	1·467	1·460	1·454	1·449	1·444	1·439	1·435	1·431
115°	1·563	1·554	1·546	1·539	1·532	1·527	1·522	1·517	1·513	1·509	1·506
120°	1·628	1·619	1·612	1·606	1·600	1·595	1·590	1·586	1·582	1·579	1·576
125°	1·688	1·680	1·674	1·668	1·663	1·658	1·654	1·650	1·647	1·644	1·641
130°	1·742	1·736	1·730	1·725	1·721	1·717	1·713	1·710	1·707	1·704	1·702
135°	1·792	1·786	1·781	1·777	1·773	1·770	1·767	1·764	1·761	1·759	1·757
140°	1·836	1·831	1·827	1·824	1·821	1·818	1·815	1·813	1·811	1·809	1·808
145°	1·875	1·871	1·868	1·865	1·863	1·861	1·859	1·857	1·855	1·853	1·852
150°	1·908	1·905	1·903	1·901	1·899	1·897	1·895	1·894	1·893	1·892	1·891
155°	1·936	1·934	1·933	1·931	1·930	1·929	1·928	1·927	1·926	1·925	1·924
160°	1·959	1·958	1·957	1·956	1·955	1·954	1·953	1·952	1·952	1·952	1·951
165°	1·977	1·976	1·976	1·975	1·975	1·974	1·974	1·974	1·973	1·973	1·973
170°	1·990	1·990	1·989	1·989	1·989	1·989	1·988	1·988	1·988	1·988	1·988
175°	1·998	1·997	1·997	1·997	1·997	1·997	1·997	1·997	1·997	1·997	1·997
180°	2·000	2·000	2·000	2·000	2·000	2·000	2·000	2·000	2·000	2·000	2·000

Table 2. *Conversion factors C for tangential pressures and values of relative piston velocity* $\dot{y}_P/(\omega R_0)$

For a given value of L/R_0 (connecting rod/crank radius ratio), the piston velocity $\dot{y}_P$ is obtained from the tabulated values by multiplying these by ωR_0, where $R_0 =$ crank radius, and $\omega = (2\pi/60) \times N$, N being the crankshaft speed [rev./min.].

$L/R_0 =$	3·0	3·2	3·4	3·6	3·8	4·0	4·2	4·4	4·6	4·8	5·0
Crank angle					Values of $C = \dot{y}/(\omega R_0)$						
$\phi = 0°$	0·000	0·000	0·000	0·000	0·000	0·000	0·000	0·000	0·000	0·000	0·00(
5°	0·116	0·114	0·113	0·111	0·110	0·109	0·108	0·107	0·106	0·105	0·10
10°	0·231	0·227	0·224	0·221	0·218	0·216	0·214	0·212	0·210	0·209	0·20
15°	0·343	0·337	0·332	0·328	0·325	0·322	0·319	0·316	0·313	0·311	0·30
20°	0·450	0·443	0·437	0·432	0·427	0·423	0·419	0·415	0·412	0·409	0·40(
25°	0·552	0·543	0·536	0·530	0·524	0·519	0·514	0·510	0·506	0·503	0·50(
30°	0·646	0·637	0·629	0·622	0·615	0·609	0·604	0·599	0·595	0·591	0·58'
35°	0·733	0·723	0·714	0·706	0·699	0·692	0·686	0·681	0·676	0·672	0·66
40°	0·811	0·800	0·790	0·782	0·774	0·767	0·761	0·756	0·751	0·746	0·74
45°	0·879	0·867	0·858	0·849	0·841	0·834	0·828	0·822	0·817	0·812	0·80
50°	0·936	0·925	0·915	0·906	0·898	0·891	0·885	0·880	0·875	0·870	0·86(
55°	0·982	0·971	0·962	0·953	0·946	0·939	0·933	0·928	0·923	0·918	0·91
60°	1·017	1·007	0·998	0·990	0·983	0·977	0·971	0·966	0·962	0·958	0·95
65°	1·040	1·031	1·023	1·016	1·010	1·005	1·000	0·995	0·991	0·987	0·98
70°	1·053	1·045	1·038	1·032	1·027	1·022	1·018	1·014	1·011	1·008	1·00
75°	1·054	1·048	1·043	1·038	1·034	1·030	1·027	1·024	1·021	1·019	1·01'
80°	1·045	1·041	1·037	1·034	1·031	1·029	1·027	1·025	1·023	1·021	1·02(
85°	1·027	1·025	1·023	1·021	1·020	1·019	1·018	1·017	1·016	1·015	1·01
90°	1·000	1·000	1·000	1·000	1·000	1·000	1·000	1·000	1·000	1·000	1·00(
95°	0·966	0·968	0·970	0·972	0·973	0·974	0·975	0·976	0·977	0·978	0·97
100°	0·925	0·929	0·932	0·935	0·938	0·941	0·943	0·945	0·947	0·948	0·95(
105°	0·878	0·884	0·889	0·894	0·898	0·902	0·905	0·908	0·911	0·913	0·91
110°	0·827	0·835	0·841	0·847	0·852	0·857	0·861	0·865	0·868	0·871	0·87
115°	0·772	0·781	0·789	0·796	0·803	0·808	0·813	0·817	0·821	0·825	0·82
120°	0·715	0·726	0·734	0·742	0·749	0·755	0·761	0·766	0·770	0·774	0·77
125°	0·656	0·667	0·677	0·685	0·693	0·699	0·705	0·711	0·716	0·720	0·72
130°	0·596	0·608	0·617	0·626	0·634	0·641	0·647	0·653	0·658	0·662	0·66
135°	0·536	0·547	0·557	0·566	0·573	0·580	0·586	0·592	0·597	0·602	0·60(
140°	0·475	0·486	0·495	0·504	0·511	0·518	0·524	0·530	0·535	0·539	0·54
145°	0·414	0·424	0·433	0·441	0·449	0·455	0·461	0·466	0·471	0·475	0·47
150°	0·354	0·363	0·371	0·379	0·385	0·391	0·396	0·401	0·405	0·409	0·41
155°	0·294	0·302	0·309	0·316	0·321	0·326	0·331	0·335	0·339	0·343	0·34
160°	0·234	0·241	0·247	0·252	0·257	0·261	0·265	0·269	0·272	0·275	0·27
165°	0·175	0·180	0·185	0·189	0·193	0·196	0·199	0·202	0·205	0·207	0·20
170°	0·117	0·120	0·123	0·126	0·129	0·131	0·133	0·135	0·137	0·138	0·13
175°	0·058	0·060	0·062	0·063	0·064	0·065	0·066	0·067	0·068	0·069	0·070
180°	0·000	0·000	0·000	0·000	0·000	0·000	0·000	0·000	0·000	0·000	0·000

This table also gives the 'conversion factor' $C(\phi)$ for tangential pressures, required for converting cylinder pressure diagrams into diagrams of tangential pressure.

The tangential pressure p_t is obtained by multiplying the tabulated figures by the corresponding value of the cylinder pressure p at the crank-angle position considered: $p_t = C \times p$.

Table 3. *Values of relative piston acceleration $\ddot{y}_P/(\omega^2 R_0)$*

For a given value of L/R_0 (connecting rod/crank radius ratio), the piston acceleration $\ddot{y}_P$ is obtained by multiplying the corresponding figures in this table by $\omega^2 R_0$, where R_0 = crank radius, and $\omega = (2\pi/60) \times N$, N being the crankshaft speed [rev./min.].

$L/R_0 =$	3·0	3·2	3·4	3·6	3·8	4·0	4·2	4·4	4·6	4·8	5·0
Crank angle						Values of $\ddot{y}/(\omega^2 R_0)$					
$\phi = 0°$	1·333	1·313	1·295	1·278	1·263	1·250	1·237	1·227	1·217	1·208	1·200
5°	1·323	1·303	1·285	1·269	1·255	1·242	1·230	1·220	1·211	1·202	1·193
10°	1·299	1·279	1·262	1·247	1·233	1·220	1·208	1·198	1·189	1·180	1·171
15°	1·256	1·237	1·220	1·207	1·194	1·182	1·171	1·162	1·154	1·147	1·140
20°	1·195	1·180	1·167	1·154	1·142	1·131	1·122	1·113	1·105	1·098	1·093
25°	1·120	1·107	1·096	1·085	1·075	1·067	1·059	1·052	1·045	1·039	1·034
30°	1·033	1·023	1·013	1·004	0·997	0·991	0·985	0·979	0·974	0·970	0·967
35°	0·934	0·926	0·920	0·914	0·909	0·905	0·901	0·897	0·894	0·891	0·888
40°	0·825	0·821	0·818	0·815	0·812	0·810	0·808	0·806	0·804	0·802	0·801
45°	0·707	0·707	0·707	0·707	0·707	0·707	0·707	0·707	0·707	0·707	0·707
50°	0·586	0·589	0·592	0·595	0·597	0·599	0·601	0·603	0·605	0·606	0·607
55°	0·460	0·466	0·472	0·478	0·483	0·488	0·492	0·496	0·499	0·502	0·505
60°	0·334	0·343	0·351	0·359	0·367	0·375	0·382	0·388	0·393	0·398	0·401
65°	0·208	0·221	0·233	0·244	0·253	0·262	0·269	0·276	0·283	0·289	0·294
70°	0·087	0·102	0·116	0·128	0·140	0·151	0·160	0·168	0·176	0·183	0·188
75°	−0·029	−0·012	0·003	0·017	0·031	0·042	0·052	0·062	0·071	0·079	0·086
80°	−0·140	−0·122	−0·105	−0·088	−0·074	−0·061	−0·051	−0·041	−0·031	−0·022	−0·014
85°	−0·241	−0·221	−0·203	−0·186	−0·172	−0·159	−0·148	−0·137	−0·127	−0·118	−0·109
90°	−0·334	−0·313	−0·294	−0·278	−0·264	−0·250	−0·237	−0·227	−0·216	−0·208	−0·201
95°	−0·416	−0·396	−0·376	−0·360	−0·345	−0·333	−0·322	−0·312	−0·302	−0·293	−0·285
100°	−0·487	−0·468	−0·449	−0·434	−0·420	−0·409	−0·398	−0·387	−0·378	−0·369	−0·361
105°	−0·548	−0·532	−0·515	−0·499	−0·486	−0·475	−0·465	−0·456	−0·447	−0·437	−0·432
110°	−0·597	−0·582	−0·567	−0·554	−0·544	−0·534	−0·524	−0·515	−0·507	−0·498	−0·494
115°	−0·637	−0·625	−0·613	−0·602	−0·592	−0·583	−0·577	−0·569	−0·561	−0·555	−0·551
120°	−0·666	−0·655	−0·646	−0·638	−0·632	−0·625	−0·620	−0·615	−0·610	−0·605	−0·600
125°	−0·688	−0·680	−0·675	−0·670	−0·664	−0·659	−0·655	−0·651	−0·648	−0·645	−0·642
130°	−0·701	−0·697	−0·693	−0·690	−0·688	−0·686	−0·684	−0·683	−0·681	−0·679	−0·677
135°	−0·707	−0·707	−0·707	−0·707	−0·707	−0·707	−0·707	−0·707	−0·707	−0·707	−0·707
140°	−0·708	−0·711	−0·714	−0·717	−0·720	−0·723	−0·723	−0·726	−0·727	−0·729	−0·731
145°	−0·706	−0·711	−0·717	−0·724	−0·730	−0·734	−0·737	−0·741	−0·745	−0·748	−0·751
150°	−0·699	−0·708	−0·718	−0·727	−0·735	−0·741	−0·747	−0·752	−0·757	−0·762	−0·766
155°	−0·692	−0·705	−0·718	−0·728	−0·737	−0·746	−0·753	−0·760	−0·766	−0·772	−0·777
160°	−0·684	−0·700	−0·714	−0·727	−0·738	−0·748	−0·758	−0·766	−0·772	−0·780	−0·786
165°	−0·677	−0·696	−0·712	−0·725	−0·738	−0·749	−0·760	−0·767	−0·774	−0·780	−0·792
170°	−0·672	−0·692	−0·710	−0·724	−0·738	−0·750	−0·760	−0·768	−0·774	−0·780	−0·797
175°	−0·668	−0·686	−0·706	−0·724	−0·738	−0·750	−0·762	−0·772	−0·782	−0·792	−0·799
180°	−0·666	−0·685	−0·705	−0·723	−0·738	−0·750	−0·763	−0·773	−0·783	−0·794	−0·801

The values in this table can also be used to determine the inertia torque of reciprocating parts by the Tabulation Method, which is described on p. 276.

2·14 Conversion from cylinder-pressure diagram to tangential-pressure diagram

In torsional vibration investigations, it is necessary to know the magnitude of the engine torque acting on the crankshaft. This can be determined from a $p\phi$ diagram representing cylinder pressure against crank angle, either by calculation or by graphical constructions. The two methods are described below.

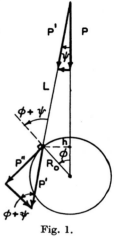

Fig. 1.

Calculation method

Using the notation of Fig. 1, with

$$L = \text{connecting rod length [in.]},$$
$$R_0 = \text{crank radius [in.]},$$
$$A = \text{piston area [in.}^2\text{]},$$
$$p = \text{cylinder pressure [Lb./in.}^2\text{]},$$
$$\phi = \text{crank angle [deg.]},$$
$$P, P', P'' = \text{forces [Lb.]},$$
$$T = \text{torque [Lb.in.]},$$
$$p_t = \text{tangential pressure [Lb./in.}^2\text{]},$$

the tangential pressure p_t on the crankpin can be calculated from the cylinder pressure for the same crank angle position by means of the relation

$$\frac{T}{AR_0} \equiv p_t \cong p \times \left\{ \sin\phi \left[1 + \frac{R_0}{L}\cos\phi \right] \right\} \cong p \times C(\phi) \quad \text{[Lb./in.}^2\text{]}.$$

The 'conversion function' $C(\phi)$ is the same as the function for relative piston velocity, which is tabulated on p. 258 for 5-degree intervals. The torque is, then $T = AR_0 \times p_t$ [Lb.in.].

Derivation. From Fig. 1, the component $P' = p'A$ of the piston force $P = pA$ is obtained as $P' = P/\cos\psi$. As

$$P'' = P'\sin(\phi + \psi) \quad \text{and} \quad h = R_0\sin\phi = L\sin\psi,$$

these relations give

$$T = R_0 P'' = R_0 \frac{pA}{\cos\psi}\sin(\phi + \psi) = R_0 Ap[\tan\psi\,\cos\phi + \sin\phi]$$

and

$$\tan\psi = \left(\frac{R_0}{L}\right) \times \sin\phi \Big/ \sqrt{\left\{1 - \left(\frac{R_0}{L}\right)^2\sin^2\phi\right\}}$$

$$\cong \frac{R_0}{L}\sin\phi\left[1 + \frac{R_0^2}{2L^2}\sin^2\phi + \frac{3R_0^4}{8L^4}\sin^4\phi + \ldots\right]$$

hence $\quad T \cong R_0 Ap\sin\phi\left[1 + \cos\phi\left\{\frac{R_0}{L} + \frac{R_0^3}{2L^3}\sin^2\phi + \frac{3R_0^5}{8L^5}\sin^4\phi + \ldots\right\}\right],$

[260]

and, neglecting in this approximation the powers of R_0/L higher than the first, we obtain the expression given at the beginning of this section.

The diagram obtained is of the type shown in Fig. 2c.

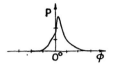 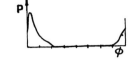 

Fig. 2a. $p\phi$ diagram.   Fig. 2b. $p\phi$ diagram with T.D.C.   Fig. 2c. p or T/AR diagram
                               at the beginning of the graph.   obtained from Fig. 2b.

Graphical method

To obtain reliable values, the drawing should be of a fairly large size (at least 15 in.). The procedure is as follows:

(1) Draw a circle of radius R_0 corresponding to the crank radius (see Fig. 3);

(2) Draw a diameter BD and erect a line, through the circle centre O, perpendicular to this diameter;

(3) For any value of crank angle ϕ considered, draw the corresponding radius OE, and join E with a point A on the line of action perpendicular to BD, in such a way that $\overline{EA} = L$, where L = length of connecting rod;

(4) Extend the line $\overline{EA}$ until it intersects the diameter BD at a point C;

(5) Then, if p = cylinder pressure corresponding to the crank-angle position ϕ, the tangential pressure p_t is obtained as

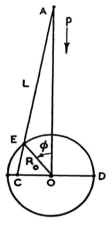

Fig. 3.

$$\frac{T}{AR_0} \equiv p_t = \frac{r}{R_0} \times p = \frac{\overline{OC}}{\overline{OB}} \times p \quad [\text{Lb./in.}^2].$$

The construction is then repeated for various values of ϕ to obtain the complete diagram of Fig. 2c.

Derivation. Referring to both Fig. 3 and Fig. 1, we see that

$$\overline{OC} = r = \overline{OA} \tan \psi = \tan \psi \times (R_0 \cos \phi + L \cos \psi) = R_0 \cos \phi \tan \psi + L \sin \psi.$$

As $L \sin \psi = R_0 \sin \phi$, we have $r = R_0(\cos \phi \tan \psi + \sin \phi)$, so that

$$T = Apr = ApR_0(\cos \phi \tan \psi + \sin \phi),$$

which is the same expression as that for $T = R_0 P''$ obtained for the 'calculation method'.

Note. Obviously, it does not matter whether the pressures are taken from a $p\phi$ or a pV diagram, provided the correct crank-angle positions are used.

Tangential pressure diagram for an opposed-piston engine

The calculation is as follows:

For the crank angle ϕ considered, the angles ψ_1 and ψ_2 are calculated by means of the relations (see Fig. 4)

$$\sin \psi_1 = (R_1/L_1)\sin \phi$$

and
$$\sin \psi_2 = (R_2/L_2)\sin \phi.$$

From these the values of $\cos \psi_1$ and $\cos \psi_2$ are determined. We then calculate the torque arms

(1) for the lower piston:

$$r_1 = \overline{OA}\tan \psi_1 = (L_1 \cos \psi_1 + R_1 \cos \phi) \times \tan \psi_1,$$

and (2) for the upper piston:

$$r_2 = \overline{OB}\tan \psi_2 = (L_2 \cos \psi_2 - R_2 \cos \phi) \times \tan \psi_2.$$

The total torque T [Lb.ft.] is then

$$T = pA(r_1 + r_2) = T_1 + T_2,$$

where p is the cylinder pressure at the crank angle ϕ, and A the area of the cylinder bore. T_1 and T_2 are the contributions to the total torque of the lower and the upper piston, respectively.

If there are two different crank radii, R_1 and R_2, in order to obtain the tangential pressure the practice is to take the value of the mean radius $R_0 = \frac{1}{2}(R_1 + R_2)$, so that finally

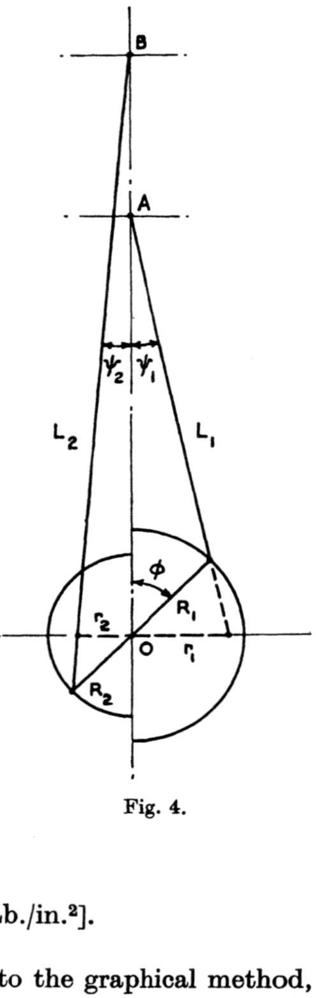

Fig. 4.

$$\frac{T}{AR_0} \equiv p_t = p \times \frac{r_1 + r_2}{\frac{1}{2}(R_1 + R_2)} \quad [\text{Lb./in.}^2].$$

This step-by-step calculation is preferable to the graphical method, and it can be conveniently tabulated.

2·15 Fourier analysis of tangential-pressure curve

The $T\phi$ curve of torque plotted against crank angle can be expressed as a Fourier series

$$T = |C_0| + |C_1|\sin(\omega t + \epsilon_1) + |C_2|\sin(2\omega t + \epsilon_2) + \ldots = \sum_{n=0}^{\infty} |C_n|\sin(n\omega t + \epsilon_n).$$

This series expresses the fact that the value of the torque T, at any instant t, can be obtained geometrically by adding together the corre-

[262]

sponding amplitudes of a number of sinusoidal curves. $|C_0|$ is the mean transmitted torque and $|C_1|$, $|C_2|$, ... are the *torque harmonics* which vary sinusoidally with frequencies which are integer multiples of the number of working cycles per minute N^* of one engine cylinder, represented by $\omega = (2\pi/60) \times N^*$ [rad./sec.]. The phases of these sine functions, relative to the T.D.C. position ($\phi = 0$), are denoted by $\epsilon_1, \epsilon_2, \ldots$.

Similarly, since $p_t = T/(AR_0)$, where $A = $ piston area and $R_0 = $ crank radius, the tangential pressure can be expressed by a Fourier series

$$p_t = |T_0| + |T_1| \sin(\omega t + \epsilon_1) + \ldots + |T_n| \sin(n\omega t + \epsilon_n) + \ldots,$$

where $|T_n| = |C_n|/(AR_0)$, that is, the *harmonic component of tangential pressure* [Lb./in.²] is equal to the corresponding torque harmonic divided by the piston area and the crank radius.

Work done by torque harmonics. The reason for using these individual harmonics is that, in an engine system vibrating at a frequency equal to n times the number of working cycles per unit time of one cylinder, the vibratory work performed by a repetitive non-sinusoidal torque is solely the work done by the nth torque harmonic.

In other words, if $\theta = \theta_0 \sin n\omega t$ is the shaft vibration at the position subjected to the torque $T = |C_0| + \sum_m |C_m| \sin(m\omega t + \epsilon_m)$, the work done in one cycle of the vibration is

$$U = U_n = \pi |C_n| \theta_0 \sin \epsilon_n = \pi |T_n| AR_0 \theta_0 \sin \epsilon_n,$$

and all the other terms U_m ($m \neq n$) give a zero result.

Derivation. The work per vibration cycle of period $\tau = 2\pi/(m\omega)$ performed by the mth torque harmonic can be written as

$$U_m = \int_0^\tau C_m \dot\theta \, dt = \int_0^{2\pi/m\omega} AR_0 |T_m| \sin(m\omega t + \epsilon_m) \times \theta_0 n\omega \cos n\omega t \, dt$$

$$= \theta_0 AR_0 |T_m| n\omega \int_0^{2\pi/m\omega} \{\cos \epsilon_m \sin m\omega t \cos n\omega t$$
$$+ \sin \epsilon_m \cos m\omega t \cos n\omega t\} \, dt,$$

and, noting that

$$\sin m\omega t \cos n\omega t = \tfrac{1}{2}[\sin(m+n)\omega t + \sin(m-n)\omega t],$$
$$\cos m\omega t \cos n\omega t = \tfrac{1}{2}[\cos(m+n)\omega t + \cos(m-n)\omega t],$$

the integral is easily evaluated. The results are:

if $m \neq n$: $\qquad U_m = 0$;

if $m = n$: $\qquad U_m = U_n = \pi |T_n| AR_0 \theta_0 \sin \epsilon_n$;

so that $\qquad U = \sum_{k=0}^{\infty} U_k = 0 + 0 + \ldots + U_n + 0 + \ldots = U_n.$

Thus, in order to determine the vibratory work developed at a frequency n times greater than the number of working cycles per minute of an engine cylinder, it is necessary to know the value of the nth torque harmonic, or the corresponding component of the tangential pressure.

In view of the fact that the values of the harmonics vary as functions of the i.m.e.p., it is usual practice to analyse four or five tangential-pressure diagrams and plot the values of each harmonic against i.m.e.p. from zero load to full load, as shown in Fig. 1.

In four-stroke cycle engines, the working cycle covers 720 crank-angle degrees, i.e. two crankshaft revolutions. Relatively to the engine speed, the 'harmonic number' n^* should then be divided by two to obtain the 'order number' n of the critical speed, so that, as shown in Fig. 1, half-order

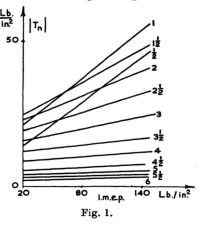

Fig. 1.

components are obtained. Thus, the following definitions are used:

$$\text{Critical speed order number} = n = \frac{F}{N} = \frac{\text{natural frequency}}{\text{running speed}} \quad \begin{array}{l}\text{[vibrations} \\ \text{per rev]},\end{array}$$

$$\text{Harmonic number} = n^* = \frac{F}{N^*} = \frac{\text{natural frequency}}{\text{no. of working cycles per min.}}$$

[vibrations per working cycle].

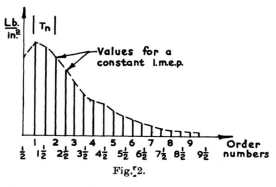

Fig. 2.

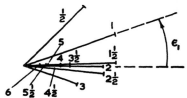

Fig. 3. Polar plot of $|T_n|$-values with their corresponding phase angles ϵ_n.

For two-stroke cycle engines, $n = n^*$. The 'critical speed order number' is generally used to designate tangential-pressure components. There are still some published data in the technical literature, however, which refer to the 'harmonic number', and it is advisable to ascertain which is used in each case. Throughout this handbook, only the 'critical speed order number' is employed.

Other ways of plotting tangential pressure components are shown in Figs. 2 and 3.

2·151 Procedures used for Fourier-harmonic analysis

The Fourier harmonics of a periodic curve can be determined by a number of methods, some of which are listed below:

Mechanical methods:

Harmonic analyser (Stanley-Mader type): a planimetering instrument of this type is in use at the B.I.C.E.R.A. Laboratory;

Michelson-Stratton harmonic analyser and synthesizer (this fairly large apparatus is described in *Phil. Mag.*, January, 1898).

Graphical methods:

Circle-and-radii method (for description, see Eagle, A., *A Practical Treatise on Fourier's Theorem*, p. 95 (Longmans, Green and Co. London, 1925)).

Tabulation methods:

These are extensively used and their accuracy depends on the number of ordinates employed; a calculating machine is an advantage in this connexion. The following list covers tables ranging from 12 to 72 ordinates:

12 ordinates: Runge, C. and König, H., *Numerisches Rechnen*, pp. 218–33;

24 ordinates: Hütte, *Des Ingenieurs Taschenbuch*, vol. 1;

36 ordinates: Runge, C., *Z. Phys. Math.* vol. 48, p. 449;

48 ordinates: Ker Wilson, W., *Practical Solution of Vibration Problems*, vol. 2, pp. 637–49 (Chapman and Hall Ltd. 1942);

48 ordinates: Manley, R. G., *Waveform Analysis*, pp. 202–8 (Chapman and Hall, Ltd. 1945);

48 ordinates: Den Hartog, J. P., *Mechanical Vibrations*, 2nd ed., pp. 25–30 (McGraw Hill Book Co. 1940);

72 ordinates: Stansfield, R. S., 'The measurement of torsional vibrations', *Proc. Instn Mech. Engrs, Lond.*, vol. 148, no. 5 (1942), pp. 187–93.

The last-mentioned tabulated schedule is considered sufficiently extensive to obtain reliable values for 24 harmonics, i.e. up to the 12th order for four-stroke cycle single-acting engines.

The 72-ordinate schedule is recommended as a standard method for the harmonic analysis of tangential-pressure diagrams by the Torsional Vibration Panel of B.I.C.E.R.A. In view of the importance of having this schedule used as widely as possible, permission has been obtained from its author for its reproduction in the following pages of this section.

Note. The tabulations are generally based on the following relations for determining the coefficients a_n and b_n of an empirical function $y(\phi)$ with a 360° period:

$$\frac{a_0}{2} \simeq \frac{1}{N}\sum_{k=1}^{N} y_k, \quad a_n \simeq \frac{2}{N}\sum_{k=1}^{N} y_k \cos\left[n\,\frac{360°}{N}k\right], \quad b_n \simeq \frac{2}{N}\sum_{k=1}^{N} y_k \sin\left[n\,\frac{360°}{N}k\right],$$

where N = number of ordinates, $k = 1, 2, 3, \ldots$ = ordinate number, and n = number of the harmonic component. For a 720° period, the 360° should be placed by 720° in the above expressions. The approximations are obtained from $\phi = k \times 360°/N = 2\pi k/N$, and $d\phi \simeq \Delta\phi = 360°/N$, applied to the usual equations for a_n and b_n.

2·152 72-ordinate harmonic analysis schedule

The following schedule for harmonic analyses was first published by R. Stansfield.† The schedule has been arranged by J. G. Withers on the basis of data prepared by the late R. P. G. Denman.

The torque diagram or tangential-pressure diagram to be analysed is usually plotted with its origin at the T.D.C. position. The ordinate at the origin is designated Y_0, and ordinates at 5, 10, 15, 20, ..., 355° for a two-stroke cycle (or at 10, 20, 30, 40, ..., 710° for a four-stroke cycle) after this are designated $Y_1, Y_2, Y_3, Y_4, \ldots, Y_{71}$. The values of these ordinates must be filled in, in the appropriate spaces in Table 1. Sums and differences of pairs of ordinates in the same columns are taken, and form the quantities S and D. These latter terms are then treated in the same manner to give L and M, and P and R terms. Finally, the L and R terms are similarly added and subtracted to obtain terms N and T, and V and W.

The harmonic coefficients can now be calculated by filling in Tables 2–5. Each term given on these sheets must be multiplied by the factor on the extreme left in the same row. Totals are then made for each column, with due regard to signs, and entered in the spaces S_c and S_e, and D_c and D_e. Then

$$\left.\begin{aligned} A_{(n)} &= S_{cn} + S_{en} \\ A_{(36-n)} &= S_{cn} - S_{en} \end{aligned}\right\} \text{sine terms} \quad \text{and} \quad \left.\begin{aligned} B_{(n)} &= D_{cn} + D_{en} \\ B_{(36-n)} &= D_{cn} - D_{en} \end{aligned}\right\} \text{cosine terms,}$$

excepting
$$B_0 = \tfrac{1}{2}(D_{c0} + D_{e0}), \quad A_{18} = S_{c18},$$
$$B_{36} = \tfrac{1}{2}(D_{c0} - D_{e0}), \quad B_{18} = D_{c18}.$$

The above are the coefficients of the series

$$f(x) = A_1 \sin 2\pi nx + A_2 \sin 4\pi nx + \ldots + A_{35} \sin 70\pi nx$$
$$+ B_0 + B_1 \cos 2\pi nx + B_2 \cos 4\pi nx + \ldots + B_{36} \cos 72\pi nx.$$

† Stansfield, R., 'The measurement of torsional vibrations', *Proc. Inst. Mech. Engrs, Lond.*, vol. 148, no. 5, pp. 175–93.

Checks. On account of the large number of operations to be carried out, it is to be expected that mistakes will occur. Therefore it is wise to make checks as indicated below:

(1) S_c terms. Check for D_1; if an error, check for D_9.

$$\tfrac{1}{4}D_1 = S_{c1}\sin 5° + S_{c2}\sin 10° + \ldots + S_{c17}\sin 85° + \tfrac{1}{2}S_{c18}\sin 90°.$$

This checks all S_c terms.

$$\tfrac{1}{4}D_9 = (S_{c1} + S_{c3} - S_{c5} - S_{c7} + S_{c9} + S_{c11} - S_{c13} - S_{c15} + S_{c17})\sin 45°$$
$$+ S_{c2} - S_{c6} + S_{c10} - S_{c14} + \tfrac{1}{2}S_{c18}.$$

This checks all S_c terms except 4, 8, 12 and 16.

(2) S_e terms. Check for D_2; if an error, check for D_{18}.

$$\tfrac{1}{4}D_2 = (S_{e1} + S_{e17})\sin 10° + (S_{e2} + S_{e16})\sin 20° + (S_{e3} + S_{e15})\sin 30° + \ldots$$
$$+ (S_{e8} + S_{e10})\sin 80° + S_{e9}.$$

This checks all S_e terms, except S_{e18}.

$$\tfrac{1}{4}D_{18} = S_{e1} - S_{e3} + S_{e5} - S_{e7} + S_{e9} - S_{e11} + S_{e13} - S_{e15} + S_{e17}.$$

This checks all odd-numbered S_e terms.

(3) D_c terms. Check for S_0; if an error, check for S_{18}.

$$\tfrac{1}{2}S_0 = \tfrac{1}{2}D_{c0} + D_{c1} + D_{c2} + \ldots + D_{c17} + \tfrac{1}{2}D_{c18}.$$

This checks all D_c terms.

$$\tfrac{1}{4}S_{18} = \tfrac{1}{2}D_{c0} - D_{c2} + D_{c4} - D_{c6} + - \ldots + D_{c16} - \tfrac{1}{2}D_{c18}.$$

This checks all even-numbered D_c terms.

(4) D_e terms. Check for S_1; if an error, check for S_9.

$$\tfrac{1}{4}S_1 = \tfrac{1}{2}D_{e0} + D_{e1}\sin 85° + D_{e2}\sin 80° + \ldots + D_{e17}\sin 5°.$$

This checks all D_e terms.

$$\tfrac{1}{4}S_9 = \tfrac{1}{2}D_{e0} - D_{e4} + D_{e8} - D_{e12} + D_{e16}$$
$$+ (D_{e1} - D_{e3} - D_{e5} + D_{e7} + D_{e9} - D_{e11} - D_{e13} + D_{e15} + D_{e17})\sin 45°.$$

This checks all D_e terms, except 2, 6, 10 and 14.

2·16   Inertia force due to reciprocating parts

If $W_{\text{recip.}}$ = weight [Lb.] of the reciprocating parts in an engine cylinder (i.e. weights of: upper end of connecting rod + piston + rings + gudgeon pin),

$\omega = (2\pi/60) \times N$, N being the crankshaft speed [rev./min.],

L = connecting rod length [in.], and

R_0 = crank radius [in.],

Table 1. 72-ordinate schedule: Y ordinates

Ordinates Y	Y_0	Y_1	Y_2	Y_3	Y_4	Y_5	Y_6	Y_7	Y_8	Y_9	Y_{10}	Y_{11}	Y_{12}	Y_{13}	Y_{14}	Y_{15}	Y_{16}	Y_{17}	Y_{18}
		Y_{71}	Y_{70}	Y_{69}	Y_{68}	Y_{67}	Y_{66}	Y_{65}	Y_{64}	Y_{63}	Y_{62}	Y_{61}	Y_{60}	Y_{59}	Y_{58}	Y_{57}	Y_{56}	Y_{55}	Y_{54}
Sums S	S_0	S_1	S_2	S_3	S_4	S_5	S_6	S_7	S_8	S_9	S_{10}	S_{11}	S_{12}	S_{13}	S_{14}	S_{15}	S_{16}	S_{17}	S_{18}
Differences D	D_0	D_1	D_2	D_3	D_4	D_5	D_6	D_7	D_8	D_9	D_{10}	D_{11}	D_{12}	D_{13}	D_{14}	D_{15}	D_{16}	D_{17}	D_{18}
Ordinates Y	Y_{19}	Y_{20}	Y_{21}	Y_{22}	Y_{23}	Y_{24}	Y_{25}	Y_{26}	Y_{27}	Y_{28}	Y_{29}	Y_{30}	Y_{31}	Y_{32}	Y_{33}	Y_{34}	Y_{35}	Y_{36}	
	Y_{53}	Y_{52}	Y_{51}	Y_{50}	Y_{49}	Y_{48}	Y_{47}	Y_{46}	Y_{45}	Y_{44}	Y_{43}	Y_{42}	Y_{41}	Y_{40}	Y_{39}	Y_{38}	Y_{37}		
Sums S	S_{19}	S_{20}	S_{21}	S_{22}	S_{23}	S_{24}	S_{25}	S_{26}	S_{27}	S_{28}	S_{29}	S_{30}	S_{31}	S_{32}	S_{33}	S_{34}	S_{35}	S_{36}	
Differences D	D_{19}	D_{20}	D_{21}	D_{22}	D_{23}	D_{24}	D_{25}	D_{26}	D_{27}	D_{28}	D_{29}	D_{30}	D_{31}	D_{32}	D_{33}	D_{34}	D_{35}	D_{36}	
S	S_0	S_1	S_2	S_3	S_4	S_5	S_6	S_7	S_8	S_9	S_{10}	S_{11}	S_{12}	S_{13}	S_{14}	S_{15}	S_{16}	S_{17}	S_{18}
	S_{36}	S_{35}	S_{34}	S_{33}	S_{32}	S_{31}	S_{30}	S_{29}	S_{28}	S_{27}	S_{26}	S_{25}	S_{24}	S_{23}	S_{22}	S_{21}	S_{20}	S_{19}	

†Sums L	L_0	L_1	L_2	L_3	L_4	L_5	L_6	L_7	L_8	L_9	L_{10}	L_{11}	L_{12}	L_{13}	L_{14}	L_{15}	L_{16}	L_{17}	L_{18}
Differences M	M_0	M_1	M_2	M_3	M_4	M_5	M_6	M_7	M_8	M_9	M_{10}	M_{11}	M_{12}	M_{13}	M_{14}	M_{15}	M_{16}	M_{17}	M_{18}
D	D_0	D_1	D_2	D_3	D_4	D_5	D_6	D_7	D_8	D_9	D_{10}	D_{11}	D_{12}	D_{13}	D_{14}	D_{15}	D_{16}	D_{17}	D_{18}
	D_{36}	D_{35}	D_{34}	D_{33}	D_{32}	D_{31}	D_{30}	D_{29}	D_{28}	D_{27}	D_{26}	D_{25}	D_{24}	D_{23}	D_{22}	D_{21}	D_{20}	D_{19}	
Sums P	P_0	P_1	P_2	P_3	P_4	P_5	P_6	P_7	P_8	P_9	P_{10}	P_{11}	P_{12}	P_{13}	P_{14}	P_{15}	P_{16}	P_{17}	P_{18}
Differences R	R_0	R_1	R_2	R_3	R_4	R_5	R_6	R_7	R_8	R_9	R_{10}	R_{11}	R_{12}	R_{13}	R_{14}	R_{15}	R_{16}	R_{17}	R_{18}
L L_0	L_1	L_2	L_3	L_4	L_5	L_6	L_7	L_8	L_9	R_0	R_1	R_2	R_3	R_4	R_5	R_6	R_7	R_8	R_9
L_{18}	L_{17}	L_{16}	L_{15}	L_{14}	L_{13}	L_{12}	L_{11}	L_{10}		R_{18}	R_{17}	R_{16}	R_{15}	R_{14}	R_{13}	R_{12}	R_{11}	R_{10}	
Sums N_0	N_1	N_2	N_3	N_4	N_5	N_6	N_7	N_8	N_9	V_0	V_1	V_2	V_3	V_4	V_5	V_6	V_7	V_8	V_9
Differences T_0	T_1	T_2	T_3	T_4	T_5	T_6	T_7	T_8	T_9	W_0	W_1	W_2	W_3	W_4	W_5	W_6	W_7	W_8	W_9

† $L_0 = S_0 + S_{36}$, etc.

[269]

Table 2. *72-ordinate schedule: sine terms*

Multiplier	A_1 and A_{35}		A_5 and A_{31}		A_7 and A_{29}		A_{11} and A_{25}		A_{13} and A_{23}		A_{17} and A_{19}	
0·0024	P_1		P_7		P_5		$-P_{13}$		$-P_{11}$		P_{17}	
0·0048		P_2		$-P_{14}$		$-P_{10}$		$-P_{10}$		$-P_{14}$		P_2
0·0072	P_3		P_{15}		P_{15}		P_3		$-P_3$		$-P_{15}$	
0·0095		P_4		$-P_8$		$-P_{16}$		P_{16}		P_8		$-P_4$
0·01175	P_5		P_1		P_{11}		P_7		P_{17}		P_{13}	
0·0139		P_6		P_6		$-P_6$		$-P_6$		P_6		P_6
0·0159	P_7		$-P_{13}$		P_1		$-P_{17}$		$-P_5$		$-P_{11}$	
0·01785		P_8		P_{16}		P_4		$-P_4$		$-P_{16}$		$-P_8$
0·0196	P_9		$-P_9$		$-P_9$		P_9		$-P_9$		P_9	
0·0213		P_{10}		P_2		P_{14}		P_{14}		P_2		P_{10}
0·0227	P_{11}		P_5		$-P_{17}$		P_1		P_{13}		$-P_7$	
0·02405		P_{12}		$-P_{12}$		P_{12}		$-P_{12}$		P_{12}		$-P_{12}$
0·0252	P_{13}		P_{17}		$-P_7$		$-P_{11}$		P_1		P_5	
0·0261		P_{14}		$-P_{10}$		P_2		P_2		$-P_{10}$		P_{14}
0·0268	P_{15}		P_3		P_3		P_{15}		$-P_{15}$		$-P_3$	
0·02735		P_{16}		P_4		$-P_8$		P_8		$-P_4$		$-P_{16}$
0·0277	P_{17}		$-P_{11}$		P_{13}		$-P_5$		P_7		P_1	
0·0278		P_{18}		P_{18}		$-P_{18}$		$-P_{18}$		P_{18}		P_{18}
Sums	S_{c1}	S_{e1}	S_{c5}	S_{e5}	S_{c7}	S_{e7}	S_{c11}	S_{e11}	S_{c13}	S_{e13}	S_{c17}	S_{e17}
Components	A_1	A_{35}	A_5	A_{31}	A_7	A_{29}	A_{11}	A_{25}	A_{13}	A_{23}	A_{17}	A_{19}

For $n=1$ to 17: $\quad A_{(n)}=S_{c(n)}+S_{e(n)}, \quad A_{18}=S_{c18},$
$$A_{(36-n)}=S_{c(n)}-S_{e(n)}, \quad B_{18}=D_{c18}.$$

Table 3. 72-ordinate schedule: sine terms

Multiplier	A_2 and A_{34}		A_4 and A_{32}		A_8 and A_{28}		A_{10} and A_{26}		A_{14} and A_{22}		A_{16} and A_{20}	
0·0048	V_1						$-V_7$		$-V_5$			
0·0095		V_2	W_1	W_8	$-W_5$	W_4		$-V_4$		$-V_8$	$-W_7$	W_2
0·0139	V_3						V_3		$-V_3$			
0·01785		V_4	W_7	W_2	W_1	$-W_8$		V_8		V_2	W_5	$-W_4$
0·0213	V_5						V_1		V_7			
0·02405		V_6	W_3	W_6	W_3	$-W_6$		$-V_6$		V_6	$-W_3$	W_6
0·0261	V_7						$-V_5$		V_1			
0·02735		V_8	W_5	W_4	$-W_7$	W_2		V_2		$-V_4$	W_1	$-W_8$
0·0278	V_9						V_9		$-V_9$			
Sums	S_{c2}	S_{s2}	S_{c4}	S_{s4}	S_{c8}	S_{s8}	S_{c10}	S_{s10}	S_{c14}	S_{s14}	S_{c16}	S_{s16}
Components	A_2	A_{34}	A_4	A_{32}	A_8	A_{28}	A_{10}	A_{26}	A_{14}	A_{22}	A_{16}	A_{20}

Multiplier	A_3 and A_{33}		A_6 and A_{30}		A_9 and A_{27}		A_{12} and A_{24}		A_{15} and A_{21}		A_{18}	B_{18}
0·0072	C_1								C_5			
0·0139			C_2	U_1					C_2			
0·0196	C_3					F_1			$-C_3$			
0·02405			C_4	U_2			E_1	E_2	$-C_4$			
0·0268	C_5								C_1			
0·0278			C_6	U_3		F_2			C_6		G_1	Z_0
Sums	S_{c3}	S_{s3}	S_{c6}	S_{s6}	S_{c9}	S_{s9}	S_{c12}	S_{s12}	S_{c15}	S_{s15}	S_{c18}	S_{s18}
Components	A_3	A_{33}	A_6	A_{30}	A_9	A_{27}	A_{12}	A_{24}	A_{15}	A_{21}	A_{18}	B_{18}

$C_1 = P_1 + P_{11} - P_{13} =$
$C_2 = P_2 + P_{10} - P_{14} =$
$C_3 = P_3 + P_9 - P_{15} =$
$C_4 = P_4 + P_8 - P_{16} =$
$C_5 = P_5 + P_7 - P_{17} =$
$C_6 = P_6 - P_{18}$ =

$U_1 = V_1 + V_5 - V_7$
$U_2 = V_2 + V_4 - V_8$
$U_3 = V_3 - V_9$
$F_1 = P_1 + P_3 + P_9 + P_{11} + P_{17} - (P_5 + P_7 + P_{13} + P_{15}) =$
$F_2 = P_2 + P_{10} + P_{18} - (P_6 + P_{14})$ =

$E_1 = W_1 + W_7 - W_5$ =
$E_2 = W_3 + W_8 - W_4$ =
$G_1 = V_1 + V_6 + V_9 - (V_3 + V_7) =$
$Z_0 = (T_0 + T_4 + T_8) - (T_2 + T_6) =$

[271]

Table 4. *72-ordinate schedule: cosine terms*

Multiplier	B_1 and B_{35}		B_5 and B_{31}		B_7 and B_{29}		B_{11} and B_{25}		B_{13} and B_{23}		B_{17} and B_{19}	
0·0024		M_{17}	M_{11}			$-M_{13}$	M_5			$-M_7$	M_1	
0·0048	M_{16}			$-M_4$	M_8		M_8			$-M_4$	M_{16}	
0·0072		M_{15}	M_3			$-M_3$		$-M_{15}$		$-M_{15}$		$-M_3$
0·0095	M_{14}			$-M_{10}$	M_2			$-M_2$	M_{10}			$-M_{14}$
0·01175		M_{13}	M_{17}			$-M_7$		$-M_{11}$	M_1		M_5	
0·0139	M_{12}		M_{12}		M_{12}		M_{12}		M_{12}		M_{12}	
0·0159		M_{11}		$-M_5$		$-M_{17}$	M_1			$-M_{13}$		$-M_7$
0·01785	M_{10}		M_2			$-M_{14}$	M_{14}			$-M_2$		$-M_{10}$
0·0196		M_9		$-M_9$	M_9			$-M_9$		$-M_9$		M_9
0·0213	M_8		M_{16}			$-M_4$		$-M_4$	M_{16}		M_8	
0·0227		M_7	M_{13}		M_1			$-M_{17}$	M_5			$-M_{11}$
0·02405	M_6		$-M_6$		$-M_6$		M_6		M_6		$-M_6$	
0·0252		M_5	M_1		M_{11}		M_7		M_{17}		M_{13}	
0·0261	M_4		$-M_8$		$-M_{16}$		$-M_{16}$		$-M_8$		M_4	
0·0268		M_3	M_{15}			$-M_{15}$	$-M_3$		$-M_3$			$-M_{15}$
0·02735	M_2		M_{14}		M_{10}			$-M_{10}$		$-M_{14}$	$-M_2$	
0·0277		M_1		$-M_7$		$-M_5$	M_{13}		M_{11}		M_{17}	
0·0278	M_0		M_0		M_0		M_0		M_0		M_0	
Sums	D_{c1}	D_{e1}	D_{c5}	D_{e5}	D_{c7}	D_{e7}	D_{c11}	D_{e11}	D_{c13}	D_{e13}	D_{c17}	D_{e17}
Components	B_1	B_{35}	B_5	B_{31}	B_7	B_{29}	B_{11}	B_{25}	B_{13}	B_{23}	B_{17}	B_{19}

For $n=1$ to 17: $\quad B_{(n)} = D_{c(n)} + D_{e(n)}, \quad B_0 = \tfrac{1}{2}(D_{c0} + D_{e0}),$

$\qquad\qquad\qquad\quad B_{(36-n)} = D_{c(n)} - D_{e(n)}, \quad B_{36} = \tfrac{1}{2}(D_{c0} - D_{e0}).$

Table 5. 72-ordinate schedule: cosine terms

Multiplier	B_2 and B_{34}		B_4 and B_{32}		B_8 and B_{28}		B_{10} and B_{26}		B_{14} and B_{22}		B_{16} and B_{20}	
0·0048	T_8		N_4	$-N_5$	N_2	N_7	$-T_2$		T_4		N_8	N_1
0·0095		T_7					$-T_5$		T_1			
0·0139	T_6		$-N_6$	N_3	$-N_6$	$-N_3$	T_6		T_6		$-N_6$	$-N_3$
0·01785		T_5					T_1		$-T_7$			
0·0213	T_4		N_2	$-N_7$	N_8	N_1	T_8		$-T_2$		N_4	N_5
0·02405		T_3					$-T_3$		$-T_3$			
0·0261	T_2		$-N_8$	N_1	$-N_4$	$-N_5$	$-T_4$		$-T_8$		$-N_2$	$-N_7$
0·02735		T_1					T_7		T_5			
0·0278	T_0		N_0	$-N_9$	N_0	N_9	T_0		T_0		N_0	N_9
Sums	D_{c2}	D_{e2}	D_{c4}	D_{e4}	D_{c8}	D_{e8}	D_{c10}	D_{e10}	D_{c14}	D_{e14}	D_{c16}	D_{e16}
Components	B_2	B_{34}	B_4	B_{32}	B_8	B_{28}	B_{10}	B_{26}	B_{14}	B_{22}	B_{16}	B_{20}

Multiplier	B_3 and B_{33}		B_6 and B_{30}		B_9 and B_{27}		B_{12} and B_{24}		B_{15} and B_{21}		B_0 and B_{36}	
0·0072		H_5							H_1			
0·0139	H_4		J_2				$-K_2$	K_1	H_4			
0·0196		H_3				X_1				$-H_3$		
0·02405	H_2		J_1							$-H_2$		
0·0268		H_1							H_5			
0·0278	H_0		J_0		X_0		K_0	$-K_3$	H_0		Q_0	Q_1
Sums	D_{c3}	D_{e3}	D_{c6}	D_{e6}	D_{c9}	D_{e9}	D_{c12}	D_{e12}	D_{c15}	D_{e15}	D_{c0}	D_{e0}
Components	B_3	B_{33}	B_6	B_{30}	B_9	B_{27}	B_{12}	B_{24}	B_{15}	B_{21}	B_0	B_{36}

$$H_0=M_0-M_{12} \qquad = \qquad J_0=T_0-T_6 \qquad = \qquad K_0=N_0+N_6 \qquad =$$
$$H_1=M_1-(M_{11}+M_{13})= \qquad J_1=T_1-(T_5+T_7) \qquad = \qquad K_1=N_1+N_5+N_7 \qquad =$$
$$H_2=M_2-(M_{10}+M_{14})= \qquad J_2=T_2-(T_4+T_8) \qquad = \qquad K_2=N_2+N_4+N_8 \qquad =$$
$$H_3=M_3-(M_9+M_{15})= \qquad X_0=M_0+M_6+M_{10}-(M_4+M_{12})= \qquad K_3=N_3+N_9 \qquad =$$
$$H_4=M_4-(M_8+M_{16})= \qquad X_1=M_1+M_7+M_9+M_{15}+M_{17} \qquad Q_0=N_0+N_2+N_4+N_6+N_8=$$
$$H_5=M_5-(M_7+M_{17})= \qquad -(M_3+M_5+M_{11}+M_{13}) \qquad = \qquad Q_1=N_1+N_3+N_5+N_7+N_9=$$

the inertia force $P_{\text{recip.}}$ due to these reciprocating parts at any crank-angle position $\phi = \omega t$ [rad.] is

$$P_{\text{recip.}} = -\frac{W_{\text{recip.}}}{g} \times \ddot{y}_P \cong -\frac{W_{\text{recip.}}}{386\cdot 4}\,\omega^2 R_0\left[\cos\phi + \frac{R_0}{L}\cos 2\phi\right] \quad [\text{Lb.}],$$

$$= -2\cdot 84 \times 10^{-5} \times N^2 R_0 W_{\text{recip.}} \times \left[\cos\phi + \frac{R_0}{L}\cos 2\phi\right] \quad [\text{Lb.}],$$

in accordance with the relations for piston acceleration $\ddot{y}_P$ derived in section 2·13. Numerical values of $\ddot{y}_P/(\omega^2 R_0)$, which is termed the 'inertia factor', are given in Table 3 of section 2·13 in steps of 5 degrees for various values of the ratio R_0/L.

Metric units. With all linear dimensions in centimetres, and weights in Kg., the inertia force $P_{\text{recip.}}$ can be obtained in Kg. by replacing 386·4 by 981·0, or in the alternative expression by replacing the constant $-2\cdot 84 \times 10^{-5}$ by $-1\cdot 1186 \times 10^{-5}$.

2·161   *Tangential-pressure components of inertia torque*

The inertia torque $T_{(I)}$ due to the reciprocating parts can be related to a corresponding tangential pressure $p_{t(I)}$, which is obtained by dividing the expression for $T_{(I)}$ by the crank radius R_0 and the piston area A. This gives $p_{t(I)}$ in terms of the crank angle position ϕ:

$$\frac{T_{(I)}}{AR_0} \equiv p_{t(I)} \cong \frac{W_{\text{recip.}}}{gA}\,\omega^2 R_0\left\{\frac{R_0}{4L}\sin\phi - \tfrac{1}{2}\sin 2\phi - \frac{3R_0}{4L}\sin 3\phi\right\} \quad [\text{Lb./in.}^2],$$

where W = weight of reciprocating parts [Lb.], $g = 386\cdot 4$ [in./sec.^{2}], R_0 = crank radius [in.], L = length of connecting rod [in.], and $\omega = (2\pi/60) \times N$, N being the crankshaft speed [rev./min.].

Thus the components consist of sine terms only, which can be written as follows:

$$S_{1(I)} = +\frac{W_{\text{recip.}}}{gA}\,\omega^2 R_0 \times \frac{R_0}{4L},$$

$$S_{2(I)} = -\frac{W_{\text{recip.}}}{gA}\,\omega^2 R_0 \times \tfrac{1}{2}.$$

and $$S_{3(I)} = -\frac{W_{\text{recip.}}}{gA}\,\omega^2 R_0 \times \frac{3R_0}{4L}.$$

The + or − sign should not be omitted when adding one of these components to the corresponding sine component for gas pressure.

Derivation. Considering Fig. 1, we can write

$$\overline{OC} = r = \overline{OA}\tan\psi = \tan\psi\,(R_0\cos\phi + L\cos\psi),$$

where $$L = \overline{AE},$$

or $$r = R_0\cos\phi\tan\psi + L\sin\psi,$$

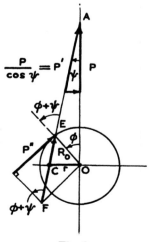

Fig. 1.

[274]

and, since $L \sin \psi = R_0 \sin \phi$,

$$r = R_0[\cos \phi \tan \psi + \sin \phi].$$

If P is the inertia force of the reciprocating masses, the corresponding torque is

$$T = rP = R_0 P \times [\cos \phi \tan \psi + \sin \phi].$$

Another way of obtaining this expression is as follows:

$$T = \overline{OE} \times P'' = R_0 \times \overline{FE} \times \sin (\phi + \psi) = R_0 P' \sin (\phi + \psi)$$

$$= R_0 P \frac{\sin (\phi + \psi)}{\cos \psi} = R_0 P[\sin \phi + \cos \phi \tan \psi].$$

Therefore
$$rP = R_0 P'' = T.$$

As $P = -\dfrac{W_{\text{recip.}}}{g} \times \ddot{y}_P$ (where $\ddot{y}_P$ = piston acceleration and $g = 386 \cdot 4$ in./sec.²), using the results obtained in section 2·13, we have:

$$\ddot{y}_P \cong \omega^2 R_0 \left[\cos \phi + \frac{R_0}{L} \cos 2\phi \right],$$

$$r = \tan \psi (R_0 \cos \phi + L \cos \psi) = \tan \psi \left[R_0 \cos \phi + L \sqrt{\left\{ 1 - \left(\frac{R_0}{L} \right)^2 \sin^2 \phi \right\}} \right]$$

or $\quad r \cong \tan \psi \left[R_0 \cos \phi + L \left(1 - \frac{R_0^2}{2L^2} \sin^2 \phi \right) \right] \cong \tan \psi \left[R_0 \cos \phi + \left(L - \frac{R_0^2}{4L} \right) \right.$

$$\left. + \frac{R_0^2}{4L} \cos 2\phi \right],$$

since $\sin^2 \phi = \frac{1}{2}(1 - \cos 2\phi)$. Furthermore

$$\tan \psi = \sin \psi / \sqrt{(1 - \sin^2 \psi)} \cong \frac{R_0}{L} \sin \phi \times \left[1 + \frac{R_0^2}{2L^2} \sin^2 \phi \right].$$

Therefore, with these three expressions for $\tan \psi$, $R_0 \cos \phi + L \cos \psi$ and $\ddot{y}_P$, we obtain

$$T_{(I)} \cong \frac{R_0}{L} \sin \phi \times \left[1 + \frac{R_0^2}{2L^2} \sin^2 \phi \right] \times \left[R_0 \cos \phi + \left(L - \frac{R_0^2}{4L} \right) + \frac{R_0^2}{4L} \cos 2\phi \right]$$

$$\times \frac{-W_{\text{recip.}}}{g} \times \omega^2 R_0 \left[\cos \phi + \frac{R_0}{L} \cos 2\phi \right].$$

Disregarding all powers of R_0/L greater than the first, this gives

$$T_{(I)} \cong -\frac{W_{\text{recip.}}}{g} \omega^2 R_0^2 \left\{ \cos^2 \phi + \cos \phi \left(\frac{L}{R_0} - \frac{R_0}{4L} \right) + \frac{R_0}{4L} \cos \phi \cos 2\phi \right.$$

$$\left. + \frac{R_0}{L} \cos \phi \cos 2\phi + \cos 2\phi \right\} \times \frac{R_0}{L} \sin \phi$$

$$\cong -\frac{W_{\text{recip.}}}{g} \omega^2 R_0^2 \left\{ \cos^2 \phi + \frac{L}{R_0} \cos \phi + \cos 2\phi \right\} \times \frac{R_0}{L} \sin \phi.$$

As $\cos^2\phi = \tfrac{1}{2}(1+\cos 2\phi)$ and $\sin\phi\cos 2\phi = \tfrac{1}{2}(\sin 3\phi - \sin\phi)$,

we have

$$T_{(I)} \cong -\frac{W_{\text{recip.}}}{g}\,\omega^2 R_0^2\left\{\frac{R_0}{2L}\sin\phi + \frac{R_0}{2L}\sin\phi\cos 2\phi + \tfrac{1}{2}\sin 2\phi + \frac{R_0}{L}\sin\phi\cos 2\phi\right\}$$

and

$$T_{(I)} \cong +\frac{W_{\text{recip.}}}{g}\,\omega^2 R_0^2\left[\frac{R_0}{4L}\sin\phi - \tfrac{1}{2}\sin 2\phi - \frac{3R_0}{4L}\sin 3\phi\right].$$

This shows that the torque due to inertia forces of the reciprocating masses is a trigonometrical series of sine terms only. The complete series can be regarded as the equivalent of a Fourier series for the inertia torque. By dividing $T_{(I)}$ by AR_0 we obtain the usual approximation for the tangential pressure components due to inertia, which is given at the beginning of this section.

Note. The complete expansion results in the following series:

$$\frac{T_{(I)}}{AR_0} \equiv p_{t(I)} = \frac{W_{\text{recip.}}}{gA}\,\omega^2 R_0\left\{\frac{R_0}{4L}\sin\phi - \left(\frac{1}{2}+\frac{R_0^4}{32L^4}+\dots\right)\sin 2\phi\right.$$

$$-\left(\frac{3R_0}{4L}+\frac{9R_0^3}{32L^3}+\dots\right)\sin 3\phi - \left(\frac{R_0^2}{4L^2}+\frac{R_0^4}{8L^4}+\dots\right)\sin 4\phi$$

$$\left.+\left(\frac{5R_0^3}{32L^3}+\dots\right)\sin 5\phi + \left(\frac{3R_0^4}{32L^4}+\dots\right)\sin 6\phi+\dots\right\}.$$

However, the additional terms with higher powers of R_0/L are not generally required, particularly since a comparable accuracy is not obtained in determining the tangential pressure components due to gas pressure. This remark also applies to the components due to the dead weight of the reciprocating parts.

For the corrections for the connecting rod couple, which are required in some cases, the reader is referred to the tabulated values given by Ker Wilson.[†]

Tabulation method

The inertia torque curve can be determined as a function of the crank angle ϕ by tabulation, with the aid of the relations

$$T_{(I)} = r \times \frac{-W_{\text{recip.}}}{g}\,\ddot{y}_P \cong r \times \frac{-W_{\text{recip.}}}{g} \times \omega^2 R_0 \times \left[\cos\phi + \frac{R_0}{L}\cos 2\phi\right],$$

using the numerical values for the 'inertia factor'

$$\ddot{y}/(\omega^2 R_0) \cong \cos\phi + \frac{R_0}{L}\cos 2\phi$$

given in Table 3 of section 2·13.

† Ker Wilson, W., *Practical Solution of Torsional Vibration Problems*, vol. 1, 2nd ed. (rev.), pp. 445–8 (Chapman and Hall, Ltd., London, 1942).

The tabulation is as follows:

(1)	(2)	(3)	(4)	(5)	(6)	(7)
ϕ	$\cos\phi + \dfrac{R_0}{L}\cos 2\phi$ (from Table 3)	$\dfrac{R_0}{L}\sin\phi$	$R_0\cos\phi + L$	$r \cong (3)\times(4)$	$(2)\times r$	$T_{(I)} = -\omega^2 R_0 \dfrac{W_{\text{recip.}}}{386\cdot 4}\times(6)$ [Lb.in.]
0°						
5°						
10°						
...						

where $W_{\text{recip.}}$ = weight [Lb.] of the reciprocating masses, $g = 386\cdot 4$ [in./sec.²], and $\omega = (2\pi/60)\times N$, N being the crankshaft speed [rev./min.]. In some cases a graphical construction is used to determine the values of the torque arm r.

The $T\phi$ curve of inertia torque against crank angle is then drawn to a suitable scale and its harmonics (consisting of sine terms only) are obtained either by means of a 72-ordinate schedule or by means of a harmonic analyser.

These torque harmonics are then divided by AR_0 (where A = piston area and R_0 = crank radius) to obtain the corresponding tangential pressure components due to inertia torque.

Metric units. See Note on p. 274.

2·17 Sine, cosine, and resultant components

The resultant tangential components $|T_n|$ due to gas pressure can be expressed by means of their corresponding sine and cosine 'zero-phase' components as follows:

$$|T_n| = \sqrt{(S_n^2 + C_n^2)} \quad \text{[Lb./in.}^2\text{]}$$

and vice versa. In the above equation S_n = sine component and C_n = cosine component.†

If the inertia-torque component is also taken into account, the nth order resultant tangential-pressure component is obtained as

$$|T_n| = \sqrt{\{(S_G + S_I)_n^2 + C_{Gn}^2\}},$$

where the subscripts G and I are used to denote gas pressure and inertia terms, respectively. In most applications, only the first three components of inertia-torque require consideration, and for higher orders the resultant due to gas pressure only is used.

Derivation. The Fourier series for tangential pressure is

$$\frac{T}{AR_0} \equiv p_t = |T_0| + |T_1|\sin(\omega t + \epsilon_1) + \dots + |T_n|\sin(n\omega t + \epsilon_n) + \dots,$$

† *Metric units.* With torque expressed in Kg.cm., area in cm.² and length in cm., the resultant tangential-pressure components are obtained in Kg./cm.². It may be noted that 1 Kg./cm.² = 14·22 Lb./in.².

where ϵ_n = phase angle of the nth order component relative to the T.D.C. position, and $\omega = (2\pi/60) \times N^*$, N^* being the number of working cycles per minute of one cylinder. Considering for instance the nth term, this may be rewritten as

$$|T_n| \sin(n\omega t + \epsilon_n) = |T_n| \cos\epsilon_n \sin n\omega t + |T_n| \sin\epsilon_n \cos n\omega t$$

$$= S_n \sin n\omega t + C_n \cos n\omega t,$$

with the relations

$$S_n = |T_n| \cos\epsilon_n, \quad C_n = |T_n| \sin\epsilon_n, \quad \tan\epsilon_n = C_n/S_n$$

and

$$|T_n| = \sqrt{(S_n^2 + C_n^2)}.$$

The Fourier series is thus converted into sums of sine and cosine terms:

$$p_t = |T_0| + C_1 \cos\omega t + C_2 \cos 2\omega t + \ldots + S_1 \sin\omega t + S_2 \sin 2\omega t + \ldots.$$

As the components of tangential pressure due to inertia torque are all sine terms, they can be added to the sine terms due to gas pressure in order to obtain the resultant components as indicated above.

2·18 Curves of tangential-pressure components

Tangential-pressure components are derived from either theoretical or experimental diagrams of torque or cylinder pressure.

It is possible to construct theoretical cylinder-pressure diagrams with the aid of thermodynamic charts taking account of the properties of air-and-fuel mixtures before and after combustion, including effective changes in specific heats and dissociation. However, such diagrams are of limited value for torsional-vibration investigations, and it is preferable, wherever possible, to base design-stage calculations on reliable experimental data obtained from a range of engines of a similar class.

The following curves give the tangential-pressure components derived from cylinder-pressure diagrams which were determined from tests of various classes of engines. Apart from two or three exceptions, the most accurate procedures at present available were used for tests and evaluations, i.e. pressure indication by the Stansfield point-by-point method or by Farnboro' equipment operated by the Standard-B.I.C.E.R.A. spark-control unit, and harmonic analysis with the Stansfield 72-ordinate schedule or the Stanley (Mader-type) harmonic analyser. All the cylinder-pressure diagrams, except in two cases, were taken and evaluated by member firms of B.I.C.E.R.A. or by the Association's laboratory.

The lines of the tangential pressure components plotted against i.m.e.p. are mean values based on at least three engine types of the class of engines considered, and departures from this rule have only been made in the cases of opposed-piston engines and turbo-charged engines, for which further test results are not at present available. It should be noted that the deviations between the mean lines and the evaluated test points increase from about 10 % for low-order components up to 100 %, in some cases, for the highest-order components. This is to be expected, since the fluctuations in cylinder-pressure diagrams vary from cylinder to cylinder and even from one engine cycle to another.

Accordingly, the curves shown in the following Figs. 1–34 are drawn to scales which are adequate for practical applications without attempting to present the results with greater accuracy than can reasonably be warranted.

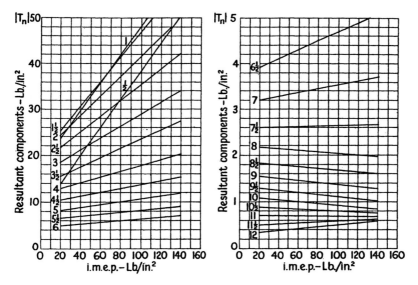

Note. These are mean lines based on tests of six different engine types.

Maximum deviations from mean-line values:

orders	per cent.
$\frac{1}{2}$ to 6	16
$6\frac{1}{2}$ to 12	15 to 100

For conversion into metric units: 1 Lb./in.² = 0·07037 Kg./cm.².

Figs. 1 and 2. Resultant harmonic components of tangential pressure, due to gas pressure only. Four-stroke cycle single-acting compression-ignition engines, with running speeds less than 1000 rev./min.

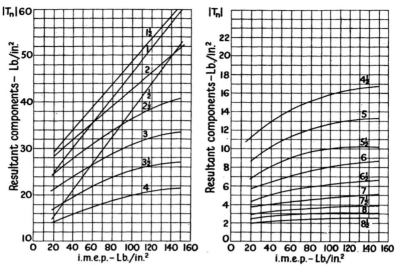

Note. These are mean lines based on tests of five different engine types.

Maximum deviations from mean-line values: orders $\frac{1}{2}$ to $2\frac{1}{2}$: 20%; 3 to $6\frac{1}{2}$: 10%; 7 to 11: over 20%; $11\frac{1}{2}$ and 12: 20%.

For conversion into metric units: 1 Lb./in.^{2} = 0·07037 Kg./cm.^{2}.

Figs. 3 and 4. Resultant harmonic components of tangential pressure, due to gas pressure only. Four-stroke cycle single-acting compression-ignition engines, with running speeds above 1000 rev./min.

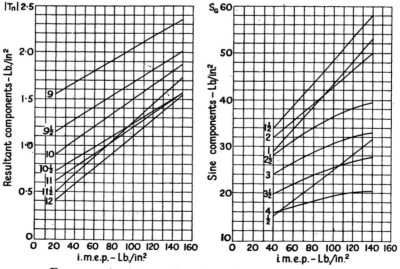

For conversion into metric units: 1 Lb./in.^{2} = 0·07037 Kg./cm.^{2}.

Figs. 5 and 6. Resultant components for orders 9 to 12, and sine terms of harmonic components of tangential pressure, due to gas pressure only. Four-stroke cycle single-acting compression-ignition engines, with running speeds above 1000 rev./min.

[280]

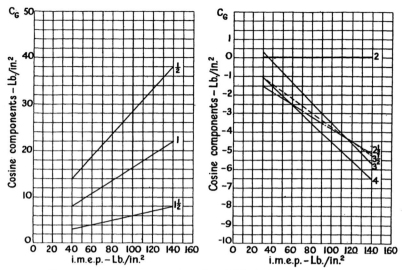

For conversion into metric units: 1 Lb./in.² = 0·07037 Kg./cm.².

Figs. 7 and 8. Cosine terms of harmonic components of tangential pressure, due to gas pressure only. Four-stroke cycle single-acting compression-ignition engines, with runnings speeds above 1000 rev./min.

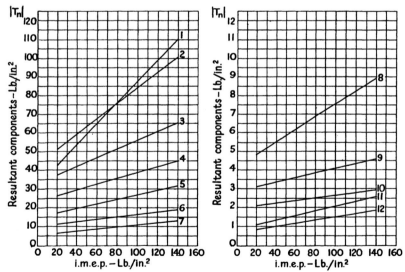

Note. These are mean lines based on tests of three different engine types.

Maximum deviations from mean-line values: orders 1 to 5: 10%; 6 to 9: 25%; 10 and 11: 60%; 12: 80%.

For conversion into metric units: 1 Lb./in.² = 0·07037 Kg./cm.².

Figs. 9 and 10. Resultant harmonic components of tangential pressure, due to gas pressure only. Two-stroke cycle single-acting compression-ignition engines, with running speeds less than 1000 rev./min.

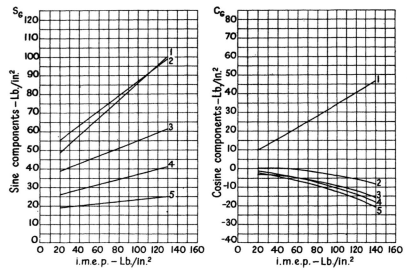

For conversion into metric units: 1 Lb./in.² = 0·07037 Kg./cm.².

Figs. 11 and 12. Sine terms (S_G) and cosine terms (C_G) of harmonic components of tangential pressure, due to gas pressure only. Two-stroke cycle single-acting compression-ignition engines, with runnings speeds less than 1000 rev./min.

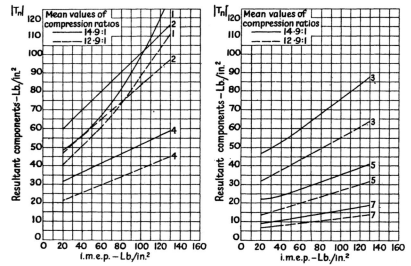

Note. These lines are based on tests of two different engine types, with three different compression ratios A, B and C. A: 15·5:1, B: 14·3:1, C: 12·9:1.

Two sets of curves are given, viz.

Curves for compression ratio of $14·9 = \frac{1}{2}(15·5 + 14·3)$; and curves for compression ratio of 12·9.

Maximum deviations of test points from these lines are as follows: orders 1, 2, 4, 6 and 8: 20%; 3, 5 and 7: 27%; 9 and 10: 32%: 11 and 12: 59%.

For conversion into metric units: 1 Lb./in.² = 0·07037 Kg./cm.².

Figs. 13 and 14. Resultant harmonic components of tangential pressure, due to gas pressure only. Two-stroke cycle single-acting compression-ignition engines, with running speeds above 1000 rev./min.

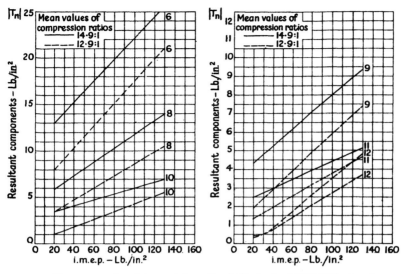

For conversion into metric units: 1 Lb./in.² = 0·07037 Kg./cm.².

Figs. 15 and 16. Resultant harmonic components of tangential pressure, due to gas pressure only. Two-stroke cycle single-acting compression-ignition engines, with running speeds above 1000 rev./min.

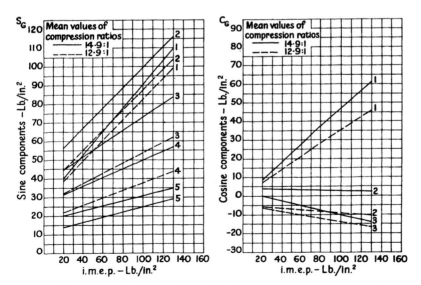

For conversion into metric units: 1 Lb./in.² = 0·07037 Kg./cm.².

Figs. 17 and 18. Sine terms (S_G) and cosine terms (C_G) of harmonic components of tangential pressure, due to gas pressure only. Two-stroke cycle single-acting compression-ignition engines, with runnings speeds above 1000 rev./min.

[283]

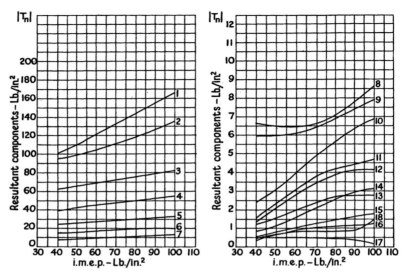

Note. The lines in Figs. 19 to 24 are based on tangential-pressure diagrams derived from cylinder-pressure diagrams by means of the relations for opposed-piston engines given in section 2·14.

For conversion into metric units: 1 Lb./in.² = 0·07037 Kg./cm.².

Figs. 19 and 20. Resultant harmonic components of tangential pressure, due to gas pressure only. Two-stroke cycle Doxford opposed-piston engine.

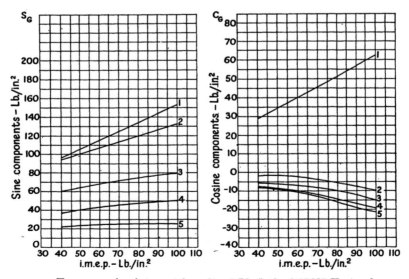

For conversion into metric units: 1 Lb./in.² = 0·07037 Kg./cm.².

Figs. 21 and 22. Sine terms (S_G) and cosine terms (C_G) of harmonic components of tangential pressure, due to gas pressure only. Two-stroke cycle Doxford opposed-piston engine.

[284]

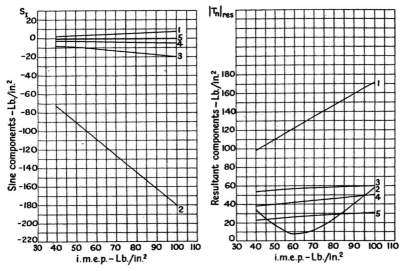

For conversion into metric units: 1 Lb./in.² = 0·07037 Kg./cm.².

Figs. 23 and 24. Sine terms (S_I) due to inertia forces and resultant components $(|\,T_n\,|_{res.})$ of tangential pressure including contributions of gas pressure and inertia forces. Two-stroke cycle Doxford opposed-piston engine.

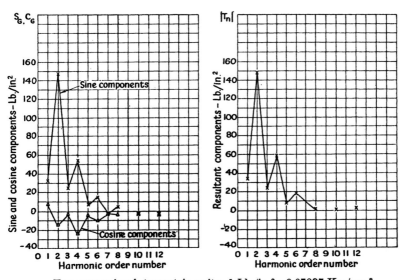

For conversion into metric units: 1 Lb./in.² = 0·07037 Kg./cm.².

Figs. 25 and 26. Sine terms (S_G), cosine terms (C_G) and resultant components of tangential pressure on one crank, due to gas pressure only. Two-stroke cycle double-acting Fullagar opposed-piston engine. At an i.m.e.p. of 87 Lb./in.².

[285]

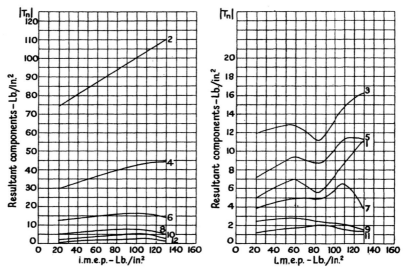

For conversion into metric units: 1 Lb./in.² = 0·07037 Kg./cm.².

Figs. 27 and 28. Resultant harmonic components of tangential pressure, due to gas pressure only. Two-stroke cycle M.A.N. double-acting engine, with maximum pressure of 596 Lb./in.² (42 Kg./cm.²) at full load.

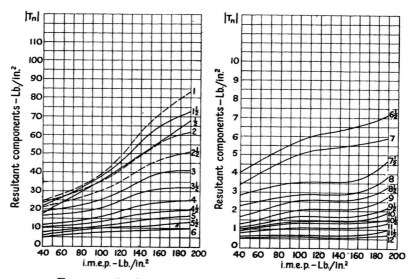

For conversion into metric units: 1 Lb./in.² = 0·07037 Kg./cm.².

Figs. 29 and 30. Resultant harmonic components of tangential pressure, due to gas pressure only. Four-stroke cycle single-acting compression-ignition engine, with Büchi turbo-charger. Running speed less than 1000 rev./min.

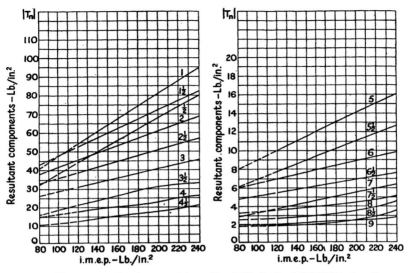

For conversion into metric units: 1 Lb./in.² = 0·07037 Kg./cm.².

Figs. 31 and 32. Resultant harmonic components of tangential pressure, due to gas pressure only. Four-stroke cycle single-acting compression-ignition engine, with Napier turbo-charger. Running speed 600 rev./min.

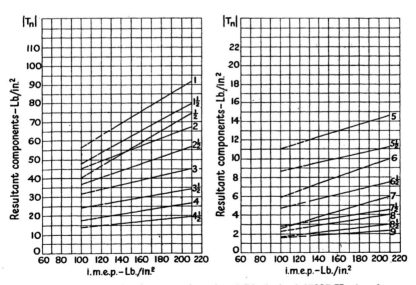

For conversion into metric units: 1 Lb./in.² = 0·07037 Kg./cm.².

Figs. 33 and 34. Resultant harmonic components of tangential pressure, due to gas pressure only. Four-stroke cycle single-acting compression-ignition engine, with Napier turbo-charger. Running speed 750 rev./min.

[287]

2·19 Factors affecting the magnitude of tangential-pressure components

The magnitude of tangential-pressure components due to gas pressure depends on the shape of the cylinder-pressure diagram, which is the result of a variety of factors. These factors and their effects have been considered by the Torsional Vibration Panel of B.I.C.E.R.A. and are summarized below:

(1) In general, it is considered that the tangential-pressure components $|T_n|$ of two-stroke cycle engines are about twice as great as the four-stroke cycle components of the same critical order number.

(2) High-speed engines have higher $|T_n|$ values than slow-speed engines.

(3) For most order numbers, the $|T_n|$ values increase with i.m.e.p. and maximum pressure.

(4) They also tend to increase for higher values of compression ratio and compression pressure. (Thus the $|T_n|$ values for compression-ignition engines are about twice as high as the corresponding figures for electrical-ignition petrol engines.)

(5) Pressure-charging results in an increase in $|T_n|$ values, which may be offset, however, by the use of a lower compression ratio than for the same engine operating as an atmospherically charged unit. For turbo-charged engines, the harmonic components show low values of $|T_n|$ at low i.m.e.p.'s, and increase somewhat more rapidly at higher i.m.e.p. levels than the corresponding components for atmospherically charged engines.

(6) In some cases it has been found that pre-chamber engines may have $|T_n|$ values about 10 % lower than those of direct-injection engines.

(7) Investigations which are being carried out by the B.I.C.E.R.A. Torsional Vibration Panel indicate that, by advancing the injection timing, the tangential-pressure components for a given value of i.m.e.p. may be increased (with corresponding increases in vibration amplitudes).

In view of these considerations, it is not surprising that it is found difficult to assign accurate values of harmonic components to any type of engine, unless these values are based on tests of the particular engine considered. The inaccuracy is greatest for high-order numbers and may easily reach 100 % in some cases. It is fortunate that these divergences occur mainly for low-value components, but their range of variation is one of the main reasons why formulae for predicted vibration amplitudes and stresses can only give very approximate estimated values.

2·191 *Tangential-pressure due to gas pressures in multi-cylinder engines*

For multi-cylinder engines, it is advisable to take if possible cylinder-pressure diagrams from all the cylinders, in order to verify that the

[288]

tangential pressures are similar. As an alternative, the following simple test, used at the B.I.C.E.R.A. Laboratory, and first suggested by members of the Torsional Vibration Panel, can be carried out:

(1) Run the engine at full load on a major critical.

(2) Indicate the torsional vibration of the forward end of the crankshaft on an oscilloscope screen with the aid of suitable equipment and a time-sweep arrangement controlled by the camshaft speed or the crankshaft speed, and consider the unfiltered diagram only.

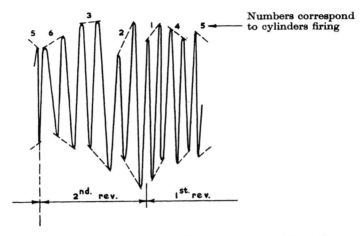

Fig. 1. Unfiltered torsional vibration diagram of 6th order critical. Test engine on ¾ load 69·5 Lb./in.² b.m.e.p. at 1173 rev./min. Typical oscillogram for a 4-stroke cycle engine obtained with a camshaft-operated time sweep. Firing order 145632. All cylinders operating.

Maximum amplitude of corresponding impulses:

1	2	3	4	5	6
6·4	6·8	6·0	5·6	5·3	4·8 cm.

(3) Relate the peak of each successive vibration impulse to the firing of the corresponding cylinder, for instance by cutting off the fuel supply to each cylinder, for one cylinder at a time, by raising the corresponding fuel pump plunger off its cam while the engine is kept at constant load. (The reduction in maximum pressure of the individual cylinders results in a marked reduction in amplitude of the corresponding impulses.)

(4) Photograph or draw on tracing paper the vibration diagram of the engine running on the major critical with all cylinders firing.

(5) The peaks of the successive impulses should follow the same sequence as the firing order (see Fig. 1).

(6) The amplitudes of the impulses should decrease gradually from the forward-end cylinder impulse to that of the cylinder nearest to the node. If the diagram does not conform to this condition, but shows, for instance, one or two impulses very much greater (or very much smaller)

than they should be according to their cylinder number, the cylinders are not all operating satisfactorily.

(7) In such a case, it is advisable to check the injection nozzles, the working pressures of the injector nozzle springs, the fuel delivery to each cylinder, the design of the manifolds, the piston rings and other components associated with the cylinders investigated.

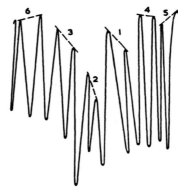

Fig. 2. Nozzle of cylinder 2 cut out. Unfiltered diagram of 6th order critical. Same load and engine speed as in Fig. 1.

Although the amplitudes of the pulses corresponding to no. 2 cylinder have dropped, the total amplitude of the diagram has remained unchanged. Filtered diagrams for the test engine confirmed this result.

It should be noted that, if the engine rack setting and speed are kept constant, any reduction in amplitude of the impulses of one cylinder is offset by changes in the impulses from the other cylinders.

This procedure can be used for rapid check tests of series-produced engines in order to verify that their vibration impulses are similar to those of a prototype.

Typical diagrams obtained at the B.I.C.E.R.A. Laboratory by means of the above method are shown in Figs. 1 and 2.

2·2 PHASE-ANGLE DIAGRAMS AND PHASE-VECTOR SUMS

NOTATION

Sym-bol	Brief definition	Typical units	Sym-bol	Brief definition	Typical units
A	area	in.², cm.²	U	energy, work	in.Lb., Kg.cm.
F	natural frequency	vib./min.	α	angle; crank angle	deg., rad.
h	order of harmonic	vib./working cycle	β	phase angle	deg., rad.
h	parameter	—	β	angle between phase vector sums of Vee engines	deg., rad.
J	moment of inertia	Lb.in.sec.², Kg.cm.sec.²	γ	phase angle	deg., rad.
K	shaft stiffness	Lb.in./rad., Kg.cm./rad.	Δ	relative amplitude (from Holzer table)	rad.
M	dynamic magnifier	—	$\overrightarrow{\Sigma\Delta}$	phase vector sum	—
n	critical speed order number	vib./rev.	ϵ	phase angle	deg., rad.
R	radius	in., cm.	θ	vibration amplitude	deg., rad.
t	time	sec., min.	σ	firing interval (Vee engines)	deg.
T	torque	Lb.in., Kg.cm.	ϕ	Vee angle	deg.
$\lvert T_n \rvert$	nth order component of tangential pressure	Lb./in.², Kg./cm.²	ω	natural phase velocity of vibration	rad./sec.

Subscripts are used in the text to define more precisely. Their meaning is fully explained in their context in each instance.

2·21 Resultant excitation torque† of a multi-cylinder engine

Torsional vibration in reciprocating engine installations is mainly due to excitation by the harmonics of the gas-pressure torque and the inertia torque (for other types of excitation, see section 1·35).

2·211 Combination of torques due to individual cylinders

Although all the masses of a system (without damping) vibrate in phase, i.e. reach their maximum amplitudes simultaneously, the phasing of the excitation torques of the individual cylinders is related to the firing sequence and the angles between the various cranks. The resultant torque is therefore obtained by adding the torque vectors together, taking due account of their phase angles, on the basis of the following considerations:

(1) The phase angles of the torque vectors depend on the phase of the harmonics relative to the firing dead-centre positions of their respective

† Alternative designations: resultant harmonic torque, resultant input torque.

cranks. However, as this phase is the same for all the torque harmonics, it can be disregarded.

(2) The firing dead-centre positions are not reached simultaneously. When piston no. 1 has fired, its crank moves away from its T.D.C. position,[†] and the next crank comes up towards T.D.C. During this interval of time, the phase of the harmonic torque due to crank no. 1 varies at a frequency which is also the natural frequency of the engine system. Therefore, this phase variation between the harmonic torques of consecutive cranks, which depends on the vibration frequency as well as on the crank-angle spacing, determines the phase angle between the harmonic torque vectors (see Fig. 1).

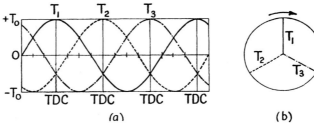

Fig. 1. Phase angles between sinusoidally-varying torques (a) and corresponding representation by means of 'rotating vectors' (b).

In addition, it is convenient to represent not the torque vectors themselves, but their 'reduced values', i.e. their contribution to the torsional amplitude at mass no. 1, so that one writes

$$\vec{T}_{1\,\text{red.}} = \frac{\pi\vec{T_1}\theta_1}{\pi\vec{T_1}\theta_1} = \vec{\Delta}_1 = 1, \quad \vec{T}_{2\,\text{red.}} = \frac{\pi\vec{T_2}\theta_2}{\pi\vec{T_1}\theta_1} = \vec{\Delta}_2, \quad \vec{T}_{3\,\text{red.}} = \frac{\pi\vec{T_3}\theta_3}{\pi\vec{T_1}\theta_1} = \vec{\Delta}_3, \quad \text{etc.}$$

The procedures used for obtaining these phase-vector diagrams are described in the following (with illustrative examples).

In this connexion, it is useful to recall the following definitions:

(1) Critical speed order number

$$= n = \frac{F}{N_c} = \frac{\text{natural frequency of engine [vib./min.]}}{\text{critical running speed [rev./min.]}} = \left[\frac{\text{vib.}}{\text{rev.}}\right],$$

so that $n \times N_c = F$.

(2) Peak value of nth order torque excitation from one engine cylinder

$$= T^{(n)} = |T_n| A R_0 \quad [\text{Lb.in.}], [‡]$$

where $|T_n| = n$th order component of tangential pressure [Lb./in.²],

$A = $ piston area [in.²],

$R_0 = $ crank radius [in.].

$T^{(n)}$ is also termed the 'harmonic torque per cylinder'.

[†] T.D.C. = 'firing dead centre'; in vertical in-line engines the 'top dead centre'.
[‡] *Metric units.* The torque $T^{(n)}$ is obtained in Kg.cm. if $|T_n|$ is expressed in Kg./cm.², A in cm.² and R_0 in centimetres.

(3) Relative amplitude of the mth mass

$$\Delta_m = \frac{\theta_m}{\theta_1} = \frac{\text{vibration amplitude of } m\text{th mass}}{\text{vibration amplitude of mass no. 1}},$$

with $\Delta_1 = $ unity. These relative amplitudes are determined from Holzer tables.

With these expressions, the *resultant excitation torque* of an m-cylinder engine is obtained as

$$T^* = |\, T_n \,|\, A R_0 \overrightarrow{\Sigma\Delta}_{(n)} \quad \text{[Lb.in.]},$$

under resonance conditions. The expression $\overrightarrow{\Sigma\Delta}_{(n)}$ is known as the *phase-vector sum* of the engine excitation, for the nth order vibration.

2·212 *Phase-vector sums of two- and four-stroke cycle engines*

The value of the phase-vector sum $\overrightarrow{\Sigma\Delta}_{(n)}$ is obtained as follows:

(1) Draw a 'phase-angle diagram' (or 'star diagram') for all the engine cylinders, with vectors of unit length $\overrightarrow{01}$, $\overrightarrow{02}$, $\overrightarrow{03}$, ..., as explained on p. 296;

(2) Multiply $\overrightarrow{01}$ by Δ_1 $(= \text{unity})$, $\overrightarrow{02}$ by Δ_2, $\overrightarrow{03}$ by Δ_3, etc., and determine the resultant of the vectors thus obtained, which is the 'phase-vector sum'. The resultant can be obtained either by geometrical construction or by numerical evaluation of the x-axis and y-axis components.

Note. In fact, to evaluate the resultant torque, only the absolute value $|\,\overrightarrow{\Sigma\Delta}_{(n)}\,|$, i.e. the length of the resultant, is required.

Details of the derivation of this procedure are given on p. 303.

EXAMPLE 1. It is required to determine the value of $\overrightarrow{\Sigma\Delta}$ for a three-cylinder engine for the 4th order critical. The relative amplitudes, obtained from a Holzer tabulation, are: $\Delta_1 = 1\cdot00$, $\Delta_2 = 0\cdot85$ and $\Delta_3 = 0\cdot5$.

(a) *Graphical method.* The phase-angle diagram (determined as shown on p. 296) is given in Fig. 2a. The relative-amplitude diagram of the engine is shown in Fig. 2b. We construct the modified star diagram from Fig. 2a by multiplying $\overrightarrow{01}$ by $\Delta_1 = 1\cdot00$, $\overrightarrow{02}$ by Δ_2, etc., and this gives Fig. 2c.

The vectors of Fig. 2c are added together geometrically to obtain Fig. 2d. The resultant, drawn from point O towards the tip of the last vector $(\overrightarrow{03} \times \Delta_3)$, is $\overrightarrow{\Sigma\Delta}$.

Taking Δ_1 as unity, we find that $\overrightarrow{\Sigma\Delta}$ has the length $|\,\overrightarrow{\Sigma\Delta}\,| = 0\cdot430$. Its phase angle relative to $\overrightarrow{\Delta_1}$ is $\beta_{\text{res.}} \cong 40°$, as determined by means of a protractor.

Note. To obtain sufficient accuracy, the diagram of Fig. 2d should be drawn at least twice as large as the accompanying figures. In practice, Fig. 2d is obtained directly from Fig. 2a, using the numerical values for Δ_1, Δ_2, etc., without drawing Figs. 2b and 2c.

(b) *Evaluation method.* This gives more accurate results than method (a), and with some practice it can be used fairly rapidly.

Δ	β	sin β	Δ sin β	cos β	Δ cos β
$\Delta_1 = 1 \cdot 000$	0°	0	0	$+1 \cdot 0$	$+1 \cdot 000$
$\Delta_2 = 0 \cdot 850$	120°	$+0 \cdot 866$	$+0 \cdot 735$	$-0 \cdot 500$	$-0 \cdot 425$
$\Delta_3 = 0 \cdot 500$	240°	$-0 \cdot 866$	$-0 \cdot 433$	$-0 \cdot 500$	$-0 \cdot 250$
			$\Sigma\Delta \sin \beta = +0 \cdot 302$		$\Sigma\Delta \cos \beta = +0 \cdot 325$

$$|\overrightarrow{\Sigma\Delta}| = \sqrt{\{(0 \cdot 302)^2 + (0 \cdot 325)^2\}} = 0 \cdot 443.$$

$$\beta_{res.} = \tan^{-1} \frac{0 \cdot 302}{0 \cdot 325} = \tan^{-1} 0 \cdot 928 = 42° \ 40'.$$

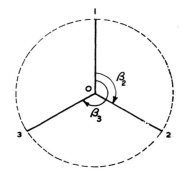

Fig. 2a.

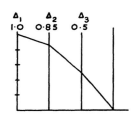

Fig. 2b.

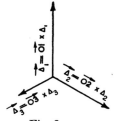

Fig. 2c.

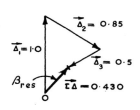

Fig. 2d.

Note. If some of the Holzer-table amplitudes have negative values, these should be plotted in the negative direction of the lines of action of the corresponding vectors in the construction of the vector diagram for $|\overrightarrow{\Sigma\Delta}|$.

EXAMPLE 2. It is required to determine the relative vector sum $\overrightarrow{\Sigma\Delta}$ for a six-cylinder engine, with the firing order 1 3 5 6 4 2, for the 5th order of vibration. The phase-angle diagram for this order is given in Fig. 3a, and the relative amplitudes of the engine inertias are: $\Delta_1 = +1 \cdot 000$, $\Delta_2 = +0 \cdot 770$, $\Delta_3 = +0 \cdot 390$, $\Delta_4 = -0 \cdot 040$, $\Delta_5 = -0 \cdot 500$ and $\Delta_6 = -0 \cdot 880$.

The details of the graphical construction are shown in Fig. 3c. The results are:

$$|\overrightarrow{\Sigma\Delta}| = 0 \cdot 210 \quad \text{and} \quad \beta_{res.} = 195°.$$

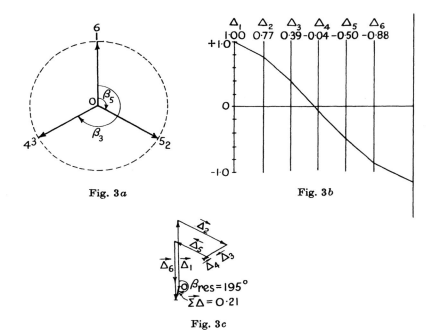

Fig. 3a

Fig. 3b

Fig. 3c

The same rule regarding negative amplitude values applies with the tabulation method. The detailed calculation for this example is as follows:

Cyl. no.	Δ	β	$\sin \beta$	$\Delta \sin \beta$		$\cos \beta$	$\Delta \cos \beta$	
1	1·000	0°	0	0		+1·000	+1·000	
2	0·770	120°	+0·866		+0·666	−0·500	−0·385	
3	0·390	240°	−0·866	−0·338		−0·500	−0·195	
4	−0·040	240°	−0·866		+0·035	−0·500		+0·020
5	−0·500	120°	+0·866	−0·433		−0·500		+0·250
6	−0·880	0°	0	0		+1·000	−0·880	
				−0·771	+0·701		−1·460	+1·270
				$\Sigma\Delta \sin \beta = -0\cdot070$			$\Sigma\Delta \cos \beta = -1\cdot460 + 1\cdot270$	
							$= -0\cdot190$	

$$|\overrightarrow{\Sigma\Delta}| = \sqrt{\{(-0\cdot070)^2 + (-0\cdot190)^2\}} = 0\cdot2025.$$

$$\beta_{\text{res.}} = \tan^{-1}\left(\frac{-0\cdot070}{-0\cdot190}\right) = \tan^{-1}\{0\cdot368\} + 180° = 20°\ 12' + 180° = 200°\ 12'.$$

It is also of interest to note that for this engine the phase-vector sums for the 6th and 5½th orders are $\overrightarrow{\Sigma\Delta}_{(6)} = 0\cdot740$ and $\overrightarrow{\Sigma\Delta}_{(5\frac{1}{2})} = 2\cdot77$. The subject of the relative importance of $\overrightarrow{\Sigma\Delta}$-values is discussed in sections 2·25 and 2·26.

Note regarding two-stroke cycle engines

The procedure for the determination of $\overrightarrow{\Sigma\Delta}$ is exactly the same as for four-stroke cycle engines. However, with two-stroke cycle engines there are no 'half-orders' of vibration, hence there are no half-order diagrams. The other phase-angle diagrams (see section 2·22) are directly applicable.

The reason for this is as follows. If we define

$$\text{order of harmonic} = h = \frac{\text{natural frequency [vib./min.]}}{\text{number of working cycles per minute}}$$

$$= \text{number of vibrations per working cycle,}$$

and

$$\text{critical speed order number} = n = \frac{\text{natural frequency}}{\text{engine speed [rev./min.]}} =$$

$$= \text{number of vibrations per revolution,}$$

it follows immediately that $n = h$ for two-stroke cycle engines, and $n = h/2$ for four-stroke cycle engines, and as harmonics are integer numbers, there are only integer critical order numbers for two-stroke cycle engines.

As regards the relative magnitude of these orders, it should be remembered that, for instance, the $2\frac{1}{2}$ order of a four-stroke cycle engine corresponds to the 5th harmonic of the torque in the working cycle, and hence corresponds to the 5th order critical speed of a two-stroke cycle engine.

2·22 Phase-angle diagrams of excitation torques

2·221 *Construction of diagrams*

The procedure for the construction of a phase-angle diagram will be described for the case of a 6-cylinder four-stroke in-line engine taken as an example. (The same procedure applies, of course, to two-stroke cycle engines.)

Firing order: 1 3 5 6 4 2. The diagram giving the angles between the various cranks, as seen when looking along the crankshaft axis from crank 1 to crank 2, is shown in Fig. 1. Crank 1 is assumed to be in its firing dead-centre (T.D.C.) position.

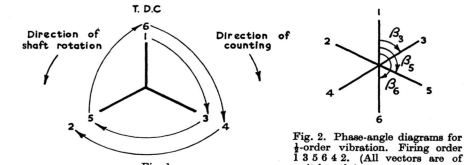

Fig. 1.

Fig. 2. Phase-angle diagrams for $\frac{1}{2}$-order vibration. Firing order 1 3 5 6 4 2. (All vectors are of unit length.)

The cylinders are counted in the clockwise direction if the shaft rotation is anticlockwise in the diagram.

We consider first a vibration due to a torque component of order $\frac{1}{2}$.

(*a*) When piston 1 is at T.D.C., it fires.

(*b*) When piston 1 is at the position 5, piston 3 is at T.D.C. and fires.

While the crank has thus rotated through 120°, the phase of its half-order excitation torque has moved through $\frac{1}{2} \times 120° = 60°$. Hence the torque vibration of crankthrow 3 lags 60° behind the crank rotation and when piston 3 is at T.D.C. the vibration is 60° before T.D.C. We draw the torque phase-vector 3 at this position in Fig. 2.

(c) When crank 1 is at position 3 in Fig. 1, piston 5 is at T.D.C. and fires. The excitation torque 5 has only gone through $\frac{1}{2} \times 240° = 120°$. Hence the vibration of 5 is at $240° - 120° = 120°$ before T.D.C. We represent the vibration phase vector at this position in Fig. 2.

(d) *Generally*, if crank 1 has gone through α degrees, anticlockwise, the torque vibration vector of the crank initially α degrees before T.D.C. has moved through $\beta = $ T.D.C. $- n\alpha$ degrees, clockwise (n being the order number of the vibration), when the crank initially at α reaches T.D.C. (where T.D.C. = firing dead centre for no. 1 crank).

Therefore: When crank 1 is at position 6 ($\alpha = 360°$), in the present example, the torque vibration vector of 6 is at

$$\beta = \text{T.D.C.} - \tfrac{1}{2} \times 360° = -180°.$$

When crank 1 is at position 4 ($\alpha = 480°$), the torque vibration vector of 4 is at $\beta = $ T.D.C. $- \frac{1}{2} \times 480° = -240°$ (counted clockwise). Finally, when crank 1 is at position 2 ($\alpha = 600°$), the torque vibration vector of 2 is at $\beta = $ T.D.C. $- \frac{1}{2} \times 600° = -300°$.

These results are included in Fig. 2.

Note 1. The rule given under (d) applies to all in-line engines with one piston per cylinder. Further rules, for Vee engines, are given on p. 304.

Note 2. It is, of course, equally possible to determine phase-angle diagrams by rotating the vector of each crank anticlockwise, through an angle $n\alpha$, where α = angle of vector before T.D.C. and n = order of vibration. (For instance, in Fig. 2, the position of crank 3 is obtained from $\frac{1}{2} \times 120° = 60°$, anticlockwise, relative to the position of this crank in Fig. 1.)

2·222 *Examples of phase-angle diagrams for two-stroke cycle engines*

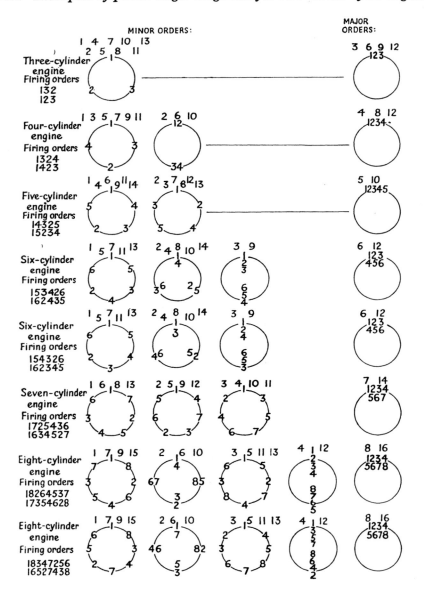

2·223 *Examples of phase-angle diagrams for four-stroke cycle engines*

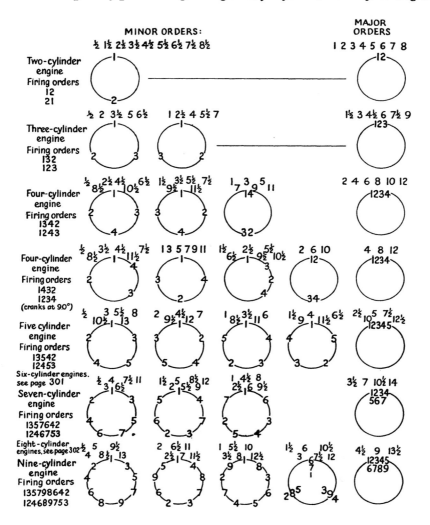

Note. The firing sequence is shown by the ½-order diagram, and the crank sequence by the 1st order, for four-stroke cycle single-acting engines; both are shown by the 1st order in two-stroke cycle single-acting engines.

Examples Of Phase-Angle Diagrams For Four-Stroke Cycle Engines

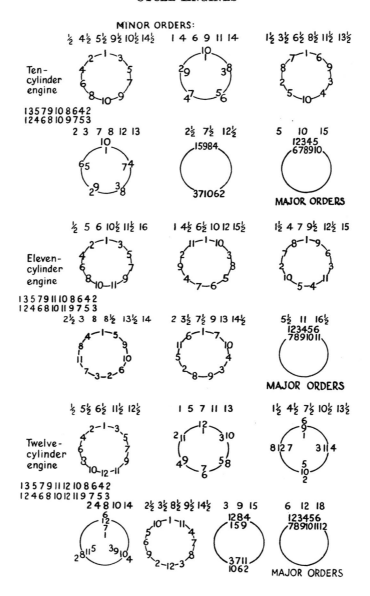

MINOR ORDERS:

Ten-cylinder engine

1 3 5 7 9 10 8 6 4 2
1 2 4 6 8 10 9 7 5 3

Eleven-cylinder engine

1 3 5 7 9 11 10 8 6 4 2
1 2 4 6 8 10 11 9 7 5 3

Twelve-cylinder engine

1 3 5 7 9 11 12 10 8 6 4 2
1 2 4 6 8 10 12 11 9 7 5 3

MAJOR ORDERS

EXAMPLES OF PHASE-ANGLE DIAGRAMS FOR FOUR-STROKE CYCLE ENGINES

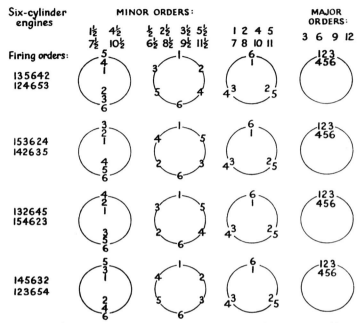

Six-cylinder engines

Firing orders:
135642
124653

153624
142635

132645
154623

145632
123654

All the phase-angle diagrams given above can be represented by the following four diagrams:

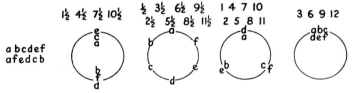

a b c d e f
a f e d c b

Note. Additional information on the subject of firing sequences and phase-angle diagrams can be found in the following references:

Ker Wilson, W., *Practical solution of torsional vibration problems*, vol 1, pp. 471–588 (Chapman and Hall, 2nd ed. (reprinted), 1942, (firing orders and discussion of their effects).

Scheuermeyer, M., *Einfluss der Zündfolgen auf die Drehschwingungen von Reihenmotoren* (Influence of firing order on the torsional vibrations of in-line engines), p. 69. (Dissertation, T. H. München, 1932; and Werft–Reederei–Hafen, 1 March, 1933.

Schlaefke, K., "Die Einfluss des V-Winkels auf die Kurbelwellen–Drehschwingungen von V-Motoren" (Influence of Vee-angle on torsional vibrations of Vee-engines), *Z. V.D.I.*, vol. 80, 1936, pp. 1253–1254.

Schrön, H., *Die Zündfolge* (The firing order) (R. Oldenbourg, Munich and Berlin, 1938).

EXAMPLES OF PHASE-ANGLE DIAGRAMS FOR FOUR-STROKE CYCLE ENGINES

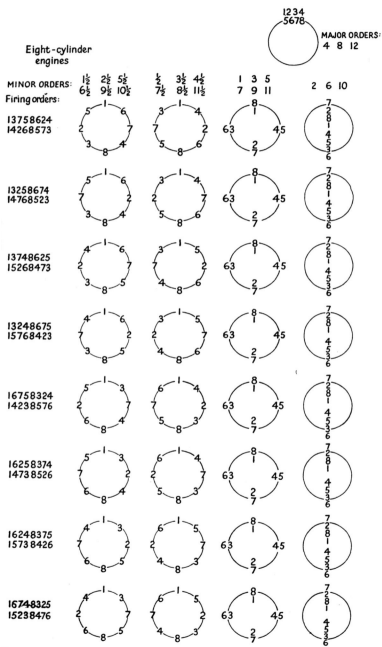

Eight-cylinder engines

MAJOR ORDERS: 4 8 12

MINOR ORDERS: 1½ 2½ 5½ 6½ 9½ 10½   ½ 3½ 4½ 7½ 8½ 11½   1 3 5 7 9 11   2 6 10

Firing orders:

13758624
14268573

13258674
14768523

13748625
15268473

13248675
15768423

16758324
14238576

16258374
14738526

16248375
15738426

16748325
15238476

Note. When the phase-angle diagrams for all major and minor orders have been determined for a given firing order, the corresponding diagrams for a different firing order can be obtained directly by simple substitution.

For instance, for an eight-cylinder engine of which the initial firing order 1 3 7 5 8 6 2 4 is changed to 1 3 2 5 8 6 7 4, the only substitutions required in the diagrams are: 7 replaced by 2, and 2 replaced by 7. In doing this, it is assumed, of course, that the crankshaft arrangement is unchanged and that only the firing sequence is altered.

The entire set of phase-angle diagrams of the preceding page can be represented, therefore, by the five diagrams given below:

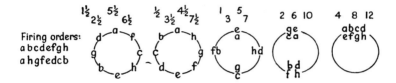

Firing orders:
a b c d e f g h
a h g f e d c b

2·23   Derivation of method for determining phase-vector sums

The work input per cycle of the mth engine cylinder is [see section 2·1]

$$U_{(m)} = \pi T^{(m)} \theta_m \sin \epsilon_m \quad \text{[in.Lb./cycle],}\dagger \tag{1}$$

where $T^{(m)}$ and θ_m are the peak values of periodic torque [Lb.in.] and vibration amplitude [rad.], and ϵ_m is the phase lag of θ_m relative to $T^{(m)}$.

The total work input of all the cylinders per vibration cycle is therefore

$$U = \Sigma U_{(m)} = \pi \Sigma (T^{(m)} \theta_m \sin \epsilon_m). \tag{2}$$

The values of the individual phase angles ϵ_m between $T^{(m)}$ and θ_m are not known, but we do know the angles β_m of the torques $T^{(m)}$ of each cylinder relative to the torque $T^{(1)}$ of cylinder no. 1.

We may also assume a value γ for the phase angle between $T^{(1)}$ and θ_1. Therefore, since $\beta_1 = 0$,

$$\sin \epsilon_1 = \sin \gamma.$$

Moreover, as all the individual masses (the 'inertias per line') vibrate with the same phase as θ_1, we can write

$$\sin \epsilon_2 = \sin (\beta_2 + \gamma), \quad \ldots\ldots \ldots\ldots\ldots\ldots,$$

$$\sin \epsilon_3 = \sin (\beta_3 + \gamma), \quad \sin \epsilon_m = \sin (\beta_m + \gamma),$$

and as all the $T^{(m)}$ have the same absolute value, eq. (2) may be rewritten as

$$U = \pi T^{(m)} \Sigma \theta_m \sin (\beta_m + \gamma). \tag{2'}$$

† Or Kg.cm./cycle if $T^{(m)}$ is in Kg.cm.

Introducing relative amplitudes, $\Delta_m = \theta_m / \theta_1$ (as obtained from Holzer tables), so that $\theta_1 = \Delta_1 \times \theta_1$, $\theta_2 = \theta_1 \times \Delta_2$, ..., we have

$$U = \pi T^{(m)} \theta_1 \Sigma \Delta_m \sin (\beta_m + \gamma). \tag{3}$$

The engine can be regarded as equivalent to a one-mass system, with a work input

$$U = \pi T^* \theta_1 \sin \epsilon^*, \tag{4}$$

which becomes $U = \pi T^* \theta_1$ at resonance (since then $\epsilon^* = 90°$). Therefore

$$T^* \sin \epsilon^* = T^{(m)} \Sigma \Delta_m \sin (\beta_m + \gamma), \tag{5}$$

where γ is unknown, while the β_m-values are the known phase angles of the cylinder torques relative to the periodic torque of cylinder no. 1.

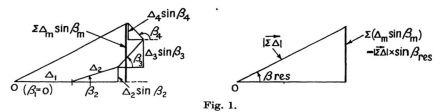

Fig. 1.

Since the Δ_m-values are also known, we can construct a vector diagram as shown in Fig. 1, assuming $\gamma = 0$ for the time being. Fig. 1 shows that

$$T^{(m)} \Sigma (\Delta_m \sin \beta_m) \equiv T^{(m)} \, | \, \overrightarrow{\Sigma \Delta} \, | \sin \beta_{\text{res.}}$$

Now eq. (5) is valid for any value of γ, in particular for $\gamma = 0$, when $\sin \epsilon^* = \sin \beta_{\text{res.}}$. Therefore, the resultant torque is

$$T^* = T^{(m)} \times | \, \overrightarrow{\Sigma \Delta} \, |,$$

where $T^{(m)} = | \, T_m \, | \, A R_0 =$ vibratory torque (of mth order) due to one cylinder, and $\overrightarrow{\Sigma \Delta}$ is the resultant vector in the right-hand diagram of Fig. 1. Its value is obtained as indicated previously in section 2·212 (p. 293).

2·24   Phase-vector sums for Vee engines

The method for obtaining the phase-vector sum of a Vee engine of Vee-angle ϕ_V is as follows:

The resultants $\overrightarrow{\Sigma \Delta}_A$ and $\overrightarrow{\Sigma \Delta}_B$, for banks A and B, are determined separately. The resultant $\overrightarrow{\Sigma \Delta}_V$ for the entire engine is obtained by geometrical addition of these two vectors:

$$\overrightarrow{\Sigma \Delta}_V = \overrightarrow{\Sigma \Delta}_A + \overrightarrow{\Sigma \Delta}_B. \tag{1}$$

The angle between these vectors is

$$\beta = n \times \sigma,$$

where n = critical speed order number, and σ = firing interval between two cylinders acting on the same crankpin.

Usually, the firing interval between corresponding cylinders of the two banks is:

(1) $\sigma = \phi_V$ for two-stroke cycle engines;

(2) $\sigma = 360° \pm \phi_V$ for four-stroke cycle engines with an *even* number of cylinders in each bank;

(3) $\sigma = \phi_V$ for four-stroke cycle engines with an *odd* number of cylinders in each bank and *even* firing intervals;

(4) $\sigma = \phi_V$ or $\sigma = 360° \pm \phi_V$ for four-stroke cycle engines with an *odd* number of cylinders in each bank and *uneven* firing intervals.

Provided that (as is usual) both banks are similar and have the same firing order, it is, therefore, only necessary to determine the value of one resultant, for instance $\overrightarrow{\Sigma\Delta}_A$ for bank A, and the resultant vector sum for the entire engine is obtained as

$$|\overrightarrow{\Sigma\Delta}_V| = 2\,|\overrightarrow{\Sigma\Delta}_A|\cos\frac{n\sigma}{2}. \qquad (2)$$

The resultant torque T^* of the engine, for the nth order of vibration, is then determined from

$$T^* = |T_n|\,AR_0\,|\overrightarrow{\Sigma\Delta}_V| \quad \text{[Lb.in.],†}$$

where $|T_n|$ = nth order component of tangential pressure [Lb./in.²], A = area of one piston [in.²], R_0 = crank radius [in.] and $|\overrightarrow{\Sigma\Delta}_V|$ = absolute value of resultant vector sum obtained by means of eq. (1) or eq. (2).

Note. The use of a firing interval $\sigma = 360° - \phi_V$ gives the following results for the case of a ten-cylinder four-stroke cycle engine with $\phi_V = 40°$, as compared with a firing interval $\sigma = \phi_V$ for the same engine:

With $\sigma = 360° - \phi_V$, it is found that in this example

(a) consecutive firing is eliminated, with a consequent reduction in bearing load;

(b) certain vibration orders are reduced (notably the $7\tfrac{1}{2}$th order);

(c) the crank angles between firing cylinders are alternately 112° and 32° (with a firing interval $\sigma = 40°$ between the banks, these become 40° and 104°).

Thus, although the 40° firing interval gives slightly less uneven crank angles between firing cylinders, this slight disadvantage appears to be outweighed by the advantages given under (a) and (b) above.

---

† *Metric units.* T is obtained in Kg.cm. if A is expressed in cm.² and R_0 in centimetres.

2·241   *Effect of Vee angle on* $\overrightarrow{\Sigma\Delta_V}$

†Considering again eq. (2) of section 2·24, with $\sigma = 360° + \phi_V$:

$$|\overrightarrow{\Sigma\Delta_V}| = 2\,|\overrightarrow{\Sigma\Delta_A}|\cos\left\{\frac{n(360° + \phi_V)}{2}\right\}, \qquad (2)$$

where $\overrightarrow{\Sigma\Delta_V} =$ resultant phase-vector sum for the entire engine, $\overrightarrow{\Sigma\Delta_A} =$ resultant vector for one bank, $n =$ critical order number, and $\phi_V =$ Vee angle, it is seen that it is possible to cancel out $\overrightarrow{\Sigma\Delta_V}$ when the cosine is zero. This occurs when

$$|\overrightarrow{\Sigma\Delta_V}| = 0: \quad \phi_V = \left[\frac{2h+1}{n} - 2\right] \times 180° \quad \left(\text{with } \frac{2h+1}{n} \geqslant 2\right), \qquad (3)$$

where $h =$ any integer (1, 2, 3, 4, etc.) such that $(2h+1)/n$ is not less than 2, and $n =$ order of vibration considered.

However, although a particular Vee angle may eliminate $\overrightarrow{\Sigma\Delta}$ for a certain value of n, it may also result in values equal to $\overrightarrow{\Sigma\Delta_A}$, or even greater, for other orders. The corresponding relations are:

$$|\overrightarrow{\Sigma\Delta_V}| = 2\,|\overrightarrow{\Sigma\Delta_A}|: \quad \phi_V = \left[\frac{2h}{n} - 2\right] \times 180° \quad \left(\text{with } \frac{2h}{n} \geqslant 2\right), \qquad (4)$$

$$|\overrightarrow{\Sigma\Delta_V}| = |\overrightarrow{\Sigma\Delta_A}|: \begin{cases} \phi_V = \left[\dfrac{3h+2}{n} - 3\right] \times 120° & (\text{with } (3h+2)/n \geqslant 3), \quad (5a) \\[2ex] \phi_V = \left[\dfrac{3h+1}{n} - 3\right] \times 120° & (\text{with } (3h+1)/n \geqslant 3). \quad (5b) \end{cases}$$

Derivation. Equation (3) is obtained from eq. (2) as follows:

$$\frac{n}{2}(360° + \phi_V) = 90°,\ 270°,\ \ldots$$

or $\qquad n(360° + \phi_V) = 180°,\ 540°,\ \ldots = (2h+1) \times 180°,$

hence $n\phi_V = (2h+1) \times 180° - n \times 360°$, resulting in eq. (3).

Equation (4) is obtained similarly, from $\dfrac{n}{2}(360° + \phi_V) = 0°,\ 180°,$ $360°,\ \ldots.$

For eqs. (5a) and (5b), we have:

$$\cos\left\{\frac{n}{2}(360° + \phi_V)\right\} = \pm\tfrac{1}{2}, \quad \text{when} \quad \frac{n}{2}(360° + \phi_V) = 60° \times [1, 4, 7, 10, \ldots],$$

and when $\qquad \dfrac{n}{2}(360° + \phi_V) = 120° \times [1, 2\tfrac{1}{2}, 4, 5\tfrac{1}{2}, \ldots],$

that is, when $\qquad n(360° + \phi_V) = (3h+1) \times 120°$

and when $\qquad n(360° + \phi_V) = (3h+2) \times 120°,$

which give the expressions for ϕ_V for $|\overrightarrow{\Sigma\Delta_V}| = |\overrightarrow{\Sigma\Delta_A}|$.

† Similar reasonings apply for $\sigma = 360° - \phi_V$ and $\sigma = \phi_V$.

Table giving values of the ratio $\left|\dfrac{\overrightarrow{\Sigma\Delta_V}}{\overrightarrow{\Sigma\Delta_A}}\right| = \dfrac{\text{resultant for Vee-engine}}{\text{resultant for one bank}}$ for different Vee angles ϕ_V and critical order numbers n

Critical order number n	$\Sigma\Delta_V/\Sigma\Delta_A = 0$	$\Sigma\Delta_V/\Sigma\Delta_A = 1\cdot0$	$\Sigma\Delta_V/\Sigma\Delta_A = 2\cdot0$
	† Vee angles $\phi_V = \sigma - 360°$ and $\phi_V = 360° - \sigma$		
2	90°	60° 120°	0° 180°
2½	0° 144°	24° 120° 168°	72°
3	60° 180°	40° 80° 160°	0° 120°
3½	0° 103°	17° 86° 120°	51½°
4	45° 135°	30° 60° 120° 150°	0° 90° 180°
4½	0° 80° 160°	13½° 66½° 93½° 145½° 173°	40° 120°
5	36° 108° 180°	24° 48° 96° 120° 168°	0° 72° 144°
5½	0° 65° 130°	11° 54½° 76½° 120° 142°	33½° 97½° 162½°
6	30° 90° 150°	20° 40° 80° 100° 140° 160°	0° 60° 120° 180°
6½	0° 55½° 111° 166°	9° 46° 65° 101½° 120½° 157° 176°	27½° 82½° 137½°
7	25½° 76½° 128° 180°	17° 34° 68½° 85½° 119½° 136½° 171°	0° 51° 102° 153°
7½	0° 48° 96° 144°	8° 40° 56° 88° 104° 136° 152°	24° 72° 120° 168°
8	22½° 67½° 112½° 157½°	15° 30° 60° 75° 105° 120° 150° 165°	0° 45° 90° 135° 180°
8½	0° 42½° 85° 127½° 170°	7° 35½° 49½° 78° 92° 120½° 134½° 163° 177°	21° 63½° 106° 148½°
9	20° 60° 100° 140° 180°	13½° 26½° 53½° 66½° 93½° 106½° 133½° 146½° 173½°	0° 40° 80° 120° 160°
9½	0° 38° 76° 114° 152°	6½° 31½° 44½° 69½° 82½° 107½° 120½° 145½° 158½°	19° 57° 95° 143° 171°
10	18° 54° 90° 126° 162°	12° 24° 48° 60° 84° 96° 120° 132° 156° 168°	0° 36° 72° 108° 144° 180°

Note. In the above table, the values of Vee angle are rounded to $\frac{1}{2}°$. The values are calculated by means of eqs. (2) to (5b), and can be plotted in the form of curves of n against ϕ_V, if necessary. The values given above apply to engines in which both banks are similar and have the same firing order.

Example of use of table. Comparison of values of $|\overrightarrow{\Sigma\Delta_V}|/|\overrightarrow{\Sigma\Delta_A}|$ obtained for a twelve-cylinder Vee engine with Vee angles of 30°, 45°, 60° and 90°.

30° Vee: no 6th order; 7½th between 1·0 and 2·0; 9th between 1·0 and 2·0; 4½th between 1·0 and 2·0; 3rd between 1·0 and 2·0.

45° Vee: 6th, 9th, 4½th between 1·0 and 2·0; 3rd and 7½th between 0 and 1·0.

60° Vee: no 3rd order; 4½th and 7½ between 1·0 and 2·0; 6th order 2·0 (maximum value); no 9th order.

90° Vee: no 6th order; no 10th order; 3rd and 9th between 1·0 and 2·0; 4½th and 7½th between 0 and 1·0.

2·242 Vee engines with articulated connecting rods

For Vee engine arrangements incorporating master and articulated connecting rods, the $\overrightarrow{\Sigma\Delta_V}$-values can generally be calculated in the same

† For $\phi_V = \sigma$, the values for the integer orders are the same; for half orders, add $(180°/n)$ to the values in the above table (n being the order of vibration considered). Also, for $\Sigma\Delta_V/\Sigma\Delta_A = 2$, all half-orders include $\phi_V = 0°$.

[307]

way as for Vee engines with connecting rods operating directly on the crankpin (side-by-side or fork-and-blade type arrangements).

However, it should be noted that there is one important exception, viz. the $\overrightarrow{\Sigma\Delta_V}$-values do not cancel out for certain Vee angles. (Equation (3) of p. 306 is not valid in this case.) An investigation by Kimmel[†] has shown that the critical order numbers n determined by

$$n = \frac{h}{\phi_V} \times 180°$$

(where $\phi_V = $ Vee angle, and $h = 1, 3, 5, 7, \dots$) can correspond to large values of $\overrightarrow{\Sigma\Delta_V}$ for Vee engines with articulated connecting rods, instead of giving zero values as in the case of Vee engines with both connecting rods directly coupled. The possibility of appreciable vibration amplitudes for these order numbers should therefore be considered in dealing with articulated systems, both at the design stage (see Kimmel's calculations[†]) and in subsequent torsiograph tests.

2·25 Effect of changes in firing order
In general, the firing order is selected in order to obtain

(i) a crankshaft arrangement giving adequate balancing of inertia forces and couples,[‡]

(ii) low $\overrightarrow{\Sigma\Delta}$-values for criticals of certain orders occurring in the speed range of the engine ($\overrightarrow{\Sigma\Delta}$ being the relative vector sum for each critical order considered, see section 2·212),

(iii) low bearing loads due to gas and inertia forces transmitted to the main bearings, and

(iv) a suitable exhaust pipe arrangement on a turbo-charged engine.

In some cases, particularly with long multi-throw crankshafts, failures of middle bearings are experienced. It may be possible to overcome this trouble (a) by using a different firing order, (b) by changing the conditions as regards inertia balance, with a non-uniform distribution of balance weights, and (c) by means of a combination of (a) and (b).

A convenient method of determining the resultant bearing load of a middle bearing is to consider load conditions with either of the pistons adjacent to the middle bearing at its T.D.C. position, taking account of carry-over moments due to the other cylinders by means of Clapeyron's equation of three moments (or other suitable methods).[§]

[†] Kimmel, A., 'Effect of connecting-rod arrangement on the torsional vibration behaviour of Vee engines', *M.T.Z.*, vol. 3 (1941), pp. 18–22.

[‡] Den Hartog, J. P., *Mechanical Vibrations* (McGraw-Hill, New York and London). Ker Wilson, W., *Balancing of Oil Engines* (Griffin, London). Ker Wilson, W., *Practical Solution of Torsional Vibration Problems* (Chapman and Hall, Ltd.). Laugharne Thornton, G., *Mechanics Applied to Vibrations and Balancing* (Chapman and Hall, Ltd.). Sharp, A., *Balancing of Engines* (Longmans, London). Toft, L. and Kersey, A. T. J., *Theory of Machines* (Pitman, London).

[§] Conway, H. D., *Aircraft Strength of Materials* (Chapman and Hall, Ltd.).

In addition, comparative tests of engines with different firing orders can be made, with indication of journal movement obtained by means of two capacitive or electromagnetic pick-ups inserted at 45° on either side of the vertical in each of the journal shells under investigation. The bearing movement diagram is altered considerably by torsional vibration. A typical oscillograph recording obtained with torsional vibration conditions is shown in Fig. 1.

Fig. 1. Indicated diagram of transverse movement of a journal centre in a multi-cylinder engine, with torsional vibration. (Diagram recorded at the B.I.C.E.R.A. Laboratory.)

2·251 *Comparison of effects of two different firing orders*

As an example, we shall consider the effects of different firing orders on a six-cylinder slow-speed four-stroke cycle engine, with the firing orders 1 2 4 6 5 3 and 1 5 3 6 2 4.

From a Holzer table, the relative amplitudes Δ_1, Δ_2, ..., of the six-cylinder inertias have been determined with the following values:

Cyl. no.	1	2	3	4	5	6
Δ	1·0000	0·9472	0·8442	0·6971	0·5131	0·3021

The phase-angle diagrams for the two firing orders are as follows:

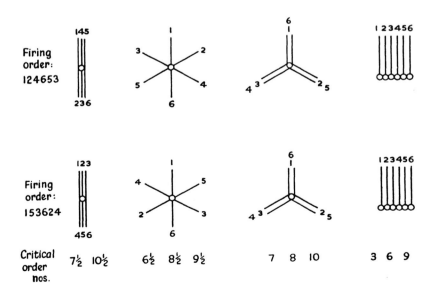

We first determine the phase-vector sums for the various orders:

(a) Firing order 1 2 4 6 5 3: 6 and 9: $\Delta_1 + \Delta_2 + ... + \Delta_6 = 4\cdot3037$; $7\frac{1}{2}$, $10\frac{1}{2}$: $\Delta_1 + \Delta_4 + \Delta_5 - (\Delta_2 + \Delta_3 + \Delta_6) = 2\cdot2102 - 1\cdot9464 = +0\cdot2368$.

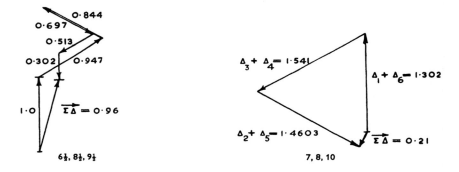

(b) Firing order 1 5 3 6 2 4: 6 and 9: $\Delta_1 + \Delta_2 + ... + \Delta_6 = 4\cdot3037$; $7\frac{1}{2}$, $10\frac{1}{2}$: $\Delta_1 + \Delta_2 + \Delta_3 - (\Delta_4 + \Delta_5 + \Delta_6) = 2\cdot8714 - 1\cdot5123 = 1\cdot3591$.

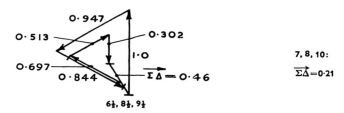

The results are included in columns 3 and 5 of the table on p. 311.

It is seen from this table that (1) the 'major criticals' 6 and 9 have the same $\overrightarrow{\Sigma\Delta}$-values for both firing orders; (2) the firing order 1 2 4 6 5 3 gives small $\overrightarrow{\Sigma\Delta}$-values for the 8, $7\frac{1}{2}$ and 7, but a comparatively large value for the $8\frac{1}{2}$ and $6\frac{1}{2}$; (3) the firing order 1 5 3 6 2 4 gives a comparatively large $7\frac{1}{2}$, but smaller values for the $8\frac{1}{2}$ and $6\frac{1}{2}$.

Relative values of resultant torque. As the resultant torque of the engine for each order of vibration is given by

$$T^* = |T_n| A R_0 \overrightarrow{\Sigma\Delta}_n,$$

where $|T_n| =$ component of tangential pressure [Lb./in.²] for the critical order n, $A =$ piston area [in.²], and $R_0 =$ crank radius [in.] (see section 2·211), we can calculate $|T_n| \overrightarrow{\Sigma\Delta}_n$ directly (since A and R_0 are constants) in order to obtain the relative values of resultant torque.

For the slow-speed four-stroke cycle engine of this example, the

$|T_n|$-values (at, say, 100 Lb./in.^{2} i.m.e.p.) are taken from the curves of $|T_n|$-components in Fig. 1 of section 2·18. The results can be tabulated:

n	$|T_n|$	$\overrightarrow{\Sigma\Delta}$ 124653	$|T_n|\overrightarrow{\Sigma\Delta}$	$\overrightarrow{\Sigma\Delta}$ 153624	$|T_n|\overrightarrow{\Sigma\Delta}$
9	1·4	4·30	6·0	4·30	6·0
8½	1·7	0·96	1·63	0·46	0·79
8	2·1	0·21	0·45	0·21	0·45
7½	2·6	0·27	0·70	1·36	3·55
7	3·5	0·21	0·74	0·21	0·74
6½	4·6	0·96	4·4	0·46	2·12
6	6·5	4·30	27·9	4·30	27·9

The frequency $F = 3950$ vib./min. being also obtained from the Holzer tables, we can plot the 'relative torques' $|T_n|\overrightarrow{\Sigma\Delta}_n$ against engine speed, as shown in Fig. 1.

Test results of this same engine are shown in Fig. 2.

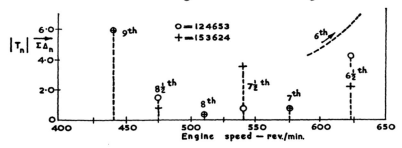

Fig. 1.

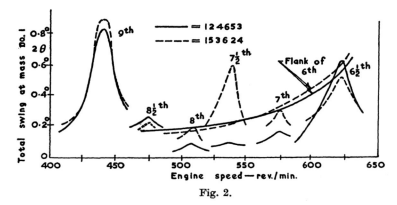

Fig. 2.

2·26 Major and minor criticals

Phase-angle diagrams can be classified into two main types (see Fig. 1, p. 312):

(*a*) diagrams in which all the vectors are in phase (i.e. point in the same direction); and

(*b*) diagrams in which the vectors have different phase angles.

The critical running speeds associated with phase-angle diagrams of type (*a*) are called 'major criticals'; all others are 'minor criticals'.

For any critical considered, the vibratory torque input of the engine depends on the phase-vector sum $\overrightarrow{\Sigma\Delta}$, obtained by multiplying each of the unit vectors of the phase-angle diagram by its corresponding relative amplitude Δ, determined from a Holzer table (see p. 293).

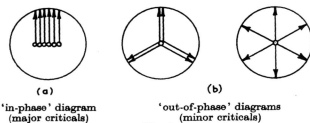

(a)

'in-phase' diagram
(major criticals)

(b)

'out-of-phase' diagrams
(minor criticals)

Fig. 1.

If the node of the vibrating system is at one end of the engine crankshaft, all the Δ-values have the same sign, and the major criticals are also the most dangerous, since they correspond to the highest values of $\overrightarrow{\Sigma\Delta}$.

If the node is approximately half-way along the crankshaft, the major criticals may be less important than some of the minor orders. In all cases, therefore, the phase-vector sum $\overrightarrow{\Sigma\Delta}$ is ultimately the decisive factor.

Note. Examples of systems with nodes near the main flywheel and at the middle of the crankshaft are given on pp. 294 and 295.

2·27 Energy input and energy dissipation

(1) *Energy input.* As already shown in connexion with the derivation of phase-vector sums (see p. 303), the work input per vibration cycle obtained from a sinusoidally varying component of the engine torque is

$$U_{in} = \pi T^* \theta_1 = \pi \, |\, T_n \,|\, A R_0 \overrightarrow{\Sigma\Delta} \times \theta_1 \quad \text{[in.Lb./cycle]}, \qquad (1)$$

where $\theta_1 =$ vibration amplitude [rad.] at cylinder no. 1, $\overrightarrow{\Sigma\Delta} =$ phase-vector sum for the nth order critical considered, $R_0 =$ crank radius [in.], $A =$ piston area [in.²], and $|\, T_n \,| = n$th order tangential-pressure component [Lb./in.²].

EXAMPLE. Six-cylinder four-stroke cycle engine running on the 6th order critical. $|\, T_6 \,| = 6{\cdot}0$ Lb./in.², bore = 5 in., stroke = 6 in., $\overrightarrow{\Sigma\Delta} = 4{\cdot}100$, measured total swing $2\theta = 1{\cdot}2°$. The work input is

$$U_{in} = \pi \times 6 \times \frac{\pi}{4} \times 25 \times \frac{6}{2} \times 4{\cdot}1 \times \frac{1{\cdot}2}{2 \times 57{\cdot}3} = 47{\cdot}5 \quad \text{[in.Lb./cycle]}.$$

(2) *Energy dissipation.* The energy is dissipated by damping during each vibration cycle. It is usual to assume that the damping torque T_d is a proportion of the engine inertia torque, using the relation

$$T_d = J_E \omega^2 \theta_1 / M \quad \text{[Lb.in.]}, \qquad (2)$$

where $M =$ dynamic magnifier of the system (see section 2·4), $J_E = \Sigma(J_{\text{cyl.}} \Delta^2_{\text{cyl.}}) + J_{MF} \Delta^2_{MF} =$ equivalent moment of inertia of the engine [Lb.in.sec.^{2}], obtained by multiplying the J-values for the engine cylinders and the main flywheel by their corresponding Holzer-table amplitudes squared, and $\omega = (2\pi/60) \times F$ [rad./sec.], F being the natural frequency of the vibration [vib./min.]. In this formula θ_1 is again the vibration amplitude at cylinder no. 1, expressed in radians.

The energy dissipation per vibration cycle is therefore

$$U_d = \pi T_d \theta_1 = \pi J_E \omega^2 \theta_1^2 / M \quad \text{[in.Lb./cycle].} \qquad (3)$$

EXAMPLE. Further data for the six-cylinder engine already considered are: $J_E = 2·65$ Lb.in.sec.^{2}, $F = 10,200$ vib./min., $\theta_1 = 0·6°$ and $M = 21·88$. The energy dissipated by damping is

$$U_d = \pi \times 2·65 \times \left(\frac{2\pi}{60} \times 10,200\right)^2 \times \left(\frac{0·6}{57·3}\right)^2 \times \frac{1}{21·88} = 47·5 \quad \text{[in.Lb./cycle].}$$

Note. The energy equilibrium requires that $U_d = U_{in}$, as shown in the above example. In stress-prediction calculations, however, it is necessary to estimate a value for M, then determining θ_1 from eqs. (1) and (3); although this estimation of M is based on experience and statistical data, it may still be inaccurate and result in an appreciable error in the expected value of θ_1 [see section 2·4, Prediction of vibration amplitudes and stresses].

Metric units. The above equations (1), (2) and (3) are directly applicable in metric units, with, for instance, linear dimensions in cm., torques in Kg.cm., inertias in Kg.cm.sec.^{2}, and energies in cm.Kg.

2·28   Static deflexion ('equilibrium' amplitude)

For the definition of static deflexion, the engine is regarded as equivalent to a one-mass system, consisting of an equivalent inertia J_E and a shaft of stiffness K_E, coupled to an infinite mass.

If the torque $T^* = |T_n| A R_0 \Sigma \vec{\Delta}$ were applied statically, the mass J_E would have a corresponding static deflexion

$$\theta_s = T^*/K_E \quad \text{[radians].}$$

As the stiffness is determined from the relation $K_E = \omega^2 J_E$ (where $\omega =$ natural phase velocity of the engine system), we may substitute this value in the above equation, in order to obtain the 'equivalent static deflexion' of the system:

$$\theta_s = \frac{T^*}{J_E \omega^2} = \frac{|T_n| A R_0 \Sigma\vec{\Delta}}{J_E \omega^2} \quad \text{[radians].} \qquad (4)$$

[313]

This deflexion is also at cylinder no. 1. It is related to the vibration amplitude θ_1 at this same position by means of the dynamic magnifier M:

$$M = \theta_1/\theta_s,$$

so that θ_1 can be estimated by assuming a value for M and using equation (4):

$$\pm \theta_1 = M \times |T_n| \, AR_0 \overrightarrow{\Sigma\Delta} \times 57\cdot3/[J_E\omega^2] \quad \text{[degrees]}. \tag{5}$$

Note. The expression for θ_s can also be determined by equating the energy input to the energy dissipated per cycle, using equations (1) and (3), and assuming that $M = 1$ under non-resonant conditions. The derivation is thus based on a consideration of energy equilibrium, and θ_s is therefore also sometimes called the 'equilibrium amplitude'.†

It is found useful, in practice, to reduce the engine to an equivalent one-mass system, following the above procedure. Further information regarding statistical values for dynamic magnifiers, and the limitations of this method, will be found in section 2·4 (Prediction of vibration amplitudes and stresses).

† The assumption made in previous literature that θ_s is obtained as the *effective* deflexion of a multi-mass system at zero frequency is unfounded (see, for instance, the curves of vibration amplitude against excitation frequency for a two-mass system in section 2·45), since even the 'node position' varies with frequency, as can be seen from Holzer tables.

2·3 EVALUATION OF STRESSES FROM VIBRATION MEASUREMENTS

NOTATION

Symbol	Brief definition	Typical units
a	distance	in., cm.
a	length of side	in., cm.
A	area	in.^{2}, cm.^{2}
A	parameter	Lb.in.^{2}, Kg.cm.^{2}
b	distance	in., cm.
B	parameter	Lb.in.^{2}, Kg.cm.^{2}
C	relative groove depth	—
C	parameter	Lb.in.^{2}, Kg.cm.^{2}
d	diameter	in., cm.
D	diameter	in., cm.
e	distance	in., cm.
e	eccentricity	in., cm.
E	modulus of elasticity	Lb./in.^{2}, Kg./cm.^{2}
F	vibration frequency	vib./min.
F	parameter	Lb.in.^{2}, Kg.cm.^{2}
g	acceleration due to gravity	in./sec.^{2}, cm./sec.^{2}
G	modulus of rigidity	Lb./in.^{2}, Kg./cm.^{2}
h	distance	in., cm.
h	weld thickness	in., cm.
H	distance	in., cm.
I	second moment of area	in.^{4}, cm.^{4}
J	mass moment of inertia	Lb.in.sec.^{2}, Kg.cm.sec.^{2}
K	shaft stiffness	Lb.in./rad., Kg.cm./rad.
L	length	in., cm.
L	thickness	in., cm.
m	diameter ratio	—
M	parameter (in formulae for shrink fits)	Lb.in.^{2}, Kg.cm.^{2}
N	shaft speed	rev./min.
N	number (of bolts, splines, etc.)	—
p	direct stress	Lb./in.^{2}, Kg./cm.^{2}
p	pressure	Lb./in.^{2}, Kg./cm.^{2}
P	pressure force	Lb., Kg.
$P*$	force per inch of weld	Lb./in.
q	shear stress	Lb./in.^{2}, Kg./cm.^{2}
$q*$	shear stress per degree	Lb./(in.^{2} deg.), Kg./(cm.^{2} deg.)
Q	shear force	Lb., Kg.
r	radius, fillet radius	in., cm.
r	speed ratio	—
R	shaft radius	in., cm.
s	distance	in., cm.
S	stress concentration factor	—
t	depth of groove	in., cm.
T	torque	Lb.in., Kg.cm.
v	velocity	in./sec., cm./sec.
w	specific weight	Lb./in.^{3}, Kg./cm.^{3}
x, y	distances	in., cm.
Z	sectional modulus	in.^{3}, cm.^{3}
α	angle	rad., deg.
γ	shear strain	—
Δ	Holzer-table amplitude	rad.
δ	difference in diameters	in., cm.
θ	vibration amplitude	deg., rad.
λ	relative length, relative eccentricity	—
μ	frictional coefficient	—
ρ	radius of curvature	in., cm.
σ	Poisson's ratio	—
ϕ	angle	rad., deg.
ω	phase velocity	rad./sec.

Subscripts are used in the text to define more precisely. Their meaning is fully explained in their context in each instance.

2·31 Evaluation of stresses from vibration-amplitude diagrams and Holzer tables

Preliminary remarks. The dynamic stresses of a vibrating system can be determined in two manners:

(1) The maximum strain at the node of the shafting is measured by suitable means (e.g. resistance-wire strain gauges) and the torsional stress q is evaluated by means of the relation

$$q = G\gamma \quad [\text{Lb./in.}^2],\dagger$$

where G = modulus of rigidity [Lb./in.^{2}] and γ = torsional strain.

(2) The deflexion θ_1 [rad.] at the free end of the system is measured by means of torsiograph equipment and the stress at the node is determined from the relation

$$q = \frac{1}{Z}\Sigma(J\omega^2\Delta) \times \theta_1 \quad [\text{Lb./in.}^2],$$

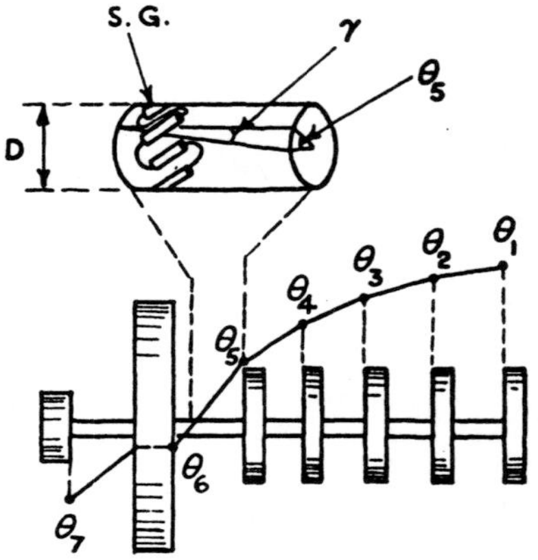

where Z = modulus of shaft cross-section [in.^{3}] and $\Sigma(J\omega^2\Delta)$ = maximum inertia torque at the node [Lb.in.] per radian at mass no. 1 ($\Delta_1 = 1$), as deter-

Fig. 1. Evaluation of stresses from strain measurements with strain gauges (S.G.) and from vibration amplitude measurements (amplitude θ_1 at mass no. 1).

mined by the Holzer table for the particular system.

As the node is not always accessible for measurements, and the use of strain gauges on rotating shaft systems is a technique requiring additional instrumentation, the second method is generally employed for stress determination, as described in the following section. [For details of instrumentation and measuring techniques, see section 4.]

2·311 *Stresses at resonant speeds*

Procedure for stress evaluation

The information required for the evaluation of stresses from vibration-amplitude measurement is as follows:

    (i) Vibration-amplitude diagram;

    (ii) Holzer tabulation; and

    (iii) Sectional modulus of shafting at each section.

The sectional modulus is given by

$$Z = \frac{\pi}{16}D^3 \quad [\text{in.}^3] \quad \text{for a solid shaft,}$$

† *Metric units.* The formulae given in this section are also valid in metric units, for instance with: stress q and modulus of rigidity G expressed in Kg./cm.^{2}, sectional modulus Z in cm.^{3}, torques T and $\Sigma(J\omega^2\Delta)$ in Kg.cm., inertias J in Kg.cm.sec.^{2}, shaft stiffnesses K in Kg.cm./rad., and linear dimensions in centimetres.

[316]

and
$$Z = \frac{\pi}{16} \times \frac{D^4 - d^4}{D} \quad [\text{in.}^3]$$

for a hollow shaft of outer diameter D and inner diameter d.

The sectional modulus should be determined for the smallest diameter shaft occurring in each shaft portion, i.e. it should be based on crankpin diameter if this is smaller than the journal diameter (and vice versa).

For a six-cylinder engine taken as an example, the vibration-amplitude diagram and the Holzer table are given below.

The Holzer table has an additional column (8), which gives the *stress in each shaft section for* $\pm 1°$ *amplitude at the free end of the system*. This is termed the 'stress per degree' and is denoted in the following by q^*.

The values of q^* [Lb./in.^{2} deg.] in column (8) are obtained from those of the total torque $\Sigma(J\omega^2\Delta)$ by means of the relations

$$q^* = \frac{1}{Z}\Sigma(J\omega^2\Delta) \times \frac{12}{57\cdot3} \quad \text{if} \quad \Sigma(J\omega^2\Delta) \text{ is in Lb.ft.,}$$

$$q^* = \frac{1}{Z}\Sigma(J\omega^2\Delta) \times \frac{1}{57\cdot3} \quad \text{if} \quad \Sigma(J\omega^2\Delta) \text{ is in Lb.in.}$$

Then, if $q^*_{\text{max.}}$ is the greatest value obtained, the stress in the corresponding shaft portion is

$$\pm q_{\text{max.}} = \pm \theta_1 \times q^*_{\text{max.}} \quad [\text{Lb./in.}^2],$$

where $\pm\theta_1 =$ measured vibration amplitude [degrees], i.e. measured 'half total swing' at the free end of the engine crankshaft.

Holzer table with column for stress per degree

For crankshaft: $Z_c = \pi D^3/16 = \pi(5\cdot5118)^3/16 = 32\cdot88$ in.^{3}, $\quad 12/[57\cdot3 \times Z_c] = 6\cdot36 \times 10^{-3}$
For shaft between cylinder no. 6 and flywheel: $Z_F = \pi(5\cdot430)^3/16 = 31\cdot42$ in.^{3},
$\quad 12/[57\cdot3 \times Z_F] = 6\cdot67 \times 10^{-3}$.
For generator shaft: $Z_G = \pi \times (7)^3/16 = 67\cdot35$ in.^{3}, $\quad 12/[57\cdot3 \times Z_G] = 3\cdot11 \times 10^{-3}$.

$F = 3940$ vib./min., $\quad \omega^2 = 0\cdot170 \times 10^6$ rad.^{2}/sec.^{2}

Mass no.	J [Lb.ft.sec.^{2}] (1)	$\frac{J\omega^2}{10^6}$ [Lb.ft.] (2)	Δ_m [rad.] (3)	$\frac{J\omega^2\Delta_m}{10^6}$ [Lb.ft.] (4)	$\frac{\Sigma J\omega^2\Delta_m}{10^6}$ [Lb.ft.] (5)	$\frac{K}{10^6}$ [Lb.ft./rad.] (6)	Δ_{sh} [rad.] (7)	q^* [see footnote] [Lb./(in.^{2} deg.)] (8)
1	1·8	0·306	1·0000	0·306	0·306	5·8	0·0528	1950
2	1·8	0·306	0·9472	0·290	0·596	5·8	0·1030	3800
3	1·8	0·306	0·8442	0·258	0·854	5·8	0·1471	5400
4	1·8	0·306	0·6971	0·213	1·067	5·8	0·1840	6800
5	1·8	0·306	0·5131	0·157	1·224	5·8	0·2110	7820
6	1·8	0·306	0·3021	0·092	1·316	3·92	0·3360	8778 (q^*_{max})
F.W.	148·0	25·15	−0·0339	−0·853	0·463	28·2	0·0164	1440
Gen.	49·0	8·33	−0·0503	−0·419	+0·044	—	—	—

Note. Column (8) = $\{12/[57\cdot3 \times Z]\} \times$ column (5), using the values $6\cdot36 \times 10^{-3}$, $6\cdot67 \times 10^{-3}$ and $3\cdot11 \times 10^{-3}$ already calculated.

For the 9th order critical, the peak amplitude is $\theta_1 = \pm\,0{\cdot}45°$ (from Fig. 2). Therefore the stress at the node is

$$q_{max.} = \pm\,0{\cdot}45\ \text{deg.} \times 8778\,\frac{\text{Lb.}}{\text{in.}^2\,\text{deg.}} = \pm\,3950 \quad [\text{Lb./in.}^2],$$

where $8778 = q^*_{max.}$ is taken from the Holzer table given above.

It is convenient to tabulate the peak stresses for the various criticals as follows:

Engine speed [rev./min.]	$N =$	450	475	510	540	580	624
Critical order no.	$n =$	9	$8\frac{1}{2}$	8	$7\frac{1}{2}$	7	$6\frac{1}{2}$
Peak amplitude [deg.]	$\pm\,\theta =$	0·45°	0·11°	0·085°	0·3°	0·14°	0·25°
Peak stress (at node) [Lb./in.²]	$\pm\,q_{max.} =$	3950	965	745	2635	1230	2195

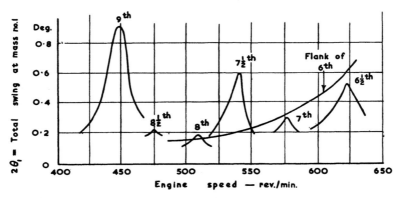

Fig. 2. Diagram of vibration amplitudes plotted against engine speed.†
From torsiograph test at 85·5 Lb./in.² i.m.e.p.

Remark on 'nominal stresses'. It should be noted that the stresses obtained by the above evaluation are 'nominal stresses', i.e. stresses calculated for a plain shaft portion. The nominal-stress values are the figures usually considered.

The effective stresses are somewhat higher, since they include stress concentration factors for fillets, grooves, oil holes, etc. The evaluation of effective stresses is dealt with in a subsequent part of this section.

Stress evaluation from 'filtered' and 'unfiltered' diagrams. The diagram of Fig. 2 was obtained by means of an electrical pick-up, and the pick-up output voltage was passed through a filter or frequency analyser before being indicated on an oscilloscope screen.

The vibration amplitudes recorded without a filter are the result of the combination of the various peak and flank amplitudes of the individual criticals. The 'resultant stresses' can therefore be evaluated

† Alternative designation: vibration spectrum.

from these unfiltered diagrams in the same way as from filtered recordings. The unfiltered diagram may, however, include contributions due to different modes of vibration, so that a check by harmonic-analysis methods may be required if the unfiltered vibration records are not supplemented by a filtered torsiograph test.

Unfiltered readings require correction for cyclic speed variation if the natural frequency is low compared with the frequency of the cyclic speed variation. In such cases, filtered readings are an advantage.

2.3111 *Stresses from vibration amplitudes measured at the driven-machine end or at an intermediate position along the shafting*

1. Close-coupled systems, or systems with negligible driven-machine damping

The most convenient method is to evaluate the corresponding vibration amplitudes at mass no. 1 from recordings taken at other positions. The stresses can then be determined from θ_1-values and a Holzer table as previously described.

The relation between the relative amplitude (i.e. the Holzer-table amplitude) Δ_M at the driven machine and its true amplitude θ_M is

$$\theta_M = \Delta_M \times \theta_1, \quad \text{hence} \quad \theta_1 = \theta_M / \Delta_M.$$

Thus the corresponding value of θ_1 is obtained by dividing the measured value θ_M, recorded at the driven-machine end, by the Holzer-table amplitude Δ_M.

The conversion is the same if the vibration amplitude θ_k is recorded at the position of an intermediate mass J_k, with a Holzer-table amplitude Δ_k:

$$\theta_1 = \theta_k / \Delta_k.$$

Note. If a clutch is included in the system, the measurements taken at the driven-machine end of the system may not give sufficiently correct indications of the vibration in the engine part of the system, so that measurements should also be taken ahead of the clutch.

This remark also applies to geared systems, for which it is useful to have recordings from the individual branches.

2. Systems with long line-shafting or appreciable driven-machine damping

For ship-propulsion installations, with long line-shafting and appreciable propeller damping in the one-node frequency vibration (the 'propeller mode'), the evaluation of stresses from vibration measurements at intermediate points on the shafting should preferably take account of phase-angle shifts due to propeller damping.

An appropriate evaluation can be made by means of Archer's method. A numerical example of recordings evaluated by this method is given in detail in section 4·11 (p. 598). Values of propeller damping coefficients for use in this method can be derived from the charts in section 2·422. For other driven machines with appreciable damping, the damping coefficients can be assessed from the values and formulae given in section 2·421.

2·3112 *Expressions used for stress evaluation*

There are various expressions which can be used for stress evaluation. Those most frequently employed are listed below:

$$q_{node} = q^*_{node}\theta_1 = q^*_{node}\frac{\theta_M}{\Delta_M}$$

$$= \frac{\theta_M}{\Delta_M} \times \frac{\Sigma(J\omega^2\Delta)_{node}}{57\cdot3 \times Z_{node}} = K_n \times (\Delta_{n-1} - \Delta_n) \times \frac{\theta_n}{\Delta_n} \times \frac{1}{57\cdot3 \times Z_{node}},$$

all the θ-values (half total swing) being expressed in degrees in these relations. The symbols are those used in the Holzer table of p. 317.

2·3113 *Stress-distribution diagrams*

It is assumed that the stress is constant in the shafting between each pair of successive masses. It is then possible to plot a broken line representing stress distribution over the length of the shafting in the vibrating system.

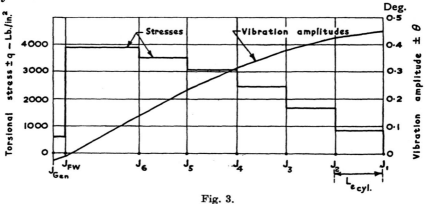

Fig. 3.

The stress-distribution curve of the six-cylinder system previously considered in this section is given in Fig. 3 for the 9th order. The stress values are obtained by multiplying each q^*-value in column (8) of the Holzer table (p. 317) by $\theta_1 = 0\cdot45°$. The vibration amplitudes are determined by multiplying the values in column (3) by θ_1.

[320]

2·312 *Stresses at non-resonant speeds*

Stress evaluation for flank amplitudes and Holzer tables with gas torques

In some cases it is desired to run an engine on the flank of a resonance curve of a certain critical, and it is then necessary to evaluate the torsional vibration stresses for this condition.

To fix the ideas, we consider the six-cylinder engine taken as an example earlier in this section and propose to determine the nominal stress at the node when the engine is running at 433 rev./min. with a 'flank amplitude' $\theta_1 = \pm 0.2°$ obtained from torsiograph measurements (see Fig. 4).

If the peak stress is already fairly low (as in the present example) an approximate value of the flank stress can be obtained directly from the ratio of measured flank and peak vibration amplitudes, using the relation

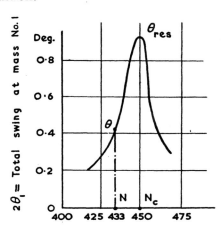

Fig. 4.

$$q = q_{\text{res.}} \times \frac{\theta}{\theta_{\text{res.}}} \quad [\text{Lb./in.}^2],$$

and since $\theta_{\text{res.}} = 0.45°$, $\theta = 0.2°$ and $q_{\text{res.}} = 3950$ Lb./in.^{2} (from peak stress evaluation, see p. 318), we have

$$q = 3950 \times 0.2/0.45 = 1755 \quad [\text{Lb./in.}^2].$$

If the peak stress is a high value and the flank position is fairly close to the resonant speed, it is necessary to evaluate the flank stress more accurately. This is achievable by the following procedure:

(1) Calculate the ratio of flank speed to critical speed

$$r = N/N_c = 433/450 = 0.9622.$$

(2) Take the natural frequency $F_{\text{res.}}$ as determined by a Holzer table for the system, and obtain the forced frequency F from

$$F = rF_{\text{res.}} \quad [\text{vib./min.}].$$

In the present example, this gives $0.9622 \times 3940 = 3791$ vib./min. [Do not use the measured natural frequency, since the Holzer table gives zero residual torque for the calculated frequency of 3940 vib./min., but not for the observed frequency.]

(3) Evaluate a Holzer tabulation for the forced frequency F (3791 vib./min.), using the measured flank amplitude ($\theta = 0.2°$) instead of $\Delta_1 = 1$ radian. The tabulation is given below. *Note.*

$$\theta = 0.2° \times 0.01745 \text{ rad./deg.} = 0.00349 \text{ rad.}$$

$F = 3791$ vib./min., $\omega = 397$ rad./sec., $\omega^2 = 0.1571 \times 10^6$ (rad./sec.)^2

Mass no.	J [Lb.ft. sec.^{2}]	$\dfrac{J\omega^2}{10^6}$ [Lb.ft./rad.]	Δ [rad.]	$\dfrac{J\omega^2\Delta}{10^6}$ [Lb.ft.]	$\dfrac{\Sigma(J\omega^2\Delta)}{10^6}$ [Lb.ft.]	$\dfrac{K}{10^6}$ [Lb.ft./rad.]	Δ_{sh} [rad.]
1	1·8	0·2837	0·003490	0·000990	0·000990	5·8	0·000170
2	1·8	0·2837	0·003320	0·000942	0·001932	5·8	0·000333
3	1·8	0·2837	0·002987	0·000845	0·002777	5·8	0·000478
4	1·8	0·2837	0·002409	0·000683	0·003460	5·8	0·000596
5	1·8	0·2837	0·001813	0·000514	0·003974	5·8	0·000684
6	1·8	0·2837	0·001129	0·000320	0·004294	3·92	0·001096
F.W.	148·0	23·326	0·000033	0·000770	0·005064	28·2	0·000180
Gen.	49·0	7·729	−0·000147	−0·001136	+0·004928	—	—

(4) Evaluate a 'gas-torque tabulation', as detailed below, denoting the gas torque due to each cylinder by $T = T_0 \times 10^6$.

$F = 3791$ vib./min., $\omega = 397$ rad./sec., $\omega^2 = 0.1571 \times 10^6$ (rad./sec.)^2

Mass no.	J [Lb.ft. sec.^{2}]	$\dfrac{J\omega^2}{10^6}$ [Lb.ft./rad.]	Δ (due to gas torque) [rad.]	$\dfrac{J\omega^2\Delta}{10^6}$ [Lb.ft.]	$\dfrac{T}{10^6}$ (gas torque) [Lb.ft.]	$\dfrac{\Sigma(J\omega^2\Delta+T)}{10^6}$ [Lb.ft.]	$\dfrac{K}{10^6}$ [Lb.ft./rad.]	Δ_{sh} [rad.]
1	1·8	0·2837	—	—	T_0	T_0	5·8	$0·17241T_0$
2	1·8	0·2837	$-0·17241T_0$	$-0·04891T_0$	T_0	$1·95109T_0$	5·8	$0·33639T_0$
3	1·8	0·2837	$-0·50980T_0$	$-0·14463T_0$	T_0	$2·80636T_0$	5·8	$0·48386T_0$
4	1·8	0·2837	$-0·99366T_0$	$-0·28190T_0$	T_0	$3·62446T_0$	5·8	$0·62491T_0$
5	1·8	0·2837	$-1·61857T_0$	$-0·45919T_0$	T_0	$4·16527T_0$	5·8	$0·71815T_0$
6	1·8	0·2837	$-2·33672T_0$	$-0·66293T_0$	T_0	$4·50234T_0$	3·92	$1·14856T_0$
F.W.	148·0	23·326	$-3·48528T_0$	$-81·29764T_0$	—	$-76·79530T_0$	28·2	$-2·72324T_0$
Gen.	49·0	7·729	$-0·76204T_0$	$-5·88981T_0$	—	$-82·68511T_0$	—	—

[*Note.* In line 2: $1·95109T_0 = (T_0 - 0·04891T_0) + T_0$.]

(5) Equate the sum of the residual torques to zero, in order to obtain the value of T_0:

$\Sigma(J\omega^2\Delta)$ (from previous table) + $\Sigma(J\omega^2\Delta+T)$ (from above table) = 0.

In the present example $\Sigma(J\omega^2\Delta) = +0.004928 \times 10^6$ from the Holzer table and $\Sigma(J\omega^2\Delta+T) = -82.68511 \times T_0 \times 10^6$; therefore

$$0.004928 - 82.68511T_0 = 0, \quad T_0 = 5.95 \times 10^{-5},$$

and
$$T = T_0 \times 10^6 = 59.5 \text{ Lb.ft.}$$

(6) Determine the maximum value of total torque at the node, by taking the sum of the corresponding values from both tables:

From the total torques in lines 6 of the two tables:

$$0.004294 \times 10^6 + 4.50234 \times 59.5 = 4294 + 267.9 = 4561.9 \quad \text{[Lb.ft.]}.$$

(7) The stress at the node is then obtained by dividing this total torque by the modulus of the corresponding shaft cross-section: $q = T/Z$. For the example, $Z_F = 31.42$ in.^{3}, so that

$$q = 4562 \times 12/31.42 = 1740 \quad \text{[Lb./in.}^2\text{]}.$$

The estimated value obtained as a first approximation from

$$q = q_{\text{res.}} \times \theta/\theta_{\text{res.}} \quad \text{was} \quad 1755 \, \text{Lb./in.}^2.$$

The value of 1740 Lb./in.^{2} gives a somewhat more accurate estimate. It should be noted, however, that the above forced-frequency calculation is also an approximation, which assumes that the damped curve is closely similar to the undamped resonance curve. This is true to some extent, and the two curves become practically identical at some distance away from the resonance point.

In many cases, either of the two methods of assessing stresses on flanks of criticals gives sufficient information. In border-line cases, the determination by means of Holzer tables with damping gives further accuracy in the results. [For details of this procedure, see 'Stress prediction by tabulation methods', section 2·44.]

The computations for gas torque, which were given in a separate table to explain the required procedure, are in fact part of the Holzer tabulation of the system with a forced frequency and an impressed torque at each engine mass. This follows directly from the differential equations of the system. Therefore, all the operations in the two previous tables can be included in a single tabulation, as indicated below.

Forced frequency $F = 3791$ vib./min., $\omega = 397$ rad./sec., $\omega^2 = 0.1571 \times 10^6$ (rad./sec.)^2,
$\Delta_1 = 0.2° = 0.00349$ rad. (measured flank amplitude), $T = T_0 \times 10^6$ Lb.ft. = gas torque per cylinder.

Mass no.	J [Lb.ft.sec.^{2}]	$\dfrac{J\omega^2}{10^6}$ [Lb.ft./rad.]	Δ [rad.]		$\dfrac{J\omega^2\Delta}{10^6}$ [Lb.ft.]	
1	1·8	0·2837	0·003490	—	0·000990	—
2	1·8	0·2837	0·003320	$-0.17241T_0$	0·000942	$-0.04891T_0$
3	1·8	0·2837	0·002987	$-0.50980T_0$	0·000845	$-0.14463T_0$
4	1·8	0·2837	0·002409	$-0.99366T_0$	0·000683	$-0.28190T_0$
5	1·8	0·2837	0·001813	$-1.61857T_0$	0·000514	$-0.45919T_0$
6	1·8	0·2837	0·001129	$-2.33672T_0$	0·000320	$-0.66293T_0$
F.W.	148·0	23·326	0·000033	$-3.48528T_0$	0·000770	$-81.29764T_0$
Gen.	49·0	7·729	-0.000147	$-0.76204T_0$	-0.001136	$-5.88981T_0$

Mass no.	Gas torque $\dfrac{T}{10^6}$ [Lb.ft.]	Total torque $\dfrac{\Sigma(J\omega^2\Delta + T)}{10^6}$ [Lb.ft.]		$\dfrac{K}{10^6}$ [Lb.ft./rad.]	$\dfrac{1}{K} \times \Sigma(J\omega^2\Delta + T) = \Delta_{sh}$ [rad.]	Shaft deflexion
1	T_0	0·000990	$+T_0$	5·8	0·000170	$+0.17241T_0$
2	T_0	0·001932	$+ 1.95109T_0$	5·8	0·000333	$+0.33639T_0$
3	T_0	0·002777	$+ 2.80636T_0$	5·8	0·000478	$+0.48386T_0$
4	T_0	0·003460	$+ 3.62446T_0$	5·8	0·000596	$+0.62491T_0$
5	T_0	0·003974	$+ 4.16527T_0$	5·8	0·000684	$+0.71815T_0$
6	T_0	0·004294	$+ 4.50234T_0$	3·92	0·001096	$+1.14856T_0$
F.W.	—	0·005064	$- 76.79530T_0$	28·2	0·000180	$-2.72324T_0$
Gen.	—	$+0.004928$	$-82.68511T_0$	—	—	—

Gas torque: $+0.004928 - 82.68511T_0 = 0$, $T_0 = 0.004928/82.68511 = 5.95 \times 10^{-5}$, $T = T_0 \times 10^6 = 59.5$ Lb.ft.

Total torque at node (line 6): $0.004294 \times 10^6 + 4.50234 \times 59.5 = 4294 + 267.9 = 4561.9$ Lb.ft. $= T_{\text{total}}$.

Stress at node: $q = T_{\text{total}}/Z_F = 4562/31.42 = 1740$ Lb./in.^{2}.

Note. The above procedure applies for major orders. Details of the procedure for minor orders are included in the evaluation of forced-frequency conditions of systems with damping, given in section 2·44. In most applications, large flank amplitudes are due to major criticals only, and an evaluation in the manner indicated above is usually sufficient.

<h2>2·32 Stress concentration factors, fatigue strength factors, and permissible torques</h2>

Stress concentration factors

The nominal shear stress of a component having been determined, it is then possible to assess the effective maximum stress

$$q_{max.} = Sq_{nom.},$$

using values for the 'stress concentration factor' S applying to the particular shape and type of loading of the component.

In general, the S-values obtained by theoretical analysis, photoelastic or analogy methods are somewhat higher than those determined from strain measurements on metal specimens subjected to static loadings. This, to some extent, is so because there is no strain gauge of practically zero gauge length so as to indicate true peak-stress values and also because for very sharp notches there is a local reduction in stress due to the plasticity of the material when the yield point is exceeded.

The classification societies (see section 2·5) require only the nominal stress values, but it is assumed that in all cases the component is designed to have fairly low S-values. The stress concentration factors for various types of shafts and components are given in the following section, mainly for torsional loading. The values given are based on the latest data available.

Fatigue strength factors

The presence of a stress raiser has generally the effect of reducing the fatigue strength of a component, i.e. the stress above which failure is likely to occur under dynamic loading. The reduction in fatigue strength can be assessed by means of the factor

$$S_F = \frac{\text{fatigue strength of specimen without stress raiser}}{\text{fatigue strength of specimen with stress raiser}}.$$

For low and medium carbon steels, the S_F-values are usually considerably lower than the S-values derived from static tests. For alloy steels with high tensile strength, the S_F-values are closer to the S-values.

Permissible torque

For some components, it is necessary to calculate the maximum permissible torque that can be transmitted. Torque formulae for certain applications are included in this section, as well as other details regarding design features.

[324]

Shafts of simple cross-section

Circle with concentric hole: $q_{max.} = T \left/ \dfrac{\pi}{16} \times \dfrac{D^4 - d^4}{D} \right. = \text{max. stress at outer}$ periphery, with $T = \text{torque}$.

Circle with eccentric hole: See p. 326.

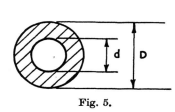

Fig. 5.

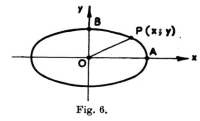

Fig. 6.

†*Ellipse:* $\qquad\qquad \overline{OA} = a, \quad \overline{OB} = b.$

Stress at any point $P(x, y)$:

$$q(x, y) = T \times \frac{\sqrt{(b^4 x^2 + a^4 y^2)}}{\dfrac{\pi}{2} a^3 b^3}.$$

Max. stress at B:

$$q_{max.} = q_B = T \times \frac{1}{\dfrac{\pi}{2} ab^2}.$$

†*Square:* $\qquad\qquad a = \text{length of side.}$

Max. stress at A:

$$q_{max.} = q_A = 4 \cdot 808 T / a^3.$$

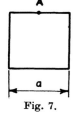

Fig. 7.

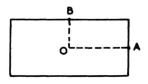

Fig. 8.

†*Rectangle:* $\qquad\qquad OA = a, \quad OB = b.$

Max. stress at B:

$$q_{max.} = q_B = \lambda \frac{T}{ab^2},$$

where λ is a value taken from the following table:

$a/b =$	1·0	1·2	1·4	1·6	1·8	2·0	2·5	3·0	4·0	5·0
$\lambda =$	0·60	0·57	0·55	0·53	0·52	0·508	0·485	0·468	0·443	0·429

(B. de St Venant.)

† For the derivations of these expressions, see, for instance: Wang, C. T., *Applied Elasticity,* pp. 85 and 90 (McGraw-Hill Book Co., Inc. 1953).

Equilateral triangle: a = length of side.

Max. stress at A:

$$q_{max.} = q_A = \frac{20T}{a^3}.$$

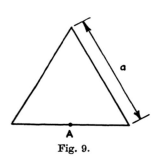

Fig. 9.

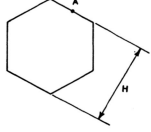

Fig. 10.

Hexagon: Max. stress at A:

$$q_{max.} = q_A = \frac{5\cdot3T}{H^3}.$$

Shaft with eccentric hole

Let e = eccentricity, d = diameter of hole, D = outer diameter of shaft. For a shaft with a concentric hole ($e = 0$), the torsional stress is given by

$$q_{nom.} = \frac{T \times (D/2)}{\frac{\pi}{32} \times (D^4 - d^4)} \quad [\text{Lb./in.}^2].$$

According to T. S. Wilson† the stress concentration factor for a shaft with an eccentric hole is

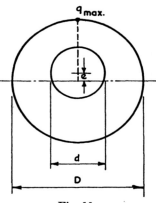

Fig. 11.

$$S = \frac{q_{max.}}{q_{nom.}} = 1 + \frac{4m^2}{1-m^2}\lambda + \frac{32m^2\lambda^2}{(1-m^2)(1-m^4)} + \frac{48m^2(1+2m^2+3m^4+2m^6)\lambda^3}{(1-m^2)(1-m^4)(1-m^6)}$$

$$+ \frac{64m^2(2+12m^2+19m^4+28m^6+18m^8+14m^{10}+3m^{12})\lambda^4}{(1-m^2)(1-m^4)(1-m^6)(1-m^8)} + \ldots,$$

where $m = d/D$ and $\lambda = e/D$.

Curves of $q_{max.}/q_{nom.}$ as a function of e/D and d/D, calculated by Wilson, are given in the following graph.

Note. The curves in Fig. 12 are for values of e/D between zero and two-fifths of the geometrical limit $(e/D)_{max.} = \frac{1}{2}(1-m)$.

† Wilson, T. S., 'The eccentric circular tube', *Aircr. Engng*, March 1942, pp. 76–9.

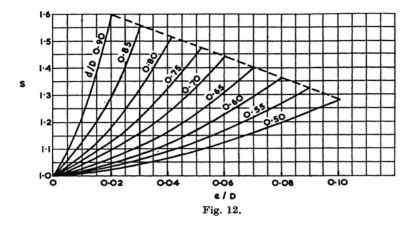

Fig. 12.

Stepped cylindrical shafts

(1) Sonntag's formula for torsional stress concentration factor:

$$S = \frac{q_{\text{max.}}}{q_{\text{nom.}}} = \left(\frac{d}{D}\right)\left(1\cdot5 + 3\frac{r}{d}\right)\frac{1+4\frac{r}{d}}{1+6\frac{r}{d}} + \left(1+\frac{1}{12\frac{r}{d}}\right)\left[1 - \frac{d}{D} - 2\frac{r}{d}\times\frac{d}{D}\right],$$

where d = diameter of small shaft portion, D = diameter of large shaft portion, r = fillet radius, and $q_{\text{nom.}} = T\big/\left(\frac{\pi}{16}d^3\right)$, T being the applied torque. The formula is applicable to shafts for which $d/D \leqslant 1\big/\left[1 + 2\frac{r}{d}\right]$.

A simpler version of the above formula is

$$S = 1 + \frac{d}{12r}\left[1 - \frac{1+2\frac{r}{d}}{\frac{D}{d}\left(1+6\frac{r}{d}\right)}\right].$$

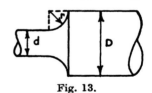

Fig. 13.

[See Sonntag, R., *Z. angew. Math. Mech.* vol. 9, 1929, pp. 1–22.]

(2) Thum and Bautz values for torsional stress concentration factor:
These values, obtained experimentally, are substantially in agreement with the values calculated by Sonntag's formula. The results are given in the following table:

$r/d =$	0·025	0·05	0·075	0·10	0·20	0·30	0·40	0·50
$D/d = 1\cdot5$ $S =$	2·5	2·15	1·92	1·80	1·50	1·35	1·25	1·25
$D/d = 3\cdot0$ $S =$	3·75	2·8	2·45	2·25	1·75	1·47	1·35	1·30

[See Thum, A. and Bautz, W., *Z. Ver. dtsch. Ing.* vol. 79 (1935), pp. 1303–6.]

(3) Experimental values of Weigand:

[See Weigand, A., *Luftfahrtforschung*, July 1943, pp. 217–9.] The following results were obtained using a Lehr-Greinacher strain gauge [described in *Forschung a. d. Geb. d. Ingenieurwesens*, p. 66, 1936], with a gauge length of 1·3 mm.

The figures are somewhat lower than those of Sonntag, Thum and Bautz, but agree with Jacobsen's measured values

$$S = q_{max.}/q_{nom.}.$$

	$r/d =$	0·10	0·12	0·14	0·16	0·18	0·20	0·22	0·24	0·25
$d/D = 0·5$	$S =$	1·46	1·42	1·38	1·34	1·31	1·28	1·25	1·23	1·22
0·6		1·425	1·375	1·34	1·30	1·27	1·24	1·22	1·21	1·208
0·7		1·38	1·33	1·29	1·26	1·23	1·21	1·19	1·18	1·18
0·8		1·32	1·28	1·245	1·22	1·18	1·165	1·15	1·14	1·14
0·9		1·235	1·20	1·18	1·15	1·13	1·115	1·09	1·08	1·08

Note. For *flanged* sections, one can use $D =$ pitch circle diameter of bolts, and $d =$ shaft diameter.

(4) Jacobsen's values:

[See Jacobsen, L. S., 'Torsional stress concentrations in shafts of circular cross-section and variable diameter', *Trans. Amer. Soc. Mech. Engrs*, vol. 47, 1925, p. 619.]

The fillet stresses of stepped shafts were determined by Jacobsen using an electrical analogy method, in which the stress function and the 'angle of twist' function are equivalent to streamlines and constant-potential lines obtained by measurements on 'hollow razor blade' specimens.

Jacobsen's values are in good agreement with those obtained with strain gauges by Weigand in the range covered by both investigators. They are lower than the corresponding values obtained by Sonntag's formula, which is based on a hydrodynamic analogy.

$$S = q_{max.}/q_{nom.}, \quad \text{where} \quad q_{nom.} = T \Big/ \left(\frac{\pi}{16} d^3\right).$$

	$r/d =$	0·01	0·02	0·03	0·04	0·06	0·08	0·10	0·20
$D/d = 1·05$	$S =$	1·45	1·28	1·23	1·18	1·14	1·12	1·11	1·07
1·10		1·95	1·56	1·44	1·35	1·27	1·22	1·18	1·12
1·20		2·40	1·95	1·77	1·65	1·50	1·40	1·34	1·23
2·00		3·00	2·25	1·95	1·84	1·68	1·53	1·45	1·28

Note. The stress concentration factors for stepped shafts are also applicable to crank-throws within certain limits. For this purpose $d =$ diameter of crankpin or journal, and $D =$ web width, r being the fillet radius.

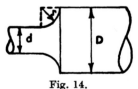

Fig. 14.

In view of the importance of this application, the fatigue strength factors S_F for torsional loading, and corresponding factors for bending, are given in the following tables.

Fatigue strength factors S_F for stepped shafts in torsion

$r/d =$		0·02	0·04	0·05	0·1	0·15	0·2	0·3	0·4
$D/d = 1\cdot8$	$S_F =$	—	—	—	1·42	1·30	1·20	1·15	1·13
$D/d = 1\cdot4$	$S_F =$	—	—	1·55	1·30	1·22	1·17	1·13	1·11
$D/d = 1\cdot2$	$S_F =$	—	—	1·37	1·25	1·17	1·13	1·10	1·08
$D/d = 1\cdot16$	$S_F =$	1·80	1·40	1·35	1·23	1·16	1·12	—	—

[Reference: Dolan, D. J., *Bull. Univ. Illinois, Engng Exp. Sta.* no. 293 (1937). Bühler, H. and Scheil, E., *Arch. Eisenhüttenw.* Jan. 1933, p. 283.]

Note.

$$S_F = \frac{\text{fatigue strength of plain shaft of diameter } d}{\text{fatigue strength of stepped shaft with smaller-portion diameter } d}.$$

Bending stress concentration factors S for stepped shafts

$r/d =$		0·02	0·03	0·04	0·05	0·1	0·15	0·2	0·25	0·3	0·4	Ref.
$D/d = 3\cdot0$	$S =$	—	3·5	3·1	2·85	2·25	1·95	1·75	1·60	1·48	1·33	T. and B.
	$S =$	—	—	2·9	2·65	1·98	1·71	1·55	1·43	1·35	1·26	F. and T.D.
$D/d = 1\cdot5$	$S =$	—	2·98	2·72	2·47	1·80	1·55	1·45	1·37	1·30	1·22	T. and B.
	$S =$	2·55	2·40	2·25	2·15	1·80	1·65	1·60	1·47	1·38	1·27	F. and T.D.

[Reference: Thum, A. and Bautz, W., *Z. Ver. dtsch. Ing.* vol. 79 (1935), pp. 1303–6. Frocht, M. M., *J. Appl. Mech.* June 1935, pp. A 67–8. Timoshenko, S. and Dietz, W., *Trans. Amer. Soc. Mech. Engrs*, vol. 47 (1925), pp. 199–237.]

Fatigue strength factors S_F for stepped shafts in bending

$r/d =$		0·05	0·08	0·1	0·15	0·2	0·3	0·4
$D/d = 1\cdot8$	$S_F =$	—	1·78	1·68	1·55	1·43	1·30	1·25
$D/d = 1\cdot4$	$S_F =$	1·8	1·68	1·62	1·49	1·39	1·25	1·20
$D/d = 1\cdot2$	$S_F =$	1·72	1·62	1·57	1·43	1·33	—	—

[Reference: Horger, O. J. *et alii*, *Trans. Amer. Soc. Mech. Engrs*, vol. 67 (1945), p. 149. Thum, A. and Bruder, E., *Dtsch. Kraftfahrtforsch.* 1940, no. 1, p. 1, and 1938, no. 11, p. 1.]

Shaft with transverse hole

Approximate formulae of G. Seeger (*Technik, Berl.*, vol. 3, no. 7 (July 1948), pp. 309–11).

(1) Sectional modulus of hollow shaft in torsion:

$$Z_h \cong \left[1 - 0\cdot9 \frac{d}{D} \right] \times \frac{\pi}{16} D^3 \quad [\text{in.}^3],$$

where $D =$ shaft diameter, $d =$ hole diameter. This formula is sufficiently accurate if $d/D < 0\cdot5$.

(2) Second polar moment of area of cross-section:

$$I_p \cong \left[1 - 0\cdot9 \frac{d}{D} \right] \times \frac{\pi}{32} D^4 \quad [\text{in.}^4].$$

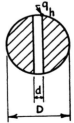

Fig. 15.

The exact expression is

$$I_p = \frac{\pi}{32} D^4 \left[1 - \frac{2}{\pi} \sin^{-1} (d/D) \right] - \frac{d}{48} \times (D^2 + 2d^2) \times \sqrt{(D^2 - d^2)} \quad [\text{in.}^4].$$

(3) Stress concentration ratio S:
On the basis of strain gauge measurements, the following values were
determined by Seeger:

$d/D=$	0·05	0·1	0·2	0·4	0·6
$S=$	2·0	2·4	2·6	2·75	2·9

where $S = q_{max.}/q_h$, q_h being determined from the relation

$$q_h = T/Z_h \quad [\text{Lb./in.}^2], \dagger$$

where $T =$ torque [Lb.in.], and $Z_h =$ sectional modulus [in.^{3}] obtained
from the formula given above.
(4) Fatigue strength ratio S_f:
Results of tests by Armbruster, Cornelius and Bollenrath, Dolan,
Herold, Houdremont, Lehr, and Seeger give the following mean values:

$d/D=$	0·0	0·05	0·1	0·2	0·3	0·35
$S_f=$	1·00	1·25	1·50	1·75	1·85	1·90

where $S_f =$ ratio of torsional fatigue strength of solid shaft to that of the
shaft with a transverse oil hole.

Shafts with circumferential groove
Complete semicircle groove section
The nominal torsional stress at the bottom of the groove is

$$q_{nom.} = \frac{T}{\dfrac{\pi}{16} d^3} \quad [\text{Lb./in.}^2].$$

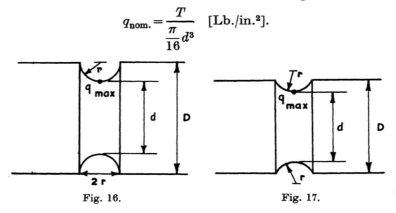

Fig. 16.   Fig. 17.

According to H. Okubo (*J. Amer. Soc. Mech. Engrs*, March 1952,
pp. 16–18), the stress concentration at this position is

$$\frac{q_{max.}}{q_{nom.}} = 2\frac{d}{D}.$$

† Or in metric units: q_h is in Kg./cm.^{2} if T is in Kg.cm. and Z_h in cm.^{3}.

Circular arc groove section

According to W. Herold (*Z. Ver. dtsch. Ing.* 1937, pp. 505 et seq.), the stress concentration

$$q_{max.}/q_{nom.} = q_{max.} \bigg/ \left[T \bigg/ \frac{\pi}{16} d^3 \right]$$

at the bottom of the groove can be determined from the values given in the following table:

$r/d =$	0·01	0·1	0·2	0·3	0·4
		Stress Concentration Factor $q_{max.}/q_{nom.}$			
$d/D = 0·7$	3·0	2·2	1·95	1·85	—
$d/D = 0·9$	2·3	1·65	1·40	—	—

Elliptical groove section

According to Okubo, if

$$q_{nom.} = \frac{T}{\dfrac{\pi}{16} d^3} = \frac{T}{\dfrac{\pi}{16} (D - 2b)^3},$$

the stress concentration factor is determined as

$$\frac{q_{max.}}{q_{nom.}} = \left(1 + \frac{b}{a} \right) \left(1 - \frac{2b}{D} \right).$$

For $a = b$, this reduces to Okubo's expression for the semicircle.

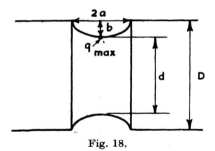

Fig. 18.

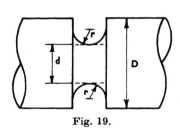

Fig. 19.

Circumferential grooves with parallel sides

Torsional stress concentration factors S

	$r/d =$	0·005	0·01	0·02	0·04	0·06	0·08	0·10
$D/d = 1·4$	$S =$	5·5	3·3	2·7	2·45	2·35	2·25	2·20
1·2		4·5	2·85	2·4	2·20	2·10	1·95	1·90
1·1		3·75	2·45	2·10	1·80	1·70	1·65	1·60
1·02		2·35	1·75	1·45	1·30	1·25	1·20	1·18
1·01		1·70	1·40	1·30	1·20	1·15	1·10	1·08

$$q_{max.} = S \times q_{nom.} = S \times T \bigg/ \left(\frac{\pi}{16} d^3 \right).$$

[Reference: Thum, A. and Bautz, W., *Z. Ver. dtsch. Ing.* vol. 79 (1935), pp. 1303–6.]

Keys and keyways

Some bibliographic references are given below:

Föppl, O. and Meyer, W. *Schweiz. Bauztg*, vol. 105, no. 14 (1935).

Gibson, W. H. H. *J. Inst. Engrs Aust.* vol. 10 (1938).

Korn, A. H. *Prod. Engng*, April 1943, pp. 242–3.

Leven, M. M. *Proc. Soc. Exp. Stress Anal.* vol. 7, no. 2 (1950), p. 141.

Okubo, H. *J. Appl. Mech.* Dec. 1950, pp. 359–62, and *Quart. J. Mech.* June 1950, pp. 162–7.

Solakian, A. G. and Karelitz, G. B. *Trans. Amer. Soc. Mech. Engrs*, 15 June 1931, vol. 54, no. 11, pp. 97–123.

Axial grooves in shafts of circular cross-section

U-*section grooves*

For both types of grooves shown in Figs. 20 and 21, the stress concentration factor can be determined by means of Neuber's formula[†] as

$$S = \frac{q_{max.}}{q_{nom.}} = \left[1 + \sqrt{\left(\frac{t}{r} \right)} \right] \times C,$$

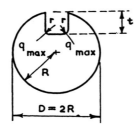

Fig. 20. Groove with flat bottom portion.

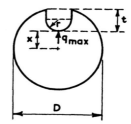

Fig. 21. Groove with round bottom portion.

where $q_{nom.} = T \times (16/\pi D^3) =$ nominal stress at the periphery of the unnotched cross-section, $t =$ depth of groove, $r =$ groove radius, and $C = [1 + (2x/R)]/3$, for a cautious estimate, with

$$x = R - t \quad \text{and} \quad R = D/2 = \text{shaft radius.}$$

Circular-arc grooves

The type of groove considered by Rossbach[‡] is a circular groove intersecting the shaft periphery at right angles (Fig. 22), i.e. a groove in accordance with the relation

$$\overline{OA}^2 = r^2 + R^2 = a^2,$$

† Neuber, H., *Investigation of Stress Concentrations (Kerbspannungslehre)*, (Springer, Berlin, 1937), p. 133.

‡ Rossbach, H. F., 'Torsion of notched circular cross-sections', *Ingen.-Arch.* vol. 10 (1939), pp. 142–52.

the point A being the centre of the groove circle. The values of the stress concentration factor are as follows (with

$$S = q_{max.}/q_{nom.},$$

and $q_{nom.} = T \times (16/\pi D^3) =$ nominal stress at the periphery of the un-notched cross-section):

$\overline{OB}/R = (a-r)/R =$	0·99	0·9	0·8	0·7	0·6	0·5	0·4	0·3	0·2
$S =$	2·00	1·88	1·74	1·63	1·50	1·38	1·26	1·16	1·06

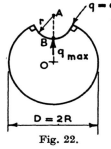

Fig. 22.

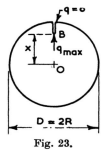

Fig. 23.

Another case also evaluated by Rossbach is that of an extremely narrow slot, with a groove radius of curvature (Fig. 23) also approaching to zero. For $x/R = 0.75$, the stress concentration factor obtained is then $S = 3.0$.

[For an explanation of the zero shear stress at the upper corners, see, for instance: Den Hartog, J.P., *Advanced strength of materials*, McGraw-Hill, 1952, pp. 2–3.]

Splined shafts

Stiffness. The stiffness is usually based on the root diameter d, so that in terms of a reference diameter D_e, the equivalent length of a splined shaft is

$$L_e = L \times \frac{D_e^4}{d^4}.$$

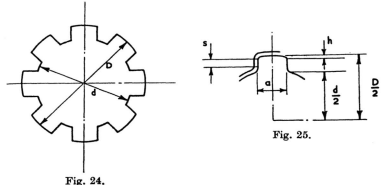

Fig. 24.

Fig. 25.

Torque-carrying capacity. This is determined from the relation

$$T = N_s \times \frac{D_m}{2} \times sLp \quad \text{[Lb.in.]},$$

where T = total transmitted torque [Lb.in.], N_s = number of splines, $D_m = \frac{1}{2}(D+d)$ = mean diameter [in.], $s = \frac{1}{2}(D-d) - 2h$ = contact distance [in.], L = length of spline [in.], and p = permissible side pressure [Lb./in.^{2}].

For a conservative calculation, in cases where radii are not apparent at the corners, s should not be taken greater than half the height of the spline, i.e.

$$s \leqslant \tfrac{1}{4}(D-d).$$

The value of p depends on the properties of the material, but it is also limited by shear-stress considerations. Equating the shear force Q to the side-pressure force P, we obtain

$$Q = qaL = psL = P,$$

and if the spline width is $a \simeq 4s$, this reduces to $p = 4q$.

Assuming a permissible nominal shear stress $q = 3000$ Lb./in.^{2},

$$p = 4 \times 3000 = 12{,}000 \quad [\text{Lb./in.}^2],$$

and this value can be used in the above equation for torque.†

Maximum shear stress. This can be calculated from Neuber's formula for U-section grooves [see p. 332]. Stress-relieving fillets at the spline roots are shown in the two figures given below [from W. Herold, *Z. Ver. dtsch. Ing.* 1 May 1937, pp. 505–9].

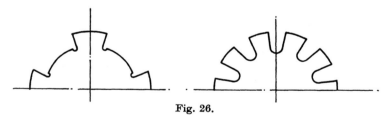

Fig. 26.

Circular shaft with diametrically opposite flat sides

The maximum stress at the mid-point of the flat sides is determined by means of the Carter and Oliphint formulae:

Approximate formula:

$$q_{\text{max.}} = \frac{GR\theta}{L} \times \frac{1 - \frac{3}{4}\alpha^2}{1 - \dfrac{2\alpha}{\pi} + \left(\dfrac{\alpha}{\pi}\right)^2} \quad [\text{Lb./in.}^2],$$

† *Note.* As data on permissible values of side pressures and shear stress for splined shafts are not available, it is suggested that, where possible, the following limits should not be exceeded:

$$p \leqslant 15{,}000 \text{ Lb./in.}^2, \quad q \leqslant 4000 \text{ Lb./in.}^2,$$

and that fillets or grooves be used at the spline roots to reduce stress concentrations. [See also McCain, G. L., 'Engineering of involute splines', *S. A. E. Quart. Trans.* October 1951, pp. 494–531.]

where G = modulus of rigidity [Lb./in.^{2}], R = shaft radius [in.], α = angle AOB [rad.], and L = shaft length [in.]. θ is the angle of twist [rad.] of the shaft of length L.

To obtain a value for θ, it is necessary to know the stiffness K, since $\theta = T/K$, where T = torque [Lb.in.]. For small α-values, the stiffness is approximately that of a circular shaft of radius OB. For large α-values, it is approximately equal to the stiffness of the circumscribed rectangle (shown dotted in the figure). More accurate values for stiffness can be estimated by the Griffith and Taylor method (see section 1·2).

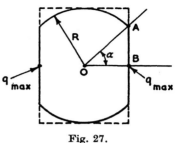

Fig. 27.

Exact formula:

$$q_{\text{max.}} = \frac{GR\theta}{L} \times \frac{(1-\alpha^2)\left(\dfrac{\alpha^2}{4}+1\right)}{\left(1-\dfrac{2\alpha}{\pi}+\dfrac{\alpha^2}{\pi^2}\right)} \quad [\text{Lb./in.}^2].$$

[Reference: Carter, W. J. and Oliphint, J. B., 'Torsion of a circular shaft with diametrically opposite flat sides', *Amer. Soc. Mech. Engrs*, Paper no. 52-S-2.]

Circular shaft with one flat side

The maximum stress is at the mid-point of the flat side. The stress concentration factor

$$S = \frac{q_{\text{max.}}}{T\left/\dfrac{\pi}{16}D_0^3\right.}$$

can be determined for the corresponding value of α (angle AOB) from the graph given below.

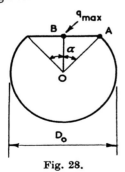

Fig. 28.

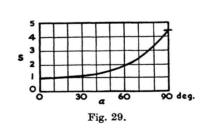

Fig. 29.

[Reference: Weigand, A., 'The problem of torsion in prismatic members of circular segmental cross-section', *Luftfahrtforschung*, vol. 20, no. 12 (8 Feb. 1944), pp. 333–40; and *Tech. Memor. Nat. Adv. Comm. Aero.*, Wash., no. 1148 (1948).]

Welded shaft sections

The shear stress is usually calculated by means of the Batho-Bredt formula

$$q = \frac{2T}{\pi D_m^2 h} \quad [\text{Lb./in.}^2],$$

where T = applied torque [Lb.in.], h = effective thickness of weld [in.], and $D_m = D - h$ = mean diameter of welded ring.

The 'force per inch of weld' (along the periphery) is

$$P^* = \frac{P}{\pi D_m} = qh = \frac{2T}{\pi D_m^2} \quad [\text{Lb./in.}].$$

Note. The Batho-Bredt formula can be applied to thin-walled cylinders or prisms of any shape. The formula in such cases is

$$q = \frac{T}{2Ah} \quad [\text{Lb./in.}^2],$$

where A is the full area of the cross-section (shown shaded in the accompanying figure). The 'force per inch of weld' is then $P^* = T/(2A)$.

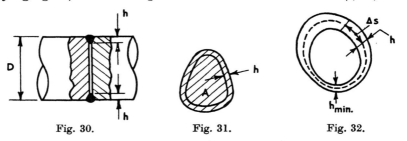

Fig. 30.   Fig. 31.   Fig. 32.

For cross-sections of varying wall thickness, the formulae for shear stress q, polar second moment of area I_p, and modulus of section Z, are as follows:†

$$q_{\text{max.}} = \frac{T}{2Ah_{\text{min.}}} \quad [\text{Lb./in.}^2],$$

$$I_p = 4A^2 \bigg/ \Sigma \frac{\Delta s}{h} \quad [\text{in.}^4],$$

$$Z = 2Ah_{\text{min.}} \quad [\text{in.}^3],$$

where A = area enclosed within perimeter $\Sigma \Delta s$ along the line of mean thickness. For I_p, the sum is taken over the entire perimeter.

Note. The expression for I_p applied to a thin circular section gives

$$I_p = 4\pi^2 \left(\frac{1}{2} \times \frac{D+d}{2}\right)^4 \bigg/ \left[\frac{\pi(D+d)/2}{(D-d)/2}\right] = \frac{\pi}{32}(D^2 - d^2) \times \frac{D^2 + 2Dd + d^2}{2},$$

† These formulae are also valid in metric units, with the linear dimensions in cm., torques in Kg.cm., and stresses and moduli of rigidity in Kg./cm.².

which is a sufficiently accurate approximation to $\frac{\pi}{32}(D^2-d^2)\times(D^2+d^2)$ for small values of $D-d$. (The error is less than $0\cdot3\%$ if $(D-d)<0\cdot1\times D$.)

See also: Boyd, G. M., 'Effects of residual stresses in welded structures', *Brit. Welding J.* Dec. 1954, p. 560. Vreedenburgh, C. G. J., 'New principles for the calculation of welded joints', *Int. Shipb. Progr.* vol. 1, no. 4 (1954), p. 200.

Shrink-fit assemblies

For the purpose of stiffness calculations, the assembly can be regarded as a shaft with integral coupling, if the tangential stress of the shrink fit is up to the yield point of the material or even somewhat higher.

The maximum (vibratory + steady) torque which can be transmitted by a shrink fit is determined from

$$T_{\max.}=\frac{\pi}{2}\mu p_r L d^2 \quad \text{[Lb.in.]},$$

where $\mu=0\cdot20-0\cdot25$ for steel-on-steel (dry), $p_r=$ radial pressure [Lb./in.²], $L=$ axial length of mating surfaces [in.] and $d=$ diameter of shaft [in.].

The stress distribution of a good shrink fit can only be impaired by the addition of a key. The mating surfaces should be absolutely dry and clean, since the presence of oil on steel can reduce the value of μ to $0\cdot10$, or less.

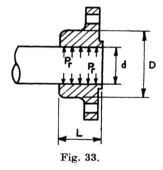

Fig. 33.   Fig. 34.

Similar materials, solid shaft

If $\delta=$ difference in diameter of the two members before heating, the radial pressure due to the shrink fit will be

$$p_r=\frac{E\delta}{2d}\left[1-\left(\frac{d}{D}\right)^2\right] \quad \text{[Lb./in.²]},$$

where $E=$ modulus of elasticity [Lb./in.²], $d=$ shaft diameter, and $D=$ effective outer diameter of outer member, so that the torque-carrying capacity is given by

$$T_{\max.}=\frac{\pi}{4}E\delta\mu Ld\left[1-\left(\frac{d}{D}\right)^2\right] \quad \text{[Lb.in.]}.$$

The tangential stress p_t at the interface is given by

$$p_t = \frac{E\delta}{2d}\left[1+\left(\frac{d}{D}\right)^2\right] \quad [\text{Lb./in.}^2].$$

The value of δ is based on the tangential stress, and it is customary to go up to the yield point, $p_t \cong p_{YP}$, or even slightly higher.

Similar materials, hollow shaft

The torque-carrying capacity is in this case based on the following formulae for radial and tangential stresses:

$$p_r = \frac{\dfrac{\delta E}{2d}}{\dfrac{(b^2-a^2)}{(b^2-1)(1-a^2)}},$$

where $a = d_H/d$ and $b = D/d$, and

$$p_t = p_r\frac{b^2+1}{b^2-1}.$$

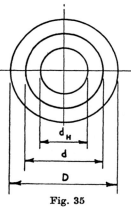

Fig. 35

Dissimilar materials

With the subscripts d = outer diameter of shaft, and D = outer diameter of outer member, and denoting Poisson's ratio by σ [$0.25 \leqslant \sigma \leqslant 0.3$], the value of the diametral interference δ for a given radial stress p_r at the interface is obtainable from

$$\delta = \frac{dp_r M}{E_d E_D(d^2-d_H^2)(d^2-D^2)}.$$

To obtain the value of M, first calculate

$$\alpha_d = 1-\sigma_d, \quad \beta_d = 1+\sigma_d,$$
$$\alpha_D = 1-\sigma_D, \quad \beta_D = 1+\sigma_D,$$

and determine

$$A = \alpha_d E_D - \alpha_D E_d, \quad B = \alpha_d E_D + \beta_D E_d,$$
$$C = \alpha_D E_d + \beta_d E_D, \quad F = \beta_d E_D - \beta_D E_d,$$

then

$$M = Ad^4 - Bd^2D^2 + Cd_H^2 d^2 - Fd_H^2 D^2.$$

Note. It is good practice to make the length of shaft used as a mating surface slightly greater (e.g. 0.020 in.) in diameter than the remainder of the shaft; this facilitates inspection and detection of any possible cracks on the peripheral edges of the shrink-fit assembly.

REFERENCES

Salmon, E. H. *Materials and Structures*, vol. 1, pp. 38–45.
Southwell, R. V. *An Introduction to the Theory of Elasticity*, pp. 403–8.
Kienzle, W. and Heiss, A. 'Force fits', *Werkstattstechnik*, vol. 32, no. 21 (1938), pp. 468–73.

Trock, B. 'Holding power of shrink fits', *Iron Age*, November 1953, p. 113.

Thomson, A. S. T., Scott, A. W. and Moir, C. M. 'Shrink-fit investigations on simple rings and on full-scale crankshaft webs'. Paper presented to the Institution of Mechanical Engineers on 2 April 1954; also in *Engineer*, 9 April 1954, pp. 531–33.

Hazelett, R. 'Increasing the holding power of press fits', *Prod. Engng*, April 1955, pp. 209–11.

Griffith and Taylor formulae for maximum shear stress

(1) *Shaft cross-sections with no re-entrant angles:*

$$q_{max.} = \frac{2R}{1 + \left(\frac{\pi R^2}{A}\right)^2} \times \left[1 + 0.15\left\{\left(\frac{\pi R^2}{A}\right)^2 - \frac{R}{\rho}\right\}\right] \times \frac{T}{I_{p\,(eff.)}} \quad [\text{Lb./in.}^2],$$

where R = radius of largest inscribed circle in cross-section [in.], A = area of cross-section [in.²], ρ = radius of curvature of the boundary [in.] (at point where ρ has its greatest value), T = applied torque [Lb.in.], $I_{p\,(eff.)}$ = effective polar second moment of area of cross-section [in.⁴], determined by Griffith and Taylor method described in section 1·2 (Shaft stiffness).

(2) *Shaft cross-sections with re-entrant angles:*

$$q_{max.} = \frac{2R}{1 + \left(\frac{\pi R^2}{A}\right)^2} \left[1 + \left\{0.272 \log_{10}\left(1 - \frac{R}{\rho}\right) - 0.238 \frac{R}{\rho}\right\} \times \tanh \frac{2\alpha}{\pi}\right]$$
$$\times \frac{T}{I_{p\,(eff.)}} \quad [\text{Lb./in.}^2],$$

where the symbols $R, A, T, I_{p\,(eff.)}$ are the same as defined above, ρ = radius of curvature (negative for re-entrant angles, positive for non re-entrant angles), and α = angle [in radians] turned through by the tangent in turning around the re-entrant portion.

Note. For I-sections, and other compounded sections, the formulae may be applied to each component part considered separately.

[Reference: Griffith, A. A. and Taylor, G. I., *Rep. Memor. Adv. Comm. Aero., Lond.*, nos. 333, 334 and 392.]

EXAMPLE. Cruciform cross-section. R = radius of largest inscribed circle; ρ = radius of fillet at root, with 'negative' curvature, hence its value should be entered with a minus sign in the second equation, so that the expression in the curved brackets becomes

$$\left\{0.272 \log_{10}\left(1 + \frac{R}{|\rho|}\right) + 0.238 \frac{R}{|\rho|}\right\};$$

moreover, $\alpha = 90°$, so that $\tanh(2\alpha/\pi) = \tanh 1$.

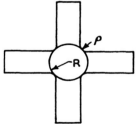

Fig. 36.

Flywheels and coupling flanges

Permissible stresses in flywheels

For large cast-iron flywheels, it is recommended that the peripheral velocity should not be above 100 ft./sec., corresponding to a hoop stress of about 1060 Lb./in.^{2}. For built-up flywheels, the normally recommended value is even lower, viz. 80 ft./sec.

However, for one-piece flywheels of steel or high-grade cast iron, higher calculated stresses are permissible, and values typical of current practice are between 140 and 175 ft./sec.

The hoop stress is determined from the formula

$$p = \frac{w}{g} \times v^2 \quad [\text{Lb./in.}^2],$$

where w = specific weight of material [Lb./in.^{3}], $g = 386$ in./sec.^{2} and v = peripheral velocity [in./sec.]. For steel $w = 0.283$ Lb./in.^{3}, and $w/g = 7.32 \times 10^{-4}$ Lb.sec.^{2}/in.^{4}. For other materials, see tabulated values at the beginning of this handbook.

Flywheels should be designed with large junction radii at changes in section. In most flywheels, it is necessary to machine a hole at the centre; but this should be avoided where possible in flywheels for which the hoop stress is already on the high side, since it results in a local increase in stress of about 100 %. The amount of non-symmetrical overhang of the rim should be reduced as much as possible, to reduce bending stresses. The combined centrifugal and bending stresses at the junction of the bolting flange and the centre disk should not exceed a safe limit (for steel and high-grade irons, a limit of about 6500 Lb./in.^{2} is considered acceptable).

Coupling flange bolts

The calculated shear stress in these bolts can be 25,000–30,000 Lb./in.^{2}.†
If N_B = number of bolts, D_{PC} = pitch circle diameter [in.], D_B = bolt diameter [in.], and T = total transmitted torque [Lb.in.], the bolt stress is

$$q = \frac{8}{\pi} \times \frac{T}{N_B D_{PC} D_B^2} \quad [\text{Lb./in.}^2].$$

Coupling flange thickness

The thickness L_{CF} of the coupling flange is usually $L_{CF} \cong 1.5 D_B$ for cast iron, and $L_{CF} \cong D_B$ for steel.

† *Note.* Lloyd's Register allow only 3,875 Lb./in.^{2} for intermediate shaft coupling bolts of 28/32 Ton/in.^{2} U.T.S. as stress due to the transmitted torque, which allows for the possible additional torques due to torsional vibration.

For further details on rotating disks and built-up flywheels, see for instance:

Den Hartog, J. P. *Advanced Strength of Materials*, 1st ed. p. 49 (McGraw-Hill Book Co., Inc. 1952).

Heusinger, W. *Forsch. IngWes.* vol. 9 (1938), pp. 197 and 309; vol. 13 (1942), p. 209.

ten Bosch, N. *Berechnung der Maschinenelemente* (3rd ed. (revised)), pp. 350–3 (Springer: Berlin, 1953).

Wang, C. T. *Applied Elasticity*, p. 63 (McGraw-Hill Publishing Company Ltd. 1953).

Disk clutch

The length of clutch shaft on which the centre plates are situated in its driving condition is regarded as having zero equivalent length.

If any slip occurs, the clutch isolates, to some extent, the driving end of the system as regards torsional vibration (see Notes on Clutches in section 1·2, p. 129).

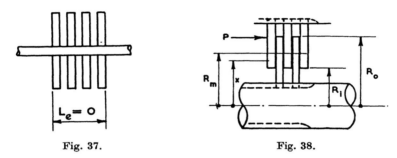

Fig. 37.   Fig. 38.

The spring pressure P [Lb.] required in order to transmit the total torque [steady + vibratory torque] is obtained as

$$P = \frac{T}{\frac{4}{3}\mu N_f \left[R_m - \dfrac{R_o R_i}{4 R_m} \right]} \quad \text{[Lb.]},$$

where N_f = number of friction surfaces, μ = friction coefficient, R_m = mean radius $= \frac{1}{2}(R_o + R_i)$ [in.], R_o = outer radius of centre plates [in.], R_i = inner radius of outer plates [in.], T = total torque [Lb.in.].

For steel or cast-iron surfaces $\mu \cong 0\cdot1$ with oil; however, for dry and clean surfaces, μ can be as high as $0\cdot25$–$0\cdot30$.

The derivation of the above equation is as follows:

$$T = \frac{P}{\pi(R_o^2 - R_i^2)} \mu N_f \int_{R_i}^{R_o} 2\pi x\, dx\, x = \tfrac{2}{3}\mu N_f \frac{R_o^3 - R_i^3}{R_o^2 - R_i^2} P,$$

and as $\qquad (R_o^3 - R_i^3)/(R_o^2 - R_i^2) = [(R_o + R_i)^2 - R_o R_i]/2R_m,$

this gives the above equation for P.

The thermal torque capacity of disk clutches is considered in an article by A. F. Gagne, Jr., in *Prod. Engng*, December 1953, pp. 182–7.

[341]

Gear-tooth fillets

(1) *Lewis formulae* (see Fig. 39):

$$p_{(A)} = + \frac{6x_P P \cos(\alpha - \phi)}{Lb^2} - \frac{P \sin(\alpha - \phi)}{Lb},$$

$$p_{(B)} = - \frac{6x_P P \cos(\alpha - \phi)}{Lb^2} - \frac{P \sin(\alpha - \phi)}{Lb} \quad \text{[Lb./in.}^2\text{]}.$$

The first term on the right-hand side represents the stress due to bending. This nominal stress should be multiplied by a stress concentration factor S for bending loads [taken, for instance, from Neuber's nomographs (ref. see p. 344)] to obtain a value for the effective stress.—A discussion of the limitations of the Lewis formulae is given in the paper by Heywood mentioned below.

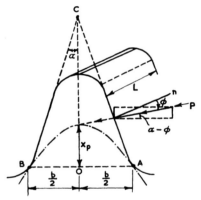

Fig. 39. Gear tooth subjected to a force P.

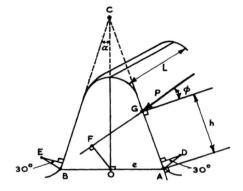

Fig. 40. $\overline{OA} = e$, $\overline{OF} = a$, $\overline{AG} = h$, $\overline{AD} = \overline{BE} = r$.

(2) *Heywood's formula* (see Fig. 40):

$$p_{\text{fillet}} = S \times \left[\frac{1 \cdot 5a}{e^2} + \sqrt{\left(\frac{0 \cdot 36}{he} \right) \left\{ 1 + \frac{\sin \phi}{4} \right\}} \right] \times \frac{P}{L} \quad \text{[Lb./in.}^2\text{]},$$

where

$$S = 1 + 0 \cdot 26 \times \left(\frac{e}{r} \right)^{0 \cdot 7}$$

$$= \text{stress concentration factor.}$$

The point A is obtained as the intersection of the fillet contour (of radius r) with a line AD drawn at 30° from D, the centre of the fillet radius.

[See Heywood, R. B., 'Tensile fillet stresses in loaded projections', *Proc. Instn Mech. Engrs, Lond.*, vol. 159 (1948) (W.E.P. 45).]

Crankwebs

Shear stress in web due to torsion about its greatest principal axis

    (1) Rectangular cross-section: see 'Rectangle', p. 325.

    (2) Elliptical cross-section: see 'Ellipse', p. 325.

    (3) Circular cross-section with flat sides: see p. 335.

Stresses in normal external fillets

    Torsion: see pp. 327 and 328 (in particular, the values of Weigand and Jacobsen); also the data on p. 329.

    Bending: see p. 329.

    In order to apply these results to crankweb fillets, one may write r = fillet radius, d = crankpin (or journal) diameter, and D = web width (regarding the web as a large-diameter shaft portion).

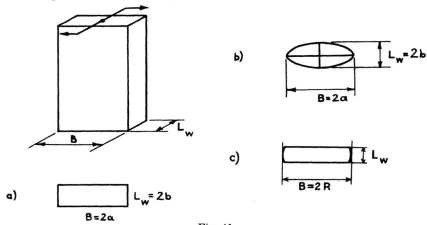

Fig. 41.

Stresses in recessed fillets

    Complete information on this subject is not available.

    In an investigation by Simonetti (for reference, see p. 345), it was found that for $r/d \cong 1/23$ and $d/D \cong 2\cdot0$ the stresses in the recessed fillets corresponded to stress concentration factors

$$S_{(t)} = 1\cdot6\text{--}1\cdot7 \quad \text{for torsion,}$$

and $\qquad S_{(b)} = 2\cdot3\text{--}2\cdot4 \quad$ for bending.

    Comparing with the corresponding values for normal fillets, one finds $S_{(t)} = 1\cdot68\text{--}1\cdot84$ (from table on p. 328) and $S_{(b)} = 1\cdot80\text{--}2\cdot85$ (or a mean value of $2\cdot3$, from the table for S-values on p. 329). In the absence of further data on recessed fillets, it

Fig. 42. r = fillet radius.
d = shaft diameter.
D = web width.

would appear possible, therefore, to estimate their S-values from those of normal fillets.

Bending fatigue strength of crankwebs

The bending fatigue strength S_F of webs increases approximately as a parabolic curve in relation to the ultimate tensile strength of the material. At low U.T.S. values, $S_F \cong \frac{1}{6} \times$ U.T.S.; for very high U.T.S. values, $S_F \cong \frac{1}{12} \times$ U.T.S., for carbon steel and for rectangular webs with little or no overlap.

ADDITIONAL BIBLIOGRAPHY

Bunyan, T. W. and Attia, H. H. Effect of fillet radii on torsional fatigue strength of marine shafting. Paper no. 1170, *Institution of Engineers and Shipbuilders in Scotland*, 7 April 1953, pp. 425–92.

Cox, H. L. Four studies in the theory of stress concentration. *Rep. Memor. Aero. Res. Comm., Lond.*, (R. & M. 2704/ARC Monogr.), 58 pp. (1953).

Calvert, N. G. Impact torsion experiments and experiments on the effect of rate of testing on the criterion of failure of certain mild steels. *Inst. Mech. Engrs* (1954).

Gilbert, G. N. J. Fatigue properties of cast irons. *J. Res. Brit. Cast Iron Ass.* vol. 5, no. 3 (Dec. 1953), pp. 94–8.

Gough, H. J. Engineering steels under combined static and cyclic stresses. *Proc. Instn Mech. Engrs, Lond.*, vol. 160, no. 4 (1949), pp. 417–40.

Hetényi, M. *Handbook of Experimental Stress Analysis* (John Wiley and Sons, Inc., 1950). (Bibliography on fatigue, pp. 26–7.)

Heywood, R. B. *Designing by Photoelasticity* (Chapman and Hall, 1952). [Stress concentration and fatigue factors; also: Bibliography, particularly refs. nos. 450 to 826.]

Ker Wilson, W. The strength of large crankshafts. *Gas Oil Pwr*, October 1951, pp. 241–3.

Lehr, E., Echhardt, and Heurich. *Dynamic Strain Measurements on Crankshafts.* B.I.C.E.R.A. Translation no. 48/T 2.

Lehr, E. and Greinacher. *Forsch. IngWes.* vol. 7 (1936), p. 66.

Lehr, E. and Reuf. Fatigue strength of large crankshafts. *M.T.Z.* no. 11/12 (1943), pp. 349–57.

Lipson, C., Noll, G. C. and Clock, L. S. *Stress and Strength of Manufactured Parts* (McGraw-Hill Book Co. Inc. 1950). [Stress concentration and fatigue strength factors.]

Martinaglia, L. Structural durability of large crankshafts. *Sulzer Tech. Rev.* no. 2 (1943), pp. 19–28; *Gas Oil Pwr*, March 1946.

Matthaes, K. Torsional fatigue strength of crankshafts. *Z. Ver. dtsch. Ing.* 1 Dec. 1954, pp. 1154–6.

Neuber, H. *Theory of Notch Stresses (Kerbspannungslehre)* (J. W. Edwards and Co., Ann Arbor, Mich. 1946).

Osgood, W. R. *Residual Stresses in Metals and Metal Construction* (Reinhold Publishing Corp. 1954).

Phillips, C. E. and Heywood, R. B. The size effect in fatigue of plain and notched steel specimens loaded under reversed direct stress. *Proc. Instn Mech. Engrs, Lond.*, vol. 165 (1951) (W.E.P. no. 65).

Peterson, R. E. *Stress Concentration Design Factors* (John Wiley and Sons, Inc. 1953). (Nomograms.)

Roark, R. J. *Formulas for Stress and Strain* (McGraw-Hill Book Co., Inc., 3rd ed. 1954).

Sankey, G. O. Plastic models for vibration analysis. *Proc. Soc. Exp. Stress Anal.* vol. 11, no. 2 (1954), pp. 81–90.

Siebel, E. and Stieler, M. Non-uniform stress-distribution with alternating loads. *Z. Ver. dtsch. Ing.* 11 Feb. 1955, pp. 121–6.

Simonetti, M. G. Large crankshafts, stress measurement and calculation. *Proc. 1st Paris Congress on Internal Combustion Engines.* [Gives S_F values for various types of normal and recessed fillets.]

Smithells, C. J. *Metals Reference Book* (Butterworth's Scientific Publications, London, 1949). [Fatigue strength of cast irons and steels, pp. 566–76.]

Timoshenko, S. Stress concentration and fatigue of metals. *Proc. Instn Mech. Engrs, Lond.,* vol. 157 (1947) (W.E.P. Issue 28).

Three Keys to Satisfaction. Published by Climax Molybdenum Company, New York, N.Y. [Comprises sections on: design of components; materials; and treatments.]

Work, C. E. and Dolan, T. J. The influence of temperature and the rate of strain on the properties of metals in torsion. *Bull. Univ. Illinois Exp. Sta.* series no. 420, vol. 51, no. 24 (Nov. 1953).

2·4 PREDICTION OF VIBRATION AMPLITUDES AND STRESSES†

NOTATION

Symbol	Brief definition	Typical units
a	engine constant	—
a	specific damping coefficient	—
A	area	in.², cm.²
A_Q	torque coefficient	—
b	engine constant	—
c	engine constant	—
c	damping coefficient	Lb.in.sec./rad., Kg.cm.sec./rad.
C	cyclic speed variation	—
C	engine constant	—
d	diameter	in., cm.
D	diameter	in., cm.
$\mathscr{D}$	denominator	—
e	$=2\cdot71828\ldots$ (base of natural logarithms)	—
E	e.m.f.	volt
f	torque coefficient	—
F	vibration frequency	vib./min.
G	modulus of rigidity	Lb./in.², Kg./cm.²
h	harmonic order number	—
H	power	h.p.
J	mass moment of inertia	Lb.in.sec.², Kg.cm.sec.²
k	engine constant, brake constant	—
K	shaft stiffness	Lb.in./rad., Kg.cm./rad.
K	parameter	—
L	length	in., cm.
m	critical speed order number	—
M	dynamic magnifier	—
n	exponent	—
N	running speed	rev./min.
$N_{\text{cyl.}}$	number of cylinders	—
P^*	propeller pitch	ft., m.
q	shear stress	Lb./in.², Kg./cm.²
r	ratio	—
r	exponent	—
R	radius	in., cm.
s	fractional slip	—
S	parameter	—
t	time	sec.
T	torque	Lb.in., Kg.cm.
$\lvert T_m \rvert$	harmonic component of tangential pressure	Lb./in.², Kg./cm.²
U	work, energy	in.Lb., cm.Kg.
V	volume	in.³, cm.³
x	distance	in., cm.
x	specific torque	Lb.in./rad., Kg.cm./rad.
Z	sectional modulus	in.³, cm.³
Z	displacement impedance	Lb.in./rad., Kg.cm./rad.
α	angle	rad., deg.
α	damping coefficient per unit volume	Lb.sec./in.², Kg.sec./cm.²
γ	damping ratio	—
δ	logarithmic decrement	—
Δ	Holzer-table amplitude	rad.
θ	vibration amplitude	deg., rad.
λ	ratio	—
ϕ	phase angle	deg., rad.
ψ	phase angle	deg., rad.
ω	phase velocity	rad./sec.

Complex numbers are denoted by $a+jA$, $b+jB$, ..., $x+jX$, $y+jY$, etc., in calculations relating to Holzer tables with damping.

Subscripts are used in the text to define more precisely. Their meaning is fully explained in the context in each instance.

† All the systems considered in detail in the present Section are without dampers or detuners. Systems with these vibration-reducing devices are dealt with separately, in Section 3.

2·40   Total shear stress

The shear stress q_{total} occurring in a shaft is the sum of the mean transmission stress q_{mt} and the alternating vibration stress $\pm q_{vib.}$:

$$q_{total} = q_{mt} \pm q_{vib.}$$

The following sections 2·41–2·45 give methods for the assessment of $q_{vib.}$

The mean transmission stress in a shaft transmitting the total power H_b [b.h.p.] of an engine is obtained from the relation

$$q_{mt} = \frac{3·208 \times 10^5}{N} \times \frac{D}{D^4 - d^4} \times H_b \quad [\text{Lb./in.}^2] \tag{1}$$

for hollow shafts of outer diameter D and inner diameter d [in.], H_b denoting the engine power [b.h.p.] at the engine speed N [rev./min.] considered. (For solid shafts, the same formula applies, with $d = 0$.)

Metric units. If D and d are in centimetres, and H_b is in metric horse-power units [1 h.p. (Brit.) = 1·0139 h.p. (metric)], the stress q_{mt} is obtained in Kg./cm.² if the numerical constant in eq. (1) is replaced by $3·648 \times 10^5$.

(*Note.* $3·648 \times 10^5 = 3·208 \times 10^5 \times (2·54)^3 / [14·22 \times 1·0139]$.)

2·41   Systems depending mainly on engine damping

2·411   *Vibration amplitudes at resonance: one-node frequency conditions*

This section deals with the prediction at the design stage of torsional vibration amplitudes and stresses occurring under one-node frequency conditions in systems of the following types:

> engines coupled to generators or alternators,
>
> engines coupled to water brakes, and
>
> engines coupled to compressors or pumps,

in all cases the engines being without dampers or detuning couplings.

The calculations for systems with propeller damping are detailed in section 2·422. The calculations for systems with dampers or detuners are given in section 3.

The estimation of stress in systems depending mainly on engine damping can give only approximate results, since the damping action depends on a large number of factors, which still require extensive investigation (see section 2·421).

It has been found that, for the engine systems mentioned above, there are three formulae for stress prediction which generally give smallest errors, viz.: the provisional B.I.C.E.R.A. formula, Ker Wilson's square-root formula, and Shannon's formula. For further reference and particularly in order to show the various aspects from which the problem of stress determination has been considered, eleven further formulae for stress estimation are also included.

A discussion of the various aspects of this subject is given on p. 384.

Vee engines. The formulae given on pp. 348–61 are also applicable to Vee engines, provided that $\overline{\Sigma\Delta}$ represents the resultant vector sum for both banks, and that $J_{\text{cyl.}}$ includes all the inertias per line of cylinders, i.e. two pistons, two connecting rods, etc.

Note on relative accuracy of measurements. It should be borne in mind that measured values are also subject to error, owing to the inevitable limitations of the equipment used (see section 4, Instrumentation). It can be said that, in practice, it is difficult to reduce the error in measurement below 0·05° total swing. This means that the percentage measuring error is of the order of

5% for vibrations of 1° total swing,

10% for vibrations of 0·5° total swing,

and 50% for vibrations of 0·1° total swing.

In view of this, the accuracy of a method of calculation should not be judged by comparing calculated values with measured total swings of less than 0·2°.

1. Provisional B.I.C.E.R.A. formula

The following provisional formula is recommended by the B.I.C.E.R.A. Torsional Vibration Panel for the estimation of torsional vibration amplitudes, the stress being evaluated by the method described in section 2·3. This formula applies to major orders of one-node close-coupled systems, without damper, detuner or propeller, and either with or without a front-end pulley or flywheel. A system is regarded as 'close-coupled' when its node is situated in the crankshaft portion between the last cylinder and the main flywheel.

Vibration amplitude at mass no. 1:

$$\pm\,\theta_1 = C_1 C_2 C_3 \times (1+a)\,bc_1 c_2 \quad \text{degrees},$$

where

$$C_1 = \frac{6 \cdot 62 \times 10^4 \times (\overrightarrow{\Sigma\Delta})^2 \times D_{\text{min.}}}{N_{\text{cyl.}} \times \Sigma J\omega^2 \Delta_{\text{(at node)}}} \times \sqrt{\left(\frac{G}{G_{\text{steel}}}\right)};$$

$$C_2 = \sqrt{\left(\frac{A R_0 \,|\, T_m\,|}{\Sigma(\Delta_{\text{cyl.}}^2)}\right)}, \quad C_3 = \sqrt{\left(\frac{K_F + K_M}{K_F} \times \frac{\Sigma J_{\text{cyl.}}}{J_F + J_M}\right)};$$

and

$a = 0$ for five- and seven-cylinder engines.

$a = 0 \cdot 33$ for all other engines,

[348]

$b = 0.66$ for engines with no balance weights,

$b = 1.00$ for engines with balance weights on all crankthrows (where $J_{\text{cyl.}}$ with balance weights equals to about $1.3 J_{\text{cyl.}}$ without these, using otherwise an intermediate value for b),

$c_1 = 1.0$ for engines with no clutch or slip-type coupling,

$c_1 = N_{\text{cyl.}}/(2+m)$ for engines with clutch or slip-type coupling (where $m =$ critical order number = vibration frequency [vib./min.] divided by engine speed [rev./min.]),

$c_2 = 1.0$ for engines with no auxiliary drive at forward end,

$c_2 = 0.8$ for engines with auxiliary drive at forward end of crankshaft,

the notation used being as follows (see Figs. 1 and 2):

$(\overrightarrow{\Sigma\Delta})^2$ phase-vector sum squared (from Holzer table and phase-angle diagram),

$D_{\text{min.}}$ crankpin diameter or journal diameter [in.], using whichever is the smaller of the two,

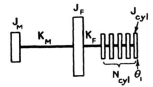 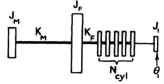

Fig. 1. Engine system without front-end flywheel or pulley.   Fig. 2. Engine system with front-end flywheel or pulley.

$N_{\text{cyl.}}$ number of cylinders,

$\Sigma J \omega^2 \Delta$ maximum inertia torque [Lb.in.], at node, from Holzer table with $\Delta_1 = 1$ radian,

G modulus of rigidity of crankshaft material [Lb./in.^{2}],

G_{steel} 11.8×10^6 Lb./in.^{2},

$|T_m|$ mth order resultant component [Lb./in.^{2}] of tangential pressure (values in section 2.1),

A area of one piston [in.^{2}],

R_0 crank radius [in.],

$\Sigma(\Delta_{\text{cyl.}}^2)$ sum of squared Holzer-table amplitudes, for cylinder masses only,

K_F shaft stiffness between last cylinder and flywheel [Lb.in./rad.],

J_F main-flywheel inertia [Lb.in.sec.²],

K_M shaft stiffness between main flywheel and driven machine [Lb.in./rad.],

J_M inertia of driven machine [Lb.in.sec.²],

$\Sigma J_{\text{cyl.}}$ total engine cylinder inertia ($= N_{\text{cyl.}} \times J_{\text{cyl.}}$ for identical cylinder inertias) [Lb.in.sec.²].

The nominal stress at the node is then determined from

$$q_N = \theta_1 q_N^* \quad [\text{Lb./in.}^2],$$

where $q_N^* =$ stress at node [Lb./in.²deg.] per degree amplitude at mass no. 1 (see section 2·3) for the shaft portion of diameter $D_{\min}$. [see above].

NUMERICAL EXAMPLE. Five-cylinder engine fitted with auxiliary flywheel at forward end and coupled to a water brake. Balance weights for all cylinders. Other data, together with Holzer table, given below. Determination of θ_{AF} for the 5th order critical.

Holzer table

$F = 2904$ vib./min., $\omega = 304\cdot0$ rad./sec., $\omega^2 = 0\cdot092 \times 10^6$

Mass no.	J [Lb.ft. sec.²]	$\dfrac{J\omega^2}{10^6}$ [Lb.ft./ rad.]	Δ [rad.]	$\dfrac{J\omega^2\Delta}{10^6}$ [Lb.ft.]	$\dfrac{\Sigma J\omega^2\Delta}{10^6}$ [Lb.ft.]	$\dfrac{K}{10^6}$ [Lb.ft./ rad.]	Δ_{sh} [rad.]	$\Delta^2_{\text{cyl.}}$ [rad.]	q^* [Lb./ in.²deg.]
AF	11·0	1·012	1·000	1·012	1·012	3·8	0·266	—	—
1	7·03	0·647	0·734	0·429	1·441	12·6	0·114	0·540	—
2	6·35	0·584	0·620	0·362	1·803	12·6	0·143	0·385	—
3	6·35	0·584	0·477	0·279	2·082	12·6	0·165	0·227	—
4	6·35	0·584	0·312	0·182	2·264	12·6	0·180	0·097	—
5	7·03	0·647	0·132	0·085	2·349	11·4	0·206	0·018	9900
F	278·0	25·576	−0·074	−1·893	0·456	3·15	0·145	—	—
B	25·4	2·337	−0·219	−0·512	−0·056	—	—	—	—

$$\Sigma(\Delta^2_{\text{cyl.}}) = 1\cdot267$$

5th order: $\overrightarrow{\Sigma\Delta} = 2\cdot275$.

$$\begin{cases} N_{\text{cyl.}} = 5, \quad D_c = D_{\min.} = 6\cdot25 \text{ in.,} \\ \Sigma J\omega^2\Delta_{(\text{node})} = 2\cdot349 \times 10^6 \text{ Lb.ft.,} \\ \qquad\qquad\quad = 2\cdot817 \times 10^7 \text{ Lb.in.,} \\ G = 11\cdot8 \times 10^6 \text{ Lb./in.}^2. \end{cases}$$

Therefore

$$C_1 = \frac{6\cdot62 \times 10^4 \times (2\cdot275)^2 \times 6\cdot25}{5 \times 2\cdot817 \times 10^7} \times 1$$

$$= 0\cdot01520.$$

$$\begin{cases} A = 83\cdot6 \text{ in.}^2, \quad R_0 = 7\cdot25 \text{ in.,} \\ |T_m| = 10\cdot5 \text{ Lb./in.}^2 \text{ (5th order),} \\ \Sigma(\Delta^2_{\text{cyl.}}) = 1\cdot267, \end{cases}$$

so that

$$C_2 = \sqrt{\left(\frac{10\cdot5 \times 83\cdot6 \times 7\cdot25}{1\cdot267} \right)} = 70\cdot87.$$

As the J- and K-values appear in C_3 as non-dimensional ratios, they need not be converted to Lb.in.sec.^{2} and Lb.in./rad. units, respectively.

$$\begin{cases} K_F = 11\cdot4 \times 10^6, \quad K_M = 3\cdot15 \times 10^6 \,\text{Lb.ft./rad.}, \\ \Sigma J_{\text{cyl.}} = 2 \times 7\cdot03 + 3 \times 6\cdot35 = 33\cdot11 \,\text{Lb.ft.sec.}^2, \quad J_F = 278\cdot0 \,\text{Lb.ft.sec.}^2, \\ J_M = 25\cdot4 \,\text{Lb.ft.sec.}^2. \end{cases}$$

Therefore
$$C_3 = \sqrt{\left/ \left(\frac{11\cdot4 + 3\cdot15}{11\cdot4} \times \frac{33\cdot11}{278\cdot0 + 25\cdot4} \right) \right.} = 0\cdot373.$$

For a five-cylinder engine, $a = 0$.
For an engine with balance weights on all cylinders, $b = 1$.
No clutch, $c_1 = 1\cdot0$.
No auxiliary drive at front end, $c_2 = 1\cdot0$.
Hence
$$(1 + a)\, b c_1 c_2 = 1\cdot0.$$

Vibration amplitude at mass no. 1 (i.e. at auxiliary flywheel 'AF') at forward end:

$$\pm\, \theta_{AF} = \pm\, C_1 C_2 C_3 \times (1 + a)\, b c_1 c_2,$$

$$\pm\, \theta_{AF} = 0\cdot01520 \times 70\cdot87 \times 0\cdot373 \times 1\cdot0 = 0\cdot402^\circ.$$

Total swing $= 2\theta_{AF} = 0\cdot804^\circ$.
Stress at node (i.e. in shafting between cylinder no. 5 and main flywheel):

$$q_N = q_N^* \theta_{AF} = 9900 \times 0\cdot402 = 3980 \quad [\text{Lb./in.}^2].$$

Note. For two-node systems, and for one-node systems which are not of the 'close-coupled' type, see section 2·412, p. 362 and section 2·45, pp. 432–50.

Metric units. In C_1, replace $6\cdot62 \times 10^4$ by $2\cdot8 \times 10^4$ in order to obtain $\pm\, \theta_1$ in degrees, when $D_{\text{min.}}$ is in cm., and $\Sigma J \omega^2 \Delta$ in Kg.cm. Then $A R_0 |T_m|$ should also be in Kg.cm. in C_2.

2. Ker Wilson's square-root formula

[Ker Wilson, W., *Practical Solution of Torsional Vibration Problems*, vol. 2, pp. 121–2.]

Vibration amplitude at mass no. 1: $\quad \pm\, \theta_1 = 500\sqrt{(\theta_s / q_N^*)}$ [degrees],

where $\quad q_N^* = $ stress at node [Lb./in.^{2}deg.] per degree amplitude at mass no. 1,

$\theta_s = $ equivalent static deflexion at mass no. 1 [deg.], this being obtained (see section 2·28) from the relation

$$\theta_s = \frac{57\cdot3 \times |T_m|\, A R_0 \overrightarrow{\Sigma \Delta}}{\omega^2 \Sigma(J \Delta^2)} \quad \text{[degrees]},$$

where $|T_m| = m$th order resultant component of tangential pressure [Lb./in.^{2}],

$A = $ piston area [in.^{2}], $\qquad R_0 = $ crank radius [in.],

$\overrightarrow{\Sigma \Delta} = $ phase-vector sum, $\qquad \omega = (2\pi/60) \times F$,

$F = $ natural frequency [vib./min.], and

$\Sigma(J \Delta^2) = $ equivalent engine inertia [Lb.in.sec.^{2}].

Note. The equivalent engine inertia is determined as

$$\Sigma(J\Delta^2) = \Sigma(J_{\text{cyl.}}\,\Delta^2_{\text{cyl.}}) + J_F\Delta^2_F \quad [\text{Lb.in.sec.}^2],$$

where $J_{\text{cyl.}}$ = inertia per cylinder (see section 1·1),

   J_F = flywheel inertia,

   $\Delta^2_{\text{cyl.}}$ = Holzer-table amplitude squared, for each cylinder inertia,

   Δ^2_F = Holzer-table amplitude squared, for the main flywheel.

The stress at the node q_N is obtained by means of either of the relations

$$q_N = \theta_1 q_N^* \quad \text{or} \quad q_N = 500\sqrt{(q_N^*\theta_s)} \quad [\text{Lb./in.}^2],$$

where q_N^* = stress at node per degree amplitude at mass no. 1 [Lb./in.^{2}deg.].
(For *Vee engines*, $J_{\text{cyl.}}$ = moment of inertia per line, i.e. of two cylinders.)

NUMERICAL EXAMPLE. Determination of vibration amplitude of a five-cylinder engine for the $7\frac{1}{2}$th order critical, by means of the data given below:

$$|T_m| = 2\cdot7\,\text{Lb./in.}^2\,(7\tfrac{1}{2}\text{th order}), \quad A = 176\cdot7\,\text{in.}^2, \quad R_0 = 10\,\text{in.}, \quad \overrightarrow{\Sigma\Delta} = 3\cdot15,$$

$$\Sigma(J\Delta^2) = 1200\,\text{Lb.in.sec.}^2, \quad \omega^2 = 6\cdot85 \times 10^4\,(\text{rad./sec.})^2, \quad q_N^* = 11{,}650\,\text{Lb./in.}^2\text{deg.}$$

$$\theta_s = \frac{57\cdot3 \times 2\cdot7 \times 176\cdot7 \times 10 \times 3\cdot15}{6\cdot85 \times 10^4 \times 1\cdot2 \times 10^3} = 0\cdot0105^\circ,$$

$$\pm\,\theta_1 = 500\sqrt{(0\cdot0105/11{,}650)} = 500 \times \frac{1}{\sqrt{(1\cdot1095 \times 10^6)}} = 0\cdot475^\circ,$$

$$q_N = q_N^*\theta_1 = 11{,}650 \times 0\cdot475 = 5530\,\text{Lb./in.}^2.$$

Metric units. The coefficient of θ_s is unchanged (i.e. 57·3), when $|T_m|$ is in Kg./cm.^{2}, A in cm.^{2}, R_0 in cm., and J in Kg.cm.sec.^{2}.

If q_N^* is in Kg./cm.^{2}, the numerical factor 500 should be replaced by 132·6 to obtain $\pm\,\theta_1$ in degrees, and q_N in Kg./cm.^{2}.

3. Shannon's formula

[Shannon, J. F., 'Damping influences in torsional oscillation', *Proc. Instn Mech. Engrs, Lond.*, vol. 131 (December 1935), pp. 387–435.]

Vibration amplitude at mass no. 1: $\pm\,\theta_1 = M \times \theta_s$ [degrees],

where θ_s = equivalent static deflexion [deg.], calculated as already shown
for Ker Wilson's formula, and

$$M = 58\cdot2 - 48\cdot3\,\frac{\Sigma\,|\Delta_{\text{cyl.}}|}{N_{\text{cyl.}}} = \text{dynamic magnifier},$$

$N_{\text{cyl.}}$ being the number of cylinders and $\Sigma\,|\Delta_{\text{cyl.}}|$ the arithmetical sum of the absolute values of the Holzer-table amplitudes for the cylinder inertias only.

Note. $\Sigma\,|\Delta_{\text{cyl.}}|$ is thus the same for all orders (major or minor) of a given mode of vibration.

The stress at the node is obtained from the stress per degree q_N^* by means of the relation

$$q_N = \theta_1 q_N^* \quad [\text{Lb.}/\text{in.}^2].$$

NUMERICAL EXAMPLE. Determination of predicted amplitude for a four-cylinder engine by means of the following data for the 10th order critical:

$$|T_m| = 1{\cdot}6 \, \text{Lb.}/\text{in.}^2 \,(10\text{th order}), \quad A = 16{\cdot}0 \, \text{in.}^2, \quad R_0 = 2{\cdot}75 \, \text{in.}, \quad \overrightarrow{\Sigma\Delta} = 2{\cdot}953,$$

$$\Sigma(J\Delta^2) = 1{\cdot}45 \, \text{Lb.in.sec.}^2, \quad \omega^2 = 1{\cdot}71 \times 10^6, \quad q_N^* = 10{,}050 \, \text{Lb.}/(\text{in.}^2\text{deg.}),$$

$$\theta_s = \frac{1{\cdot}6 \times 16{\cdot}0 \times 2{\cdot}75 \times 2{\cdot}953}{1{\cdot}45 \times 1{\cdot}71 \times 10^6} \times 57{\cdot}3 = 0{\cdot}0048°,$$

$$\Sigma\Delta_{\text{cyl.}}/N_{\text{cyl.}} = 2{\cdot}953/4 = 0{\cdot}736,$$

$$M = 58{\cdot}2 - 0{\cdot}736 \times 48{\cdot}3 = 22{\cdot}6, \quad \pm\theta_1 = 22{\cdot}6 \times 0{\cdot}0048 = 0{\cdot}109°,$$

$$q_N = q_N^* \theta_1 = 10{,}050 \times 0{\cdot}109 \cong 1100 \, \text{Lb.}/\text{in.}^2.$$

Metric units. With Kg. and cm. replacing Lb. and in., respectively, the above formulae are also valid in metric units, without alteration of their numerical factors.

4. Archer's method

[Archer, S., Communication received in August 1948.]

For 2-node marine and 1- and 2-node close-coupled systems
Vibration amplitude at mass no. 1: $\pm\theta_1 = M\theta_s$ [degrees]

with
$$\theta_s = \frac{57{\cdot}3 \times |T_m| A R_0 \overrightarrow{\Sigma\Delta}}{\omega^2 \Sigma(J\Delta^2)} \quad [\text{degrees}].$$

$\Sigma(J\Delta^2)$ includes all the inertias in the system (i.e. it includes those of the driven machine). The dynamic magnifier M is obtained from the following tabulated values:

θ_s (°) =	0·0003	0·0005	0·0007	0·0008	0·001	0·0014	0·0018	0·002
M =	50	42	38	36	33·5	30·2	28	27
θ_s (°) =	0·0024	0·0026	0·003	0·004	0·005	0·0055	0·0057	
M =	25·6	25	24·2	22·8	21·7	21·2	21·0	
θ_s (°) =	0·006	0·01	0·02	0·03	0·04	0·05	0·057	
M =	20·9	19·4	16·3	13·8	12·4	11·4	11·0	

Stress at node $\qquad q_N = \theta_1 \times q_N^* \quad [\text{Lb.}/\text{in.}^2],$

where $q_N^* = $ nodal stress per degree amplitude at mass no. 1 [Lb./(in.^{2}deg.)].

Metric units. The above expressions for θ_s and θ_1 are also valid in metric units, for instance with linear dimensions in Kg., tangential pressures $|T_m|$ in Kg./cm.^{2}, and moments of inertia J in Kg.cm.sec.^{2}.

5. Bibby's formula

(a) *Original Bibby formula:*

Stress at node: $\qquad q_N = k \times \sqrt{(|T_m| A R_0 \overrightarrow{\Sigma\Delta}/D_j^3)} \quad [\text{Lb.}/\text{in.}^2],$

Vibration amplitude at mass no. 1:

$$\pm\,\theta_1 = q_N/q_N^* \quad \text{[degrees]}.$$

In these equations $q_N^* = $ nominal stress at node [Lb./in.^{2}deg.] per degree amplitude at mass no. 1 of the system, as determined by means of a Holzer table, $k = 1100$, and $D_j = $ journal diameter [in.].

The expression for engine torque in the square root should be in Lb.in.

(b) *Modified Bibby formula:*

$$q_N = k \times \sqrt{\left(\frac{|\,T_m\,|\,A R_0 \overrightarrow{\Sigma\Delta}}{D_j^3} \times \lambda\right)} \quad \text{[Lb./in.}^2\text{]},$$

with

$$k = 1340 - 78\cdot5m,$$

where $m = $ critical order number (vibration frequency [vib./min.] divided by running speed [rev./min.]), and $\lambda = L_e/L_{e(N)}$ as shown in Fig. 3.

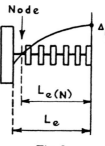

Fig. 3.

NUMERICAL EXAMPLE. Predicted stress value for the 9th order critical of a six-cylinder engine, on the basis of the following values:

$$|\,T_m\,| = 1\cdot38 \text{ Lb./in.}^2 \text{ (9th order)}, \quad A = 128\cdot7 \text{ in.}^2,$$

$$R_0 = 7\cdot29 \text{ in.},$$

$$\overrightarrow{\Sigma\Delta} = 4\cdot087, \quad \lambda = 1\cdot08, \quad D_j = 9\cdot06 \text{ in.}$$

(a) With constant k-factor:

$$q_N = \sqrt{\left(\frac{1\cdot38 \times 128\cdot7 \times 7\cdot29 \times 4\cdot087 \times 1\cdot08}{743\cdot68}\right)} \times 1100 = 3050 \quad \text{[Lb./in.}^2\text{]}.$$

(b) With variable k-factor:

$$k = 1340 - 78\cdot5 \times 9 = 633\cdot5,$$

$$q_N = \frac{633\cdot5}{1100} \times 3050 = 1755 \quad \text{[Lb./in.}^2\text{]}.$$

Metric units. The expression $|\,T_m\,|\,A R_0/D_j^3$ has the dimensions of stress. If it is evaluated in Kg./cm.^{2}, the nodal stress q_N will also be in Kg./cm.^{2} in the original formula if the factor 1100 is replaced by 4147 (which is equal to $1100 \times \sqrt{(14\cdot22)}$).

For the modified formula, the expression for k in metric units is

$$k = 5049 - 296m.$$

6. Den Hartog's formula

[Den Hartog, J. P., *Mechanical Vibrations*, 2nd ed. p. 254.]

Vibration amplitude at mass no. 1:

$$\pm\,\theta_1 = 4\cdot8 \times 10^{-3} \times \frac{L_1}{D_c^{3\cdot3}} \times \left[\frac{T_{\text{res.}}}{C}\right]^{\frac{1}{1\cdot3}} \quad \text{[degrees]},$$

where L_1 = true length of one crankthrow [in.] from cylinder centre to cylinder centre,

D_c = crankpin diameter [in.],

$T_{res.} = |T_m| AR_0 \overrightarrow{\Sigma\Delta}$ = resultant excitation torque [Lb.in.], and

$C = \Sigma(\Delta_{sh}^{2\cdot3})$ = sum of 2·3 powers of relative angles of twist for the various shaft sections between the cylinders only.

Note. $\Delta_{sh} = \Sigma(J\omega^2\Delta)/K$ is obtained from the last column of the Holzer table, for each of the shaft portions between cylinders.

The stress at the node is determined from $q_N = \theta_1 q_N^*$ [Lb./in.^{2}], where q_N^* = nodal stress per degree.

NUMERICAL EXAMPLE. Determination of 9th order critical for a three-cylinder engine on the basis of the data given below:

$$L_1 = 10\cdot375 \text{ in. (length of one crankthrow)}, \quad D_c = 3\cdot0 \text{ in.},$$

$$T_{res.} = |T_m| AR_0\overrightarrow{\Sigma\Delta} = 2\cdot1 \times 16 \times 2\cdot75 \times 1\cdot952 = 180\cdot4 \text{ Lb.in. (9th order)};$$

relative shaft twists Δ_{sh} between engine masses from the following table:

	Auxil. flywheel		Cyl. no. 1		Cyl. no. 2		Cyl. no. 3		Main flywheel
Δ_{sh}	0·111		0·208		0·299		0·424		
$\Delta_{sh}^{2\cdot3}$	—		0·0269		0·0626				—

Therefore $\quad C = \Sigma(\Delta^{2\cdot3}) = 0\cdot0269 + 0\cdot0626 = 0\cdot0895,$

$$\pm\theta_1 = 4\cdot8 \times 10^{-3} \times \frac{10\cdot375}{(3\cdot0)^{3\cdot3}} \times \left\{\frac{180\cdot4}{0\cdot0895}\right\}^{\frac{1}{1\cdot3}} = 4\cdot8 \times 10^{-3} \times 0\cdot2645 \times \{2\cdot011 \times 10^3\}^{\frac{1}{1\cdot3}}$$

$$= 1\cdot27 \times 10^{-3} \times 3\cdot42 \times 10^2 \cong 0\cdot43°.$$

Metric units. If L and D are in centimetres and the torque $T_{res.}$ is in Kg.cm., in the expression for $\pm\theta_1$ the numerical constant $4\cdot8 \times 10^{-3}$ should be replaced by $3\cdot47 \times 10^{-2}$. The relation for q_N is not affected by the change to metric units.

7. Dorey's method

[Dorey, S. F., 'Strength of marine engine shafting', *Trans. N.-E. Cst Instn Engrs Shipb.* vol. 50 (1939).]

Vibration amplitude at mass no. 1:

$$\pm\theta_1 = 57\cdot3 \times U_{in}/U_d \quad \text{[degrees]},$$

where $U_{in} = \pi |T| AR_0\overrightarrow{\Sigma\Delta} = \pi \times$ resultant input torque [in.Lb.], and

$U_d = \Sigma(LR^2 q_{rad.}^2 \times 10^{-8}).$

In the expression for the dissipated energy U_d, the factors used are as follows:

L = actual length [in.] of each shaft portion,

R = actual radius [in.] of each shaft portion,

$q_{\text{rad.}} = \Sigma(J\omega^2\Delta)/Z$ [Lb./in.^{2}rad.] = stress per radian amplitude at mass no. 1,

Z being the modulus of the corresponding section, and $\Sigma(J\omega^2\Delta)$ the total torque, from a Holzer table, for the corresponding section. The torque for intermediate shaft sections is obtained by simple interpolation between two successive $\Sigma(J\omega^2\Delta)$-values, as indicated in the following example.

NUMERICAL EXAMPLE. Calculations for a five-cylinder engine coupled to a generator are given below.

Journal diameter $D_j = 10\cdot5$ in.	$Z_j = \pi D_j^3/16 = 227$ in.^{3}.
Crankpin diameter $D_c = 9\cdot0$ in.	$Z_c = \pi D_c^3/16 = 143$ in.^{3}.
Generator shaft diameter $D_s = 11\cdot875$ in.	$Z_s = \pi D_s^3/16 = 329$ in.^{3}.

In the following, $R = \frac{1}{2}D_j$, $\frac{1}{2}D_c$ or $\frac{1}{2}D_s$, L = length of corresponding shaft portion [in.], $q_{\text{rad.}}$ = stress for a vibration amplitude, $\Delta_1 = 1$ rad. at the free end [Lb./in.^{2}rad.] and T_{mean} = mean torque = $\Sigma(J\omega^2\Delta)$ for journal portions (from Holzer table); for crankpin portions the values taken are mean values of the torques for two adjacent journals. Also j. = journal and c. = crankpin.

	$T_{\text{mean}}/10^6$ [Lb.in./rad.]	R^2 [in.^{2}]	L [in.]	$q_{\text{rad.}}/10^4$ [Lb./in.^{2}rad.]	$q_{\text{rad.}}^2 \times 10^{-8}$	$10^{-8}LR^2q_{\text{rad.}}^2$
No. 1 j.	26·8	27·56	9·0	11·8	140	$0\cdot3 \times 10^5$
No. 1 c.	37·1†	20·25	7·75	26·0	680	$1\cdot1 \times 10^5$
No. 2 j.	47·4	27·56	9·0	20·9	440	$1\cdot1 \times 10^5$
No. 2 c.	56·4	20·25	7·75	39·4	1550	$2\cdot4 \times 10^5$
No. 3 j.	65·4	27·56	9·0	28·8	830	$2\cdot1 \times 10^5$
No. 3 c.	72·6	20·25	7·75	50·8	2580	$4\cdot0 \times 10^5$
No. 4 j.	79·8	27·56	9·0	35·1	1230	$3\cdot0 \times 10^5$
No. 4 c.	84·7	20·25	7·75	59·3	3520	$5\cdot5 \times 10^5$
No. 5 j.	89·5	27·56	9·0	39·4	1550	$3\cdot8 \times 10^5$
No. 5 c.	92·1	20·25	7·75	64·5	4160	$6\cdot5 \times 10^5$
No. 6 j.	94·7	27·56	26·5	41·7	1740	$12\cdot7 \times 10^5$
Gen.	33·5	35·25	21·0	10·2	100	$0\cdot7 \times 10^5$
						$\Sigma = \overline{43\cdot2 \times 10^5}$

† $37\cdot1 = \frac{1}{2}(26\cdot8 + 47\cdot4)$.

Hence $\qquad U_d = 43\cdot2 \times 10^5$ in.Lb./cycle for $\Delta_1 = 1$ rad.

$$U_{in} = \pi\,|\,T_m\,|\,AR_0\,\overrightarrow{\Sigma\Delta} = \pi \times 2\cdot7 \times 176\cdot7 \times 10 \times 3\cdot147$$

$$= 47{,}100 \text{ in.Lb./cycle for } \Delta_1 = 1 \text{ rad. } (7\tfrac{1}{2} \text{ order});$$

$$= \pi \times 1\cdot0 \times 176\cdot7 \times 10 \times 3\cdot147$$

$$= 17{,}410 \text{ in.Lb./cycle for } \Delta_1 = 1 \text{ rad. (10th order).}$$

Therefore we have

$7\frac{1}{2}$ order: $\pm\,\theta_1 = U_{in}/U_d = 47,100/43 \cdot 2 \times 10^5 = 0 \cdot 0109$ rad. ($= 0 \cdot 625°$),

$\qquad\qquad q_N = \theta_1 q_{\text{rad.}(N)} = 0 \cdot 0109 \times 64 \cdot 5 \times 10^4 = 7030$ Lb./in.^{2},

10th order: $q_N = 2600$ Lb./in.^{2}.

8. Draminsky's formulae

[Draminsky, P., *Daempningen ved torsionssvingninger i krumtapaksler* (published by Nyt Nordisk Forlag Arnold Busck, Copenhagen, 1947); also 'Crankshaft damping', *Proc. Instn Mech. Engrs*, Lond., 1949.] The formula for $\pm\,\theta_1$ is also valid in metric units, with $T_{\text{res.}}$ in Kg.cm., and J in Kg.cm.sec.^{2}.

Vibration amplitude at mass no. 1:

$$\pm\,\theta_1 = M \times \frac{T_{\text{res.}}}{\omega^2 \Sigma (J\Delta^2)} \times 57 \cdot 3 \quad \text{[degrees]},$$

where $T_{\text{res.}} = |\,T_m\,|\,AR_0\,\overrightarrow{\Sigma\Delta} =$ resultant excitation torque [Lb.in.],

$\qquad \omega =$ natural phase velocity [rad./sec.],

$\qquad \Sigma (J\Delta^2) =$ sum of all inertias multiplied by their Holzer-table amplitudes squared [Lb.in.sec.^{2}],

$\qquad M =$ dynamic magnifier, determined by one of the following relations:

Engines with main flywheel

Two- to five-cylinder engines:

$$M = \frac{100}{0 \cdot 16 bmr + 0 \cdot 6},$$

engines with six or more cylinders:

$$M = \frac{100}{0 \cdot 6 + [0 \cdot 004 N_{\text{cyl.}}^2 + 0 \cdot 06] \times bmr},$$

where $b = 1 \cdot 0$ for engines without balance weights,

$\qquad b = 0 \cdot 7$ for engines with balance weights,

$\qquad m =$ critical order number (natural frequency [vib./min.] divided by engine speed [rev./min.]),

$\qquad r = \Sigma(\Delta_{\text{cyl.}}^2 J_{\text{cyl.}})/\Sigma(\Delta^2 J)$; the sum in the numerator is for cylinder inertias only, whereas that in the denominator includes all the masses in the system, and

$N_{\text{cyl.}} =$ number of cylinders.

Engines without main flywheel

Engines with less than ten cylinders:

$$M = \frac{100}{0 \cdot 1 bmr + 0 \cdot 6},$$

[357]

engines with ten or more cylinders:

$$M = \frac{100}{0\cdot6 + \left[0\cdot004\left(\dfrac{N_{\mathrm{cyl.}}}{2}\right)^2 + 0\cdot06\right] \times bmr}.$$

NUMERICAL EXAMPLE. Seven-cylinder single-acting four-stroke cycle engine, with main flywheel. $\omega^2 = 0\cdot576 \times 10^6$, $\Sigma(J\Delta^2) = 22\cdot80$ Lb.in.sec.^{2} (all inertias). No balance weights; 7th order excitation torque: $T_{\mathrm{res.}} = 2184$ Lb.in. Equivalent inertia of cylinder masses only: $\Sigma(\Delta^2_{\mathrm{cyl.}} J_{\mathrm{cyl.}}) = 19\cdot26$ Lb.in.sec.^{2}. Determination of vibration amplitude for 7th order critical:

$$57\cdot3 \times T_{\mathrm{res.}}/\{\omega^2\Sigma(J\Delta^2)\} = \frac{57\cdot3 \times 2\cdot184 \times 10^3}{0\cdot576 \times 10^6 \times 22\cdot80} = \theta_s = 9\cdot61 \times 10^{-3} \text{ deg.}$$

$$b = 1\cdot0, \quad m = 7 \text{ (critical order number)}, \quad r = 19\cdot26/22\cdot80 = 0\cdot844.$$

Therefore $$M = \frac{100}{0\cdot6 + [0\cdot004 \times 49 + 0\cdot06] \times 1 \times 7 \times 0\cdot844} = 47\cdot3,$$

and $$\pm\,\theta_1 = M\theta_s = 47\cdot3 \times 0\cdot00961 = 0\cdot455°.$$

9. Ker Wilson's linear formula

Vibration amplitude at mass no. 1:

$$\pm\,\theta_1 = 70 \times \theta_s \quad [\text{degrees}].$$

Stress at node: $\qquad q_N = \theta_1 q_N^* \quad [\text{Lb./in.}^2]$,

where θ_s = equivalent static deflexion [deg.], and

q_N^* = node stress per degree amplitude at mass no. 1 [Lb./in.^{2}deg.], as defined previously for Ker Wilson's square-root formula.

Metric units. The above formulae remain unchanged when metric units are used. The stress q_N is in Kg./cm.^{2} if q_N^* is in Kg./cm.^{2}deg.

10. Lewis-Rowett formula

[Lewis, F. M., 'Torsional vibration in the diesel engine', *Trans. Soc. Nav. Archit.*, *N.Y.*, vol. 23 (November 1925), p. 109. Rowett, F. E., 'Elastic hysteresis in steel', *Proc. Roy. Soc.* vol. 89 (1914).]

Vibration amplitude at mass no. 1:

$$\pm\,\theta_1 = k \times \left[\frac{T_{\mathrm{res.}}}{C}\right]^{\frac{1}{1\cdot3}} \quad [\text{degrees}],$$

where $T_{\mathrm{res.}} = |T_m| A R_0 \overrightarrow{\Sigma\Delta}$ = resultant excitation torque [Lb.in.],

$k = 8\cdot0 \times 10^{-3}$, and

C = value obtained by one of the following expressions:

$$C = \Sigma\left\{\frac{D^{4\cdot3}}{L^{1\cdot3}} \times \Delta^{2\cdot3}_{sh}\right\} \quad \text{or} \quad C \simeq \frac{D_e^{4\cdot3}}{L_e^{1\cdot3}} \Sigma(\Delta^{2\cdot3}_{sh}).$$

In the first expression, the D and L refer to the actual diameters and lengths [in.] of all shaft portions in the system, and Δ_{sh} is the corresponding shaft twist obtained from the last column of a Holzer table. In the second expression, D_e is the diameter of the equivalent shaft, and L_e the total equivalent length of all shafting in the system.

Note. The webs are not taken into account in dealing with actual lengths L, but are included in the expression based on the equivalent length L_e.

Metric units. When $T_{res.}$ is in Kg.cm., and D and L are in cm., the numerical factor $8 \cdot 0 \times 10^{-3}$ in $\pm \theta_1$ should be replaced by $k = 6 \cdot 0 \times 10^{-2}$.

11. Lundberg's formula

[Lundberg, S., 'Beräkning av torsionssvängningar i motorenaxelledningar', *Trans. Chalmers Univ. Technology, Gothenburg*, no. 35 (1944); N. J. Gumperts Bokhandel A.-B., Gothenburg.]

Vibration amplitude at mass no. 1:

$$\pm \theta_1 = \frac{1600 \times |T_m| \overrightarrow{\Sigma\Delta}}{kcR_0 \omega \Sigma(\Delta^2)} \quad \text{[degrees]},$$

where $|T_m| = m$th order resultant component of tangential pressure [Lb./in.²],

$\overrightarrow{\Sigma\Delta}$ = relative vector sum,

R_0 = crank radius [in.],

ω = natural phase velocity [rad./sec.],

$\Sigma(\Delta^2)$ = sum of Holzer amplitudes squared for entire system, and

c = damping coefficient obtained from the table below.

Factor k: For engines coupled to generators: $k = 1$.
For engine systems with long line shafts: $k = 1 \cdot 2$.

Table of damping coefficients c

Note. $\lambda = L_e/L_{e(N)} =$ length ratio, as shown in Fig. 4.

$\begin{cases} h = \text{harmonic order number [natural frequency [vib./min.]} \\ \quad \text{divided by number of working cycles per min.],} \\ h = \text{critical order number of two-stroke cycle single-acting} \\ \quad \text{engines,} \\ h = 2 \times \text{critical order number for four-stroke cycle single-acting} \\ \quad \text{engines.} \end{cases}$

$h = 1$	2	4	6	8	10	12	14	16	18	20	22
				Damping coefficients c							
$\lambda = 1{\cdot}00$ 5·0	4·7	4·0	3·6	3·2	2·9	2·65	2·4	2·2	2·0	1·95	1·90
1·25 5·6	5·3	4·4	3·9	3·5	3·2	2·9	2·65	2·4	2·2	2·15	2·1
1·50 6·2	5·8	4·9	4·3	3·8	3·5	3·25	3·0	2·8	2·6	2·5	2·4
1·75 6·8	6·3	5·4	4·7	4·2	3·9	3·6	3·3	3·1	2·9	2·75	2·6
2·00 7·5	7·0	6·0	5·2	4·6	4·2	3·9	3·6	3·3	3·2	3·0	2·85

NUMERICAL EXAMPLE. Six-cylinder four-stroke cycle engine coupled to water brake with flexible shaft.

$R_0 = 3{\cdot}35$ in., $\overrightarrow{\Sigma\Delta} = 3{\cdot}83$, $\omega = 1{\cdot}104 \times 10^3$ rad./sec.,

$\Sigma(\Delta^2) = 3{\cdot}03$, $|T_9| = 1{\cdot}55$ Lb./in.^{2},   and $\lambda = 1{\cdot}12$.

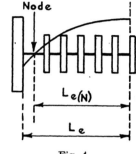

$m = 9$ (critical order number), hence $h = 2m = 18$; $k = 1{\cdot}2$ (flexible brake shaft), and $c = 2{\cdot}1$ by interpolation in the above table. Therefore

$$\pm\,\theta_1 = 1600 \times \frac{1{\cdot}55 \times 3{\cdot}83}{1{\cdot}2 \times 2{\cdot}1 \times 3{\cdot}35 \times 1104 \times 3{\cdot}03} = 0{\cdot}33^\circ.$$

Metric units. With $|T_m|$ in Kg./cm.^{2}, and R_0 in cm., in the above formula for $\pm\,\theta_1$ the factor 1600 should be replaced by $5{\cdot}88 \times 10^4$.

Fig. 4.

12. Mitchell's formula

[Mitchell, R. W. S., 'The design office problem in the estimation of the torsional resonance characteristics of small marine diesel-propulsion units', *Proc. Instn Mech. Engrs, Lond.*, vol. 149, no. 4 (September 1943), pp. 133–42.]

Vibration amplitude at mass no. 1:

$$\pm\,\theta_1 = 3{\cdot}5\,\frac{|T_m|\,AR_0\,\overrightarrow{\Sigma\Delta}}{k \times \omega J_{\text{cyl.}}^{0{\cdot}8} \times \Sigma(\Delta^2)}\quad\text{[degrees]},$$

where $|T_m| = m$th order resultant component of tangential pressure [Lb./in.^{2}],

$A = $ piston area [in.^{2}],

$R_0 = $ crank radius [in.],

$\overrightarrow{\Sigma\Delta} = $ phase-vector sum,

$k = 1 + 270 \times C$, where $C = $ cyclic speed variation (e.g. for $C = 0{\cdot}01$, $k = 3{\cdot}7$),

$\omega = $ natural phase velocity [rad./sec.],

$J_{\text{cyl.}}^{0{\cdot}8} = $ inertia per line [Lb.in.sec.^{2}] raised to 0·8th power,

$\Sigma(\Delta^2) = $ sum of Holzer-table amplitudes squared, for *all* the inertias in the system.

Note. Methods for determining cyclic speed variation are described in section 2·6.

Metric units. With A in cm.^{2}, R in cm., $|T_m|$ in Kg./cm.^{2} and J in Kg.cm.sec.^{2}, the numerical coefficient 3·5 in the above formula for $\pm\theta_1$ should be replaced by 3·36.

13. Shannon's linear formula

Vibration amplitude at mass no. 1:

$$\pm\theta_1 = 50 \times \theta_s \quad \text{[degrees]},$$

where $\qquad \theta_s = 57\cdot3 \times |T_m| \, AR_0\overrightarrow{\Sigma\Delta}/[\omega^2\Sigma(J\Delta^2)] \quad \text{[degrees]},$

the value of $\Sigma(J\Delta^2)$ including all masses in the system.

Metric units. The above formula for θ_s is also valid in metric units, with $|T_m|\, AR_0 =$ torque in Kg.cm., and $\omega^2\Sigma(J\Delta^2)$ also in Kg.cm.

14. Zdanowich's method

[Zdanowich, R. W., Communication received in 1950.]

Vibration amplitude at mass no. 1:

$$\pm\theta_1 = M_0 \times \frac{V}{1000} \times \theta_s \quad \text{[degrees]},$$

where $\theta_s = 57\cdot3 \times |T_m|\, AR_0\overrightarrow{\Sigma\Delta}/[\omega^2\Sigma(J\Delta^2)]$ degrees; the value of $\Sigma(J\Delta^2)$ including only the engine masses in the system, i.e. the engine inertias including that of the main flywheel,

$V = 2N_{\text{cyl.}} \times AR_0$ [in.^{3}] = total swept volume of engine; and

$M_0 =$ dynamic-magnifier value determined, according to the vibration frequency of the system, from the following table:

Natural frequency	vib./min.	1080	1200	1440	1680	1920	2160	2400	2640	2880
	c./sec.	18	20	24	28	32	36	40	44	48
M_0		0·4	0·5	0·8	1·2	1·8	2·5	3·2	4·1	5·0
Natural frequency	vib./min.	3120	3360	3600	3840	4080	4320	4560	4800	5040
	c./sec.	52	56	60	64	68	72	76	80	84
M_0		6·0	7·1	8·2	9·4	10·7	11·9	13·1	14·4	15·8
Natural frequency	vib./min.	5280	5520	5760	6000	6240	6480	6720	6960	7200
	c./sec.	88	92	96	100	104	108	112	116	120
M_0		17·2	18·6	20·0	21·5	23·0	24·5	26·0	27·5	29·0

NUMERICAL EXAMPLE. Estimation of 8th order critical of an eight-cylinder engine on the basis of the following values: bore 10·25 in., stroke 14·5 in., frequency 2610 vib./min., equivalent static amplitude $\theta_s = 0\cdot0156°$.

From the tabulated values:

$$M_0 = 4\cdot0.$$

Swept volume (all cylinders):

$$V = 8 \times \frac{\pi}{4}(10\cdot25)^2 \times 14\cdot5 = 9580 \text{ in.}^3.$$

Vibration amplitude:

$$\pm\,\theta_1 = 4 \cdot 0 \times 9 \cdot 58 \times 0 \cdot 0156 = 38 \cdot 3 \times \theta_s = 0 \cdot 597° \cong 0 \cdot 6°.$$

Note. For frequencies above 40 c./sec., the M_0-values can be obtained from the following relation:

$$M_0 \cong 0 \cdot 25 \times \left(\frac{F}{10}\right)^{1 \cdot 921}, \quad \text{where} \quad F = \text{frequency in cycles/sec.}$$

Metric units. The equation for θ_s remains unchanged for evaluations in metric units, with $|\,T_m\,|\,A R_0 \overrightarrow{\Sigma\Delta}$ and $\omega^2\Sigma(J\Delta^2)$ both in Kg.cm. If the total swept volume V is in cm.^{3}, the equation for θ_1 becomes

$$\pm\,\theta_1 = M_0 \times \frac{V}{16,390} \times \theta_s \quad \text{[degrees]}.$$

2·412 Vibration amplitudes at resonance: two-node frequency conditions

The damping actions under two-node frequency vibrations are usually different from those occurring with one-node vibrations.

However, the prediction of vibration amplitudes and stresses for the two-node mode is usually made with the same formulae as for the one-node mode, i.e. by means of the formulae and methods given in section 2·411. Some of these formulae have been specifically indicated as applicable to two-node conditions, for instance, those of Ker Wilson (p. 351), Shannon (p. 352), Archer (p. 353), Dorey (p. 355) and Mitchell (p. 360).

These formulae, and the expressions given in section 2·45, pp. 432–50, can also be used for estimating the one-node vibration amplitudes of *systems which are not 'close-coupled'*, e.g. systems with the node in the driven-machine shafting after the main flywheel, or systems in which the node is near the middle of the crankshaft.

2·413 Vibration amplitudes at non-resonant speeds

Three methods of estimating vibration amplitudes and stresses on the flanks of criticals are indicated below:

(1) Zero-damping method

If $N_c = $ critical running speed [rev./min.],

$N = $ speed on flank of critical [rev./min.],

$\theta_s = $ equivalent static amplitude [deg.],

then the flank amplitude $\theta_{1(N)}$ is obtained by means of the relation

$$\pm\,\theta_{1(N)} = \frac{\theta_s}{1 - \left(\dfrac{N}{N_c}\right)^2} \quad \text{[degrees]}, \tag{1}$$

where $\theta_s = 57\cdot3 \times |T_m|\, AR_0 \overrightarrow{\Sigma\Delta}/[\{\Sigma(J_{\text{cyl.}}\Delta^2_{\text{cyl.}}) + J_F\Delta^2_F\}\,\omega^2]$ is the equivalent static amplitude, calculated as shown in the example of p. 352 in section 2·411. This method applies to points fairly remote from the resonance peak, i.e. to points below $N/N_c = 0\cdot85$ or above $N/N_c = 1\cdot15$.

(2) *Finite-damping method*

The vibration amplitude at the non-resonant speed N is obtained from

$$\pm\,\theta_{1(N)} = \frac{1}{\sqrt{\left\{\dfrac{1}{\theta_s^2}\left[1 - \left(\dfrac{N}{N_c}\right)^2\right]^2 + \dfrac{1}{\theta_{1\,\text{res.}}^2} \times \left(\dfrac{N}{N_c}\right)^2\right\}}}\qquad\text{[degrees]},\qquad(2)$$

where θ_s is the equivalent static amplitude calculated as mentioned above, and $\theta_{1\,\text{res.}}$ = vibration amplitude [deg.] at resonance peak (speed N_c) calculated by one of the formulae in section 2·411. Damped resonance curves, for values of $M = \theta_{1\,\text{res.}}/\theta_s$ between 5 and 100, are given in section 2·45 (Analytical Methods). The stress at the non-resonant speed is estimated in both methods from

$$q_{(N)} = q_{\text{res.}} \times \frac{\theta_{1(N)}}{\theta_{1\,\text{res.}}}\qquad\text{[Lb./in.}^2\text{]}.\qquad(3)$$

Note. If the tangential-pressure component $|T_m|$ varies at different speeds, for instance, if $|T_m|_N = \left(\dfrac{N}{N_c}\right)^2 \times |T_m|_{N_c}$ (propeller law); equations (1) and (2) should be correspondingly modified. For the propeller law, the relations are

$$(1')\quad \pm\,\theta_{1(N)} = \frac{\theta_s}{\left(\dfrac{N_c}{N}\right)^2 - 1},\quad\text{and}\quad (2')\quad \theta_{1(N)} = \frac{1}{\sqrt{\left\{\dfrac{1}{\theta_s^2}\left[\left(\dfrac{N_c}{N}\right)^2 - 1\right]^2 + \dfrac{N_c^2}{N^2\,\theta_{1\,\text{res.}}^2}\right\}}},$$

respectively, $\theta_{1\,\text{res.}}$ and θ_s being obtained with $|T_m|_{N_c}$. The stress is then estimated as previously.

(3) *Tabulation method*

This method consists in evaluating a Holzer table for the forced frequency corresponding to the non-resonant engine speed considered. Details are given on pages 534 (zero damping) and 413 (finite damping). For a general discussion of flank amplitudes, see pages 434 and 450.

2·42   Damping factors

Note. The damping is assumed to be of viscous nature, i.e. proportional to the vibration velocity $\dot{\theta}$ [rad./sec.]. One can then write

$$T_d = c\dot{\theta},$$

where T_d is the damping torque [Lb.in.] and c the 'damping coefficient' [Lb.in./(rad./sec.) = Lb.in.sec./rad.] related to the system considered.

For simplicity of calculation, it is convenient to assume that the damping torque is linearly proportional to $\dot\theta$ and not a more complicated function thereof; as for instance $T_d = c_1 \dot\theta^{1/2}$ or $T_d = c_1 \dot\theta^2$. Thus, when a system is investigated in regard to vibratory conditions, it is convenient to determine its 'equivalent viscous damping coefficient' c. This is obtained as indicated in the following example.

EXAMPLE. An engine system is characterized by the following equivalent one-mass values:

Excitation torque	$T_0 = 1500$ Lb.in.,
Equivalent inertia	$J_e = \Sigma(J\Delta^2) = 2\cdot4$ Lb.in.sec.^{2},
Natural phase velocity	$\omega = 1\cdot12 \times 10^3$ rad./sec.,
Equivalent shaft stiffness	$K_e = \omega^2 J_e = 1\cdot25 \times 10^6 \times 2\cdot4 = 3\cdot0 \times 10^6$ Lb.in./rad.,
Vibration amplitude	$\pm\,\theta_1 = 0\cdot6°$. (*Note.* $1° \cong 0\cdot0175$ rad.)

It is required to determine its dynamic magnifier M and its damping coefficient c The relations are as follows:

$$M = \frac{\theta_1}{T_0/K_e} = \frac{J_e\omega}{c} = \frac{K_e}{\omega c}$$

(for derivation, see: Analytical Methods, section 2·45, p. 429). With the numerical values, one obtains directly

$$M = \frac{0\cdot0175 \times 0\cdot6}{1500/(3 \times 10^6)} = 21\cdot0,$$

hence
$$c = \frac{J_e\omega}{M} = \frac{2\cdot4 \times 1\cdot12 \times 10^3}{21\cdot0} = 128\cdot0 \quad \text{[Lb.in.sec./rad.]}.$$

Conversely, if T_0, ω and c had been specified with the above values, and it were required to determine $\pm\,\theta_1$, this is obtained as

$$\pm\,\theta_1 = \frac{T_0}{\omega c} = \frac{1500}{1120 \times 128} = 1\cdot05 \times 10^{-3} \text{ rad.} = 0\cdot6°,$$

and if K_e were also given,

$$M = K_e/\omega c = 3\cdot0 \times 10^6/(1120 \times 128) = 21\cdot0.$$

Metric units. If T_d is in Kg.cm., the damping coefficient c is obtained in Kg.cm.sec./rad.

2·421 *Damping factors of engine parts and driven machines*

This section gives numerical values and formulae for the estimation of damping coefficients of individual parts.

The practice hitherto has been to use a separate damping coefficient mainly for the calculation of propeller damping (see section 2·422), and the effects of the various parts of the system have been included in parameters figuring in the expression for the dynamic magnifier of an entire system. In many cases, however, there remain appreciable differences between predicted and measured values. The individual

[364]

damping actions of parts of a vibrating system will thus require detailed consideration in future work on the calculation of vibration amplitudes and stresses.

The following empirical values of damping coefficients can be used in three different manners:

(1) Expressed as reciprocals of dynamic magnifiers (see p. 364), they can be used in conjunction with equivalent static amplitudes in 'one-mass system' calculations.

(2) The coefficients can be employed in Holzer tables with damping (see section 2·44, Tabulation Methods) or for computations with electronic machines for the solution of system determinants or matrices.

(3) They can be used in the formulae for equivalent two-, three- and four-mass systems, with damping, given in section 2·45.

The values given in this section are based on the work of various investigators in different countries. The accuracy with which they were derived depends on whether the investigation was based on the assumption of an equivalent one-mass system, and other assumptions on the type of damping. Although individual coefficients at present may give only qualitative indications, they are also of use as reference values for further work on this subject.

Use of dynamic magnifiers derived from damping coefficients

To illustrate the use of damping coefficients by the first of the three methods mentioned above, it is convenient to see how a damping factor is determined in detail by correlation of an engine test with calculations.

Consider an engine with its flywheel, running on a critical at no-load, without a brake or driven machine. Let $\theta_1 =$ measured amplitude at cylinder no. 1.

The calculated value of equivalent static amplitude is obtained as

$$\theta_s = \frac{57 \cdot 3 \times |T_m| \, AR_0 \, \overrightarrow{\Sigma\Delta}}{\omega^2 \Sigma(J_{\text{cyl.}} \Delta^2_{\text{cyl.}})}$$

for the critical order considered. Then the 'engine dynamic magnifier' M_E is determined simply from

$$M_E = \theta_1/\theta_s.$$

The numerator of the expression for θ_s corresponds to the resultant torque, ω is the natural phase velocity, and $\Sigma(J_{\text{cyl.}} \Delta^2_{\text{cyl.}})$ is the 'single mass' of the equivalent one-mass system, which consists of cylinder inertias only, multiplied by their corresponding Holzer-table amplitudes squared. (The flywheel is the 'infinite mass' to which the shaft carrying the 'single mass' is clamped.)

Then, for a one-mass system, as shown in section 2·45, the 'engine

[365]

damping coefficient' can be determined from the dynamic magnifier, by means of the relation
$$c_E = \omega \Sigma (J_{\text{cyl.}} \Delta^2_{\text{cyl.}})/M_E.$$

If $J_{\text{cyl.}}$ is the same value for all cylinders, we have

$$\Sigma (J_{\text{cyl.}} \Delta^2_{\text{cyl.}}) = J_{\text{cyl.}} \Sigma (\Delta^2_{\text{cyl.}}),$$

and we may obtain the 'damping coefficient per cylinder', $c_{\text{cyl.}}$, from

$$c_{\text{cyl.}} = c_E / \Sigma (\Delta^2_{\text{cyl.}}).$$

If the torque is expressed in Lb.in., and the inertias are in Lb.in.sec.², the values for damping coefficients are obtained in Lb.in.sec./rad. (In metric units these quantities are expressed in Kg.cm., Kg.cm.sec.² and Kg.cm.sec./rad., respectively.)

Reverting to prediction of vibration amplitudes, we have

$$\theta_1 = \frac{57\cdot3 T_{\text{res.}}}{\omega \Sigma (c_{\text{cyl.}} \Delta^2_{\text{cyl.}})} = \frac{57\cdot3 T_{\text{res.}}}{\omega c_E},$$

or $\qquad \theta_1 = \theta_s M_E,$ where $\qquad M_E = \dfrac{\omega \Sigma \{ J_{\text{cyl.}} \Delta^2_{\text{cyl.}} \}}{\Sigma \{ c_{\text{cyl.}} \Delta^2_{\text{cyl.}} \}} = \dfrac{\omega J_{\text{cyl.}}}{c_{\text{cyl.}}},$

all give the same value of vibration amplitude at mass no. 1.

The next step is to consider the engine operating with a driven machine of inertia J_{DM}. The measured amplitude at mass no. 1 is $\theta_{1(DM)}$. The damping coefficient of the driven machine is represented by

$$c_{DM} = \omega J_{DM}/M_{DM},$$

where Δ_M is the Holzer amplitude of the driven machine. c_{DM} is obtained from the 'total damping coefficient of the system', c_T:

$$c_T = \omega \{ \Sigma (J_{\text{cyl.}} \Delta^2_{\text{cyl.}}) + J_{DM} \Delta^2_{DM} \}/M_T, \qquad (1)$$

where $M_T = \theta_{1(DM)}/\theta_s$, with θ_s denoting the same expression as previously.

We may write

$$c_T = c_{DM} \Delta^2_{DM} + c_E = \frac{\omega J_{DM} \Delta^2_{DM}}{M_{DM}} + \frac{\omega \Sigma (J_{\text{cyl.}} \Delta^2_{\text{cyl.}})}{M_E}, \qquad (2)$$

so that from equations (1) and (2)

$$c_D = c_E \left\{ \frac{\theta_1}{\theta_{1(DM)}} \times \left[1 + \frac{J_{DM} \Delta^2_{DM}}{\Sigma (J_{\text{cyl.}} \Delta^2_{\text{cyl.}})} \right] - 1 \right\}.$$

This enables c_{DM} to be obtained by comparative measurements of θ_1 and $\theta_{1(DM)}$. Details of the derivation of this expression are given at the end of this subsection.

Reverting to the prediction of vibration amplitudes, we can calculate $\theta_{1(DM)}$ of the engine with driven machine from

$$\theta_{1(DM)} = \frac{57\cdot3 T_{\text{res.}}}{\omega \{ \Sigma (c_{\text{cyl.}} \Delta^2_{\text{cyl.}}) + c_{DM} \Delta^2_{DM} \}} \quad \text{[degrees]}, \qquad (3a)$$

or, using dynamic magnifiers M_E (for the bare engine) and M_{DM} (for the driven machine), from

$$\theta_{1(DM)} = \frac{57 \cdot 3 T_{\text{res.}}}{\omega^2 \left\{ \dfrac{\Sigma(J_{\text{cyl.}} \Delta^2_{\text{cyl.}})}{M_E} + \dfrac{J_{DM} \Delta^2_{DM}}{M_{DM}} \right\}} \quad \text{[degrees]}, \tag{3b}$$

or more simply from

$$\theta_{1(DM)} = M_T \times \frac{57 \cdot 3 T_{\text{res.}}}{\omega^2 \, \Sigma(J_{\text{cyl.}} \Delta^2_{\text{cyl.}})} = M_T \times \theta_s, \tag{3c}$$

where

$$M_T = M_E \times \frac{1}{1 + \dfrac{M_E}{M_{DM}} \times \dfrac{J_{DM} \Delta^2_{DM}}{\Sigma(J_{\text{cyl.}} \Delta^2_{\text{cyl.}})}}.$$

This also gives the relation for amplitudes of vibration with and without the driven machine

$$\theta_1 / \theta_{1(DM)} = M_E / M_T.$$

For 'one-mass system' evaluations based on θ_s, the values of damping coefficients given in the following pages can, in most cases, be converted into expressions for dynamic magnifiers using the relation

$$M_{DM} = \omega J_{DM} / c_{DM},$$

and then substituted in equations (3c) to determine the vibration amplitude.

Derivation of expression for c_{DM}. Using $M_E = \theta_1 / \theta_s$ and $M_T = \theta_{1(DM)} / \theta_s$, together with the relation $c_E = \omega \Sigma (J_{\text{cyl.}} \Delta^2_{\text{cyl.}}) \, \theta_s / \theta_1$, we obtain from equation (1) divided by c_E

$$\frac{c_T}{c_E} = \frac{\theta_1}{\theta_{1(DM)}} \times \left[1 + \frac{J_{DM} \Delta^2_{DM}}{\Sigma(J_{\text{cyl.}} \Delta^2_{\text{cyl}})} \right] = 1 + \frac{c_{DM}}{c_E} \Delta^2_{DM},$$

and c_{DM} is thereby determined.

2·4211 General remarks on damping coefficients and dynamic magnifiers of engine parts and driven machines, with numerical values

Alternators

In most cases, the alternator produces a damping action which is regarded as negligible ($c_A = 0$ and $M_A = \infty$). However, when damper windings are used, the damping may be quite considerable, and may be estimated as $c_{DM} \cong 0 \cdot 5 \, \omega J_{MD}$. [Evaluated from Wennerberg's results; for reference, see p. 496.]

Bearing damping

(1) For bearings with a diametral clearance of about $0 \cdot 001$ in., per inch of journal diameter, the damping coefficient per pair of journal

bearings can be represented by the following expression for the 'dynamic magnifier per cylinder':

$$M_{\text{cyl.}} = \frac{100}{0 \cdot 2bm + 0 \cdot 6},$$

where $b = 1 \cdot 0$ for engines without balance weights, $b = 0 \cdot 5$ for engines with balance weights, and $m = $ critical order number (i.e. frequency divided by running speed).

The corresponding expression for multi-cylinder engines is

$$M_E = \frac{100}{0 \cdot 2bm\Sigma(\Delta_{\text{cyl.}}^2) + 0 \cdot 6},$$

where $\Sigma(\Delta_{\text{cyl.}}^2) = $ sum of Holzer-table amplitudes for cylinder inertias only. The term $0 \cdot 6$ represents foundation and shaft hysteresis damping.

[Draminsky, P., 'Crankshaft damping', *Proc. Instn Mech. Engrs, Lond.*, (1949).]

(2) For calculations using damping coefficients, the 'damping coefficient per cylinder' is

$$c_{\text{cyl.}} = \omega J_{\text{cyl.}}/M_{\text{cyl.}} \quad [\text{Lb.in.sec./rad.}],$$

where $\omega = $ natural phase velocity of system [rad./sec.], $J_{\text{cyl.}} = $ inertia per cylinder (per two cylinders for Vee engines) [Lb.in.sec.^{2}] and $M_{\text{cyl.}}$ the first expression given above. The 'total engine damping coefficient' is

$$c_E = \omega\Sigma(J_{\text{cyl.}}\Delta_{\text{cyl.}}^2)/M_E,$$

where $\Delta_{\text{cyl.}} = $ Holzer-table amplitude corresponding to each cylinder.

(3) Owing to the forces acting on the various crankthrows and the effects of the relative crank angles, the journal centres perform cycloidal-type movements, which are associated with damping and bearing constraints. [See also: Oil Viscosity, p. 374.] Regarding the use of $c_{\text{cyl.}}$ as a 'shaft damping' coefficient, see section 2·45, Engine with flywheel and front-end pulley.

Belt drives

Belt drives (particularly Vee belts with appreciable initial tension) can introduce damping into a system. For low-power drives this is not important, but an allowance for damping should be made (on the basis of comparative tests) for drives taking a power of the order of 0·1 % to 0·5 % or more of the total engine power. See also: Chain-driven Auxiliaries.

Brakes, eddy-current

The damping coefficient can be estimated by means of the expression given for electromagnetic couplings. See Couplings, Electromagnetic.

[368]

Brakes, hydraulic

The damping coefficient of water brakes is approximately

$$c_B = 10^6 \times H_b/N^2 \quad \text{[Lb.in.sec./rad.]},$$

where H_b = brake power [b.h.p.], and N = running speed [rev./min.]. The associated dynamic magnifier is

$$M_B = \omega J_B/c_B,$$

where ω = natural phase velocity of entire system [rad./sec.], J_B = brake inertia [Lb.in.sec.^{2}].

Note. In the curved portion of the brake characteristic, the maximum power $H_{max.}$ that can be absorbed by a two-row bucket type brake can be expressed as

$$H_{max.} = kN^n \quad \text{[h.p.]}$$

at any speed N. In this expression k = proportionality factor, and $n \cong 2 \cdot 75$.

Chain-driven auxiliaries

If an engine has chain-driven auxiliaries at the forward end of the crankshaft, for water pumps or lubricating oil pumps, etc., the drive may have an appreciable damping effect.

With front-end auxiliaries absorbing a power of $0 \cdot 5$–$1 \cdot 0 \%$ of the total engine power, the vibration amplitude $\pm \theta_1$ may be reduced by as much as 30 %, particularly if chain slack is allowed.

Auxiliaries driven from points close to the node of the engine system, i.e. near the main flywheel, do not have an appreciable damping effect.

Clutches

The fractional slip is a value generally between $0 \cdot 01$ and $0 \cdot 05$ (i.e., slip of 1–5%). Clutch slip reduces high-amplitude criticals. If M = dynamic magnifier of a system without a slipping clutch, and M_C = magnifier with a slipping clutch, then (approximately)

$$M_C = \frac{N_{cyl.}}{2+m} \times M, \quad \text{with} \quad m \geqslant N_{cyl.},$$

where $N_{cyl.}$ = total number of engine cylinders, and m = critical order number [vibration frequency divided by running speed].

The damping action requires careful determination in a system tested with and without the clutch, taking into account the lubrication conditions and spring loading required to maintain the required amount of slip.

Couplings, Bibby

The detuning action of Bibby couplings causes a reduction in vibration amplitude between 3 : 1 and 5 : 1. Thus, if M_C = dynamic magnifier of engine system with a coupling of this type, and $M =$ magnifier without the coupling, an approximate relation is

$$M_C = 0.25M.$$

For further details of Bibby couplings, see section 1·24.

Couplings, electromagnetic

These generally have a high damping coefficient, which can be estimated from

$$c_C = 6 \times 10^5 \times \frac{1}{s} \times \frac{H_b}{N^2} \quad \text{[Lb.in.sec./rad.]},$$

where $s =$ fractional slip (0·01–0·03), and $H_b =$ transmitted power [h.p.] at N [rev./min.].

If $J_C =$ total inertia of the coupling masses [Lb.in.sec.^{2}], the associated dynamic magnifier is

$$M_C = \omega J_C / c_C.$$

[Metz and Ericsson, *Proc. Inst. Mar. Engrs*, 9 November 1937.]

Couplings, fluid

An approximate expression for the damping coefficient is

$$c_C = 6 \times 10^5 \times \frac{m^2 + 1}{m^2} \times \frac{H_b}{N^2} \quad \text{[Lb.in.sec./rad.]},$$

where $N =$ rotary speed of coupling [rev./min.], $H_b =$ power transmitted [h.p.], and $m =$ critical order number (i.e. vibration frequency divided by running speed). As for electromagnetic couplings, the dynamic magnifier is given by

$$M_C = \omega J_C / c_C.$$

[For further information on fluid couplings, see section 1·24.]

[Sinclair, H., *Engineering*, 24 April 1938.]

Couplings, rubber

The damping coefficient varies with the vibration frequency and should be based on corresponding dynamic tests. For a hollow-sandwich coupling, made of natural rubber plus 'filler' materials, which was dynamically tested in torsion at the B.I.C.E.R.A. Laboratory, the following values of α (= damping coefficient c divided by volume V of rubber) were obtained at different vibration frequencies F:

F [vib./min.]	200	400	600	1,000	1,500	3,000	8,000	10,000	12,000
$\alpha = c/V$ [Lb.sec./in.^{2}]	16	7	4·5	2·7	2·2	1·6	0·6	0·4	0·38

Thus, if $c = \alpha V$ [Lb.in.sec./rad.] is the damping coefficient of the coupling, the corresponding dynamic magnifier is

$$M_C = \omega J_C / c,$$

where $J_C =$ coupling inertia [Lb.in.sec.^{2}].

[370]

Cyclic speed variation

The detuning due to cyclic speed variation is a factor which reduces the dynamic magnifier M_T of the entire engine system. Let

$$C = (N_{max.} - N_{min.})/N_{mean} = \text{cyclic speed variation,}$$

r = number of torsional vibration cycles per speed variation cycle,

then the magnifier taking account of cyclic speed variation is

$$M_{T(C)} = M_T \times \cos\left\{\frac{Cr}{\pi}\right\},$$

so that the amplitude $\pm \theta_{1(C)}$ with speed variation is related to the value $\pm \theta_1$ without this variation by

$$\theta_{1(C)} = \theta_1 \times \cos\left\{\frac{Cr}{\pi}\right\} = \theta_s \times M_{T(C)},$$

θ_s being the equivalent static amplitude.

Note. $r \cong F/N^*$, where

$$F = \text{natural frequency} \quad [\text{vib./min.}],$$

$$N^* = N_{cyl.} \times \tfrac{1}{2}N_{mean} \quad \text{(for four-stroke cycle engines).}$$

In many cases, the effect of cyclic speed variation is small enough to be neglected.

[Rybner, *Ingeniøren*, 1943, no. 47, and Draminsky, P., *Daempningen ved torsionssvingninger i krumtapaksler.*]

For the experimental determination of C, see section 2·6.

Cylinder damping

The 'damping coefficient per engine cylinder', $c_{cyl.}$, can be expressed as

$$c_{cyl.} = k_{cyl.}\,\omega J_{cyl.},$$

where $J_{cyl.}$ = cylinder inertia [Lb.in.sec.^{2}], ω = natural phase velocity [rad./sec.], and $k_{cyl.}$ = a factor of the order of 0·01. The total damping of the engine is then

$$c_E = k_{cyl.}\,\omega\Sigma(J_{cyl.}\Delta_{cyl.}^2) \quad [\text{Lb.in.sec./rad.}].$$

With the damping $c_{DM} = k_{DM}\,\omega J_{DM}$ due to the driven machine, the total damping coefficient of the system is obtained as

$$c_T = [k_{cyl.}\,\omega J_{cyl.}\,\Sigma(\Delta_{cyl.}^2) + c_{DM}\Delta_{DM}^2].$$

Multi-cylinder engines. The 'damping per engine cylinder' may have different values for engines with the same basic type of cylinder but with different numbers of cylinders or different crank arrangements, or operating at appreciably different loads.

Increase in amplitude with running time. Torsiograph measurements on new engines, and on engines fitted with new pistons, investigated by members of the Torsional Vibration Panel of B.I.C.E.R.A., have shown that the vibration amplitudes may increase up to 30 % of their initial value during the running-in period the first 20 hours. Generally, stationary values of amplitude are obtained thereafter, but it has been found in some cases that the amplitudes tend to increase somewhat, over a period of years.

Critical speeds and normal running speed. In some cases, the dynamic magnifiers of engine systems appear to be greater for criticals with higher order numbers. This may be due to the driven machine characteristic. For example, if a hydraulic brake has a power/speed curve $H \propto N^{2 \cdot 75}$, and a damping coefficient $c \propto H/N^2$, then $c \propto N^{0 \cdot 75}$, so that there is less brake damping at low running speeds. [$\propto$ means 'proportional to'.]

Damping due to piston ring chatter. Rig tests carried out at the B.I.C.E.R.A. Laboratory indicated that piston-ring chatter (at the torsional vibration frequency of an engine system) produces negligible damping when the piston rings are mounted on a reciprocating piston in a cylinder. With a piston vibrating about a stationary position, however, there is a noticeable amount of damping due to ring vibration. This damping increases with increased gas pressures in the cylinder.

As a reciprocating piston is 'stationary' only near its top and bottom dead-centre positions, damping due to piston ring flutter may occur only at these extreme positions, where it has little effect on torsional vibration.

'Identical' engines. Tests of identically built engines frequently show different vibration amplitudes for two or more engines of the same type. These variations may be due to different injection pressures or combustion conditions, and can be expected since it is known that cylinder-pressure diagrams show variations for different cylinders.

Driven machines

The power H absorbed by a driven machine is generally given by

$$H = \frac{T\omega_N}{6600} \quad \text{[h.p.]},$$

where T = driving torque [Lb.in.] and $\omega_N = (2\pi/60) \times N$, N being the running speed [rev./min.]. Assuming that the damping torque is proportional to the driving torque this gives a relation for the damping coefficient

$$c = kT/\omega_N \quad \text{[Lb.in.sec./rad.]},$$

where k = proportionality factor. Therefore, expressing T in terms of H, this gives

$$c = k \times 6600 \times \frac{H}{\omega_N^2} \simeq 6 \times 10^5 \times k \times \frac{H}{N^2}.$$

The value of k depends on the type of driven machine considered. The corresponding dynamic magnifier is

$$M_{DM} = \omega J_{DM}/c,$$

where J_{DM} = inertia of driven machine [Lb.in.sec.^{2}], ω = natural phase velocity of engine system [rad./sec.].

Foundation damping

Foundation damping, i.e. the damping due to vibration of the test bed or engine mounting (including the crankcase) contributes to the total damping of a system. Combined with shaft hysteresis damping, according to Draminsky, it can be represented by a dynamic magnifier

$$M_{(f+h)} \cong 167 \cong \frac{1}{0 \cdot 06}.$$

It should be noted that crankcase vibrations due to unbalanced engine forces and couples may be magnified, in some cases, by the flexibility of the test bed or engine mounting. In such cases, comparative tests (using reed vibrometers) with different engine mountings usually give adequate indications.

Gear-driven auxiliaries

The damping action obtained gives similar decreases in vibration amplitude to those observed with chain-driven auxiliaries.

Gear backlash, however, may introduce additional vibration frequencies into the system [see section 1·3]. For auxiliary masses with large inertia torques, it is advisable to take torsiograph recordings at these masses, since their vibration amplitudes may not be those indicated by Holzer tabulations.

Generators, d.c.

According to F. M. Lewis (*Trans. Soc. Nav. Archit., N.Y.*, vol. 33 (1925), p. 109), the damping coefficient of generators can be estimated as

$$c = 6 \times 10^5 \times \frac{E}{E - E_c} \times \frac{H}{N^2} \quad [\text{Lb.in.sec./rad.}],$$

where N = running speed [rev./min.], H = power absorbed [h.p.], E = total e.m.f. (terminal voltage plus internal drop), and E_c = counter e.m.f. of driven machines. With J_M = inertia of rotating mass of generator [Lb.in.sec.^{2}], ω = natural phase velocity [rad./sec.] of engine system, the corresponding dynamic magnifier is

$$M_M = \omega J_M/c.$$

Injection timing

It has been found that the dynamic magnifier of an engine may be altered by a variation of the injection timing. Tests carried out by members of the Torsional Vibration Panel and the B.I.C.E.R.A. Laboratory indicate that the vibration amplitude is decreased by retarding the injection timing from its optimum position, whereas it is increased when the injection timing is advanced 3 to 5 degrees further than the optimum setting before T.D.C.

[The 'optimum' timing is the setting which gives maximum efficiency, i.e. minimum specific fuel consumption, at the engine speed considered.]

In all these tests, the variation in indicator diagrams and torque harmonics was taken into account, so that the variation referred to is a separate effect.

Keyed-on or screwed-on assemblies

Components with keyed-on or screwed-on connexions tend to damp vibrations, but this action is sometimes associated with fretting or gradual loosening of the connected parts. Therefore, if reductions in vibration amplitudes are observed in the course of tests, it is advisable to verify the effectiveness of these connexions. These remarks also apply to some extent to shrink fits.

Oil viscosity

Torsiograph tests, carried out on an engine at the B.I.C.E.R.A. Laboratory, using lubricating oils of different viscosities (S.A.E. 10, 20 and 30), showed no appreciable variation in peak amplitudes of vibration with the various oils used. Cold-starting tests were also made, and with torsiograph measurements taken within 3 minutes after starting, the results obtained were the same as with the warm engine.

The Laboratory is pursuing these matters.

Propellers

See Propeller damping, section 2·422, pp. 375 et seq.

Pumps and compressors

Appreciable damping may be obtained when the engine system includes auxiliary drives with oil pumps or piston compressors. Comparative tests with and without direct drive of these auxiliaries are required to determine the damping effect in each case. Air damping due to rotary impeller-type auxiliaries may also occur if these are mechanically coupled to the engine system.

Shaft hysteresis

The damping due to torsional shaft hysteresis is small in the normal range of permissible stresses for engine shafting. The formulae of Lewis, Dorey and Den Hartog, assume hysteresis to be the major factor in engine damping, but also use numerical coefficients adjusted to give results of the right order. Draminsky, for the normal range of permissible shaft stresses, proposes for the hysteresis damping of mild steel a dynamic magnifier $M_h = 250$.

2·422 *Propeller damping*

1. Approximate estimation

[For detailed calculations taking account of propeller pitch, etc., see (2) Archer's method, pp. 376–384.]

An approximate value of the damping coefficient c_P of a propeller is obtained from

$$c_P = \frac{n \times T_P}{\omega_C} = \frac{n \times T_P}{\dfrac{2\pi}{60} \times N_c} \simeq 30 \times \frac{T_P}{N_c} \quad \text{[Lb.in.sec./rad.]},$$

where n = torque exponent ($\simeq 3$), T_P = propeller torque [Lb.in.] at critical speed, N_c = propeller speed [rev./min.] at critical speed, and

$$\omega_c = (2\pi/60) \times N_c.$$

If propeller damping only is considered, the vibration amplitude $\pm \theta_1$ at mass no. 1 of the engine, at any propeller service speed N_P [rev./min.] is, approximately,

$$\pm \theta_1 = 1 \cdot 91 \times \frac{N_P^2}{\omega N_c} \times \frac{|T_m| A R_0 \overrightarrow{\Sigma\Delta}}{T_P \Delta_P^2} \quad \text{[degrees]},$$

where ω = natural phase velocity of the entire system [rad./sec.], Δ_P = Holzer-table amplitude of the propeller (with $\Delta_1 = 1$ rad. at mass no. 1) and $|T_m| A R_0 \overrightarrow{\Sigma\Delta}$ = excitation torque of the engine cylinders.

Note. Propeller torque T_P [Lb.in.] is related to propeller power H_P [shaft h.p.] by the expression
$$T_P = 6 \cdot 303 \times 10^4 \times H_P / N_P,$$

where N_P = corresponding propeller speed [rev./min.], and for two different speeds N_{P_1} and N_{P_2}, the corresponding propeller powers are

$$H_{P_2} = H_{P_1} \times \left(\frac{N_{P_2}}{N_{P_1}} \right)^3.$$

Sometimes the calculation of vibration amplitude is based on the 'equivalent static amplitude' of the system:

$$\theta_s = \frac{|T_m| A R_0 \overrightarrow{\Sigma\Delta} \times 57 \cdot 3}{\omega^2 \times [\Sigma(J_{cyl.} \Delta_{cyl.}^2) + J_P \Delta_P^2]} \quad \text{[degrees]}.$$

The corresponding dynamic magnifier is

$$M_P = \frac{\omega[\Sigma(J_{cyl.} \Delta_{cyl.}^2) + J_P \Delta_P^2]}{3 \cdot 15 \times \dfrac{T_P}{\omega_P} \times \Delta_P^2},$$

or, if $\omega = m \times \omega_P$ (where m = critical speed order number, i.e. natural frequency divided by *propeller* speed), using the relation for T_P and H_P,

$$M_P = \frac{N_P^3 \times [\Sigma(J_{\text{cyl.}} \Delta_{\text{cyl.}}^2) + J_P \Delta_P^2]}{1 \cdot 8 \times 10^7 \times H_P \times \Delta_P^2}.$$

Finally,

$$\pm \theta_1 = M_P \times \theta_s \quad \text{[degrees]}.$$

Metric units. If T_P is in Kg.cm., the value of c_P is in Kg.cm.sec./rad. without changing the numerical factor 30. The expression for $\pm \theta_1$ is unchanged in metric units. As $1 \text{ Lb.in.} = 1 \cdot 155 \text{ Kg.cm.}$, and $1 \text{ h.p. (Brit.)} = 1 \cdot 0139 \text{ h.p. (metric)}$, in the equation for T_P the constant $6 \cdot 303 \times 10^4$ should be replaced by $7 \cdot 18 \times 10^4$ to obtain T_P in Kg.cm. from H_P in metric h.p. Moreover, for the dynamic magnifier M_P, the constant $2 \cdot 04 \times 10^7$ should be used (with J-values in Kg.cm.sec.^{2}) in place of $1 \cdot 8 \times 10^7$.

2. Archer's method

[Archer, S.,' Contribution to improved accuracy in the calculation and measurement of torsional vibration stresses in marine propeller shafting', *Proc. Instn Mech. Engrs, Lond.*, vol. 164 (1951), pp. 351–6. 'Torsional vibration damping coefficients for marine propellers', *Engineering*, 13 May 1955, pp. 594–8.]

This method was developed on the basis of the results obtained by Troost in open-water tests of bronze two-, three-, four- and five-bladed modern propellers. [L. Troost, *Trans. N.-E. Cst Instn Engrs Shipb.*, vol. 54, p. 321, and vol. 56, p. 91.] In his papers mentioned above, Archer also gives the mathematical expressions by means of which the results of Troost's methodical tests were converted into the curves of Figs. 2–10 in this section.

Notation

D = blade tip diameter [ft.],

$A_0 = \pi D^2 / 4$ [ft.^{2}],

A_s = propeller surface area (developed area) [ft.^{2}],

$r = A_s / A_0$ = disk-area ratio,

P^* = pitch (mean face pitch) [ft.],

P^*/D = pitch/diameter ratio,

H_R = rated engine power [h.p.],

N_R = rated speed of propeller [rev./min.],

N_C = critical speed of propeller [rev./min.],

$H_P \cong_{(R)} 0 \cdot 985 \times H_R$ = delivered shaft power at the propeller [h.p.] at rated speed N_R,

$H_{P(C)} = H_{P(R)} \times (N_C/N_R)^3$ = delivered shaft power at the propeller [h.p.], at critical speed N_C,

c_P = propeller damping coefficient [Lb.in.sec./rad.],

$T_{P(C)} = 6 \cdot 303 \times 10^4 \times H_{P(C)}/N_C$ = mean propeller torque at critical speed [Lb.in.],

$$A_Q = 9 \cdot 5 \times H_{P(R)} \Big/ \left\{ D^2 \left[\frac{N_R D}{1000} \right]^3 \right\} = \text{torque coefficient}$$

$$\text{(non-dimensional)},$$

$a = c_P N_C/T_{P(C)}$ = specific damping coefficient (non-dimensional).

Note. All the quantities indicated above are related to sea-water conditions.

Procedure

(1) Evaluate the non-dimensional quantities r, P^*/D and the torque coefficient

$$A_Q = 9 \cdot 5 \times \frac{H_{P(R)}}{D^2 \left[\dfrac{N_R D}{1000} \right]^3}.$$

(2) Referring to Figs. 2–10, determine the specific damping coefficient a.

(3) Calculate the mean propeller torque from

$$T_{P(C)} = 6 \cdot 303 \times 10^4 \times H_{P(C)}/N_C,$$

or from

$$T_{P(C)} = 6 \cdot 303 \times 10^4 \times \frac{H_{P(R)}}{N_C} \times \left(\frac{N_C}{N_R} \right)^3 = 6 \cdot 303 \times 10^4 \times \frac{H_{P(R)}}{N_R} \times \left(\frac{N_C}{N_R} \right)^2 \quad \text{[Lb.in.]}.$$

(4) Finally, evaluate the propeller damping coefficient from

$$c_P = a T_{P(C)}/N_C \quad \text{[Lb.in.sec./rad.]}.$$

It should be noted that the curves of Figs. 2–10 are for bronze propellers. For cast-iron propellers, a suggested approximation is to increase the nominal mean face pitch by about 10% (or whatever percentage experience indicates as appropriate for the higher blade thickness fractions) and treat as an equivalent bronze propeller. The reason for this is that the blade thicknesses of cast-iron propellers are generally greater than those for bronze propellers. Since the effective pitch of a propeller blade is determined both by the face and the back of the blade, this will result in a larger effective pitch for the cast-iron propeller.

NUMERICAL EXAMPLE. Four-bladed propeller with the following data: $D = 16 \cdot 406$ ft., $P^*/D = 0 \cdot 709$, $r = A_s/A_0 = 0 \cdot 45$. Rated engine power $H_R = 2300$ h.p. at $N_R = 100$ rev./min. It is required to determine the damping coefficient of the propeller at a critical speed $N_C = 49$ rev./min.

Torque coefficient

$$A_Q = 9\cdot5 \times \frac{2300 \times 0\cdot985}{(16\cdot406)^2 \times \left[\dfrac{100 \times 16\cdot406}{1000}\right]^3} = 18\cdot17.$$

Interpolations for coefficient a:

From Fig. 1a, for $r = 0\cdot40$:

$$P^*/D = 0\cdot60, \quad a = 22\cdot0,$$

$$P^*/D = 0\cdot80, \quad a = 33\cdot2,$$

therefore $\quad P^*/D = 0\cdot709, \quad a = 22 + \dfrac{0\cdot709 - 0\cdot600}{0\cdot8 - 0\cdot6} \times (33\cdot2 - 22\cdot0) = 28\cdot10.$

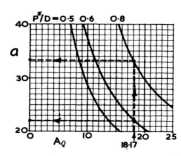
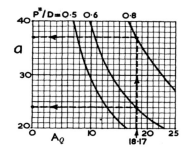

Fig. 1a. Damping curves for 4-bladed propeller with $r = 0\cdot40$.

Fig. 1b. Damping curves for 4-bladed propeller with $r = 0\cdot55$.

(These figures are reprints of parts of Figs. 6 and 7.)

From Fig. 1b, for $r = 0\cdot55$:

$$P^*/D = 0\cdot6, \quad a = 23\cdot20,$$

$$P^*/D = 0\cdot8, \quad a = 37\cdot00,$$

$$P^*/D = 0\cdot709, \quad a = 23\cdot20 + \frac{0\cdot709 - 0\cdot600}{0\cdot8 - 0\cdot6} \times (37\cdot00 - 23\cdot20) = 30\cdot72.$$

Hence, for $r = 0\cdot45$:

$$a = 28\cdot10 + \frac{0\cdot45 - 0\cdot40}{0\cdot55 - 0\cdot40} \times (30\cdot72 - 28\cdot10) = 28\cdot98 \simeq 29\cdot0.$$

Propeller torque:

$$T_{P(C)} = 6\cdot303 \times 10^4 \times \frac{2300 \times 0\cdot985}{100} \times \left(\frac{49}{100}\right)^2 = 347{,}900 \quad [\text{Lb.in.}].$$

Propeller damping coefficient:

$$c_P = a \times \frac{T_{P(C)}}{N_C} = 29 \times \frac{347{,}900}{49} = 205{,}700 \quad [\text{Lb.in.sec./rad.}].$$

This value of c_P can be used for vibration amplitude evaluations by analytical, graphical, or tabulation methods (sections 2·45, 2·43 and 2·44).

Curves for specific damping coefficients

Values of the specific damping coefficient a, for two-, three-, four- and five-bladed propellers, are plotted against A_Q in Figs. 2–10.

[378]

Curves of percentage real slip

In addition, for the convenience of those who prefer to work on the basis of slip rather than with torque coefficients, the figures also give the percentage real slip S, plotted against A_Q.

It should be noted that the S-curves, which are related to the S-scale on the right-hand side of the figures, can be used directly only to obtain A_Q-values.

To determine corresponding a-values, the procedure is as follows: (1) Starting from a given S-value on the right-hand scale of the figures,

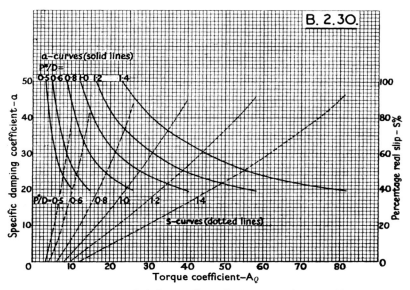

Fig. 2. Two-bladed propellers. Disk-area ratio: $r = 0.30$.

follow a horizontal line to the S-curve corresponding to the P^*/D-ratio considered. (2) At the intersection, read downwards and thus determine the corresponding A_Q-value. (3) Then proceed vertically from this value of A_Q to the a-curve with the same P^*/D ratio, and read the corresponding a-value on the left-hand scale of the figure.

Note on the derivation of the propeller damping coefficients

The torsional vibration damping coefficients a are derived from Troost's standard 'B' series of systematic model tests carried out in the experimental tank at Wageningen, Holland.†

Each model propeller had a diameter of 240 mm. (about 9·5 in.) and was tested at an immersion to the centre of the propeller equal to its diameter. The tests were

† Troost, L., 'Open water test series with modern propeller forms', Part 3. *Trans. N.-E. Cst. Instn Engrs Shipb.*, vol. 67 (1950–1), pp. 89–130.

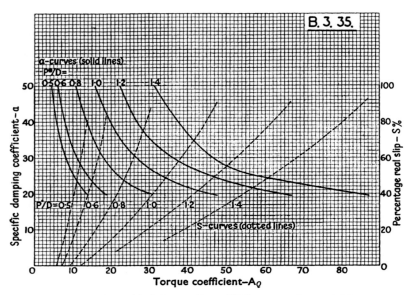

Fig. 3. Three-bladed propellers. Disk-area ratio: $r = 0.35$.

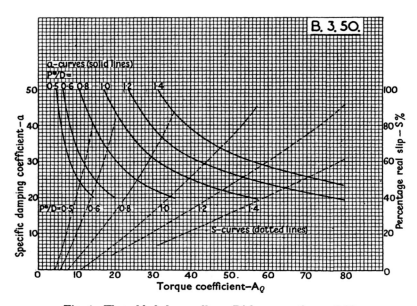

Fig. 4. Three-bladed propellers. Disk-area ratio: $r = 0.50$.

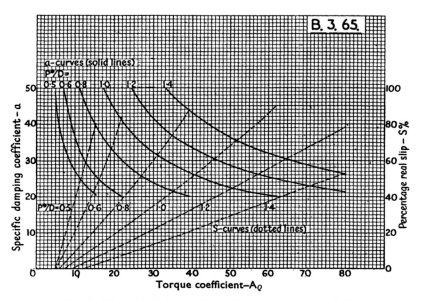

Fig. 5. Three-bladed propellers. Disk-area ratio: $r = 0.65$.

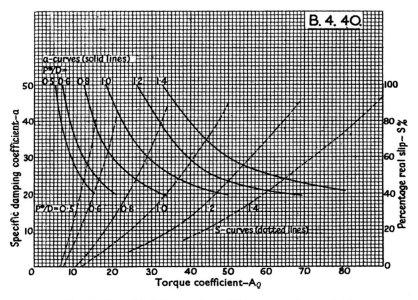

Fig. 6. Four-bladed propellers. Disk-area ratio: $r = 0.40$.

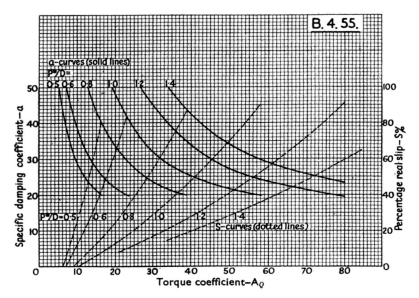

Fig. 7. Four-bladed propellers. Disk-area ratio: $r = 0.55$.

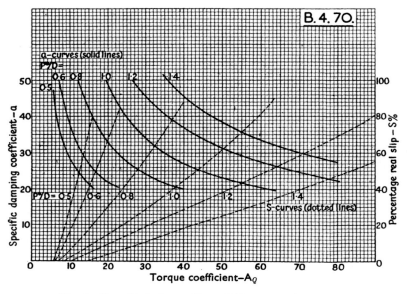

Fig. 8. Four-bladed propellers. Disk-area ratio: $r = 0.70$.

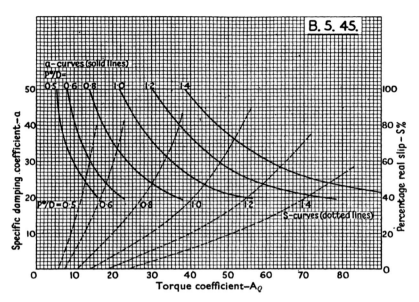

Fig. 9. Five-bladed propellers. Disk-area ratio: $r = 0.45$.

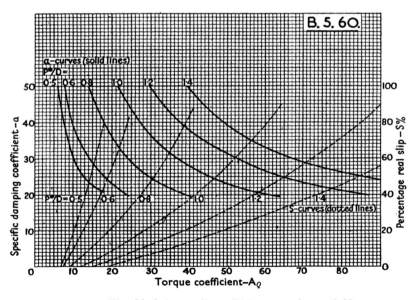

Fig. 10. Five-bladed propellers. Disk-area ratio: $r = 0.60$.

carried out in open water, i.e. not behind a model, and the revolutions for each propeller were kept constant while the speed of advance was varied over the full range of slips.

The propeller blades were of the wide-tip type with aerofoil sections over the inner part and circular-back sections over the outer part of the blade. With the exception of the four-bladed propellers, which had a pitch reduction of 20 % towards the boss, the propellers had a constant face pitch.

The following table summarizes the 'B' series:

No. of blades	Disk area ratio						
2	0·30						
3		0·35			0·50		0·65
4			0·40			0·55		0·70
5				0·45			0·60

No. of blades	Blade thickness fraction	Boss/ diameter ratio	Pitch reduction	Rake of blades	Pitch/diameter ratio $P*/D$
2	0·055	0·18	0	15°	0·5 to 1·4
3	0·05	0·18	0	15°	0·5 to 1·4
3	0·045	0·167	20 %	15°	0·5 to 1·4
5	0·04	0·167	0	15°	0·5 to 1·4

The derivation of the a-coefficients, due to Archer,† is as follows:

At a constant speed of advance V, the damping coefficient is obtained as

$$c_P = \left[\frac{\partial T_P}{\partial \omega}\right]_V = \frac{60}{2\pi}\left[2 - \lambda\frac{f'(\lambda)}{f(\lambda)}\right] = \frac{aT_P}{N},$$

where T_P = propeller torque,

ω = instantaneous angular velocity of propeller,

$\lambda = V/(ND)$ = non-dimensional speed coefficient (N being the propeller speed [rev./min.] and D the propeller diameter),

$f(\lambda) = T_P/(\rho D^5 N^2)$ = non-dimensional torque coefficient (ρ being the density of the water used in the model tests), and

$f'(\lambda) = \partial f(\lambda)/\partial \lambda$ = slope of $f(\lambda)$-curve in Troost's systematic series torque diagrams.

The non-dimensional damping coefficient a is determined from

$$a = Nc_P/T_P.$$

General note on formulae for predicted amplitudes and stresses

The various formulae which have been given in pp. 348–363 for the prediction of vibration amplitudes and stresses employ empirical coefficients. Furthermore, they are all based on a reduction of the actual multi-mass system to an equivalent one-mass system.

This one-mass system is obtained by means of two assumptions, viz.: (i) the inertias of the system can be combined to obtain an equivalent engine inertia, and (ii) the resultant torque based on a phase-vector sum represents the action of the torques due to the individual cylinders.

The limitations of one-mass system evaluations are generally realized and have already led to the use of two-mass system stress calculations for systems with propeller damping (see sections 2·422 and 2·45).

† Archer, S., loc. cit. (p. 376).

As regards the use of a single expression $\overrightarrow{\Sigma\Delta}$ to represent the combined effects of cylinder torques, it may be said that it gives fairly accurate results in many instances for major orders. It is evident with minor orders that the prediction is not sufficiently accurate, even if the minor-order torque itself has a high value due to a low-order high-amplitude excitation.

The examination of a forced-frequency tabulation shows that the action of the cylinder torques is not represented, even in a first approximation, by $\overrightarrow{\Sigma\Delta}$, although it does depend on the phase angle. The correct approach to this part of the problem is to use equivalent systems with more than one mass (and more than one torque position) for stress evaluation. This can be achieved by means of (i) graphical methods, (ii) Holzer tables with gas-pressure torques and damping, or (iii) formulae for two, three- or four-mass systems.

Methods (i) and (ii) are numerical, and are not generally suited to give a clear picture of the relations between torques, damping actions and inertias. Method (iii) shows each quantity in an explicit formula. Its development requires laboratory work to determine true damping actions and will take some time. All three types of procedure will now be considered in the following section. With time, it should be possible to decide which of these three can be regarded as the most suitable for replacing the empirical formulae now in use and which have been given in the present section. The use of more efficient means of evaluation, such as electronic computers, should help considerably in introducing more accurate methods for stress prediction.

2·43 Graphical and semi-graphical methods

2·431 *One-mass system*

The system consists of a shaft of stiffness K, clamped to an infinite mass at one end and rigidly connected at the other to a mass with a moment of inertia J. The mass J is subjected to a sinusoidally varying torque $T = T_0 \sin \omega t$ (where $\omega =$ phase velocity of excitation and $t =$ time). The vibration amplitude θ of the mass J is damped by a dashpot with a damping coefficient $c = T_0/\dot{\theta}_1$ (where $\dot{\theta}_1$ is the maximum value of $d\theta/dt$).

Using complex numbers, so that $\theta = \theta_1 e^{j\omega t} = \theta_1(\cos \omega t + j \sin \omega t)$, with $j^2 = -1$, the equation of motion gives

$$-\omega^2 J \theta_1 + j\omega c \theta_1 + K\theta_1 = T_0.$$

In order to determine the 'polygon of forces', i.e. the torque-vector diagram, it is convenient to re-write the above equation as follows:

$$+\omega^2 J - j\omega c - K + \frac{T_0}{\theta_1} = 0.$$

The torque vectors can now be drawn directly, using a suitable scale:

(1) Draw a pair of rectangular axes, and denote the real axis by $+\mathscr{R}$, and the imaginary axis by $+\mathscr{I}$, as shown in Fig. 2.

(2) Draw $+\omega^2 J$ in the $+\mathscr{R}$ direction, then $-j\omega c$ downwards (the length of this vector being ωc), and draw $-K$ in the $-\mathscr{R}$ direction.

(3) Close the polygon by means of a vector T_0/θ as shown in Fig. 3.

(4) Measure the length of the vector T_0/θ, and express this value in Lb.in./rad.† by means of the scale already used for the other vectors.

(5) As T_0 is a known value, the amplitude θ of the vibration can be determined by calculating

$$\theta = T_0/(T_0/\theta)$$

and the phase angle ϕ between T_0 and θ is obtained either directly from the diagram or by evaluating the relation [derived in section 2·45]

$$\tan\phi = \frac{\omega c}{K - \omega^2 J}.$$

The procedure will now be illustrated by a numerical example.

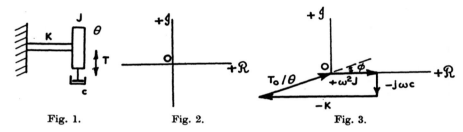

Fig. 1.   Fig. 2.   Fig. 3.

EXAMPLE. An engine is reduced to an equivalent one-mass system (see sections 2·3 and 2·41) with the following values:

Excitation torque $T_0 = 1100$ Lb.in., shaft stiffness $K = 3 \times 10^6$ Lb.in./rad., inertia $J = 3$ Lb.in.sec.², $\omega_0^2 = K/J = 10^6$ (rad./sec.)², and damping coefficient $c = 180$ Lb.in.sec./rad.

It is required to determine by vector diagrams the vibration amplitude θ of J, for the following ratios $\omega/\omega_0 = 0·2$, 0·8, 1·0, 1·2 and 3·0. The detailed procedure is given below.

$\omega/\omega_0 = 0·2$:   Phase velocity $\omega = 0·2 \times 10^3$ rad./sec., $\omega^2 = 0·04 \times 10^6$.

                Inertia torque $J\omega^2 = 0·12 \times 10^6$ Lb.in./rad.

                Damping torque $c\omega = 180 \times 200 = 0·036 \times 10^6$ Lb.in./rad.

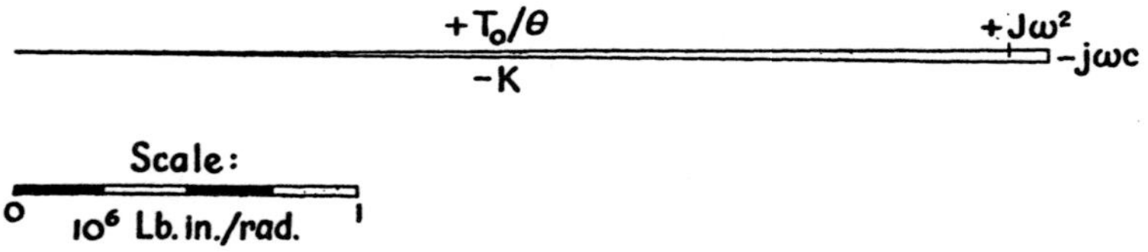

$+T_0/\theta$ $+J\omega^2$
—— $-j\omega c$
                —K

Scale:

○————————————|
   10^6 Lb.in./rad.

Fig. 4. 'STIFFNESS REGION.' T_0 is mainly resisted by K.

Evaluation: $\theta \cong T_0/K = 1·1 \times 10^3/3 \times 10^6 \cong 4 \times 10^{-4}$ rad.

            $\tan\phi = 0·036/[3 - 0·12] = 0·0125$, $\phi \cong 0°\,43'$.

† Or Kg.cm./rad. in metric units.

$\omega/\omega_0 = 0\cdot8$: $\omega^2 = 0\cdot64 \times 10^6$, $J\omega^2 = 1\cdot92 \times 10^6$, $c\omega = 180 \times 800 = 0\cdot144 \times 10^6$.
Scale: As above.

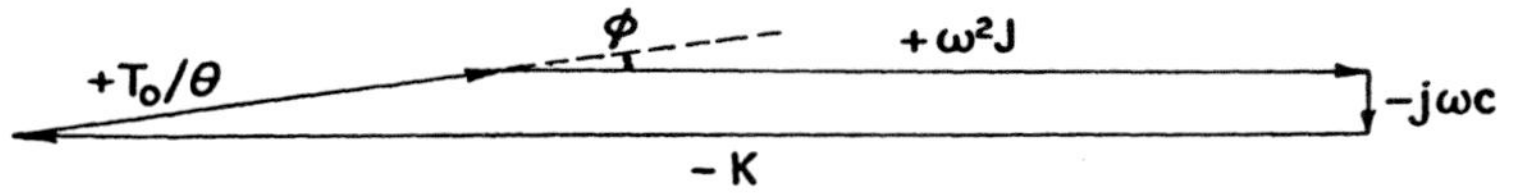

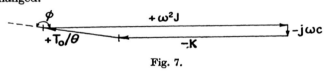

Fig. 5.

Evaluation: $T_0/\theta = 1\cdot06 \times 10^6$, $\theta = T_0/(1\cdot06 \times 10^6) = \dfrac{1\cdot1 \times 10^3}{1\cdot06 \times 10^6} = 1\cdot04 \times 10^{-3}\,\text{rad}$.

$$\tan\phi = 0\cdot144/[3 - 1\cdot92] = 0\cdot133, \quad \phi \cong 7^\circ\,36'.$$

$\omega/\omega_0 = 1\cdot0$: $\omega_0^2 = 10^6$, $J\omega_0^2 = 3 \times 10^6$, $c\omega_0 = 180 \times 10^3 = 0\cdot18 \times 10^6$.
Scale: unchanged.

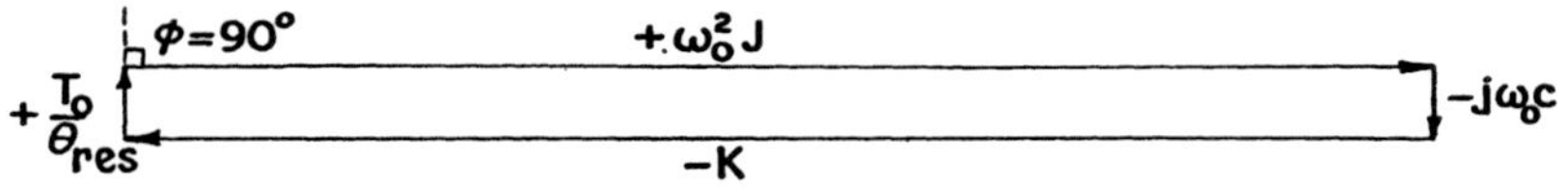

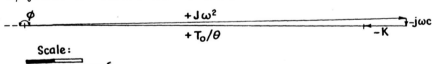

Fig. 6. 'RESONANCE REGION.' T_0 is only opposed by damping.

$$\theta_{\text{res.}} = T_0/c\omega_0 = 1\cdot1 \times 10^3/0\cdot18 \times 10^6 = 6\cdot2 \times 10^{-3}\,\text{rad.} \cong \pm 0\cdot35^\circ.$$
$$\phi = 90^\circ.$$

$\omega/\omega_0 = 1\cdot2$: $\omega^2 = 1\cdot44 \times 10^6$, $J\omega^2 = 4\cdot32 \times 10^6$, $c\omega = 180 \times 1200 = 0\cdot216 \times 10^6$.
Scale: unchanged.

Fig. 7.

$$T_0/\theta = 1\cdot35 \times 10^6, \quad \theta = T_0/1\cdot35 \times 10^6 = 1\cdot1 \times 10^3/1\cdot35 \times 10^6 \cong 8\cdot2 \times 10^{-4}\,\text{rad}.$$

$\omega/\omega_0 = 3\cdot0$: $\omega^2 = 9 \times 10^6$, $J\omega^2 = 27 \times 10^6$, $c\omega = 180 \times 3000 = 0\cdot54 \times 10^6$.

Scale:

Fig. 8. 'INERTIA REGION.' T_0 mainly resisted by inertia torque.

$$T_0/\theta = 23\cdot84 \times 10^6, \quad \theta = T_0/23\cdot84 \times 10^6 = 1\cdot1 \times 10^3/23\cdot84 \times 10^6 = 4\cdot6 \times 10^{-5}\,\text{rad}.$$
$$\tan\phi = 0\cdot54/[3\cdot0 - 27\cdot0] = -0\cdot54/24\cdot0 = -0\cdot0225;$$
$$\phi = 180^\circ - 1^\circ\,18' = 178^\circ\,42'.$$

Metric units. The above example is also valid in metric units, without altering the numerical values, if the 'Lb.' is replaced by 'Kg.' throughout, and 'in.' is replaced by 'cm.'.

2·432   Two-mass system

The system consists of two masses of moments of inertia J_1 and J_2, connected by shafting of stiffness K. The damping at each mass is represented by the dashpots with damping coefficients c_1 and c_2, and mass J_2 is subjected to a sinusoidally-varying torque $T = T_0 \sin \omega t$.

The vector diagrams for the torques acting on each mass will first be considered separately, and then combined in a single diagram.

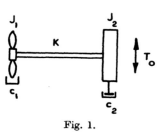

Fig. 1.

The procedure is similar to that used for the one-mass system and will be illustrated by means of a numerical example, which is self-explanatory to a certain extent. To give a clear idea of the construction of the diagrams, the values used for the input torque T_0 and the damping coefficients c_1 and c_2 are much greater than those normally occurring in engine systems.

Mass no. 1.   Numerical values:

$$\left. \begin{array}{c} \omega^2 J_1 = 2 \\ K = 3 \\ c_1 \omega = 0\cdot8 \end{array} \right\} \quad \begin{array}{c} \text{(All torques [Lb.in.] are} \\ \text{divided by } 10^6.) \end{array}$$

Equation of motion:

$$J_1 \ddot{\theta}_1 + c_1 \dot{\theta}_1 + K\theta_1 - K\theta_2 = 0.$$

With $\theta_2/\theta_1 = \Delta_2$, this gives

$$+ \omega^2 J_1 - j\omega c_1 - K + K\Delta_2 = 0.$$

For details of diagram construction, see 'One-mass system'. The diagram for mass no. 1 is obtained as shown in Fig. 2a:

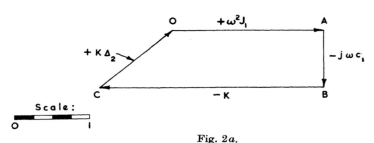

Fig. 2a.

The diagram $OABCO$ gives

$$\overline{CO} = K\Delta_2 = 1\cdot28.$$

Hence
$$\Delta_2 = 1\cdot28/K = 0\cdot43.$$

[388]

Mass no. 2.

$$\left.\begin{array}{c} \omega^2 J_2 = 1\cdot2 \\ K = 3\cdot0 \\ c_2\omega = 0\cdot2 \\ T_0 = 0\cdot0063 \end{array}\right\}$$ so that, with $\begin{cases} \omega^2 J_2\Delta_2 = 0\cdot52, \\ c_2\omega\Delta_2 = 0\cdot086. \end{cases}$
$\Delta_2 = 0\cdot43$

The equation of motion is

$$K(\theta_2 - \theta_1) + J_2\ddot{\theta}_2 + c_2\dot{\theta}_2 = T_0,$$

and this corresponds to the equation

$$-K\Delta_2 + K + J_2\omega^2\Delta_2 - j\omega c_2\Delta_2 + \frac{T_0}{\theta_1} = 0.$$

The diagram is shown in Fig. 2b:

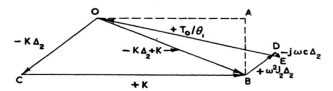

Fig. 2b. Note that $\overline{BD}$ is parallel to $\overline{CO}$, and that $\overline{DE}$ is perpendicular to $\overline{BD}$ (in the clockwise direction).

The diagram *OCBDEO* gives

$$\overline{EO} = T_0/\theta_1 = 2\cdot53.$$

Therefore $\pm\theta_1 = T_0/2\cdot53 = 0\cdot0063/2\cdot53 = 2\cdot5 \times 10^{-3}\,\text{rad.},$

$$\pm\theta_2 = \Delta_2\theta_1 = 0\cdot43 \times 2\cdot5 \times 10^{-3} = 1\cdot075 \times 10^{-3}\,\text{rad.}$$

Total deflexion of shaft θ_{sh}. From the vector diagram of Fig. 2c, we have

$$2\cdot14/K = \Delta_{sh} = 2\cdot14/3 = 0\cdot71, \quad \theta_{sh} = \Delta_{sh} \times \theta_1 = 1\cdot78 \times 10^{-3}\,\text{rad.}$$

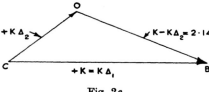

Fig. 2c.

[From the equation of motion of mass no. 1, it is seen that

$$\overrightarrow{K} = K\overrightarrow{\theta_1}/|\theta_1| = K\overrightarrow{\Delta_1}, \quad \text{with} \quad \Delta_1 = 1.$$

Therefore, $+K$ is in the direction of Δ_1.]

Combined diagrams. Referring again to Figs. 2a and 2b, one sees that for the construction of the latter the lines $\overline{OC}$ and $\overline{CB}$ are not needed, provided that the polygon of Fig. 2a is replaced by the triangle OAB, closed by the line $\overline{BO}$. The combined diagram, replacing these two figures, is shown in Fig. 2d.

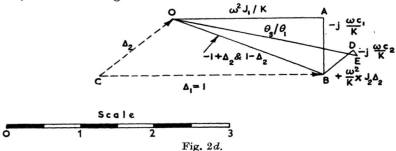

Fig. 2d.

It should be noted that all the values in the diagram can be expressed as non-dimensional quantities, if we divide the torques by $K\theta_1$ and the vibration amplitudes by θ_1. The two equations determining the diagram are then as follows:

$$\omega^2 \frac{J_1}{K} - j\frac{\omega c_1}{K} - 1 + \Delta_2 = 0,$$

which gives Δ_2 directly, and

$$-\Delta_2 + 1 + \omega^2 \frac{J_2 \Delta_2}{K} - j\frac{\omega c_2 \Delta_2}{K} + \frac{\theta_s}{\theta_1} = 0,$$

where $\theta_s = T_0/K$. These non-dimensional quantities are indicated in Fig. 2d.

Torques acting at both masses. If there is a torque T_1 acting on mass J_1, as well as a torque T_2 acting on J_2, the procedure for obtaining the vibration amplitudes is as follows (see Fig. 3):

1. Assume $T_1 = 0$ and construct the diagrams, first for J_1, then for J_2, as shown above. This gives the vibration amplitudes $\theta_{1(2)}$ and $\theta_{2(2)}$ due to the action of the torque T_2.

2. Assume $T_2 = 0$ (and $T_1 \neq 0$) and construct the diagrams, first for J_2, then for J_1, in the same manner as previously. This determines the vibration amplitudes $\theta_{1(1)}$ and $\theta_{2(1)}$ due to T_1.

3. Finally, the resultant amplitudes are obtained by vector addition of the two contributions:

$$\theta_1 = \theta_{1(1)} + \theta_{1(2)}, \qquad \theta_2 = \theta_{2(1)} + \theta_{2(2)},$$

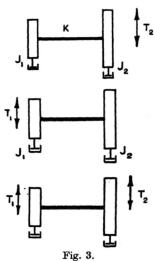

Fig. 3.

[390]

that is, by means of diagrams such as that of Fig. 2c, but using the θ-values instead of $K\Delta$-values ('Principle of superposition').

Note. The above applies when T_1 and T_2 are in phase. With torques in antiphase: $\pm\theta_1 = \theta_{1(1)} - \theta_{1(2)}$ and $\pm\theta_2 = \theta_{2(1)} - \theta_{2(2)}$.

For torques with a mutual phase angle ϕ, the vectors $\theta_{1(2)}$ and $\theta_{2(2)}$ should be rotated from their initial positions (determined by the diagrams) through this angle ϕ, before being compounded with $\theta_{1(1)}$ and $\theta_{2(1)}$, respectively, in order to obtain the resultant values.

Remark on damping coefficients. These diagrams are based on the use of damping coefficients c_1 and c_2, where c_1 refers to the driven machine and c_2 represents engine damping.

The value of c_2 can be evaluated as follows: Let M = dynamic magnifier of engine, obtained from tests or from statistical data; then

$$c_2 = \omega_0 \Sigma(J_{\text{cyl.}}\Delta^2_{\text{cyl.}})/M \quad \text{[Lb.in.sec./rad.]},$$

where ω_0 = natural phase velocity [rad./sec.] of engine system, and $\Sigma(J_{\text{cyl.}}\Delta^2_{\text{cyl.}})$ = sum of cylinder inertias multiplied by their corresponding Holzer-table amplitudes squared [Lb.in.sec.^{2}]. (In metric units, with $J_{\text{cyl.}}$ in Kg.cm.sec.^{2}, c is in Kg.cm.sec./rad.)

The value of c_1 depends on the driven machine (generator, water brake, propeller, etc.). For a preliminary assessment, a trial value of $c_1 = 0.2c_2$ may be used, except in the case of propellers. For numerical values of c_1 for various types of driven machines, see section 2·42.

The engine masses in these graphical methods for two- or three-mass systems are also represented by $J_2 = \Sigma(J_{\text{cyl.}}\Delta^2_{\text{cyl.}})$. The two-mass diagrams, as well as the three-mass diagrams given below, are convenient means of assessing the vibration amplitudes on the flanks of the resonance curve, using corresponding values of ω for the forced frequency $F = (60/2\pi) \times \omega$ [vib./min.]. For resonant conditions, the natural phase velocity ω_0 should be determined first, from formulae or by means of the usual Holzer tables.

2·433   Three-mass system

The method is simply an extension of the procedure used for the two-mass system. The details are indicated in the following numerical example.

The system considered consists of a driven machine of inertia J_1, a flywheel J_2, and an engine with a moment of inertia represented by J_3. These masses are coupled together by means of shaft lengths of stiffness K_1 and K_2, respectively (see Fig. 4). It is assumed that damping is present only in the driven machine (c_1) and in the engine (c_3). The

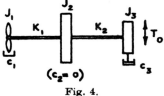

Fig. 4.

example will be based on the following numerical values (all torques [Lb.in.] being divided by 10^6):

Mass no. 1	Mass no. 2	Mass no. 3
$\omega^2 J_1 = 2$	$\omega^2 J_2 = 4$	$\omega^2 J_3 = 1 \cdot 2$
$K_1 = 3$	$K_1 = 3$	$K_2 = 2$
$c_1 \omega = 0 \cdot 8$	$K_2 = 2$	$c_3 \omega = 0 \cdot 2$
	$c_2 = 0$	$T_0 = 0 \cdot 0245$

Mass no. 1. The diagram (Fig. 5a) is constructed by means of the equation

$$\omega^2 J_1 - j\omega c_1 - K_1 + K_1 \Delta_2 = 0.$$

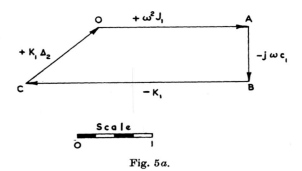

Fig. 5a.

The polygon $OABCO$ gives

$$\overline{CO} = K_1 \Delta_2 = 1 \cdot 28,$$

from which

$$\Delta_2 = 1 \cdot 28 / K_1 = 1 \cdot 28 / 3 = 0 \cdot 43.$$

Mass no. 2. The diagram (see Fig. 5b) is constructed by means of the relation

$$K_1 - K_1 \Delta_2 + \omega^2 J_2 \Delta_2 - j\omega c_2 \Delta_2 - K_2 \Delta_2 + K_2 \Delta_3 = 0.$$

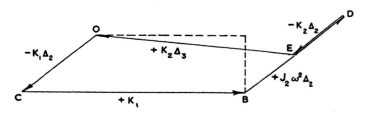

Fig. 5b.

[392]

As Δ_2 is a known value, we have:

$$K_1 \Delta_2 = 1 \cdot 28,$$

$$\omega^2 J_2 \Delta_2 = 4 \times 0 \cdot 43 = 1 \cdot 72,$$

$$\omega c_2 \Delta_2 = 0,$$

$$K_2 \Delta_2 = 2 \times 0 \cdot 43 = 0 \cdot 86.$$

The polygon $OCBDEO$ gives

$$\overline{EO} = K_2 \Delta_3 = 2 \cdot 74,$$

hence $\qquad \Delta_3 = 2 \cdot 64 / K_2 = 2 \cdot 74 / 2 = 1 \cdot 37.$

Mass no. 3. Having determined the value of Δ_3, we evaluate the quantities related to mass no. 3, and obtain

$$\omega^2 J_3 \Delta_3 = 1 \cdot 2 \times 1 \cdot 37 = 1 \cdot 64, \quad c_3 \omega \Delta_3 = 0 \cdot 2 \times 1 \cdot 37 = 0 \cdot 27.$$

The equation for this mass is

$$+ K_2 \Delta_2 - K_2 \Delta_3 + \omega^2 J_3 \Delta_3 - j \omega c_3 \Delta_3 + \frac{T_0}{\theta_1} = 0,$$

and the corresponding diagram is shown in Fig. 5c.

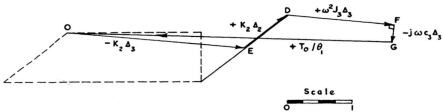

Fig. 5c. Note that $\overline{DF}$ is parallel to $\overline{OE}$, and that $\overline{FG}$ is perpendicular to $\overline{DF}$ (in the clockwise direction).

From the polygon $OEDFGO$, we have:

$$\overline{GO} = T_0 / \theta_1 = 4 \cdot 9.$$

Therefore

$$\pm \theta_1 = T_0 / 4 \cdot 9 = 0 \cdot 0245 / 4 \cdot 9 = 0 \cdot 005 \text{ rad.},$$

$$\pm \theta_2 = \Delta_2 \theta_1 = 0 \cdot 43 \times 0 \cdot 005 = 2 \cdot 15 \times 10^{-3} \text{ rad.},$$

$$\pm \theta_3 = \Delta_3 \theta_1 = 1 \cdot 37 \times 0 \cdot 005 = 6 \cdot 85 \times 10^{-3} \text{ rad.}$$

Diagram of vibration amplitudes. The diagram of the vibration amplitudes of the damped system at the forced frequency considered can be obtained by replotting from a common origin O vectors in the directions of Δ_1, Δ_2 and Δ_3, as determined from the above diagrams. The result is shown in Fig. 6.

[393]

The Holzer table for the undamped system does not give exactly the same relative amplitudes. Details of the tabulation are given below:

Mass no.	$J\omega^2/10^6$	Δ	$J\omega^2\Delta/10^6$	$\Sigma/10^6$	$K/10^6$	Δ_{sh}
1	2·00	1·00	2·00	2·00	3·00	0·67
2	4·00	0·33	1·32	3·32	2·00	1·66
3	1·20	−1·33	−1·60	+1·72	—	—

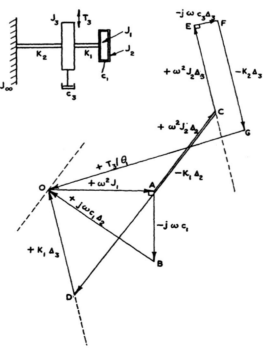

Fig. 6.

Further examples. The diagrams in Fig. 7 and 8 show how this graphical method can be applied to systems including a viscous fluid damper or a fluid flywheel. The equations, by means of which the diagrams are constructed, are also indicated.

2·434   *System with damper*

Fig. 7. $J_1 =$ inner mass of damper. $J_2 =$ damper casing. $c_1 =$ damping coefficient of damper. $J_3 =$ engine inertia. $J_\infty =$ main flywheel inertia (regarded as infinite). $T_3 =$ excitation torque. K_1, $K_2 =$ shaft stiffnesses. $c_3 =$ damping coefficient of engine.

($\Delta_1 = 1$) Equations

$$\omega^2 J_1 - j\omega c_1 + j\omega c_1 \Delta_2 = 0$$

$$-j\omega c_1 \Delta_2 + j\omega c_1 + \omega^2 J_2 \Delta_2 - K_1 \Delta_2 + K_1 \Delta_3 = 0$$

$$K_1 \Delta_2 - K_1 \Delta_3 + \omega^2 J_3 \Delta_3 - j\omega c_3 \Delta_3 - K_2 \Delta_3 + \frac{T_3}{\theta_1} = 0.$$

Diagram construction

$$\overline{OA} + \overline{AB} + \overline{BO} = 0 \quad (\overline{BO} \text{ gives } \Delta_2)$$

$$\overline{OB} + \overline{BA} + \overline{AC} + \overline{CD} + \overline{DO} = 0 \quad (\overline{AC} \text{ perpendicular to } \overline{OB} \text{ in anti-}$$
$$\text{clockwise direction: } \overline{DO} \text{ gives } \Delta_3)$$

$$\overline{OD} + \overline{DC} + \overline{CE} + \overline{EF} + \overline{FG} + \overline{GO} = 0 \quad (\overline{CE} \text{ parallel to } \overline{DO}; \, \overline{GO} \text{ gives } T_3/\theta_1).$$

From $\overline{GO} = +T_3/\theta_1$, it follows that $\theta_1 = T_3/\overline{GO}$, $\theta_2 = \Delta_2 \theta_1$ and $\theta_3 = \Delta_3 \theta_1$. Directions of amplitude vectors Δ_1, Δ_2 and Δ_3 are parallel to $\overrightarrow{OA}$, $\overrightarrow{AC}$ and $\overrightarrow{CE}$, respectively.

2·435 *System with fluid coupling*

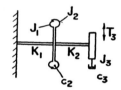

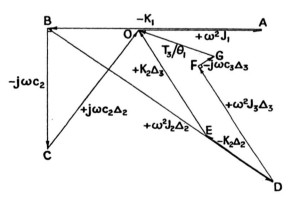

Fig. 8. J_1=driven portion of coupling. J_2=driving portion of coupling. J_3=engine inertia. c =damping coefficient of coupling. T_3=excitation torque. K_1, K_2=shaft stiffnesses. c_3=damping coefficient of engine.

($\Delta_1 = 1$) Equations

$$\omega^2 J_1 - K_1 - j\omega c_2 + j\omega c_2 \Delta_2 = 0$$

$$-j\omega c_2 \Delta_2 + j\omega c_2 + \omega^2 J_2 \Delta_2 - K_2 \Delta_2 + K_2 \Delta_3 = 0$$

$$-K_2 \Delta_3 + K_2 \Delta_2 + \omega^2 J_3 \Delta_3 - j\omega c_3 \Delta_3 + \frac{T_3}{\theta_1} = 0$$

Diagram construction

$$\overline{OA} + \overline{AB} + \overline{BC} + \overline{CO} = 0 \quad (\overline{CO} \text{ determines } \Delta_2)$$

$$\overline{OC} + \overline{CB} + \overline{BD} + \overline{DE} + \overline{EO} = 0 \quad (\overline{BD} \text{ perpendicular to } \overline{OC} \text{ in anti-clockwise direction; } EO \text{ gives } \Delta_3)$$

$$\overline{OE} + \overline{ED} + \overline{DF} + \overline{FG} + \overline{GO} = 0 \quad (\overline{DF} \text{ parallel to } \overline{EO}; \overline{GO} \text{ gives } T_3/\theta_1).$$

The directions of the amplitude vectors are given by the inertia torques.

Note. These examples are given only as illustrations of the procedure. For a detailed treatment of dampers, see section 3; for fluid couplings, see section 2·421 and section 1·247.

2·436 *System with torque acting on intermediate mass*

In the systems hitherto considered, the torque was applied at one of the end-masses J_1 or J_3. If the torque acts on the intermediate mass J_2 (see Fig. 9), the procedure, to be employed is as follows:

(*a*) Construct the diagram for mass no 1, using the equation

$$+\omega^2 J_1 - j\omega c_1 - K_1 + K_1 \Delta_2 = 0.$$

This gives the values of $K_1 \Delta_2$ and $\Delta_2 = |\theta_2/\theta_1|$, as well as the direction of θ_2 relative to θ_1 (see Fig. 10*a*).

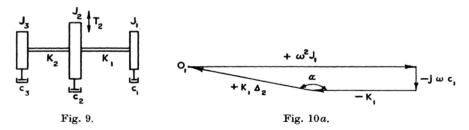

Fig. 9.   Fig. 10*a*.

(*b*) Construct the diagram for mass no. 3 by means of

$$+\omega^2 J_3 - j\omega c_3 - K_2 + K_2 \Delta_2^* = 0.$$

This gives $K\Delta_2^*$ and $\Delta_2^* = |\theta_2/\theta_3|$, as well as the direction of θ_3 relative to θ_2 (Fig. 10*b*).

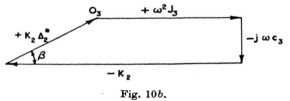

Fig. 10*b*.

Evaluate from Figs. 10a and 10b the ratio

$$\frac{K_1\Delta_2}{K_2\Delta_2^*} = \frac{K_1}{K_2} \times \frac{(\theta_2/\theta_1)}{(\theta_2/\theta_3)} = \frac{K_1}{K_2} \times \frac{|\theta_3|}{|\theta_1|} = r,$$

and determine

$$\Delta_3 = \left|\frac{\theta_3}{\theta_1}\right| = r \times \frac{K_2}{K_1}.$$

(c) Construct finally the diagram of the intermediate mass J_2, on which the torque is applied, by using the equation

$$K_1 - K_1\Delta_2 + \omega^2 J_2\Delta_2 - j\omega c_2\Delta_2 - K_2\Delta_2 + K_2\Delta_3 + \left(\frac{T_2}{\theta_1}\right) = 0,$$

taking account of the phase angles α and β between Δ_1 and Δ_2, and Δ_2 and Δ_3, respectively. The resultant (see Fig. 10c) gives the value of $x = T_2/\theta_1$, hence $\theta_1 = T_2/x$, $\theta_2 = \Delta_2\theta_1$ and $\theta_3 = \Delta_3\theta_1$.

A numerical example is given at the end of this section.

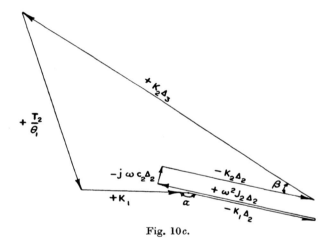

Fig. 10c.

Note. The procedure can be applied to a system with more than three masses.

In this case, the diagrams are constructed from the end-masses, up to the masses on either side of the intermediate mass subjected to external torque. After relating the amplitudes on either side, the diagram of the intermediate mass is built up and gives T/θ_1.

If the system has more than one intermediate mass subjected to external torque, the above procedure is employed first for T_A acting on J_A, then repeated for T_B acting on J_B, etc. (Principle of Superposition). The resultant vibration amplitudes are

$$\theta_{1\,\text{res.}} = \theta_{1A} + \theta_{1B} + \ldots, \quad \theta_{2\,\text{res.}} = \theta_{2A} + \theta_{2B} + \ldots, \quad \text{etc.}$$

NUMERICAL EXAMPLE. We consider a three-mass system with the torque T_2 acting on the intermediate mass J_2. The numerical values related to each mass are given below. [All torques in Lb.in. are divided by 10^6.]

Mass no. 1   Mass no. 2   Mass no. 3

$\omega^2 J_1 = 6\cdot0$ $\omega^2 J_2 = 3\cdot0$ $\omega^2 J_3 = 1\cdot0$

$\omega c_1 = 0\cdot2$ $\omega c_2 = 0\cdot1$ $\omega c_3 = 0\cdot2$

$K_1 = 2\cdot5$ $K_1 = 2\cdot5$ $K_2 = 2\cdot0$

              $K_2 = 2\cdot0$

              $T_2 = 0\cdot0154$

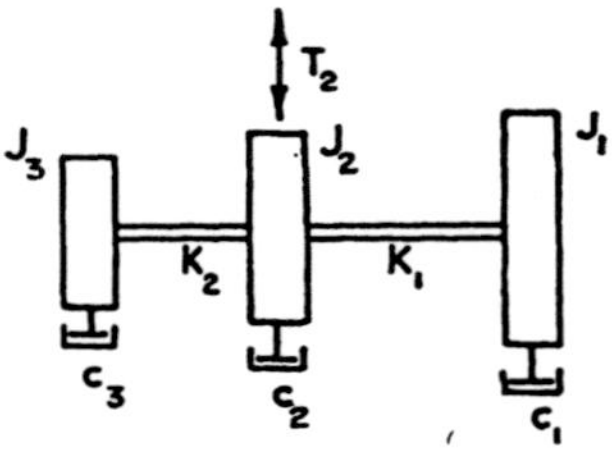

Fig. 11.

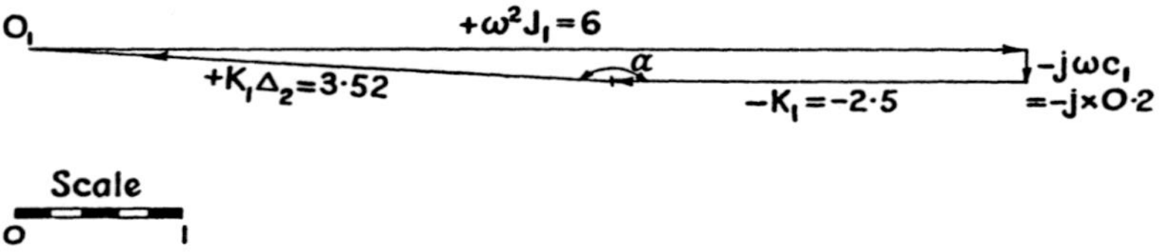

Fig. 12. Diagram of mass no. 1 (based on equation: $\omega^2 J_1 - j\omega c_1 - K_1 + K_1 \Delta_2 = 0$).

This gives $\theta_2/\theta_1 = \Delta_2 = 3\cdot52/K_1 = 3\cdot52/2\cdot5 = 1\cdot41.$

$0\cdot2/3\cdot52 = 0\cdot0568 = \sin(180° - \alpha), \quad 180° - \alpha = 3° 15', \quad \alpha = 176° 45'.$

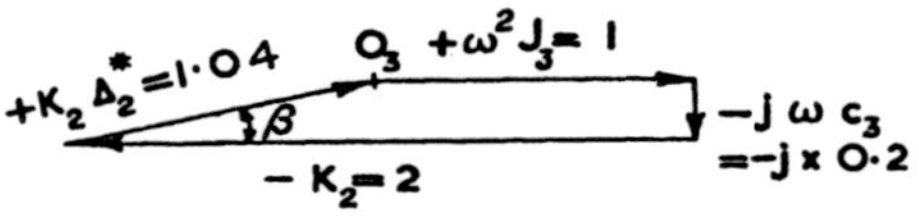

Fig. 13. Diagram of mass no. 3 (based on equation: $\omega^2 J_3 - j\omega c_3 - K_2 + K_2 \Delta_2^* = 0$).
[In this diagram, $+K_2$ is in the direction of $\Delta_3^* = 1$.]
$\sin\beta = 0\cdot2/1\cdot04 = 0\cdot1924, \quad \beta = 11° 6'.$

This gives $\dfrac{K_1(\theta_2/\theta_1)}{K_2(\theta_2/\theta_3)} = \dfrac{3\cdot52}{1\cdot04} = \dfrac{K_1}{K_2} \times \dfrac{\theta_3}{\theta_1} = 3\cdot38,$

hence $\Delta_3 = \theta_3/\theta_1 = 3\cdot38 \times K_2/K_1 = 2\cdot7.$

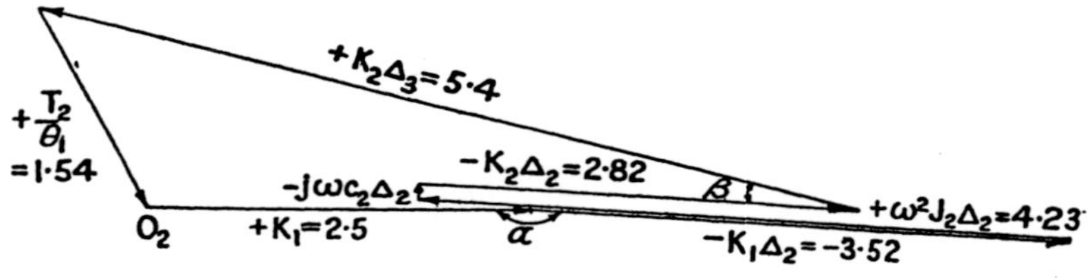

Fig. 14. Diagram of mass no. 2 (from equation:

$$K_1 - K_1 \Delta_2 + \omega^2 J_2 \Delta_2 - j\omega c_2 \Delta_2 - K_2 \Delta_2 + K_2 \Delta_3 + \left(\frac{T_2}{\theta_1}\right) = 0).$$

$\omega^2 J_2 \Delta_2 = 3 \times 1\cdot41 = 4\cdot23, \quad \omega c_2 \Delta_2 = 0\cdot1 \times 1\cdot41 = 0\cdot14,$

$K_2 \Delta_2 = 2 \times 1\cdot41 = 2\cdot82, \quad K_2 \Delta_3 = 2 \times 2\cdot7 \quad = 5\cdot4.$

[398]

This diagram gives $T_2/\theta_1 = 1.54$. Therefore

$$\pm \theta_1 = T_2/1.54 = 0.01\,\text{rad.,} \qquad \pm \theta_2 = \Delta_2 \times \theta_1 = 0.014\,\text{rad.,} \qquad \pm \theta_3 = \Delta_3 \theta_1 = 0.027\,\text{rad.}$$

Moreover, from the vibration amplitudes diagram, the shaft deflexions are $\theta_{12} = 0.024$ and $\theta_{23} = 0.0136\,\text{rad.}$

Fig. 15. Vibration amplitudes diagram. The directions of θ_1, θ_2 and θ_3 are those of $+\Delta_1$, $+\Delta_2$ and $+\Delta_3$ in the diagram of mass no. 2, with $+\Delta_1$ parallel to $+K_1$.

The Holzer tabulation for the system without damping is given below.

Mass no.	$J\omega^2/10^6$	Δ	$J\omega^2\Delta/10^6$	$\Sigma/10^6$	$K/10^6$	Δ_{sh}
1	6·0	1·00	6·00	6·00	2·5	2·4
2	3·0	−1·40	−4·20	1·80	2·0	0·9
3	1·0	−2·30	−2·30	−0·50	—	—

General remarks on semi-graphical methods

In the previous pages, a basic procedure has been described which is applicable to a wide variety of systems. The details were worked out at the B.I.C.E.R.A. Laboratory with a view to obtaining a method which would be simple to explain and to apply. After becoming familiar with its main features, it is easy to introduce non-dimensional ratios and other variants in the sequence of construction of the diagrams.

The accuracy of semi-graphical methods depends on the size of the diagrams and on the extent to which computations are carried out. In the extreme case, all the diagrams could be replaced by computations of the x- and y-axis components of the various vectors. The diagrams then serve mainly as a check and for a visualization of the results.

Various methods using diagram constructions have been evolved for the solution of vibration problems and some are listed in the bibliography given below.

BIBLIOGRAPHY

Archer, S. Contribution to improved accuracy in the calculation and measurement of torsional vibration stresses in marine propeller shafting. *Proc. Instn Mech. Engrs, Lond.*, vol. 164, no. 3 (1951), pp. 351–66. [Diagrams used in conjunction with analytical formulae, see section 2·45.]

Bishop, R. E. D. and Welbourn, D. B. The problem of the dynamic vibration absorber. *Engineering*, 19 December 1952, p. 796.

Bishop, R. E. D. On the graphical solution of transient vibration problems. *Proc. Instn Mech. Engrs, Lond.*, vol. 168, no. 10 (1954), pp. 229–322.

Cole, E. B. *Theory of Vibrations for Engineers* (Crosby, Lockwood and Son, Ltd., London, 1950), pp. 182–9.

Den Hartog, J. P. *Mechanical Vibrations*, 2nd ed. (McGraw-Hill Book Co. 1940), pp. 61–70.

Draminsky, P. *Daempningen ved torsionssvingninger i krumtapaksler (Damping of Torsional Vibrations in Crankshafts)*, (Nyt Nordisk Forlag Arnold Busck, Copenhagen, 1947), pp. 14–30.

Everitt, W. L. *Communication Engineering* (McGraw-Hill Book Co. 1937), pp. 53–74.

Hansen, H. M. and Chenea, P. F. *Mechanics of Vibration* (John Wiley and Sons, Inc. 1952), pp. 9–14, 82–6 and 138–84.

Hansen, S. Forced vibrations in two-mass systems. *Premier Congrès International des Moteurs à Combustion Interne*, Paris, May 1951, vol. 1, pp. 345–61.

Inglis, Sir C. *Applied Mechanics for Engineers* (Cambridge University Press, 1951), p. 311.

Kurzemann, W. Semi-graphical method for the study of damped three-mass systems. *Engrs Digest*, November 1953, pp. 421–3.

Marquard, E. *Schwingungsdynamik des schnellen Strassenfahrzeugs (Dynamics of Vibration of High-speed Vehicles)* (W. Girardet Verlag, Essen, 1952), pp. 98–100.

Myklestad, N. O. *Vibration Analysis* (McGraw-Hill Book Co. 1944), pp. 106–9.

Robertson, D. Vector methods of studying mechanical vibrations. *Engineer*, vol. 61 (1931), pp. 230–1, 256–7, 288–9 and 314–5.

Thomson, W. T. *Mechanical Vibrations* (George Allen and Unwin, Ltd.), pp. 62–71.

Trendlenburg, F. *Einführung in die Akustik (Introduction to Acoustics)* (Springer-Verlag, Berlin, 1950), p. 27.

2·44 Tabulation methods for systems with gas torque and damping

2·441 *The Holzer method*

The tabulation method for systems with gas torque and damping was first used by Holzer† and subsequently developed by Spaetgens.‡ It has since been extended somewhat in order that it may be used for determining both vibration amplitudes and stresses, under resonant and non-resonant conditions.

The method is a step-by-step evaluation based on the equations of motion of the system, as can be seen from the derivation of the method, which is outlined on p. 411. It gives an exact treatment, which is more convenient than graphical methods, and, particularly for the determination of large flank amplitudes of minor orders, the evaluation is comparatively more rapid and more easily checked at any stage than by any other methods hitherto developed.

In the following, the method is described with the aid of numerical examples, viz.: two-mass system, pp. 401–410; eight-mass system: minor order, pp. 413–424; major order, pp. 413–420 and 424–426.

The evaluations for a representative multi-mass system are given with full details, so as to enable a standard procedure to be followed in practical work.

† Holzer, H., *Calculation of Torsional Vibrations (Die Berechnung der Drehschwingungen)*, (Springer, Berlin, 1922); in particular pp. 111, 153, 157 and 181.
‡ Spaetgens, T. W., 'Holzer method for forced-damped torsional vibrations', *J. Appl. Mech.* March 1950, pp. 57–63.

Note. Although this method gives a detailed and correct treatment of multi-mass systems, it should be borne in mind that, like all other methods, it depends on a knowledge of the various damping factors in an engine system for the prediction of vibration amplitudes and stresses. These are still uncertain, except in special cases (systems with marine propellers, viscous-fluid dampers, etc.). See sections 2·421, 2·422 and 3·3.

2·442 Holzer tabulation for systems with gas torque and damping

Basic procedure

Details of the procedure will be given as follows:

(a) Tabulation,

(b) Vibration amplitude at mass no. 1,

(c) Vibration amplitudes at other masses,

(d) Shaft deflexions,

(e) Stresses,

(f) Vector diagram of vibration amplitudes and torques.

In general, only (a), (b), (d) and (e) are required, but the other two calculations help to complete the picture of the vibration conditions. The advantages of the method are mainly apparent with multi-mass systems, which can be treated correctly in this manner. In work of this kind, an electrically operated desk-type calculating machine will be found useful.

For convenience, the procedure will be described by considering a simple example of a two-mass system, with the following values (see also Fig. 1):†

Inertias: $J_1 = 480$ Lb.in.sec.^{2},

$J_2 = 800$ Lb.in.sec.^{2}.

Shaft stiffness: $K = 3 \times 10^6$ Lb.in./rad.

Smallest shaft diameter: $D = 6·0$ in.

Gas torque: $T_1 = 6000$ Lb.in.

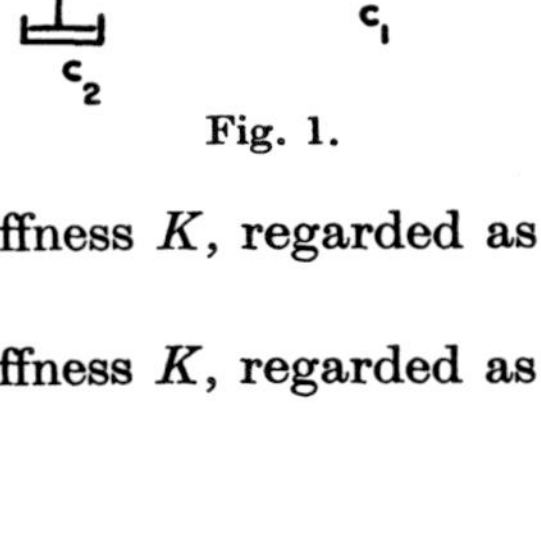

Fig. 1.

Dynamic magnifiers:

$M_1 = 80$ (magnifier related to mass J_1 and stiffness K, regarded as a one-mass system).

$M_2 = 40$ (magnifier related to mass J_2 and stiffness K, regarded as a one-mass system).

(a) *Tabulation.* The procedure is as follows:

(1) Determine the natural frequency of the system without damping, by means of an ordinary Holzer table.

† *Metric units.* This numerical example is also valid in metric units, without any changes in the numerical values, if 'Lb.' is replaced by 'Kg.' and 'in.' by 'cm.'. Thus $J_1 = 480$ Kg.cm.sec.^{2}, $T_1 = 6000$ Kg.cm., $c_1 = 600$ Kg.cm.sec./rad., etc.

Details of this table are given below.

$F = 955$ vib./min., $\omega = 100$ rad./sec., $\omega^2 = 0.01 \times 10^6$.

J	$\dfrac{J\omega^2}{10^6}$	Δ	$\dfrac{J\omega^2\Delta}{10^6}$	$\dfrac{\Sigma J\omega^2\Delta}{10^6}$	$\dfrac{K}{10^6}$	Δ_{sh}
480	4.80	1.00	4.80	4.80	3.0	1.60
800	8.00	-0.60	-4.80	0.00	—	—

(2) Determine the damping coefficients associated with each mass:

$$c_1 = \frac{J_1\omega}{M_1} \quad \text{and} \quad c_2 = \frac{J_2\omega}{M_2}.$$

The damping coefficients are

$$c_1 = \frac{480 \times 100}{80} = 600 \quad \text{[Lb.in.sec./rad.]},$$

$$c_2 = \frac{800 \times 100}{40} = 2000 \quad \text{[Lb.in.sec./rad.]}.$$

(3) Express the gas torque as

$$T/10^6 = T' + jT''$$
$$= T_0 \cos\phi + jT_0 \sin\phi,$$

where $j^2 = -1$ and $\phi = $ unknown phase angle between T and the vibration amplitude θ_1 at mass no. 1.

The reduced gas torque is

$$T_0 = T/10^6 = 6000/10^6 = 0.006,$$

and its components are

$$T' = T_0 \cos\phi = 0.006 \cos\phi,$$

and
$$T'' = T_0 \sin\phi = 0.006 \sin\phi,$$

where ϕ will be determined later.

(4) Evaluate the quantities

$$J_1\omega^2 - j\omega c_1 \quad \text{and} \quad J_2\omega^2 - j\omega c_2.$$

The evaluation gives:

$$J_1\omega^2 - j\omega c_1 = 480 \times 0.01 \times 10^6 - j100 \times 600$$
$$= (4.8 - j0.06) \times 10^6,$$

$$J_2\omega^2 - j\omega c_2 = 800 \times 0.01 \times 10^6 - j100 \times 2000$$
$$= (8.0 - j0.2) \times 10^6.$$

(5) Prepare a table as shown below and fill in with the numerical values already available:

$$F = 955 \text{ vib./min.}, \quad \omega = 100 \text{ rad./sec.}, \quad \omega^2 = 0.01 \times 10^6 \text{ (rad./sec.)}^2.$$

(1) Mass no.	(2) $\dfrac{J\omega^2 - j\omega c}{10^6}$ [Lb.in./rad.]	(3) θ [rad.]	(4) $\dfrac{J\omega^2 - j\omega c}{10^6} \times \theta$ [Lb.in.]	(5) $\dfrac{T}{10^6}$ [Lb.in.]	(6) $\dfrac{\Sigma[J\omega^2\theta - j\omega c\theta + T]}{10^6}$ [Lb.in.]	(7) $\dfrac{K}{10^6}$ [Lb.in./rad.]	(8) $\theta_{sh} = \dfrac{(6)}{(7)}$ [rad.]
1	$4.8 - j0.06$	θ_1	$4.8\theta_1 - j0.06\theta_1$	$T' + jT''$	$4.8\theta_1 + T' + j(-0.06\theta_1 + T'')$	3.0	—
2	$8.0 - j0.20$	—	—	—	—	—	—

Note. In column (5) it is sufficient to write $T' + jT''$, bearing in mind that the corresponding numerical values will be used after the tabulation is completed.

(6) Evaluate column (8), line 1, and columns (3) and (4), line 2 [using the multiplication rule for complex numbers:

$$(a + jA)(b + jB) = (ab - AB) + j(Ab + aB)$$
$$= c + jC].$$

Evaluation:
$$\theta_{sh} = [(4.8\theta_1 + T') + j(-0.06\theta_1 + T'')] \times \frac{1}{3.0}$$

$$= (1.6\theta_1 + 0.33T') + j(-0.02\theta_1 + 0.33T'').$$

Column (3), line 2:

$$\theta_2 = \theta_1 - \theta_{sh}$$
$$= (\theta_1 - 1.6\theta_1 - 0.33T') + (0j - j[-0.02\theta_1 + 0.33T'']),$$
$$\theta_2 = (-0.6\theta_1 - 0.33T') + j(0.02\theta_1 - 0.33T'').$$

Column (4), line 2:
$$a = 8.0,$$
$$A = -0.20,$$
$$b = -0.6\theta_1 - 0.33T',$$
$$B = 0.02\theta_1 - 0.33T'';$$

hence $c = (ab) - (AB) = (-4.8\theta_1 - 2.64T') - (-0.004\theta_1 + 0.066T'')$
$$= -4.796\theta_1 - 2.64T' - 0.066T''.$$

$$jC = j[(Ab) + (aB)] = j[(+0.12\theta_1 + 0.066T') + (0.16\theta_1 - 2.64T'')]$$
$$= j[0.28\theta_1 + 0.066T' - 2.64T''].$$

We thus have the number $c + jC$ of column (4), line 2.

(7) Add the contributions of columns (4) and (5), line 2, to the terms of column (6) of the previous line, and write the results in column (6), line 2.

In this example, there are no gas torque contributions on line 2.

26-2

Hence, the summation for column (6) of line 2 is as follows:

$$\Sigma = (4 \cdot 8\theta_1 + T') + j(-0 \cdot 06\theta_1 + T'') + c + jC,$$

or, written in full,

$$\Sigma = [4 \cdot 8\theta_1 + T' - 4 \cdot 796\theta_1 - 2 \cdot 64T' - 0 \cdot 066T'']$$
$$+ j[-0 \cdot 06\theta_1 + T'' + 0 \cdot 28\theta_1 - 2 \cdot 64T'' + 0 \cdot 066T']$$
$$= [0 \cdot 004\theta_1 - 1 \cdot 64T' - 0 \cdot 066T''] + j[0 \cdot 22\theta_1 - 1 \cdot 64T'' + 0 \cdot 066T']$$
$$= d + jD.$$

The tabulation is now complete, and resultant table is given below.

$F = 955$ vib./min., $\omega = 100$ rad./sec., $\omega^2 = 0 \cdot 01 \times 10^6$ (rad./sec.)^2.

(1) Mass no.	(2) $\dfrac{J\omega^2 - j\omega c}{10^6}$ [Lb.in./rad.]	(3) θ [rad.]	(4) $\dfrac{J\omega^2 - j\omega c}{10^6} \times \theta$ [Lb.in.]	(5) $\dfrac{T}{10^6}$ [Lb.in.]
1 {	$4 \cdot 8$ $-j0 \cdot 06$	θ_1 —	$4 \cdot 8\theta_1$ $-j0 \cdot 06\theta_1$	T' $+jT''$
2 {	$8 \cdot 0$ $-j0 \cdot 02$	$-0 \cdot 6\theta_1 - 0 \cdot 33T'$ $+j[0 \cdot 02\theta_1 - 0 \cdot 33T'']$	$-4 \cdot 796\theta_1 - 2 \cdot 64T'$ $-0 \cdot 066T''$ $+j[0 \cdot 28\theta_1$ $+0 \cdot 066T' - 2 \cdot 64T'']$	— —

(1) Mass no.	(6) $\dfrac{\Sigma[J\omega^2\theta - j\omega c\theta + T]}{10^6}$ [Lb.in.]	(7) $\dfrac{K}{10^6}$ [Lb.in./rad.]	(8) $\theta_{sh} = \dfrac{(6)}{(7)}$ [rad.]
1 {	$4 \cdot 8\theta_1 + T'$ $+j[-0 \cdot 06\theta_1 + T'']$	$3 \cdot 0$ —	$1 \cdot 6\theta_1 + 0 \cdot 33T'$ $+j[-0 \cdot 02\theta_1 + 0 \cdot 33T'']$
2 {	$0 \cdot 004\theta_1 - 1 \cdot 64T' - 0 \cdot 066T''$ $+j[0 \cdot 22\theta_1 - 1 \cdot 64T'' + 0 \cdot 066T']$	— —	— —

(b) *Vibration amplitude at mass no. 1.* For equilibrium conditions, the resultant torque acting on the last mass must be equal to zero ($\Sigma = 0$). This gives two equations, viz.: $d = 0$ and $D = 0$. The procedure is therefore as follows:

(1) Solve both these equations for θ_1, equate the results, and substitute in these the expressions for T' and T'' already calculated.

From these, determine the phase angle ϕ of the gas torque T of mass no. 1 relative to θ_1.

The equations are:

$$d \equiv 0 \cdot 004\theta_1 - 1 \cdot 64T' - 0 \cdot 066T'' = 0,$$

and

$$D \equiv 0 \cdot 22\theta_1 + 0 \cdot 066T' - 1 \cdot 64T'' = 0.$$

These give:

$$\theta_1 = \frac{1}{0 \cdot 004}[1 \cdot 64T' + 0 \cdot 066T''] \quad \text{and} \quad \theta_1 = \frac{1}{0 \cdot 22}[-0 \cdot 066T' + 1 \cdot 64T''].$$

Hence
$$-0{\cdot}066T' + 1{\cdot}64T'' = \frac{0{\cdot}22}{0{\cdot}004}[1{\cdot}64T' + 0{\cdot}066T'']$$

$$= 90{\cdot}1T' + 3{\cdot}63T'';$$

this gives
$$[-0{\cdot}066 - 90{\cdot}1]\,T' = [-1{\cdot}64 + 3{\cdot}63]\,T''$$

or
$$-90{\cdot}166T_0\cos\phi = +1{\cdot}99T_0\sin\phi.$$

Consequently
$$\tan\phi = -\frac{90{\cdot}166}{1{\cdot}99} = -45{\cdot}3,$$

$$\tan(180° - \phi) = +45{\cdot}3,$$

and
$$180° - \phi = 88°\,44'; \quad \text{so that} \quad \underline{\phi = 91°\,16'.}$$

(2) Evaluate $\cos\phi$ and $\sin\phi$.

Substitute these values in either of the resultant-torque equations ($d = 0$ or $D = 0$) and thus determine the vibration amplitude θ_1.

From $\phi = 91°\,16'$, we have:†

$$\sin[91°\,16'] = \sin[90° + 1°\,16'] = \cos[1°\,16'] = 0{\cdot}99976,$$

$$\cos[91°\,16'] = -\sin[1°\,16'] = -0{\cdot}02211.$$

Introducing these values in $d = 0$, we obtain (with $T' = 0{\cdot}006\cos\phi$ and $T'' = 0{\cdot}006\sin\phi$)
$$0{\cdot}004\theta_1 - 1{\cdot}64T' - 0{\cdot}066T'' = 0,$$

therefore
$$0{\cdot}004\theta_1 = 0{\cdot}006[1{\cdot}64 \times \cos\phi - 0{\cdot}066 \times \sin\phi]$$

and
$$\theta_1 = \frac{0{\cdot}006}{0{\cdot}004}[1{\cdot}64 \times (-0{\cdot}02211) - 0{\cdot}066 \times 0{\cdot}99976]$$

$$= 1{\cdot}5 \times [-0{\cdot}0363 + 0{\cdot}0660]$$

$$= 1{\cdot}5 \times 0{\cdot}0297 = 0{\cdot}0445\ \text{rad.,}$$

or
$$\pm\,\theta_1 = 57{\cdot}3 \times 0{\cdot}0445 = \underline{2{\cdot}55°.}$$

The same value should be obtained by means of the second equation $D = 0$:
$$0{\cdot}22\theta_1 + 0{\cdot}066T' - 1{\cdot}64T'' = 0,$$

$$\theta_1 = \frac{0{\cdot}006}{0{\cdot}22}[-0{\cdot}066 \times (-0{\cdot}02211) + 1{\cdot}64 \times 0{\cdot}99976]$$

$$= 0{\cdot}0273[+0{\cdot}00146 + 1{\cdot}64] = 0{\cdot}0273 \times 1{\cdot}6415$$

$$= 0{\cdot}0448\ \text{rad.,}$$

or
$$\pm\,\theta_1 = 57{\cdot}3 \times 0{\cdot}0448 = \underline{2{\cdot}57°.}$$

† For ready reference: $\sin(90° \pm \phi) = \cos\phi$, $\cos(90° - \phi) = \sin\phi$, $\cos(90° + \phi) = -\sin\phi$.

We take the mean value, 2·56°, for further computations.

The agreement is within 0·02°, or $100 \times 0.02/2.56 = 0.8 \%$.

Besides giving a partial check of the calculation, this indicates the order of accuracy of the evaluation, which was made on a small-size cylindrical slide rule (of 60 inches developed length) without more than normal care in setting up.

(c) *Vibration amplitude at mass no. 2*

(1) Substitute the values of θ_1 (in radians), T' and T'' in the complex-number expression in column (3) of line 2.

The values to be used are:

$$\theta_1 = 0.0447 \text{ rad.},$$

$$T' = 0.006 \cos \phi = 0.006 \times (-0.0211) = -0.00013,$$

$$T'' = 0.006 \sin \phi = 0.006 \times 0.99976 = +0.006$$

and the expression for θ_2 taken from the table on the previous tabulation is

$$\theta_2 = [-0.6\theta_1 - 0.33T'] + j[0.02\theta_1 - 0.33T''].$$

We can thus substitute directly and obtain

$$\theta_2 = [-0.6 \times 0.0447 - 0.33 \times (-0.00013)]$$

$$+ j[0.02 \times 0.0447 - 0.33 \times 0.006]$$

$$= [-0.02682 + 0.00004] + j[0.000894 - 0.00198]$$

$$= -0.02678 - j0.001086$$

$$\cong -0.02678 - j0.0011 = b + jB.$$

(2) Calculate the absolute value $|\theta_2|$ of θ_2, and the internal phase angle ψ_2 between θ_2 and θ_1.

$$|\theta_2| = \sqrt{(b^2 + B^2)}$$

$$= \sqrt{\{(0.02678)^2 + (0.0011)^2\}} = 10^{-2} \times \sqrt{\{(2.678)^2 + (0.11)^2\}}$$

$$= 10^{-2} \times \sqrt{(7.17 + 0.01)} = 0.0268 \text{ rad.},$$

or

$$\pm \theta_2 = 57.3 \times 0.0268 = \underline{1.54°}.$$

The phase angle ψ_2 is obtained as follows.

We calculate $\tan \alpha_2 = \dfrac{B}{b} = \dfrac{-0.0011}{-0.02678} = +0.0410,$

and since both b and B are negative, α_2 must† be in the 3rd quadrant; therefore

$$\alpha_2 - 180° = \tan^{-1}[0{\cdot}0410] = 2° \, 21',$$

or

$$\alpha_2 = 182° \, 21'.$$

Now, since α_2 is greater than $180°$, it is the 'external' angle. The 'internal' phase angle is therefore

$$\psi_2 = 360° - \alpha_2 = \underline{177° \, 39'}.$$

(d) *Shaft deflexion.* To obtain the shaft deflexion θ_{sh}, substitute the numerical values of θ_1, T' and T'' in the expression obtained in column (8), line 1, of the table.

Determine the absolute value $|\theta_{sh}|$.

We have

$$\theta_{sh} = [1{\cdot}6\theta_1 + 0{\cdot}33T''] + j[-0{\cdot}02\theta_1 + 0{\cdot}33T'']$$
$$= [1{\cdot}6 \times 0{\cdot}0447 + 0{\cdot}33 \times (-0{\cdot}00013)]$$
$$\qquad\qquad + j[-0{\cdot}02 \times 0{\cdot}0447 + 0{\cdot}33 \times 0{\cdot}006]$$
$$= [0{\cdot}071520 - 0{\cdot}000043] + j[-0{\cdot}000894 + 0{\cdot}00198]$$
$$= 0{\cdot}071516 + j0{\cdot}001086.$$

The absolute value is

$$|\theta_{sh}| = 10^{-2} \times \sqrt{\{(7{\cdot}152)^2 + (0{\cdot}109)^2\}} = 10^{-2} \times \sqrt{(51{\cdot}15 + 0{\cdot}01)}$$
$$= 0{\cdot}07184 \, \text{rad.},$$

and

$$\pm \theta_{sh} \simeq 57{\cdot}3 \times 0{\cdot}0718 \simeq \underline{4{\cdot}11°}.$$

(e) *Stresses*

(1) Determine the vibratory shaft torque from the relation

$$T_{sh} = K\theta_{sh} \quad [\text{Lb.in.}],$$

where θ_{sh} is the twist deflexion of the shaft with a stiffness K.

As, in this example,

$$K = 3 \times 10^6 \, \text{Lb.in./rad.} \quad \text{and} \quad \theta_{sh} = 0{\cdot}07184 \, \text{rad.},$$

---

† For ready reference: Let $r = |B/b| = $ absolute value of the ratio B/b. Then the 'internal phase angle' ψ of the vector $b + jB$ relative to the positive real axis can be determined by means of the following table:

	Sign of			
	b	B		
I	+	+	$\alpha = \tan^{-1} r$	$\psi = \tan^{-1} r$
II	−	+	$\alpha = 180° - \tan^{-1} r$	$\psi = 180° - \tan^{-1} r$
III	−	−	$\alpha = 180° + \tan^{-1} r$	$\psi = 180° - \tan^{-1} r$
IV	+	−	$\alpha = 360° - \tan^{-1} r$	$\psi = \tan^{-1} r$

In the numerical example given above, sign of b is −, sign of B is −, hence (case III) we have

$$\psi = 180° - \tan^{-1} r = 180° - \tan^{-1}[0{\cdot}0403] = 180° - 2° \, 18' = 177° \, 42'.$$

the vibratory shaft torque is

$$T_{sh} = 3 \times 7 \cdot 184 \times 10^4 = 215{,}520 \text{ Lb.in.}$$

(2) Determine the nominal torsional stress q from the relation

$$q = T_{sh}/Z \quad [\text{Lb./in.}^2],$$

where Z = lowest value of sectional modulus [in.³] in the shaft portion considered.

The smallest diameter is $D = 6 \cdot 0$ in., and the corresponding sectional modulus is

$$Z = \frac{\pi}{16} D^3 = 42 \cdot 42 \text{ in.}^3.$$

Therefore the nominal stress is

$$q = 215{,}520/42 \cdot 42 = 5080 \text{ Lb./in.}^2.$$

(3) Determine the 'point of minimum amplitude' which corresponds to the 'node position' in systems without damping.†

† In damped multi-mass systems, there is no true 'node', i.e. no shaft position at which the vibration amplitude is constantly zero, but there is a corresponding point, which may be termed the 'point of minimum amplitude' (see Fig. 2).

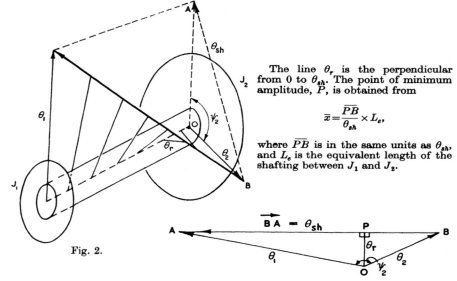

The line θ_r is the perpendicular from 0 to θ_{sh}. The point of minimum amplitude, P, is obtained from

$$\bar{x} = \frac{\overline{PB}}{\theta_{sh}} \times L_e,$$

where $\overline{PB}$ is in the same units as θ_{sh}, and L_e is the equivalent length of the shafting between J_1 and J_2.

Fig. 2.

Fig. 3. Diagram for determining the minimum amplitude θ_r.

Note. The absence of a true node, and the procedure for determining the point of minimum amplitude, were indicated by S. Archer, 'Contribution to increased accuracy in the calculation and measurement of torsional vibration stresses in marine propeller shafting', *Proc. Instn Mech. Engrs, Lond.*, vol. 164 (1951), pp. 351–66.

The 'node position' $\bar{x}$ and the 'minimum amplitude' $\theta_{\text{red.}}$ can be conveniently determined by a diagram. To construct Fig. 4a,

(1) draw a horizontal line for $\theta_1 = 2 \cdot 56°$,

(2) draw $\theta_2 = 1 \cdot 54°$, at an angle $\psi = 177° \, 39'$ relative to θ_1,

(3) join the tips of the vectors θ_1 and θ_2 so as to obtain θ_{sh}, and erect a perpendicular to θ_{sh} through the origin. This gives θ_r and $\bar{x}$. (In this example, $L_e = 500$ in.)

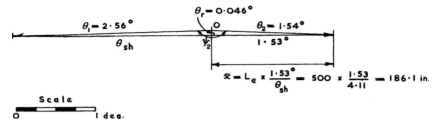

$$\bar{x} = L_e \times \frac{1 \cdot 53°}{\theta_{sh}} = 500 \times \frac{1 \cdot 53}{4 \cdot 11} = 186 \cdot 1 \text{ in.}$$

Scale

0   1 deg.

Fig. 4a.

Note. The values of $|\theta_r|$ and $\bar{x}$ can be calculated, where further accuracy is required, by means of the following relations:

$$|\theta_r| = \frac{|\theta_1| A_2}{|\theta_{sh}|}$$

and

$$\bar{x} = -\frac{a_2 a_{sh} + A_2 A_{sh}}{|\theta_{sh}|} \times \frac{L_e}{|\theta_{sh}|},$$

where $\theta_1 = a_1 + jA_1$, $\theta_2 = a_2 + jA_2$ and $\theta_{sh} = a_{sh} + jA_{sh}$.

$|\theta_r|$ is obtained from Fig. 4b as follows:

     $A_{sh} = A_2$ and $a_{sh} = a_2 + |\theta_1|$.

     Area $\triangle OAB = \frac{1}{2} |\theta_r| \, |\theta_{sh}|$,

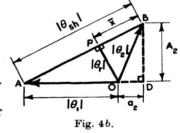

but this area is also equal to $\frac{1}{2} |\theta_1| A_2$. Hence $|\theta_r| = |\theta_1| A_2 / |\theta_{sh}|$.

The expression for $\bar{x}$ is obtained by vector algebra.

Fig. 4b.

We have seen (see p. 406):

     $|\theta_1| = 0 \cdot 0447$ rad.,

     $\theta_2 = -0 \cdot 0268 - j0 \cdot 0011 = a_2 + jA_2$,

     $\theta_{sh} = +0 \cdot 071516 + j0 \cdot 001086 = a_{sh} + jA_{sh}$,

and $|\theta_{sh}| = 0 \cdot 07184$ rad.,

[409]

so that

$$|\theta_r| = \frac{|\theta_1| A_2}{|\theta_{sh}|} = \frac{0\cdot0447 \times 0\cdot0011}{0\cdot07184} = 0\cdot00684\,\mathrm{rad.} = \underline{0\cdot040^\circ},$$

$$\bar{x} = \frac{-(-0\cdot0268) \times 0\cdot071516 - (-0\cdot0011) \times 0\cdot001086}{0\cdot07184} \times \frac{500}{0\cdot07184}$$

$$= \frac{0\cdot019140}{7\cdot184} \times \frac{500}{0\cdot07184} = \underline{185\cdot4\,\mathrm{in.}}$$

(f) Vector diagram

The lengths and directions of all the vibration-amplitude and torque vectors are known. The vector diagram of the complete system can therefore be drawn directly.

The values obtained from evaluations (a) to (d) are:

Torque angle (relative to θ_1): $\phi_1 = 91^\circ\,16'$.

$$|\theta_1| = 2\cdot56^\circ, \quad \psi_1 = 0^\circ,$$
$$|\theta_2| = 1\cdot54^\circ, \quad \psi_2 = 177^\circ\,39',$$
$$|\theta_{sh}| \simeq 4\cdot11^\circ.$$

Therefore, the vector diagram can be constructed as shown in Fig. 5.

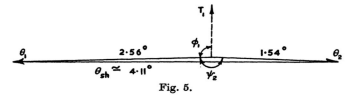

Fig. 5.

It is thus evident, from the above example, that the complete evaluation of the results of a Holzer tabulation with damping and gas torque can give all the quantities required for an analysis of the system.

The shaft deflexion can be determined correctly, by simple substitution of the values of θ_1, T' and T'' in the expression obtained for θ_{sh}. This value may be appreciably less than the twist determined from forced-frequency tabulations without damping, particularly under non-resonant conditions. The nominal stress is then reduced correspondingly, and this fact indicates the value of this method for practical work.

Basic theory of Holzer method for systems with gas torque and damping

The Holzer tabulation method with damping and excitation torque is based on considerations similar to those described for tabulations without damping (see section 1·3211, p. 161). It is also an extension of the 'dual' tabulation for forced frequencies without damping, given in section 2·312, pp. 321—324. It will be convenient to consider its derivation on the basis of a two-mass system taken as an example.

Two-mass system. The equations of motion of a two-mass system of the type illustrated in the figure are:

$$J_1 \frac{d^2\theta_1'}{dt^2} + c_1 \frac{d\theta_1'}{dt} + K(\theta_1' - \theta_2') = T = T_0\, e^{j(\omega t + \phi)}, \qquad (1)$$

$$J_2 \frac{d^2\theta_2'}{dt^2} + c_2 \frac{d\theta_2'}{dt} + K(\theta_2' - \theta_1') = 0, \qquad (2)$$

where $J_1, J_2 =$ mass moments of inertia [Lb.in.sec.^{2}], $\theta_1', \theta_2' =$ vibration amplitudes [rad.], $c_1, c_2 =$ damping coefficients [Lb.in.sec./rad.], $K =$ shaft stiffness [Lb.in./rad.] and $T_0\, e^{j(\omega t + \phi)} \equiv T_0\{\cos(\omega t + \phi) + j\sin(\omega t + \phi)\} =$ excitation torque [Lb.in.] acting with a phase velocity ω [rad./sec.] on mass J_1. Moreover, $\phi =$ value of phase angle (still to be determined) between maximum amplitudes of T and θ_1'.

Substitution in eqs. (1) and (2) of the expressions $\theta_1' = \theta_1\, e^{j\omega t}$ and $\theta_2' = \theta_2\, e^{j\omega t}$, where $\theta_1, \theta_2 =$ maximum values of θ_1' and θ_2', respectively, gives

$$-\omega^2 J_1 \theta_1 + j\omega c_1 \theta_1 + K(\theta_1 - \theta_2) = T_0\, e^{j\phi}, \qquad (1')$$

$$-\omega^2 J_2 \theta_2 + j\omega c_2 \theta_2 + K(\theta_2 - \theta_1) = 0. \qquad (2')$$

For the gas torque, one may write $T_0\, e^{j\phi} = T_0 \cos\phi + jT_0 \sin\phi$, and after rearrangement of the terms, eqs. (1') and (2') become

$$[+\omega^2 J_1 \theta_1 + T_0 \cos\phi] + j[-\omega c_1 \theta_1 + T_0 \sin\phi] = K(\theta_1 - \theta_2), \qquad (1'')$$

$$[+\omega^2 J_2 \theta_2] \qquad\qquad + j[-\omega c_2 \theta_2] \qquad\qquad = K(\theta_2 - \theta_1). \qquad (2'')$$

Then

$$\left[\omega^2 \frac{J_1 \theta_1}{K} + \frac{T_0}{K}\cos\phi\right] + j\left[-\frac{\omega c_1}{K}\theta_1 + \frac{T_0}{K}\cos\phi\right] = \theta_1 - \theta_2 = \theta_{sh} \qquad (1''')$$

and $\theta_2 = \theta_1 - \theta_{sh} = \theta_1 - \left[\omega^2 \dfrac{J_1 \theta_1}{K} + \dfrac{T_0}{K}\cos\phi\right] - j\left[-\dfrac{\omega c_1}{K}\theta_1 + \dfrac{T_0}{K}\sin\phi\right].$

Note that θ_1, which is to be determined, is assumed to be a positive value situated on the real axis. The expression for θ_2, substituted in eq. (2''), gives

$$\omega^2 J_2 \theta_2 + j[-\omega c_2 \theta_2]$$
$$= (\omega^2 J_2 - j\omega c_2) \times \left\{\left[\theta_1 - \omega^2 \frac{J_1 \theta_1}{K} - \frac{T_0}{K}\cos\phi\right] - j\left[-\frac{\omega c_1}{K}\theta_1 + \frac{T_0}{K}\sin\phi\right]\right\}$$
$$= \left\{\omega^2 J_2 \theta_1 - \omega^4 \frac{J_1 J_2}{K}\theta_1 - \omega^2 \frac{J_2}{K}T_0 \cos\phi + \omega^2 \frac{c_2 c_1}{K}\theta_1 - \omega c_2 \frac{T_0}{K}\sin\phi\right\}$$
$$- j\left\{\omega c_2 \theta_1 - \omega^3 \frac{J_1 c_2}{K}\theta_1 - \omega \frac{c_2 T_0}{K}\cos\phi - \omega^3 \frac{J_2 c_1}{K}\theta_1 + \omega^2 \frac{J_2 T_0}{K}\sin\phi\right\}. \qquad (2''')$$

The condition must also be fulfilled that the total torque at the end-mass J_2 is zero; and, since $K(\theta_2 - \theta_1)$ appears with opposite signs in eqs. (1″) and (2″), it means that the sum of the left-hand side expressions in these two equations must be zero. Accordingly,

$$0 = \{[\omega^2 J_1 \theta_1 + T_0 \cos \phi] + j[-\omega c_1 \theta_1 + T_0 \sin \phi]\} + \{[\omega^2 J_2 \theta_2] + j[-\omega c_2 \theta_2]\},$$

or, with use of eq. (2‴),

$$0 = \left\{ \omega^2 J_1 \theta_1 + T_0 \cos \phi + \omega^2 J_2 \theta_1 - \omega^4 \frac{J_1 J_2}{K} \theta_1 - \omega^2 \frac{J_2}{K} T_0 \cos \phi \right.$$
$$\left. + \omega^2 \frac{c_2 c_1}{K} \theta_1 - \omega c_2 \frac{T_0}{K} \sin \phi \right\}$$
$$+ j \left\{ -\omega c_1 \theta_1 + T_0 \sin \phi - \omega c_2 \theta_1 + \omega^3 \frac{J_1 c_2}{K} \theta_1 + \omega \frac{c_2}{K} T_0 \cos \phi \right.$$
$$\left. + \omega^3 \frac{J_2 c_1}{K} \theta_1 - \omega^2 \frac{J_2}{K} T_0 \sin \phi \right\}.$$

This is a complex-number equation of the form $a + jA = 0$. It is equal to zero only if $a = 0$ and $A = 0$ separately, so that two equations (one for each series of terms in the curved brackets) are obtained, each of which must equal zero. Therefore

$$\left(\omega^2 J_1 + \omega^2 J_2 - \omega^4 \frac{J_1 J_2}{K} + \omega^2 \frac{c_1 c_2}{K} \right) \theta_1 = -\left(1 - \omega^2 \frac{J_2}{K} \right) T_0 \cos \phi + \omega c_2 \frac{T_0}{K} \sin \phi, \quad (3)$$

$$\left(-\omega c_1 - \omega c_2 + \omega^3 \frac{J_1 c_2}{K} + \omega^3 \frac{J_2 c_1}{K} \right) \theta_1 = -\left(1 - \omega^2 \frac{J_2}{K} \right) T_0 \sin \phi - \omega c_2 \frac{T_0}{K} \cos \phi, \quad (4)$$

or, more simply, $\qquad\qquad c\theta_1 = b \qquad\qquad\qquad (3')$

and $\qquad\qquad\qquad\qquad e\theta_1 = d. \qquad\qquad\qquad (4')$

These equations are solvable numerically, first by eliminating θ_1, so that

$$\theta_1 = b/c \quad \text{and} \quad \theta_1 = d/e \quad \text{give} \quad be = cd.$$

Then all the terms with $T_0 \cos \phi$ are grouped together, and those with $T_0 \sin \phi$ are grouped similarly. The result is an equation of the form

$$fT_0 \cos \phi = gT_0 \sin \phi,$$

from which $\tan \phi$ can be determined. Thus ϕ, $\sin \phi$ and $\cos \phi$ are also determined, and when the latter two are introduced in eq. (3), the value of θ_1 can be computed directly. The same value of θ_1 is obtained if the values for $\sin \phi$ and $\cos \phi$ are used in conjunction with eq. (4).

Systems with a large number of masses. For multi-mass systems, the procedure follows the same lines as for the simple two-mass system. The terms with K-factors are the 'coupling terms', which are successively eliminated, in order to obtain a complex-number equation in θ_1. The tabulation arrangement provides a systematic procedure for computation.

For 'minor orders', i.e. when the excitation torques of the individual cylinders are not in phase, it is necessary to take account also of their respective phase angles, and the manner in which this is carried out is described in the following numerical example.

Tabulation for vibration amplitudes and stresses of major and minor orders

A single tabulation is sufficient for all major and minor orders of an engine system. It should also be noted that this tabulation is in fact an extension of the forced-frequency tabulations with gas torque given on pp. 321–324 of section 2·3. For large flank amplitudes of minor orders, there is no alternative method so far known which provides more rapid results.

As an example, in the following we shall determine the vibration amplitudes for the $6\frac{1}{2}$th and 9th order criticals of a 6-cylinder 4-stroke cycle engine with flywheel and driven machine. The engine data (see Fig. 6) are as follows:

Bore: 240 mm. (9·449 in.).

Stroke: 300 mm. (11·812 in.).

Inertias: $J_1 = J_2 = \ldots = J_6 = J_{cyl.} = 1·8$ Lb.ft.sec.^{2},

$J_7 = J_F = 148·0$ Lb.ft.sec.^{2},

$J_8 = J_G = 49·0$ Lb.ft.sec.^{2}.

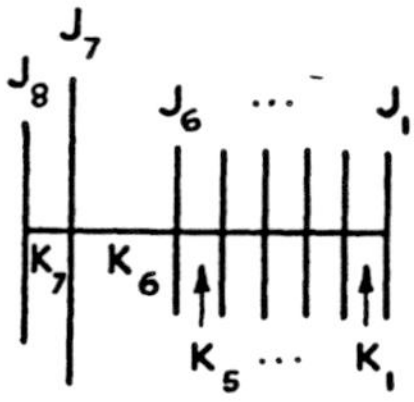

Fig. 6.

Stiffnesses: $K_1 = K_2 = \ldots = K_5 = K_{cyl.} = 5·8 \times 10^6$ Lb.ft./rad.,

$K_6 = K_F = 3·92 \times 10^6$ Lb.ft./rad.,

$K_7 = K_G = 28·2 \times 10^6$ Lb.ft./rad.

Damping coefficients. From previous measurements, it was found that engine systems of this type have a dynamic magnifier $M \cong 49$. This magnifier is associated with an engine damping coefficient (see section 2·42, pp. 367–368):

$$c_E = \omega_0 \Sigma (J_{cyl.} \Delta^2_{cyl.})/M$$

$$= 412·6 \times 6·208/49 = 52·27 \quad [\text{Lb.ft.sec./rad.}].$$

In the above equation, $\omega_0 =$ natural phase velocity of the system [rad./sec.], $J_{cyl.} =$ moment of inertia per line [Lb.ft.sec.^{2}], and $\Delta_{cyl.} =$ corresponding relative amplitude, taken from an ordinary Holzer table without damping, given on p. 415. In this example it is assumed that the

[413]

contribution of the driven machine to the overall damping can be neglected. The damping coefficient per cylinder, $c_{\text{cyl.}}$, is therefore

$$c_{\text{cyl.}} = c_E / \Sigma \Delta_{\text{cyl.}}^2$$
$$= 52 \cdot 27 / 3 \cdot 45 = 15 \cdot 2 \quad [\text{Lb.ft.sec./rad.}].$$

Gas torques. The excitation torque per cylinder, for the $6\frac{1}{2}$th and the 9th order, respectively, is obtained in the usual manner:

$6\frac{1}{2}$th: $\quad T_{\text{cyl.}} = |T_{6\frac{1}{2}}| A R_0 = 4 \cdot 6 \times 70 \cdot 1205 \times 5 \cdot 906$
$$= 1905 \text{ Lb.in} = 158 \cdot 8 \text{ Lb.ft.}$$

9th: $\quad T_{\text{cyl.}} = |T_9| A R_0 = 1 \cdot 38 \times 70 \cdot 1205 \times 5 \cdot 906$
$$= 571 \cdot 5 \text{ Lb.in.} = 47 \cdot 63 \text{ Lb.ft.}$$

In the above expressions $|T_{6\frac{1}{2}}| = 4 \cdot 6 \text{ Lb./in.}^2$, and $|T_9| = 1 \cdot 38 \text{ Lb./in.}^2$, are the components of tangential pressure due to gas torque, for an engine of this class, determined for 90 Lb./in.^{2} i.m.e.p. (from Fig. 2 of section 2·1 of this Handbook), $A = 70 \cdot 1205 \text{ in.}^2 = $ piston area, and $R_0 = 5 \cdot 906 \text{ in.}$ (150 mm.) = crank radius.

Phase angles of torques. The firing order is 1 5 3 6 2 4. No 'phase-vector' summation is required, but it is necessary to have the phase-angle diagrams. For the $6\frac{1}{2}$th (minor) order we consider Fig. 7, which is taken from section 2·22.

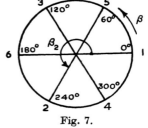

Fig. 7.

The torque due to cylinder no. 1 has a phase angle ϕ_1 relative to the vibration θ_1 of mass J_1 (θ_1 is not shown in Fig. 7). We therefore resolve T_1 into its sine and cosine components (using complex numbers), and divide it by 10^6 for convenience, so as to obtain

$$T_{\text{cyl.(no. 1)}}/10^6 = T_1/10^6 = T_0 \cos \phi + j T_0 \sin \phi = a + jA.$$

The torques from the other cylinders have phase angles β_2, β_3, $\cdots$ relative to T_1. Fig. 7 gives the relations

$$T_2/10^6 = T_0 \cos(\phi_1 - 240°) + j T_0 \sin(\phi_1 - 240°) = b + jB,$$
$$T_3/10^6 = T_0 \cos(\phi_1 - 120°) + j T_0 \sin(\phi_1 - 120°) = c + jC,$$
$$T_4/10^6 = T_0 \cos(\phi_1 - 300°) + j T_0 \sin(\phi_1 - 300°) = d + jD,$$
$$T_5/10^6 = T_0 \cos(\phi_1 - 60°) + j T_0 \sin(\phi_1 - 60°) = e + jE,$$
$$T_6/10^6 = T_0 \cos(\phi_1 - 180°) + j T_0 \sin(\phi_1 - 180°) = f + jF.$$

For the Holzer table with gas torques and damping, it is convenient to use simply $a + jA$, $b + jB$, etc., the values of these expressions being determined after completing the tabulation.

[414]

Furthermore, as $a+jA$, $b+jB$, etc., can be related to any phase-angle diagram, the tabulation is valid for all order numbers. For the 9th (major) order, the evaluation is simplified since in this case

$$a+jA = b+jB = \ldots,$$

that is $\qquad a=b=c=\ldots, \quad A=B=C=\ldots.$

Holzer tabulation without gas torques or damping. This is given below. For ease of reference, the columns are denoted by (1), (2), ..., and the horizontal rows by the letters p, q, r, ... Thus, for instance, $q(3)$ refers to the value in the 2nd row of column (3).

$F = 3940$ vib./min., $\omega = 412 \cdot 6$ rad./sec., $\omega^2 = 0 \cdot 17024 \times 10^6$ (rad./sec.)^2.

Columns	(1)	(2)	(3)	(4)	(5)	(6)	(7)	(8)	Additional columns used for evaluation of damping	
		J	$\dfrac{J\omega^2}{10^6}$		$\dfrac{J\omega^2\Delta}{10^6}$	$\dfrac{\Sigma}{10^6}$	$\dfrac{K}{10^6}$		Δ^2	$J\Delta^2$
Rows	Mass no.	[Lb.ft. sec.²]	[Lb.ft./rad.]	Δ [rad.]	[Lb.ft.]	[Lb.ft.]	[Lb.ft./ rad.]	$\Delta_{\bullet\bullet}$ [rad.]	[rad.²]	[Lb.in.sec.²]
p	1	1·8	0·30643	1·00000	0·30643	0·30643	5·8	0·05283	1·00000	1·800
q	2	1·8	0·30643	0·94717	0·29024	0·59667	5·8	0·10287	0·89713	1·615
r	3	1·8	0·30643	0·84430	0·25872	0·85539	5·8	0·14748	0·71284	1·283
s	4	1·8	0·30643	0·69682	0·21353	1·06892	5·8	0·18430	0·48556	0·874
t	5	1·8	0·30643	0·51252	0·15705	1·22597	5·8	0·21137	0·26268	0·473
u	6	1·8	0·30643	0·30115	0·09228	1·31825	3·92	0·33629	0·09065	0·163
v	7	148·0	25·195	−0·03514	−0·88535	0·43290	28·2	0·01535	3·44890 =	6·208 =
w	8	49·0	8·3417	−0·05049	−0·42117	+0·01173	—	—	$\Sigma(\Delta^2_{cyl.})$	$(\Sigma J_{cyl.}\Delta^2_{cyl.})$

Combined expressions for inertia torque and damping torque

The Holzer table without damping gives for masses nos. 1–6 the inertia torque $J\omega^2 = 0 \cdot 30463 \times 10^6$ Lb.ft./rad. As the phase velocity is $\omega = 412 \cdot 6$ rad., the damping torque is

$$-jc_{cyl.} \times \omega = -j15 \cdot 2 \times 412 \cdot 6 = -j6271 \cdot 5 \quad \text{[Lb.ft./rad.]}.$$

The symbol $-j$ indicates the phase-angle relation between damping torque and the torque $+J\omega^2$. As both have to be multiplied by the vibration amplitude θ of the corresponding mass, it is convenient to write both torques as a combined expression

$$(J\omega^2 - j\omega c)/10^6 = [0 \cdot 30463 - j0 \cdot 00627] \times 10^6 \quad \text{[Lb.ft./rad.]}$$

for use in column (3) of the Holzer table with damping. We now proceed to calculate the expressions indicated in this table. It is useful to have a desk-type electrically driven calculating machine for this work and to organize the intermediate computations on separate sheets (for instance as shown in the following pages) to minimize error risk and facilitate checking.

The only operations requiring special attention are those of column (3), which involve the multiplication of complex numbers, say $x+jX$ and $y+jY$, in accordance with the relation

$$(x+jX)(y+jY) = [(xy)-(XY)] + j[(Xy)+(xY)] = z+jZ.$$

Examples showing how this operation may be organized for a large number of terms are shown on p. 418.

[415]

$F = 3940$ vib./min., $\omega = 412.6$ rad./sec., $\omega^2 = 0.17024 \times 10^6$ (rad./sec.)^2

Columns	(1)	(2) $\dfrac{J\omega^2 - j\omega c}{10^6}$	(3)	(4) $\dfrac{J\omega^2 - j\omega c}{10^6} \times \theta$
Rows	Mass no.	[Lb.ft./rad.]	θ [rad.]	[Lb.ft.]
p	1	0.30643 $-j0.00627$	θ_1	$0.30643\theta_1 - j0.00627\theta_1$
q	2	0.30643 $-j0.00627$	$0.94717\theta_1 - 0.17241a$ $+j[0.00108\theta_1 - 0.17241A]$	$0.29025\theta_1 - 0.05283a - 0.00108A$ $+j[-0.00561\theta_1 + 0.00108a - 0.05283A]$
r	3	0.30643 $-j0.00627$	$0.84429\theta_1 - 0.33572a + 0.00019A$ $-0.17241b + j[-0.00313\theta_1 + 0.00019a$ $-0.33572A - 0.17241B]$	$0.25869\theta_1 - 0.10287a - 0.00204A$ $-0.05283b - 0.00108B + j[-0.00625\theta_1$ $+0.00204a - 0.10287A + 0.00108b$ $-0.05283B]$
s	4	0.30643 $-j0.00627$	$0.69681\theta_1 - 0.48129a + 0.00073A$ $-0.33572b + 0.00019B - 0.17241c$ $+j[0\theta_1 - 0.00073a - 0.48129A$ $-0.00019b - 0.33578B - 0.17241C]$	$0.21352\theta_1 - 0.14748a - 0.00280A$ $-0.10287b - 0.00204B - 0.05283c$ $-0.00108C + j[-0.00437\theta_1 + 0.00280a$ $-0.14748A + 0.00204b - 0.10287B$ $+0.00108c - 0.05283C]$
t	5	0.30643 $-j0.00627$	$0.51286\theta_1 - 0.60143a + 0.00175A$ $-0.48129b + 0.00075B - 0.33572c$ $+0.00019C - 0.17241d + j[+0.00388\theta_1$ $-0.00175a - 0.60143A - 0.00073b$ $-0.48129B - 0.00019c - 0.33572C$ $-0.17241D]$	$0.15716\theta_1 - 0.18431a - 0.00323A$ $-0.14748b - 0.00280B - 0.10287c$ $-0.00204C - 0.05283d - 0.00108D$ $+j[-0.00203\theta_1 + 0.00323a - 0.18431A$ $+0.00280b - 0.19748B + 0.00204c$ $-0.10287C + 0.00108d - 0.05283D]$
u	6	0.30643 $-j0.00627$	$0.30182\theta_1 - 0.68979a + 0.00333A$ $-0.60143b + 0.00175B - 0.48129c$ $+0.00073C - 0.33572d + 0.00019D$ $-0.17241e + j[+0.00811\theta_1 - 0.0033a$ $-0.68979A - 0.00175b - 0.60143B$ $-0.00073c - 0.48129C - 0.00019d$ $-0.33572D - 0.17241E]$	$0.09254\theta_1 - 0.21139a - 0.00330A$ $-0.18431b - 0.00323B - 0.14714c$ $-0.00280C - 0.10287d - 0.00204D$ $-0.05283e - 0.00108E + j[0.00060\theta_1$ $+0.00330a - 0.21139A + 0.00323b$ $-0.18431B + 0.00280c - 0.14748C$ $+0.00204d - 0.10287D + 0.00108e$ $-0.05283E]$
v	7	$25.195 + j \times 0$	$-0.03404\theta_1 - 0.76661a + 0.00651A$ $-0.73217b + 0.00408B - 0.65905c$ $+0.00224C - 0.55110d + 0.00099D$ $-0.41404e + 0.00028E - 0.25510f$ $+j[+0.01421\theta_1 - 0.00651a - 0.76661A$ $-0.00408b - 0.73217B - 0.00224c$ $-0.65905C - 0.00099d - 0.55110D$ $-0.00028e - 0.41404E - 0.25510F]$	$-0.85764\theta_1 - 19.31474a + 0.16402A$ $-18.44702b + 0.10280B - 16.60476c$ $+0.05644C - 13.88496d + 0.02494D$ $-10.43174e + 0.00705E - 6.4274f$ $+j[+0.35802\theta_1 - 0.16402a - 19.31474A$ $-0.10280b - 18.44702B - 0.05644c$ $-16.60476C - 0.02494d - 13.88496D$ $-0.00705e - 10.43174E - 6.42724F]$
w	8	$8.3417 + j \times 0$	$-0.05031\theta_1 - 0.09237a + 0.00114A$ $-0.09620b + 0.00076B - 0.09494c$ $+0.00045C - 0.08867d + 0.00022D$ $-0.07771e + 0.00007E - 0.06264f$ $+j[+0.00236\theta_1 - 0.00114a - 0.09237A$ $-0.00076b - 0.09620B - 0.00045c$ $-0.09494C - 0.00022d - 0.08867D$ $-0.00007e - 0.07171E - 0.06264F]$	$-0.41967\theta_1 - 0.77052a + 0.00951A$ $-0.80247b + 0.00634B - 0.79196c$ $+0.00375C - 0.73966d + 0.00184D$ $-0.64823e + 0.00058E - 0.52252f$ $+j[+0.01969\theta_1 - 0.00951a - 0.77052A$ $-0.00634b - 0.80247B - 0.00375c$ $-0.79196C - 0.00184d - 0.73966D$ $-0.00058e - 0.64823E - 0.52252F]$

The equations for the torques acting on the last mass are:

(i) $0.03928\theta_1 - 19.78414a + 0.16108A - 18.73696b + 0.09999B + 0.05427C - 13.78032d$
$+ 0.02366D - 10.13280e + 0.00655E - 5.94976f = 0,$

(ii) $0.35378\theta_1 - 0.16108a - 19.78414A - 0.09999b - 18.73696B - 0.05427c - 16.69990C - 0.02366D$
$- 13.78032D - 0.00655e - 10.13280E - 5.94976F = 0.$

six-cylinder engine with flywheel and generator

$$F = 3940 \text{ vib./min.}, \quad \omega = 412\cdot6 \text{ rad./sec.}, \quad \omega^2 = 0\cdot17024 \times 10^6 \text{ (rad./sec.)}^2$$

(5) $\dfrac{T}{10^6}$ [Lb.ft.]	(6) $\dfrac{\Sigma[J\omega^2\theta - j\omega c\theta + T]}{10^6}$ [Lb.ft.]	(7) $\dfrac{K}{10^6}$ [Lb.ft./rad.]	(8) $\theta_{sh} = \dfrac{(6)}{(7)}$ [rad.]
$a+jA$	$0\cdot30643\theta_1 + a + j[-0\cdot00627\theta_1 + A]$	$5\cdot8$	$0\cdot05283\theta_1 + 0\cdot17241a + j[-0\cdot00108\theta_1 + 0\cdot17241A]$
$b+jB$	$0\cdot59668\theta_1 + 0\cdot94717a - 0\cdot00108A + b + j[-0\cdot01188\theta_1 + 0\cdot00108a + 0\cdot94717A + B]$	$5\cdot8$	$0\cdot10288\theta_1 + 0\cdot16331a - 0\cdot00019A + 0\cdot17241b + j[-0\cdot00205\theta_1 + 0\cdot00019a + 0\cdot16331A + 0\cdot17241B]$
$c+jC$	$0\cdot85537\theta_1 + 0\cdot84430a - 0\cdot00312A + 0\cdot94717b - 0\cdot00108B + a + j[-0\cdot01813\theta_1 + 0\cdot00312a + 0\cdot84430A + 0\cdot00108b + 0\cdot94717B + C]$	$5\cdot8$	$0\cdot14748\theta_1 + 0\cdot14557a - 0\cdot00554A + 0\cdot16331b - 0\cdot00019B + 0\cdot17241c + j[-0\cdot00313\theta_1 + 0\cdot00054a + 0\cdot14557A + 0\cdot00019b + 0\cdot16331B + 0\cdot17241C]$
$d+jD$	$1\cdot06689\theta_1 + 0\cdot69682a - 0\cdot00592A + 0\cdot84430b - 0\cdot00312B + 0\cdot94717c - 0\cdot00108C + d + j[-0\cdot02550\theta_1 + 0\cdot00592a + 0\cdot69682A + 0\cdot00312b + 0\cdot84430B + 0\cdot00108c + 0\cdot94717C + D]$	$5\cdot8$	$0\cdot18395\theta_1 + 0\cdot12014a - 0\cdot00102A + 0\cdot14557b - 0\cdot00054B + 0\cdot16331c - 0\cdot00019C + 0\cdot17241d + j[-0\cdot00388\theta_1 + 0\cdot00102a + 0\cdot12014A + 0\cdot00054b + 0\cdot16331B + 0\cdot00019c + 0\cdot16331C + 0\cdot17241D]$
$e+jE$	$1\cdot22405\theta_1 + 0\cdot51251a - 0\cdot00915A + 0\cdot69682b - 0\cdot00592B + 0\cdot84430c - 0\cdot00312C + 0\cdot94717d - 0\cdot00108D + e + j[-0\cdot02453\theta_1 + 0\cdot00915a + 0\cdot51251A + 0\cdot00592b + 0\cdot69682B + 0\cdot00312c + 0\cdot84430C + 0\cdot00108d + 0\cdot94717D + E]$	$5\cdot8$	$0\cdot21104\theta_1 + 0\cdot08836a - 0\cdot00158A + 0\cdot12014b - 0\cdot00102B + 0\cdot14557c - 0\cdot00054C + 0\cdot16331d - 0\cdot00019D + 0\cdot17241e + j[-0\cdot00423\theta_1 + 0\cdot00158a + 0\cdot08836A + 0\cdot00102b + 0\cdot12014B + 0\cdot00054c + 0\cdot14557C + 0\cdot00019d + 0\cdot16331D + 0\cdot17241E]$
$f+jF$	$1\cdot31659\theta_1 + 0\cdot30112a - 0\cdot01245A + 0\cdot51251b - 0\cdot00915B + 0\cdot69682c - 0\cdot00592C + 0\cdot84430d - 0\cdot00312D + 0\cdot94717e - 0\cdot00108E + f + j[-0\cdot02393\theta_1 + 0\cdot01245a + 0\cdot30112A + 0\cdot00915b + 0\cdot51251B + 0\cdot00592c + 0\cdot69682C + 0\cdot00312d + 0\cdot84430D + 0\cdot00108e + 0\cdot94717E + F]$	$3\cdot92$	$0\cdot33586\theta_1 + 0\cdot07682a - 0\cdot00318A + 0\cdot13074b - 0\cdot00233B + 0\cdot17776c - 0\cdot00151C + 0\cdot21538d - 0\cdot00080D + 0\cdot24163e - 0\cdot00028E + 0\cdot25510f + j[-0\cdot00610\theta_1 + 0\cdot00318a + 0\cdot07682A + 0\cdot00233b + 0\cdot13074B + 0\cdot00151c + 0\cdot17776C + 0\cdot00080d + 0\cdot21538D + 0\cdot00028e + 0\cdot24163E + 0\cdot25510F]$
—	$+0\cdot45895\theta_1 - 19\cdot01362a + 0\cdot15157A - 17\cdot93449b + 0\cdot09365B - 15\cdot90794c + 0\cdot05052C - 13\cdot04066d + 0\cdot02182D - 9\cdot48457e + 0\cdot00597E - 5\cdot42724f + j[+0\cdot33409\theta_1 - 0\cdot15157a - 19\cdot01362A - 0\cdot09365b - 17\cdot93449B - 0\cdot05052c - 15\cdot90794C - 0\cdot02182d - 13\cdot04066D - 0\cdot00597e - 9\cdot48457E - 5\cdot42724F]$	$28\cdot2$	$+0\cdot01627\theta_1 - 0\cdot67424a + 0\cdot00537A - 0\cdot63597b + 0\cdot0332B - 0\cdot56411c + 0\cdot00179C - 0\cdot46243d + 0\cdot00077D - 0\cdot33633e + 0\cdot00021E - 0\cdot19246f + j[+0\cdot01185\theta_1 - 0\cdot00537a - 0\cdot67424A - 0\cdot0332b - 0\cdot63597B - 0\cdot00179c - 0\cdot56411C - 0\cdot00077d - 0\cdot46243D - 0\cdot00021e - 0\cdot33633E - 0\cdot19246F]$
—	$+0\cdot03928\theta_1 - 19\cdot78414a + 0\cdot16108A - 18\cdot73696b + 0\cdot09999B - 16\cdot69990c + 0\cdot05427C - 13\cdot78032d + 0\cdot02366D - 10\cdot13280e + 0\cdot00655E - 5\cdot94976f + j[+0\cdot35378\theta_1 - 0\cdot16108a - 19\cdot78414A - 0\cdot09999b - 18\cdot73696B - 0\cdot05427c - 16\cdot69990C - 0\cdot02366d - 13\cdot78032D - 0\cdot00655e - 10\cdot13280E - 5\cdot94976F]$	—	—

In general, it is not necessary to use more than three or four significant figures after the decimal point. The table above has been calculated with greater detail for reference purposes; it also indicates the symmetry of the numerical coefficients and their association with new factors in the course of the calculation.

Detailed multiplications for column (4) in the tabulation

Multiplication rule: $(x+jX)(y+jY)=[(xy)-(XY)]+j[(Xy)+(xY)]$

q (4) $(0\cdot30643-j0\cdot00627)\times\{0\cdot94717\theta_1-0\cdot17241a+j[0\cdot00108\theta_1-0\cdot17241A]\}$
$$=[(0\cdot29024\theta_1-0\cdot05283a)-(-0\cdot00001\theta_1+0\cdot00108A)]$$
$$+j[(-0\cdot00594\theta_1+0\cdot00108a)+(0\cdot00033\theta_1-0\cdot05283A)]$$

r (4) $x=0\cdot30643$
        $X=-0\cdot00627$

$y=$	y_1	$+y_2$	$+y_3$	$+y_4$
	$0\cdot84429\theta_1$	$-0\cdot33572a$	$+0\cdot00019A$	$-0\cdot17241b$
$Y=$	Y_1	$+Y_2$	$+Y_3$	$+Y_4$
	$-0\cdot00313\theta_1$	$-0\cdot00019a$	$-0\cdot33572A$	$-0\cdot17241B$
$(xy)=$	$0\cdot25872\theta_1$	$-0\cdot10287a$	$+0\cdot00006A$	$-0\cdot05283b$
$-(XY)=$	$-0\cdot00003\theta_1$	$-0\cdot00000a$	$-0\cdot00210A$	$-0\cdot00108B$
$(xy)-(XY)=$	$+0\cdot25869\theta_1$	$-0\cdot10287a$	$-0\cdot00204A$	$-0\cdot05283b-0\cdot00108B$
$(Xy)=$	$-0\cdot00529\theta_1$	$+0\cdot00210a$	$-0\cdot00000A$	$+0\cdot00108b$
$+(xY)=$	$-0\cdot00096\theta_1$	$-0\cdot00006a$	$-0\cdot10287A$	$-0\cdot05283B$
$(Xy)+(xY)=$	$-0\cdot00625\theta_1$	$+0\cdot00204a$	$-0\cdot10287A$	$+0\cdot00108b-0\cdot05283B$

Similar calculations occur for $s(4)$ and $t(4)$. The operations for $u(4)$ are given in the following table. Lines v and w are simpler, since no damping coefficients are associated with the last two masses in this example. Checking is facilitated by the symmetry of many of the terms, in the coefficients of a, A, b, B, etc.

Detailed multiplications for u(4) in the tabulation

Multiplication rule: $(x+jX)(y+jY)=[(xy)-(XY)]+j[(Xy)+(xY)]$

$u(4)$ $x=\ \ 0\cdot30643$
        $X=-0\cdot00627$

$y=$	y_1	$+y_2$	$+y_3$	$+y_4$	$+y_5$
	$0\cdot30182\theta_1$	$-0\cdot68979a$	$+0\cdot00333A$	$-0\cdot60143b$	$+0\cdot00175B$
$Y=$	Y_1	$+Y_2$	$+Y_3$	$+Y_4$	$+Y_5$
	$0\cdot00811\theta_1$	$-0\cdot00333a$	$-0\cdot68979A$	$-0\cdot00175b$	$-0\cdot60143B$
$(xy)=$	$+0\cdot09249\theta_1$	$-0\cdot21137a$	$+0\cdot00102A$	$-0\cdot18430b$	$+0\cdot00054B$
$-(XY)=$	$+0\cdot00005\theta_1$	$-0\cdot00002a$	$-0\cdot00432A$	$-0\cdot00001b$	$-0\cdot00377B$
$(xy)-(XY)=$	$+0\cdot09254\theta_1$	$-0\cdot21139a$	$-0\cdot00330A$	$-0\cdot18431b$	$-0\cdot00323B$
$(Xy)=$	$-0\cdot00189\theta_1$	$+0\cdot00432a$	$-0\cdot00002A$	$+0\cdot00377b$	$-0\cdot00001B$
$+(xY)=$	$+0\cdot00249\theta_1$	$-0\cdot00102a$	$-0\cdot21137A$	$-0\cdot00054b$	$-0\cdot18430B$
$(Xy)+(xY)=$	$+0\cdot00060\theta_1$	$+0\cdot00330a$	$-0\cdot21139A$	$+0\cdot00323b$	$-0\cdot18430B$

$y=$	y_6	$+y_7$	$+y_8$	$+y_9$	$+y_{10}$
	$-0\cdot48129c$	$+0\cdot00073C$	$-0\cdot33572d$	$+0\cdot00019D$	$-0\cdot17241e$
$Y=$	$+Y_6$	$+Y_7$	$+Y_8$	$+Y_9$	$+Y_{10}$
	$-0\cdot00073c$	$-0\cdot48129C$	$-0\cdot00019d$	$-0\cdot33572D$	$-0\cdot17241E$
$(xy)=$	$-0\cdot14748c$	$+0\cdot00022C$	$-0\cdot10287d$	$+0\cdot00006D$	$-0\cdot05283e$
$-(XY)=$	$-0\cdot00000c$	$-0\cdot00302C$	$-0\cdot00000d$	$-0\cdot00210D$	$-0\cdot00108E$
$(xy)-(XY)=$	$-0\cdot14748c$	$-0\cdot00280C$	$-0\cdot10287d$	$-0\cdot00204D$	$-0\cdot05283e-0\cdot00108E$
$(Xy)=$	$+0\cdot00302c$	$-0\cdot00000C$	$+0\cdot00210d$	$-0\cdot00000D$	$+0\cdot00108e$
$+(xY)=$	$-0\cdot00022c$	$-0\cdot14748C$	$-0\cdot00006d$	$-0\cdot10287D$	$-0\cdot05283E$
$(Xy)+(xY)=$	$+0\cdot00280c$	$-0\cdot14748C$	$+0\cdot00204d$	$-0\cdot10287D$	$+0\cdot00108e-0\cdot05283E$

We may now re-write eqs. (i) and (ii), obtained from the Holzer table with damping and gas torques (see p. 416), as follows:

$$-\theta_1 = \frac{1}{0 \cdot 03928} \begin{bmatrix} -19 \cdot 78414a \\ +\ 0 \cdot 16108A \\ -18 \cdot 73696b \\ +\ 0 \cdot 09999B \\ -16 \cdot 69990c \\ +\ 0 \cdot 05427C \\ -13 \cdot 78032d \\ +\ 0 \cdot 02366D \\ -10 \cdot 13280e \\ +\ 0 \cdot 00655E \\ -\ 5 \cdot 94976f \end{bmatrix} = \frac{1}{0 \cdot 35378} \begin{bmatrix} -\ 0 \cdot 16108a \\ -19 \cdot 78414A \\ -\ 0 \cdot 09999b \\ -18 \cdot 73696B \\ -\ 0 \cdot 05427c \\ -16 \cdot 69990C \\ -\ 0 \cdot 02366d \\ -13 \cdot 78032D \\ -\ 0 \cdot 00655e \\ -10 \cdot 13280E \\ -\ 5 \cdot 94976F \end{bmatrix}$$

and these give, with $0 \cdot 03928/0 \cdot 35378 = 0 \cdot 11103$ used for multiplying each term in the square brackets on the right-hand side,

$$\begin{bmatrix} -19 \cdot 78414a \\ +\ 0 \cdot 16108A \\ -18 \cdot 73696b \\ +\ 0 \cdot 09999B \\ -16 \cdot 69990c \\ +\ 0 \cdot 05427C \\ -13 \cdot 78032d \\ +\ 0 \cdot 02366D \\ -10 \cdot 13280e \\ +\ 0 \cdot 00655E \\ -\ 5 \cdot 94976f \end{bmatrix} = \begin{bmatrix} -0 \cdot 01788a \\ -2 \cdot 19663A \\ -0 \cdot 01110b \\ -2 \cdot 08036B \\ -0 \cdot 00603c \\ -1 \cdot 85419C \\ -0 \cdot 00263d \\ -1 \cdot 53003D \\ -0 \cdot 00073e \\ -1 \cdot 12504E \\ -0 \cdot 66060F \end{bmatrix}$$

By transferring the terms from the right to the left, we obtain

$$(-19 \cdot 78414 + 0 \cdot 01788)\,a = -19 \cdot 76626a,$$

and so on; so that the result is

$$\begin{bmatrix} -19 \cdot 76226a \\ +\ 2 \cdot 35771A \\ -18 \cdot 72586b \\ +\ 2 \cdot 18035B \\ -16 \cdot 69387c \\ +\ 1 \cdot 98046C \\ -13 \cdot 77769d \\ +\ 1 \cdot 55369D \\ -10 \cdot 13207e \\ +\ 1 \cdot 13159E \\ -\ 5 \cdot 94976f \\ +\ 0 \cdot 66060F \end{bmatrix} = 0. \tag{iii}$$

It should be noted that, as the expressions involving phase angles have not yet been introduced in place of the coefficients a, A, b, B, etc., the above equation is valid for all major and minor orders of the system.

We shall now determine the trigonometrical expressions for a, A, b, B, etc., by means of a phase-angle diagram, for the $6\frac{1}{2}$th (minor) order.

The corresponding (and much simpler) calculation for the 9th (major) order is given on p. 424.

Numerical values of torque components for $6\frac{1}{2}$th order

We shall now determine, for the $6\frac{1}{2}$th (minor) order, the values of the gas torque components, a, A, b, B, etc., using equation (3) and the trigonometrical relations obtained from the phase-angle diagram (see p. 414).

The phase angles β_2, β_3, ..., of the torques T_2, T_3, ... relative to torque T_1 are shown again in Fig. 8. We have already determined the torque per cylinder as

$$T_{\text{cyl.}} = |T_{6\frac{1}{2}}| \, AR_0 = 158\cdot8 \,[\text{Lb.ft.}],$$

and $\qquad T_0 = T_{\text{cyl.}}/10^6 = 0\cdot0001588.$

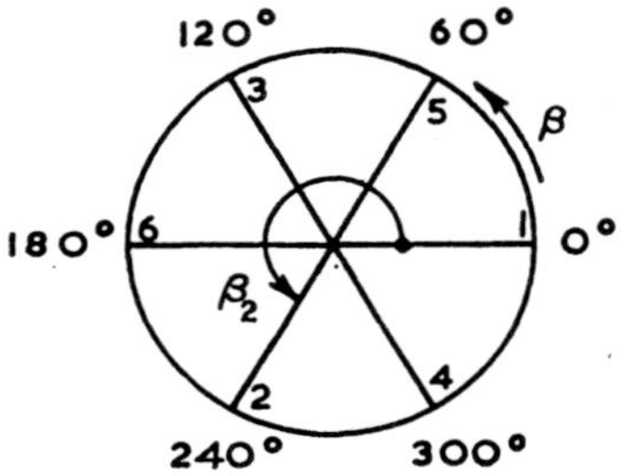

From accompanying diagram

Fig. 8.

β_5: $\quad \cos 60° = +0\cdot5, \qquad \sin 60° = +0\cdot866;$

β_3: $\quad \cos 120° = -0\cdot5, \qquad \sin 120° = +0\cdot866;$

β_6: $\quad \cos 180° = -1\cdot0, \qquad \sin 180° = 0;$

β_2: $\quad \cos 240° = -0\cdot5, \qquad \sin 240° = -0\cdot866;$

β_4: $\quad \cos 300° = +0\cdot5, \qquad \sin 300° = -0\cdot866.$

Addition theorems:

$$\cos(\phi_1 - \beta) = \cos\beta\cos\phi_1 + \sin\beta\sin\phi_1, \quad \sin(\phi_1 - \beta) = \cos\beta\sin\phi_1 - \sin\beta\cos\phi_1.$$

The evaluation is as follows:

$$T_1/10^6 = a + jA = T_0[\cos\phi_1 + j\sin\phi_1] = 0\cdot0001588[\cos\phi_1 + j\sin\phi_1],$$

where $\phi_1 =$ phase angle between T_1 and θ_1,

$$T_2/10^6 = b + jB = T_0[\cos\beta_2\cos\phi_1 + \sin\beta_2\sin\phi_1 + j(\cos\beta_2\sin\phi_1 - \sin\beta_2\cos\phi_1)]$$
$$= T_0[-0\cdot5\cos\phi_1 - 0\cdot866\sin\phi_1 + j(-0\cdot5\sin\phi_1 + 0\cdot866\cos\phi_1)],$$
$$T_3/10^6 = c + jC = T_0[-0\cdot5\cos\phi_1 + 0\cdot866\sin\phi_1 + j(-0\cdot5\sin\phi_1 - 0\cdot866\cos\phi_1)],$$
$$T_4/10^6 = d + jD = T_0[+0\cdot5\cos\phi_1 - 0\cdot866\sin\phi_1 + j(+0\cdot5\sin\phi_1 + 0\cdot866\cos\phi_1)],$$
$$T_5/10^6 = e + jE = T_0[+0\cdot5\cos\phi_1 + 0\cdot866\sin\phi_1 + j(+0\cdot5\sin\phi_1 - 0\cdot866\cos\phi_1)],$$
$$T_6/10^6 = f + jF = T_0[-\cos\phi_1 + j(-\sin\phi_1)].$$

The values of a, A, b, B, etc., can therefore be expressed in terms of the phase angle ϕ_1:

$$a = T_0[\cos\phi_1], \qquad\qquad A = T_0[\sin\phi_1],$$
$$b = T_0[-0\cdot5\cos\phi_1 - 0\cdot866\sin\phi_1], \quad B = T_0[+0\cdot866\cos\phi_1 - 0\cdot5\sin\phi_1],$$
$$c = T_0[-0\cdot5\cos\phi_1 + 0\cdot866\sin\phi_1], \quad C = T_0[-0\cdot866\cos\phi_1 - 0\cdot5\sin\phi_1],$$
$$d = T_0[+0\cdot5\cos\phi_1 - 0\cdot866\sin\phi_1], \quad D = T_0[+0\cdot866\cos\phi_1 + 0\cdot5\sin\phi_1],$$
$$e = T_0[+0\cdot5\cos\phi_1 + 0\cdot866\sin\phi_1], \quad E = T_0[-0\cdot866\cos\phi_1 + 0\cdot5\sin\phi_1],$$
$$f = T_0[-\cos\phi_1], \qquad\qquad F = T_0[-\sin\phi_1],$$

and substitution of these expressions in eq. (iii) gives

$$
\begin{bmatrix}
-19{\cdot}76226a \\
+ 2{\cdot}35771A \\
- 18{\cdot}72586b \\
+ 2{\cdot}18035B \\
- 16{\cdot}69387c \\
+ 1{\cdot}98046C \\
- 13{\cdot}77769d \\
+ 1{\cdot}55369D \\
- 10{\cdot}13207e \\
+ 1{\cdot}13159E \\
- 5{\cdot}94976f \\
+ 0{\cdot}66060F
\end{bmatrix}
= T_0 \times
\begin{bmatrix}
-19{\cdot}76226\cos\phi_1 + 0 \\
+ 0 \qquad\qquad + 2{\cdot}35771\sin\phi_1 \\
+ 9{\cdot}36293\cos\phi_1 + 16{\cdot}21659\sin\phi_1 \\
+ 1{\cdot}88818\cos\phi_1 - 1{\cdot}09018\sin\phi_1 \\
+ 8{\cdot}34694\cos\phi_1 - 14{\cdot}45689\sin\phi_1 \\
- 1{\cdot}65273\cos\phi_1 - 0{\cdot}95423\sin\phi_1 \\
- 6{\cdot}88885\cos\phi_1 + 11{\cdot}91348\sin\phi_1 \\
+ 1{\cdot}34550\cos\phi_1 + 0{\cdot}77685\sin\phi_1 \\
- 5{\cdot}06604\cos\phi_1 - 8{\cdot}77437\sin\phi_1 \\
- 0{\cdot}97996\cos\phi_1 + 0{\cdot}56580\sin\phi_1 \\
+ 5{\cdot}94976\cos\phi_1 + 0 \\
+ 0 \qquad\qquad - 0{\cdot}66060\sin\phi_1
\end{bmatrix}
= 0,
$$

so that, by summating the coefficients, we obtain

$$-7{\cdot}46053\cos\phi_1 + 5{\cdot}89416\sin\phi_1 = 0.$$

Therefore
$$\tan\phi_1 = +\frac{7{\cdot}46053}{5{\cdot}89416} = 1{\cdot}26575$$

and
$$\phi_1 \simeq 51^\circ\, 41' \quad (\text{or } 231^\circ\, 41').$$

Taking $\phi_1 \simeq 51^\circ\, 41'$, we find

$$\cos[51^\circ\, 41'] = 0{\cdot}62001 \quad \text{and} \quad \sin[51^\circ\, 41'] = 0{\cdot}78460.$$

We also evaluate:
$$0{\cdot}5\cos\phi_1 = 0{\cdot}31000, \quad 0{\cdot}866\cos\phi_1 = 0{\cdot}53693,$$
$$0{\cdot}866\sin\phi_1 = 0{\cdot}67946, \quad 0{\cdot}5\sin\phi_1 = 0{\cdot}39230.$$

By means of these values, it is now possible to determine the torque components:

$$a = T_0\cos\phi_1 = 0{\cdot}62001T_0,$$
$$A = T_0\sin\phi_1 = 0{\cdot}78460T_0,$$
$$b = T_0[-0{\cdot}31000 - 0{\cdot}67946] = -0{\cdot}98946T_0,$$
$$B = T_0[+0{\cdot}53693 - 0{\cdot}39230] = +0{\cdot}14463T_0,$$
$$c = T_0[-0{\cdot}31000 + 0{\cdot}67946] = +0{\cdot}36946T_0,$$
$$C = T_0[-0{\cdot}53693 - 0{\cdot}39230] = -0{\cdot}92923T_0,$$
$$d = T_0[+0{\cdot}31000 - 0{\cdot}67946] = -0{\cdot}36946T_0,$$
$$D = T_0[+0{\cdot}53693 + 0{\cdot}39230] = +0{\cdot}92923T_0,$$
$$e = T_0[+0{\cdot}31000 + 0{\cdot}67946] = +0{\cdot}98946T_0,$$
$$E = T_0[-0{\cdot}53693 + 0{\cdot}39230] = -0{\cdot}144463T_0,$$
$$f = T_0[(-1) \times 0{\cdot}62001] = -0{\cdot}62001T_0,$$
$$F = T_0[(-1) \times 0{\cdot}78460] = -0{\cdot}78460T_0.$$

These values can be checked by plotting the vectors $a+jA$, $b+jB$, etc., as shown in the diagram of Fig. 9.

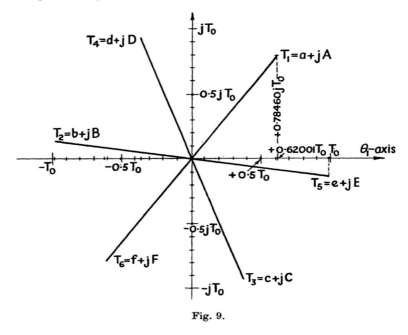

Fig. 9.

Using again equation (i), we finally obtain θ_1 as follows:

$$-\theta_1 = \frac{1}{0.03928}\begin{bmatrix} -19.78414 \times 0.62001T_0 \\ + 0.16108 \times 0.78460T_0 \\ -18.73696 \times (-0.98946)\,T_0 \\ + 0.09999 \times 0.14463T_0 \\ -16.69990 \times 0.36946T_0 \\ + 0.05427 \times (-0.92923)\,T_0 \\ -13.78032 \times (-0.36946)\,T_0 \\ + 0.02366 \times 0.92923T_0 \\ -10.13280 \times 0.98946T_0 \\ + 0.00655 \times (-0.14463)\,T_0 \\ - 5.94976 \times (-0.62001)\,T_0 \end{bmatrix} = \frac{T_0}{0.03928}\begin{bmatrix} -12.26636 \\ + 0.12638 \\ +18.53947 \\ + 0.01446 \\ - 6.16995 \\ - 0.05043 \\ + 5.09128 \\ + 0.02199 \\ -10.02600 \\ - 0.00095 \\ + 3.68891 \end{bmatrix},$$

$$-\theta_1 = \frac{T_0}{0.03928} \times [-28.51369 + 27.48249] = \frac{0.0001588}{0.03928} \times (-1.0312)$$

or

$$\theta_1 = + \frac{1.63755 \times 10^{-4}}{3.928 \times 10^{-2}} = 0.004169\ [\text{rad.}],$$

so that

$$\pm\,\theta_1 = 57.3 \times 0.004169 \cong \underline{0.24^\circ}.$$

[422]

It is advisable to check this result by means of equation (ii); this gives

$$-\theta_1 = \frac{-T_0}{0\cdot35378}\begin{bmatrix} + & 0\cdot16108 \times 0\cdot62001 \\ + & 19\cdot78414 \times 0\cdot78460 \\ + & 0\cdot09999 \times (-0\cdot98946) \\ + & 18\cdot73696 \times 0\cdot14463 \\ + & 0\cdot05427 \times 0\cdot36946 \\ + & 16\cdot69990 \times (-0\cdot92923) \\ + & 0\cdot02366 \times (-0\cdot36946) \\ + & 13\cdot78032 \times 0\cdot92923 \\ + & 0\cdot00655 \times 0\cdot98946 \\ + & 10\cdot13280 \times (-0\cdot14463) \\ + & 5\cdot94976 \times (-0\cdot78460) \end{bmatrix} = \frac{-T_0}{0\cdot35378}\begin{bmatrix} + & 0\cdot09987 \\ + & 15\cdot52264 \\ - & 0\cdot09894 \\ + & 2\cdot70993 \\ + & 0\cdot02005 \\ - & 15\cdot51805 \\ - & 0\cdot00874 \\ + & 12\cdot80509 \\ + & 0\cdot00648 \\ - & 1\cdot46551 \\ - & 4\cdot66818 \end{bmatrix}$$

$$+\theta_1 = \frac{T_0}{0\cdot35378} \times [31\cdot16406 - 21\cdot75942] = \frac{0\cdot0001588}{0\cdot35378} \times 9\cdot40464 = 0\cdot004221 \text{ [rad.]},$$

so that
$$\pm\,\theta_1 = 57\cdot3 \times 0\cdot004221 \simeq 0\cdot24^{\circ}.$$

The agreement is within 1·2 %, i.e.

$$\frac{422 - 417}{\tfrac{1}{2}(422 + 417)} \times 100 = 1\cdot2,$$

and this can be regarded as an indication of the accuracy of the entire calculation.

Shaft twist

By substituting the numerical values for θ_1, a, A, b, B, etc., in the corresponding expression in the Holzer table with gas torques and damping, the twist of any shaft portion can be determined. In most cases it is sufficient to determine shaft twist for the 'node' section of the shafting. Details of the evaluation for the present example are given below.

From $u(8)$, i.e. row u, column 8, of the tabulation, we have:

$$\begin{aligned}
\theta'_{sh(6-7)} &= 0\cdot33586\theta_1 + 0\cdot07682a - 0\cdot00318A + 0\cdot13074b - 0\cdot00233B \\
&\quad + 0\cdot17776c - 0\cdot00151C + 0\cdot21538d - 0\cdot00080D + 0\cdot24163e \\
&\quad - 0\cdot00028E + 0\cdot25510f \\
&= 0\cdot33586 \times 0\cdot0042 + 0\cdot000159\{ + 0\cdot07682 \times 0\cdot62 - 0\cdot0038 \times 0\cdot7846 \\
&\quad + 0\cdot1307 \times (-0\cdot9895) - 0\cdot0023 \times 0\cdot1446 + 0\cdot1778 \times 0\cdot3695 \\
&\quad - 0\cdot0015 \times (-0\cdot9292) + 0\cdot2154 \times (-0\cdot3695) - 0\cdot0008 \times 0\cdot9292 \\
&\quad + 0\cdot2416 \times 0\cdot9895 - 0\cdot0003 \times (-0\cdot1446) + 0\cdot2551 \times (-0\cdot620)\} \\
&= 0\cdot00141 + 0\cdot000159\{0\cdot3524 - 0\cdot3725\} = 0\cdot00141.
\end{aligned}$$

$$\begin{aligned}
\theta''_{sh(6-7)} &= -0\cdot00610\theta_1 + 0\cdot00318a + 0\cdot07682A + 0\cdot00233b + 0\cdot13074B \\
&\quad + 0\cdot00151c + 0\cdot17776C + 0\cdot00080d + 0\cdot21538D + 0\cdot00028e \\
&\quad + 0\cdot24163E + 0\cdot2551F \\
&= -0\cdot0061 \times 0\cdot0042 + 0\cdot000159\{ + 0\cdot0032 \times 0\cdot62 + 0\cdot0768 \times 0\cdot7846 \\
&\quad + 0\cdot0023 \times (-0\cdot9895) + 0\cdot1307 \times 0\cdot1446 + 0\cdot0015 \times 0\cdot3695 \\
&\quad + 0\cdot1778 \times (-0\cdot9292) + 0\cdot0008 \times (-0\cdot3695) + 0\cdot2154 \times 0\cdot9292 \\
&\quad + 0\cdot0003 \times 0\cdot9895 + 0\cdot2416 \times (-0\cdot1446) + 0\cdot2551 \times (-0\cdot7846)\} \\
&= -0\cdot000026 + 0\cdot000159\{0\cdot280 - 0\cdot402\} = -0\cdot000045.
\end{aligned}$$

Therefore $\theta_{sh} = \theta'_{sh} + j\theta''_{sh} = 0 \cdot 00141 + j[-0 \cdot 00005]$,

$|\theta_{sh}| = 10^{-4} \times \sqrt{\{(14 \cdot 1)^2 + (0 \cdot 5)^2\}} = 10^{-4} \times \sqrt{(198 \cdot 81 + 0 \cdot 25)} = \underline{0 \cdot 00141}$ [rad.].

The shaft twist $\Delta_{sh(6-7)}$ relative to the deflexion at mass no. 1 is

$$\Delta_{sh} = \frac{\theta_{sh}}{\theta_1} = \frac{0 \cdot 001411}{0 \cdot 00420} = 0 \cdot 33595.$$

Vibration amplitudes of masses J_2 to J_8

The vibration amplitudes θ_2, θ_3, ..., θ_8 are determined by direct substitutions in the expressions of column (4) of the tabulation, i.e. by the same procedure as for shaft twist.

For mass no. 2, we have

$$\theta_2 = [0 \cdot 94717\theta_1 - 0 \cdot 17241a] + j[0 \cdot 00108\theta_1 - 0 \cdot 17241A]$$

$$= [0 \cdot 94717 \times 0 \cdot 0042 - 0 \cdot 17241 \times 0 \cdot 62 \times 0 \cdot 000159]$$

$$+ j[0 \cdot 00108 \times 0 \cdot 0042 - 0 \cdot 17241 \times 0 \cdot 7846 \times 0 \cdot 000159]$$

$$= 0 \cdot 00396 - j0 \cdot 00002,$$

$$|\theta_2| = 0 \cdot 00396 \text{ [rad.]}.$$

The evaluations are similar for the vibration amplitudes at masses J_3 to J_8. The results are

$\theta_3 = 0 \cdot 00355 - j0 \cdot 00006,$ $\theta_6 = 0 \cdot 00120 - j0 \cdot 00004,$

$\theta_4 = 0 \cdot 00282 - j0 \cdot 00004,$ $\theta_7 = 0 \cdot 00017 - j0 \cdot 000003,$

$\theta_5 = 0 \cdot 00215 - j0 \cdot 00005,$ $\theta_8 = -0 \cdot 00021 - j0 \cdot 0....$

The slight irregularity in the decreasing values of the imaginary terms is not surprising, since the calculations of θ_1, a, A, etc., have been restricted to five decimal places. The accuracy of $\theta_1 = 0 \cdot 00420$ rad. is $1 \cdot 2 \%$, i.e. of the order of $\pm 0 \cdot 000002$, so that the variation of $0 \cdot 00001$ in the imaginary terms is acceptable. It should be noted that for 'inherent engine damping', as in the case of this example, the absolute values are practically equal to the real terms only (i.e. $|\theta_3| \simeq 0 \cdot 0355$, etc.) and that the amplitudes relative to mass no. 1 are practically identical with those of the Holzer table without damping (i.e. $|\theta_2/\theta_1| \simeq \Delta_2$, etc.).

The procedure, however, does not only represent a frequency determination, but also gives the predicted values of θ_1, θ_2, ..., etc., without any empirical formulae or assumptions regarding vector summations for torques, and equivalent engine inertias $\Sigma J\Delta^2$.

In systems with heavier damping than in this example, for instance, systems with propeller damping or damping obtained from a viscous-fluid damper, the relative amplitudes with and without damping may be appreciably different, particularly under non-resonant conditions. These conditions alter the value of the shaft stress and call for careful investigation.

Determination of vibration amplitude for 9th (major) order

The evaluation is fairly simple. As the 9th order is a major order, all the gas torques are in phase with one another, so that we may write

$$T_1 = T_2 = \ldots = T_6, \quad T_{cyl.} = T_0 \times 10^6 = 47 \cdot 63 \text{ [Lb.ft.]},$$

and $\quad T_1/10^6 = T_0 \cos \phi_1 + jT_0 \sin \phi_1 = a + jA = b + jB = \ldots = f + jF.$

[424]

Reverting to equation (iii) of p. 419, we see that it may now be rewritten as follows:

$$0 = \begin{bmatrix} -19 \cdot 76626a + 2 \cdot 35771A \\ -18 \cdot 72586a + 2 \cdot 18035A \\ -16 \cdot 69387a + 1 \cdot 90846A \\ -13 \cdot 77769a + 1 \cdot 55369A \\ -10 \cdot 13207a + 1 \cdot 13519A \\ -\; 5 \cdot 94976a + 0 \cdot 66060A \end{bmatrix} = -85 \cdot 04851a + 9 \cdot 79240A,$$

since $a = b = \ldots = f$, and $A = B = \ldots = F$.

Moreover, as $\qquad a = T_0 \cos \phi_1 \quad$ and $\quad A = T_0 \sin \phi_1,$

the phase angle ϕ_1 of T_1 relative to θ_1 can be obtained:

$$-85 \cdot 04851 \cos \phi_1 + 9 \cdot 79240 \sin \phi_1 = 0,$$

$$\tan \phi_1 = 85 \cdot 04851/9 \cdot 79240 = +8 \cdot 68515,$$

$$\phi_1 = 83° \, 26',$$

therefore, $\qquad \sin \phi_1 = +0 \cdot 99344 \quad$ and $\quad \cos \phi_1 = +0 \cdot 11436.$

The value of θ_1 can now be obtained from equation (i) of p. 416 as follows:

$$-\theta_1 = \frac{T_0}{0 \cdot 03928} \times \left\{ \begin{bmatrix} -19 \cdot 78414 \\ -18 \cdot 73696 \\ -16 \cdot 69990 \\ -13 \cdot 78032 \\ -10 \cdot 13280 \\ -\; 5 \cdot 94976 \end{bmatrix} \times 0 \cdot 11436 + \begin{bmatrix} +0 \cdot 16108 \\ +0 \cdot 09999 \\ +0 \cdot 05427 \\ +0 \cdot 02366 \\ +0 \cdot 00655 \end{bmatrix} \times 0 \cdot 99344 \right\},$$

$$-\theta_1 = \frac{4 \cdot 76 \times 10^{-5}}{3 \cdot 928 \times 10^{-2}} \times \{-85 \cdot 08388 \times 0 \cdot 11436 + 0 \cdot 34555 \times 0 \cdot 99344\} = \underline{0 \cdot 011375 \, [\text{rad.}]},$$

so that $\qquad \pm \theta_1 = 57 \cdot 3 \times 0 \cdot 011375 = 0 \cdot 65°.$

Checking this result by means of equation (ii), we find

$$-\theta_1 = \frac{T_0}{0 \cdot 35378} \times \left\{ \begin{bmatrix} -0 \cdot 16108 \\ -0 \cdot 09999 \\ -0 \cdot 05427 \\ -0 \cdot 02366 \\ -0 \cdot 00655 \end{bmatrix} \times 0 \cdot 11436 + \begin{bmatrix} -19 \cdot 78414 \\ -18 \cdot 73696 \\ -16 \cdot 69990 \\ -13 \cdot 78032 \\ -10 \cdot 13282 \\ -\; 5 \cdot 94676 \end{bmatrix} \times 0 \cdot 99344 \right\}$$

$$= \frac{4 \cdot 76 \times 10^{-4}}{3 \cdot 5378} \times \{-0 \cdot 34565 \times 0 \cdot 11436 - 85 \cdot 08390 \times 0 \cdot 99344\} = \underline{1 \cdot 1366 \times 10^{-2} \, [\text{rad.}]},$$

so that $\qquad \pm \theta_1 = 57 \cdot 3 \times 0 \cdot 011366 = 0 \cdot 65°.$

The agreement is within $0 \cdot 1 \%$, i.e.

$$\frac{11375 - 11366}{\frac{1}{2}(11375 + 11366)} \times 100 = 0 \cdot 09,$$

and this can be regarded as an indication of the accuracy of the calculation.

Finally, we shall evaluate the shaft twist $\theta_{sh(6-7)}$ between masses J_6 and J_7. The expression, taken from $u(8)$ in the tabulation, gives

$$\theta'_{sh} = 0.33586\theta_1 + 4.76 \times 10^{-5} \left\{ \begin{bmatrix} +0.07682 \\ +0.13074 \\ +0.17776 \\ +0.21538 \\ +0.24163 \\ +0.25510 \end{bmatrix} \times 0.11436 - \begin{bmatrix} +0.00381 \\ +0.00233 \\ +0.00151 \\ +0.00080 \\ +0.00028 \end{bmatrix} \times 0.9934 \right\}$$

$$= 0.33586 \times 1.1372 \times 10^{-2} + 4.76 \times 10^{-5} \times 1.0887 = \underline{+0.003871} \text{ [rad.]},$$

$$\theta''_{sh} = -0.00610 \times 1.1372 \times 10^{-2} \left\{ \begin{array}{l} \\ +4.76 \times 10^{-5} \end{array} \right. \left\{ \begin{bmatrix} +0.00318 \\ +0.00233 \\ +0.00151 \\ +0.00180 \\ +0.00028 \end{bmatrix} \times 0.11436 + \begin{bmatrix} +0.07682 \\ +0.13074 \\ +0.17776 \\ +0.21538 \\ +0.24163 \\ +0.25510 \end{bmatrix} \times 0.9934 \right\}$$

$$= -6.9369 \times 10^{-5} + 5.1894 \times 10^{-5}$$

$$= -1.75 \times 10^{-5},$$

therefore $\quad\quad \theta_{sh} = \theta'_{sh} + j\theta''_{sh} = 0.003871 + j[-0.000018] \quad$ [rad.].

The shaft twist relative to the deflexion of mass no. 1 is

$$\theta_{sh}/\theta_1 \simeq 0.003871/0.001372 \simeq \underline{0.34046}.$$

Note. The tabulation for multi-mass systems with damping makes it possible to allow for damping due to auxiliary drives at the corresponding points in the system. For instance, if, in the above example, a pump drive produces additional damping at mass no. 1, the value of the damping coefficient c_1 related to this mass can be increased slightly until the corresponding reduction in vibration amplitude is obtained.

Systems with damping acting between two masses ('shaft' damping) can be dealt with by using the expression

$$K_{12} + j\omega c_{12}$$

instead of K_{12} in column (7) of the tabulation. The subscript '12' indicates that c_{12} is related to the movement of two consecutive masses, e.g. J_1 and J_2. As the reciprocal of $K + j\omega c$ is

$$\frac{1}{K + j\omega c} = \frac{K - j\omega c}{K^2 + (\omega c)^2},$$

the value of the shaft deflexion θ_{sh} in column (8) of the tabulation is obtained by multiplying the expression for Σ (the total torque) by this reciprocal value. [For an example, see Dampers, p. 535.]

2·443 Other tabulation methods for systems with gas torques and damping

Instead of having all the gas torques T_1, T_2, T_3, ... of the individual cylinders in a single table, as in the Holzer tables with damping given in the previous pages, it is also possible to prepare separate tables, each of which contains only one T-value (e.g. $T_1 = T_1' + jT_1''$) in the column for gas torque, all the other torques being left equal to zero.

For each table the evaluation can be carried out so as to obtain the 'corresponding deflexion' at mass 1, i.e.

$$\theta_{1(1)} \quad \text{for the table with} \quad T_1 = T_1' + jT_1'',$$

$$\theta_{1(2)} \quad \text{for the table with} \quad T_2 = T_2' + jT_2'',$$

$$\text{etc.}$$

The resultant deflexion θ_1 for all the torques is then obtained by geometrical addition, that is, by a diagram construction in which the θ-contributions are phased in relation to their corresponding torque vectors. This possibility has already been employed in semi-graphical methods (see section 2·43, pp. 358–389). The extent of the calculations for each table is thus reduced, but the number of tables required is then equal to the number of excitation torques (i.e. the number of engine cylinders).

A tabulation method of this type has been developed by Giraudeau.† The method differs somewhat, however, from the Holzer-table method and is more closely related to the semi-graphical method mentioned above, in that for each applied-torque position tables are calculated from both ends of the system. The number of tables is thus equal to twice the number of engine cylinders in the system.

Furthermore, instead of using 'mass damping' (i.e. $j\omega c_1 \theta_1$, etc.), Giraudeau employs 'shaft damping' (i.e. $j\omega c_{12}(\theta_1 - \theta_2)$), which is probably more representative of true damping conditions in close-coupled systems (see section 2·45, p. 446). The term for damping torque is therefore added to the corresponding term for shaft torque (see Note on p. 426) and the expressions obtained are of the form

$$\frac{K_{12}'}{\omega_0^2}(\theta_1 - \theta_2) = \left(\frac{K_{12}}{\omega^2 R_0^2} + j\frac{c_{12}}{\omega^2 R_0^2}\right)(\theta_1 - \theta_2),$$

where $K =$ shaft stiffness, $\omega =$ phase velocity, $R_0 =$ crank radius, and $c_{12} =$ 'shaft damping' coefficient.

Giraudeau's method is also of interest because it makes use of the reciprocal theorem to reduce some of the calculation work. However, the amount of computation required is still at least as great as by Holzer's method.

† Giraudeau, A., *Vibrating Systems with a Number of Viscous-damping Points; Application to Torsional Vibrations of Line Shafting*, 31 pp. (Ass. Technique Maritime et Aéronautique, Session 1954).

2·45 Formulae based on analytical methods for calculations of systems with damping

In previous sections of the handbook, various procedures have been given for the assessment of vibration amplitudes, viz. Empirical formulae (section 2·41), Graphical methods (section 2·43) and Tabulation methods (section 2·44).

Comparing these, it may be said that the empirical formulae are the simplest; however, most of them assume that the engine system can be reduced to an equivalent one-mass system, which is not always possible (e.g. in the case of marine propulsion systems). The graphical methods, and the tabulation methods with damping, take full account of the complete system, but, in order to do this, they require vector-polygon constructions or complex-number calculations which increase in complexity with the number of masses and excitation torques.

The formulae based on analytical methods outlined in this section are the exact theoretical formulae for peak and flank amplitudes of various types of systems. These formulae can be evaluated fairly rapidly (for instance, the vibration amplitude at mass no. 1 of a three-mass system can be evaluated directly, whereas with vectors or complex number tabulations the entire system has to be calculated). Being algebraic expressions, they can also be used for comparisons and to assess the trend of amplitude variations when altering the value of one of the elements in a vibrating system. Finally, the analysis provides indications of anti-resonant conditions, in a manner which illustrates the value of analytical treatment.

The formulae given in the following are only for one-mass, two-mass and three-mass systems. However, it is possible to develop formulae for systems with larger numbers of masses, particularly if some of the inertias and shaft stiffnesses have identical values, which is frequently the case in engine systems. Formulae for peak and flank amplitudes, shaft twists and dynamic magnifiers are given for all systems mentioned below:

One-mass system.

Two-mass system with 'mass damping' (i.e. engine and propeller damping).

Two-mass system with 'shaft damping' (i.e. damping along shafting between two masses).

Three-mass systems:

Engine with flywheel and front-end pulley (comparison with two-mass system without pulley) (shaft damping).

Engine with flywheel and front-end pulley (mass damping).

Two-cylinder engine with main flywheel (with use of relative amplitudes from ordinary Holzer tables).

Engine with flywheel, and driven machine (mass damping).

Transient vibrations: free decaying vibrations of a one-mass system (coupling tests).

[428]

2·451 One-mass system with excitation torque

Multi-mass systems have hitherto frequently been reduced to an equivalent one-mass system for the estimation of vibration amplitude. The principal expressions for a one-mass system (Fig. 1) are as follows:

Let T = sinusoidally-varying excitation torque [Lb.in.],†

$\quad K$ = shaft stiffness [Lb.in./rad.],

$\quad J$ = inertia of mass [Lb.in.sec.²],

$\quad c = T/\dot\theta$ = damping coefficient [Lb.in.sec./rad.],

$\quad \omega$ = phase velocity of excitation torque [rad./sec.],

$\quad \omega_0$ = natural phase velocity of system [rad./sec.],

$\quad \theta$ = vibration amplitude of mass J [rad.].

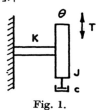

Fig. 1.

The static deflexion (at zero frequency) is

$$\theta_s = T_0/K \quad \text{[rad.]}.$$

Dynamic magnifier at resonance:

$$M = \theta_1/\theta_s, \quad \text{where} \quad \theta_1 = \text{peak amplitude};$$

other relations for M are:

$$M = J\omega_0/c = K/(\omega_0 c).$$

Peak amplitude at resonance:

$$\theta_1 = \frac{T_0}{K} \times M = \frac{T_0}{\omega_0^2 J} \times M = \frac{T_0}{\omega_0 c}. \quad \text{[rad.]}.$$

Flank amplitude at non-resonant frequency:

$$\theta_F = \frac{\theta_1}{\sqrt{\left\{ M^2\left[1 - \frac{\omega^2}{\omega_0^2}\right]^2 + \left(\frac{\omega}{\omega_0}\right)^2 \right\}}}.$$

A graph of θ_F/θ_1 plotted against ω/ω_0 is given for various M-values in Fig. 3.

Derivation. The equation of motion is

$$J\ddot\theta + c\dot\theta + K\theta = T.$$

With complex numbers, the excitation torque is represented by

$$T = T_0\, e^{j\omega t} \equiv T_0\{\cos \omega t + j \sin \omega t\},$$

---

† *Metric units.* The formulae in this section are also valid in metric units, with linear dimensions expressed in cm. and forces in Kg. One thus obtains T in Kg.cm., K in Kg.cm./rad., J in Kg.cm.sec.², and c in Kg.cm.sec./rad.

and the amplitude is $\theta = \theta^* e^{j\omega t}$, where $j^2 = -1$, hence the equation of motion may be rewritten as

$$[-\omega^2 J + j\omega c + K]\theta^* = T_0 = Z\theta^*,$$

and as the absolute values only are required: $|T_0| = |Z||\theta^*|$, hence with $|Z| = \sqrt{\{[K - \omega^2 J]^2 + (\omega c)^2\}}$

$$\theta_F = T_0/\sqrt{\{[K - \omega^2 J]^2 + (\omega c)^2\}}.$$

Introducing $\omega_0^2 = K/J$, one obtains the above formula for θ_F, using $\theta_s = T_0/K$, $M = K/\omega_0 c$ and $\theta_1 = M\theta_s$.

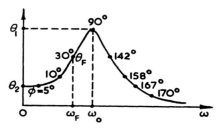

Fig. 2. Resonance curve of a one-mass system, showing variations of amplitude θ and phase angle ϕ as functions of the phase velocity ω of the excitation torque.

The phase angle ϕ between the applied torque T and θ_F is given by

$$\tan\phi = \omega c/[K - \omega^2 J] = M \times \frac{\omega}{\omega_0} \Big/ \left[1 - \left(\frac{\omega}{\omega_0}\right)^2\right],$$

so that it is apparent that the phase angle ϕ of the vibration is $0°$ at zero frequency, and that θ_F lags $90°$ behind T at resonance (see Fig. 2). At still higher frequencies the lag increases and ϕ tends towards $180°$.

Experimental determination of one-mass system dynamic magnifier

The value of M can be determined experimentally as follows:
(1) Measure the peak amplitude θ_1 at the critical speed N_c;
(2) measure a flank amplitude θ_F at a speed N;
(3) evaluate the relation

$$M = \frac{\sqrt{\{(\theta_1/\theta_F)^2 - (N/N_c)^2\}}}{1 - (N/N_c)^2}.$$

This equation assumes that the excitation torque T is the same at both speeds.

Derivation. The above expression is obtained by solving for M the formula for the flank amplitude θ_F given on the previous page. This leads to $M^2[1 - (\omega/\omega_0)^2]^2 + (\omega/\omega_0)^2 = (\theta_1/\theta_F)^2$ and thus gives M. The ratio of the phase velocities is equal to that of the running speeds at which the torque is produced.

[430]

Flank-amplitude curves for equivalent one-mass systems

Fig. 3 gives curves of relative flank amplitudes θ_F/θ_1 plotted against speed ratio N/N_c (this being also the frequency ratio) for values of dynamic magnifiers M between 5 and 100.

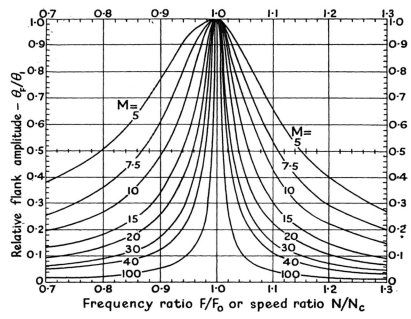

Fig. 3. Flank-amplitude curves for one-mass systems.

EXAMPLE. $N/N_c = 1\cdot1$, $M = 15$, $\theta_1 = \pm 0\cdot64°$. The flank value $\theta_F/\theta_1 = 0\cdot3$ from the above diagram, therefore, is: $\theta_F = 0\cdot3\theta_1 \cong 0\cdot19°$.

Remarks on use of $e^{j\omega t}$

The expression $e^{j\omega t}$, which was first employed in electrical calculations, is extremely handy for all types of vibration problems.

Its geometrical meaning can be appreciated from the relation

$$e^{j\omega t} \equiv \cos \omega t + j \sin \omega t,$$

which is valid because the MacLaurin series for $e^{j\omega t}$ is equal to the sum of the corresponding series for $\cos \omega t$ and $j \cos \omega t$, where $j = \sqrt{-1}$. By writing

$$\cos \omega t = \frac{x}{r} \quad \text{and} \quad \sin \omega t = \frac{y}{r},$$

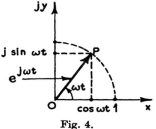

Fig. 4.

where $r^2 = x^2 + y^2$, one sees that, as shown in Fig. 4, $e^{j\omega t}$ is a vector of unit length, in polar co-ordinates, with $\overline{OP} = 1$ and an angle $\phi = \omega t$ relative to the x-axis.

Thus a 'rotating vector' is simply a vector in polar co-ordinates, in the complex-number plane. If an excitation torque varies sinusoidally as a function of time, it can be represented by

$$T = \mathscr{R}[T_0\, e^{j\omega t}] = T_0 \cos \omega t,$$

where $\mathscr{R}[e^{j\omega t}]$ denotes the real part of $e^{j\omega t}$, that is, $\cos \omega t$, which is the projection of the rotating T-vector on the x-axis. The sinusoidally varying angular motion of a mass J_1 can similarly be written as

$$\theta_1 = \mathscr{R}[\theta'_{01}\, e^{j(\omega t + \alpha_1)}] = \mathscr{R}[\theta_{01}\, e^{j\omega t}],$$

where $\theta_{01} = \theta'_{01} e^{j\alpha_1}$, α_1 being the phase angle of θ_1 relative to T, which is determined, where necessary, after evaluating the absolute value $|\theta_1|$ (i.e. the 'length') of θ_1. For relations between θ_1 and T, the restriction to the real components is unnecessary, and is generally omitted.

In the evaluation of systems vibrating under steady-state conditions (that is, with constant values for peak amplitudes), the use of $e^{j\omega t}$ as a substitution in differential equations results in an appreciable shortening of the computation work, particularly for multi-mass systems with damping.

Finally, it may be noted that the symbols j and i are both employed to denote $\sqrt{-1}$, and there is no special distinction between the use of i or j as an 'operator' or as an imaginary number, except that i is used to denote electric current so that it is convenient to write j for $\sqrt{-1}$. In this handbook, j is employed, in conformity with electrical engineering practice.

2·452 *Two-mass system with engine damping and propeller damping*

The system considered consists of two inertias J_1 and J_2, connected together by a shaft of stiffness K, and each provided with a dashpot (damping coefficients c_1 and c_2 in Fig. 5). The input torque T_1 acts on the engine inertia J_1.

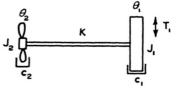

Fig. 5.

The evaluation can be based on any of the following consistent sets of units:

Inertias	Lb.in.sec.^{2}	Lb.ft.sec.^{2}	Kg.cm.sec.^{2}
Stiffnesses	Lb.in./rad.	Lb.ft./rad.	Kg.cm./rad.
Damping coefficients	Lb.in.sec./rad.	Lb.ft.sec./rad.	Kg.cm.sec./rad.
Torque	Lb.in.	Lb.ft.	Kg.cm.

(a) *Resonant conditions.* At resonance, the vibration amplitudes (in radians) are given by the following expressions:

Mass no. 1:

$$\theta_1 = \frac{T_1}{\omega} \times \frac{\sqrt{\left\{\left(\dfrac{J_2}{J_1}\right)^2 + \left(\dfrac{\omega c_2}{K}\right)^2\right\}}}{\sqrt{\left\{\left(c_2\dfrac{J_1}{J_2}\right)^2 + \left(c_1\dfrac{J_2}{J_1}\right)^2 + 2c_1 c_2 + \left(\dfrac{\omega c_1 c_2}{K}\right)^2\right\}}},$$

Mass no. 2:

$$\theta_2 = \theta_1 \times \cfrac{1}{\sqrt{\left\{\left(\dfrac{J_2}{J_1}\right)^2 + \left(\dfrac{\omega c_2}{K}\right)^2\right\}}},$$

Shaft twist:

$$\theta_{sh} = |\theta_1 - \theta_2| = \theta_1 \times \cfrac{\sqrt{\left\{\left(1 + \dfrac{J_2}{J_1}\right)^2 + \left(\dfrac{\omega c_2}{K}\right)^2\right\}}}{\sqrt{\left\{\left(\dfrac{J_2}{J_1}\right)^2 + \left(\dfrac{\omega c_2}{K}\right)^2\right\}}}.$$

The damping coefficient c_1 can be obtained from the dynamic magnifier M_1 of the engine without a driven machine, by means of the relation $c_1 = J_1 \omega / M_1$. The propeller damping is determined by one of the methods indicated in section 2·422. The above formulae are, of course, also valid when either c_1 or c_2 is zero. Moreover, J_2 and c_2 may refer to driven machines of various types.

In addition, in certain applications it is useful to consider the vibratory shaft torque. This is obtained directly as

$$T_{sh} = K \times \theta_{sh}.$$

Since in a two-mass system, $(J_1 + J_2)/J_1 = \omega^2 J_2/K$, this torque can also be written as follows:

$$T_{sh} = \frac{T_1 \times \sqrt{(\omega^2 J_2^2 + c_2^2)}}{\sqrt{\{(c_2 J_1/J_2)^2 + (c_1 J_2/J_1)^2 + 2c_1 c_2 + (\omega c_1 c_2/K)^2\}}}.$$

(b) *Non-resonant conditions.* When ω is not the natural phase velocity of the system, it is necessary to evaluate first the expression:

$$\mathscr{D} = \omega \sqrt{\{\omega^2 [\omega^2 J_1 J_2 - K(J_1 + J_2) - c_1 c_2]^2 + [c_2(K - \omega^2 J_1) + c_1(K - \omega^2 J_2)]^2\}}.$$

The vibration amplitudes and the shaft twist are then obtained from

$$\theta_1 = \frac{T_1}{\mathscr{D}} \times \sqrt{\{(K - \omega^2 J_2)^2 + (\omega c_2)^2\}}, \qquad \theta_2 = \frac{T_1}{\mathscr{D}} \times K,$$

$$\theta_{sh} = |\theta_1 - \theta_2| = \frac{T_1}{\mathscr{D}} \times \sqrt{\{(\omega^2 J_2)^2 + (\omega c_2)^2\}},$$

while the shaft torque is given by $T_{sh} = K\theta_{sh}$.

(c) *Variation of vibration amplitudes with running speed.* It is of interest to consider the shape of the resonance curves obtained when the two-mass system is excited by torques at various frequencies, i.e. corresponding to various running speeds. For this, it is convenient to consider separately the cases for which either $c_2 = 0$ or $c_1 = 0$.

System with engine damping	System with driven-machine damping
$c_1 \neq 0, \quad c_2 = 0.$	$c_1 = 0, \quad c_2 \neq 0.$
Peak-amplitude values (at resonance)	Peak-amplitude values (at resonance)

$$\theta_1 = T_1/(\omega c_1)$$

$$\theta_2 = \frac{T_1}{\omega c_2} \times \frac{J_2}{J_1}$$

$$\theta_2 = \theta_1 \times J_1/J_2$$

$$\theta_1 = \theta_2 \times \sqrt{\{(J_2/J_1)^2 + (\omega c_2/K)^2\}}$$

$$\theta_{sh} = \theta_1 \times [1 + (J_1/J_2)]$$

$$\theta_{sh} = \theta_2 \times \sqrt{\{[1 + (J_2/J_1)]^2 + (\omega c_2/K)^2\}}$$

'Anti-resonant' values — 'Anti-resonant' values

(a) When $\omega^2 = K/J_1$,   (a) When $\omega^2 = K/J_1$,

$$\theta_1 = \frac{T_1}{K} \times \frac{\left(1 - \dfrac{J_2}{J_1}\right)}{\sqrt{\left\{1 + \dfrac{c_1^2}{KJ_1}\left(1 - \dfrac{J_2}{J_1}\right)^2\right\}}}$$

$$\theta_1 = \frac{T_1}{K} \times \sqrt{\left\{\left(\frac{J_2}{J_1} - 1\right)^2 + \frac{c_2^2}{KJ_1}\right\}}$$

$$\theta_2 = \theta_1/[1 - (J_2/J_1)]$$

$$\theta_2 = T_1/K$$

(b) When $\omega^2 = K/J_2$,   (b) When $\omega^2 = K/J_2$,

$$\theta_1 = 0$$

$$\theta_2 = \frac{T_1}{K} \bigg/ \sqrt{\left\{1 + \frac{c_2^2}{KJ_2}\left(\frac{J_1}{J_2} - 1\right)^2\right\}}$$

$$\theta_2 = T_1/K$$

$$\theta_1 = \theta_2 c_2/\sqrt{(KJ_2)}$$

The resonance curves for both cases are indicated in Figs. 6 and 7.

Note. Regarding the values of c_1 and c_2, see the remarks on p. 433.

In Figs. 6 and 7 the vibration amplitudes increase towards infinity as the ω-values become increasingly small. Referring to the foregoing formulae for non-resonant conditions (p. 433), it is seen that this rise in θ-values is due to the expression for $\mathscr{D}$, which decreases proportionally to ω^2 as $\omega \to 0$.

The physical explanation is that the ascending portion of the θ-curves, in the range of ω-values between the lowest 'anti-resonant' frequency and zero, represents the *cyclic speed variation* of the system.

Whereas a one-mass system, rigidly clamped at one end of its shafting (see Fig. 1 of this section), has a *static deflexion* when torque is applied at zero frequency, a two-mass system without external clamping points (Fig. 5) tends to 'roll' in the direction of applied torque, when torque is applied to one of its masses at very low frequency. As the frequency of the torque excitation increases, the inertia torque builds up gradually and this reaction results in shaft twist, i.e. torsional vibration. This

[434]

behaviour is typical of all multi-mass systems without clamping constraints. (For further details on cyclic speed variation, see section 2·6.)

Figs. 6 and 7 also show that the 'anti-resonant' values of ω determine the 'low flank-amplitude' positions between zero and the resonant frequency. If $\sqrt{(K/J_1)}$ or $\sqrt{(K/J_2)}$ (whichever has the lower value of the two) is very close to $\omega_{\text{res.}}$, the resonance curve has a steep flank on the left-hand side; if the 'anti-resonant' value is not close to $\omega_{\text{res.}}$, the resonance flank is wider. Therefore, in some cases, it may be possible *to alter the flank shape* without varying the natural frequency of the system.

System with engine damping   System with driven-machine damping

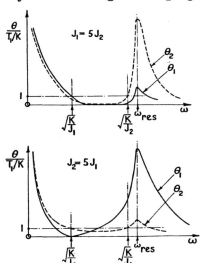 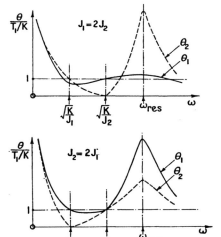

Fig. 6. System with engine damping. Curves of vibration amplitudes, showing peak values and 'anti-resonant' values.

Fig. 7. System with driven-machine damping. Curves of vibration amplitudes, showing peak values and 'anti-resonant' values.

In a two-mass system, this may be achieved: (*a*) by altering the value of K to a new value K', in which case the inertias must be altered to new values J_1' and J_2', such that

$$\omega^2/K' = (1/J_1') + (1/J_2'),$$

where ω is the constant value of resonant phase velocity; or (*b*) by retaining the original K-value and altering only the inertias so as to have new values J_1'' and J_2''; in this case, the modified values must fulfil the condition

$$\omega^2/K = (1/J_1'') + (1/J_2'') = (1/J_1) + (1/J_2).$$

28-2

2·453 *Outline of method used to obtain formulae*

For the two-mass system with engine and driven-machine damping, shown in Fig. 5, the equations of motion are

$$J_1\ddot{\theta}_1 + c_1\dot{\theta}_1 + K(\theta_1 - \theta_2) = T_1,$$

$$J_2\ddot{\theta}_2 + c_2\dot{\theta}_2 + K(\theta_2 - \theta_1) = 0.$$

Substituting in these

$$T_1 = T_1^* e^{j\omega t}, \quad \theta_1 = \theta_1^* e^{j\omega t} \quad \text{and} \quad \theta_2 = \theta_2^* e^{j\omega t},$$

we obtain the characteristic equations:

$$(K - \omega^2 J_1 + j\omega c_1)\,\theta_1^* \qquad - K\theta_2^* \qquad = T_1^*$$

$$- K\theta_1^* \qquad + (K - \omega^2 J_2 + j\omega c_2)\,\theta_2^* = 0.$$

From these, the following determinants are obtained:

$$\mathscr{D} \equiv \mathrm{Det}_s = \begin{vmatrix} K - \omega^2 J_1 + j\omega c_1 & -K \\ -K & K - \omega^2 J_2 + j\omega c_2 \end{vmatrix}, \quad \begin{matrix} \mathrm{Det}_1 = T_1^*(K - \omega^2 J_2 + j\omega c_2), \\ \mathrm{Det}_2 = T_1^* K. \end{matrix}$$

Hence

$$\mathscr{D} = -K(\omega^2 J_2 + \omega^2 J_1) + \omega^4 J_1 J_2 - \omega^2 c_1 c_2 + j\omega\{c_1(K - \omega^2 J_2) + c_2(K - \omega^2 J_1)\}.$$

With resonance occurring when $\omega^4 J_1 J_2 - \omega^2 K(J_1 + J_2) = 0$, one may write

$$\omega^2 J_1 = K + K(J_1/J_2) \quad \text{or} \quad K - \omega^2 J_1 = -KJ_1/J_2,$$

and

$$\omega^2 J_2 = K + K(J_2/J_1) \quad \text{or} \quad K - \omega^2 J_2 = -KJ_2/J_1.$$

Thus, under resonance conditions,

$$\mathscr{D}_{\mathrm{res.}} = -j\omega\{c_1(J_2/J_1) + c_2(J_1/J_2)\}K - \omega^2 c_1 c_2,$$

and

$$|\mathscr{D}_{\mathrm{res.}}| = \omega K \sqrt{\{(c_1 J_2/J_1)^2 + (c_2 J_1/J_2)^2 + 2c_1 c_2 + (\omega c_1 c_2/K)^2\}}.$$

The vibration amplitudes are therefore

$$|\theta_1^*| = |\mathrm{Det}_1|/|\mathscr{D}_{\mathrm{res.}}|, \quad |\theta_2^*| = |\mathrm{Det}_2|/|\mathscr{D}_{\mathrm{res.}}|,$$

and the shaft twist† is determined as

$$|\theta_{sh}^*| = |\theta_1^* - \theta_2^*| = |\mathrm{Det}_1 - \mathrm{Det}_2|/|\mathscr{D}_{\mathrm{res.}}|.$$

For simplicity, the asterisks and the double bars denoting absolute values have been omitted in the formulae set out in the previous pages.

The procedure used has been given in detail in the above example. The analysis of systems with more than two masses can be carried out in a

---

† It should be noted that

$$|\theta_{sh}^*| = |T_1^*\{K - \omega^2 J_2 + j\omega c_2 - K\}|/|D_{\mathrm{res.}}| = |T_1^*(-\omega^2 J_2 + j\omega c_2)|/|D_{\mathrm{res.}}|$$
$$= T_1^* \sqrt{\{(\omega J)^2 + (\omega c_2)^2\}}/|D_{\mathrm{res.}}|,$$

and that this is not the same as $|\theta_1| - |\theta_2|$. The latter only represents the shaft twist if there is no damping and if θ_1 and θ_2 are both vibrating in phase (i.e. at a forcing frequency well below resonance).

similar manner, so that in the following only the resultant formulae will generally be indicated.

Additional formulae for vector diagram construction (Archer's method, used in section 4, p. 600). Let

$$\theta_1 = (a + jA)/\mathscr{D} \quad \text{and} \quad \theta_2 = (b + jB)/\mathscr{D},$$

at resonance, so that

$$a = K - \omega^2 J_2 = -KJ_2/J_1, \quad A = \omega c_2, \quad b = +K, \quad B = 0.$$

Disregarding $\mathscr{D}$ which is common to both amplitude vectors, one sees that θ_1 will be situated in the second quadrant, whereas θ_2 is on the real axis. By vector geometry, the internal angle ψ between θ_2 and θ_1 is therefore obtainable from

$$\tan \psi = \frac{bA - aB}{ab + AB} = \frac{K\omega c_2}{-K^2 J_2/J_1} = -\frac{\omega c_2}{K} \times \frac{J_1}{J_2}.$$

Now

$$+ \omega c_2 J_1/KJ_2 = -\tan \psi = \tan(180° - \psi),$$

so that

$$\psi = 180° - \tan^{-1}\left[\frac{\omega c_2}{K} \times \frac{J_1}{J_2}\right].$$

Furthermore,

$$\left|\frac{\theta_2}{\theta_1}\right| = \frac{K}{|(K - \omega^2 J_2) + j\omega c_2|} = \frac{1}{\sqrt{\left\{\left(\frac{J_2}{J_1}\right)^2 + \left(\frac{\omega c_2}{K}\right)^2\right\}}} = \frac{1}{\frac{J_2}{J_1}\sqrt{\left\{1 + \left(\frac{\omega c_2 J_1}{KJ_2}\right)^2\right\}}}$$

$$= \frac{J_1/J_2}{\sqrt{(1 + \tan^2 \psi)}} = \frac{J_1/J_2}{|\sec \psi|}.$$

Taking $\theta_1/|\theta_1| = $ unit vector, we can thus plot $\theta_2/|\theta_1|$ at an angle ψ to $\theta_1/|\theta_1|$, and by joining the tips of these vectors we obtain

$$\theta_{sh}/|\theta_1| = (\theta_1 - \theta_2)/|\theta_1|.$$

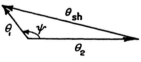

If $|\theta_1|$ (or $|\theta_2|$) is determined by torsiograph measurements, we thus have θ_2 (or θ_1) and θ_{sh} from the vector diagram.

Dynamic magnifier of two-mass system

In one-mass systems with the shaft clamped at one end, the vibration amplitude θ_1 at mass no. 1 is equal to the shaft twist θ_{sh}, so that the dynamic magnifier may be written as

$$M = \frac{\theta_1}{T_1/K} \tag{1}$$

or

$$M = \frac{\theta_{sh}}{T_1/K}, \tag{2}$$

where $T_1 = $ input torque and $K = $ shaft stiffness.

In two-mass systems without clamping points, θ_1 is no longer equal to the shaft twist, so that only eq. (2) is valid.

Using the expression already derived for θ_{sh} at resonance, the dynamic magnifier is obtained directly as

$$\frac{\theta_{sh}}{T_1/K} = M = \frac{\sqrt{\left\{(1+\lambda)^2 + \left(\frac{\omega c_2}{K}\right)^2\right\}}}{\frac{\omega}{K}\sqrt{\left\{\left(\frac{c_2}{\lambda}\right)^2 + (\lambda c_1)^2 + 2c_1 c_2 + \left(\frac{\omega c_1 c_2}{K}\right)^2\right\}}}, \qquad (3)$$

where $\lambda = J_2/J_1$. The damping coefficients c_1 and c_2 can be related to one-mass system dynamic magnifiers, as follows. When the inertia J_2 is clamped, the engine system consists of J_1 and K, and has a dynamic magnifier

$$M_1 = J_1 \omega_1/c_1, \quad \text{so that} \quad c_1^2 = \frac{J_1^2 \omega_1^2}{M_1^2} = \frac{K J_1}{M_1^2},$$

since $\omega_1^2 = K/J_1$. Similarly, when the engine end is clamped, the driven-machine system (consisting of J_2 and K) has a dynamic magnifier

$$M_2 = J_2 \omega_2/c_2, \quad \text{so that} \quad c_2^2 = \frac{J_2^2 \omega_2^2}{M_2^2} = \frac{K J_2}{M_2^2},$$

since $\omega_2^2 = K/J_2$.

Substituting these expressions for c_1 and c_2 in eq. (3) and using the relation: $\omega^2/K = (1/J_1) + (1/J_2)$, we obtain from

$$M = \frac{\sqrt{\left\{\left(1 + \frac{J_2}{J_1}\right)^2 + \left(\frac{1}{J_1} + \frac{1}{J_2}\right)\frac{J_2}{M_2^2}\right\}}}{\sqrt{\left\{\frac{\omega^2}{K}\frac{c_2^2}{K}\left(\frac{J_1}{J_2}\right)^2 + \frac{\omega^2}{K}\frac{c_1^2}{K}\left(\frac{J_2}{J_1}\right)^2 + 2\frac{\omega^2}{K}\frac{c_1 c_2}{K} + \frac{\omega^4}{K^2}\left(\frac{c_1 c_2}{K}\right)^2\right\}}}$$

the expression

$$M = \frac{\sqrt{\left\{(1+\lambda) + \frac{1}{M_2^2}\right\}}}{\sqrt{\left\{\frac{1}{\lambda^2 M_2^2} + \frac{\lambda}{M_1^2} + \frac{2}{\sqrt{\lambda}\, M_1 M_2} + \frac{1+\lambda}{\lambda} \times \frac{1}{M_1^2 M_2^2}\right\}}}.$$

Fig. 8 shows curves of M plotted against $\lambda = J_2/J_1$, for various values of M_1 and M_2 used as parameters.

The curves of Fig. 8a, based on the theoretical formula for the dynamic magnifier M of a two-mass system, give rise to the following remarks. It is an advantage to make good use of the damping available, so as to obtain a low value of M for the entire system, or, more specifically, an M which will be lower than, or at most equal to, the lower of the two values M_1 and M_2 considered.

From Fig. 8a it appears that this can be achieved by using, if possible,
(1) with predominantly driven-machine damping ($M_2 = 10$), J_2/J_1 values less than unity;

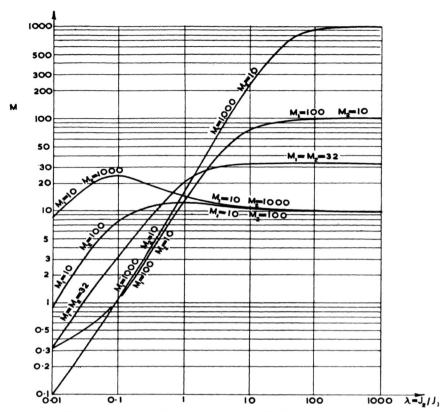

Fig. 8a. Curves of dynamic magnifier M of two-mass system.

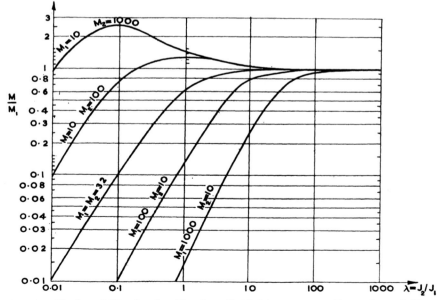

Fig. 8b. Replot of Fig. 8a, for M-values divided by corresponding values of M_1.

(2) with engine damping approximately equal to that of the driven machine ($M_1 = M_2 = 32$), J_2/J_1 values less than 10;

(3) with predominantly engine damping ($M_1 = 10$), J_2/J_1 values greater than unity if the driven-machine damping is negligible ($M_2 = 1000$), or J_2/J_1 values less than unity if the driven-machine damping is not negligible ($M_2 = 100$).

Fig. 8b is a replot of Fig. 8a, arranged so as to facilitate interpolation.

Note. The above applies to systems in which the node is situated in the shafting between the driven machine and the main flywheel of the engine.

Experimental determination of dynamic magnifier of two-mass system

(a) *From measurements of θ_1*

$$M = \frac{\theta_1}{(T_1/K)} \times \frac{\sqrt{\left((1+\lambda)^2 + \dfrac{1+\lambda}{M_2^2}\right)}}{\sqrt{\left(\lambda^2 + \dfrac{1+\lambda}{M_2^2}\right)}},$$

where θ_1 = resonant amplitude at mass no. 1 [rad.], T_1 = excitation torque [Lb.ft.], K = shaft stiffness [Lb.ft./rad.] and $\lambda = J_2/J_1$. This equation assumes that the magnifier M_2 of the driven-machine system is known.

(b) *From measurements of θ_1 and θ_2*

$$M = \frac{\theta_2}{(T_1/K)} \times \sqrt{\left\{\left(\frac{\theta_1}{\theta_2}\right)^2 + 1 + \lambda\right\}},$$

where θ_1 and θ_2 are vibration amplitudes at resonance, measured at both ends of the system (i.e. at J_1 and J_2).

Derivations. The formula under (a) is obtained from the relation between θ_1 and θ_{sh} at resonance (see p. 433) after substituting

$$\frac{\omega^2}{K} \times \frac{c_2^2}{K} = \left(\frac{1}{J_2} + \frac{1}{J_1}\right) \times \frac{J_2}{M_2^2}$$

in the expression for M on p. 438.

The formula under (b) is derived as follows:

$$\theta_1/\theta_2 = \sqrt{\{(KJ_2/J_1)^2 + (\omega c_2)^2\}}/K, \quad \text{hence} \quad \omega^2 c_2^2 = K^2\left[\left(\frac{\theta_1}{\theta_2}\right)^2 - \lambda^2\right];$$

furthermore

$$M = \frac{\theta_2}{T_1/K} \times \sqrt{\left\{J_2^2\left(\frac{1}{J_2} + \frac{1}{J_1}\right)^2 + \left[\left(\frac{\theta_1}{\theta_2}\right)^2 - \lambda^2\right]\right\}},$$

and this gives the above relation.

2·454 *Two-mass system with shaft damping*

The system considered (Fig. 9) consists of two masses of inertias J_1 and J_2, a shaft of stiffness K, and a dashpot with a damping coefficient c, connected to the two masses. The excitation torque T_1 acts on the inertia J_1.

[440]

The vibration amplitudes and the shaft twist are determined by the following relations:

(a) *Resonant conditions*

$$\theta_1 = \frac{T_1}{\omega c} \times \frac{J_1 J_2}{(J_1+J_2)^2} \times \sqrt{\left\{\left(\frac{J_2}{J_1}\right)^2 + \left(\frac{\omega c}{K}\right)^2\right\}},$$

$$\theta_2 = \frac{T_1}{\omega c} \times \frac{J_1 J_2}{(J_1+J_2)^2} \times \sqrt{\left\{1 + \left(\frac{\omega c}{K}\right)^2\right\}}.$$

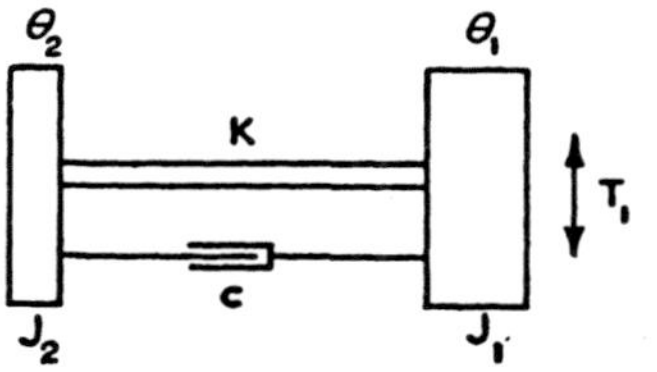

Fig. 9.

Shaft twist:

$$\theta_{sh} = \frac{T_1 J_2}{\omega c (J_1+J_2)};$$

Shaft torque:

$$T_{sh} = K\theta_{sh};$$

Dynamic magnifier:

$$M = \frac{\theta_{sh}}{T_1/K} = \frac{KJ_2}{\omega c (J_1+J_2)} = \frac{J_2 \omega}{c} \times \frac{J_1 J_2}{(J_1+J_2)^2}.$$

In these formulae, the damping coefficient may be replaced by the following expressions based on the dynamic magnifiers M_1 and M_2 of the sub-systems K/J_1 and K/J_2:

$$c = \sqrt{(KJ_1)}/M_1 = \sqrt{(KJ_2)}/M_2.$$

All these equations are derived from the equations of motion of the system, viz.

$$J_1\ddot{\theta}_1 + c(\dot{\theta}_1 - \dot{\theta}_2) + K(\theta_1 - \theta_2) = T_1 e^{j\omega t},$$

$$J_2\ddot{\theta}_2 + c(\dot{\theta}_2 - \dot{\theta}_1) + K(\theta_2 - \theta_1) = 0,$$

using the method previously described for the two-mass system with 'mass damping', i.e. damping actions concentrated at masses J_1 and J_2.

(b) *Non-resonant conditions*

$$\mathscr{D} = \omega^2 \sqrt{\{[\omega^2 J_1 J_2 - K(J_1+J_2)]^2 + (\omega c)^2 \times [J_1 + J_2]^2\}},$$

$$\theta_1 = \frac{T_1}{\mathscr{D}} \times \sqrt{\{(K - \omega^2 J_2)^2 + (\omega c)^2\}},$$

$$\theta_2 = \frac{T_1}{\mathscr{D}} \times \sqrt{\{K^2 + (\omega c)^2\}},$$

$$\theta_{sh} = |\theta_1 - \theta_2| = \frac{T_1}{\mathscr{D}} \times \omega^2 J_2.$$

2·455 Three-mass systems

1. Engine with front-end pulley (shaft damping)

(a) *Engine without pulley.* The two-mass system consists of a main flywheel J_3 and the sum of the cylinder inertias $\Sigma J_{cyl.} = J_2$, connected together by a shaft of stiffness K_2 (see Fig. 10).

[441]

The excitation torque T_E acts on J_2, and between this mass and the main flywheel there is 'shaft damping', represented by the damping coefficient c. The stiffness K_2 is obtained from

$$K_2 = \frac{\omega_E^2}{\dfrac{1}{J_3}+\dfrac{1}{J_2}} = \frac{\omega_E^2}{\dfrac{1}{J_3}+\dfrac{1}{\Sigma J_{\text{cyl.}}}},$$

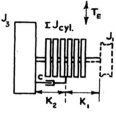

where ω_E is the natural phase velocity of the engine-and-flywheel portion of the system. The vibration amplitude at J_2 is

$$\theta_2 \simeq \frac{T_E}{\omega_E c} \times \frac{J_3^2}{(J_2+J_3)^2}$$

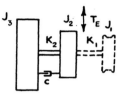

as already determined for two-mass systems with shaft damping (see p. 441), assuming that the term $(c\omega/K_2)$ is small enough to be neglected, which is usually the case.

(b) *Engine with pulley.* The three-mass system, which includes the pulley inertia J_1 and the shafting of stiffness K_1 between J_1 and J_2, has a natural phase velocity ω_* which can be obtained from

Fig. 10. Engine with front-end pulley. It is assumed that the belt stiffness is so low that the effect of other auxiliary masses may be neglected.

$$\omega_*^4 - \left\{ K_1\left(\frac{1}{J_1}+\frac{1}{J_2}\right) + K_2\left(\frac{1}{J_2}+\frac{1}{J_3}\right) \right\} \omega_*^2 + \frac{K_1 K_2}{J_1 J_2 J_3}(J_1+J_2+J_3) = 0.$$

Its vibration amplitudes and shaft twists at resonance are

$$\theta_1^* = \frac{T_E^*}{\omega_* c} \times \frac{\sqrt{\left\{\left(1-\dfrac{\omega_*^2 J_3}{K_2}\right)^2 + \left(\dfrac{\omega_* c}{K_2}\right)^2\right\}}}{\mathscr{D}_{\text{res.}}}, \quad \theta_2^* = \theta_1^* \times \left(1-\frac{\omega_*^2 J_1}{K_1}\right),$$

$$\theta_3^* = \frac{T_E^*}{\omega_* c} \times \frac{\left(1-\dfrac{\omega_*^2 J_1}{K_1}\right)\sqrt{\left\{1+\left(\dfrac{\omega_* c}{K_2}\right)^2\right\}}}{\mathscr{D}_{\text{res.}}},$$

$$\theta_{sh1}^* = |\,\theta_1^* - \theta_2^*\,| = \theta_1^* \times \frac{\omega_*^2 J_1}{K_1},$$

$$\theta_{sh2}^* = |\,\theta_2^* - \theta_3^*\,| = \frac{T_E^*}{\omega_* c \mathscr{D}_{\text{res.}}} \times \left(1-\frac{\omega_*^2 J_1}{K_1}\right) \times \frac{\omega_*^2 J_3}{K_2},$$

where
$$\mathscr{D}_{\text{res.}} = \frac{\omega_*^4}{K_1 K_2}(J_1 J_2 + J_1 J_3) - \frac{\omega_*^2}{K_2}(J_1+J_2+J_3).$$

In the above, an asterisk is used to denote values pertaining to the three-mass system.

By means of the expressions for θ_1^* and θ_2, it is possible to assess the

[442]

effect of the front-end pulley as regards amplitudes of vibration. This will be discussed in the following with the aid of a numerical example.

(c) *Effect of front-end pulley.* The effect of the front-end pulley can be assessed by means of the ratio

$$\frac{\theta_1^*}{\theta_2} = \frac{\text{vibration amplitude at first mass of three-mass system}}{\text{vibration amplitude at first mass of two-mass system}}.$$

Using the formulae for θ_1^* and θ_2^*, and noting that ω_* is the one-node phase velocity, the amplitude ratio may be written as

$$\frac{\theta_1^*}{\theta_2} \simeq \frac{T_E^*}{T_E} \times \frac{\omega_E}{\omega_*} \times \frac{\left(1 - \dfrac{\omega_*^2 J_3}{K_2}\right) \times \dfrac{(J_2 + J_3)^2}{J_3^2}}{\left[\dfrac{\omega_*^4 J_1}{K_1 K_2}(J_2 + J_3) - \dfrac{\omega_*^2}{K_2}(J_1 + J_2 + J_3)\right]},$$

where ω_E and ω_* are determined from the frequency formulae for two-mass and three-mass systems, whereas T_E^* and T_E are based on phase vector sums $\overrightarrow{\Sigma\Delta}^*$ and $\overrightarrow{\Sigma\Delta}$ (for the relative amplitudes at the engine cylinders see section 2·2) taken from corresponding Holzer tables. Thus, one may write

$$T_E^*/T_E = \overrightarrow{\Sigma\Delta}^*/\overrightarrow{\Sigma\Delta}.$$

NUMERICAL EXAMPLE. Considering the six-cylinder engine with the inertia and stiffness values indicated in Fig. 11, it is required to assess the effect on vibration amplitudes of the addition of a front-end pulley J_1 equal to 1·5 or 6·0 Lb.in.sec.². The measured frequency of the engine without the pulley is $F_E \cong 10{,}000$ vib./min.†

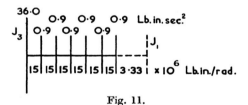

Fig. 11.

The first calculations are to determine K_1 and K_2, as follows:

$$\omega_E = \left(\frac{2\pi}{60}\right) \times F_E = 1047 \text{ rad./sec.}, \qquad \omega_E^2 = 1·094 \times 10^6 \text{ rad.}^2/\text{sec.}^2,$$

hence

$$K_2 = \omega_E^2 \Big/ \left[\frac{1}{36·0} + \frac{1}{(6 \times 0·9)}\right] = 5·13 \times 10^6 \text{ Lb.in./rad.}$$

† *Metric units.* The example is also valid in metric units, with the same numerical values, when 'Kg.' is used instead of 'Lb.', and 'cm.' instead of 'in.'. Then

$$K_1 = 1·98 \times 10^6 \text{ Kg.cm./rad.}, \qquad J_3 = 36·0 \text{ Kg.cm.sec.}^2, \quad \text{etc.}$$

The stiffness of the entire length of shafting from J_3 to J_1 is obtained from

$$\frac{1}{K_{\text{total}}} = \left[\frac{6}{15 \cdot 0} + \frac{1}{3 \cdot 33}\right] \times 10^{-6} = [0 \cdot 4 + 0 \cdot 3] \times 10^{-6} = 0 \cdot 7 \times 10^{-6},$$

so that

$$\frac{1}{K_1} = \frac{1}{K_{\text{total}}} - \frac{1}{K_2} = \left[0 \cdot 7 - \frac{1}{5 \cdot 13}\right] \times 10^{-6} = 0 \cdot 505 \times 10^{-6},$$

and

$$K_1 = 10^6 / 0 \cdot 505 = 1 \cdot 98 \times 10^6 \, \text{Lb.in./rad.}$$

The three-mass system thus has the following constants

$$J_1 = 1 \cdot 5 \text{ or } 6 \cdot 0 \, \text{Lb.in.sec.}^2, \quad K_1 = 1 \cdot 98 \times 10^6 \, \text{Lb.in./rad.},$$

$$J_2 = 5 \cdot 4 \, \text{Lb.in.sec.}^2, \quad\quad\quad K_2 = 5 \cdot 13 \times 10^6 \, \text{Lb.in./rad.}$$

$$J_3 = 36 \cdot 0 \, \text{Lb.in.sec.}^2,$$

Calculations for $J_1 = 1 \cdot 5 \, \text{Lb.in.sec.}^2$:

$$\omega_*^4 - 10^6 \left\{ \left(\frac{1}{1 \cdot 5} + \frac{1}{5 \cdot 4}\right) \times 1 \cdot 98 + \left(\frac{1}{5 \cdot 4} + \frac{1}{36}\right) \times 5 \cdot 13 \right\} \omega_*^2 + 10^{12} \times \frac{1 \cdot 98 \times 5 \cdot 13}{1 \cdot 5 \times 5 \cdot 4 \times 36}$$
$$\times (1 \cdot 5 + 5 \cdot 4 + 36) = 0,$$

$$\omega_*^4 - 10^6 \{1 \cdot 686 + 1 \cdot 088\} \omega_*^2 - 1 \cdot 495 \times 10^{12} = 0,$$

$$\omega_*^2 = \tfrac{1}{2}\{2 \cdot 774 - \sqrt{[(2 \cdot 774)^2 - 4 \times 1 \cdot 495]}\} \times 10^6 = 0 \cdot 732 \times 10^6,$$

therefore $\omega_* = 856 \, \text{rad./sec.}$ and $F_* = \dfrac{60}{2\pi}\omega_* = 9 \cdot 55 \omega_* = 8175 \, \text{vib./min.}$

From Holzer tabulations, we have $\overrightarrow{\Sigma\Delta} = 3 \cdot 78$ and $\overrightarrow{\Sigma\Delta}^* = 2 \cdot 7$. Further preliminary calculations are as follows:

$$1 - \frac{\omega_*^2 J_3}{K_2} = 1 - \frac{0 \cdot 732 \times 36}{5 \cdot 13} = -4 \cdot 133, \quad\quad \frac{(J_2 + J_3)^2}{J_3^2} = \frac{(5 \cdot 4 + 36)^2}{36^2} = 1 \cdot 322,$$

$$\omega_*^2 \frac{J_1}{K_1} = \frac{0 \cdot 732 \times 1 \cdot 5}{1 \cdot 98} = 0 \cdot 554, \quad\quad \omega_*^2 \frac{J_1}{K_2} = \frac{0 \cdot 732 \times 1 \cdot 5}{5 \cdot 13} = 0 \cdot 214,$$

$$\omega_*^2 \frac{J_2}{K_2} = \frac{0 \cdot 732 \times 5 \cdot 4}{5 \cdot 13} = 0 \cdot 770, \quad\quad \text{hence}$$

$$\frac{\omega_*^2}{K_2}(J_1 + J_2 + J_3) = 0 \cdot 214 + 0 \cdot 770 + 5 \cdot 13$$

$$\omega_*^2 \frac{J_3}{K_2} = \frac{0 \cdot 732 \times 36}{5 \cdot 13} = 5 \cdot 133, \quad\quad\quad\quad = 6 \cdot 114.$$

hence

$$\omega_*^4 \frac{J_1}{K_1}\left(\frac{J_2 + J_3}{K_2}\right) = 0 \cdot 554 \times (0 \cdot 770 + 5 \cdot 133)$$
$$= 3 \cdot 271.$$

The amplitude ratio is

$$\frac{\theta_1^*}{\theta_2} = \frac{2 \cdot 7}{3 \cdot 78} \times \frac{1 \cdot 047}{0 \cdot 856} \times \frac{(-4 \cdot 133) \times 1 \cdot 322}{[3 \cdot 271 - 6 \cdot 114]} = 1 \cdot 67,$$

that is, θ_1^* is 67 % greater than θ_2.

[444]

Calculations for $J_1 = 6\cdot0$ Lb.in.sec.^{2}:

$$\omega_*^4 - 10^6\left\{\left(\frac{1}{6\cdot0} + \frac{1}{5\cdot4}\right) \times 1\cdot98 + \left(\frac{1}{5\cdot4} + \frac{1}{36}\right) \times 5\cdot13\right\}\omega_*^2 + \frac{1\cdot98 \times 5\cdot13}{6 \times 5\cdot4 \times 36}$$
$$\times (6\cdot0 + 5\cdot4 + 36) = 0,$$

$$\omega_*^4 - 1\cdot785 \times 10^6 \times \omega_*^2 + 0\cdot414 = 0,$$

$$\omega_*^2 = 0\cdot274 \times 10^6,$$

$$\omega_* = 523\cdot5 \text{ rad./sec.} \quad \text{and} \quad F_* = 4995 \text{ vib./min.}$$

From Holzer tabulations. $\overrightarrow{\Sigma\Delta} = 3\cdot78$ and $\overrightarrow{\Sigma\Delta^*} = 1\cdot7$. Further preliminary calculations:

$$1 - \frac{\omega_*^2 J_3}{K_2} = 1 - \frac{0\cdot274 \times 1\cdot5}{5\cdot13} = -0\cdot922, \qquad \frac{(J_2 + J_3)^2}{J_3^2} = 1\cdot322, \quad \text{as above,}$$

$$\omega_*^2 \frac{J_1}{K_1} = \frac{0\cdot274 \times 6\cdot0}{1\cdot98} = 0\cdot830, \qquad\qquad \omega_*^2 \frac{J_1}{K_2} = \frac{0\cdot274 \times 6\cdot0}{5\cdot13} = 0\cdot320,$$

$$\omega_*^2 \frac{J_2}{K_2} = \frac{0\cdot274 \times 5\cdot4}{5\cdot13} = 0\cdot288, \qquad \text{hence}$$

$$\omega_*^2 \frac{J_3}{K_2} = \frac{0\cdot274 \times 36\cdot0}{5\cdot13} = 1\cdot922, \qquad \frac{\omega_*^2}{K_2}(J_1 + J_2 + J_3) = 0\cdot320 + 0\cdot288 + 1\cdot922$$
$$= 2\cdot530.$$

hence

$$\omega_*^4 \frac{J_1}{K_1}\left(\frac{J_2 + J_3}{K_2}\right) = 0\cdot83 \times (0\cdot288 + 1\cdot922)$$
$$= 1\cdot835.$$

The amplitude ratio is

$$\frac{\theta_1^*}{\theta_2} = \frac{1\cdot7}{3\cdot78} \times \frac{1\cdot047}{0\cdot5235} \times \frac{(-0\cdot922) \times 1\cdot322}{[1\cdot835 - 2\cdot530]} = 1\cdot57,$$

that is, θ_1^* is 57 % greater than θ_2.

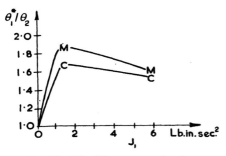

Fig. 12. M—M = measured values.
C—C = calculated values.

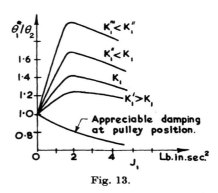

Fig. 13.

Comparison with measured results. An engine of the type indicated in Fig. 11 was tested with and without front-end flywheel inertias at the B.I.C.E.R.A.

Laboratory. The results are compared with calculated values in the following table and are also illustrated in Fig. 12.

	ω^* meas.	ω^* calc.	θ_1^*/θ_2 meas.	θ_1^*/θ_2 calc.
$J_1=0$	1047	—	1	
1·5	860	856	1·87	1·67
6·0	563	524	1·60	1·57
Lb.in.sec.²	rad./sec.		deg.	

The calculated values of θ_1^*/θ_2 are in substantial agreement with the measured results and also show the same trend. It should be noted that:

(1) the value of the stiffness K_1 is a decisive factor for the value of the amplitude increase, i.e. the more flexible the front-end shafting is, the greater the initial increase will be in θ_1^*/θ_2 (see Fig. 13);

(2) tests indicate that the effect is somewhat more pronounced for major than for minor criticals; and

(3) a decrease in amplitude may be obtained (i.e. $\theta_1^*/\theta_2<1$) if appreciable damping occurs at the pulley position (see Fig. 13).

These effects due to the addition of a front-end flywheel have also been recorded with various engine systems investigated by members of the B.I.C.E.R.A. Torsional Vibration Panel.

Remarks on 'shaft damping' and 'mass damping'

Calculations similar to those given above were made on the assumption that the system had 'mass damping' (Fig. 14b), that is, a damping action

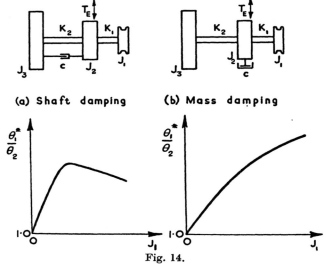

(a) Shaft damping   (b) Mass damping

Fig. 14.

concentrated at the engine mass J_2. It was found that the trend of the θ_1^*/θ_2-values thus obtained was not in agreement with the trend of the test results.

On the other hand, 'mass damping' is operative in systems with long intermediate shafting between various masses (e.g. in marine propulsion systems). Therefore, it appears that

(a) close-coupled systems can be treated by assuming that 'shaft damping' occurs between the engine cylinders and the main flywheel; and

(b) loosely coupled systems can be regarded as having 'mass damping' at the equivalent engine mass and at the driven-machine mass.

A physical explanation for the decreasing trend of θ_1^*/θ_2 after the initial rise may be that, with increasing values of J_1, there is more relative motion, and hence more damping action, between masses J_2 and J_3 of the system.

2. Engine with front-end pulley (mass damping)†

In this system (Fig. 15), $J_1 =$ pulley inertia, $J_2 = \Sigma J_{cyl.} =$ sum of engine cylinder inertias, and $J_3 =$ inertia of main flywheel.

The stiffness K_2 is obtained from the relation

$$K_2 = \frac{\omega_E^2}{\dfrac{1}{J_3} + \dfrac{1}{J_2}} = \frac{\omega_E^2}{\dfrac{1}{J_3} + \dfrac{1}{\Sigma J_{cyl.}}},$$

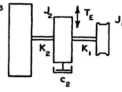

where ω_E is the natural phase velocity of the engine-and-flywheel portion of the system. Having determined K_2, its corresponding equivalent length of shaft is obtained from

$$L_{e2} = \frac{\dfrac{\pi}{32} G D_e^4}{K_2},$$

Fig. 15.

and K_1 is determined as the stiffness of the remaining portion of shafting, i.e. of $L_{e1} = L_{e\,total} - L_{e2}$. (In the above $G =$ modulus of rigidity, and $D_e =$ reference diameter of the equivalent shafts.)

At resonance, the exact expressions for vibration amplitudes and shaft twists are as follows:

$$\theta_1 = \frac{T_E/(\omega c_2)}{\dfrac{\omega^2 J_1}{K_1} - 1}, \quad \theta_2 = \frac{T_E}{\omega c_2}, \quad \theta_3 = \frac{T_E/(\omega c_2)}{\dfrac{\omega^2 J_3}{K_2} - 1},$$

$$\theta_{sh1} = |\theta_1 - \theta_2| = \frac{T_E/(\omega c_2)}{\dfrac{K_1}{\omega^2 J_1} - 1}, \quad \theta_{sh2} = |\theta_2 - \theta_3| = \frac{T_E/(\omega c_2)}{1 - \dfrac{K_2}{\omega^2 J_3}},$$

and the corresponding dynamic magnifiers are $M_{sh1} = \theta_{sh1} \times K_1/T_E$ and $M_{sh2} = \theta_{sh2} \times K_2/T_E$.

Introducing relative amplitudes Δ_1, Δ_2, etc., from Holzer tables at the resonant frequency, the amplitude equations may be rewritten as

$$\theta_1 = \theta_2/\Delta_2, \quad \theta_2 = T_E/(\omega c_2), \quad \theta_3 = \theta_2 \Delta_3/\Delta_2.$$

† This case is included for completeness and as a reference for further work on systems with front-end flywheels or pulleys.

Note. With damping (represented by a damping coefficient c_1) at mass J_1, the expressions become more complicated. Thus, the vibration amplitude at J_1 is then

$$\theta_1 = \frac{T_E}{\mathscr{D}_{\text{res.}}} \times \left(1 - \frac{\omega^2 J_3}{K_2}\right),$$

where

$$\mathscr{D}_{\text{res.}} = \omega \times \sqrt{\left\{\omega^2 \left[\frac{c_1 c_2}{K_1} - \frac{\omega^2 c_1 c_2 J_3}{K_1 K_2}\right]^2 + \left[(c_1 + c_2) - \omega^2 \left\{c_2 \left(\frac{J_3}{K_2} + \frac{J_1}{K_1}\right) + c_1 \left(\frac{J_2 + J_3}{K_1} + \frac{J_3}{K_2}\right)\right\} + \frac{\omega^4 J_3}{K_1 K_2} \{c_1 J_2 + c_2 J_1\}\right]^2\right\}}.$$

3. Two-cylinder engine with main flywheel

The system comprises two cylinder inertias J_1 and J_2, a main flywheel J_3, and two shaft portions of stiffnesses K_1 and K_2. The engine damping is represented by two dashpots c_1 and c_2 at the cylinder positions (see Fig. 16).

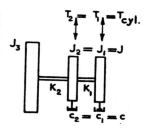

Fig. 16.

The exact expressions for vibration amplitudes and shaft twists, valid for both one-node and two-node vibration conditions, are given below. Only in-phase torques, corresponding to 'major criticals' will be considered:

$$\theta_1 = \frac{T_{\text{cyl.}}}{\mathscr{D}_{\text{res.}}} \times \sqrt{\left\{\left[2 - \omega^2 \left(\frac{J + J_3}{K_1} + \frac{2J_3}{K_2}\right) + \omega^4 \frac{J_3 J}{K_1 K_2}\right]^2 + \left(\frac{\omega c}{K_1}\right)^2 \times \left[1 - \frac{\omega^2 J_3}{K_2}\right]^2\right\}},$$

$$\theta_2 = \frac{T_{\text{cyl.}}}{\mathscr{D}_{\text{res.}}} \times \sqrt{\left\{\left[2 - \omega^2 \left(\frac{J}{K_1} + \frac{2J_3}{K_2}\right) + \omega^4 \frac{J_3 J}{K_1 K_2}\right]^2 + \left(\frac{\omega c}{K_1}\right)^2 \times \left[1 - \frac{\omega^2 J_3}{K_2}\right]^2\right\}},$$

$$\theta_3 = \frac{T_{\text{cyl.}}}{\mathscr{D}_{\text{res.}}} \times \sqrt{\left\{\left(2 - \frac{\omega^2 J}{K_1}\right)^2 + \left(\frac{\omega c}{K_1}\right)^2\right\}},$$

where $\omega =$ phase velocity at resonance, obtained from the frequency formula for three-mass systems (see section 1·3) or from a Holzer tabulation without damping, and

$$\mathscr{D}_{\text{res.}} = \omega c \sqrt{\left\{\left(\frac{\omega c}{K_1}\right)^2 \times \left[1 - \frac{\omega^2 J_3}{K_2}\right]^2 + \left[2 - \omega^2 \left(\frac{2J + J_3}{K_1} + \frac{2J_3}{K_2}\right) + 2\omega^4 \frac{J J_3}{K_1 K_2}\right]^2\right\}}.$$

For one-node frequency conditions, the terms preceded by the factor $(\omega c/K_1)^2$ can generally be neglected. The shaft twists and their corresponding dynamic magnifiers are

$$\theta_{sh\,1} = |\theta_1 - \theta_2| = \frac{T_{\text{cyl.}}}{\mathscr{D}_{\text{res.}}} \times \frac{J_3 \omega^2}{K_1}, \qquad M_{sh\,1} = \frac{\theta_{sh\,1}}{T_{\text{cyl.}}/K_1},$$

$$\theta_{sh\,2} = |\theta_2 - \theta_3| = \frac{T_{\text{cyl.}}}{\mathscr{D}_{\text{res.}}} \times \sqrt{\left\{\left[\frac{2\omega^2 J_3}{K_2} - \omega^4 \frac{J J_3}{K_1 K_2}\right]^2 + \left(\frac{\omega c}{K_1}\right)^2 \times \left(\frac{\omega^2 J_3}{K_2}\right)^2\right\}},$$

$$M_{sh\,2} = \frac{\theta_{sh\,2}}{2T_{\text{cyl.}}/K_2}.$$

Approximate formulae. Using the relative amplitudes $\Delta_1 = \theta_1/\theta_1$, $\Delta_2 = \theta_2/\theta_1$, etc., from an ordinary Holzer tabulation, disregarding the terms with $(\omega c/K_1)$, and noting that $\Delta_1 - \Delta_2 = J\omega^2/K_1$, so that

$$1 - (\omega^2 J/K_1) = \Delta_2, \quad J_3\omega^2/K_2 = (\Delta_1 - \Delta_2) \times (J_3/J) \times (K_1/K_2),$$

the expression for $\mathscr{D}_{\mathrm{res.}}$ reduces to

$$\mathscr{D}_{\mathrm{res.}} \cong \omega c \left[2\Delta_2 + \frac{J_3}{J}\left(\Delta_2 - 1 + \frac{K_1}{K_2}\Delta_2^2 \right) \right],$$

so that $\qquad \theta_3 \cong \dfrac{T_{\mathrm{cyl.}}}{\omega c} \times (1 + \Delta_2) \Big/ \left[2\Delta_2 + \dfrac{J_3}{J}\left(\Delta_2 - 1 + \dfrac{K_1}{K_2}\Delta_2^2 \right) \right],$

and as $\Delta_3 J_3 = -(\Delta_1 J_1 + \Delta_2 J_2) = -J(1 + \Delta_2)$ and $\theta_3 = \theta_1 \times \Delta_3$, we finally obtain

$$\theta_1 \cong \frac{\theta_3}{\Delta_3} \cong \frac{(T_{\mathrm{cyl.}}/\omega c)}{1 - \left(\Delta_2 + \dfrac{K_1}{K_2}\Delta_2^2 \right) - 2\dfrac{J}{J_3}\Delta_2}.$$

4. Engine with main flywheel and driven machine (loose-coupled)

In the system considered (Fig. 17), J_1 is the sum of all the cylinder inertias and K_1 is the shaft stiffness between these and the main flywheel J_2. To obtain K_1, use is made of the relation

$$K_1 = \frac{\omega_E^2}{\dfrac{1}{J_1} + \dfrac{1}{J_2}} = \frac{\omega_E^2}{\dfrac{1}{J_2} + \dfrac{1}{\Sigma J_{\mathrm{cyl.}}}},$$

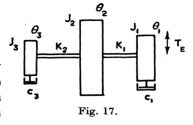

Fig. 17.

where ω_E is the natural phase velocity of the engine-and-flywheel portion of the system. The entire system also comprises a driven machine of inertia J_3, and its corresponding shaft of stiffness K_2. The excitation torque T_E acts on J_1 and the damping action is represented by dashpots with damping coefficients c_1 and c_3.

The exact expressions for vibration amplitudes and shaft twists, valid for both one- and two-node resonant-frequency conditions, are as follows:

$$\theta_1 = \frac{T_E}{\mathscr{D}_{\mathrm{res.}}} \times \sqrt{\left\{ \left[1 - \omega^2\left(\frac{J_2 + J_3}{K_1} + \frac{J_3}{K_2} \right) + \omega^4\frac{J_2 J_3}{K_1 K_2} \right]^2 \right.}$$
$$\left. \overline{ + \left(\frac{\omega c_3}{K_2} \right)^2 \times \left[1 + \frac{K_2}{K_1} - \frac{\omega^2 J_2}{K_1} \right]^2 \right\}},$$

$$\theta_2 = \frac{T_E}{\mathscr{D}_{\mathrm{res.}}} \times \sqrt{\left\{ \left[1 - \omega^2\frac{J_3}{K_2} \right]^2 + \left(\frac{\omega c_3}{K_2} \right)^2 \right\}}, \quad \theta_3 = \frac{T_E}{\mathscr{D}_{\mathrm{res.}}}.$$

where ω = phase velocity at resonance, obtained from the frequency formula for three-mass systems (see section 1·3) or from a Holzer tabulation without damping, and

$$\mathscr{D}_{\text{res.}} = \omega \sqrt{\left\{\omega^2 \left[\frac{J_2\omega^2}{K_1} - 1 - \frac{K_2}{K_1}\right]^2 \left(\frac{c_1 c_3}{K_2}\right)^2 + \left[c_1 + c_3 - \omega^2 c_1\left(\frac{J_2+J_3}{K_1} + \frac{J_3}{K_2}\right)\right.\right.}$$
$$\left.\left. - \omega^2 c_3\left(\frac{J_1+J_2}{K_2} + \frac{J_1}{K_1}\right) + \omega^4 \frac{J_2}{K_1 K_2}(c_1 J_3 + c_3 J_1)\right]^2\right\}}.$$

In some cases, the driving-machine damping is negligible ($c_1 \cong 0$). The shaft twists and their corresponding dynamic magnifiers are

$$\theta_{sh\,1} = |\theta_1 - \theta_2| = \frac{T_E}{\mathscr{D}_{\text{res.}}} \times \sqrt{\left\{\omega^4\left[\frac{J_2 J_3}{K_1 K_2}\omega^2 - \frac{J_2 + J_3}{K_1}\right]^2\right.}$$
$$\left. + \left(\frac{\omega c_3}{K_1}\right)^2 \times \left[1 - \frac{\omega^2 J_2}{K_2}\right]^2\right\},$$

$$\theta_{sh\,2} = |\theta_2 - \theta_3| = \frac{T_E}{\mathscr{D}_{\text{res.}}} \times \sqrt{\left\{\left(\frac{\omega^2 J_3}{K_2}\right)^2 + \left(\frac{\omega c_3}{K_2}\right)^2\right\}},$$

$$M_{sh\,1} = \frac{\theta_{sh\,1}}{T_E/K_1}, \qquad M_{sh\,2} = \frac{\theta_{sh\,2}}{T_E/K_2}.$$

Approximate formulae. Using the relative amplitudes Δ_1, Δ_2, etc., from an ordinary Holzer tabulation, and disregarding the terms with c_3, the expression for θ_3 can be written as

$$\theta_3 \cong \frac{T_E/(\omega c_1)}{1 - (1-\Delta_2) \times \left[\frac{J_2+J_3}{J_1} + \frac{J_3 K_1}{J_1 K_2}\right] + (1-\Delta_2)^2 \times \frac{J_2 J_3}{J_1^2} \times \frac{K_1}{K_2}},$$

and θ_1 is obtained from the relation

$$\theta_1 \cong \frac{-\theta_3 J_3}{J_1 + J_2 \Delta_2} \cong \frac{\theta_3}{\Delta_3}.$$

These approximate formulae in conjunction with Holzer tables, may be used to assess the effect on θ_1 of various values for the inertias J_2 and J_3 of the flywheel and the driven machine.

2·456 *Shape of resonance curves*

The resonance curve of a one-mass system is of the type shown in Fig. 18.

The resonance curves of engine systems are generally of the types shown in Figs. 19 and 20, and it is only in certain cases that parts of the curve contours of Figs. 19 and 20 are comparable to the curve in Fig. 18. θ_{st} is the amplitude of vibration obtained with a torque T acting on the shaft stiffness K at zero frequency ('static' or 'equilibrium' amplitude).

In certain applications, it is necessary to operate an engine on the flank of a large-amplitude critical. By altering the slope of this flank, it may be possible to reduce the stresses at the operating speed considerably.

This is achievable as indicated by the following considerations. Let

J_M = driven machine inertia [Lb.in.sec.²],

J_F = flywheel inertia [Lb.in.sec.²],

K_M = driven-machine stiffness (between J_M and J_F) [Lb.in./rad.],

K_F = flywheel shaft stiffness (between J_F and the last cylinder centre) [Lb.in./rad.].

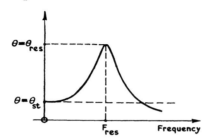

Fig. 18.

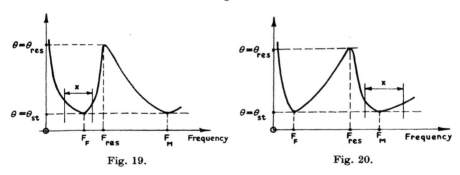

Fig. 19.   Fig. 20.

In the range of frequencies on either side of $F_{res.}$ (one-node frequency), the system is characterized by two 'anti-resonant' frequencies (using electrical engineering terminology), namely,

$$F_M = 9.55 \sqrt{(K_M/J_M)} \quad \text{and} \quad F_F = 9.55 \sqrt{\{(K_M + K_F)/J_F\}} \quad \text{[vib./min.]}.$$

F_F is always the lower value of the two if the flywheel inertia is large and K_M is not much greater than K_F. In some systems, however, J_F is smaller than J_M, in which case F_M has the lower value.

The 'low-flank amplitude' (L.F.A.) point in the vicinity of the frequencies x in the speed range is the important value (in Fig. 19 it is point F_F, in Fig. 20 it is point F_M). It can be shown by calculations for damped multi-mass systems that the closer this point is to $F_{\text{res.}}$, the steeper the flank will be. Therefore it may be said that:

(1) if the critical is above the speed range, its flank on the left-hand side (see Fig. 19) can be made steeper, so as to reach more rapidly low-amplitude values, by modifying the system so as to raise the L.F.A. frequency on the low side of $F_{\text{res.}}$ to a value closer to $F_{\text{res.}}$;

(2) if the critical is below the speed range (Fig. 20), its flank can be steepened, and therefore lowered, by reducing the L.F.A. frequency (which is on the high side of $F_{\text{res.}}$) to a value closer to $F_{\text{res.}}$.

It should also be noted that engines with a front-end pulley (or front-end flywheel) J_P, and a shaft stiffness K_P between J_P and the first cylinder centre, have a further anti-resonant frequency

$$F_P = 9 \cdot 55 \sqrt{(K_P/J_P)} \quad \text{[vib./min.]},$$

which may be close to the frequencies associated with the speed range of the engine. In such a case, the value F_P determines a further L.F.A. position, which may be suitably located to reduce flank amplitudes, in the same way as F_F or F_M.

Finally, it may be mentioned that the values of the anti-resonant frequencies of the engine cylinders, viz. $F'_{\text{cyl.}} = 9 \cdot 55 \times \sqrt{(K_{\text{cyl.}}/J_{\text{cyl.}})}$ and $F''_{\text{cyl.}} = 9 \cdot 55 \sqrt{(2K_{\text{cyl.}}/J_{\text{cyl.}})}$ (where $J_{\text{cyl.}} = $ inertia per cylinder and $K_{\text{cyl.}} = $ crankthrow stiffness) are usually so far above the frequency of the highest mode of vibration considered that they cannot be used to alter the shape of resonance curves.

Behaviour of systems at anti-resonant frequencies

The following is an outline of the considerations leading to the above conclusions on anti-resonant effects.

Two-mass system. The two-mass system has a natural phase velocity ω which is obtained from $\omega^2_{\text{res.}} = (K/J_1) + (K/J_2)$.

From the formulae given on p. 434 for a system with engine damping, it is seen that

when $\omega = \omega_1 = \sqrt{(K/J_1)}$:

$$\theta_2 \cong T_1/K \quad \text{and} \quad \theta_1 = [1 - (J_2/J_1)]\,\theta_2;$$

also when $\omega = \omega_2 = \sqrt{(K/J_2)}$:

$$\theta_1 = 0 \quad \text{and} \quad \theta_2 \cong T_1/K,$$

where T_1/K is the low amplitude of vibration due to the excitation torque of the engine in the absence of resonance. Thus ω_1 and ω_2 correspond to low-amplitude, i.e. anti-resonant conditions. As $\omega^2_{\text{res.}} = \omega_1^2 + \omega_2^2$, both ω_1 and ω_2 are values lower than $\omega_{\text{res.}}$.

[452]

An approximate physical explanation of anti-resonant effects is as follows:

When the excitation frequency of a system is such that the inertia torque of one mass is equal (and opposite) to the shaft torque, the amount of torque remaining for the excitation of the second mass is extremely small, so that the second mass tends to have a very low vibration amplitude. This, in turn, usually indicates that the first mass will also vibrate at a fairly low amplitude, hence the entire system is in an 'anti-resonant' condition.

General case of n-mass systems. Similar considerations apply for systems with a large number of masses (see Fig. 21):

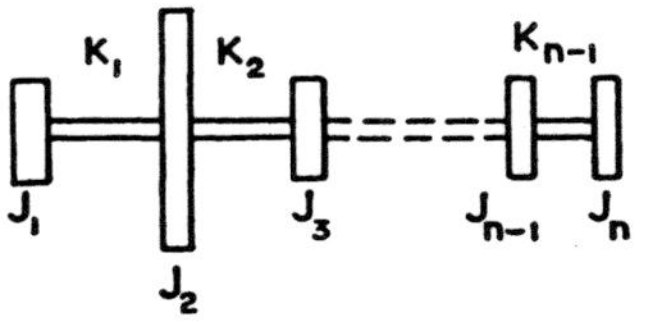

When $\omega = \omega_1 = \sqrt{(K_1/J_1)}$, J_1 may be regarded as vibrating with shaft K_1 while J_2 (and the remainder of the system) is clamped.

Fig. 21.

When $\omega = \omega_2 = \sqrt{\{(K_1 + K_2)/J_2\}}$, J_2 may be regarded as vibrating between K_1 and K_2, while J_1 and J_3 (with the remainder of the system) are clamped. Similar conditions occur for $\omega_3 = \sqrt{\{(K_2 + K_3)/J_3\}}$, etc. The last mass J_n has no shaft beyond it, so that the equation for its anti-resonant value is $\omega_n = \sqrt{(K_{n-1}/J_n)}$.

It should be noted that in systems with more than two masses, the one-node resonant frequency is flanked by anti-resonant points on either side. For instance, considering a three-mass system with

$$K_1 = 30 \times 10^6 \text{ Lb.in./rad}, \qquad J_1 = 10 \text{ Lb.in.sec.}^2,$$
$$K_2 = 10 \times 10^6 \text{ Lb.in./rad.}, \qquad J_2 = 60 \text{ Lb.in.sec.}^2,$$
$$J_3 = 20 \text{ Lb.in.sec.}^2,$$

the anti-resonant values are

$$\omega_1 = \sqrt{(K_1/J_1)} = 1732, \qquad \omega_2 = \sqrt{\{(K_1 + K_2)/J_2\}} = 819$$

and
$$\omega_3 = \sqrt{(K_2/J_3)} = 707 \text{ rad./sec.},$$

whereas the natural phase velocities are obtained from

$$\omega_{\text{res.}}^2 = \frac{1}{2}\left[\omega_1^2 + \omega_2^2 + \omega_3^2 \mp \sqrt{\left\{(\omega_1^2 + \omega_2^2 + \omega_3^2)^2 - 4\omega_1^2\omega_3^2\left(1 + \frac{J_1 + J_3}{J_2}\right)\right\}}\right],$$

which gives $\omega_{\text{res.}}^{\text{I}} = 800$ and $\omega_{\text{res.}}^{\text{II}} = 1880 \text{ rad./sec.}$ In fact, all $\omega_{\text{res.}}$-values except that of the highest mode of vibration have anti-resonant values on both sides.

In general, only the anti-resonant values close to the $\omega_{\text{res.}}$ of the mode of vibration considered have an effect on the flank steepness of the corresponding resonance curve.

Finally, it may be mentioned that if two anti-resonant values coincide (multiple anti-resonance) they still correspond to an anti-resonant

[453]

frequency of the system. For instance, in the case of a two-mass system, let $\omega_2 = \omega_1$. Then

$$\omega_{\text{res.}}^2 = \omega_1^2 + \omega_2^2 = 2\omega_1^2,$$

which indicates that the resonant frequency is $\sqrt{2}$ times higher than the anti-resonant frequency.

2·457 *Remarks on various analytical methods*

In evolving formulae for vibration amplitudes of multi-mass systems, the calculations may be shortened by making use of the principle of superposition and the reciprocal theorem. The superposition principle (see section 1·3, p. 239) states that if $\theta_{1(1)}$ is the amplitude at mass 1 due to a torque T_1 acting on this mass, and $\theta_{1(2)}$ the amplitude at mass 1 due to a torque T_2 acting on mass 2, then the total amplitude at mass 1 is

$$\theta_1 = \theta_{1(1)} + \theta_{1(2)}.$$

The reciprocal theorem states that for equal torques $\theta_{2(1)} = \theta_{1(2)}$, i.e. that the amplitude at mass 2 due to a torque T_1 acting at mass 1 is equal to the amplitude already calculated for mass 1 under the effect of a torque T_2 of equal magnitude acting at mass 2. Both theorems are useful for reducing determinant calculations.

Furthermore, the determinants of systems with a large number of masses can be expanded into pairs of sub-determinants by means of Laplace's expansion, which is more rapid than the normal procedure for large numbers of elements. If determinants are evaluated numerically (after introducing a numerical value for ω) before taking the ratios $D_1/D_{\text{res.}} = \theta_1$, etc., the condensation method of Gauss reduces some of the computation work.

Among other methods for obtaining analytical expressions, particular mention should be made of the following:

the 'Admittance method' (or 'Receptance method'), in which the mechanical admittance of the vibration amplitude at any mass is obtained by adding together a certain number of terms, such as K, $\omega^2 J_2$, $\omega^2 J_1$, etc., and then dividing these by the expanded expression for the system determinant;

the 'Mobility method', in which $j\omega\theta/T$ is used as a basis for calculations; and

the 'Impedance method', using one of the electromechanical analogies given in section 1·3 (p. 236), the system being calculated as an electrical circuit.

BIBLIOGRAPHY

Southwell, R. V. *An Introduction to the Theory of Elasticity* (Clarendon Press, Oxford, 1936). (Superposition principle and reciprocal theorem.)
Aitken, A. C. *Determinants and matrices* (Oliver and Boyd, 1951). (Laplace expansion and condensation method.)

Frazer, R. A., Duncan, W. J. and Collar, A. R. *Elementary Matrices* (Cambridge University Press, 1952).

Duncan, W. J. Mechanical admittances and their applications to oscillation problems. *Rep. Memor. Aero. Res. Comm., Lond.*, no. 2000 (1947).

Hansen, H. M. and Chenea, P. F. *Mechanics of Vibration* (J. Wiley and Sons, Inc.; Chapman and Hall, Ltd, London). (Mobility methods.)

2·458 *Transient vibrations*

The results of the following considerations are used in evaluating recordings of decaying vibrations, for instance, of couplings investigated on test rigs (see section 1·2).

Free damped vibration of a one-mass system

The system consists of a mass of moment of inertia J, supported by a shaft of stiffness K and provided with a dashpot giving 'viscous damping', i.e. a damping torque T_D proportional to the velocity of vibration $\dot{\theta}$, so that we may write $T_D = c\dot{\theta}$, where c = damping coefficient.

The system is twisted to obtain an initial deflexion θ_0 and is then suddenly released. It is proposed to determine the instantaneous amplitude of the decaying vibration.

The equation of motion is

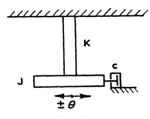

$$J\ddot{\theta} + c\dot{\theta} + K\theta = 0,$$

which expresses the fact that inertia torque, viscous torque and shaft torque are in equilibrium at any instant.

Fig. 22.

Using the relations $\omega_0^2 = K/J$ (where ω_0 = natural phase velocity of the undamped system) and $\gamma = c/(2J\omega_0)$ = damping ratio (the reason for this will be apparent later), we may divide the above equation by J and obtain

$$\ddot{\theta} + 2\gamma\omega_0\dot{\theta} + \omega_0^2\theta = 0. \tag{1}$$

By means of the substitution $\theta = K e^{rt}$, we have the characteristic equation

$$r^2 + 2\gamma\omega_0 r + \omega_0^2 = 0,$$

with the roots

$$r_{1,2} = -\gamma\omega_0 \pm \sqrt{(\gamma^2 - 1)}\,\omega_0 \quad \text{if} \quad \gamma \geqslant 1,$$

$$r_{1,2} = -\gamma\omega_0 \pm i\sqrt{(1 - \gamma^2)}\,\omega_0 \quad \text{if} \quad \gamma < 1 \quad (i^2 = -1).$$

We consider only the latter case ($\gamma < 1$). It should be noted that when $\gamma = 1$, there is only one real root. Thus $\gamma = 1$ or $c = 2J\omega_0$ is a 'critical' condition. It can be shown that, for $\gamma = 1$, the system no longer oscillates but returns to its zero or undeflected position following a non-periodic decay curve. Vibrations of periodic type occur only if $\gamma < 1$.

The general solution of eq. (1) is

$$\theta = K_1 e^{r_1 t} + K_2 e^{r_2 t} = e^{-\gamma\omega_0 t}[K_1 e^{i\omega_D t} + K_2 e^{-i\omega_D t}],$$

where $\omega_D = \omega_0 \sqrt{(1-\gamma^2)}$ = natural phase velocity of the damped system. Introducing the constants $C = \frac{1}{2}(K_1 + K_2)$ and $S = \frac{1}{2i}(K_1 - K_2)$ and using the relation

$$e^{\pm i\omega_D t} = \cos \omega_D t \pm i \sin \omega_D t,$$

we may rewrite the general solution as

$$\theta = e^{-\gamma\omega_0 t}[C \cos \omega_D t + S \sin \omega_D t], \tag{2}$$

where C and S have to be determined from the initial conditions. At the beginning of the test, let t (time) equal zero. If at $t = 0$ the vibration amplitude is θ^* and the velocity of vibration is $\dot{\theta}_{(t=0)} = \omega^*$, we can make use of these conditions in eq. (2) as follows:

$t = 0:\quad \theta = \theta^*:\quad$ eq. (2) gives $\theta^* = C,$

$t = 0:\quad \dot{\theta} = \omega^*:\quad$ eq. (2) gives

$$\frac{d\theta}{dt} = \dot{\theta} = -\gamma\omega_0 e^{-\gamma\omega_0 t} [C \cos \omega_D t + S \sin \omega_D t]$$

$$+ e^{-\gamma\omega_0 t} [-\omega_D C \sin \omega_D t + \omega_D S \cos \omega_D t],$$

$$\dot{\theta}_{(t=0)} = \omega^* = -\gamma\omega_0 C + \omega_D S = -\gamma\omega_0 \theta^* + \omega_D S,$$

hence

$$S = \frac{\omega^*}{\omega_D} + \frac{\gamma\omega_0}{\omega_D}\theta^* = \frac{\omega^*}{\omega_D} + \frac{\gamma}{\sqrt{(1-\gamma^2)}}\theta^*,$$

since $\omega_D = \omega_0 \sqrt{(1-\gamma^2)}$. We thus have eq. (2) in the form

$$\theta = e^{-\gamma\omega_0 t}\left[\theta^* \cos \omega_D t + \left(\frac{\omega^*}{\omega_D} + \frac{\gamma}{\sqrt{(1-\gamma^2)}}\theta^*\right) \sin \omega_D t\right]. \tag{3}$$

In our experiment, the system is released with zero velocity ($\omega^* = 0$). Therefore, equation (3) reduces to

$$\theta = \theta^* e^{-\gamma\omega_0 t}\left[\cos \omega_D t + \frac{\gamma}{\sqrt{(1-\gamma^2)}} \sin \omega_D t\right]. \tag{3'}$$

By means of the substitutions

$$a = R \cos \epsilon = 1, \quad b = -R \sin \epsilon = \frac{\gamma}{\sqrt{(1-\gamma^2)}},$$

the terms in the square brackets in eq. (3) can be written as $R \cos(\omega_D t + \epsilon)$. We have

$$R = \sqrt{(a^2 + b^2)} = \frac{1}{\sqrt{(1-\gamma^2)}}, \quad -\tan \epsilon = \tan(-\epsilon) = \frac{+b}{-a} = \frac{\gamma}{\sqrt{(1-\gamma^2)}},$$

so that

$$\theta = \frac{\theta^*}{\sqrt{(1-\gamma^2)}} \times e^{-\gamma\omega_0 t} \times \cos\left(\omega_D t - \tan^{-1}\frac{\gamma}{\sqrt{(1-\gamma^2)}}\right). \tag{4}$$

[456]

For small values of γ (say $\gamma \leqslant 0\cdot1$) such as occur in practice, the following approximation is obtained:

$$\theta = \theta^* \times \left(1 + \frac{\gamma^2}{2}\right) \times e^{-\gamma\omega_0 t} \times \cos(\omega_D t - \gamma), \qquad (5)$$

and the corresponding decaying vibration is shown in Fig. 23.

At $t=0$, the phase angle of the vibration is γ. The relative peaks of the decaying vibration are also shifted by an angle γ, ahead of the $0°$, $180°$, $360°$, ..., positions. (This can be verified by equating $d\theta/dt$ to zero to obtain relative maxima.)

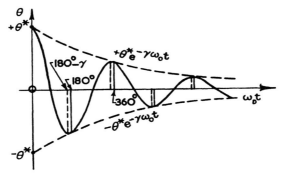

Fig. 23. The first negative peak of the vibration is at $\omega_D t = 180° - \gamma$; the point of contact with the envelope (shown dotted) is at $180°$, with a slightly reduced amplitude.

The above results can also be obtained by Laplace transforms, but the derivation for this one-mass system is not shortened appreciably.

Effect of γ on phase velocity ω_D. The relation between the natural phase velocities with and without damping, ω_D and ω_0, is

$$\omega_D = \omega_0 \sqrt{(1 - \gamma^2)},$$

which may be rewritten as

$$(\omega_D/\omega_0)^2 + \gamma^2 = 1.$$

This is the equation of a circle $y^2 + x^2 = 1$, of unit radius (Fig. 24). For small values of γ, we may write $\omega_D \cong \omega_0$.

For $\gamma < 0\cdot1$, the error is less than

$$\sqrt{(1 - (0\cdot1)^2)} - 1 \cong 1 - \tfrac{1}{2}(0\cdot01) - 1 = -0\cdot005,$$

i.e. smaller than $0\cdot5\,\%$.

Period τ_D. Owing to the damping action, the vibration takes less time to pass from a peak value to the zero position than from zero to the next peak value. However, the period of time τ_D between two consecutive zero positions (or peaks with the same sign) is constant. The following relations apply:

$$\tau_D = \frac{2\pi}{\omega_D} = \frac{2\pi}{\omega_0 \sqrt{(1 - \gamma^2)}} \cong \frac{2\pi}{\omega_0}.$$

Logarithmic decrement δ. This is defined as the natural logarithm of the ratio between peak amplitudes θ_1 and θ_2 of two consecutive cycles. If

$$\theta_1 = \theta^*(1 + \tfrac{1}{2}\gamma^2)\, e^{-\gamma\omega_0 t_1} \times \cos(\omega_D t_1 - \gamma),$$
$$\theta_2 = \theta^*(1 + \tfrac{1}{2}\gamma^2)\, e^{-\gamma\omega_0 t_2} \times \cos(\omega_D t_2 - \gamma),$$

the two cosines have the same value since $\omega_D t_2 = \omega_D t_1 + 2\pi$, and

$$\delta = \log_e(\theta_1/\theta_2) = \log_e\{e^{-\gamma\omega_0 t_1} \times e^{+\gamma\omega_0 t_2}\} = \gamma\omega_0(t_2 - t_1),$$

hence $\delta = \gamma\omega_0 \tau_D = \text{constant}$. This we may rewrite as follows, with $2\pi/\omega_D = \tau_D$:

$$\delta = \gamma\omega_0 \times \frac{2\pi}{\omega_D} = 2\pi\frac{\gamma}{\sqrt{(1-\gamma^2)}} \cong 2\pi\gamma.$$

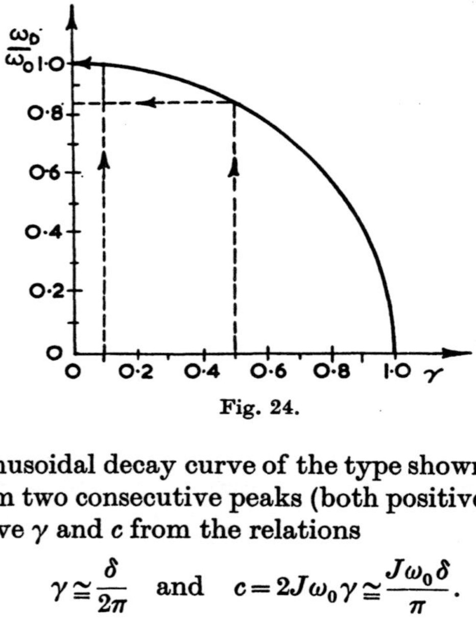

Fig. 24.

Thus, from a sinusoidal decay curve of the type shown in Fig. 23, we can determine δ from two consecutive peaks (both positive or both negative) and we then have γ and c from the relations

$$\gamma \cong \frac{\delta}{2\pi} \quad \text{and} \quad c = 2J\omega_0\gamma \cong \frac{J\omega_0\delta}{\pi}.$$

If the peaks are n cycles apart,

$$\delta = \frac{1}{n}\log_e\{\theta_1/\theta_{n+1}\}.$$

Typical numerical values are given in the following table:

$\gamma =$	0·1	0·01
$\delta =$	0·628	0·063
$\theta_1/\theta_2 =$	1·87/1·00	1·064/1·000

Energy dissipation per vibration cycle. The value of the viscous damping coefficient being known, we can express the energy dissipation per vibration cycle, occurring through viscous damping, as

$$U_{\text{visc.}} = \pi c\omega_0 \theta_n^2 = J\omega_0^2 \delta\theta_n^2.$$

[458]

2·5 RECOMMENDATIONS, RULES AND GUIDANCE NOTES OF CLASSIFICATION SOCIETIES REGARDING PERMISSIBLE TORSIONAL VIBRATION STRESSES

The following pages give extracts of rules and recommendations of the various Classification Societies on the subject of permissible torsional vibration stresses.

It should be noted that, as these rules and recommendations are periodically revised and modified, the extracts given in the following can only be regarded as indicative of the general requirements. For complete information, including subsequent additions, application should be made to the corresponding Classification Society.

The information given in this section comprises data received from the following Classification Societies: American Bureau of Shipping Bureau Veritas; Germanischer Lloyd; Lloyd's Register of Shipping Nippon Kaiji Kyōkai; Det Norske Veritas; Registro Italiano Navale.

NOTATION

Symbol	Brief definition	Typical units
d	diameter	in., mm.
D	diameter	in., mm.
N	speed	rev./min.
q	torsional stress	Lb./in.^{2}, Kg./cm.^{2}
r	speed ratio	—
λ	speed ratio	—

Subscripts are used in the text to define more precisely. Their meaning is fully explained in the context in each instance.

2·51 American Bureau of Shipping

Extracts from Rules, Section 34, paragraph 2:

Torsional critical speeds. Allowable stresses

'...Particulars to be submitted for all engines should include the type of engine, maximum continuous brake horsepower and revolutions per minute, maximum firing pressure, mean indicated pressure, critical speed data, and weights of reciprocating parts, weight and diameter of flywheel or flywheel effect for the engine.'

In addition to the above, a general provision of section 31 states that application of Rule formulas for rotating parts of machinery 'does not relieve the manufacturer from responsibility for the presence of dangerous vibrations in the installation at speeds within the operating range'.

Although no responsibility is assumed for determining when a dangerous condition exists, the American Bureau of Shipping do review torsional critical speed tabulations in order that their experience may be available to the interested parties.

The question of permissible vibratory stresses has been investigated by the American Bureau of Shipping, and their tentative conclusions may be summarized as follows:

In attempting to determine which criticals appear to be strong enough to call for corrective measures, and which are weak enough to be tolerated, the usual criteria are expressed in terms of allowable vibratory stress in the shafting. It is considered that any such criterion must be evaluated in the light of other relevant factors; for example, common troubles such as undue liner wear, bearing difficulties, and excessive maintenance costs may be aggravated by continuous operation for a major portion of the service life of the engine on a weak critical which is harmless to the shafting. In applying a criterion of allowable stress, it is to be borne in mind that

(1) vibratory stresses are seldom amenable to precise calculation, and experienced engineers may arrive at widely divergent conclusions when estimating such stresses;

(2) measured vibratory stresses are not reliable unless extraordinary precautions are taken; and

(3) conditions arising in service may substantially increase the severity of certain criticals.

Subject to the foregoing limitations, torsional vibratory stresses above ± 5000 Lb./in.^{2} in the crankshaft or tailshaft are regarded as probably dangerous, and when they occur within the operating range a barred range or suitable corrective measures should normally be provided. Stresses below ± 2000 Lb./in.^{2} are considered harmless to the shafting but not necessarily to the engine and associated equipment.

Vibratory stresses in the intermediate range from ± 2000 to ± 5000 Lb./in.^{2} present the most difficult problems, and the experience and judgement of the designer must be relied on to determine whether corrective measures are required. The proportion of the service life during which it is expected that the engine will operate on or near the critical in question is a basic consideration.

Because most marine engines are operated for the greater part of their service life within a rather narrow range of speeds, it is considered that stresses due to forced vibration at the normal operating speed may

[460]

require attention. In particular, when a strong critical is placed just above the operating range, stress due to forced vibration at normal speed is likely to be high, and should be investigated. Some operating difficulties that have been experienced may have been due in part to neglect of this consideration.

New York, 15 April 1954

2·52   Bureau Veritas

International Register for the Classification of Shipping and Aircraft

It is desired to draw attention to the experimental nature of these recommendations, which are based on tests and service experience of the past ten years. The findings are naturally subject to possible future amendments and, moreover, modifications may be permissible in certain cases. The right is reserved to depart from the general values given hereinafter in any case submitted.

Extracts from Bureau Veritas Rules, Edition 1951. Article 155, paragraph 5:

Torsional vibrations

Calculations of the torsional vibrations are, except in approved cases, to be submitted for consideration to the Administration.

These calculations should indicate, in particular:

The basic data used in the preparation of these calculations and especially the dynamic characteristics of the equivalent system;

Tables of natural frequencies for the one- and two-node frequencies and for the three-node frequency where necessary;

Vector summations of relative amplitudes with phases determined by crank angle and firing order for each mode of vibration and the various harmonics which may result in dangerous critical running speeds;

Maximum stresses at resonance caused by torsional vibration at these critical speeds.

In principle, calculations of torsional vibration should also be submitted for auxiliary engines of more than 100 b.h.p.

The results of these calculations may be shown in a graph, in which the critical speeds are indicated by their corresponding order number and mode of vibration.

In certain cases, the Administration may consider it necessary for torsiograph recordings to be taken during a trial trip.

Limits for additional stresses (torsional) in parts of
the line shafting of ships

With a view to the application of the above prescriptions of its Rules, Bureau Veritas has prepared the following graphs which take into consideration particularly the results of systematic tests carried out by subjecting to alternating stresses full-size parts of crankshafts, line shafting, etc.

For a general assessment, the results of the above tests indicate that only critical speeds for which the vibratory stresses are generally below 2·5 Kg./mm.² may be permitted for parts also subject to appreciable bending, such as crankshafts and propeller shafts, for any extensive running period. Moreover, for continuous running in the normal service range, Bureau Veritas considers it advisable not to exceed, in general, the value of 2·0 Kg./mm.² for vibratory stresses in crankshafts and propeller shafts.

With regard to the intermediate shafts, it may be noted that certain unfavourable factors, such as bending and corrosion, become practically negligible, so that the above-mentioned values may be appreciably increased.

For certain portions of the line shafting, it appears reasonable, therefore, to allow higher values for the limiting stress for conditions under which the ratio of critical speed to normal speed is not near unity, since these running ranges are in fact ranges for transient running for limited periods of time. On the other hand, from metallurgical considerations (e.g. imperfection of forging), the vibratory stress that is permissible should be reduced for shafts of large diameter.

The graphs given hereinafter take into account the foregoing considerations, as well as service experience. They indicate for the various portions of the line shafting, and for auxiliary engines, the limits of stress for transient and for continuous operating conditions. These limits are given for various shaft diameters and for various values of the ratio N_c/N_s (critical speed divided by service speed). These values are the result of recent experience in the last ten years. Bureau Veritas deems it justified to consider, at present, that they have given reasonable safety in most cases, providing also that the following remarks are taken into account in using these graphs:

(*a*) The values indicated in the graphs refer to stresses derived from actual measurements of vibration amplitude, obtained by means of torsiographs or equivalent instrumentation;

(*b*) The limiting stresses under transient conditions should not, in principle, be exceeded when passing through a critical speed, unless it be proved by means of actual measurements of vibration amplitude that the resonance values of the latter cannot be attained during operations in which the engine speed rapidly passes through the critical range;

(c) For cases where the normal running speed coincides with a critical speed, it is strongly recommended to reduce the values of limiting permissible stresses given by the graphs, by at least 50 %.

As regards electrical generating sets, the limiting stresses have been indicated in the table entitled 'Torsional vibration stresses in crankshafts of auxiliary engines' and are given below. For these sets, the permitted stresses for crankshafts under continuous operating conditions are the same as for propulsion engines. On the other hand, the permitted stress values for transient conditions are appreciably higher than those allowed for crankshafts or propulsion engines under similar conditions, in view of the fact that auxiliary engines starting at zero load pass very rapidly through the critical speed range situated below service speed, and this is so both during starting and stopping conditions.

Torsional vibration stresses in crankshafts
of auxiliary engines

U.T.S. = 41–50 Kg./mm.^{2}. [*Note.* 1 Kg./mm.^{2} = 1422 Lb./in.^{2}.]

Range of fatigue limits for transient conditions

From $N_c/N_s = 0$ to 0·92, constant limits for permissible torsional vibration stresses, as follows:

Shaft diameter [mm.]	Maximum permissible stress [Kg./mm.^{2}]
0	11·8
100	11·1
200	10·3
300	9·55
400	8·8

Range of fatigue limits for continuous running conditions

From $N_c/N_s = 0·92$ to 1·08, constant limits of permissible torsional vibration stresses, as follows:

Shaft diameter [mm.]	Maximum permissible stress [Kg./mm.^{2}]
0	1·65
100	1·55
200	1·42
300	1·30
400	1·20

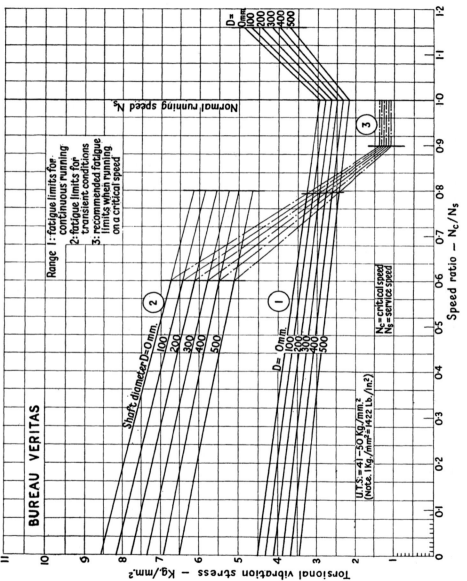

Fig. 1. Torsional vibration stresses in crankshafts.

Within the figure:

BUREAU VERITAS

Range 1: fatigue limits for continuous running
2: fatigue limits for transient conditions
3: recommended fatigue limits when running on a critical speed

Normal running speed N_s

Shaft diameter D=0mm.
100
200
300
400
500

D = 0mm.
100
200
300
400
500

D = 0mm.
100
200
300
400
500

N_c = critical speed
N_s = service speed

U.T.S.= 41–50 Kg./mm.²
(Note. 1Kg./mm²= 1422 Lb./In.²)

Torsional vibration stress — Kg./mm.²

Speed ratio — N_c/N_s

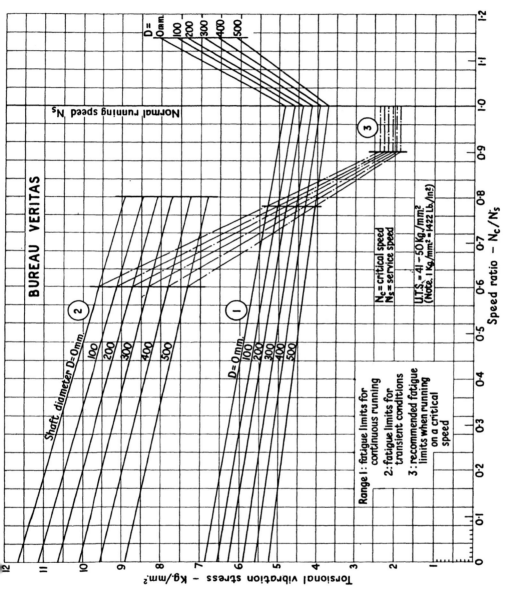

Fig. 2. Torsional vibration stresses in intermediate shafts.

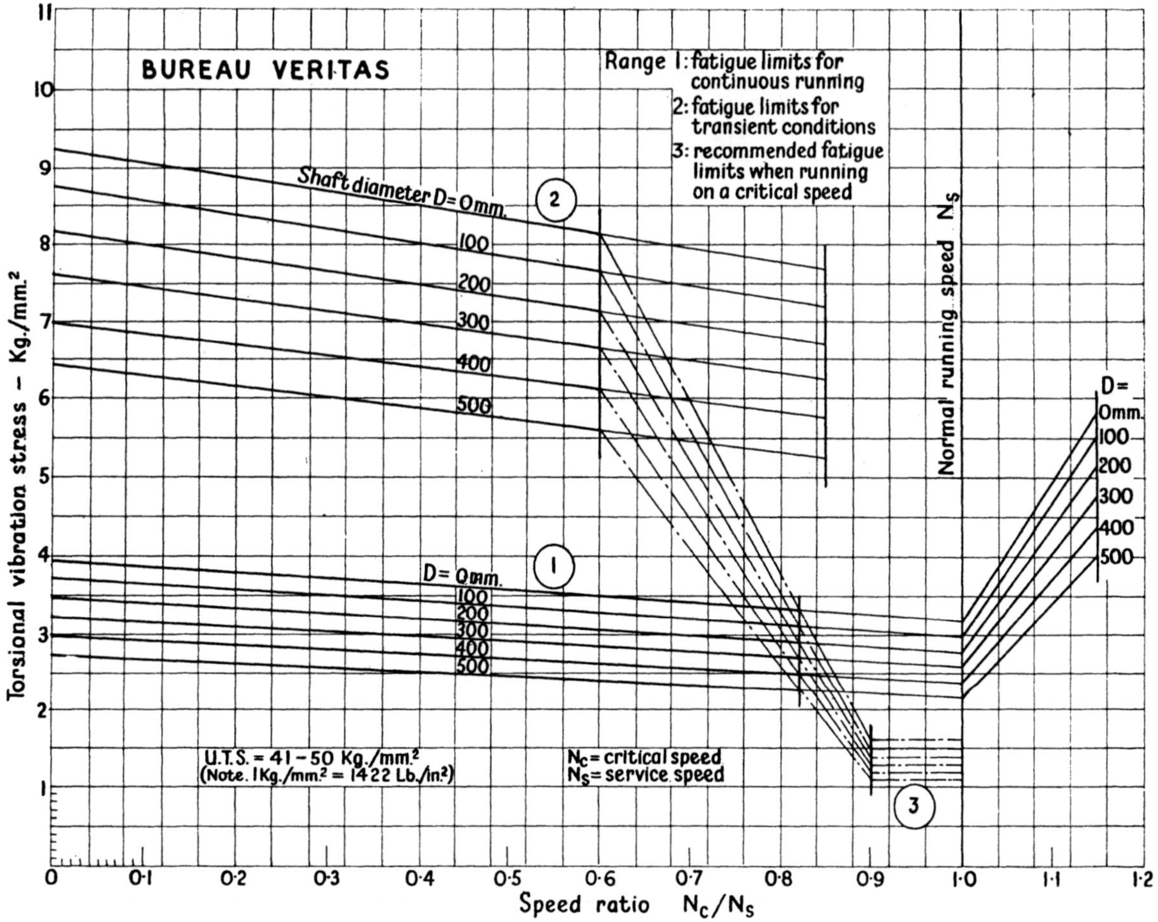

Fig. 3. Torsional vibration stresses in propeller shafts.

2·53   Germanischer Lloyd

Guidance notes relating to permissible torsional vibration stresses in critical speed range (Issue '*a*', July 1954)

For continuous running

	Crankshafts	Thrust and intermediate shafts	Propeller shafts
Permissible range		Limit value $\lambda = 0.9$. The range 'f' from $\lambda = 0.9$ to 1.05 should be free of criticals†	
Permissible stress q Kg./cm.²	$(300 - 0.18 D^{mm})$ $+ 130(1 - \lambda)$‡	$(500 - 0.28 D^{mm})$ $+ 200(1 - \lambda)$	$(300 - 0.15 D^{mm})$ $+ 160(1 - \lambda)$

For transient running

	Crankshafts	Thrust and intermediate shafts	Propeller shafts
Permissible range		Limit value $\lambda = 0.8$. No criticals should be run through in range 'g' above $\lambda = 0.8$	
Permissible stress q^* Kg./cm.²	$(730 - 0.45 D^{mm})$ $+ 200(1 - \lambda)$	$(850 - 0.48 D^{mm})$ $+ 300(1 - \lambda)$	$(660 - 0.38 D^{mm})$ $+ 270(1 - \lambda)$

† For engine installations in ships operating under special service conditions, higher critical speeds may, in agreement with the Germanischer Lloyd, also be permitted in the critical speed range from $\lambda = 0.9$ to 1.05, if the vibration stress without damping effect (damper failure) does not exceed 50 % of the value obtained from these Guidance Notes by extrapolation to $\lambda = 1.0$ and the operation of the ship is not detrimentally affected by the engine running under these conditions; wherever necessary, this should be proved by measurement.

‡ For semi- and completely built-up crankshafts, the permissible stress q may be increased by $\Delta = D^{mm}/15$ Kg./cm.². Crankshafts free-shape forged in special forging procedures which do not cut the direction of the grain in the individual throws may also be given an addition for shape.

In the above, $\lambda = N/N_{FL} = $ partial-load speed/full-load speed.

GERMANISCHER LLOYD

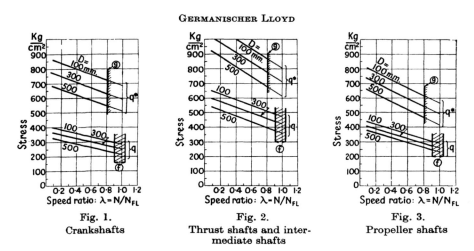

Fig. 1.
Crankshafts

Fig. 2.
Thrust shafts and intermediate shafts

Fig. 3.
Propeller shafts

The *stress evaluation* is to be based on the maximum vibration amplitude during one period, in the critical range recorded by the torsiograph.

United with the British Corporation Register

Guidance Notes

relating to the magnitude of torsional vibration stresses and position of critical speeds relative to normal service r.p.m. for main propelling and auxiliary heavy-oil engine machinery (Modification of revised Guidance Notes of 1 January 1952)—*As amended 24 August 1954.*

Having regard to the unavoidable differences in practice between calculated and measured vibration stresses, the recommended limits contained in these notes are to be taken, in the final instance, as measured stresses derived by torsiograph, or by equivalent means such as strain gauges or torsionmeters. Furthermore, the stresses are nominal values based on the plain section of the shafting neglecting stress raisers.

The stress limits given herein should not be treated as design values and are intended to be invoked only in those cases where relatively high stresses are unavoidable.

With shafts having stress concentrations of normal severity, the presence of criticals at or near the service revolutions giving nominal stresses considered permissible for continuous operation, whilst acceptable from the aspect of possible shafting failures, may nevertheless bring about many undesirable service troubles such as increased cylinder liner wear, broken oil feed pipes, wear and tear in valve-timing and blower gears, excessive noise and vibration, etc. It is therefore strongly recommended that at least over the speed range from 90 to 100 % of the service revolutions, the stresses should, wherever possible, be held to not more than 50 % of the limiting values indicated in these notes. This recommendation would also apply to auxiliary machinery over the speed range of $7\frac{1}{2}$ % above and below the rated full-load revolutions.

In general the stresses apply to the effect of a single order only, but cases will arise where account must be taken of the simultaneous effect of the flanks of adjacent orders and these may require special consideration.

It is assumed that the shafting is of 28/32 ton quality carbon steel. Where steels of higher tensile strength are employed it is not considered, at the present state of knowledge, that stress values should be increased, on account of the greater notch sensitivity to fatigue of the higher tensile steels. Special consideration, however, will be given in the case of small high tensile steel forgings where it is desired that higher fatigue stresses should be accepted.

(I) PROPELLING MACHINERY

(1) *Crankshafts*

(A) Where the critical occurs at or below the service r.p.m., vibration stresses not exceeding the values given by the following formula are considered satisfactory for continuous running, viz.

$$q_c = \pm (4400 - 70d) \times (1 \cdot 27 - 0 \cdot 27r), \tag{1}$$

where q_c = maximum value of the vibration stress for continuous running in Lb./in.^{2},

d = minimum diameter of crank pin or journal, whichever is the smaller, in inches,

r = ratio of critical r.p.m. divided by service r.p.m.

Where the vibration stresses exceed values given by formula (1), the critical speed should be avoided for continuous running. The maximum values of the vibration stresses due to such transient criticals should not exceed those given by the following formula, viz.

$$q_t = \pm 2 \cdot 35 q_c, \tag{2}$$

except as indicated under clause (7) of General, p. 473, where

q_t = maximum value of the vibration stress due to transient criticals in Lb./in.^{2}, and

q_c is as previously defined in formula (1).

Transient criticals should be arranged sufficiently removed from the service r.p.m. to ensure that at $r = 0 \cdot 85$ the stress due to the upper flank does not exceed q_c (formula (1)).

Where the vibration stresses exceed the limiting values for continuous running as given by formula (1), a notice board should be fitted at the control station stating that the engine is not to be run continuously between the following speed limits, above and below the critical speed, and the engine tachometer should be marked correspondingly, viz.

Range of engine r.p.m. to be avoided:

$$\text{from } N_c / 1 \cdot 05 \text{ to } 1 \cdot 05 N_c \text{ inclusive}, \tag{3}$$

where N_c = critical r.p.m.

(B) Where the critical occurs above the service r.p.m., the vibration stresses in crankshafts should not increase beyond the values given by the following formula at revolutions up to $1 \cdot 16$ times the service r.p.m., viz.

$$q = \pm (5 \cdot 5r - 4 \cdot 5) q_c, \tag{4}$$

where q = maximum value of the vibration stress at revolutions N in Lb./in.^{2},

q_c = value of vibration stress for continuous running as given by formula (1) for $r = 1$, in Lb./in.^{2}, and

r = ratio N/N_s = r.p.m. above service speed divided by service r.p.m.

[469]

(2) *Intermediate and thrust shafts*

(A) Where the critical occurs at or below the service r.p.m., vibration stresses not exceeding the values given by the following formula are considered satisfactory for continuous running, viz.

$$q_c = \pm (6800 - 80d) \times (1 \cdot 42 - 0 \cdot 42r), \tag{5}$$

where q_c = maximum value of the vibration stress for continuous running in Lb./in.^{2},

d = minimum shaft diameter, in inches, and

r = ratio of critical r.p.m. divided by service r.p.m.

Where the vibration stresses exceed values given by formula (5), the critical speeds should be avoided for continuous running. The maximum values of the vibration stresses due to such transient criticals should not exceed those given by the following formula, viz.

$$q_t = \pm 1 \cdot 7 q_c, \tag{6}$$

except as indicated under Clause (7) of General, p. 473, where

q_t = maximum value of vibration stress due to transient criticals, in Lb./in.^{2}, and

q_c is as previously defined in formula (5).

Transient criticals should be arranged sufficiently removed from the service r.p.m. to ensure that at $r = 0 \cdot 80$ the stress due to the upper flank does not exceed q_c (formula (5)).

Where vibration stresses exceed the limiting values for continuous running as given by formula (5), a notice board should be fitted at the control station stating that the engine is not to be run continuously between the following speed limits, above and below the critical speed, and the engine tachometer should be marked correspondingly, viz.

Range of engine r.p.m. to be avoided:

$$\text{from } 16N_c/(18 - r) \text{ to } (18 - r) N_c/16 \text{ inclusive}, \tag{7}$$

where N_c = critical r.p.m., and

r = ratio of critical r.p.m. divided by service r.p.m.

(B) Where the critical occurs above the service r.p.m., the vibration stresses in intermediate and thrust shafts should not increase beyond values given by the following formula at revolutions up to $1 \cdot 16$ times the service r.p.m., viz.

$$q = \pm (4 \cdot 2r - 3 \cdot 2) q_c, \tag{8}$$

where q = maximum value of vibration stress at revolutions N, in Lb./in.^{2},

q_c = value of vibration stress for continuous running as given by formula (5) for $r = 1$, in Lb./in.^{2}, and

r = ratio N/N_s = r.p.m. above service speed divided by service r.p.m.

[470]

(3) Screw shafts

(A) Where the critical occurs at or below the service r.p.m., vibration stresses not exceeding values given by the following formula are considered satisfactory for continuous running, viz.

$$q = \pm (4100 - 50d) \times (1 \cdot 55 - 0 \cdot 55r), \tag{9}$$

where q_c = maximum value of the vibration stress for continuous running, in Lb./in.^{2},

d = minimum shaft diameter at top of cone, in inches, and

r = ratio of critical r.p.m. divided by service r.p.m.

Where the vibration stresses exceed values given by formula (9), the critical speeds should be avoided for continuous running. The maximum values of the vibration stress due to such transient criticals should not exceed those given by the following formula, viz.

$$q_t = \pm 1 \cdot 9 q_c, \tag{10}$$

except as indicated under clause (7) of General, p. 473, where

q_t = maximum value of vibration stress due to transient criticals, in Lb./in.^{2}, and

q_c is as previously defined in formula (9).

Transient criticals should be arranged sufficiently removed from the service r.p.m. to ensure that at $r = 0 \cdot 80$ the stress due to the upper flank does not exceed q_c (formula (9)).

Where vibration stresses exceed the limiting values for continuous running as given by formula (9), a notice board should be fitted at the control station stating that the engine is not to be run continuously between the following speed limits, above and below the critical speed, and the engine tachometer should be marked correspondingly, viz.

Range of engine r.p.m. to be avoided:

$$\text{from } 16N_c/(18-r) \text{ to } (18-r)N_c/16 \text{ inclusive}, \tag{11}$$

where N_c = critical r.p.m., and

r = ratio of critical r.p.m. divided by service r.p.m.

(B) Where the critical occurs above the service r.p.m., the vibration stresses in screw shafts should not increase beyond values given by the following formula at revolutions up to $1 \cdot 16$ times the service r.p.m., viz.

$$q = \pm (4 \cdot 2r - 3 \cdot 2) q_c, \tag{12}$$

where q = maximum value of vibration stress at revolutions N, in Lb./in.^{2},

q_c = value of vibration stress for continuous running as given by formula (9) for $r = 1$, in Lb./in.^{2}, and

$r = N/N_s$ = r.p.m. above service speed divided by service r.p.m.

General

(1) *Service r.p.m.* The 'Service r.p.m.' may be defined as the maximum revolutions per minute at which the engines are intended to operate continuously at sea.

(2) *Main machinery installations for trawlers.* For trawler installations the engine revolutions when proceeding to and from the fishing grounds will be considered as the service r.p.m. The limiting values of the vibration stresses are, therefore, as given in formulae (1)–(12) inclusive. The range of revolutions used during trawling should, however, be quoted when the torsional vibration characteristics are submitted for approval and every endeavour made to ensure that the range of trawling speeds is maintained clear of critical speeds.

Within the range of revolutions used for trawling, when the vibration stress is in excess of the limiting value for continuous operation as given by formulae (1), (5) or (9), the speed ranges over which the engine should not be run continuously are given by formulae (3), (7) or (11) and should be excluded from the effective range of trawling revolutions.

(3) *Governor control.* For installations in which, under governor control, the full load to no load momentary speed rise does not exceed $7\frac{1}{2}\%$ of the full load r.p.m., the application of formulae (4), (8) and (12) may be restricted to revolutions at or below $1 \cdot 10 N_s$, provided the limits of governor control are demonstrated during engine trials on the test bed.

Where the full load to no load momentary speed rise under governor control is other than $7\frac{1}{2}\%$ of the full load r.p.m., the upper speed limit of application of formulae (4), (8) and (12) may be specially considered.

(4) *'Barred' speed ranges.* Where the above Guidance Notes state that a notice board should be fitted indicating that a range of revolutions should be avoided for continuous running, the tachometer readings should be checked against the counter readings, or by equivalent means, in the presence of the Society's Surveyors to verify that the tachometer reads correct within $\pm 2\%$ in way of the restricted range of revolutions.

In cases where the resonance curve of a transient critical vibration has been derived from torsiograph measurements, the range of revolutions to be avoided for continuous running may be taken as that over which the measured vibration stresses are in excess of those given by formulae (1), (5) or (9).

(5) *Excessive vibration stresses.* In cases where vibration stresses due to transient criticals exceed the limiting values given by formulae (2), (6) or (10), damping or detuning arrangements should be provided, or preferably the dynamic system suitably amended to reduce the magnitude of the vibration stress or remove the critical vibration from the operating range of the installation. Where dampers or detuners are fitted, it may be necessary to request that torsiograph records should be taken to verify that the vibration stress is effectively limited to values not

[472]

exceeding those given by formulae (2), (6) or (10). The use of dampers or detuners to control criticals within the range between $0.9N_s$ and $1.05N_s$ should preferably be avoided.

(6) *'Tick over' r.p.m.* When the torsional vibration characteristics of a main machinery installation are submitted for approval, the minimum r.p.m. at which the engine can operate continuously under load should be stated.

(7) *Transient criticals.* For ships whose service demands frequent manoeuvring, such as coasters, trawlers, tugs, ferryboats, hopper barges and similar ships, the maximum values of vibration stresses arising from the transient criticals should not exceed 80 % of those given by formulae (2), (6) or (10) respectively. Where, however, this recommendation cannot be met, the speed ranges between the limits of $r = 0.85$ and $r = 1.0$ for crankshafts and $r = 0.80$ and $r = 1.0$ for thrust, intermediate and screw shafts should be maintained clear of all criticals giving maximum stresses exceeding two-thirds of the values of q_c given by formulae (1), (5) or (9) respectively. For trawlers this would apply to the range of trawling speeds also. In no case, however, should q_t exceed the limiting values given by formulae (2), (6) or (10), respectively.

(8) *Gear hammer.* In installations having reversing and/or reduction gearing, or geared scavenge blowers, etc., where the vibratory torques at the gears exceed the mean transmission torques at the criticals considered, it may be necessary to impose a restricted range of revolutions in way of each critical speed at which gear hammer is reported under steady values of mean engine torque.

Further, in the event of gear hammer being detected under similar conditions at speeds other than the calculated critical speeds, torsiograph records may be requested to confirm the calculated natural frequencies.

In all cases where there is a possibility of gear hammer, the backlash should be kept to a minimum. Care should also be taken to ensure that at a critical speed the vibratory torque does not induce an excessive tooth load when taken in conjunction with that imposed by the mean engine torque.

(9) *Spare propellers.* In ocean-going ships in which a spare propeller is carried, separate calculations should be submitted for the one-node mode when its moment of inertia is different from that of the working propeller.

This procedure also applies in the case of small ships such as coasters, where it is often the practice to fit a provisional cast-iron propeller subsequently changed for one of bronze.

Particularly in the smaller installations where the moment of inertia of the propeller, as made, sometimes differs considerably from the design value, it may in certain cases be found necessary to 'swing' the propeller for purposes of verification. This would normally be done at the propeller maker's works in the presence of the Society's surveyors.

(10) *Controllable-pitch propellers.* In ships having controllable-pitch propellers care should be taken to avoid running with low pitch settings in the vicinity of major one-node critical speeds. Torsiograph records may be requested if criticals cannot be avoided when operating under these conditions.

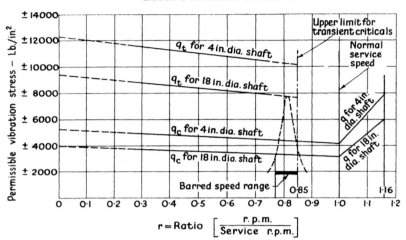

Fig. 1. Permissible vibration stresses in crankshafts.

(11) *Screw shafts.* The recommended stress limits for screw shafts apply to fully protected shafts only, i.e. to those having a continuous liner with efficient sealing arrangements against corrosion of the shaft by sea water, or alternatively, without liner and with an equivalent approved type of oil-gland. In other cases special consideration would be necessary.

The limits are also intended to apply to the minimum section of the shaft between the forward end of the propeller boss and the stern bush. For other sections special consideration would be given.

Where it is found necessary, for tuning purposes, to fit oversize shafting, the scantlings of the shaft connexions such as couplings, bolts, keys, etc., may, if desired, be based on the Rule size of shaft. In other cases where the excess size is required to reduce vibration *stress*, the coupling dimensions should be based on the actual diameter of the shaft fitted.

(II) AUXILIARY SETS

The following notes are applicable to heavy-oil engines developing 150 b.h.p., or 100 kW. and over, and driving auxiliary machinery used for essential services at sea.

[474]

The dynamic system comprising engine and driven machinery should be so designed that vibration stresses in the crankshafts and transmission shafting resulting from critical speeds do not exceed values given by the following formula within speed limits of $N_s/1\cdot075$ and $1\cdot075N_s$, N_s being the full load r.p.m., viz.

$$q'_c = \pm\,(4400 - 70d), \tag{13}$$

where $q'_c =$ maximum value of vibration stress for continuous operation within the speed range specified above, in Lb./in.^{2}, and

$d =$ shaft diameter in inches, or in the case of crankshafts, as defined for formula (1), p. 469.

LLOYD'S REGISTER OF SHIPPING

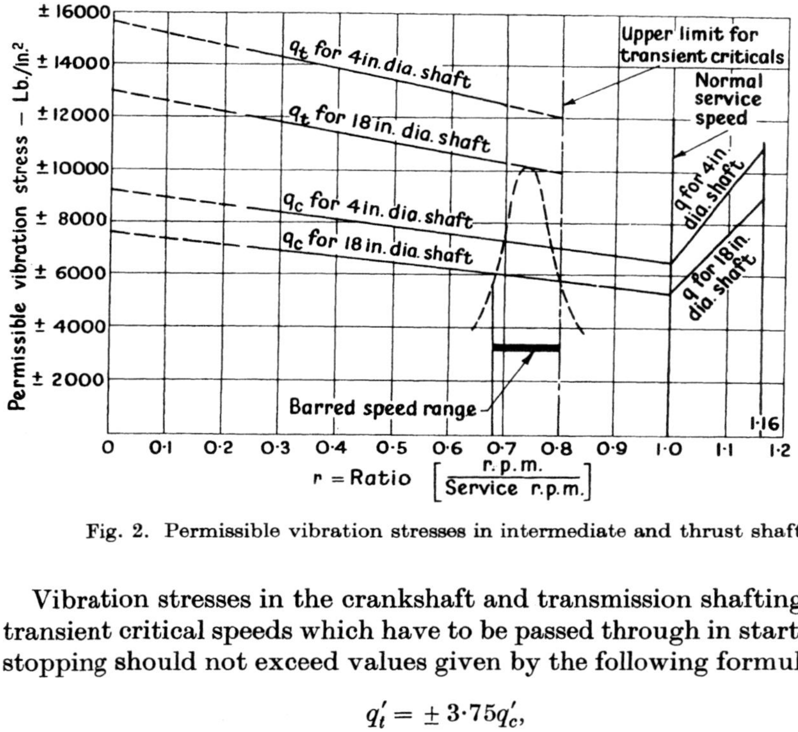

Fig. 2. Permissible vibration stresses in intermediate and thrust shafts.

Vibration stresses in the crankshaft and transmission shafting due to transient critical speeds which have to be passed through in starting and stopping should not exceed values given by the following formula, viz.

$$q'_t = \pm\,3\cdot75q'_c, \tag{14}$$

where $q'_t =$ maximum value of vibration stress due to transient criticals, in Lb./in.^{2}.

It is also recommended that the *vibration torques* applied to generator armatures should be reduced to the lowest possible value and, in any case, should be limited to not more than *twice full load engine torque* over the range $\pm\,7\frac{1}{2}\%$ on each side of the full load revolutions and not more than *six times full load engine torque* in passing through transient criticals. In the former case the vibration torque should be taken as the arithmetical sum of the torques arising from all significant major orders.

These recommendations are designed to protect the armature structure from excessive inertia torques and, in particular, the keyed connexions of core laminations to spider arms.

Experience has shown that where the vibration *stress* in armature shafts has been fully acceptable due to the stiff construction adopted, the corresponding vibratory *torques* may in some cases be sufficient to cause loosening of laminations and other defects. This is especially likely where it has been found necessary to lighten the flywheel or armature with a view to raising a major critical above full load revolutions and where the resulting flank *torques* are not sufficiently reduced.

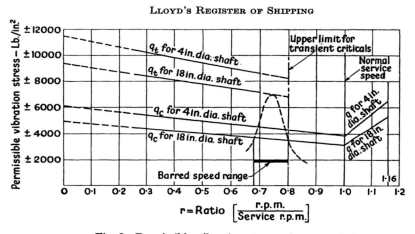

Fig. 3. Permissible vibration stresses in screw shafts.

Where these recommendations cannot be complied with, particular care should be taken with the fastenings of the electrical machine and preferably not less than double keying, or equivalent means of fastening, should be employed for the armature laminations.

Where transient criticals occur in close proximity to the limits of the speed range given by $N_s/1\cdot075$ and $1\cdot075N_s$, torsiograph tests on the installation, witnessed by the Society's surveyors, may be considered necessary to confirm that the vibratory conditions are satisfactory within this speed range. During these trials the engine manufacturer should demonstrate the effective limits of governor control within the stated limits of r.p.m.

In cases where vibration stresses due to transient criticals exceed values given by formula (14), the conditions stated under clause (5) of

[476]

General Remarks on Main Engine Installations are generally applicable, except that dampers and detuners should preferably not be used to control critical speeds occurring within the range of speeds between $N_s/1\cdot075$ and $1\cdot075N_s$ inclusive.

2·55 Nippon Kaiji Kyōkai—The Japanese Marine Corporation

Extract from Rules (Kosen Kisoku), Part 34, chapter 2, article 4:

Internal combustion engines: drawings and data

1. Calculation sheets of torsional vibrations should be submitted for consideration together with drawings of shafting and propeller. The calculation sheets should include:

(1) Natural frequency tables for one- and two-node vibrations, and for higher modes of vibration if necessary;

(2) Vector sum diagrams for all orders of vibration occurring at speeds up to 20 % above the maximum continuous running speed;

(3) Details of crank arrangement and firing order.

2. The vibratory stresses should be carefully estimated by the engine makers, taking due account of results previously obtained for engines of the same type.

3. Where dangerous stresses are likely to occur in or near the range of normal running speeds, and this is deemed necessary by the Committee, measurements of vibration amplitudes of the installation after fitting should be carried out on board ship for the purpose of verifying the calculations.

4. Similar calculations should also be submitted for internal combustion engines of 100 b.h.p. or more, driving generators or auxiliary machinery used for essential service at sea.

In addition to the above, the following *tentative rules* are applied to all Japanese motor vessels classified by the Corporation:

(1) The engine should not be run continuously in the range of speeds where the vibration stresses occurring in the crankshaft or propeller shaft exceed ± 300 Kg./cm.²

(2) The critical speeds at which the maximum stress exceeds ± 200 Kg./cm.² in the crankshaft or the propeller shaft should be marked on the engine tachometer, indicating that it is advisable to avoid continuous running in this speed range.

(3) For intermediate and thrust shafts, the permissible stresses are 1·5 times the above limits, i.e. ± 450 and ± 300 Kg./cm.², respectively.

Tokyo, 6 April 1954

*Rules relating to torsional vibration stresses and calculations
of torsional vibration characteristics*

(Extracts from the 1953 Rules for I.C. engines.) Chapter 7, section 2 A:

2. The formulas for the determination of the diameters of shafts do
not take into consideration the stresses caused by torsional vibration.
Calculations of torsional vibration characteristics, together with the
necessary drawings of the engine, shafting, propeller or other components, are to be submitted for consideration.

The calculations should comprise:

(*a*) The basic elements necessary for the preparation of these calculations;

(*b*) The tables of natural frequencies for the one- and two-node
frequencies and also higher-node frequencies if necessary;

(*c*) Particulars of the firing order and vector summation for all orders
of vibration occurring at speeds up to 15 % above the service speed;

(*d*) Particulars of calculated stresses at critical orders of importance.
The calculated stresses are to comply with the requirements given
under 4;

(*e*) Where the estimation of vibration stresses is based on actual
measurements previously obtained on similar installations, particulars
should be submitted.

3 *a*. Calculations of torsional vibration are to be submitted for all
diesel installations intended for propulsion and for all diesel aggregates
intended for essential services. Generally, submission of torsional
vibration characteristics will not be required for main and auxiliary
engines of 150 b.h.p. or less.

3 *b*. In addition to the above requirements, calculations of torsional
vibration are to be submitted for the dynamic system with the *spare
propeller*, in which case a one-node frequency table only will be required.

4. *Permissible torsional vibration stresses in shafts*

(*a*) The permissible additional vibration stresses shown in the following diagrams for main engines and the formulas given under (*h*) below
for auxiliary engines, are applicable for shafts made of steel having a
tensile breaking strength of 28–32 tons per square inch. It is not considered advisable that stress values should be increased for shafts made
of steels with a higher tensile strength. The permissible values given apply
to *uniform cylindrical sections* of the shafting and to stresses derived from
records *actually* measured by torsiograph or other equivalent means.
Ample fillets of a good finish should be provided where a change in
shaft diameter occurs and in keyways which may result in increased
stresses.

(*b*) The permissible stresses are determined for a continuous operation at the service speed from the aspect of possible failures in the shafting, but they may nevertheless cause severe strain on the engine, increased wear and excessive vibration, noise, etc. It is therefore recommended to reduce the stresses as far as possible below the limiting values given in the diagrams and by the formulas.

DET NORSKE VERITAS

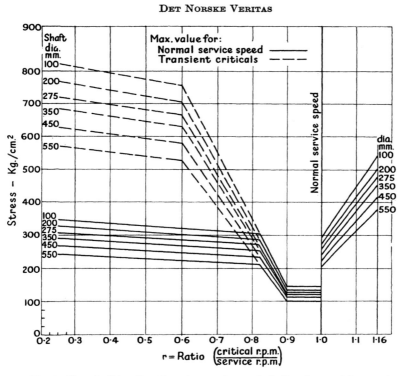

Fig. 1. Permissible vibration stresses in crankshafts of propulsion engines.

(*c*) *For ships such as coasters, trawlers, fishing vessels, tugs* and similar ships whose service demands frequent manoeuvring, the maximum value of vibration stress is not to exceed 80 % of the permissible additional stresses for transient criticals given in the diagrams.

(*d*) *The vibration stress limits for propeller shafts* apply to shafts fitted with a continuous liner and where efficient sealing arrangement between the propeller boss and the liner is provided, as well as to shafts without liner, where an approved type of oil gland is fitted between the propeller and stern tube. The stresses are to be calculated for the minimum diameter of the shaft between the forward end of the propeller boss and stern tube.

[479]

(e) 'Barred' speed ranges. Where vibration stresses due to transient criticals exceed the limiting values for continuous running (see the diagrams), the critical range is to be 'barred'.

(f) Where a critical must be avoided for continuous running, a notice board is to be fitted at the control platform indicating which ranges of revolutions are not to be run continuously. The tachometer is to be

DET NORSKE VERITAS

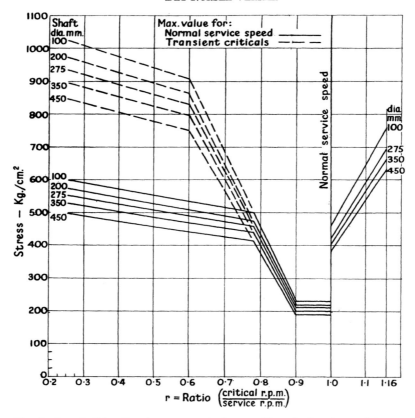

Fig. 2. Permissible vibration stresses in intermediate shafts of propulsion engines.

marked correspondingly. The 'barred' speed ranges are determined by the following formulas:

(A) For vibration stresses in crankshafts:

from $N_c/1\cdot05$ to $1\cdot05N_c$ inclusive.

(B) For vibration stresses in thrust, intermediate and screw shafts:

from $16N_c/(18-r)$ to $(18-r)\,N_c/16$ inclusive.

[480]

In these formulas, N_c = critical r.p.m. and r = ratio of critical r.p.m. divided by service r.p.m.

Where the measured vibration stresses call for a 'barred' range exceeding the limits determined by the above formulas, the 'barred' range is to be based on the torsiograms.

(g) *Control of tachometer.* Where the limits of stress required above state that a notice board is to be fitted indicating that a range of revolutions should be avoided for continuous running, the Surveyors are to

DET NORSKE VERITAS

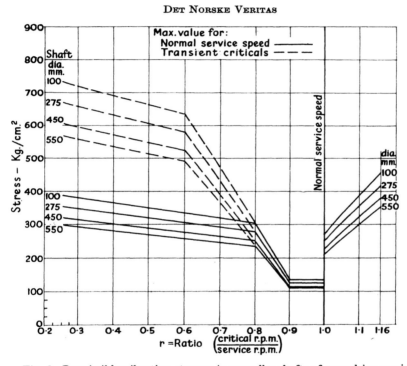

Fig. 3. Permissible vibration stresses in propeller shafts of propulsion engines.

check the tachometer readings against the counter readings and so verify that the tachometer reads correct to within ± 2 % in way of the restricted range of revolutions. Where necessary, the tachometer readings should be corrected.

(h) *Auxiliary internal combustion engines of 150 b.h.p. or more, intended for essential services at sea.* For continuous operation, the vibration stresses in the crankshafts and transmission shafting should not exceed those determined by the following formula:

$$q_{\mathrm{crit.}} = \pm\,(309 - 1 \cdot 93 d),$$

where $q_{\text{crit.}}$ = maximum permissible value of the vibration stress for continuous running [Kg./cm.²], and

d = minimum shaft diameter [cm.].

The formula is applicable within the following limits of engine r.p.m.:

$$N_s/1\cdot07 \quad \text{and} \quad 1\cdot07N_s,$$

where N_s = r.p.m. at full load.

Vibration stresses in the shafts due to transient critical speeds which have to be passed through in starting and stopping, are not to exceed values given by the following formula:

$$q_t = \pm 3\cdot75q_{\text{crit.}},$$

where q_t = maximum permissible value of vibration stress due to transient criticals (Kg./cm.²).

5. *Dampers and detuners*

Where vibration dampers or detuners are intended to be fitted, to reduce stress in the shafting, a drawing showing the construction of the dampers and their attachment to the shaft is to be submitted, together with details of operation; also particulars regarding maintenance of the dampers are to be stated. Unless specially approved, vibration dampers are not to be used where the additional stresses without a damper are greater than the permissible stresses within the range between $0\cdot9N_s$ and $1\cdot05N_s$.

6. *Torsiograph records*

In certain cases, the Administration may require torsiograms to be taken on the installation after it is fitted on board to confirm that severe vibratory conditions do not occur in the normal speed range. Generally, torsiograph records are required when the calculated stresses give a value very close to the permissible limit; for the determination of critical speeds and for the checking of vibration dampers, if fitted. The measurements are to be performed with an instrument of approved manufacture.

2·57 Registro Italiano Navale

The Registro Italiano Navale has not issued official rules for the torsional vibration stresses of engine systems.

However, full details of analyses, together with results of measurements, should be included wherever necessary.

Genova, 30 April 1954

2·6   CYCLIC SPEED VARIATION

NOTATION

Symbol	Brief definition	Typical units	Symbol	Brief definition	Typical units
A	area	in.², cm.²	P	force	Lb., Kg.
C	cyclic speed deviation	—	R	radius	in., cm.
			t	time	sec.
F	frequency	vib./min.	T	torque	Lb.in., Kg.cm.
H	power	h.p., kW.	$\lvert T_n \rvert$	component of tangential pressure	Lb./in.², Kg./cm.²
H	height	in., cm.			
J	moment of inertia	Lb.in.sec.², Kg.cm.sec.²	U	energy, work	ft.Lb., m.Kg.
			V	voltage	volts
K_n	inertia-force coefficient	—	Z	ratio of energies	—
			α	alternator constant	—
K	oscillograph constant	—	α	vibration amplitude	rad.
L	length	in., cm.	$\dot{\alpha}$	vibration velocity	rad./sec.
M	mass	Lb.sec.²/in., Kg.sec.²/cm.	δ	phase angle	deg.
			ϵ	phase angle	deg.
m	number of cycles per revolution	cycles/rev.	η	efficiency	—
			θ	vibration amplitude	rad., deg.
n	number of engine impulses per second	pulses/sec.	$\overrightarrow{\Sigma\Delta}$	phase-vector sum	—
			ϕ	angle of rotation	rad.
p	number of pole pairs	—	$\dot{\phi}$	angular velocity	rad./sec.
p	number of cycles per revolution	cycles/rev.	$\ddot{\phi}$	angular acceleration	rad./sec.²
			ω	phase velocity	rad./sec.

2·61   Introductory note

In torsiograph measurements monitored on an oscilloscope screen, cyclic speed variation† appears as a number of fairly large amplitude oscillations which reach increasingly large values as the speed of the engine is reduced down to stalling point.

Filter measurements in this case indicate an oscillation frequency which varies with engine speed. In practice, true cyclic speed variation is identifiable by this feature, and by the fact that the frequency values thus obtained do not usually correspond with any of the natural torsional-vibration frequencies of the engine system.‡

Although the irregular 'rolling' motion of the shafting occurs without producing torsional stresses in the shafting, it may have an effect on the damping of torsional vibrations (see section 2·42, Damping factors, p. 371). Moreover, high values of cyclic speed variation are not permissible in certain applications, e.g. electrical generator sets (owing to

† Alternative designations: cyclic irregularity, cyclic speed deviation.
‡ For usual multi-cylinder in-line engines, the cyclic-variation frequency is equal to the running speed multiplied by the number of firing strokes per crankshaft revolution.

the possibility of flicker), geared systems (to avoid load reversals and tooth impact) and systems fitted with 'soft' (e.g. rubber type) couplings. Usually, more stringent limits for cyclic speed variation are required for electrical purposes than for marine propulsion systems, pump drives, etc.

In view of the above considerations, it is necessary to estimate the cyclic speed variation of an installation by methods applicable at the design stage, and also to determine the exact value on the basis of measurements. Suitable methods for these requirements are given in the following, and illustrated by a numerical comparison of calculated and experimental results.

2·62   Methods of calculation

2·621   *Excess-energy method*

This is the only method for calculating cyclic speed variation that is recommended by the B.I.C.E.R.A. Torsional Vibration Panel. A torque diagram, of the type shown in Fig. 1, is required for the engine under consideration. The shafting is assumed to be rigid in torsion and 'rolling'

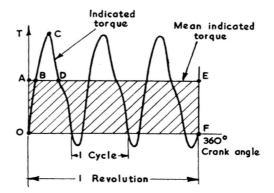

Fig. 1. Torque/crank-angle diagram.

irregularity only is considered (although in fact with torsional vibration the motion may be quite different, see section 2·45, p. 434, and section 2·64, p. 492).

The total amount of inertia of the system is

$$J = J_E + J_F + J_G \quad [\text{Lb.in.sec.}^2],$$

where J_E = engine inertia (moment of inertia per cylinder × number of cylinders), J_F = flywheel inertia, and J_G = inertia of driven machine (generator).

[484]

If ω_1 and ω_2 [rad./sec.] are the maximum and minimum angular velocities of the engine, respectively, then the excess energy, which is stored in the rotating masses, is given by

$$\Delta U = \tfrac{1}{2} J (\omega_1^2 - \omega_2^2) \quad \text{[in.Lb.]}, \tag{1}$$

this excess energy being represented by the triangular-shaped area BCD in Fig. 1.

Noting that $\omega = 2\pi N/60$, where $N =$ running speed [rev./min.], let the mean speed be assumed to be $N_0 = \tfrac{1}{2}(N_1 + N_2)$, so that the mean angular velocity is $\omega_0 = \tfrac{1}{2}(\omega_1 + \omega_2)$. Then the cyclic speed variation C is defined as

$$C = (N_1 - N_2)/N_0 = (\omega_1 - \omega_2)/\omega_0 = (\omega_1^2 - \omega_2^2)/2\omega_0^2, \tag{2}$$

and, since eq. (1) gives $\Delta U/J = \tfrac{1}{2}(\omega_1^2 - \omega_2^2)$, one obtains

$$C = \Delta U/J\omega_0^2 = \frac{\text{excess energy}}{2 \times \text{mean kinetic energy}}. \tag{3}$$

Additional notes. (A) It is sometimes required to calculate the cyclic speed variation from the brake power H_b of the engine or from the *indicated* work per revolution U_i [ft.Lb.] = area $OAEF$ in Fig. 1. The relation between these quantities is

$$H_b = (\eta_m/100) \times N_0 U_i/33,000 \quad \text{[h.p.]}, \tag{4}$$

where $\eta_m =$ mechanical efficiency of the engine (expresssed in 'per cent'). Let ΔU also be in ft.Lb. (but J still in Lb.in.sec.²); then with the ratio $\Delta U/U_i$ denoted by Z, eqs. (3) and (4) give

$$C = 3 \cdot 611 \times 10^7 \times \left(\frac{100}{\eta_m}\right) \times ZH_b/(JN_0^3). \tag{5}$$

(B) To find the moment of inertia J_F of a flywheel required to obtain a given cyclic irregularity C, from eqs. (3) or (5):

$$J_F = (\Delta U/C\omega^2) - (J_E + J_G) \quad \text{[Lb.in.sec.²]},$$

or $\qquad J_F = \left\{ 3 \cdot 611 \times 10^7 \left(\frac{100}{\eta_m}\right) \times ZH_b/(CN_0^3) \right\} - (J_E + J_G) \quad \text{[Lb.in.sec.²]}.$

Metric units. If U_i is in Kg.metres, then the brake power H_b is given by

$$H_b = (\eta_m/100) \times N_0 U_i/(60 \times 75) \quad \text{[metric h.p.]}.$$

With ΔU in Kg.m. and J in Kg.cm.sec.², eq. (5) becomes

$$C = 4 \cdot 103 \times 10^7 \times \left(\frac{100}{\eta_m}\right) \times ZH_b/(JN_0^3). \tag{5'}$$

2·622 *Formula using gas-force components*

If a torque diagram is not available, the cyclic speed variation may be assessed from the gas-force and inertia-force components. The shafting is again assumed to be rigid, which is equivalent to saying that the engine

[485]

is not running on or near to a torsional vibration critical. The procedure is as follows:

(1) Consider the phase-angle and vector-sum diagrams of the engine (see section 2·22, pp. 296 et seq.) and, treating all vectors as equal and of unit length, determine the lowest order diagram which does not give a closed polygon, i.e. which has a resultant phase vector $\overrightarrow{\Sigma\Delta}_n$ not equal to zero.

(2) Denote this 'predominant order' by n, noting that

n = number of cycles of speed variation per engine revolution,

= order of harmonic for two-stroke cycle engines,

= 'critical speed order number' (one-half the harmonic order number for four-stroke cycle engines).

(3) Determine the value of the gas-force component

$$P_G = |T_n| \, A \overrightarrow{\Sigma\Delta}_n \quad \text{[Lb.]},$$

where $|T_n|$ = nth order tangential-pressure component due to gas pressure [Lb./in.^{2}], A = area of one piston [in.^{2}], and $\overrightarrow{\Sigma\Delta}_n$ = phase-vector sum of predominant order, obtained under 1 above. (The $|T_n|$-value can be taken from the curves of section 2·18, pp. 278–287, for the i.m.e.p. value considered.)

(4) Determine the inertia-force component

$$P_I = M_{\text{recip.}} \times \omega^2 R_0 \overrightarrow{\Sigma\Delta}_n \times K_n \quad \text{[Lb.]},$$

where $M_{\text{recip.}} = W_{\text{recip.}}/g = J_{\text{recip.}}/R_0^2$ = equivalent reciprocating mass per line [Lb.sec.^{2}/in.], R_0 = crank radius [in.], $\omega = (2\pi/60) \times$ mean value of engine speed considered = mean angular velocity of engine [rad./sec.], $\overrightarrow{\Sigma\Delta}_n$ = phase-vector sum obtained under 1 above, and K_n = nth order inertia-force coefficient with the following values:

Two-Stroke Cycle Engines and Four-Stroke Cycle Engines

$$K_1 = R_0/4L$$
$$K_2 = -1/2$$
$$K_3 = -3R_0/4L$$
$$K_{4\,(\text{etc.})} \cong 0$$

In these expressions, L = length of connecting rod [in.].

(5a) If the inertia-force component P_I is zero or negligibly small, determine the cyclic speed variation as

$$C = \frac{2R_0 P_G}{n\omega^2 \Sigma J}, \tag{A}$$

where ΣJ is the sum of all the inertias in the system (i.e. engine inertias + flywheel inertia + inertia of driven machine).

[486]

(5 b) If P_I is not negligible, evaluate

$$\tan \epsilon_n = \frac{n\text{th order sine component}}{n\text{th order cosine component}}$$

from the sine and cosine component curves of gas-pressure components given in section 2·18 (or from those for the particular engine considered) and thus determine the phase-angle ϵ_n.

Finally determine the cyclic speed variation as

$$C = \frac{2R_0 \sqrt{(P_G^2 + P_I^2 + 2P_G P_I \cos \epsilon_n)}}{n\omega^2 \Sigma J}, \tag{B}$$

where $2P_G P_I$ should always be taken as a positive value.

Derivation. As $|T_n|$ is the maximum amplitude of a sinusoidal vibration, one may write

$$P_G' = P_G \sin (n\omega t + \epsilon_n), \tag{1}$$

where $n =$ order of predominant vibration, $\omega = N \times (2\pi/60)$ [rad./sec.], $N =$ mean value of the engine speed [rev./min.], and $\epsilon_n =$ phase angle relative to top dead centre (T.D.C.). (Note that $\epsilon_n \cong 0$ for $n > 1$ for four-stroke cycle engines, or for $n > 2$ for two-stroke cycle engines.)

The inertia force is

$$P_I' = P_I \sin n\omega t,$$

the phase angle being omitted, since all inertia forces are in phase or in anti-phase. The resultant force P_R is obtained by geometrical addition

$$P_R = \sqrt{(P_G^2 + P_I^2 + 2P_G P_I \cos \epsilon_n)} \tag{2}$$

and

$$P_R' = P_R \sin (n\omega t + \delta_n), \tag{2'}$$

where $\delta_n =$ phase angle of P_R relative to T.D.C.

The equation of motion of a rigid shaft system is

$$\ddot{\phi} \Sigma J = R_0 P_R' = R_0 P_R \sin (n\omega t + \delta_n), \tag{3}$$

where $\ddot{\phi} =$ angular acceleration [rad./sec.^{2}], $\Sigma J =$ sum of all moments of inertia in the system [Lb.in.sec.^{2}], $R_0 =$ crank radius [in.], and $P_R =$ resultant force [Lb.]. Therefore

$$\dot{\phi} = \int \ddot{\phi} \, dt = -R_0 P_R \cos (n\omega t + \delta_n)/[n\omega \Sigma J]. \tag{4}$$

As $\dot{\phi} = \dot{\phi}_{\text{max.}}$ when $\ddot{\phi} = 0$, eq. (3) gives $\delta_n = -n\omega t$, so that

$$\dot{\phi}_{\text{max.}} = -R_0 P_R/[n\omega \Sigma J].$$

Also, $\dot{\phi} = \dot{\phi}_{\text{min.}}$ when $\ddot{\phi} = 0$ and $\delta_n = \pi - n\omega t$, hence

$$\dot{\phi}_{\text{min.}} = +R_0 P_R/[n\omega \Sigma J].$$

Thus

$$C = \frac{1}{\omega} \times (\dot{\phi}_{\text{max.}} - \dot{\phi}_{\text{min.}}) = \frac{2R_0 P_R}{n\omega^2 \Sigma J}. \tag{5}$$

If $P_I \cong 0$, $P_R = P_G$ and the relation reduces to formula (A). If not, it is necessary to employ eq. (2) and therefore C is determined by formula (B), given above.

Note. n is the number of cycles of speed variation per engine revolution. Thus, for a six-cylinder four-stroke cycle single-acting in-line engine, with equally spaced crank angles, $n = 3$. On the other hand three cycles per engine revolution correspond to 6th order harmonics per working cycle. However, as the gas-pressure components $|T_n|$ are related to the portion of the working cycle performed in one revolution, i.e. to the critical speed order number, this 6th harmonic is denoted as the 3rd order component (critical speed order number), hence use should be made of $|T_3|$ and not $|T_6|$ in this instance. For a six-cylinder two-stroke cycle single-acting engine, $n = 6$ and also $|T_n| = |T_6|$.

Metric units. Equation (5) is also valid in metric units, with R_0 in cm., P_R in Kg., and ΣJ in Kg.cm.sec.^{2}.

2·623 *Cyclic irregularity due to torsional vibration*

Torsional vibration is also in effect a cyclic irregularity in the running speed, but it occurs with an amplitude which varies along the length of the shafting (see section 2·45, p. 434).

If an engine system is running on its nth order critical speed, and the torsional vibration amplitude (i.e. half total swing) at the driven-machine end is $\pm \theta_{DM}$ [rad.], the cyclic speed variation C_{torsion} due to torsional vibration at this position is

$$C_{\text{torsion}} = 2n\theta_{DM} \quad [\theta \text{ in rad.}]$$

(with $n = 1, 2, 3, \ldots$ for two-stroke cycle engines, and $n = \frac{1}{2}, 1, 1\frac{1}{2}, \ldots$ for four-stroke cycle engines). At any other points, e.g. at mass no. 1 of the engine, the corresponding value of C_{torsion} is obtained by means of the ratio of Holzer-table amplitudes Δ_1/Δ_{DM} as

$$C_{\text{torsion}} = 2n\theta_1 = 2n\theta_{DM} \times \Delta_1/\Delta_{DM}.$$

Thus, the *resultant* cyclic speed variation at the driven-machine end is

$$C = C_{\text{rolling}} + C_{\text{torsion}},$$

where $C_{\text{torsion}} = 2n\theta_{DM}$ and C_{rolling} is obtained by the methods of sections 2·621 or 2·622. Although there is no relation between the rolling and torsional-vibration frequencies, it is possible that the two oscillations may combine, with maximum amplitudes (beat-frequency amplitudes); hence it is suitable to take the most unfavourable case where the effects can be added arithmetically.

Derivation. At the nth order critical speed N_E [rev./min.], the torsional vibration of the driven machine can be expressed as a sine function, viz.

$$\theta'_{DM} = \theta_{DM} \sin(n\omega_E t + \epsilon_n),$$

where $\omega_E = (2\pi/60) \times N_E$ [rad./sec.]. The corresponding angular velocity is then

$$\omega(t) = \omega_E + \frac{d}{dt}\theta'_{DM} = \omega_E + \theta_{DM} n\omega_E \cos(n\omega_E t + \epsilon_n).$$

C is defined as
$$C = \frac{1}{\omega_{\text{mean}}}[\omega_{\text{max.}} - \omega_{\text{min.}}],$$

and with $\omega_E = \omega_{\text{mean}}$ this gives

$$C = \frac{1}{\omega_E}[(\omega_E + \theta_{DM} n \omega_E) - (\omega_E - \theta_{DM} n \omega_E)] = 2n\theta_{DM}.$$

2·63   Experimental determination of cyclic speed variation

2·631   *Method based on vibration amplitude*

(a) *Measurement with electrical equipment.* The pick-up is first mounted on a torsional-vibration calibrator, which is adjusted to give a sinusoidal vibration of $\pm\theta_{\text{cal.}}$ [rad.]. If a velocity-type pick-up (e.g. an S.T.C. Sunbury) is used, the output should be passed through an integrating circuit. For a displacement-type pick-up, of course, no integrating circuit is needed.

On an oscilloscope screen, the calibration amplitude $\pm\theta_{\text{cal.}}$ [rad.] corresponds to a trace of total height $H_{\text{cal.}}$ [in.].

The pick-up is then mounted on the engine†, which is run up to the mean speed N_E [rev./min.] at which the cyclic speed variation is to be measured. The oscilloscope will then show a trace of total height H_E, corresponding to a vibration amplitude $\pm\theta_{DM}$ [rad.] and the *total* speed variation of the engine is thus obtained as

$$C = 2n\theta_{DM} = 2n\theta_{\text{cal.}} \times \frac{H_E}{H_{\text{cal.}}},$$

where $n =$ order number of the predominant vibration (usually equal to the number of firing strokes per revolution). The derivation of $C = 2n\theta_{DM}$ is given in section 2·623, and it may be noted that $\theta_{DM}/\theta_{\text{cal.}} = H_E/H_{\text{cal.}}$.

(b) *Measurements with mechanical torsiographs.* The recordings of mechanical torsiographs, such as the Geiger instrument, after conversion into radians (half total swing), can be used to evaluate cyclic speed variation by means of the relation $C = 2n\theta_{DM}$.

2·632   *Method based on vibration velocity*

This method, evolved by Stansfield and Pitkethly,‡ requires the use of a pick-up giving the vibration velocity, such as the S.T.C. Sunbury torsional-vibration pick-up in which the output-signal voltage is proportional to the rate of change of flux. The signal is amplified in a 'direct' (i.e. non-integrating) circuit of the oscillograph.

The pick-up is first mounted on a torsional-vibration calibrator, which is adjusted to give a sinusoidal vibration of $\pm\alpha_{\text{cal.}}$ [rad.] and runs at a

---

† Or, if possible, preferably on the driven machine.
‡ See Stansfield, R. and Pitkethly, T. E., 'The measurement of cyclic irregularity', *Oil Engine*, January 1948, pp. 316–7.

constant speed $N_{\text{cal.}}$ [rev./min.]. This corresponds to a trace of total height $H_{\text{cal.}}$ on the oscilloscope screen.

The pick-up is then fitted to the engine, which is run up to the mean speed N_E [rev./min.] and gives on the oscillograph screen a trace with a total height H_E.

By means of these values, the cyclic speed variation is determined as

$$C = 2\alpha_{\text{cal.}}\, p_{\text{cal.}} \times \frac{H_E}{H_{\text{cal.}}} \times \frac{N_{\text{cal.}}}{N_E},$$

where $\alpha_{\text{cal.}}$ = calibrator amplitude (half total swing) [rad.], and $p_{\text{cal.}}$ = number of cycles per revolution produced by the calibrator (for a Hooke's joint or a two-peak cam $p_{\text{cal.}} = 2$; for a simple eccentric-type cam $p_{\text{cal.}} = 1$), while H_E and $H_{\text{cal.}}$ = heights of traces obtained on engine and on calibrator, respectively, N_E and $N_{\text{cal.}}$ being the corresponding speeds [rev./min.].

Derivation. If the pick-up mounted on the calibrator is vibrated sinusoidally with an amplitude $\pm \alpha_{\text{cal.}}$ [rad.] at a frequency

$$F = N_{\text{cal.}} \times p_{\text{cal.}} = (60/2\pi) \times \omega_{\text{cal.}}\, p_{\text{cal.}} \quad \text{[vib./min.]},$$

where $N_{\text{cal.}}$ = speed of calibrator motor [rev./min.], $p_{\text{cal.}}$ = number of vibrations per motor revolution (or number of peaks of calibrator cam) and $\omega_{\text{cal.}}$ = corresponding angular velocity of motor [rad./sec.], the pick-up will have an angular velocity of vibration

$$\frac{d\alpha_{\text{cal.}}}{dt} = \frac{d}{dt}[\alpha_{\text{cal.}}\sin(p_{\text{cal.}}\,\omega_{\text{cal.}}\, t)] = \alpha_{\text{cal.}}\,\omega_{\text{cal.}}\,p_{\text{cal.}} \times \cos(p_{\text{cal.}}\,\omega_{\text{cal.}}\, t)$$

with a maximum value

$$\dot{\alpha}_{\text{cal.}} = \alpha_{\text{cal.}}\, p_{\text{cal.}}\, N_{\text{cal.}} \times \frac{2\pi}{60} \quad \text{[rad./sec.]}. \tag{1}$$

This corresponds on the oscilloscope screen to a trace of total height $H_{\text{cal.}}$ [in.], in accordance with the relation

$$H_{\text{cal.}} = K\dot{\alpha}_{\text{cal.}}, \tag{2}$$

where K = oscillograph constant.

The pick-up is then mounted on the engine, which is run up to the speed N_E [rev./min.] at which the cyclic speed variation is to be measured. If the engine speed varies between $N_E + \Delta N_E$ and $N_E - \Delta N_E$ during each crankshaft revolution, the angular velocity of the engine will have a corresponding maximum variation $\dot{\alpha}_E$, giving a trace

$$H_E = K\dot{\alpha}_E \tag{3}$$

on the oscilloscope screen. As $\dot\alpha_E = (2\pi/60) \times \Delta N_E$, the cyclic speed variation can be obtained as

$$C = \frac{1}{N_E}[(N_E + \Delta N_E) - (N_E - \Delta N_E)] = \frac{2\Delta N_E}{N_E} = \frac{2}{N_E} \times \frac{60}{2\pi} \dot\alpha_E = \frac{2}{N_E} \times \frac{60}{2\pi} \times \frac{H_E}{K},$$

and as $K = H_{cal.}/\dot\alpha_{cal.}$ from eq. (2)

$$C = \frac{2}{N_E} \times \frac{60}{2\pi} \times \frac{\dot\alpha_{cal.}}{H_{cal.}} \times H_E = \frac{2}{N_E} \times \frac{60}{2\pi} \times \frac{\alpha_{cal.} \, p_{cal.} \, N_{cal.}}{H_{cal.}} \times \frac{2\pi}{60} \times H_E$$

$$= \frac{2H_E}{H_{cal.}} \times \frac{N_{cal.}}{N_E} \times \alpha_{cal.} \, p_{cal.}.$$

Note. The above relations apply irrespective of whether a stationary or a rotating type of vibration calibrator is used, since a velocity-type pick-up with seismic mass indicates only velocity variations (and not the constant rotational velocity) both on the calibrator and on the engine.

2·633 *Example of calculations and tests*

The methods described in the previous sections were applied in order to determine the cyclic speed variation of a six-cylinder four-stroke cycle in-line engine at the B.I.C.E.R.A. Laboratory. As a torque diagram was not available, it was not possible to use the excess-energy method given in section 2·621, and the value of C was calculated by means of the formula using gas-force components (section 2·622). Details of the calculation are given below for the cyclic speed variation at 500 rev./min.

Engine data. Normal running speed $N_{normal} = 1300$ rev./min. Test conditions $N_E = 500$ rev./min. and i.m.e.p. $= 100$ Lb./in.². Piston area $A = 20\cdot563$ in.², crank radius $R_0 = 3\cdot346$ in. Inertias of engine, flywheel and brake

$$\Sigma J = 6 \times J_{cyl.} + J_F + J_B = 6 \times 0\cdot87 + 44\cdot56 + 2\cdot53 = 52\cdot31 \text{ Lb.in.sec.}^2.$$

(a) Calculation

The order of the predominant vibration is $n = 3$ (three cycles per revolution), with $\overrightarrow{\Sigma\Delta_3} = 6$ (all unit vectors in phase).

The value $|T_3|$ at 100 Lb./in.² i.m.e.p. is $|T_3| = 30\cdot8$ Lb./in.² (taken from section 2·1, p. 280: curves for four-stroke cycle engines with a normal running speed above 1000 rev./min.).

The gas-force component is evaluated as

$$P_G = |T_3| \, A \overrightarrow{\Sigma\Delta_3} = 30\cdot8 \times 20\cdot563 \times 6 = 3800 \text{ Lb.}$$

The inertia-force component for the 3rd order of four-stroke cycle engines is negligible (see p. 486).

The mean angular velocity at $N_E = 500$ rev./min. is

$$\omega = N_E \times (2\pi/60) = 500/9\cdot55 = 52\cdot36 \text{ rad./sec.}$$

Therefore, the calculated cyclic speed variation is

$$C = \frac{2R_0 P_G}{n\omega^2 \Sigma J} = \frac{2 \times 3\cdot346 \times 3800}{3 \times (52\cdot36)^2 \times 52\cdot31} = \frac{1}{16\cdot92} \simeq \frac{1}{17}.$$

(b) *Test using vibration-amplitude method*

Calibration: $2\theta_{cal.} = 0 \cdot 00875$ rad. $= 0 \cdot 5°$, $n = 3$.

Calibrator motor speed $N_{cal.} = 1500$ rev./min.†

Screen reading $H_{cal.} = 13 \cdot 2$ cm.

Test: Engine speed $N_E = 500$ rev./min.

Screen reading $H_E = 24 \cdot 7$ cm.

Result: $C = 2n\theta_{cal.} \times H_E/H_{cal.} = 3 \times 0 \cdot 00875 \times \dfrac{24 \cdot 7}{13 \cdot 2}$

$$= \frac{1}{20}.$$

(c) *Test using vibration-velocity method*

Calibration: $2\alpha_{cal.} = 0 \cdot 0175$ rad.

Number of cycles per cam revolution $p_{cal.} = 1$.

Calibrator motor speed	$N_{cal.} = 1500$	1800 rev./min.
Screen reading	$H_{cal.} = 3 \cdot 1$	4·2 cm.
Test: Engine speed	$N_E = 500$	600 rev./min.
Screen reading	$H_E = 3 \cdot 1$	2·9 cm.

Results, with $C = 2p_{cal.}\,\alpha_{cal.}\,H_E\,N_{cal.}/(N_E\,H_{cal.})$:

$C = \frac{1}{16}$ at 500 rev./min. and $C = 1/27 \cdot 5$ at 600 rev./min.

2·64 Total cyclic speed variation at higher speeds

The 'true' cyclic irregularity $C_{rolling}$ at a higher speed can be obtained by extending the hyperbolic curve of $C_{rolling}$, obtained for several speeds in the lower speed range. This is done by numerical extrapolation or graphically by plotting the straight line of $\log C$ against the engine speed N on log-log paper.

The total cyclic speed variation at the higher speed is then

$$C_{total} = C_{rolling} + rC_{torsion},$$

where $r = \left| \dfrac{\Delta_{DM}}{\Delta_1} \right| = \dfrac{\text{Holzer amplitude at driven-machine end}}{\text{Holzer amplitude at free end of engine}}$,

Δ_{DM} being determined from a Holzer table, in which $\Delta_1 = 1$ rad.

---

† *Note.* The best results are obtained if the calibrator is adjusted to oscillate at n times the engine speed, using a Hooke's joint or other rotary type of calibrator. If the frequency of the cyclic speed variation is low, it may be necessary to fit the pick-up with an additional inertia ring to increase its moment of inertia.

The vibration-amplitude method is equally applicable when a Geiger torsiograph is used.

EXAMPLE. If at a higher speed of, say, 1200 rev./min., $C_{\text{rolling}} = 0\cdot01$, $C_{\text{torsion}} = 0\cdot05$ and $|\Delta_{DM}| = |-0\cdot5| = +0\cdot5$ rad., then

$$C_{\text{total}} = 0\cdot01 + \frac{0\cdot5}{1\cdot0} \times 0\cdot05 = 0\cdot035 \cong \tfrac{1}{28}.$$

It is thus apparent that the cyclic irregularity due to torsional vibration may give an important contribution to the total cyclic speed variation of the engine system.

2·65   Flicker and alternators in parallel

The cyclic speed variation C of an engine-generator set gives rise to voltage variations which may cause flickering of electric lights. The voltage V at any instant is proportional to speed and hence $(V_{\text{max.}} - V_{\text{min.}})/V_{\text{mean}} = C$, the methods for finding C having already been given in sections 2·62–2·64. Perceptible flicker depends, however, both on the value of C and on the frequency at which the voltage varies, since the sensitivity of the eye to variation in brightness of lamps depends partly on frequency.

As shown in section 2·62, a mechanical angular deviation θ_M [rad.] occurring m times per shaft revolution corresponds to a cyclic irregularity $C = 2m\theta_M$. (Usually, $m =$ number of firing strokes or working cycles per revolution.) If $p =$ number of pole pairs of the generator, it is apparent that there will be p cycles of voltage alternations per revolution, so that the mechanical deviation θ_M will correspond to an electrical deviation θ_E which is p times larger:

$$\theta_E = p\theta_M.$$

Moreover, the electrical frequency F_E [cycles/min.] of the voltage is p times the shaft speed N [rev./min.]:

$$F_E = pN,$$

so that, with $p = F_E/N$, one has

$$\theta_E = \theta_M F_E/N.$$

The *limits for flicker* are laid down in British Standard 649 : 1949. If $n =$ number of engine impulses per second,† it is required that $C \leqslant \tfrac{1}{75}$ for engines with one or two cylinders; and the limits of cyclic irregularity for engines with more than two cylinders are

$$n < 10 \quad C \leqslant 1/150,$$

$$10 \leqslant n \leqslant 20 \quad C \leqslant n/1500,$$

$$n > 20 \quad C \leqslant 1/75.$$

Furthermore, for *parallel operation* of a.c. generators, the angular

† Usually, one first considers the lowest value of n, viz. $n = mN/60$ [firing strokes per second].

deviation θ_E should not exceed $\pm 2 \cdot 5°$ or $\pm 0 \cdot 04363$ rad. This means that, since $\theta_E = p\theta_M$, the further requirement in this instance is

$$\theta_E = C_E p/(2m) \leqslant 0 \cdot 04363,$$

where C_E denotes the usual 'total cyclic speed deviation' C (taking account of torsional vibration) multiplied by a dynamic magnifier M if the system is running on the flank of an electromechanical resonance, that is,

$$C_E = MC, \quad \text{where} \quad M = \frac{1}{[1 - (F_H/60n_1)^2]},$$

F_H being the oscillation frequency [vib./min.] evaluated as shown in the following (p. 495), while $n_1 =$ number of firing strokes per second of one cylinder only (for both single- and multi-cylinder engines).

If C is specified, the J-value (of rotor and flywheel) effectively required to obtain it will be

$$J = MJ_0,$$

where $J_0 =$ value of inertia derived directly from eq. (5) of p. 487. This is so because C-values are inversely proportional to inertias, hence $M = C_E/C = J/J_0$ so that $J = MJ_0$.†

For d.c. machines in parallel, the restriction for θ_E is not applied, and only the above limits for flicker are required.

Finally, *electrical resonance* must also be avoided. The engine-generator set, operating as an electromechanical system, has an oscillation frequency F_H which is the subject of the following Note.

Note on the natural frequency of oscillation of engine-driven alternators

An alternator connected to a network has an inherent synchronizing torque which is produced whenever the rotor departs from its normal angular position for a given load, and which acts in a direction necessary to restore the original state of equilibrium. Synchronizing power is expressed in terms of kilowatts per electrical radian, and is analogous to torsional stiffness in a mechanical sense. It will be seen, therefore, that an alternator can be made to oscillate about a mean position at a frequency dependent upon the total inertia of its rotor and prime mover and the strength of the synchronizing torque, if impulses are imposed at that frequency.

This oscillation involves a variation of load angle and hence of load, and it follows that it will be accompanied by a corresponding oscillation of power superimposed on whatever steady load is being carried by the alternator, and is objectionable for the following reasons.

If the power oscillation is very large, the pull-out power of the alternator may be exceeded, causing it to drop out of synchronism. Even if the alternator does not drop out, the power swing may be large enough to operate the overload or reverse power relays on the circuit breaker. If excessive, the power pulsation may cause mechanical damage to the alternator or prime mover. Although these effects will only be noticeable if the power oscillation appreciably exceeds 100 %, a relatively

† Sometimes J is calculated from the expression $J = J_0 + \Delta J$, where $\Delta J =$ additional inertia (for specified C) required to compensate that magnification. The resultant numerical value for J is the same as that obtained directly from $J = MJ_0$.

small oscillation is undesirable, as the swinging of ammeters and wattmeters will make it difficult to observe the load being carried. Severe power pulsation may cause flickering of lights.

According to B.S. 649, the onus is on the electrical designer to ensure that such conditions will not prevail, and he requires to know from the engine designer:

(1) the total inertia of engine and flywheel;
(2) the number of firing impulses per minute from one cylinder; and
(3) the number of firing impulses per minute from all cylinders.

This last item is not used for the calculation of resonance with the natural frequency of oscillation, but it is usual to supply the information so that an estimate of cyclic irregularity may be made.

The oscillation frequency F_H can be estimated from the following formula:

$$F_H = \frac{21,000}{N} \times \sqrt{\left(\frac{\alpha F_E H_E}{J}\right)} \quad \text{[vib./min.]},$$

where N = running speed [rev./min.], $\alpha \cong 0.04$ for three-phase, and $\alpha \cong 0.07$ for single-phase alternators,† $F_E = pN$ = electrical frequency [cycles/min.], H_E = rating of alternator [kilowatts], and J = inertias of flywheel + rotor [Lb.in.sec.²].

It may be noted that the electromechanical system has an inertia J and an 'equivalent electrical stiffness' (or 'synchronizing rigidity') due to the magnetic linkage between the stator and the rotor field, and proportional to $\alpha F_E H_E / N^2$. At resonance, this stiffness is equal to the inertia torque per radian, that is, to $J F_H^2 \times (2\pi/60)^2$.

The above equation for F_H is valid for an alternator in parallel with an infinite network, or for a system formed by a number of identical alternators running in parallel with an infinite network (the system frequency then has the same F_H-value as a single unit).

For systems with dissimilar alternators in parallel: Let F_{H_1}, F_{H_2}, ..., denote the oscillation frequencies and J_1, J_2, ..., the inertias of the individual units. Then the combined system will have resultant frequencies $F_{res.}$, determined by the relation

$$\frac{J_1 F_{H_1}^2}{F_{res.}^2 - F_{H_1}^2} + \frac{J_2 F_{H_2}^2}{F_{res.}^2 - F_{H_2}^2} + \dots + \frac{J_n F_{Hn}^2}{F_{res.}^2 - F_{Hn}^2} = 0,$$

that is, there will be $n-1$ natural frequencies $F_{res.}^{I}$, $F_{res.}^{II}$, ..., for the n-mass system, and none of the engine impulse frequencies should coincide with these. (For the derivation of this expression, see section 1·3215, p. 190.)

Torque harmonics. In addition to the possible 'restricted' speed ranges due to resonance with the firing-stroke frequencies, it may be noted that one should also avoid running at speeds at which resonance due to torque harmonics (major or minor orders) might occur.

Normally, on receipt of this information, the electrical designer checks that these frequencies are at least 20% away from the natural frequency of the alternator and other alternators which are to run in parallel, and, if this is not the case, he requests the engine maker to modify the engine flywheel accordingly.

The experience of the engine makers represented on the B.I.C.E.R.A. Torsional Vibration Panel is that this 20% limit, which is the condition usually demanded, is not always practicable, particularly when matching new engine-alternator equipment with sets of various makes.

The 20% limit, therefore, should not be regarded as rigid, but should be complied with as far as possible.

† $\alpha = 0.05$ according to B.E.A.M.A. rules (see Jackson's paper in Bibliography at the end of this section). Fitchett and Holland indicate $\alpha \cong 0.03$, and also give curves showing the variation of α with load and power factor.

Derivation of expression for C_E. Consider a one-mass system as shown in Fig. 2, in which K_E = equivalent (electrical) stiffness and J = inertia. It can be shown (see section 2·45, p. 430) that, when the inertia is acted upon by a sinusoidal torque $T = T_0 \sin \omega t$, the system vibrates with an amplitude

$$\theta = \frac{T_0/K_E}{1 - (\omega/\omega_H)^2},$$

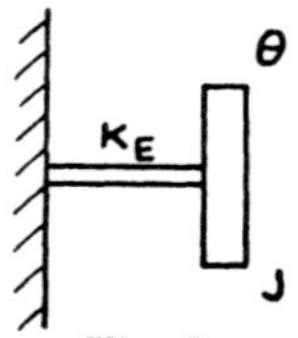

where $\omega_H^2 = K_E/J$. Now, owing to cyclic irregularity, the inertia is subjected to a vibratory torque $T_0 = -J\omega^2\theta_M$, where $\theta_M = C/2m$. (This torque is similar to that of an out-of-balance weight type exciter.) Substituting this expression, we obtain

Fig. 2.

$$\theta = \frac{-\omega^2 J\theta_M/K_E}{1 - (\omega/\omega_H)^2} = \frac{-(\omega/\omega)^2\theta_M}{1 - (\omega/\omega_H)^2}.$$

The dynamic magnifier is, therefore,

$$\frac{\theta}{\theta_M} = M = \frac{1}{1 - (\omega_H/\omega)^2}.$$

With $\omega_H = 2\pi F_H/60$ and $\omega = 2\pi n_1$, we have $\omega_H/\omega = F_H/(60n_1)$ and, since $\theta/\theta_M = C_E/C$, this gives the equation for C_E indicated on p. 494.

BIBLIOGRAPHY

British Standards Institution, B.S. no. 649 (1949).

Bradbury, C. H. *The Parallel Operation of Diesel Alternator Sets*, A.E.S.D. Tech. Publ. (1930–1).

Everest. Some factors in the parallel operation of alternators, *J. Instn Elect. Engrs*, vol. 50 (May 1913), p. 526.

Fitchett, F. and Holland, W. R. *The Design and Construction of Alternators for Coupling to Diesel Engines*, Paper S-150 read before the Diesel Engine Users' Association, 28 February 1939, pp. 5–24.

Jackson, P. *The Vibration of Oil Engines*, Paper S-115 read before the Diesel Engine Users' Association, 26 April 1933, pp. 13–17.

Ker Wilson, W. *Practical Solution of Torsional Vibration Problems*, vol. 2 (1941), pp. 601–33.

Kleiner, A. Cyclic irregularity in diesel-electric sets. *Sulzer Tech. Rev.* no. 3/4 (1947); *Schweiz. Bauztg*, 16 August 1947, p. 443.

Laible, T. Parallel operation of synchronous machines with non-uniform driving torque. Parts 1 and 2. *Bull. Oerlikon*, nos. 311 and 312, August and October 1955, pp. 45–58 and 67–74.

Ogle, H. R. *The Parallel Operation of Alternators*, parts 1 and 2. Reprint from *The Allen Engng Rev.* no. AP 7002 (1956).

Putnam, H. E. Synchronizing power in synchronous machines. *Trans. Instn Elect. Engrs*, vol. 45, p. 1116.

Simons. Das Flackern des Lichtes in elektrischen Beleuchtungsanlagen. *Elektrotech. Z.* (1917).

Stone, M. Parallel operation of a.c. generators, *Trans. Amer. Inst. Elect. Engrs*, June 1932, p. 332.

Wennerberg, J. *A.S.E.A. Journal*, September 1931, pp. 118–121. (Use of damper windings.)

PART 3

DESIGN AND OPERATION OF VARIOUS DEVICES FOR LIMITING VIBRATION

3·1   THE TUNING DISK† WITHOUT DAMPING

NOTATION

Sym-bol	Brief definition	Typical units	Sym-bol	Brief definition	Typical units
J	moment of inertia	Lb.in.sec.², Kg.cm.sec.²	Δ	Holzer-table amplitude	rad.
K	stiffness	Lb.in./rad., Kg.cm./rad.	θ	vibration amplitude	deg., rad.
N	speed	rev./min.	ω	phase velocity	rad./sec.
T	torque	Lb.in., Kg.cm.			

Subscripts are used in the text to define more precisely. Their meaning is fully explained in the context in each instance.

(1) Where a resonance peak occurs at a speed N_C in the running range of an engine, it is possible to reduce the torsional-vibration amplitude at this speed by mounting, at the front end of the crankshaft, a disk of suitable moment of inertia.

Let ω_0 = natural phase velocity [rad./sec.] of engine system without the tuning disk,

K_{TD} = stiffness [Lb.in./rad.] of shaft portion between cylinder no. 1 and tuning disk,

J_{TD} = inertia [Lb.in.sec.²] of tuning disk.

To obtain the amplitude reduction indicated above, it is necessary that the values of K_{TD} and J_{TD} should be in accordance with the relation

$$K_{TD}/J_{TD} = \omega_0^2.$$

When this relation holds, the sub-system K_{TD} and J_{TD} is said to be 'tuned' to the original frequency of the engine system. Usually, the value of K_{TD} is already fixed by the crankshaft design, within certain limits, unless an extension shaft is used at the front end of the crankshaft. In such cases, the 'tuning' is obtainable only by choosing a suitable value for J_{TD}.

(2) Although the tuning disk eliminates the peak at the engine speed N_C, it also adds a further possible mode of vibration to the engine system, so that with the tuning disk two peaks are obtained, on either side of the original resonance curve. If the value of K_{TD} is sufficiently

† Alternative designations. Tuning flywheel, harmonic balancer, dynamic vibration absorber.

high, these two new peaks may occur at speeds outside (i.e. above and below) the running range.

As regards the peak values of the two resonance curves, these may be approximately equal to that of the original single-peak amplitude. However, they may also be greater, or smaller, in some cases. These considerations are fully discussed in section 2·45 under 'Engine with front-end pulley', pp. 441–447.

(3) If K_{TD}/J_{TD} is correctly chosen, the 'anti-resonant frequency' will coincide with the original resonant frequency of the system, and the amplitude reduction will be situated at the peak of the original resonance curve (Fig. 1). If $\sqrt{(K_{TD}/J_{TD})}$ has a value not quite equal to ω_0, the amplitude reduction will occur on one of the flanks of the original resonance curve (Fig. 2).

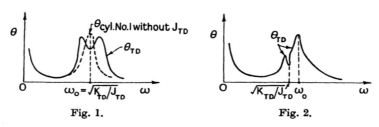

Fig. 1.   Fig. 2.

In theory, without damping, the amplitude θ_1 at cylinder no. 1 is zero (although θ_{TD} is not zero) at the anti-resonant phase velocity $\omega_0 = \sqrt{(K_{TD}/J_{TD})}$, from the analysis of two- and three-mass systems.

In practice, however, there is damping in the engine system, and θ_1 decreases to a low value, which can be determined from the Holzer-table amplitudes Δ_{TD} and Δ_1 of a forced-frequency tabulation, when θ_{TD} is known from measurements. *In general, θ_1 will always be smaller than θ_{TD}.*

(4) The height of the anti-resonant amplitude depends on the extent to which the two peaks can be separated. The 'tuning' of K_{TD} and J_{TD} to obtain the required K_{TD}/J_{TD} ratio ensures that the anti-resonance will occur at the original resonance peak. While keeping this ratio constant, however, both K_{TD} and J_{TD} can be varied together.

It can be shown that if K_{TD} is a very small value (and hence J_{TD} is also very small) only a minute separation is obtained at the resonance peak (Fig. 3). This is known as a 'loose coupling' condition.

With K_{TD}/J_{TD} maintained constant, as K_{TD} is increased, the two peaks begin to separate (Fig. 4), and as a result the common flank position, i.e. the anti-resonant amplitude, diminishes. Beyond a certain value of K_{TD}, for the same ratio, the two peaks, and their flanks can be clearly discriminated (Fig. 5) and this is known as a 'close' or 'rigid' coupling condition.

Note. These phenomena also occur with various K/J-values on the driven-machine side of the system. When there is a very soft coupling between the driven machine and the main flywheel, one speaks of the engine as being 'isolated' from the driven parts. In fact, this loosely coupled condition may correspond to two nearly coincident peaks which can be determined by tests, as well as by usual Holzer tabulations if the latter are worked out with a sufficiently large number of figures to very small values of residual torque ($\Sigma J\omega^2\Delta$).

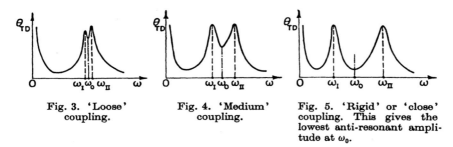

Fig. 3. 'Loose' coupling.

Fig. 4. 'Medium' coupling.

Fig. 5. 'Rigid' or 'close' coupling. This gives the lowest anti-resonant amplitude at ω_0.

It should also be noted that if more than one large-amplitude critical occurs in the running range of the original system, each of these resonance curves will be split into two peaks when a 'medium'- or 'close'-coupled tuning disk is used, so that the number of peaks in the running range will be correspondingly doubled. For variable-speed engines, therefore, the undamped tuning disk is not sufficient to deal with all possible cases. However, if the tuning disk is damped (as in the case of a rubber-type damper), the peak amplitudes will also be reduced.

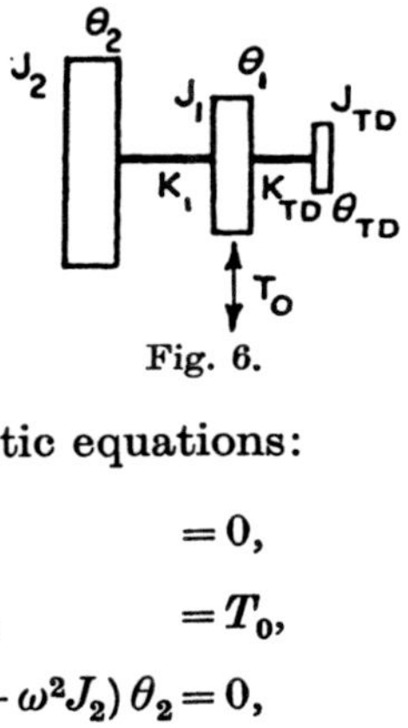

Fig. 6.

Derivation of 'tuning' ratio. Consider, as an example, a three-mass system of the type shown in Fig. 6. Its equations of motion give the characteristic equations:

$$(K_{TD}-\omega^2 J_{TD})\,\theta_{TD} - K_{TD}\theta_1 = 0,$$
$$-K_{TD}\theta_{TD}+(K_{TD}+K_1-\omega^2 J_1)\,\theta_1 - K_1\theta_2 = T_0,$$
$$-K_1\theta_1 \qquad\qquad + (K_1-\omega^2 J_2)\,\theta_2 = 0,$$

where θ = vibration amplitude and T_0 = torque. Hence, using determinants (see section 2·45), one obtains

$$\begin{vmatrix} K_{TD}-\omega^2 J_{TD} & -K_{TD} & 0 \\ -K_{TD} & K_{TD}+K_1-\omega^2 J_1 & -K_1 \\ 0 & -K_1 & K_1-\omega^2 J_2 \end{vmatrix} = \mathrm{Det}_s,$$

[501]

$$\begin{vmatrix} 0 & -K_{TD} & 0 \\ T_0 & K_{TD}+K_1-\omega^2 J_1 & -K_1 \\ 0 & -K_1 & K_1-\omega^2 J_2 \end{vmatrix} = \text{Det}_{TD},$$

$$\begin{vmatrix} K_{TD}-\omega^2 J_{TD} & 0 & 0 \\ -K_{TD} & T_0 & -K_1 \\ 0 & 0 & K_1-\omega^2 J_2 \end{vmatrix} = \text{Det}_1,$$

$$\begin{vmatrix} K_{TD}-\omega^2 J_{TD} & -K_{TD} & 0 \\ -K_{TD} & K_{TD}+K_1-\omega^2 J_1 & T_0 \\ 0 & -K_1 & 0 \end{vmatrix} = \text{Det}_2.$$

These give directly the vibration amplitudes at the three masses, viz.

$$\theta_{TD}=T_0 K_{TD}(K_1-\omega^2 J_2)/\text{Det}_s, \quad \theta_1=T_0(K_{TD}-\omega^2 J_{TD})(K_1-\omega^2 J_2)/\text{Det}_s,$$

and
$$\theta_2=T_0 K_1(K_{TD}-\omega^2 J_{TD})/\text{Det}_s.$$

Therefore θ_1 and θ_2 are both zero, when $\omega^2=K_{TD}/J_{TD}$.

Let $\omega_0^2=K_1\left(\dfrac{1}{J_1}+\dfrac{1}{J_2}\right)$. If $\omega^2=\omega_0^2$, then the phase velocity at which $\theta_1=\theta_2=0$ will be the natural phase velocity of the original two-mass system. Therefore the condition for obtaining minimum θ-values at J_1 and J_2 is that $K_{TD}/J_{TD}=\omega_0^2$.

Det_s can be reduced to a simpler form by adding column 1 to column 2 and also column 3 to column 2, so that

$$\text{Det}_s = \begin{vmatrix} K_{TD}-\omega^2 J_{TD} & -\omega^2 J_{TD} & 0 \\ -K_{TD} & -\omega^2 J_1 & -K_1 \\ 0 & -\omega^2 J_2 & K_1-\omega^2 J_2 \end{vmatrix}$$
$$= -\omega^2\{(K_{TD}-\omega^2 J_{TD})(J_1 K_1-\omega^2 J_1 J_2+J_2 K_1)+K_{TD}J_{TD}(K_1-\omega^2 J_2)\}.$$

Hence, when $\omega^2=\omega_0^2$,

$$\theta_{TD}=T_0 K_{TD}(K_1-\omega^2 J_2)/\text{Det}_s = -T_0/(\omega_0^2 J_{TD}) = -T_0/K_{TD},$$

which shows that under 'tuned' conditions, as stated under 4 above, the lower K_{TD} is, the higher the anti-resonant amplitude θ_{TD} will be.

As regards θ_1, in theory it is still zero when both J_{TD} and K_{TD} approach zero value while K_{TD}/J_{TD} is kept constant. However, in practice, K_{TD} must be sufficient to avoid excessive torsional stress in this shaft portion resulting from large θ_{TD}-values, and this sets a limit to the minimum value of J_{TD}.

Size of tuning disk. In practice, the tuning disk inertia J_{TD} is usually given a suitable value in the range $0.1 J_E$ to $0.3 J_E$, where J_E represents

[502]

the equivalent engine inertia $\Sigma J_{\text{cyl.}} \Delta_{\text{cyl.}}^2$ (denoted by J_1 in the above equations). The equivalent engine-shaft stiffness is evaluated as

$$K_E = K_1 = \omega_0^2/[(1/J_E) + (1/J_F)],$$

where J_F = inertia of main flywheel. Torsiograph tests on a prototype engine system are advisable, in order to verify that suitable anti-resonant conditions are obtained at the engine speed considered.

Resonant frequencies. By equating to zero the above expression for Det_s, and substituting $\omega_0^2 = K_{TD}/J_{TD}$, one obtains

$$\omega^4 - \omega^2 \left(\frac{K_{TD} + K_1}{J_1} + \omega_0^2 + \frac{K_1}{J_2} \right) + K_1 \left(\frac{\omega_0^2}{J_2} + \frac{\omega_0^2}{J_1} + \frac{K_{TD}}{J_1 J_2} \right) = 0.$$

This equation gives two resonant phase velocities, ω_I and ω_{II}. For very large values of J_2 (which represents the main flywheel), it can be shown that

$$\omega_I^2 \omega_{II}^2 = \omega_0^4,$$

that is, that the original phase velocity $\omega_0 = \sqrt{(\omega_I \omega_{II})}$. As one of the two new resonance values, for instance ω_I, is reduced towards zero, the other will tend to reach very high values, as indicated above under 4.

Note. Figs. 3, 4 and 5 on p. 501 refer to systems in which $\sqrt{(K_{TD}/J_{TD})}$ is equal to the natural phase velocity ω_0 of the original engine system (tuning ratio of unity). For different values of the tuning ratio, the resonant peaks ω_I and ω_{II} may, of course, be far apart, even with 'loose' coupling conditions.

3·2 THE TUNING DISK WITH DAMPING

The above designation includes all types of dampers which have their 'seismic' member elastically connected to the crankshaft and for which the damping can be regarded as proportional to vibration velocity, e.g. rubber dampers, and dampers of the viscous-shear type (with silicone fluid) which have their inner mass elastically connected to the outer casing by means of springs or rubber-like elements.

It can be stated as a general rule that all such types of damper require 'tuning', i.e. adaptation to the main engine system, but that, in contrast to tuning disks without damping, those with damping do not necessarily fulfil all requirements when tuned to the natural frequency of the original system.

Tuning disks with damping are used when it is required not only to eliminate vibration at a particular speed, but when it is also necessary to reduce peak amplitudes throughout the speed range. Thus more stringent requirements are formulated and the tuning disk with damping requires correspondingly a consideration of a wider range of conditions.

NOTATION

Symbol	Brief definition	Typical units	Symbol	Brief definition	Typical units
A	area	in.^{2}, cm.^{2}	L	length or thickness	in., cm.
A	parameter	Lb.in., Kg.cm.	M	dynamic magnifier	—
b	parameter	rad.^{2}/sec.^{2}	q	shear stress	Lb./in.^{2},
c	damping coefficient	Lb.in.sec./			Kg./cm.^{2}
		rad.	R	radius	in., cm.
		Kg.cm.sec./	t	time	sec., min.
		rad.	T	torque	Lb.in., Kg.cm.
d	parameter	rad.^{2}/sec.^{2}	W	weight	Lb., Kg.
D	diameter	in., cm.	γ	damping ratio	—
$\mathscr{D}$	denominator	—	Δ	Holzer-table amplitude	rad.
F	vibration frequency	vib./min.			
G	modulus of rigidity	Lb./in.^{2},	θ	vibration amplitude	deg., rad.
		Kg./cm.^{2}	μ	inertia ratio	—
j	$\sqrt{-1}$	—	$\overrightarrow{\Sigma\Delta}$	phase-vector sum	—
J	moment of inertia	Lb.in.sec.^{2},	ϕ	'loss angle'	deg.
		Kg.cm.sec.^{2}	ω	phase velocity	rad./sec.
K	stiffness	Lb.in./rad.,			
		Kg.cm./rad.			

Subscripts are used in the text to define more precisely. Their meaning is fully explained in their context in each instance.

3·21 Three-mass system formulae

The inherent engine damping may be regarded as negligible in comparison with the damping in the tuning disk system.

The engine system with damper and flywheel is reduced to a three-mass system of the type shown in Fig. 1, as follows:

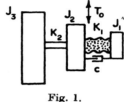

Let $J_1'' =$ moment of inertia [Lb.in.sec.^{2}]† of the 'seismic' member of the damper,

$K_1 =$ stiffness of material between the two members of the damper [Lb.in./rad.],

$c =$ damping coefficient of material between the two members of the damper [Lb.in.sec./rad.],

$T_0 =$ resultant excitation torque [Lb.in.] of the engine

$$(|T_n| A R_0 \overrightarrow{\Sigma\Delta}),$$

for the critical order considered, using a phase-vector sum $\overrightarrow{\Sigma\Delta}$ determined from a Holzer table for the engine with damper,

$J_2 =$ inertia of damper portion attached to crankshaft + sum of all cylinder inertias multiplied by their Holzer-table amplitudes squared $[J_2 = J_1'' + \Sigma(J_{\text{cyl.}} \Delta_{\text{cyl.}}^2)]$,

$K_2 = \omega_0^2 / \left[\dfrac{1}{J_2} + \dfrac{1}{J_3} \right]$, where $\omega_0 =$ natural phase velocity of engine- and main-flywheel system (without the damper),

$J_3 =$ inertia of main flywheel.

Where necessary, the values of K_1 and c may be verified experimentally by rig tests of the damper, which can be tested in the same way as couplings (see section 1·2, p. 104, and section 2·45, p. 455).

The equivalent three-mass system thus obtained has two natural phase velocities, ω_{I} and ω_{II}, which can be determined directly from

$$\omega_{\text{I, II}}^2 = \tfrac{1}{2}[b \mp \sqrt{(b^2 - 4d)}] \quad [\text{rad./sec.}]^2,$$

where

$$b = K_1 \left(\frac{1}{J_1''} + \frac{1}{J_2} \right) + K_2 \left(\frac{1}{J_2} + \frac{1}{J_3} \right) \quad \text{and} \quad d = \frac{K_1 K_2}{J_1'' J_2 J_3} (J_1'' + J_2 + J_3).$$

(a) *Resonant conditions.* The formulae for peak amplitudes for 'engines with front-end pulley', given on p. 441 of section 2·45, can be applied directly, after altering the subscripts to suit the arrangement

† The formulae in this section are also valid in metric units, for instance with J-values in Kg.cm.sec.^{2}, K in Kg.cm./rad., c in Kg.cm.sec./rad. and T in Kg.cm.

considered above in Fig. 1. Thus, substituting either ω_I or ω_{II}, one obtains the peak amplitude and the shaft twists as follows:

Amplitude at damper mass J_1'':

$$\theta_1 = \frac{T_0}{\omega c \mathscr{D}_{\text{res.}}} \times \left(1 - \frac{\omega^2 J_3}{K_2}\right) \times \sqrt{\left\{1 + \left(\frac{\omega c}{K_1}\right)^2\right\}};$$

Amplitude at engine mass J_2:

$$\theta_2 = \theta_3 \times \left(1 - \frac{\omega^2 J_3}{K_2}\right);$$

Amplitude at main flywheel J_3:

$$\theta_3 = \frac{T_0}{\omega c \mathscr{D}_{\text{res.}}} \times \sqrt{\left\{\left(1 - \frac{\omega^2 J_1''}{K_1}\right)^2 + \left(\frac{\omega c}{K_1}\right)^2\right\}};$$

Twist† of damper material (K_1):

$$\theta_{sh\,1} = |\theta_1 - \theta_2| = \frac{T_0}{\omega c \mathscr{D}_{\text{res.}}} \times \left(1 - \frac{\omega^2 J_3}{K_2}\right) \times \frac{\omega^2 J_1''}{K_1};$$

Twist† of crankshaft (K_2):

$$\theta_{sh\,2} = |\theta_2 - \theta_3| = \theta_3 \times \frac{\omega^2 J_3}{K_2}.$$

In these formulae:

$$\mathscr{D}_{\text{res.}} = \frac{\omega^2}{K_1}\left[\frac{\omega^2 J_3}{K_2}(J_1'' + J_2) - (J_1'' + J_2 + J_3)\right].$$

(b) *Anti-resonant condition.* When

$$\omega^2 = \omega_0^2 = K_1/J_1'' = K_2[(1/J_2) + (1/J_3)],$$

it will usually be sufficient to evaluate the amplitude at the damper mass J_1'':

Approximate formulae:

$$\theta_1 \cong T_0/(J_1'' \omega_0^2), \quad \theta_2 \cong 0.$$

Exact formula:

$$\theta_1 = \frac{T_0}{\mathscr{D}_{\text{a.r.}}} \times \left(1 - \frac{\omega_0^2 J_3}{K_2}\right) \times \sqrt{\left\{1 + \left(\frac{\omega_0 c}{K_1}\right)^2\right\}},$$

where

$$\mathscr{D}_{\text{a.r.}} = \omega_0^2 \sqrt{\left\{J_1''^2 \times \left(1 - \frac{\omega_0^2 J_3}{K_2}\right)^2 + \left(\frac{\omega_0 c}{K_1}\right)^2 \times \left[J_1'' + J_2 + J_3 - \frac{\omega_0^2 J_3}{K_2}(J_1'' + J_2)\right]^2\right\}}.$$

(c) *Flank amplitudes.* For non-resonant conditions, with vibration

---

† It may be recalled that these amplitude differences are obtained from the complex-number expressions for θ_1, θ_2 and θ_3, by subtracting real terms and imaginary terms separately, before taking the square root for the absolute value.

[506]

amplitudes of intermediate values, use may be made of the following relation:

Amplitude at damper mass J_1'':

$$\theta_1 = \frac{T_0}{\mathscr{D}_{\text{n.r.}}} \times \left(1 - \frac{\omega^2 J_3}{K_2}\right) \times \sqrt{\left\{1 + \left(\frac{\omega c}{K_1}\right)^2\right\}},$$

where

$$\mathscr{D}_{\text{n.r.}} = \omega^2 \sqrt{\left\{\left[\left(1 - \frac{\omega^2 J_1''}{K_1}\right)\left(J_2 + J_3 - \frac{\omega^2 J_2 J_3}{K_2}\right) + J_1''\left(1 - \frac{\omega^2 J_3}{K_2}\right)\right]^2\right.}$$
$$\left. + \left(\frac{\omega c}{K_1}\right)^2 \left[J_1'' + J_2 + J_3 - \frac{\omega^2 J_3}{K_2}(J_1'' + J_2)\right]^2\right\}.$$

Results obtainable with these formulae will now be illustrated by the following numerical example.

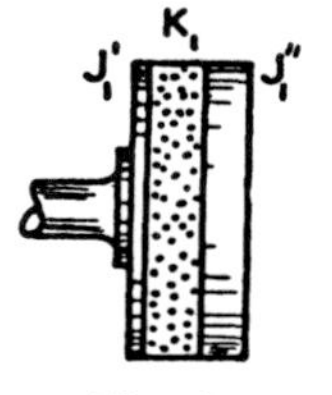

NUMERICAL EXAMPLE. A rubber-type damper with a 'seismic' mass of inertia $J_1'' = 120$ Lb.in.sec.^{2} and a stiffness $K_1 = 1 \cdot 08 \times 10^7$ Lb.in./rad. is 'tuned' to the natural phase velocity $\omega_0 = 300$ rad./sec. of an engine system, consisting of the equivalent engine inertia $J_2 = 480$ Lb.in.sec.^{2} (including the inertia J_1' of the damper member attached to the crankshaft), the main flywheel inertia $J_3 = 800$ Lb.in.sec.^{2} and the equivalent crankshaft stiffness

Fig. 2.

$$K_2 = \omega_0^2/[(1/J_2) + (1/J_3)] = 2 \cdot 7 \times 10^7 \text{ Lb.in./rad.}$$

The damper has a damping coefficient $c = 4000$ Lb.in.sec./rad. and the excitation torque of the engine is $T_0 = 12,000$ Lb.in. It is required to determine the peak amplitudes of vibration at the resonant phase velocities ω_{I} and ω_{II}, and the minimum amplitude at the anti-resonant phase velocity ω_0.

Natural phase velocities of system with damper

$$b = 10 \cdot 8 \times 10^6 \times \left(\frac{1}{120} + \frac{1}{480}\right) + 27 \cdot 0 \times 10^6 \times \left(\frac{1}{480} + \frac{1}{800}\right) = 2 \cdot 025 \times 10^5,$$

$$d = \frac{27 \cdot 0 \times 10 \cdot 8 \times 10^{12}}{120 \times 480 \times 800} \times (120 + 480 + 800) = 88 \cdot 593 \times 10^8,$$

$$\omega_{\text{I}}^2 = \tfrac{1}{2}[b - \sqrt{(b^2 - 4d)}] = 63,900, \quad \omega_{\text{II}}^2 = \tfrac{1}{2}[b + \sqrt{(b^2 - 4d)}] = 138,500.$$

$$\omega_{\text{I}} = 253 \text{ rad./sec.}, \quad F_{\text{I}} = (60/2\pi)\,\omega_{\text{I}} = 9 \cdot 55 \times 253 = 2416 \text{ vib./min.},$$

$$\omega_{\text{II}} = 372 \text{ rad./sec.}, \quad F_{\text{II}} = 3552 \text{ vib./min.},$$

$$\omega_{\text{I}}\omega_{\text{II}} = 253 \times 372 = 94,120 \simeq \omega_0^2.$$

One-node conditions

$$T_0/(\omega_{\text{I}} c) = 12 \times 10^3/(253 \times 4000) = 0 \cdot 0119.$$

$$\omega_{\text{I}}^2 J_3/K_2 = 6 \cdot 39 \times 10^4 \times 800/27 \times 10^6 = 1 \cdot 893.$$

$$\mathscr{D}_{\text{res.}} = \frac{\omega_{\text{I}}^2}{K_1}\left[\frac{\omega_{\text{I}}^2 J_3}{K_2}(J_1'' + J_2) - (J_1'' + J_2 + J_3)\right]$$
$$= \frac{6 \cdot 39 \times 10^4}{10 \cdot 8 \times 10^6}[1 \cdot 893 \times 600 - 1400] = -1 \cdot 562.$$

$$1 - (\omega_{\text{I}}^2 J_3/K_2) = -0 \cdot 893.$$

$$\dagger \omega_I c/K_1 = 253 \times 4000/10\cdot8 \times 10^6 = 0\cdot0937, \quad (\omega_I c/K_1)^2 = 0\cdot00878.$$

$$\theta_1 = \frac{0\cdot0119}{-1\cdot562} \times (-0\cdot893) \times \sqrt{(1 + 0\cdot00878)} = 0\cdot00683 \, \text{rad.} = +0\cdot39^\circ.$$

$$\omega_I^2 J_1''/K_1 = 6\cdot39 \times 10^4 \times 120/10\cdot8 \times 10^6 = 0\cdot710; \quad 1 - (\omega_I^2 J_1''/K_1) = 0\cdot290.$$

$$\theta_3 = \frac{0\cdot0119}{-1\cdot562} \times \sqrt{\{(0\cdot29)^2 + 0\cdot00878\}} = -0\cdot00232 \, \text{rad.} = -0\cdot133^\circ.$$

$$\theta_2 = \theta_3 \times [1 - (\omega_I^2 J_3/K_2)] = -0\cdot133 \times (-0\cdot893) = +0\cdot12^\circ.$$

$$\theta_{sh\,2} = |\,\theta_2 - \theta_3\,| = \theta_3 \times \omega_I^2 J_3/K_2 = -0\cdot133 \times 1\cdot893 = -0\cdot25^\circ.$$

$$\theta_{sh\,1} = |\,\theta_1 - \theta_2\,| = \frac{0\cdot0119}{-1\cdot562} \times (-0\cdot893) \times 0\cdot710 = +0\cdot00483 \, \text{rad.} = +0\cdot28^\circ.$$

Two-node conditions

Calculations similar to those detailed above, but with $\omega_{II} = 372$ rad./sec., give the values:

$$\theta_1 = +0\cdot11^\circ, \quad \theta_2 = -0\cdot06^\circ, \quad \theta_3 = +0\cdot02^\circ,$$

$$\theta_{sh\,1} = 0\cdot16^\circ, \quad \theta_{sh\,2} = 0\cdot08^\circ.$$

Anti-resonant condition

$$\theta_1 \cong T_0/(\omega_0^2 J_1'') \cong 1\cdot2 \times 10^4/[9 \times 10^4 \times 120] \cong 0\cdot0011 \, \text{rad.} \cong 0\cdot064^\circ.$$

The exact formula gives the same value for θ_1, since the damping is fairly small.

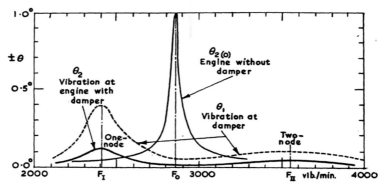

Fig. 3. Reduction in amplitudes obtained with rubber-type damper
(tuned to engine frequency F_0).

Note. A suitable bonded-rubber damper for this example could be designed to the following specification: outer diameter $D = 30$ in., thickness of rubber $L = 2\cdot208$ in., rigidity modulus of rubber compound $G = 300$ Lb./in.^{2}, thickness of seismic mass $L_1'' = 2\cdot2$ in. (cast iron), weight of seismic mass $W_1'' = 405$ Lb. Then $K_1 = GI_p/L = 10\cdot8 \times 10^6$ Lb.in./rad., and $J_1 = 120$ Lb.in.sec.^{2}. The shear stress due

$\dagger$ In rubber technology, it is usual to write $\omega_I c/K_1 = \tan \phi$, where $\phi = \tan^{-1} (\omega_I c/K_1)$ is the 'loss angle' of the rubber. Thus, ϕ is a measure of the energy dissipation of a rubber-type material. In regard to the engine system, the criterion is $\gamma = c/2J_1'' \omega_0$, where $\gamma = $ damping ratio [see section 2·45, p. 458]. In the above example, the loss angle is $\phi = \tan^{-1} [0\cdot0937] = 5^\circ 21'$, and the damping ratio is $\gamma = c/2J_1'' \omega_0 = 0\cdot055$.

to $\theta_{sh\,1}=0\cdot28°$ is $q=G\times(D/2)\times\theta_{sh\,1}/L\cong23$ Lb./in.^{2}, and the shear stress due to the dead weight W_1'' is negligible. If necessary, an oil-resisting synthetic rubber may be used, or the damper may be provided with a suitable coating protecting it against oil action.

The results of the calculation are represented in Fig. 3. These show that

(1) low peak values θ_2 are obtained at the engine mass J_2;

(2) the peak values of amplitude θ_1 at the damper mass J_1'' are not of the same height for the one-node and two-node frequencies F_I and F_{II}.

Note. Where gearwheels for auxiliary drives are mounted at the free end of the crankshaft in the original engine system, it may be an advantage to retain these at this position, or to mount them on the rigidly coupled member J_1' of the damper, since the vibration at this position is considerably smaller than at the seismic-mass end of the damper (see Fig. 4).

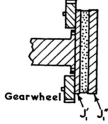

Fig. 4.

Heat dissipation of rubber dampers

In tests of engines with rubber dampers, it is advisable to measure the temperature of the rubber, in order to verify that no overheating occurs and that a fairly stable working temperature is obtained. For assessments at the design stage, it is recommended, on the basis of investigations by Zdanowich and Moyal,† that for safe operation the power dissipated per unit volume of rubber material should not exceed

$$H/V = 0\cdot009\,\frac{\text{B.Th.U.}}{\text{sec.in.}^3} = 0\cdot0126\,\frac{\text{h.p.}}{\text{in.}^3}.$$

A mean value for H can be obtained from the relation

$$H = c\omega^2\theta_{sh\,1}^2/13{,}200 \quad \text{[h.p.]},$$

where $\theta_{sh\,1}$ is determined from the corresponding formula on p. 506. [For the derivation of this expression, see p. 555, section 3·3.]

3·22 'Optimum' values of damper performance

The results of the numerical example given above show a low amplitude of vibration at the engine mass. However, the maximum vibration of the damper mass J_1'' was only $0\cdot28°$, and it may be asked whether a somewhat smaller damper vibrating at greater amplitude could be employed to obtain the same θ_2-value at the engine mass.

In theory this is possible in most cases. In practice one has to consider also the fact that a large damper inertia may be required to bring the peak of a critical below the engine speed range (e.g. to reduce gear chatter even when the crankshaft stresses are low) and that frequently

† Zdanowich, R. W. and Moyal, J. E., 'Some practical applications of rubber dampers for the suppression of torsional vibrations in engine systems', *Proc. Instn Mech. Engrs, Lond.*, vol. 153 (1945), pp. 61–82. [Detailed treatment of systems with rubber dampers by the effective-inertia method.]

the damping coefficient of the damper material is not accurately known or cannot be altered without some difficulty.

Nevertheless, it will be an advantage to have a clear idea of 'optimum' working conditions of 'tuned' dampers. The first point to be noted is that the three-mass formulae, given under 3·21, determine the exact values of resonance peaks only for dampers of 'moderate' damping capacity, i.e. with damping ratios $\gamma = c/(2J_1'' \omega_0)$ less than 0·1. For tuning disks provided with 'heavy' damping (e.g. $\gamma > 0·1$, using, for instance, silicone fluid between the elastically coupled seismic mass J_1'' and its coupling disk J_1') they are no longer valid, since they are based on the assumption that peak-amplitude resonance is at the same frequency with and without damping, and this is sufficiently correct only for small values of damping ratios.

However, the general formula for 'non-resonant amplitudes', also given under 3·21, is valid for all cases. It gives not only flank amplitudes but also the correct peak amplitudes with 'heavy' damping, provided that the right values of phase velocity are used, and these can be determined by step-by-step evaluations, using the ω-values without damping as first approximations.

The possibilities with the 'tuned' damper are as follows:

(a) damper with 'tuning ratio ω_1/ω_0 equal to unity', i.e. damper tuned to the original phase velocity ω_0 of the engine. The one- and two-node peak amplitudes are at different heights, usually with $\theta_{2\mathrm{I}} > \theta_{2\mathrm{II}}$;

(b) damper with 'optimum tuning ratio'; it will be shown that when a ratio ω_1/ω_0 lower than unity is used, it is possible to adjust the one- and two-node peak amplitudes to equal values;

(c) damper with tuning ratio below 'optimum tuning'; the one-node and two-node peak amplitudes are at different heights, usually with $\theta_{2\mathrm{I}} < \theta_{2\mathrm{II}}$.

Moreover, for all three cases, provided that sufficient damping is available, the damper may be designed for 'minimum-amplitude conditions', that is, in such a way that the one- and two-node vibrations at the engine mass will be as close as possible to the corresponding values of equivalent static deflexion ('equilibrium amplitudes') for the size of damper considered.

The minimum peak amplitudes are obtained by considering the resonance curves of the engine-and-damper system with zero damping ($c = 0$) and infinite damping ($c = \infty$), respectively. When $c = \infty$, the seismic mass J_1'' of the damper is solidly locked to the engine mass J_2. The system is then again a one-mass system, with a phase velocity $\omega_\infty = \sqrt{\{K_2/(J_2 + J_1'')\}}$, which is lower than the phase velocity $\omega_0 = \sqrt{(K_2/J_2)}$ of the original system. (For simplicity, it is assumed in the following that the flywheel inertia J_3 is infinitely large, so that the engine alone is treated as a single-mass system, see Fig. 5a.)

[510]

As shown in Fig. 5 b, the flank amplitudes of the curves with $c = 0$ and $c = \infty$ have equal values at two points marked I and II. These common points are 'fixed points' of the system, that is, any other resonance curves will pass through I and II for any c-values between zero and infinity.† (Large but finite values of c have the effect of increasing the value of ω_{I} and decreasing that of ω_{II}.)

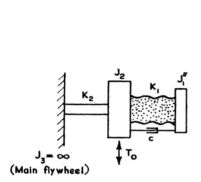

Fig. 5 a.

Fig. 5 b. Common points I and II of the resonance curves with zero and infinite damping. $M_{\mathrm{I}} = \theta_{2\mathrm{I}}/(T_0/K)$, and $M_{\mathrm{II}} = \theta_{2\mathrm{II}}/(T_0/K_2)$. The original resonance curve of the engine without a damper is denoted only by a dotted vertical line at ω_0, since ω_0 is no longer a natural phase velocity of the system when it is fitted with a damper.

By choosing suitable values of damping, the *peak* amplitudes $\theta_{2\mathrm{I}}$ and $\theta_{2\mathrm{II}}$ of the engine mass can be located at these minimum points, and the natural phase velocities are then ω_{I} and ω_{II} ('common-point phase velocities').

It should be noted that, by increasing the ratio $\mu = J_1''/J_2$, point I will be lowered as the flanks of the curves of $\omega_{\mathrm{I}(c=0)}$ and ω_{∞} are separated. However, with large μ-values, the left-hand flank of the ω_{∞}-curve comes somewhat closer again to the right-hand flank of the $\omega_{\mathrm{I}(c=0)}$-curve, so that the common-point amplitude is raised correspondingly.

The 'optimum' conditions for the *damper with a tuning ratio*

$$\omega_1/\omega_0 = \sqrt{(K_1/J_1'')}/\sqrt{(K_2/J_2)}$$

of unity can be estimated from the charts of Figs. 6 a and 6 b for any chosen value of $\mu = J_1''/J_2$. The derivation of these curves is given at the end of this section.

Similarly, the optimum conditions for the *damper with an 'optimum tuning' ratio* $\omega_1/\omega_0 = 1/(1 + \mu)$ are indicated in Figs. 7 a and 7 b. The formulae on which these curves are based are given on pp. 516–518.

† The vibration amplitude at points I and II is independent of damping, as will be shown in the derivation of these results. But the damping may be chosen in such a way as to give decreasing amplitudes on either side of points I and II, which thus become peak-amplitude values.

[511]

Example of use of Figs. 6a *and* 6b. If it is decided to use a μ-value of 0·4, the chart of Fig. 6a shows that the optimum value of the damping ratio is $\gamma = 0\cdot4$, the corresponding natural phase velocities are $\omega_{\mathrm{I}} = 0\cdot77\omega_0$ and $\omega_{\mathrm{II}} = 1\cdot18\omega_0$, and the peak amplitudes at the engine mass

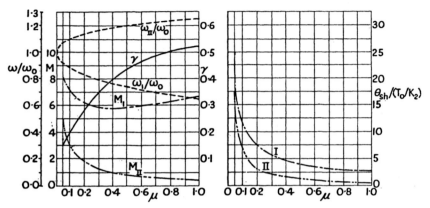

Fig. 6a. Minimum-amplitude parameters for dampers with *tuning ratio of unity*.

Fig. 6b. Twist of damper material for dampers with *tuning ratio of unity*.

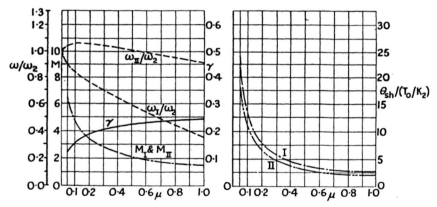

Fig. 7a. Minimum-amplitude parameters for dampers with '*optimum tuning*' ratio.

Fig. 7b. Twist of damper material for dampers with '*optimum tuning*' ratio.

J_2 are $\theta_{2\,\mathrm{I}} = 5\cdot8T_0/K_2$ and $\theta_{2\,\mathrm{II}} = 1\cdot05T_0/K_2$, these amplitude values being the lowest attainable with the μ-ratio chosen.

Note. $M = \theta_2/(T_0/K_2)$ and $\gamma = c/(2J_1''\omega_0)$.

Furthermore, the chart of Fig. 6b shows that for $\mu = 0\cdot4$ the twist in the damper material reaches the peak amplitudes $\theta_{sh\,\mathrm{I}} = 4\cdot5T_0/K_2$ for the one-node, and $\theta_{sh\,\mathrm{II}} = 2\cdot0 \times T_0/K_2$ for the two-node condition.

Note. For Fig. 6b, $\theta_{sh} = |\theta_1 - \theta_2| =$ amplitude at J_1'' minus amplitude at J_2.

[512]

Finally, it may be mentioned that, as all the curves in these figures represent non-dimensional ratios, they are valid irrespective of the (British or metric) units used.

Example of use of Figs. 7a and 7b. If it is decided to use a μ-value of 0·4, the chart of Fig. 7a shows that the optimum value of the damping ratio is $\gamma = 0\cdot22$, the corresponding natural phase velocities are $\omega_I = 0\cdot64\omega_2$ and $\omega_{II} = 1\cdot02\omega_2$, and the peak amplitudes at the engine mass J_2 are $\theta_{2I} = \theta_{2II} = 2\cdot4T_0/K_2$, these amplitudes being the lowest values attainable with the μ-value chosen.

Note. $M = \theta_2/(T_0/K_2)$ and $\gamma = c/(2J_1''\omega_2)$.

Furthermore, the chart of Fig. 7b shows that for $\mu = 0\cdot4$ the twist in the damper material reaches the peak amplitudes $\theta_{sh\,I} = 4\cdot6T_0/K_2$ for the one-node, and $\theta_{sh\,II} = 3\cdot75 \times T_0/K_2$ for the two-node condition.

Note. For Fig. 7b, $\theta_{sh} = |\theta_1 - \theta_2| =$ amplitude at J_1'' minus amplitude at J_2.

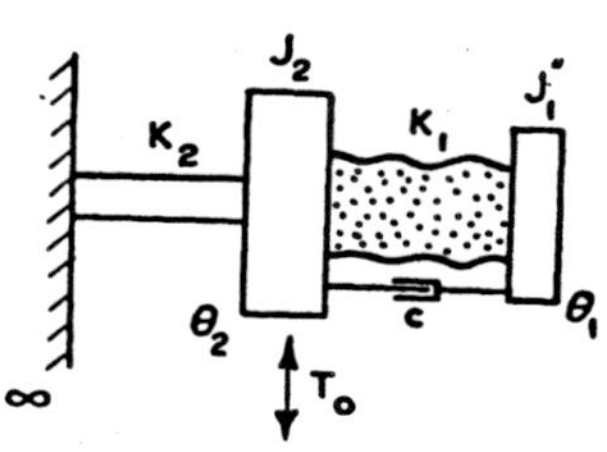

Derivation. For the system of Fig. 8, the equations of motion give the characteristic equations

$$(K_1 - \omega^2 J_1'' + j\omega c)\,\theta_1 - (K_1 + j\omega c)\,\theta_2 = 0,$$

$$-(K_1 + j\omega c)\,\theta_1 \\ + (K_1 + K_2 - \omega^2 J_2 + j\omega c)\,\theta_2 = T_0.$$

Fig. 8.

Therefore, the system determinant is

$$\mathrm{Det}_s = \begin{vmatrix} K_1 - \omega^2 J_1'' + j\omega c & -K_1 - j\omega c \\ -K_1 - j\omega c & K_1 + K_2 - \omega^2 J_2 + j\omega c \end{vmatrix}$$

$$= [K_1 K_2 - \omega^2(J_1'' K_1 + J_1'' K_2 + J_2 K_1) + \omega^4 J_1'' J_2] + j\omega c[K_2 - \omega^2(J_1'' + J_2)]$$

$$= \qquad\qquad C \qquad\qquad\qquad + jD;$$

moreover, $\qquad\qquad \mathrm{Det}_1 = T_0(K_1 + j\omega c)$

and $\qquad\qquad \mathrm{Det}_2 = T_0(K_1 - \omega^2 J_1'' + j\omega c).$

Therefore, the vibration amplitudes θ_1, θ_2 and the twist $\theta_{sh} = \theta_1 - \theta_2$ of the damper material give the following relations:

$$\left|\frac{\theta_1}{T_0}\right|^2 = \frac{K_1^2 + (c\omega)^2}{C^2 + D^2}, \quad \left|\frac{\theta_2}{T_0}\right|^2 = \frac{(K_1 - \omega^2 J_1'')^2 + (\omega c)^2}{C^2 + D^2}, \quad \left|\frac{\theta_{sh}}{T_0}\right|^2 = \frac{(\omega^2 J_1'')^2}{C^2 + D^2}.$$

When $c = 0$, the natural phase velocities $\omega_{I\,(c=0)}$ and $\omega_{II\,(c=0)}$ are determined by $C = 0$; when $c = \infty$, the natural phase velocity is $\omega_\infty = \sqrt{\{K_2/(J_2 + J_1'')\}}$.

Let $\qquad\qquad (K_1 - \omega^2 J_1'') = A, \quad \omega c = B,$

then $\qquad\qquad |\theta_2|^2 = T_0^2 \times \dfrac{A^2 + B^2}{C^2 + D^2}.$

When $(A/B)^2 = (C/D)^2 = Z^2$ say, this equation becomes

$$|\theta_2|^2 = T_0^2 \times \frac{A^2 + \dfrac{A^2}{Z^2}}{C^2 + \dfrac{C^2}{Z^2}} = T_0^2 \times \frac{A^2}{C^2},$$

that is, at ω-values giving $(A/B)^2 = (C/D)^2$, the amplitude θ_2 is *independent of damping* and has *fixed values* at these phase velocities. This is the case when

$$\pm A/B = C/D \quad \text{or} \quad \pm AD = BC,$$

that is, when

$$\pm (K_1 - \omega^2 J_1'')\, \omega c [K_2 - \omega^2 (J_1'' + J_2)]$$
$$= \omega c [K_1 K_2 - \omega^2 (J_1'' K_1 + J_1'' K_2 + J_2 K_1) + \omega^4 J_1'' J_2],$$

where ωc can be cancelled on both sides. Solving for ω^2, in order to obtain the fixed-point values, one finds that the $+$ sign merely gives $\omega = 0$. Taking the $-$ sign, one obtains

$$-2 K_1 K_2 + 2\omega^2 [J_1'' K_1 + J_1'' K_2 + J_2 K_1] - \omega^4 (J_1''^2 + 2 J_1'' J_2) = 0,$$

or, after dividing throughout by $J_1'' J_2$ and substituting $\mu = J_1''/J_2$, $\omega_1^2 = K_1/J_1''$ and $\omega_2^2 = K_2/J_2$:

$$\frac{2}{2+\mu}\, \omega_1^2 \omega_2^2 - \frac{2}{2+\mu} [\omega_1^2 (1+\mu) + \omega_2^2]\, \omega^2 + \omega^4 = 0. \tag{1a}$$

This is a quadratic equation in ω^2, which can be written

$$(\omega^2 - \omega_\mathrm{I}^2)\,(\omega^2 - \omega_\mathrm{II}^2) = 0.$$

Therefore the sum of its two roots, i.e. $\omega_\mathrm{I}^2 + \omega_\mathrm{II}^2$, is equal to the coefficient of the term in ω^2 with a negative sign prefixed, so that

$$\omega_\mathrm{I}^2 + \omega_\mathrm{II}^2 = +\frac{2}{2+\mu} [\omega_1^2 (1+\mu) + \omega_2^2]. \tag{1b}$$

Note. Let $\omega_\mathrm{I} < \omega_\mathrm{II}$. Then it can be shown with these equations that, as ω_1 varies from 0 to ∞, ω_I^2 varies from 0 to $\omega_2^2/(1+\mu)$, and ω_II^2 varies from $2\omega_2^2/(2+\mu)$ down to $\omega_2^2/(1+\mu)$ and then rises towards ∞.

The amplitudes of vibration which are independent of damping can be obtained by substituting the one-node or two-node phase velocity, ω_I or ω_II, in the equation for θ_2.

For these fixed points, where c may have any value, it is convenient to simplify the expression for θ_2 by taking $c = \infty$. Then, after dividing both the numerator and the denominator of

$$|\theta_2/T_0|^2 = (A^2 + B^2)/(C^2 + D^2)$$

by $(c\omega)$ and taking $c \to \infty$ we have

$$\left|\frac{\theta_2}{T_0}\right|^2 = \frac{(\pm 1)^2}{[K_2 - \omega^2(J_1'' + J_2)]^2} = \left(\frac{\pm B}{D}\right)^2$$

or

$$\left[\frac{\theta_2}{T_0}\right]_{\mathrm{I}} = \frac{-1}{K_2 - \omega_{\mathrm{I}}^2(J_1'' + J_2)} = \frac{M_{\mathrm{I}}}{K_2} \tag{2a}$$

and

$$\left[\frac{\theta_2}{T_0}\right]_{\mathrm{II}} = \frac{+1}{K_2 - \omega_{\mathrm{II}}^2(J_1'' + J_2)} = \frac{M_{\mathrm{II}}}{K_2}. \tag{2b}$$

(Physically, the minus sign for the one-node flank means that it is after the one-node resonance without damping, whereas the plus sign for the two-node flank indicates that it is ahead of its undamped resonance peak.)

Thus we have the phase velocities, and the amplitudes, of the fixed points. It remains to see how they can be made into peak-amplitude positions. Although the c-values do not affect the fixed points themselves, they definitely determine the values of flank amplitudes in their vicinity and, in particular, can make these have higher or lower values than at the fixed points.

In principle, the right value of c could be obtained by equating to zero the derivative $\partial \, | \, \theta_2^2/T_0^2 \, | / \partial \omega^2$, in order to obtain a horizontal tangent. To avoid laborious calculations, however, the damping coefficient may be estimated from values in the neighbourhood of ω_{I} (or ω_{II}) as follows:

With the ratios already employed, and $2\gamma = c/J_1'' \omega_2$, the non-dimensional expression for θ_2 may be rewritten as†

$$M^2 = \left|\frac{\theta_2}{T_0/K_2}\right|^2 = \frac{\left[\left(\frac{\omega_1}{\omega_2}\right)^2 - \left(\frac{\omega}{\omega_2}\right)^2\right]^2 + 4\left(\frac{\omega}{\omega_2}\right)^2 \gamma^2}{\left\{\left[\left(\frac{\omega_1}{\omega_2}\right)^2 - \left(\frac{\omega}{\omega_2}\right)^2\right]\left[1 - \left(\frac{\omega}{\omega_2}\right)^2\right] - \mu\left(\frac{\omega}{\omega_2}\right)^2 \left(\frac{\omega_1}{\omega_2}\right)^2\right\}^2}$$

$$+ 4\left(\frac{\omega}{\omega_2}\right)^2 \gamma^2 \left[1 - \left(\frac{\omega}{\omega_2}\right)^2 - \mu\left(\frac{\omega}{\omega_2}\right)^2\right]^2. \tag{3}$$

Using the correct numerical values of $M^2 = M_{\mathrm{I}}^2$ (or M_{II}^2), as derived above and *approximately* correct values for $\omega^2 = \omega_{\mathrm{I}}^2$ (or ω_{II}^2), for instance, $\omega^2 = 1 \cdot 01 \omega_{\mathrm{I}}^2$ (or $0 \cdot 99 \omega_{\mathrm{II}}^2$), we may evaluate the above equation and solve it for γ. The computation is fairly straightforward for the particular cases considered below, and gives γ-values which correspond to peak amplitudes at the fixed points.

† To use $2\gamma = c/J_1'' \omega_2$, we write

$$B^2 = (c\omega)^2 = K_1^2(c\omega/K_1)^2 = K_1 \times \left(\frac{\omega c}{\omega_1^2 J_1''}\right)^2 = K_1^2 \times \left(\frac{\omega}{\omega_1}\right)^2 \left(\frac{\omega_2}{\omega_1}\right)^2 \left(\frac{c}{\omega_2 J_1''}\right)^2 = K_1^2 \times \left(\frac{\omega}{\omega_1}\right)^2 \left(\frac{\omega_2}{\omega_1}\right)^2 \times 4\gamma^2.$$

The expressions for A, C and D are easily reduced and finally both the numerator and the denominator of the resultant fraction are multiplied by $(\omega_1/w_2)^4$ in order to obtain the above expression.

It is of interest to compare the results for three main types of tuned dampers, viz. 'hard' dampers (tuning ratio of unity), 'medium' dampers ('optimum' tuning ratio) and 'soft' dampers (tuning ratio ω_1/ω_2 of about 0·5). These are given in the following.

Formulae for tuned dampers

Damper with tuning ratio equal to unity (Figs. 6a and 6b)

$$\omega_1/\omega_2 = 1, \quad \omega_1 = \omega_2 = \omega_0.$$

Eq. (1a) becomes

$$\frac{2}{2+\mu}\,\omega_2^4 - 2\omega_2^2\omega^2 + \omega^4 = 0,$$

hence

$$\omega_{\mathrm{I}}^2 = \omega_2^2\left[1 - \sqrt{\left(\frac{\mu}{2+\mu}\right)}\right],$$

$$\omega_{\mathrm{II}}^2 = \omega_2^2\left[1 + \sqrt{\left(\frac{\mu}{2+\mu}\right)}\right].$$

Substituting these values in eq. (2a) or (2b), respectively, we find

$$M_{\mathrm{I}} = \left[\frac{\theta_2}{T_0/K_2}\right]_{\mathrm{I}} = \frac{1}{\mu - \sqrt{\left(\frac{\mu}{1+\mu}\right)}\,(1+\mu)},$$

$$M_{\mathrm{II}} = \left[\frac{\theta_2}{T_0/K_2}\right]_{\mathrm{II}} = \frac{1}{\mu + \sqrt{\left(\frac{\mu}{1+\mu}\right)}\,(1+\mu)}.$$

Inserting corresponding values of μ, M_{I} (or M_{II}) and ω_{I}^2 (or ω_{II}^2) in eq. (3), and simplifying with $\omega_1/\omega_2 = 1$, we solve for the corresponding value of γ. This is sufficient to obtain a *peak* amplitude, in the close vicinity of ω_{I} (or ω_{II}).

Note. From $\omega_1/\omega_2 = 1$, it follows that

$$K_1/K_2 = \mu.$$

A typical result is shown in Fig. 9a.

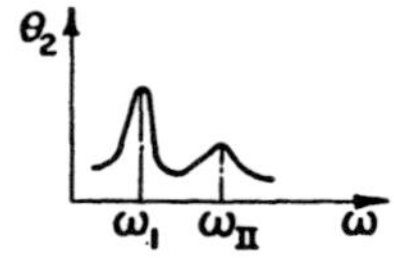

Fig. 9a.

Damper with 'optimum tuning' ratio (Figs. 7a and 7b)

Equal peak values of θ_2 are obtained if eqs. (2a) and (2b) are equated to each other, that is, if

$$-K_2 + \omega_{\mathrm{II}}^2(J_1'' + J_2) = K_2 - \omega_{\mathrm{I}}^2(J_1'' + J_2)$$

or

$$2K_2/(J_1'' + J_2) = \omega_{\mathrm{I}}^2 + \omega_{\mathrm{II}}^2$$

$$= 2\omega_2^2/(1+\mu).$$

Using eq. (1b)

$$\frac{2}{2+\mu}\,[\omega_{\mathrm{I}}^2(1+\mu) + \omega_2^2] = \frac{2}{1+\mu}\,\omega_2^2,$$

[516]

hence
$$\left(\frac{\omega_1}{\omega_2}\right) = \frac{1}{1+\mu}.$$

With this tuning ratio, eq. (1a) becomes

$$\frac{2\omega_2^4}{(2+\mu)(1+\mu)} - \frac{2}{2+\mu}\,\omega_2^2\omega^2 + \omega^4 = 0,$$

hence
$$\omega_{\mathrm{I}}^2 = \frac{\omega_2^2}{1+\mu}\left[1 - \sqrt{\left(\frac{\mu}{2+\mu}\right)}\right],$$

$$\omega_{\mathrm{II}}^2 = \frac{\omega_2^2}{1+\mu}\left[1 + \sqrt{\left(\frac{\mu}{2+\mu}\right)}\right].$$

Substituting these values in eq. (2a) or (2b), respectively, we find

$$M_{\mathrm{I}} = M_{\mathrm{II}} = \left[\frac{\theta_2}{T_0/K_2}\right]_{\mathrm{I,II}} = \sqrt{\left(\frac{2+\mu}{\mu}\right)}.$$

The corresponding γ-values are calculated by means of eq. (3). From $\omega_1/\omega_2 = 1/(1+\mu)$, it follows that

$$K_1/K_2 = \mu/(1+\mu)^2.$$

A typical result is shown in Fig. 9b

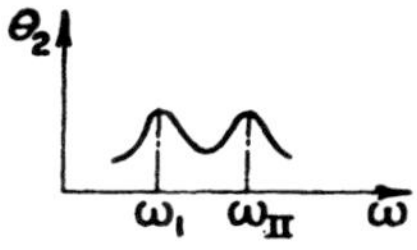

Damper with tuning ratio equal to 0·5

$$\omega_1/\omega_2 = 0{\cdot}5 \quad \text{or} \quad \omega_1^2 = \omega_2^2/4.$$

Eq. (1a) becomes

$$\frac{\omega_2^4}{2(2+\mu)} - \frac{5+\mu}{2(2+\mu)}\,\omega_2^2\omega^2 + \omega^4 = 0.$$

Hence
$$\omega_{\mathrm{I}}^2 = \frac{(1+\mu)}{2(2+\mu)}\,\omega_2^2,$$

$$\omega_{\mathrm{II}}^2 = \frac{2}{2+\mu}\,\omega_2^2.$$

Substituting these values in eq. (2a) or (2b), respectively,

$$M_{\mathrm{I}} = \frac{2(2+\mu)}{3-\mu^2},$$

$$M_{\mathrm{II}} = \frac{2+\mu}{\mu}.$$

The corresponding γ-values are calculated by means of eq. (3). From $\omega_1^2/\omega_2^2 = 1/4$, it follows that

$$K_1/K_2 = \frac{\mu}{4}.$$

A typical result is shown in Fig. 9c.

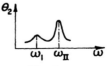
Fig. 9b.

Fig. 9c.

[517]

Twist of damper material. For all three cases θ_{sh} may be determined exactly from $|\theta_{sh}/T_0|^2 = (\omega^2 J_1'')^2/[C^2 + D^2]$, or estimated by means of the relations used for Figs. 6b and 7b:

$$\left[\frac{\theta_{sh}}{T_0/K_2}\right]_I = \sqrt{\left(\frac{M_I}{2\gamma\mu(\omega_I/\omega_2)}\right)} \quad \text{and} \quad \left[\frac{\theta_{sh}}{T_0/K_2}\right]_{II} = \sqrt{\left(\frac{M_{II}}{2\gamma\mu(\omega_{II}/\omega_2)}\right)}.$$

The derivation is as follows: equating the work input $\pi T_0 \theta_2$ to the energy dissipation $\pi c\omega\theta_{sh}^2$ in the damper, and using $2\gamma = c/\omega_2 J_1'' = c/\omega_2\mu J_2$, we have

$$\theta_{sh}^2 = T_0\theta_2/c\omega = T_0\theta_2/[2\gamma\omega\omega_2\mu J_2] = (T_0/K_2) \times (K_2/J_2) \times \theta_2/[2\gamma\mu\omega\omega_2]$$

so that

$$\theta_{sh}^2/(T_0/K_2)^2 = \{\theta_2/(T_0/K_2)\}/[2\gamma\mu(\omega/\omega_2)],$$

which gives the above formula.

General remarks on tuned dampers

(1) *Relative peak amplitudes*

Dampers with a tuning ratio of unity produce a one-node peak higher than the two-node peak because, with a fixed value of the damping coefficient c, the damping torque increases with the phase velocity, that is, it varies as $c\omega$. As $\omega_I < \omega_{II}$, the one-node vibration is less damped, and therefore reaches a higher value, than two-node vibration.

This result can be altered by varying the tuning ratio to a value less than unity. The anti-resonant point is then lower than the original ω_0-value of the system without a damper (see p. 500, Fig. 2), and, as the left-hand portion of the resonance curve is of lower value, the damping torque due to $c\omega_I$ will be sufficient to reduce it to the same amplitude as the two-node peak acted upon by $c\omega_{II}$. With a still lower tuning ratio, the one-node peak can even be made smaller than the two-node peak, as shown in Fig. 9c.

(2) *Crankshaft stresses*

It is worth emphasizing that the criterion for the effect of tuned dampers used in engine systems is the reduction which they provide in *crankshaft stresses*, and not merely the vibration amplitude reduction. If both one- and two-node criticals occur in the engine speed range, as the stress per degree $q_{deg.}$ is usually much greater for two-node than for one-node vibration conditions, it is an advantage in such cases to use a damper with a tuning ratio of unity, since then $\theta_I > \theta_{II}$ while $q_{deg.\,I} < q_{deg.\,II}$, so that it may be possible to obtain for both peaks a stress $q_{max.}$ such that

$$q_{max.} \cong \theta_I q_{deg.\,I}] \cong \theta_{II} q_{deg.\,II}.$$

'Optimum stress' conditions, therefore, should be the governing consideration in the selection of dampers for engine applications, and it is advisable to adopt the practice of checking the crankshaft stresses for engines with dampers even where large amplitude-reductions have been attained.

BIBLIOGRAPHY

The theory of tuned dampers has been the subject of extensive investigations, although the applications considered are not necessarily related to engine systems, and a list of literature references is given below.

Ormondroyd, J. and Den Hartog, J. P. *Trans. Amer. Soc. Mech. Engrs*, vol. 50, no. 7 (1928), p. 9.

Holzer, H. *Stodola-Festschrift* (Zurich, 1929), p. 234.

Hahnkamm, E. *Ann. Phys.* 5th series, vol. 14 (1932), p. 683; *Z. angew. Math. Mech.* vol. 13 (1933), p. 183; *Ingen.-Arch.* vol. 4 (1933), p. 192; *Schiffbautechn. Ges.* Versammlung, Berlin, Nov. 1935.

Zdanowich, R. W. and Moyal, J. E. *Proc. Instn Mech. Engrs, Lond.*, vol. 153, no. 3 (1945), pp. 61–82.

Arnold, R. N. *Proc. Instn Mech. Engrs, Lond.*, vol. 157, no. 25 (1947), pp. 1–19.

3·3 THE UNTUNED VISCOUS-SHEAR DAMPER†

Dampers of this type (see Fig. 1) consist of an annular seismic mass enclosed in a casing. The peripheral and lateral gaps between these two members are filled with a viscous fluid, e.g. silicone fluid.

The damper is 'untuned', since there is no elastic coupling member between the seismic mass and the casing; the seismic mass is acted upon only by the viscous torque transmitted by the fluid.

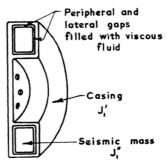

Fig. 1.

Unlike 'tuned' dampers, the untuned viscous-shear damper does not introduce an additional resonance, but lowers the value of the natural frequency of the engine system while reducing the vibration amplitudes.

The viscous-shear type damper has been extensively investigated, and the information now available on this subject will be given as follows:

(1) Two-mass system charts and formulae;
(2) Three-mass system formulae;
(3) Tabulations for common-point frequency and optimum damping;
(4) Design procedure to obtain required damping coefficient;
(5) Further information on design features and operating experience;
(6) Derivation of relations used in design procedure.

NOTATION

Symbol	Brief definition	Typical units	Symbol	Brief definition	Typical units
A	area	in.^{2}, cm.^{2}	F	vibration frequency	vib./min.
A	damper parameter	—	F	damper parameter	—
B	damper parameter	—	g	acceleration due to	in./sec.^{2},
c	damping coefficient	Lb.in.sec./rad.,		gravity	cm./sec.^{2}
		Kg.cm.sec./	H	power	h.p.
		rad.	j	$=\sqrt{-1}$	—
C	damper parameter	—	J	mass moment of	Lb.in.sec.^{2},
D	diameter	in., cm.		inertia	Kg.cm.sec.^{2}
D	damper parameter	—	k	stiffness ratio	—
$\mathscr{D}$	denominator	—	K	stiffness	Lb.in./rad.,
E	damper parameter	—			Kg.cm./rad.

† Alternative designations: Lanchester damper, viscous-fluid damper, untuned damped vibration absorber.

L	length, thickness	in., cm.		W	weight	Lb., Kg.
M	dynamic magnifier	—		x	frequency ratio	—
M_b	bending moment	Lb.in., Kg.cm.		Z	sectional modulus	in.^{3}, cm.^{3}
n	number of studs, bolts, etc.	—		α	thermal expansion coefficient	1/°C.
n	shear rate exponent	—		α	stress coefficient	Lb./in.^{4}, Kg./cm.^{4}
N	shaft speed	rev./min.				
p	direct stress, pressure	Lb./in.^{2}, Kg./cm.^{2}		β	thermal expansion coefficient	1/°C.
P	force	Lb., Kg.		β	phase angle	deg.
q	shear stress	Lb./in.^{2}, Kg./cm.^{2}		γ	damping ratio	—
				δ	damper clearance	in., cm.
r	rate of shear	sec.$^{-1}$		Δ	Holzer-table amplitude	rad.
R	radius	in., cm.				
t	time	sec., min.		ϵ	phase angle	deg.
t	temperature	°C., °F.		η	coefficient	—
T	temperature	°C., °F.		η_r	shear rate coefficient	—
T	torque	Lb.in., Kg.cm.		θ	vibration amplitude	rad., deg.
$|T_n|$	nth order component of tangential pressure	Lb./in.^{2}, Kg./cm.^{2}		θ	vibration velocity	rad./sec.
				λ	inertia ratio	—
				μ	inertia ratio	—
U	work or energy per vibration cycle	in.Lb./cycle, cm.Kg./cycle		ν	kinematic viscosity	in./sec.^{2}, cm./sec.^{2}
v	velocity	in./sec., cm./sec.		ρ	mass-density	Lb.sec.^{2}/in.^{4}, Kg.sec.^{2}/cm.^{4}
V	volume	in.^{3}, cm.^{3}		$\overrightarrow{\Sigma\Delta}$	phase-vector sum	—
w	specific weight	Lb./in.^{2}, Kg./cm.^{2}		ϕ	phase angle	deg.
				ω	phase velocity	rad./sec.

Subscripts are used in the text to define more precisely. Their meaning is fully explained in their context in each instance.

3·31 Two-mass system charts and formulae†

The vibration amplitudes obtainable with an engine fitted with an untuned viscous-shear damper depend in the first instance on the ratio $\mu = J_1''/J_2$ of the seismic-mass inertia J_1'' to the equivalent engine inertia $J_2 = \Sigma(J_{\text{cyl.}} \Delta_{\text{cyl.}}^2)$ (where $J_{\text{cyl.}}$ = moment of inertia per cylinder line and $\Delta_{\text{cyl.}}$ = Holzer-table amplitude of each $J_{\text{cyl.}}$); they also depend on the damping ratio $\gamma = c/2J_1''\omega_2$, where c = damping coefficient of damper and $\omega_2 = \sqrt{(K_2/J_2)}$ = natural phase velocity of the original engine system.

The following charts and formulae can be used to estimate the vibration conditions of an engine fitted with a viscous-shear type damper. They are derived with the aid of two simplifying assumptions, viz.

(i) the main flywheel J_3 is infinitely large;

(ii) the damper is coupled to the engine inertia J_2 by means of an infinitely stiff shaft portion between the damper casing J_1' and the centre line of cylinder no. 1 (see Figs. 2a and 2b).

† In all the following considerations, the 'inherent damping' in the engine system is not taken into account, since it will normally be negligible in comparison with the damping provided by the damper.

In practice J_3 is usually larger than J_2 without being infinite, and the stub shaft portion with the half crankthrow to the centre of cylinder no. 1 has a greater stiffness than the entire crankshaft from cylinder no. 1 to the main flywheel J_3. Therefore the following charts and formulae, which are exact for a two-mass system, also give approximate indications for engine systems, and may be used as a basis for practical work.

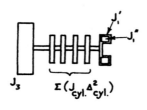

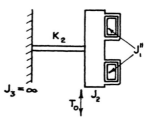

Fig. 2a. Engine system with damper.   Fig. 2b. Equivalent system.

The engine system with damper is reduced to an equivalent two-mass system as indicated in Figs. 2a and 2b.

Equivalent system.† Let

ω_2 = natural phase velocity of original engine system [rad./sec.],

$J_2 = \Sigma(J_{\mathrm{cyl.}}\,\Delta_{\mathrm{cyl.}}^2)$ = equivalent engine inertia [Lb.in.sec.²],

$K_2 = \omega_2^2 J_2$ = equivalent crankshaft stiffness [Lb.in./rad.],

J_1'' = inertia of seismic mass of damper [Lb.in.sec.²], and

T_0 = excitation torque ($|\,T_n\,|\,AR_0\overrightarrow{\Sigma\Delta}$) [Lb.in.].

Then the natural phase velocity ω_D of the system with damper, the vibration amplitude θ_2 at the damper casing (and at the engine mass J_2), and the 'optimum' damping ratio γ may be obtained from the charts of Figs. 3a and 3b.

Example of use of Figs. 3a and 3b. If it is decided to use a μ-value of 0·4, the chart of Fig. 3a shows that the corresponding natural phase velocity is $\omega_D = 0\cdot92\omega_2$, with a damping ratio $\gamma = c/2J_1''\omega_2 = 0\cdot385$ and an amplitude at the damper casing given by $M = \theta_2/(T_0/K_2) = 6\cdot0$, that is, $\theta_2 = 6\cdot0 \times T_0/K_2\,\mathrm{rad.}$

Fig. 3b shows that the vibration amplitude of the seismic mass J_1'' relative to the damper casing J_1' is $\theta_{sh} = 4\cdot2 \times T_0/K_2\,\mathrm{rad.}$, for a μ-value of 0·4.

---

† It should be noted that Figs. 3a and 3b (and the corresponding formulae) do not take account of the damper casing inertia J_1'. After deciding on a μ-value (and hence on a first estimated value for J_1'), a more detailed estimation, also using the above charts, can be made by assuming $J_1' \cong 0\cdot5J_1''$ and evaluating $J_2 = J_1' + \Sigma(J_{\mathrm{cyl.}}\,\Delta_{\mathrm{cyl.}}^2)$, with a new phase velocity ω_2 and new Δ-values of relative amplitudes obtained *from a Holzer table for the engine system with the damper casing J_1' as mass no. 1.*

Metric units. The graphs of Fig. 3a and 3b represent non-dimensional quantities and are therefore also valid in metric units (with 'Lb.' replaced by 'Kg.', and 'in.' by 'cm.').

Alternative method

The following method of estimating the vibration amplitude θ_2 at the damper casing is also commonly used:

(1) Calculate (with θ_2 as the unknown factor) the work input of the engine as

$$U_{in} = \pi T_0 \times \theta_2 \quad \text{[in.Lb./cycle]},$$

where $T_0 = $ excitation torque $= |T_n| A R_0 \overrightarrow{\Sigma\Delta}$ [Lb.in.], the phase-vector sum $\overrightarrow{\Sigma\Delta}$ being taken from a Holzer table for the engine with the damper, assuming $J_D = J_1' + \frac{1}{2}J_1''$ (where $J_1' = $ casing inertia and $J_1'' = $ inertia of inner mass [Lb.in.sec.^{2}]).

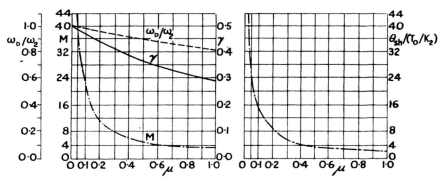

Fig. 3a. Minimum-amplitude parameters for viscous-shear type dampers.

Fig. 3b. Vibration amplitude of seismic mass relative to damper casing.

(2) Calculate the energy dissipated by the damper as

$$U_D = \frac{\pi}{2} J_1'' \omega^2 \times \theta_2^2 \quad \text{[in.Lb./cycle]},$$

where $J_1'' = $ assumed value for inner mass of damper [Lb.in.sec.^{2}], and $\omega = $ natural phase velocity [rad./sec.] of engine system with damper, from Holzer table with $J_D = J_1' + \frac{1}{2}J_1''$ as mass no. 1.

(3) Equate the work input to the energy dissipated

$$U_{in} = \pi T_0 \theta_2 = \frac{\pi}{2} J_1'' \omega^2 \theta_2^2 = U_D$$

and thus evaluate the vibration amplitude θ_2 at the damper casing as

$$\theta_2 = 2T_0/(\omega^2 J_1'') \quad \text{[rad.]},$$

or

$$\pm\, \theta_2 = 2 \times 57 \cdot 3 T_0/(\omega^2 J_1'') \quad \text{[degrees]}.$$

(4) Determine the damping coefficient of the damper as

$$c = \omega J_1'' \quad [\text{Lb.in.sec./rad.}].$$

Then follow the design procedure given in section 3·34 (see p. 537).

Note. The derivation of the expressions used above is considered on pp. 556 and 557.

Further design procedure

The values estimated from the charts in Figs. 3a and 3b, or by means of the alternative energy method, are sufficient in many cases (particularly when a large μ-ratio is chosen), and the next step is then to design the damper so as to obtain the required value for the damping coefficient c.

Thus, at this stage, the following detailed analyses may be omitted for such purposes, and the design procedures to be employed can be taken directly from section 3·34 (see p. 537).

Derivation of two-mass system formulae used for curves of Figs 3a and 3b

These curves are based on the equations previously considered on pp. 513 and 514, for the particular case where $K_1 = 0$ (i.e. where there is no elastic coupling member between the damper masses J_1'' and J_1'). In this case the system determinant becomes

$$\text{Det}_s = -\omega^2 J_1'' K_2 + \omega^4 J_1'' J_2 + j\omega c [K_2 - \omega^2 (J_1'' + J_2)],$$

while the determinants for θ_1 and θ_2 (i.e. for the amplitudes at J_1'' and J_2) are

$$\text{Det}_1 = T_0 j\omega c \quad \text{and} \quad \text{Det}_2 = T_0(-\omega^2 J_1'' + j\omega c).$$

Moreover, the relative amplitude between J_1'' and J_1' is determined by

$$\text{Det}_1 - \text{Det}_2 = + T_0 \omega^2 J_1''.$$

Therefore, we may write

$$\mathscr{D}^2 = \frac{1}{\mu^2 K_2^2} |\,\text{Det}_s\,|^2 = (x - x^2)^2 + 4\gamma^2 x [1 - x(1 + \mu)]^2,$$

where $x = (\omega/\omega_2)^2 = \omega^2/(K_2/J_2)$, so that finally,

$$\left| \frac{\theta_1}{T_0/K_2} \right|^2 = 4\gamma^2 x / \mathscr{D}^2 \quad \text{(where } \theta_1 = \text{amplitude at } J_1''),$$

$$\left| \frac{\theta_2}{T_0/K_1} \right|^2 \equiv M^2 = (x^2 + 4\gamma^2 x)/\mathscr{D}^2 \quad (\theta_2 = \text{amplitude at } J_2),$$

$$\left| \frac{\theta_{sh}}{T_0/K_2} \right|^2 = \left| \frac{\theta_1 - \theta_2}{T_0/K_2} \right|^2 = x^2/\mathscr{D}^2 \quad (\theta_{sh} = \text{relative amplitude between } J_1'' \text{ and } J_1').$$

[524]

These formulae are valid for both *peak and flank* amplitudes.

Eq. (1 a) of p. 514 gives, for $\omega_1^2 = K_1/J_1'' = 0$,

$$-\frac{2}{2+\mu}\,\omega_2^2\omega^2 + \omega^4 = 0,$$

which has two roots, viz.

$$\omega_I^2 = 0 \quad \text{and} \quad \omega_{II}^2 = \frac{2}{2+\mu}\,\omega_2^2 = \omega_D^2.$$

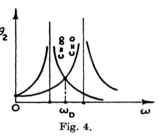

Fig. 4.

Thus, there is only one fixed point (apart from $\omega = 0$) as shown in Fig. 4.

The amplitude at J_2, at the fixed-point phase velocity ω_D, is obtained from eq. (2 b) of p. 515 as follows:

$$\left[\frac{\theta_2}{T_0/K_2}\right] \equiv M = \frac{+1}{1-(\omega_D/\omega_2)^2 \times (1+\mu)}, \quad \text{and with} \quad (\omega_D/\omega_2)^2 = 2/(2+\mu),$$

$$M = \frac{2+\mu}{\mu}.$$

To make θ_2 a peak amplitude at ω_D, the damping ratio γ must be determined from the condition that θ_2 has a horizontal tangent at ω_D. For the untuned damper, the expression for θ_2^2 is much simpler than for the tuned damper, and its derivative is obtained fairly easily, as follows:

$$\left|\frac{\theta_2}{T_0/K_2}\right|^2 = \frac{x^2+4\gamma^2 x}{(x-x^2)^2 + 4\gamma^2 x[1-x(1+\mu)]^2} = \frac{x+C}{x^3+Ax^2+Bx+C}, \quad (*)$$

where
$$A = 4\gamma^2(1+\mu)^2 - 2, \quad B = 1-8\gamma^2(1+\mu), \quad C = 4\gamma^2,$$

hence
$$0 = \frac{\partial}{\partial\omega^2}\left|\frac{\theta_2}{T_0/K_2}\right|^2 = \frac{1}{\omega_2^2}\frac{\partial}{\partial x}\left|\frac{\theta_2}{T_0/K_2}\right|^2,$$

using eq. (*) one obtains

$$0 = 2x^3 + x^2(A+3C) + 2ACx - C + BC.$$

Introducing $\omega = \omega_D$, that is, $x = (\omega_D/\omega_2)^2 = 2/(2+\mu)$, we find

$$4\gamma^4(1+\mu)(2+\mu)^2 + 2\mu(2+\mu)\gamma^2 - 1 = 0$$

which has one real and one imaginary root for γ^2; the real value gives

$$\gamma = \frac{1}{\sqrt{\{2(1+\mu)(2+\mu)\}}}.$$

This value has been used for evaluating the γ-curve in Fig. 3 a.

The above formulae were obtained in a number of investigations, by Ormondroyd and Den Hartog,[†] Hahnkamm,[†] Carter,[‡] Neugebauer,[‡] O'Connor,[‡] and Lewis.[‡]

[†] Loc. cit. p. 519.
[‡] Carter, B. C., *Tech. Rep. Aero. Res. Comm.*, R. & M. 1053, February 1926, H.M.S.O. Neugebauer, F., *Tech. Mech. Thermo-Dynam.* vol. 1 (1930), pp. 137 and 184. O'Connor, B. E., *S. A. E. Quart. Trans.* vol. 1 (1947), pp. 87–97. Lewis, F. M., *A.S.M.E. Paper*, no. 55-APM-3 (1955), 6 pp.

For the seismic-mass amplitude θ_1 and the relative amplitude $\theta_{sh} = |\theta_1 - \theta_2|$, as functions of the casing amplitude θ_2, useful expressions can be obtained as follows, from the formulae of p. 524:

$$\left|\frac{\theta_1}{\theta_2}\right|^2 = \frac{4\gamma^2 x/\mathscr{D}^2}{(x^2 + 4\gamma^2 x)/\mathscr{D}^2} = \frac{1}{\dfrac{x}{4\gamma^2} + 1} = \frac{1}{\dfrac{2}{2+\mu} \times \dfrac{2(1+\mu)(2+\mu)}{4} + 1} = \frac{1}{2+\mu};$$

$$\left|\frac{\theta_{sh}}{\theta_2}\right|^2 = \frac{x^2/\mathscr{D}^2}{(x^2 + 4\gamma^2 x)/\mathscr{D}^2} = \frac{1}{1 + (4\gamma^2/x)} = \frac{1+\mu}{2+\mu}.$$

The results are, therefore,

$$\theta_2/(T_0/K_2) = M = (2+\mu)/\mu, \qquad |\theta_1/\theta_2| = 1/\sqrt{(2+\mu)},$$

$$|\theta_{sh}/\theta_2| = \sqrt{(1+\mu)}/\sqrt{(2+\mu)}.$$

This last expression was used to evaluate the curve of Fig. 3b, p. 523. Note that $|\theta_{sh}/\theta_2|$ can never exceed unity. One can also use

$$\left|\frac{\theta_{sh}}{T_0/K_2}\right| = \left|\frac{\theta_{sh}}{\theta_2}\right| \times M = \frac{\sqrt{(1+\mu)}\,(2+\mu)}{\sqrt{(2+\mu)}\,\mu} = \frac{\sqrt{\{(1+\mu)(2+\mu)\}}}{\mu} = \frac{1}{\sqrt{2}\,\mu\gamma}.$$

3·32 Three-mass system formulae

These formulae give a more accurate assessment of the optimum phase velocity ω_D and the vibration amplitudes $\theta_{1''}$ at the damper casing and θ_2 at the engine mass.

Equivalent system

Referring to Fig. 5, let

$\omega_2 =$ natural phase velocity [rad./sec.] of engine system without damper, as determined by a Holzer table,

$J_2 = \Sigma(J_{\text{cyl.}} \Delta_{\text{cyl.}}^2) =$ equivalent engine inertia [Lb.in.sec.^{2}], from Holzer table for system without damper (the summation extending only over the cylinder inertias),

$K_2 = \omega_2^2 J_2 =$ equivalent crankshaft stiffness [Lb.in./rad.],

$J_1'' =$ inertia of seismic mass of damper [Lb.in.sec.^{2}],

$J_1' =$ inertia of damper casing [Lb.in.sec.^{2}],

$K_1 =$ stiffness of shafting between J_1' and cylinder centre no. 1 [Lb.in./rad.],

$T_0 =$ excitation torque $(|T_n| A R_0 \overrightarrow{\Sigma\Delta})$ [Lb.in.] of engine without damper.

Note. Although the flywheel inertia J_3 has a finite value which is used in the Holzer table mentioned above, the three-mass formulae are derived with the assumption that $J_3 = \infty$, i.e. that the flywheel inertia is so much larger than J_2 that the node can be situated at J_3.

Furthermore, it is convenient to introduce the following parameters:

$$\text{seismic mass ratio: } \mu = J_1''/J_2, \quad \text{casing ratio: } \lambda = J_1'/J_1'',$$

$$\text{stiffness ratio: } k = K_1/K_2.$$

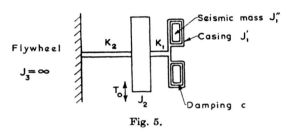

Fig. 5.

Then, for specified values of μ, λ and k, the *optimum phase velocity* ω_D of the engine system with damper can be estimated from the chart of Fig. 6, which was obtained by evaluating the following formula:

$$-\frac{A}{k}x^2 + \left[2 + A\left(1 + \frac{1}{k}\right) \right] x - 2 = 0,$$

where

$$A = (2\lambda + 1)\mu \quad \text{and} \quad x = \omega_D^2/\omega_2^2.$$

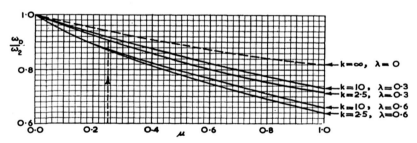

Fig. 6. Optimum phase velocity ω_D of engine with damper, taking account of casing inertia J_1' and damper shaft stiffness K_1.

Example of use of Fig. 6. Assuming that it is decided to fit a damper with a seismic mass ratio $\mu = 0.25$ and a casing ratio $\lambda = 0.333$ to a stub shaft with a stiffness ratio $k = 2.85$, one can estimate from the above figure that the optimum value of phase velocity will be $\omega_D \simeq 0.908\omega_2$. (Direct evaluation of the formula gives $\omega_D = 0.9048\omega_2$.)

For $k = \infty$ and $\lambda = 0$, one obtains $A = \mu$ and the equation for ω_D^2/ω_2^2 reduces to that previously obtained by considering the two-mass system. The above equation, however, is quadratic in x, and only the root smaller

[527]

than unity has been used in Fig. 6. The other root gives the common-point value for two-node vibration conditions, i.e. with a second node situated in the damper stub shaft K_1.

Usually, the inertia of the casing is smaller than that of the seismic mass $(0\cdot2 < \lambda < 0\cdot7)$ and the stiffness of the damper stub shaft is greater than the equivalent crankshaft stiffness $(3\cdot0 < k < 12\cdot0)$, so that the curves of Fig. 6 cover the range of values usually considered.

Vibration amplitudes

The following formulae are valid for both *peak and flank amplitudes*:
Amplitude at damper casing:

$$\theta_{1'} = \frac{T_0}{K_2} \times M_{1'} = \frac{T_0}{K_2} \times \frac{\sqrt{(x^2 + 4\gamma^2 x)}}{\sqrt{(C^2 + D^2)}} \quad [\text{rad.}].$$

Amplitude at position of cylinder no. 1 of engine:

$$\theta_2 = \frac{T_0}{K_2} \times M_2 = \frac{T_0}{K_2} \times \frac{\sqrt{(x^2 E^2 + 4\gamma^2 x F^2)}}{\sqrt{(C^2 + D^2)}} \quad [\text{rad.}],$$

where $x = \omega_D^2/\omega_2^2$,

$$C^2 = x^2 \left[1 - x(1 + \lambda\mu) + \frac{\lambda\mu}{k} x(x - 1) \right]^2,$$

$$D^2 = 4\gamma^2 x \left[1 - x(1 + \lambda\mu + \mu) + \frac{\mu x}{k}(1 + \lambda)(x - 1) \right]^2,$$

$$E^2 = \left(\frac{\lambda x}{k} - 1 \right)^2,$$

$$F^2 = \left[1 - \frac{\mu}{k}(1 + \lambda) x \right]^2.$$

In these equations, $\gamma = c/(2\omega_2 J_1'') = $ non-dimensional damping ratio, c being the damping coefficient of the damper (which is determined by γ), and $T_0 = |T_n| A R_0 \overrightarrow{\Sigma\Delta} = $ excitation torque [Lb.in.] of the engine, from the Holzer table without a damper.

Moreover, since it has been assumed that $J_3 = \infty$, θ_2 is also the shaft twist θ_{sh} of the entire crankshaft; therefore, the vibratory shaft torque is $T_{sh} = K_2 \theta_2$ [Lb.in.] and the corresponding stress at the node is

$$q = K_2 \theta_2/Z \quad [\text{Lb./in.}^2],$$

where $Z = $ sectional modulus [in.^{3}] of the smallest crankshaft cross-section in the vicinity of the node.

Optimum value of damping ratio γ

This can be taken from the corresponding curve for the two-mass system (Fig. 3a, p. 523). One can verify whether the γ-value thus obtained is sufficiently close to the optimum value by evaluating the

[528]

above expression for $\theta_{1'}$ for one or two other γ-values. This has been done in the following example.

Metric units. All the above formulae are also valid in metric units, for instance, with T in Kg.cm., K in Kg.cm./rad., c in Kg.cm.sec./rad., J in Kg.cm.sec.², q in Kg./cm.² and Z in cm.³.

NUMERICAL EXAMPLE. An engine has a natural frequency $F = 2332$ vib./min. without a damper and its equivalent inertia (determined from a Holzer table) is $J_2 = \Sigma(J_{cyl.}\, \Delta_{cyl.}^2) = 480$ Lb.in.sec.². Its excitation torque for the critical speed order number considered is $T_0 = 12,000$ Lb.in.

Assuming that it is decided to use a damper with a seismic mass J_1'' of 120 Lb.in.sec.², and a casing inertia J_1' of 40 Lb.in.sec.², and that the stiffness K_1 of the shaft portion between the damper casing and cylinder no. 1 is 81×10^6 Lb.in./rad., it is required to determine the optimum phase velocity ω_D of the system with damper, the peak amplitude at the casing and the optimum damping ratio.

(a) The natural phase velocity of the engine without the damper is

$$\omega_2 = F \times 2\pi/60 = 2332/9 \cdot 55 = 244 \cdot 18 \text{ rad./sec.};$$

therefore

$$\omega_2^2 = 5 \cdot 9625 \times 10^4 \text{ rad.}^2/\text{sec.}^2$$

and

$$K_2 = \omega_2^2 J_2 = 5 \cdot 9625 \times 10^4 \times 480 = 28 \cdot 45 \times 10^6 \text{ Lb.in./rad.}$$

The values of the non-dimensional parameters are:

$$\mu = J_1''/J_2 = 120/480 = 0 \cdot 25, \quad \lambda = J_1'/J_1'' = 40/120 = 0 \cdot 3333,$$

$$k = K_1/K_2 = 81/28 \cdot 45 = 2 \cdot 85.$$

Hence, from Fig. 6, $\omega_D/\omega_2 = 0 \cdot 908$.

Evaluation of the frequency formula gives

$$\omega_D/\omega_2 = 0 \cdot 9048, \quad \text{or} \quad \omega_D = 0 \cdot 9048\omega_2 = 220 \cdot 934 \text{ rad./sec.}$$

From Fig. 3a, p. 523, we find $\gamma = 0 \cdot 45$.

(b) Calculation of the casing amplitude $\theta_{1'}$:

$$x = (\omega_D/\omega_2)^2 = (0 \cdot 9048)^2 = 0 \cdot 81866, \quad x^2 = 0 \cdot 67037,$$

$$x^2 + 4\gamma^2 x = 0 \cdot 67037 + 4 \times (0 \cdot 45)^2 \times 0 \cdot 81866 = 1 \cdot 33348,$$

$$C^2 = 0 \cdot 67037 \times [0 \cdot 11312 - 0 \cdot 10878]^2 = 0 \cdot 00793,$$

$$D^2 = 4 \times (0 \cdot 45)^2 \times 0 \cdot 81866 \times [-0 \cdot 09154 - 0 \cdot 01736]^2 = 0 \cdot 00787,$$

$$M_{1'}^2 = (x^2 + 4\gamma^2 x)/(C^2 + D^2) = 1 \cdot 33348/0 \cdot 0158 = 84 \cdot 4, \quad M_{1'} = 9 \cdot 187,$$

$$\theta_{1'} = M_{1'} \times T_0/K_2 = 9 \cdot 187 \times 12 \cdot 0 \times 10^3/28 \cdot 45 \times 10^6 = 0 \cdot 003875 \text{ rad.}$$

or

$$\theta_{1'} = \pm 0 \cdot 222°.$$

(c) It is necessary to verify whether the value of θ_1 thus obtained is effectively the peak amplitude. This can be done by evaluating θ_1 for other values of phase velocity, on either side of the value chosen for ω_D. The results are summarized below:

$$\omega_D/\omega_2 = \quad 0\cdot8235 \quad 0\cdot8693 \quad 0\cdot9048 \quad 0\cdot915$$

$$\pm\,\theta_{1'} = \quad 0\cdot152° \quad 0\cdot265° \quad 0\cdot222° \quad 0\cdot218°$$

$$M_{1'} = \quad 6\cdot293 \quad 10\cdot987 \quad 9\cdot1869 \quad 9\cdot0327$$

Thus, with $\gamma = 0\cdot45$, the peak amplitude is not at $\omega_D/\omega_2 = 0\cdot9048$ but at the somewhat lower value of $0\cdot8693$.

Further check calculations with $\gamma = 0\cdot50$ and $\gamma = 0\cdot40$ still leave the peak at $\omega_D/\omega_2 = 0\cdot8693$; a higher peak is obtained with $0\cdot50$ and a slightly lower one with $0\cdot40$. However, there is still some uncertainty in the value of γ, since J_3 has been assumed to be infinitely large, so that $\gamma \simeq 0\cdot45$ can be regarded as sufficiently close to the optimum value of damping.

The peak-amplitude condition occurs therefore at $\omega_D = 0\cdot8693\omega_2 = 212\cdot3\,\mathrm{rad./sec}$ and the results for the system with damper are as follows:

peak amplitude at casing: $\qquad \theta_{1'} = \pm 0\cdot265°,$

peak amplitude at cylinder no. 1: $\quad \theta_2 = \pm 0\cdot179°,$

dynamic magnifiers: $\qquad M_{1'} \simeq 11\cdot0, \quad M_2 \simeq 7\cdot4.$

Criteria of performance of viscous-shear type dampers

The main criterion is whether the vibration amplitude at cylinder no. 1, and the vibration stress at the node, are sufficiently reduced by means of the damper.

A second criterion is whether a damper with given μ, λ and k values is providing the lowest possible amplitude. The performance limit is set by the dynamic magnifier of the engine system with the damper, that is, by the calculated value

$$M_{1'} = \theta_{1'}/(T_0/K_2).$$

In the example which has just been considered, $M_{1'} = 11$. This magnifier would be obtained with the damper regardless of whether the engine without a damper initially had (owing to its inherent damping) a dynamic magnifier of, say, 22 or 44, although the reduction in amplitude in these cases is $2:1$ and $4:1$, respectively. Hence for comparisons of damper performance, the criterion is the $M_{1'}$-value that it attains in practice.

Size of viscous-shear dampers. The damper considered above, with a μ-ratio of $0\cdot25$, is smaller than those generally employed. Usually, μ is taken in the range $0\cdot4 < \mu < 1\cdot0$, in order to obtain very low dynamic magnifiers ($M_{1'} \simeq 5$), as indicated by the M-curve of Fig. 3a, p. 523.

Derivation of three-mass system formulae

The equations of motion of the system indicated in Fig. 5 are

$$(-\omega^2 J_1'' + j\omega c)\,\theta_{1''} - \qquad\qquad j\omega c\theta_{1'} \qquad\qquad = 0,$$

$$-j\omega c\theta_{1''} \quad + (K_1 - \omega^2 J_1' + j\omega c)\,\theta_{1'} - K_1\theta_2 = 0,$$

$$- K_1\theta_{1'} + (K_1 + K_2 - \omega^2 J_2)\,\theta_2 = T_0.$$

[530]

The system determinant is

$$\mathrm{Det}_s = \begin{vmatrix} -\omega^2 J_1'' + j\omega c & -j\omega c & 0 \\ -j\omega c & K_1 - \omega^2 J_1' + j\omega c & -K_1 \\ 0 & -K_1 & K_1 + K_2 - \omega^2 J_2 \end{vmatrix}$$

$$= \begin{vmatrix} -\omega^2 J_1'' + j\omega c & -j\omega c & 0 \\ -\omega^2 J_1'' & K_1 - \omega^2 J_1' & -K_1 \\ 0 & -K_1 & K_1 + K_2 - \omega^2 J_2 \end{vmatrix}$$

$$= \begin{vmatrix} -\omega^2 J_1'' & 0 & 0 \\ -\omega^2 J_1'' & K_1 - \omega^2 J_1' & -K_1 \\ 0 & -K_1 & K_1 + K_2 - \omega^2 J_2 \end{vmatrix}$$

$$+ \begin{vmatrix} j\omega c & -j\omega c & 0 \\ -\omega^2 J_1'' & K_1 - \omega^2 J_1' & -K_1 \\ 0 & -K_1 & K_1 + K_2 - \omega^2 J_2 \end{vmatrix}$$

$$= -\omega^2 J_1''[(K_1 - \omega^2 J_1')(K_1 + K_2 - \omega^2 J_2) - K_1^2]$$

$$+ j\omega c[(K_1 - \omega^2 J_1' - \omega^2 J_1'')(K_1 + K_2 - \omega^2 J_2) - K_1^2].$$

(The second determinant is obtained by adding line 1 to line 2 in the first determinant; the two determinants with only real or imaginary components by splitting up the second determinant along line 1.) Introducing the ratios

$$\lambda = J_1'/J_1'', \quad \mu = J_1''/J_2, \quad \lambda\mu = J_1'/J_2, \quad \omega_2^2 = K_2/J_2,$$

$$k = K_1/K_2, \quad (\omega/\omega_2)^2 = x \quad \text{and} \quad c = 2J_1''\omega_2\gamma,$$

so that

$$\omega c/K_2 = 2J_1''\omega_2\gamma\omega/K_2 = 2\mu\omega\gamma\omega_2 J_2/K_2 = 2\mu\gamma\omega/\omega_2 = 2\mu\gamma\sqrt{x},$$

we can write

$$\mathrm{Det}_s = -K_2^3 x\mu[(k - \lambda\mu x)(k + 1 - x) - k^2]$$

$$+ j2\mu\gamma\sqrt{x}\,K_2^3[(k - \lambda\mu x - \mu x)(k + 1 - x) - k^2];$$

expanding and rearranging within the brackets, we obtain the absolute value squared as

$$|\mathrm{Det}_s|^2 = K_2^6\mu^2 k^2 \left\{ x^2\left[1 - x(1 + \lambda\mu) + \frac{\lambda\mu x}{k}(x - 1)\right]^2 \right.$$

$$\left. + 4\gamma^2 x\left[1 - x(1 + \lambda\mu + \mu) + \frac{\mu x}{k}(1 + \lambda)(x - 1)\right]^2 \right\}.$$

The determinant for the amplitude at the damper casing is

$$\text{Det}_{1'} = \begin{vmatrix} -\omega^2 J_1'' + j\omega c & 0 & 0 \\ -j\omega c & 0 & -K_1 \\ 0 & T_0 & K_1 + K_2 - \omega^2 J_2 \end{vmatrix} \begin{aligned} &= T_0(-\omega^2 J_1 + j\omega c)\, K_1, \\ &= T_0 K_2^2 k\mu[-x + j2\gamma\sqrt{x}], \end{aligned}$$

so that

$$|\text{Det}_{1'}|^2 = T_0^2 K_2^4 k^2 \mu^2 [x^2 + 4\gamma^2 x].$$

Therefore

$$M_{1'}^2 = \frac{x^2 + 4\gamma^2 x}{C^2 + D^2} = \left| \frac{\theta_{1'}}{T_0/K_2} \right|^2,$$

where $A^2 = x^2$, $B^2 = 4\gamma x$, while C^2 and D^2 are the expressions given on p. 528. We may now write, following the same reasoning as for the two-mass system formulae,

$$M_{1'}^2 = \frac{A^2 + B^2}{C^2 + D^2},$$

which reduces to an expression independent of γ, viz. $M_{1'}^2 = A^2/C^2$, when $(A/C)^2 = (B/D)^2$, as shown on p. 514. The relation $-AD = +BC$, therefore, determines the phase velocity ω_D at which $\theta_{1'}$ is independent of damping:

$$-x \times 2\gamma\sqrt{x}\left[1 - x(1 + \lambda\mu + \mu) + \frac{\mu x}{k}(1 + \lambda)(x - 1)\right]$$

$$= 2\gamma\sqrt{x} \times x\left[1 - x\left(1 + \frac{\lambda\mu}{k}\right) + \frac{\lambda\mu x}{k}(x - 1)\right],$$

and this reduces to the quadratic equation for x given on p. 527.

Finally, the determinant for the amplitude θ_2 at the engine mass is

$$\text{Det}_2 = \begin{vmatrix} -\omega^2 J_1'' + j\omega c & -j\omega c & 0 \\ -j\omega c & K_1 - \omega^2 J_1' + j\omega c & 0 \\ 0 & -K_1 & T_0 \end{vmatrix}$$

$$= T_0\{-\omega^2 J_1'' K_1 + \omega^4 J_1' J_1'' + j\omega c(K_1 - \omega^2 J_1' - \omega^2 J_1'')\},$$

so that

$$|\text{Det}_2|^2 = T_0^2 K_2^4 \mu^2 k^2 \left[x^2\left(\frac{\lambda x}{k} - 1\right)^2 + 4\gamma^2 x\left\{1 - \frac{\mu}{k}(1 + \lambda)x^2\right\}^2\right],$$

with which one obtains $M_2^2 = (x^2 E^2 + 4\gamma^2 x F^2)/(C^2 + D^2)$ and therefore also the expression for θ_2 given on p. 528.

3·33 Tabulations for common-point frequency and optimum damping

The common-point phase velocity ω_D may be estimated by means of ordinary Holzer-table calculations for the engine system with damper, using for the 'effective damper inertia' the relation $J_D \simeq J_1' + \frac{1}{2} J_1''$ (where $J_1' =$ casing inertia and $J_1'' =$ inertia of the inner member or seismic mass).

[532]

However, this approximate value for ω_D may not be sufficiently accurate, particularly in cases where the resonance curve even with heavy damping is still so important that it is necessary to locate its peak position fairly accurately at the design stage. Such cases, for instance, may occur with highly supercharged engines operating at a very high mean effective pressure, with $1\frac{1}{2}$th, 2nd or 3rd order critical speeds near the running range.

For such cases it is possible to determine ω_D and γ with adequate accuracy by means of the following method using Holzer tables (this method was evolved at the B.I.C.E.R.A. Laboratory):

(1) Evaluate forced-frequency tables, without damping, for flank amplitudes of (a) the engine system with the damper casing only, (b) the engine with the seismic mass of the damper locked to the damper casing (i.e. with $J_D = J_1' + J_1''$).

(2) Plot these flank-amplitude curves against frequency. Their point of intersection thus obtained determines the common-point frequency and hence ω_D.

(3) Prepare three Holzer tabulations for the phase velocities ω_D, $0.95\omega_D$ and $1.05\omega_D$. Evaluate these tables without damping, beginning from the driven-machine end of the system, down to the inertia of cylinder no. 1.

(4) To evaluate the last two lines (for the casing and the seismic mass of the damper), assume a value of $\gamma = 0.5$ for the damping ratio.† In this manner obtain three values for vibration amplitude at the driven-machine end, viz. $\theta_{0.95}$, θ_1 and $\theta_{1.05}$, using the three tables.

(5) Plot these θ-values against frequency. If the peak is not at ω_D but towards $0.95\omega_D$, choose a lower value of γ (e.g. 0.45 or 0.40) and evaluate again the last two lines of the tables. (A lower γ will shift the peak towards higher ω-values.)

(6) If the peak is towards $1.05\omega_D$, choose a higher value for the damping ratio ($\gamma = 0.55$) and repeat the evaluation of the last two lines of the tables. In this way, the optimum damping ratio will be determined as that which gives a peak equal to the common-point amplitude at the common-point frequency.

In the numerical example given in the following, all the cylinder inertias have been lumped together and represented by an 'equivalent engine inertia'. However, the complex-number calculation is not more involved when all the inertias are tabulated individually. This applies to major orders. For minor orders, it is simpler to use the equivalent engine inertia and the corresponding phase vector sums.

NUMERICAL EXAMPLE. It is required to determine the optimum phase velocity, the optimum peak amplitude and the damping ratio for a system consisting of an

† Details regarding the use of γ in the tables will be apparent in the numerical example in the following part of this section.

engine with a main flywheel and a damper, the data to be considered being as follows:

Damper		Engine	
seismic mass	$J_1'' = 120$ Lb.in.sec.^{2}	damper stub shaft	$K_1 = 81 \times 10^6$ Lb.in./rad.
casing	$J_1' = 40$ Lb.in.sec.^{2}	crankshaft	$K_2 = 27 \times 10^6$ Lb.in./rad.
		engine mass	$J_2 = J_E = 480$ Lb.in.sec.^{2}
		flywheel	$J_3 = J_F = 8000$ Lb.in.sec.^{2}
		excitation torque	$T_0 = 12{,}000$ Lb.in.

It may be noted that the system without the damper has a natural phase velocity $\omega_2 = \sqrt{\{K_2[(1/J_2) + (1/J_3)]\}} = 244 \cdot 18$ rad./sec., so that for the system with damper the range of ω-values to be explored is between this limit and, approximately, $0 \cdot 8\omega_2$.

(a) *Determination of common-point amplitude and frequency.* The flanks of the resonance curve for the 'system with casing only' $(J_D = J_1')$ and the 'system with damper locked' $(J_D = J_1' + J_1'')$ were evaluated in order to obtain the points A, B, C, D and E, F, G, H, I, J, plotted in Fig. 7.

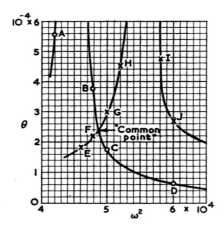

Fig. 7. Curves of flank amplitudes for the system with damper locked (points A, B, C, D) and for the system with damper casing only (points E, F, G, H, I, J). The common-point amplitude θ, for mass J_3, has a value of $0 \cdot 00024$ rad. and is situated at
$$\omega^2 = \omega_D^2 = 4 \cdot 87 \times 10^4 \text{ rad.}^2/\text{sec.}^2.$$

Two of the forced-frequency tables without damping are given below.
Forced-frequency tabulation for point 'B':

$$\omega^2 = 0 \cdot 048 \times 10^6 \text{ rad.}^2/\text{sec.}^2, \quad T' = T_0/10^6 = 0 \cdot 012$$

Mass no.	J	$\dfrac{J\omega^2}{10^6}$	θ	$\dfrac{J\omega^2\theta}{10^6}$	$\dfrac{T_0}{10^6}$	$\dfrac{\Sigma[J\omega^2\theta + T_0]}{10^6}$	$\dfrac{K}{10^6}$	θ_{sh}
3 (J_F)	8000	384·0	$1 \cdot 000\theta$	$384 \cdot 000\theta$	—	$384 \cdot 000\theta$	14·222θ	
2 (J_E)	480	23·04	$-13 \cdot 222\theta$	$-304 \cdot 635\theta$	T'	$+80 \cdot 365\theta + T'$	81·0	0·992θ $+0 \cdot 0123T'$
1 (J_D)	160	7·68	$-14 \cdot 214\theta$ $-0 \cdot 0123T'$	$-109 \cdot 164\theta$ $-0 \cdot 0945T'$	—	$-28 \cdot 799\theta$ $+0 \cdot 906T'$	—	—

$\Sigma[J\omega^2\theta + T_0] = 0$ gives $\theta_3 = 0 \cdot 906T'/28 \cdot 799 = 0 \cdot 906 \times 0 \cdot 012/28 \cdot 799 = 0 \cdot 000378$ rad. $= 0 \cdot 0237°$ at J_3. (At mass 1, the amplitude is $\theta_1 = -14 \cdot 214\theta_3 - 0 \cdot 0123T' = -0 \cdot 005373 - 0 \cdot 000148 = -0 \cdot 005521$ rad. $= -0 \cdot 316°$ at J_1.)

Forced-frequency tabulation for point 'G':

$$\omega^2 = 0.05 \times 10^6 \text{ rad.}^2/\text{sec.}^2$$

Mass no.	J	$\dfrac{J\omega^2}{10^6}$	θ	$\dfrac{J\omega^2\theta}{10^6}$	$\dfrac{T}{10^6}$	$\dfrac{\Sigma[J\omega^2\theta + T_0]}{10^6}$	$\dfrac{K}{10^6}$	θ_{sh}
3 (J_F)	8000	400·0	$1·000\theta$	$400·000\theta$	—	$400·000\theta$	27·0	$14·815\theta$
2 (J_E)	480	24·0	$-13·815\theta$	$-331·560\theta$	T'	$+68·440\theta + T'$	81·0	$0·845\theta$
								$+0·123T'$
1 (J_D)	40	2·0	$-14·660\theta$	$-29·320\theta$	—	$+39·120\theta$	—	—
			$-0·0123T'$	$-0·0246T'$		$+0·9754T'$		

$\Sigma[J\omega^2\theta + T_0] = 0$ gives $\theta_3 = -0·9754T'/39·120 = -0·9754 \times 0·012/39·120 = -0·00030$ rad. $= -0·0172°$ at J_3. (Amplitude at mass 1: $\theta_1 = -14·660\theta_3 - 0·0123T' = +0·004398 - 0·000148 = +0·004250$ rad. $= +0·144°$.)

In plotting the curves of Fig. 7, the $+$ or $-$ sign of θ is disregarded, since only absolute values are considered.

The computation procedure is substantially the same as for Holzer tables without the gas torque T_0, except that the residual torque $\Sigma[J\omega^2\theta + T_0]$ at the last mass is equated to zero and serves to determine the flank amplitude θ (see also section 2·312 for further details on forced-frequency tabulations without damping). Thus

$$\omega_D = \sqrt{(4·87 \times 10^4)} = 220·7 \text{ rad./sec.} \quad \text{and} \quad \theta_{\text{opt.}} = 0·000245 \text{ rad. at mass } J_3.$$

(b) *Determination of damping ratio γ and damping coefficient c.* It is now necessary to consider the complete system with damping between the damper casing J_1' and the seismic mass J_1'' (see Fig. 8).

To verify the position of the peak amplitude with various γ-values, three Holzer tables with damping will be required for ω^2-values in the vicinity of ω_D^2; for instance, for

$$\omega^2 = 0·048 \times 10^6, \ 0·0487 \times 10^6 \quad \text{and} \quad 0·050 \times 10^6.$$

The tabulation for $\omega^2 = 0·0487 \times 10^6$ is given below. In the column for stiffness we have $K_2/10^6$, $K_1/10^6$ and $j\omega c' = j\omega c/10^6$, this last expression representing the 'equivalent stiffness' due to the damping torque acting between J_1' and J_1''.

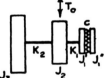

Fig. 8.

Note. $c = 2J_1''\omega_2\gamma$ [Lb.in.sec./rad.]. As γ is still undetermined, the expression $j\omega c'$ is entered without a numerical value in the tabulation.

$$\omega = 220·68 \text{ rad./sec.}, \quad \omega^2 = 0·0487 \times 10^6 \text{ rad.}^2/\text{sec.}^2$$

Mass no.	J [Lb.in. sec.²]	$\dfrac{J\omega^2}{10^6}$ [Lb.in./ rad.]	θ [rad.]	$\dfrac{J\omega^2\theta}{10^6}$ [Lb.in.]	$\dfrac{T_0}{10^6}$ [Lb.in.]	$\dfrac{\Sigma[J\omega^2\theta + T_0]}{10^6}$ [Lb.in.]	$\dfrac{K}{10^6}$ [Lb.in./ rad.]	θ_{sh} [rad.]
3	8000	389·6	$1·000\theta$	$3·896\theta$	—	$389·6\theta$	27·0	$14·330\theta$
2	480	23·376	$-13·330\theta$	$-313·940\theta$	$T'+jT''$	$+75·660\theta$	81·0	$0·934\theta$
						$+T'+jT''$		$+0·0123T'$
								$+j0·0123T''$
1'	40	1·948	$-14·364\theta$	$-27·981\theta$	—	$+47·679\theta$	$j\omega c'$	$0·9760T''/\omega c'$
			$-0·0123T'$	$-0·0240T'$		$+0·9760T'$		$-j47·679\theta/\omega c'$
			$-j0·0123T''$	$-j0·0240T''$		$+j0·9760T''$		$-j0·9760T'/\omega c'$
1''	120	5·844	$-14·3640\theta$	$-83·943\theta$	—	$-36·264\theta$	—	—
			$-0·0123T'$	$-0·0719T'$		$+0·9041T'$		
			$-0·9760T''/\omega c'$	$-5·704T''/\omega c'$		$-5·704T''/\omega c'$		
			$-j0·0123T''$	$+j278·636\theta/\omega c'$		$+j278·636\theta/\omega c'$		
			$+j0·9760T'/\omega c'$	$+j5·704T'/\omega c'$		$+5·704T'/\omega c'$		
			$+j47·679\theta/\omega c'$	$-j0·0719T''$		$+0·9041T''\}$		

Equating to zero the real and the imaginary terms of the total torque, one obtains two equations for θ (amplitude at J_3), viz.

$$\theta_3 = \frac{0\cdot9041T'}{36\cdot264} - \frac{5\cdot704T''}{36\cdot264\omega c'} = 0\cdot0250T' - 0\cdot1573T''/\omega c', \qquad (1a)$$

$$\theta_3 = -\frac{5\cdot704T'}{278\cdot636} - \frac{0\cdot9041T''\omega c'}{278\cdot636} = -0\cdot0205T' - 0\cdot00324T''\omega c', \qquad (1b)$$

which can be used with various values for $\omega c'$.

Trial solutions. Assume a value of $\omega c'$ close to the expected optimum value: for instance, let

$$10^6 \times \omega c' \equiv \omega c = \omega_D^2 J_1'' = 0\cdot0487 \times 10^6 \times 120 = 5\cdot844 \times 10^6.$$

This gives $\qquad c = 5\cdot844 \times 10^6/\omega = 26{,}482\ \text{Lb.in.sec./rad.}$

and $\qquad \gamma = c/2J_1''\omega_2 = 26{,}482/(2 \times 120 \times 244\cdot18) = 0\cdot452,$

where $\omega_2 =$ natural phase velocity of the system without the damper. With this value of $\omega c'$ equations $(1a)$ and $(1b)$ become

$$\theta_3 = 0\cdot0250T' - (0\cdot1573/5\cdot844)\,T'' = 0\cdot0250T' - 0\cdot0269T'', \qquad (2a)$$

and $\qquad \theta_3 = -0\cdot0205T' - 0\cdot00324 \times 5\cdot844T'' = -0\cdot0205T' - 0\cdot0189T''. \qquad (2b)$

Subtracting $(2b)$ from $(2a)$, we obtain

$$0 = 0\cdot0455T' - 0\cdot0080T'' \equiv [0\cdot0455T_0 \cos\phi - 0\cdot0080T_0 \sin\phi] \times 10^{-6},$$

hence $\tan\phi = 45\cdot5/8\cdot0 = +5\cdot688$, $\phi = 79°\,49'$ (phase angle between T_0 and θ_3) and $\psi = 180° - \phi = 100°\,11'$ (phase angle between T_0 and θ_2; from the Holzer table it can be seen that there is a 180° phase difference, since $\theta_2 = -13\cdot330\theta_3$). Furthermore, as

$$\cos\phi = \cos 79°\,49' = +0\cdot17679 \qquad \text{and} \quad \sin\phi = \sin 79°\,49' = +0\cdot98425,$$

$$T' = 0\cdot012 \times \cos\phi = +0\cdot00212 \quad \text{and} \qquad T'' = 0\cdot012\sin\phi = +0\cdot01181.$$

The substitution of these values in eqs. $(2a)$ and $(2b)$ gives

$$\theta_3 = -0\cdot000265\,\text{rad.} \quad \text{and} \quad \theta_3 = -0\cdot000268\,\text{rad.},$$

or a mean value of $-0\cdot0002665\,$rad. (From Fig. 7, $\theta_3 \simeq 0\cdot00024\,$rad.)

Using this result, we evaluate, by means of the corresponding expression in the table, the vibration amplitude at the damper casing as

$$\theta_{1'} = -14\cdot364\theta_3 - 0\cdot0123T' - j0\cdot0123T'' = (-3\cdot8541 - j0\cdot1452) \times 10^{-3},$$

so that its absolute value is

$$|\theta_{1'}| = 10^{-3} \times \sqrt{(14\cdot8541 + 0\cdot0211)} = 3\cdot857 \times 10^{-3}\,\text{rad.} \quad \text{or} \quad \pm\theta_{1'} = 0\cdot221°.$$

The two other Holzer tables with damping, which are not reproduced in the present text, were also evaluated in the same manner. The comparative results are

$\omega^2 =$	$0\cdot048 \times 10^6$	$0\cdot0487 \times 10^6$	$0\cdot050 \times 10^6$	$[\text{rad./sec.}]^2$
$\omega =$	$219\cdot09$	$220\cdot68$	$223\cdot61$	$[\text{rad./sec.}]$
$\theta_3 =$	$-0\cdot0002653$	$-0\cdot0002665$	$-0\cdot000119$	$[\text{rad.}]$

These results indicate that the peak amplitude is at the optimum phase velocity ω_D determined by the forced-frequency tables without damping, and that with the γ-value of 0·452 the peak amplitude at J_3 is approximately that of the common point in Fig. 7. It could be verified that $\gamma = 0·452$ is fairly close to the optimum γ-value by repeating the computation of eqs. (1 a) and (1 b) with different ωc-values for the three-phase velocities already used, but in the present example the agreement is sufficiently close to require no further verification.

Note. It is of interest to mention, however, that both ω_D and $\gamma_{\text{opt.}}$ may vary if the driven-machine portion of the system is altered, although the engine and damper are identical in all cases. This fact has been verified in torsiograph tests and indicates that optimum damping conditions depend on every inertia and stiffness of the entire system. For cases where optimum conditions are essential, the above method enables a suitable assessment to be made.

Other results obtainable from the Holzer tabulation with damping, given on p. 535, are as follows:

Amplitude at engine mass:

$$\theta_2 = -13·330\theta_3 = -13·330 \times (-0·0002665) = 0·003544 \,\text{rad.} = \pm 0·203°.$$

Static deflexion of crankshaft:

$$T_0/K_2 = 12{,}000/27 \times 10^6 = 4·444 \times 10^{-4}\,\text{rad.}$$

Dynamic magnifiers:

$$M_2 = \theta_2/(T_0/K_2) = 35·44/4·444 \simeq 8·0,$$
$$M_{1'} = \theta_{1'}/(T_0/K_2) = 38·57/4·444 \simeq 8·7.$$

Torque in crankshaft:

$$T_{sh} = K_2\theta_{sh} = 27 \times 10^6 \times 14·330 \times 0·0002665 = 103{,}140\,\text{Lb.in.}$$

3·34   Design procedure to obtain required damping coefficient

In order to obtain optimum operating conditions, it is necessary to determine the damper dimensions, the clearances between seismic mass and casing, and the nominal viscosity of the silicone fluid to be used. The following basic design procedure is recommended:

(1) Determine the necessary moment of inertia J_1'' of the seismic mass (i.e. the annular inner member) of the damper, for the required vibration amplitude $\theta_{1'}$ at the damper casing, by means of the relations between

$$\theta_{1'} = M \times T_0/K_2 \quad \text{and} \quad J_1'' = \mu J_2,$$

provided by the curve of M plotted against μ in Fig. 3 a, p. 523 (or by the more detailed methods: three-mass formulae or Holzer tables for common-point amplitude). In these equations,

M = lowest dynamic magnifier attainable with a given value of $\mu = J_1''/J_2$,

$\theta_{1'}$ = acceptable amplitude at the damper casing [rad.],

T_0 = excitation torque of engine without damper [Lb.in.],

$K_2 = \omega^2 J_2 = \omega^2 \Sigma(J_{\text{cyl.}}\Delta_{\text{cyl.}}^2)$ = equivalent crankshaft stiffness [Lb.in./ rad.] based on natural phase velocity ω and equivalent engine inertia J_2 without damper (see p. 523).

(2) Taking account of the space available at the free end of the engine, decide on suitable values for the outer radius R_o, the inner radius R_i and the thickness L of the inner mass.

For a first approximation, the value of J_1''/L can be related to values of R_o by means of the graph on p. 30, section 1·1.

In practice, there is a wide variation in the relative dimensions used, the values of R_i/R_o varying between 0·65 and 0·25 and those of L/R varying between 0·20 and 0·60.

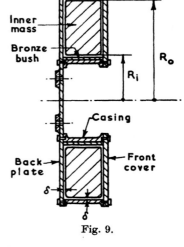

Fig. 9.

It is not possible to make a general recommendation regarding relative dimensions, since these may depend on the space available. However, it should be noted that it is necessary to provide sufficient surface area for heat dissipation. It is thus advisable to calculate the maximum power dissipated:

$$H_{\text{max.}} = 1\cdot325 \times 10^{-11} \times F^3 J_1'' \theta_1^2. \quad [\text{h.p.}],$$

where F = natural vibration frequency [vib./min.] and θ_1' = vibration amplitude with the damper [degrees] and to consider it in relation to the heat-dissipating surface. The most convenient reference area at this stage is the surface of the inner mass, which is given by

$$A = 2\pi(L + R_o - R_i)(R_o + R_i) \quad [\text{ft.}^2].$$

On the basis of tests of dampers of this type, carried out at the B.I.C.E.R.A. Laboratory, it can be recommended that for safe operation the power dissipated per unit area by such dampers should not exceed

$$H_{\text{max.}}/A = 0\cdot80 \quad [\text{h.p./ft.}^2].$$

If $H_{\text{max.}}/A$ exceeds 0·80 h.p./ft.², the relative dimensions of the damper should be altered to obtain a lower value for specific power dissipation. Alternatively, special cooling features such as fins, might be provided. [In metric units, $H_{\text{max.}}/A \leqslant 8\cdot73$ h.p. (metric)/(metre)².]

(3) Determine the damper clearances, from

$$\delta = 0\cdot010 + 0\cdot010 \times \sqrt{(D_o/10)} \quad [\text{in.}],$$

where $D_o = 2R_o$ is the outer diameter of the inner mass, in inches. This expression gives clearances in the range of values commonly used.† It is

† This formula is given only as a suggestion and it may sometimes prove more convenient to use quite different figures.

generally found satisfactory to use the same value of δ for both lateral and peripheral clearances.

(4) After this, design the damper casing and calculate its moment of inertia. Where there is no question of requiring an additional mass at the free end of the crankshaft for the purpose of modifying the vibration frequencies, the casing should have the smallest possible moment of inertia. Values of J_1'/J_1'' (casing inertia/inertia of inner mass) obtained in practice are usually between 0·35 and 0·8.

(5) Determine the optimum damping coefficient c by means of the γ-curve of Fig. 3a, p. 523, using the relation

$$c = \gamma \times 2J_1'' \omega_2 \quad \text{[Lb.in.sec./rad.]},$$

where $\omega_2 =$ natural phase velocity of the engine system with the damper casing [rad./sec.], or by trying various values of γ in the three-mass formulae or Holzer tables with damping.

Evaluate the damping coefficients c_L and c_P for the lateral and peripheral surfaces from

$$c_L = \frac{c}{1 + \dfrac{2L}{\eta_R R_o}} \quad \text{and} \quad c_P = c - c_L,$$

where the side-surface correction factor η_R is determined from the following table for various values of the mean shear rate

$$r_m = 0.49 \omega \theta_{1'} R_o / \delta \quad [\text{sec.}^{-1}]$$

and the ratio R_i/R_o:

$R_i/R_o =$	0·25	0·3	0·4	0·5	0·6	0·7	0·8
$r_m < 700$ sec.$^{-1}$ $\quad$ $\eta_R =$	1·04	1·03	1·01	0·97	0·89	0·77	0·61
$700 < r_m < 10{,}000$ sec.$^{-1}$ $\quad$ $\eta_R =$	1·11	1·10	1·08	1·03	0·94	0·81	0·58

(6) Calculate the effective viscosity ν of the silicone fluid required from the equation

$$\nu = \frac{c_P \delta}{1.448 \times 10^{-7} \times 2\pi L R_0^3} \quad \text{[centistokes]}.$$

(7) Finally, calculate the nominal viscosity† ν_0 of the silicone fluid from the effective viscosity ν by taking account of the corrections η_r for the mean shear rate and η_T for temperature, using the relation

$$\nu_0 = \nu/(\eta_r \eta_T).$$

† The term 'nominal viscosity' is used to denote the figure usually quoted by the suppliers, namely, that obtained by measurement at 25° C (77° F)., using the normal methods, such as the falling ball, which involve a very low rate of shear.

[539]

The value of η_r for the mean shear rate r_m is read off the chart of Fig. 10. The value of η_T is obtained from Fig. 11. With normal atmospheric temperatures of 59–68° F. (15–20° C.), the working temperature T may be taken as 113° F. (45° C.) for a damper outside the crankcase, and

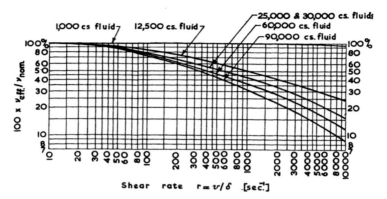

Fig. 10. Percentage variation of kinematic viscosity of silicone fluids with shear rate.

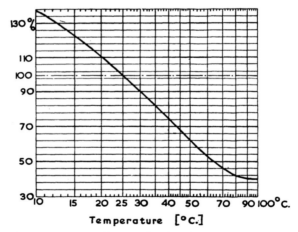

Fig. 11. Effect of temperature on kinematic viscosity of silicone fluids. This curve is applicable to silicone fluids with nominal viscosities between 12,500 and 200,000 centistokes.

167° F. (75° C.) for a damper arranged inside the crankcase, and these temperatures give values of η_T equal to 0·68 and 0·44, respectively.

Example of use of chart. A fluid with a nominal viscosity of 25,000 centistokes, subjected to a shear rate of 1000 sec.$^{-1}$, has its viscosity reduced to a fraction $\eta_r = 0·42$ of its original value; that is, its effective viscosity is $0·42 \times 25,000 = 10,500$ centistokes.

Examples of use of curves of Figs. 10 *and* 11. A fluid with an effective viscosity of 26,000 centistokes at a shear rate of 2000 sec.$^{-1}$ and a temperature of 45° C. should have a nominal viscosity

$$\nu_0 = 26,000/(0 \cdot 33 \times 0 \cdot 68) \cong 116,000 \text{ centistokes.}$$

The design procedure described above is based on the results of sections 3·31, 3·32 and 3·33 (pp. 521–37) as regards the determination of the casing amplitude $\theta_{1'}$, the inertias of the inner mass J_1'' and the casing J_1', and the damping ratio γ.

The additional details relating to H_{max}, c_P, c_L and the nominal viscosity ν_0 are derived from practical and theoretical considerations given in section 3·36 (pp. 551–8).

3·35 Further information on design features and operating experience, provided by B.I.C.E.R.A. members †

(*a*) *Materials used for dampers.* Steel has been found to be a suitable material for both the inner member and the outer casing, provided that the central location of the inner member is by a phosphor-bronze bush. Casings have also been made of aluminium alloy and inner members of tungsten, but usually steel is employed for both components.

(*b*) *Casing.* For casings assembled by means of welding, submerged electric arc welding has been used. The joints should be made in such a way that the welding material does not penetrate to the inner surfaces of the casing. No substances containing alkali should be in contact with the liquid.

For casings assembled by bolting, the studs should be spaced sufficiently close to one another to ensure adequate pressure on the joint faces. Dampers of small and medium sizes are now also assembled by machine rolling and peening the outer edge of the casing over the cover-plate, the outer edge being rolled over a soft sealing material. This eliminates the possibility of distortion or porosity of welds.

A further method of casing construction is machining from a solid piece steel; this construction has been employed for a number of medium-size dampers.

(*c*) *Surface finish.* As ferrous metals do not work at all well together with silicone fluids as lubricants, whereas other metal combinations are better in this respect, it is usual to provide the inner mass with a cadmium or zinc plating.

(*d*) *Cleaning of damper.* The use of alkaline detergents should be avoided, since these can depolymerize methyl silicones at high temperatures and thus yield methane gas, which may result in the generation of excessive pressure in the damper (risk of explosion). The suppliers

† The procedures given in this section are not necessarily descriptive of those now used by Holset and other specialist makers.

of silicone fluids recommend, therefore, that cleaning be done by degreasing in water vapour. Alternatively, the damper can be cleaned with liquid paraffin oil before filling and rinsed out with petrol or carbon tetrachloride to remove any paraffin which may remain on the damper surfaces. The former method appears preferable, however, since petrol and carbon tetrachloride are solvents which may also act on the silicone fluid.

(e) *Sealing.* To avoid leakage, joints have been made of various materials, for instance of 'hermatite compound', paper, more recently, rubber and 'gaskoid'. In some designs, soft solder is used over the filling plugs to avoid leakage. Tin soldering, however, has not always been found fully effective. Lead wire $\frac{1}{32}$ in. diameter has also been used as a jointing agent between the surfaces to be closed up. Another fully satisfactory jointing method consists in using natural rubber joint rings in recesses between the spigotted joints between the main body of the casing and its inner plate.

(f) *Leakage tests for casing.* For dampers situated outside the crankcase, a simple method is to encompass the damper with clear paper guards and run the engine for a certain period. Any leakage is then apparent from the inspection of the paper guards. Leakage can also be detected by the immersion test. For this, the damper after assembling is closed and filled with air (instead of silicone fluid) up to a pressure of 1–2 atmospheres. It is then immersed in water to verify that it is air-tight. It may be noted that leaks have also been found to occur on welded dampers owing to porosity of the weld.

(g) *Filling arrangements.* One or several fillings holes may be provided, and may be situated near, or on, the inner periphery of the casing, the outer periphery, or the middle of the flat portions of the casing.

The simplest method of filling is to coat all the inner surfaces of the damper casing and those of the inner mass with silicone fluid before assembly. After assembly, additional fluid is forced through the damper via the filling holes and discharge holes by means of a grease gun. Where discharge holes are provided, they are generally situated diametrically opposite to the filling holes on the same side of the casing.

The commercially used more elaborate method is to fill the damper under vacuum, and this is usual practice for heavy-duty dampers. This ensures that all cavities are filled with fluid free from air. This raises the question of stress caused by thermal expansion, which may be fairly high in some cases and which therefore also requires consideration at the design stage.

Spinning operation: After a first filling by either of the above methods, the damper is mounted on a shaft and rotated by a motor at a speed sufficient to allow the centrifugal force to displace the fluid from the centre to the periphery of the damper. Further filling is then carried out, either by means of a grease gun or under vacuum.

(*h*) *Loads acting on damper casing.* The casing is subjected to loads due to thermal expansion, loads required to produce a fluid-tight bolted joint, loads due to transmission of torque caused by the damper at peak amplitudes, and loads due to fluid pressure caused by centrifugal force. Thermal expansion coefficients are given under (*k*): bolt loads on circular plates can be assessed by the usual formulae for disks; it may be noted that the vibratory torque acting on the casing is $T = c\omega(\theta_{1'} - \theta_{1''})$, which can be taken from a Holzer table with damping or estimated as $T \simeq 0\cdot8\omega c\theta_{1'}$ (where $\theta_{1'}$ is the vibration amplitude of the casing); the effect of fluid pressure caused by centrifugal force is appreciable mainly in large-size dampers and will now be considered.

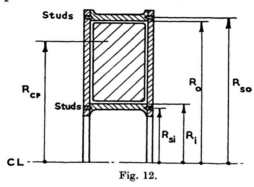

Fig. 12.

At any radius R [in.], the fluid pressure is

$$p = \frac{w}{2g}\,\omega^2 R^2 = \alpha R^2 \quad [\text{Lb./in.}^2],$$

with $\alpha = w\omega^2/2g$, where the constants are: $w =$ specific weight ($0\cdot0355$ Lb./in.^{3} for silicone fluids), $g = 386\cdot4$ in./sec.^{2} and $\omega = (2\pi/60) \times N_{\text{max.}}$, $N_{\text{max.}}$ being the maximum engine speed [rev./min.]. As the fluid may be regarded as incompressible, the force due to fluid pressure on each of the side surfaces of the casing is

$$P = \int_{R_i}^{R_o} p \times 2\pi R\,dR = \frac{\pi\alpha}{2}(R_o^4 - R_i^4) \quad [\text{Lb.}]$$

(for dampers with a fluid reservoir extending down to the centre, $R_i = 0$).

The moment of the load (on one side) about the centre line CL of the damper axis is

$$M_b = \int_{R_i}^{R_o} Rp \times 2\pi R\,dR = \frac{2\pi\alpha}{5}(R_o^5 - R_i^5) \quad [\text{Lb.in.}];$$

thus the centre of pressure is at a radius

$$R_{CP} = M_b/P = 0\cdot8(R_o^5 - R_i^5)/(R_o^4 - R_i^4) \quad [\text{in.}].$$

Stud loads and stresses: (1) Inner studs, at radius R_{si} (see Fig. 12):

$$P_i = P(R_{so} - R_{CP})/(R_{so} - R_{si}) \quad [\text{Lb.}], \qquad p_i = P_i/(n_i A_i) \quad [\text{Lb./in.}^2],$$

[543]

where $p_i =$ stress in each stud of core area A_i, n_i being the number of studs on the inner rim periphery. (2) Outer studs:

$$P_o = P - P_i \quad \text{[Lb.]}, \qquad p_o = P_o/(n_o A_o) \quad \text{[Lb./in.}^2\text{]},$$

where $n_o =$ number of studs of core area A_o on the outer rim periphery.

(*i*) *Properties of silicone fluids.*† Silicone fluids used in dampers have a specific gravity of about 0·973. Compressibility 4·5 % at 500 atm., 12·5 % at 2500 atm. Volumetric expansion, see (*k*). Boiling-point in the range 136–236° C., freezing-point − 86 to − 59° C. The effect of low temperatures on the kinematic viscosity is considerable: at 0° C., the ν-value is approximately doubled for fluids with nominal viscosities of 12,500 to 60,000 centistokes at + 25° C. For engines exposed to large variations in ambient temperature, the damper is more suitably placed inside the crankcase. Effect of shear rate, see Fig. 10 and also remarks under 'Test rigs' (*p*) given below.

(*j*) *Mixing.* To obtain a required nominal viscosity ν_0, it is possible to mix two fluids of nominal viscosities ν_1 and ν_2 (such that $\nu_1 < \nu_0 < \nu_2$). If $W_0 =$ total weight of mixture required, and W_1, $W_2 =$ weights of fluids ν_1 and ν_2, respectively, then the proportions (obtained from $W_1 + W_2 = W_0$ and $W_1 \nu_1 + W_2 \nu_2 = W_0 \nu_0$) are

$$W_1 = (\nu_2 - \nu_0) W_0/(\nu_2 - \nu_1) \quad \text{and} \quad W_2 = (\nu_0 - \nu_1) W_0/(\nu_2 - \nu_1).$$

Note. More accurate blending charts are obtainable from silicone manufacturers.

(*k*) *Expansion coefficients.* The *volumetric* expansion coefficients of silicone fluids and metals commonly used in viscous-shear dampers are as follows:

	Silicone fluids	Steel	Phosphor bronze	Aluminium alloys
$\beta =$	0·00078	0·000035	0·000054	0·000077

where $\beta \cong 3\alpha$, α being the linear expansion coefficient. Thus the volume V_t at a temperature t [° C.] is obtained from the volume V_0 at 0° C. by means of the relation $\quad V_t = V_0(1 + \beta t).$

Particularly for dampers of medium and large size, operating with appreciable power dissipation, the stresses in the casing caused by differential expansion should be considered. (See also second paragraph of (*n*).)

(*l*) *Clearances.* The clearances should be sufficiently ample to allow machining and assembling to overall tolerances of, say, 0·003 in., without

† Bibliography on properties of silicone fluids: Fitz Simmons, V. V., et al. *Trans. Amer. Soc. Mech. Engrs*, 1946, p. 365. Kauppi, T. A. and Currie, C. C., *Prod. Engng*, Feb. 1949, pp. 108–12. Georgian, G. C., *A.S.M.E. Paper*, no. 48-A-67, and *Trans. Amer. Soc. Mech. Engrs*, May 1949, pp. 389–99. Currie, C. C. and Smith, B. F., *Industr. Engng Chem.* vol. 42 (1950), p. 2457. McGregor, R. R., *Silicones and their Uses* (McGraw-Hill, 1954). Viscometers: (*a*) The SOD viscometer. Coordinating Research Council, Inc., 30 Rockefeller Plaza, New York 20, N.Y.—(*b*) British Standard 188: 1937, and Amendments: CF 4734, Jan. 1940; PD 1129, Jan. 1951; PD 1655, July 1953. (Falling-ball viscometers.)

appreciable effect on the damper effectiveness. It is found convenient to have radial and lateral clearances of the order of 0·015–0·030 in. (depending on the size of damper), but dampers have been made with radial clearances as large as 0·040 in. or as small as 0·007 in. The viscosity of the fluid is determined according to the chosen values for clearances.

(m) *Location of inner mass.* In the first designs, the clearance at the outer diameter of the inner mass was made smaller than that at the central bore, so that the inner mass was supported on its outer periphery by the casing. To avoid rubbing and pitting, phosphor-bronze rubbing pads were fitted in some cases on the outer peripheral surface of the inner mass, with successful results. However, the correct fitting of these chamfered pads was an expensive and laborious procedure.

For medium- and large-size dampers where extra precautions are required, a simpler solution was devised by providing the inner mass with a central location, in the form of a non-ferrous (e.g. phosphor-bronze) bush, fitted over the inner wall of the casing. The diametral clearance between the inner mass and the bush is of the order of a normal bearing clearance, i.e. 0·001 in. per in. of bush radius. Fully satisfactory results are obtained with this arrangement, which is now extensively used. In view of the possibility of axial vibration, this bush may also be designed with thrust surfaces for end location. It is equally possible to use a bush pressed into the mass and slightly clear of the casing.

Further alternative arrangements include the provision of lateral locating rings of bronze inserted on the inner mass. These have also proved satisfactory in operation over long periods.

It is generally considered that a fairly massive section of metal is needed immediately inside the periphery of the casing, to give good support to the bearing, in cases where the mass runs on its inner bore. It is usual for the bronze locating bush to be a tight press fit on to the inner mass of the casing, in order that expansion due to heat may not tend to free the bush.

(n) *Reservoirs of fluid.* In dampers of the non-welded type, the possibility of leakage has led to the inclusion of reservoirs of silicone fluid, in the form of internal grooves in the wall of the inner periphery of the casing. As these grooves are generally covered by the bush used for central location, the bush is provided with suitable passages between the fluid reservoir and the peripheral surface of the inner mass. Supply occurs as a result of centrifugal action during running. In view of the improved methods ensuring hermetically tight sealing of fabricated casings, it is possible that reservoirs may no longer be required for this purpose in future designs.

The provision of a fluid reservoir is not only, however, to counteract leakage. If the damper mass is guided at its inner bore and it is decided to leave an air space to allow for expansion, it is necessary to arrange a certain amount of space inside the rubbing surface, so that this surface shall always be lubricated. The fluid will flow outwards when the damper is rotating and the air space will occur in the inner groove or reservoir.

(*o*) *Performance during long periods of operation.* Viscous-shear dampers of a wide range of sizes have now proved their reliability over extensive periods of operation, running up to 10,000 hours without any reduction in effectiveness. During the first few years of development, the amplitude reductions obtained were between 2·5 : 1 and 3·5 : 1 for μ-ratios (inner mass inertia/equivalent engine inertia) of 0·75. At present, reductions of 4 : 1 to 7 : 1 are obtained for μ-ratios of 0·4–0·5; and 2·5 : 1 is achieved with a μ-value of 0·30. It should be borne in mind, however, that the amplitude reduction depends to some extent on the dynamic magnifier of the engine, so that, for instance, it is much easier to reduce the vibration of an engine with a magnifier of 40 than to achieve this if the engine magnifier is 10 or 15. (See also p. 530.)

(*p*) *Test rigs.* A first investigation of a viscous-shear damper built at the B.I.C.E.R.A. Laboratory was made in 1948, on a test rig designed as a two-mass system. This rig, with an excitation frequency equal to four times the running speed, had initially been developed for tests of a Sandner damper. The damper was mounted at one end of the system, while the second mass, of much larger inertia, was bolted at the other end to the exciter housing. The excitation torque was produced by the action of four spring-loaded guides with cam rollers, which moved radially with reciprocating motion in the exciter housing under the action of a four-peak sinusoidal cam. The torque was produced by the side thrust of the flanks of the cam on the rollers. The machine, with an additional flywheel in place of the damper, had a dynamic magnifier of 50 and produced a dynamic torque of 50,000 Lb.in., resulting in a vibration of $\pm 0.26°$ at the free end. The reduction in amplitude obtained with the first experimental viscous-shear damper was 3 : 1.

A more extensive investigation was made with a second test rig in 1951–2. A damper of smaller size (15 in. outer diameter) was used for these tests. The damper was vibrated by a radial arm oscillated by an eccentric through the medium of a connecting rod. The vibration amplitude could be altered by fitting eccentrics of different throws. The eccentric was belt-driven and the different frequencies were obtained by changing the motor speed and the pulley ratio.

The motor was mounted in a swinging frame and the torque reaction was measured by a spring-balance arrangement. This enabled the power input to be calculated. The vibration frequency was measured by taking tachometer readings at the end of the shaft carrying the eccentric. Vibration amplitudes were determined by means of a rigid metal blade mounted on the damper casing and oscillating between two adjustable insulated contacting screws connected to a battery and headphones.

Temperature measurements at four representative positions on the face and periphery of the damper were taken at suitable intervals by means of thermocouples.

Tests covering a wide range of vibration frequencies F and amplitudes θ, using fluids with nominal viscosities of 12,500 to 50,000 centistokes and various clearance values, were made with the damper and with solid masses. The first solid mass had a moment of inertia equal to that of the damper locked, whereas the other had an inertia approximately equal to that of the damper operating under optimum conditions. It was found that the frictional power absorbed by the rig system, for given values of F and θ, was not greatly affected by changing from one solid mass to the other. The relative motion of the inner mass was examined stroboscopically, the damper being fitted with a graduated glass casing for this purpose.

The results obtained in these tests served as a basis for the design procedure given in section 3·34 (pp. 537–41). In particular, it was found that

(1) the relative amplitude $\theta_{\mathrm{rel.}}$ of the inner mass was between 0·7 and 0·9 of that of the damper casing;†

(2) the temperature correction‡ was not sufficient to explain the reduced power absorption of the damper when operating at large amplitudes and high frequencies;

(3) the shear-rate correction, in addition to the temperature correction, gave good agreement between measured and calculated values of the entire range of frequencies and amplitudes investigated;

(4) in each of these tests, the damper reached equilibrium conditions only after a preliminary running period of about 2 hours; tests were carried out at normal ambient temperatures (10–20° C.) and at a higher temperature (with the casing heated to 58° C. by electric radiators before starting), and showed that the damper was capable of operating satisfactorily at temperatures which were 50° (C.) above the initial casing temperature under steady operating conditions;

(5) seizures were likely to occur if the temperature rise due to the energy dissipation by the fluid was greater than 55–60° (C.) regardless of the initial temperature of the casing; in view of this, it was inferred that for safe operation the power dissipated by dampers of this design should not exceed the values given by $H/A = 0·8$ h.p./ft.², where $A =$ total surface area of inner mass of damper.

---

† The inner mass was also observed to have a steady rotary motion inside the casing, making one complete revolution approximately every 4 min. when vibrating at 4500 vib./min. The corresponding motion per vibration cycle (i.e. $360/(4 \times 4500) = 0·02°$) was negligible. This relative rotary motion was also observed stroboscopically on a damper mounted on an engine.

‡ Taken from the curve of Fig. 11. This curve is based on published data and was also confirmed by falling-ball viscometer tests at the B.I.C.E.R.A. Laboratory.

In conclusion, it may be said that, unless the shear-rate effect is fully taken into account,† the effective viscosity of the fluid may be incorrectly estimated, and a fluid with an inadequate nominal viscosity may be chosen. The results then obtained usually show underdamped conditions. Successive trial-and-error tests with more viscous fluids may then lead to overdamped conditions. The use of the shear-rate

EXAMPLES OF DESIGNS OF VISCOUS-SHEAR DAMPERS

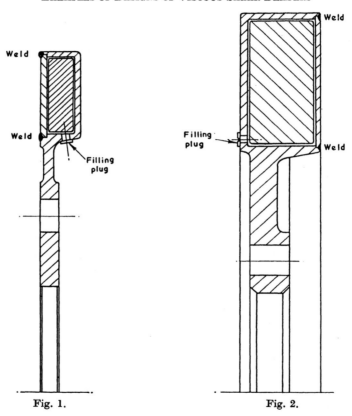

Fig. 1.   Fig. 2.

correction makes it possible to assess more closely the conditions approximating to optimum damping.

Damper proportions. For a specified value of moment of inertia, the inner mass of the damper may be designed either as a 'thin plate' with a large outer diameter or as a 'compact annulus', i.e. a very thick ring of smaller outer diameter. The first design takes up less space axially (see Figs. 1 and 2 in the following sketches given as examples); on the

† With the correction included under the integral sign for the lateral surfaces of the damper, as is the case with the procedure given in section 3·34.

[548]

other hand, where large amplitudes are expected even with the damper, the more compact design offers the possibility of reducing the shear-rate effect, since it is dependent on the damper radius (see Figs. 4, 5 and 6 given below). In practice, compromises are made between these two conditions.

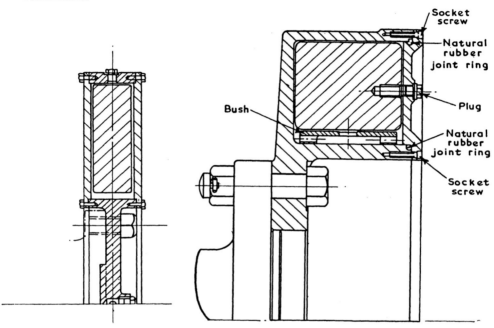

Fig. 3.

Fig. 4.

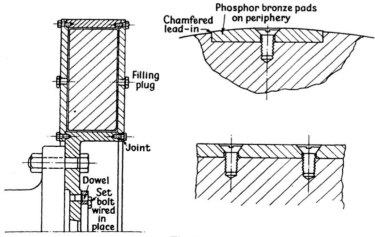

Fig. 5.

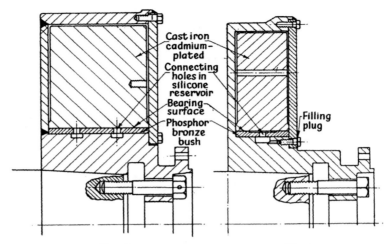

Fig. 6.   Fig. 7.

Cast iron
cadmium-
plated
Connecting
holes in
silicone
reservoir
Bearing
surface
Phosphor
bronze
bush

Filling
plug

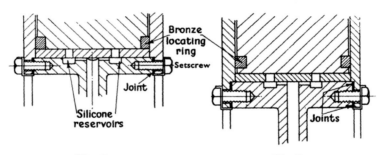

Bronze
locating
ring
Setscrew
Joint
Silicone
reservoirs
Joints

Fig. 8.   Fig. 9.

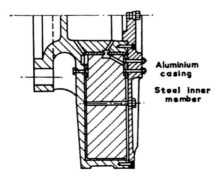

Aluminium
casing
Steel inner
member

Fig. 10.

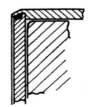

Fig. 11. Damper assembled by
machine rolling and peening of
edge of casing (Holset; Houdaille).

3·36 Derivation of relations used in design procedure

3·361 *Effect of shear rate on viscosity*

At low rates of shear, fluids generally obey Newton's law. That is, the shear stress q in a fluid of mass-density ρ and with a 'kinematic viscosity' coefficient ν (regarded as constant), between two surfaces with a clearance δ and moving with a relative velocity v, is determined by

$$q = 1\cdot55 \times 10^{-3}\nu\rho v/\delta \quad [\text{Lb./in.}^2], \tag{1}$$

where ν is expressed in centistokes (1 centistoke $= 0\cdot01$ cm.^{2}/sec. $= 0\cdot00155$ in.^{2}/sec.), ρ in Lb.sec.^{2}/in.^{4}, v in inches per second, and δ in inches.

The mass density of silicone fluids is approximately equal to that of water, i.e. $9\cdot34 \times 10^{-5}$ Lb.sec.^{2}/in.^{4}. Thus, for silicones, eq. (1) can be rewritten as

$$q = 1\cdot55 \times 10^{-3} \times 9\cdot34 \times 10^{-5} \times \nu v/\delta$$

or

$$q = 1\cdot448 \times 10^{-7} \times \nu v/\delta \quad [\text{Lb./in.}^2]. \tag{1'}\dagger$$

For high rates of shear, silicone fluids do not follow the law of eq. (1). Their kinematic viscosity decreases considerably with the rate of shear r, defined as

$$r = v/\delta \quad [\text{sec.}^{-1}]. \tag{2}$$

The information available in the first instance on this subject, in respect of silicone fluids, was a graph published by Georgian (see reference on p. 556) and subsequently confirmed by other investigators. For the purpose of the present calculations, mean values were taken for fluids with nominal viscosities ν_0 of 25,000 and 30,000 centistokes and the results were then converted so as to give relative values of viscosity as percentages of the nominal value. The resultant curve thus obtained is included in Fig. 10, p. 540. It may be noted that for 12,500 centistokes the shear-rate effect is less marked. For silicones of extremely low viscosity, i.e. of the order of 1 centistoke (which have no application in these dampers), the shear-rate effect is practically negligible.

3·362 *Viscous damping coefficient*

The damping coefficient may be defined by the relation: $T_0 = c\dot{\theta}_{r\,\text{max.}}$, where T_0 is the maximum torque during an oscillation, c is the damping coefficient, and $\dot{\theta}_{r\,\text{max.}}$ is the maximum relative angular velocity. Now $\theta_r = \theta_{ro}\sin\omega t$, where θ_r is the instantaneous relative vibration amplitude and θ_{ro} the full relative vibration amplitude between inner member and

---

$\dagger$ *Metric units.* When v and δ are in cm., and $\rho = 1\cdot019 \times 10^{-6}$ Kg.sec.^{2}/cm.^{4}, eq. (1') becomes $q = 1\cdot019 \times 10^{-8}\nu v/\delta$ [Kg./cm.^{2}].

Hence, for all the formulae in this section, metric values in Kg. and cm. are obtainable by using the numerical coefficient $1\cdot019 \times 10^{-8}$ instead of $1\cdot448 \times 10^{-7}$. (Since v/δ is in sec.$^{-1}$, it is evident that one also has $1\cdot448 \times 10^{-7}/1\cdot019 \times 10^{-8} = 14\cdot22$.)

casing, while $\omega = 2\pi F/60$, F being the vibration frequency in vib./min. The damping coefficient is therefore

$$c = T_0/(\omega\theta_{ro}) \quad \text{[Lb.in.sec./rad.]}. \tag{3}$$

The damping torque dT due to an element of area dA situated at a radius R from the axis of vibration is given by

$$dT = RqdA = RdA \times 1{\cdot}448 \times 10^{-7} \times \nu v/\delta, \tag{4}$$

using eq. (1'). For a very precise determination, the changes in the value of ν due to the variation of shear rate throughout the cycle would be taken into account by a step-by-step method. In practice, it is satisfactory to use the mean cyclic value of the shear rate, which is

$$(2/\pi) \times r_{\text{max.}} = 0{\cdot}637 v_{\text{max.}}/\delta.$$

The damping coefficient c_P for the outer periphery of the damper, with a width L and a radial clearance δ_P at the radius R_o from the centre, is therefore
$$c_P = T_0/\omega\theta_{ro} = R_o \times 1{\cdot}448 \times 10^{-7} \times \nu v A/(\delta_P \omega\theta_{ro}).$$

As $v = R\theta_{r\,\text{max.}}$ and $A = 2\pi R_o L$, this gives

$$c_P = 1{\cdot}448 \times 10^{-7} \times 2\pi\nu L R_o^3/\delta_P. \tag{5'}$$

The damping coefficient for the inner periphery is negligible if the damper is designed to obtain a fairly large area of the side faces as is usual. Where necessary, it may be taken into account by writing $R_o^3 + R_i^3$ instead of R_o^3 in eq. (5').

The coefficient c_L for the two lateral annular surfaces of the damper, with a lateral clearance δ_L and inner and outer radii R_i and R_o, is

$$c_L = 2T_0/\omega\theta_{ro} = 2 \times 1{\cdot}448 \times 10^{-7} \times \int_{R_i}^{R_o} \left(\frac{\nu}{\omega}\right) \times dR \times 2\pi R \times \omega R \times R/\delta_L$$

$$= \frac{4\pi \times 1{\cdot}448 \times 10^{-7}}{\delta_L} \int_{R_i}^{R_o} R^3 \nu \, dR.$$

From Fig. 10, p. 540, it can be shown that the kinematic viscosity coefficient ν varies with the rate of shear according to the relation

$$\nu = \nu_0 \left(\frac{10}{r}\right)^n = \nu_0 \left(\frac{10\delta_L}{\omega R\theta_{ro}}\right)^n,$$

where $\nu_0 =$ viscosity at very low shear rate, i.e. as measured by falling-ball viscometer,

$n = 0{\cdot}158$ for shear rates up to 700 sec.$^{-1}$, and

$n = 0{\cdot}424$ for shear rates from 700 to 10,000 sec.$^{-1}$

for the silicone fluids generally considered (30,000–60,000 centistoke fluids).

[552]

To take full account of the varying shear rate over the lateral surfaces of the damper, the above expression for ν is introduced under the integral sign, and the integration gives

$$c_L = \frac{4\pi \times 1 \cdot 448 \times 10^{-7} \times \nu_0}{\delta_L} \times \left(\frac{10\delta_L}{\omega\theta_{ro}}\right)^n \times \int_{R_i}^{R_o} R^{3-n}\, dR$$

$$= \frac{4\pi}{4-n} \times \frac{1 \cdot 448 \times 10^{-7} \times \nu_0}{\delta_L(\omega\theta_{ro}/10\delta_L)^n} \times [R_o^{4-n} - R_i^{4-n}].$$

But

$$\nu_0/(\omega\theta_{ro}R_o/10\delta_L)^n = \nu_P$$

is the effective viscosity at the periphery taking into account the mean rate of shear. Thus,

$$c_L = \frac{\pi}{1-(n/4)} \times 1 \cdot 448 \times 10^{-7} \times \nu_P \times \frac{R_o^4}{\delta_L} \times \left[1 - \left(\frac{R_i}{R_o}\right)^{4-n}\right].$$

The calculation is simplified by introducing an equivalent radius R_e determined as follows:

$$c_L = \pi \times 1 \cdot 448 \times 10^{-7} \times \nu_P \times \frac{R_e^4}{\delta_L}$$

$$= \frac{\pi}{1-(n/4)} \times 1 \cdot 448 \times 10^{-7} \times \nu_P \times \frac{R_o^4}{\delta_L} \times [1 - (R_i/R_o)^{4-n}].$$

Then

$$R_e^4 = R_o^4[1 - (R_i/R_o)^{4-n}]/[1 - (n/4)]$$

or

$$R_e^4 = R_o^4 \eta_R,$$

where η_R is the side surface correction factor. Values of η_R for various values of the mean shear rate r_m and the radius ratio R_i/R_o are given in the table of p. 539. Thus c_L is determined as

$$c_L = \pi \times 1 \cdot 448 \times 10^{-7} \times \nu_P \times \eta_R R_o^4/\delta_L. \tag{5''}$$

The total damping coefficient is

$$c = c_L + c_P = c_P[(c_L/c_P) + 1].$$

If the peripheral and lateral clearances are made equal, the value of $c_L/c_P = \eta_R R_o/2L$, from eqs. (5') and (5''), so that, given the required value of c from the relation $c = \gamma \times 2J_{1''}\omega_2$ (see pp. 522 and 525), c_P may be determined from

$$c_P = \frac{c}{(c_L/c_P) + 1} = \frac{c}{(\eta_R R_o/2L) + 1},$$

and then $c_L = c - c_P$. Using this result in conjunction with eq. (5''), or the value of c_P with eq. (5'), one can obtain the effective viscosity ν. The nominal viscosity is determined as $\nu_0 = \nu/(\eta_r \eta_T)$ with the aid of the graphs of Figs. 10 and 11, p. 540.

3·363 *Relative amplitude and mean shear rate*

Referring to the two-mass system formulae of p. 524 and using the subscripts I and C for the inner mass and the casing, respectively, we have

$$\text{Det}_I = T_0 j\omega c, \quad \text{Det}_C = T_0(-\omega^2 J_I + j\omega c), \quad \text{Det}_I - \text{Det}_C = T_0 \omega^2 J_I.$$

Hence the vibration amplitude of the inner mass relative to the casing amplitude is

$$\theta_r/\theta_C = (\theta_I - \theta_C)/\theta_C = \omega^2 J_I/(-\omega^2 J_I + j\omega c) = 1/[-1 + j(c/J_1\omega)];$$

taking absolute values

$$|\theta_r| = |\theta_C|/\sqrt{\{1 + (c/J_I\omega)^2\}}.$$

For optimum damping $2(1+\mu)(2+\mu) = 1/\gamma^2 = (2J_I\omega_2)^2/c^2$ so that, with $\omega^2 = 2\omega_2^2/(2+\mu)$:

$$1+\mu = \omega^2 J_I^2/c^2 \quad \text{and} \quad \omega J_I/\sqrt{(1+\mu)} = c.$$

With this expression, the relative amplitude becomes

$$|\theta_r| = \sqrt{\left(\frac{1+\mu}{2+\mu}\right)} \times |\theta_C|.$$

For $\mu = 0.5$:
$$|\theta_r| \cong 0.77 |\theta_C|.$$

The mean shear rate is obtained from

$$r_m = \frac{\omega}{\pi}\int_0^{\pi/\omega} r(t)\,dt = \frac{\omega}{\pi}\int_0^{\pi/\omega}\left(\frac{\omega R_0}{\delta}\right) \times |\theta_r|\sin\omega t\,dt = (2/\pi)\times r_{\max}.$$

$$= 0.637 \times (R_o\omega/\delta) \times |\theta_r| = 0.637 \times \sqrt{\left(\frac{1+\mu}{2+\mu}\right)} \times (R_o\omega/\delta) \times |\theta_C|.$$

For $\mu = 0.5$:
$$r_m \cong 0.49 \times (\omega R_o/\delta) \times |\theta_C|.$$

3·364 *Energy and power dissipated by the damper*

The work done by the damper per vibration cycle is, under 'optimum' conditions:

$$U_{D\,\text{opt.}} = \pi c\omega |\theta_r|^2 = \pi\omega^2 J_I \times \frac{1}{\sqrt{(1+\mu)}} \times \frac{(1+\mu)}{(2+\mu)} \times |\theta_C|^2$$

$$= \frac{\pi\sqrt{(1+\mu)}}{2+\mu}\omega^2 J_I |\theta_C|^2.$$

For $\mu = 0.5$:
$$U_{D\,\text{opt.}} = 0.49\pi\omega^2 J_I |\theta_C|^2.$$

The derivation of the expression $U_D = \pi c\omega |\theta_r|^2$, where θ_r is the relative amplitude between the damper casing and the inner mass, is given at the end of this section (p. 558).

Consider now the two cases of the damper operating under optimum and non-optimum conditions.

[554]

Let c_0 = damping coefficient obtained with a fluid at 25°C. and at zero shear rate, and

c = damping coefficient occurring under operating conditions $(c = \eta_r \eta_T c_0)$.

Then the energy dissipated by the damper per vibration cycle is†

$$U_D = \pi c \omega \mid \theta_r \mid^2 = \pi \eta_r \eta_T c_0 \omega \mid \theta_r \mid^2$$

$$= \pi \eta_r \eta_T c_0 \omega \mid \theta_C \mid^2 / [1 + (\eta_r \eta_T c_0 / J_I \omega)^2] \quad \text{[in.Lb./cycle]}.$$

If the value of $\eta_r \eta_T c_0$ is that required for optimum damping (that is, if $\eta_r \eta_T c_0 = \omega J_I / \sqrt{(1+\mu)}$), this equation reduces to

$$U_{D\,\text{opt.}} = \frac{\pi \sqrt{(1+\mu)}}{2+\mu} \omega^2 J_I \mid \theta_C \mid^2.$$

The power dissipated is

$$H_D = \frac{(60/2\pi) \times \omega}{33{,}000 \times 12} \times U_D = \frac{1}{13{,}200} \times \frac{\eta_r \eta_T c_0 \omega^2 \mid \theta_C \mid^2}{[1 + (\eta_r \eta_T c_0 / J_I \omega)^2]} \quad \text{[h.p.]}$$

under conditions which may or may not correspond to optimum damping. If optimum damping is obtained, the equation can be rewritten as

$$H_{D\,\text{opt.}} = \omega^3 J_I \mid \theta_C \mid^2 \times \sqrt{(1+\mu)} / [13{,}200 \times (2+\mu)] \quad \text{[h.p.]}.$$

If $\mid \theta_C \mid$ is in degrees, and ω is replaced by the vibration frequency $F = (60/2\pi) \times \omega$, these equations become

$$H_D = 2 \cdot 53 \times 10^{-10} \eta_r \eta_T c_0 F^2 \mid \theta_C \mid^2 / [1 + (9 \cdot 55 \eta_r \eta_T c_0 / J_I F)^2] \quad \text{[h.p.]}$$

and $\quad H_{D\,\text{opt.}} = 2 \cdot 65 \times 10^{-11} J_I F^3 \mid \theta_C \mid^2 \times \sqrt{(1+\mu)} / (2+\mu) \quad \text{[h.p.]}.$

3·365 Energy equilibrium

(1) 'Optimum' damping. The work input is $U_{in} = \pi T_0 \mid \theta_C \mid \sin \phi$, where ϕ = phase angle between $\mid \theta_C \mid$ and the excitation torque T_0. Considering the two-mass system of p. 522, one can write the equation of motion of the engine inertia J_E (which vibrates with the same amplitude as the damper casing) as follows:

$$(K - \omega^2 J_E) \theta_C + j\omega c(\theta_C - \theta_I) = T_0.$$

As $(\theta_I - \theta_C) = \theta_C \omega^2 J_I / (-\omega^2 J_I + j\omega c)$, from p. 554, it follows that

$$(K - \omega^2 J_E) \theta_C - j\omega c \theta_C / [-1 + j(c/J_I \omega)] = T_0. \tag{1}$$

† The casing and the inner mass perform sinusoidal vibrations with the same phase velocity ω but different amplitudes θ_C and θ_I. Their relative phase angle ϵ is constant. Therefore the relative amplitude $\theta_r = \theta_I - \theta_C$ is also a sinusoidal function, viz.

$$\theta_r = \sqrt{(\theta_I^2 + \theta_C^2)} \sin (\omega t - \epsilon).$$

Dividing by K throughout, and using the relations

$$\omega^2 J_E/K = \omega^2/\omega_2^2 = 2/(2+\mu), \quad c/\omega J_I = 1/\sqrt{(1+\mu)},$$

$$\omega c/K = \omega c/\omega_2^2 J_E = \omega c \bigg/ \left[\frac{2+\mu}{2}\,\omega^2 J_I \times \frac{1}{\mu}\right] = 2\mu/[(2+\mu)\sqrt{(1+\mu)}];$$

one obtains

$$T_0/K = \theta_C \times \left[1 - \frac{2}{2+\mu} + j\,\frac{2\mu}{(2+\mu)\sqrt{(1+\mu)}} \times \frac{1+j\{1/\sqrt{(1+\mu)}\}}{1+\{1/(1+\mu)\}}\right],$$

which reduces to

$$T_0/K = \theta_C[\mu^2 + j2\mu\sqrt{(1+\mu)}]/(2+\mu)^2.$$

Hence

$$\theta_C = \frac{T_0}{K} \times \frac{(2+\mu)^2[\mu^2 - j2\mu\sqrt{(1+\mu)}]}{\mu^4 + 4\mu^2(1+\mu)} = \frac{T_0}{K}\left[1 - j\,\frac{2}{\mu}\sqrt{(1+\mu)}\right].$$

Therefore, the phase angle of θ_C is obtainable from

$$\sin\phi = \frac{-\dfrac{2}{\mu}\sqrt{(1+\mu)}}{\sqrt{\left\{1 + \dfrac{4}{\mu^2}(1+\mu)\right\}}} = \frac{-2\sqrt{(1+\mu)}}{2+\mu}. \tag{2}$$

Thus, the work input is

$$U_{in} = \pi T_0\,|\,\theta_C\,| \times 2\sqrt{(1+\mu)}/(2+\mu).$$

From p. 554, the energy dissipated by the damper is (with

$$\omega^2 = \omega_2^2 \times 2/(2+\mu))$$

$$U_D = \frac{\pi\sqrt{(1+\mu)}}{2+\mu}\,\omega^2 J_I\,|\,\theta_C\,|^2 = \frac{\pi\sqrt{(1+\mu)}}{2+\mu}\,\omega^2 \mu J_E\,|\,\theta_C\,|^2 = 2\pi\mu\,\frac{\sqrt{(1+\mu)}}{(2+\mu)^2}\,K\,|\,\theta_C\,|^2,$$

so that for energy equilibrium $(U_D = U_{in})$

$$|\,\theta_C\,| = T_0(2+\mu)/K\mu,$$

and then

$$M = |\,\theta_C\,|/(T_0/K) = (2+\mu)/\mu, \tag{3}$$

in agreement with previously obtained results.

These results show that the 'phase-reducing' effect (and hence the corresponding reduction of work input, due to $\sin\phi$) is fairly small in most cases. They also indicate that, if the phase angle has to be calculated, the energy method for determining $|\,\theta_C\,|$ involves more calculations than the direct determination of $|\,\theta_C\,|$ from the equations of motion:

Numerical values of phase reduction (eq. (2)):	$\mu =$	0	0·2	0·4	0·6	0·8	1·0	2·0
	$\sin\phi =$	1·0	0·996	0·986	0·973	0·958	0·943	0·866

(2) *'Maximum' damping.*† It is of interest to compare the results obtained by the following method, which is commonly used, with those derived under (1) above.

† Georgian, G. C., 'Torsional viscous-friction dampers', *Trans. Amer. Soc. Mech. Engrs*, May 1949, pp. 389–99.

The energy dissipated by the damper is

$$U_D = \pi c \omega \, |\, \theta_r \,|^2,$$

where
$$|\, \theta_r \,| = |\, \theta_C \,| / \sqrt{\{1 + (c/J_I \omega)^2\}}.$$

From $\partial U_D / \partial c = 0$, a maximum value of U_D is obtained, when

$$c = J_I \omega. \qquad (4)$$

This gives
$$|\, \theta_r \,|_{\text{max.}} = |\, \theta_C \,| / \sqrt{2} = 0 \cdot 707 \, |\, \theta_C \,|$$

and
$$U_D = (\pi/2) \times c \omega \, |\, \theta_C \,|^2 = (\pi/2) \, J_I \omega^2 \, |\, \theta_C \,|^2.$$

Equating this to the work input, with $\sin \phi = 1$,

$$\pi T_0 \, |\, \theta_C \,| \sin \phi \cong \pi T_0 \, |\, \theta_C \,| = (\pi/2) \, J_I \omega^2 \, |\, \theta_C \,|^2,$$

one obtains the value
$$|\, \theta_C \,|_{\text{max.}} = 2 T_0 / J_I \omega^2. \qquad (5)$$

With 'optimum' damping, on the other hand, the peak amplitude at the damper casing is

$$|\, \theta_C \,|_{\text{opt.}} = \frac{T_0}{K} \times \frac{2 + \mu}{\mu} = \frac{T_0}{\omega_2^2 J_E} \times \frac{2 + \mu}{\mu} = \frac{T_0}{\omega_2^2 J_I} (2 + \mu),$$

or, since $\omega_2^2 = (2 + \mu) \, \omega^2 / 2$,

$$|\, \theta_C \,|_{\text{opt.}} = 2 T_0 / J_I \omega^2. \qquad (6)$$

This is exactly the same result as with 'maximum' damping. It is therefore necessary to consider whether the two methods differ in any other respects.

It can be seen that the requirement $\partial U_D / \partial c = 0$ given above is only a condition for obtaining maximum energy dissipation at a value $|\, \theta_C \,|$, which may be a peak or a flank amplitude of the resonance curve with damping. In fact, with $c = J_I \omega$, it is a point somewhere on the descending flank. To obtain the peak, it is necessary to determine first the horizontal tangent (see p. 525); the 'optimum' damping in its vicinity is then obtained as $c_{\text{opt.}} = \omega J_I / \sqrt{(1 + \mu)}$.

Another point is that $|\, \theta_C \,|$ is also a function of c (see eq. (1), p. 555) and should not actually be treated as a constant in $\partial U_D / \partial c$. These various considerations indicate that $c = J_I \omega$ is an approximation, valid only for small μ-values. For dampers with fairly large μ-values, $c_{\text{opt.}}$ is preferable, particularly in cases where the minimum peak amplitude is of importance.

It may be noted that it has been possible to show on engine systems fitted with viscous-shear dampers that the optimum value of damping can be reached and exceeded; in the latter case, the corresponding peak amplitude is then higher than the optimum value.

Derivation of expression for energy dissipation of damper

The expression $\pi c\omega \,|\,\theta_r\,|^2$ for the work dissipated by the damper is based on the following considerations: The casing and the inner mass vibrate with different amplitudes $|\,\theta_C\,|$ and $|\,\theta_I\,|$, but their relative phase angle β is constant. One may write

$$\theta_C = |\,\theta_C\,| \sin\omega t \quad \text{and} \quad \theta_I = |\,\theta_I\,| \sin(\omega t - \beta).$$

Therefore the relative amplitude $\theta_r = \theta_I - \theta_C$ is also sinusoidal and the energy dissipated in damping is given by the above expression. The detailed calculation is as follows:

$$\begin{aligned}
\theta_r = \theta_I - \theta_C &= |\,\theta_I\,| \sin\omega t \cos\beta - |\,\theta_I\,| \cos\omega t \sin\beta - |\,\theta_C\,| \sin\omega t \\
&= (|\,\theta_I\,| \cos\beta - |\,\theta_C\,|) \sin\omega t - |\,\theta_I\,| \sin\beta \cos\omega t \\
&= |\,\theta_r\,| \cos\epsilon \sin\omega t - |\,\theta_r\,| \sin\epsilon \cos\omega t = |\,\theta_r\,| \sin(\omega t - \epsilon).
\end{aligned}$$

Thus θ_r varies sinusoidally with time and its phase angle ϵ, relative to θ_I, is

$$\epsilon = \tan^{-1}[|\,\theta_I\,| \sin\beta/(|\,\theta_I\,| \cos\beta - |\,\theta_C\,|)].$$

The energy dissipated by the damper per vibration cycle is (with $c\dot\theta_r = $ damping torque)

$$\begin{aligned}
U_D &= 2\int_{-\epsilon/\omega}^{(\pi-\epsilon)/\omega} c\dot\theta_r \frac{d\theta_r}{dt} \, dt = 2c\int \dot\theta_r^2 \, dt \\
&= 2c\int |\,\theta_r\,|^2 \omega^2 \times \cos^2(\omega t - \epsilon) \, dt = 2c\omega^2 \,|\,\theta_r\,|^2 \int_{-\epsilon/\omega}^{(\pi-\epsilon)/\omega} \tfrac{1}{2}[1 + \cos 2(\omega t - \epsilon)] \, dt \\
&= c\omega^2 \,|\,\theta_r\,|^2 \times \left\{ \frac{\pi}{\omega} + [\sin 2(\pi - 2\epsilon) - \sin 2(-2\epsilon)] \times \frac{1}{2\omega} \right\} \\
&= \pi c\omega \,|\,\theta_r\,|^2.
\end{aligned}$$

3·4 SLIPPING-TORQUE TYPE DAMPERS

The dampers considered in the following are

the Sandner damper (with pumping chambers or gearwheels),

the Atlas damper (pumping-chamber type), and

the Lanchester damper (semi-dry friction type).

Although these dampers are considerably different in their design features, they all operate by means of a slipping torque and consequently the calculations are similar for all dampers of these types.

NOTATION

Symbol	Brief definition	Typical units	Symbol	Brief definition	Typical units
a	area	in.², cm.²	R	radius	in., cm.
A	area	in.², cm.²	t	time	sec., min.
c	damping coefficient	Lb.in.sec./ rad., Kg.cm.sec./ rad.	T	torque	Lb.in., Kg.cm.
			U	work per cycle	in.Lb./cycle, cm.Kg./cycle
J	moment of inertia	Lb.in.sec.², Kg.cm.sec.²	V	volume	in.³, cm.³
			α	damper parameter	1/(Lb.in.), 1/(Kg.cm.)
K_s	linear stiffness	Lb./in.², Kg./cm.²	β	angle	deg., rad.
			δ	deflexion	in., cm.
L	length	in., cm.	Δ	Holzer-table amplitude	rad.
N_s	number of springs	—			
p	pressure	Lb./in.², Kg./cm.²	θ	vibration amplitude	deg., rad.
			μ	inertia ratio	—
P	force	Lb., Kg.	μ	friction coefficient	—
q	shear stress	Lb./in.², Kg./cm.²	$\overrightarrow{\Sigma\Delta}$	phase-vector sum	—
			τ	time	sec., min.
r	radius	in., cm.	ω	phase velocity	rad./sec.

Subscripts are used in the text to define more precisely. Their meaning is fully explained in the context in each instance.

3·41   The Sandner damper (pumping-chamber type)

This damper (Fig. 1) consists of a hub or primary inner member, fixed to the free end of the crankshaft, and a rim or secondary inertia member (seismic mass) which rotates in unison with the hub as long as its relative acceleration lies below a predetermined value, corresponding to the operating pressure of a spring-loaded relief valve.

Beyond that value, relative movement occurs; the moment of inertia of the unlocked damper will vary, thus giving a detuning effect by changing the natural frequency of the system twice during each oscillation. At the same time vibratory energy is dissipated in the form of heat as a result of the pumping action taking place in the oil chambers arranged within the damper.

The side plates form part of the rim or seismic mass. Two sets of flat blades (shown in Fig. 4) are used to locate the rim, their arrangement being such that the torsional rigidity of the blades plays no part in the functioning of the component as a damper and does not alter the dynamic characteristics of the system. The slipping torque can be accurately set by adjustment of a spring-loaded relief valve in the central portion of the hub. The oil circulation is through the crankshaft into the hub and through radial passages into the pumping chambers.

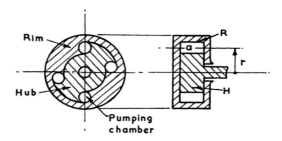

Fig. 1.

The quantity of oil in circulation is very small and oil is passed into the suction space of the valve gear after discharge through the relief valve. A non-return valve is incorporated in the oil supply.

The damper is capable of operating indefinitely on major critical speeds without wear or overheating and without any alteration occurring to the slipping torque. As an example of its effectiveness, it may be mentioned that in a typical installation consisting of a supercharged eight-cylinder four-stroke cycle engine of $17\cdot4 \times 18\cdot5$ in. (bore × stroke) coupled to a generator, the amplitude reduction obtained was 8 : 1 for an inertia ratio $\mu = J_D/J_E \simeq 0\cdot6$ (J_D and J_E being the equivalent inertias of the damper and the engine, respectively). The $3\frac{1}{2}$, 4th, $4\frac{1}{2}$ and $5\frac{1}{2}$ order criticals were completely eliminated.

Basic design procedure

(1) Determine the necessary moment of inertia J_R of the seismic mass (i.e. the rim) of the damper, for the required vibration amplitude θ_1 [rad.] at the damper hub, by means of the relation

$$J_R = \frac{\pi^2 T_0}{4\omega^2 \theta_1} \quad \text{[Lb.in.sec.}^2\text{]},$$

[560]

where $T_0 = |T_n| AR_0 \overrightarrow{\Sigma\Delta}$ = excitation torque of engine without damper [Lb.in.], and ω = natural phase velocity [rad./sec.] of engine without damper.

Note. This first value for J_R can be improved by evaluating a Holzer table with $J_D \cong 1 \cdot 5J_R$ (the hub inertia being taken as $0 \cdot 5J_R$) as mass no. 1. The tabulation gives a more representative value for ω and a better value for the phase-vector sum $\overrightarrow{\Sigma\Delta}$ for the engine system with damper. These two values can then be used in the above formula to obtain a second approximation for J_R.

(2) Taking account of the space available at the free end of the engine, decide on suitable values for the outer radius R_o (see Fig. 2), the inner radius R_i and the thickness L of the outer rim.

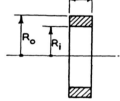

For a first approximation, the value of J_R/L can be related to values of R_o by means of the graph of p. 30, section 1·1. For typical proportions, see Fig. 4, p. 564.

(3) Calculate the maximum torque T_R, at which slipping occurs, as

Fig. 2.

$$T_{R\max.} = \frac{\sqrt{2}}{\pi} \times \omega^2\theta_1 J_R \quad \text{[Lb.in.]}.$$

(4) Determine the critical pressure p (i.e. the spring pressure of the oil valve required for optimum operation) as

$$p = T_{R\max.}/(2ra) \quad \text{[Lb./in.}^2\text{]},$$

where a = cross-sectional area [in.^{2}] of one of the pumping chambers, and r = mean radius [in.] to a from the centre of the hub. The factor 2 is used because only two of the four pumping chambers are effective in each direction.

(5) Design the hub (i.e. the inner member) in accordance with the general indications in Fig. 4. It may be noted that even for fairly large dampers of this type, the radial clearance between the hub periphery and the outer ring is only of the order of a few thousandths of an inch.

Note. If the damping of the driven machine and the hysteresis damping of the shafting are appreciable, these may also be taken into account as follows:

Let $U_{in} = \pi T_0 \theta_1$ [in.Lb./cycle] = work input of the engine,

$U_G = \pi c_G \omega \Delta_G^2 \theta_1^2$ = energy dissipation of the driven machine,

$U_H = 0 \cdot 64 \times 10^{-10} \times \Sigma(q_s^{2 \cdot 3} \times V)$ = energy dissipated by shaft hysteresis, and

$U_D = \frac{4}{\pi} \omega^2\theta_1^2 \times J_R$ = energy to be dissipated by damper.

Then for energy equilibrium

$$U_{in} = U_D + U_G + U_H;$$

hence if θ_1 is a specified value,

$$J_R = [U_{in} - U_G - U_H] \bigg/ \left(\frac{4}{\pi} \omega^2 \theta_1^2 \right) \quad \text{[Lb.in.sec.}^2\text{]}.$$

For values of the damping coefficient c_G for driven machines, see section 2·4211, p. 367. The hysteresis damping evaluation is based on the Lewis-Rowett formula, with q_s = shear stress [Lb./in.^{2}] and V = volume [in.^{3}] of each shaft section considered, see section 2·411, p. 358.

Metric units. The formulae in this section are also valid in metric units (Kg. and cm.); in the equation for U_H, however, if q is in Kg./cm.^{2} and V in cm.^{3}, the numerical coefficient is $1·53 \times 10^{-9}$ instead of $0·64 \times 10^{-10}$, the result is then in cm.Kg./cycle.

Derivation of formulae

The following theory applies to the case where the damper motion is such that slipping is continually taking place; with a lower hub velocity, however, it is possible for slipping to be intermittent.

The calculation of the Sandner damper is similar to that of the Lanchester dry-friction damper, for which the theory was developed by Den Hartog and Ormondroyd.†

Let ω = natural phase velocity of torsional vibration of the engine system without damper; $\dot{\theta}_H$ and $\dot{\theta}_R$ = angular velocities of hub and rim, respectively, due to torsional vibration only.

The motion of the hub is sinusoidal, and the motion of the rim is assumed to vary linearly with time (see Fig. 3).

Putting $t = 0$ when $\dot{\theta}_H = \dot{\theta}_{H\,\text{max.}}$, $t = t_1$ when $\dot{\theta}_H = \dot{\theta}_R > 0$, and $t = t_2$ when $\dot{\theta}_H = \dot{\theta}_R < 0$, we can then represent the motion of the rim and hub in the time interval $t_1 \leqslant t \leqslant t_2$ by

$$\left. \begin{aligned} \theta_H &= \theta_1 \sin \omega t, & \ddot{\theta}_R &= -T_R/J_R = \tan \beta, \\ \dot{\theta}_H &= \omega \theta_1 \cos \omega t, & \dot{\theta}_R &= +T_R(t_3 - t)/J_R, \\ \ddot{\theta}_H &= -\omega^2 \theta_1 \sin \omega t, & \theta_R &= T_R(t_3 - t)^2/2J_R, \end{aligned} \right\} \quad (1)$$

where $\dot{\theta} = d\theta/dt$, J_R = rim inertia, and T_R = accelerating torque acting on the rim. The time t_3 is the instant at which $\dot{\theta}_R = 0$, hence

$$t_3 = t_1 + (\tau/4) = t_1 + (\pi/2\omega),$$

since $\omega = 2\pi/\tau$, with τ = period of vibration cycle.

Slip, defined as
$$\theta_H - \theta_R = \int (\dot{\theta}_H - \dot{\theta}_R)\,dt, \quad (2)$$

occurs continuously throughout the entire cycle, and reverses its direction when $\dot{\theta}_R = \dot{\theta}_H$, at times t_1 and t_2. Slip occurs therefore twice per cycle. The energy dissipation per cycle of the damper is therefore given by

$$U_D = 2 \int_{t_1}^{t_2} T_R(\dot{\theta}_R - \dot{\theta}_H)\,dt,$$

† For bibliographic references, see p. 565.

[562]

and, using eq. (1),

$$U_D = 2T_R \int_{t_1}^{t_3} \left[\frac{T_R}{J_R}(t_3 - t) - \theta_1 \omega \cos \omega t \right] dt = 2T_R \left[\frac{T_R}{2J_R}(t_3 - t)^2 - \theta_1 \sin \omega t \right]_{t_1}^{t_3},$$

and since

$$t_3 - t_2 = -\tau/4, \quad t_3 - t_1 = +\tau/4, \quad t_2 = t_1 + (\tau/2) = t_1 + (\pi/\omega)$$

we have

$$U_D = 2T_R \left[\frac{T_R}{2J_R}\{(-\tau/4)^2 - (\tau/4)^2\} - \theta_1 \sin(\omega t_1 + \pi) + \theta_1 \sin \omega t_1 \right],$$

so that

$$U_D = 4T_R \theta_1 \sin \omega t_1. \tag{3}$$

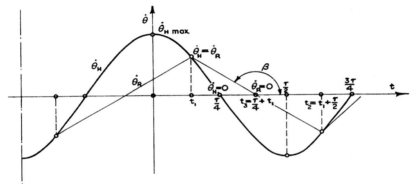

Fig. 3.

Furthermore, when $t = t_1$, $\dot{\theta}_R = \dot{\theta}_H$, or from eq. (1),

$$T_R(t_3 - t_1)/J_R = \pi T_R/(2\omega J_R) = \omega \theta_1 \cos \omega t_1,$$

hence

$$\sin \omega t_1 = \sqrt{(1 - \alpha^2 T_R^2)},$$

where $\alpha = \pi/(2\omega^2 \theta_1 J_R)$ and

$$U_D = 4T_R \theta_1 \sqrt{(1 - \alpha^2 T_R^2)}. \tag{4}$$

The maximum value of U_D is determined from

$$\partial U_D/\partial T_R = 0, \quad T_{R\,\text{max.}} = \pm 1/(\sqrt{2}\,\alpha),$$

so that

$$T_{R\,\text{max.}} = \sqrt{2}\,\omega^2 \theta_1 J_R/\pi \quad \text{and} \quad U_{D\,\text{max.}} = \frac{4}{\pi}\omega^2 \theta_1^2 J_R. \tag{5}$$

As the work input of the engine is $U_{in} = \pi T_0 \theta_1$ (where $T_0 = |T_n| AR_0 \overrightarrow{\Sigma\Delta} =$ excitation torque), the condition for energy equilibrium $(U_{in} = U_{D\,\text{max.}})$ gives J_R as a function of θ_1, as indicated in the basic design procedure (p. 560).

36-2

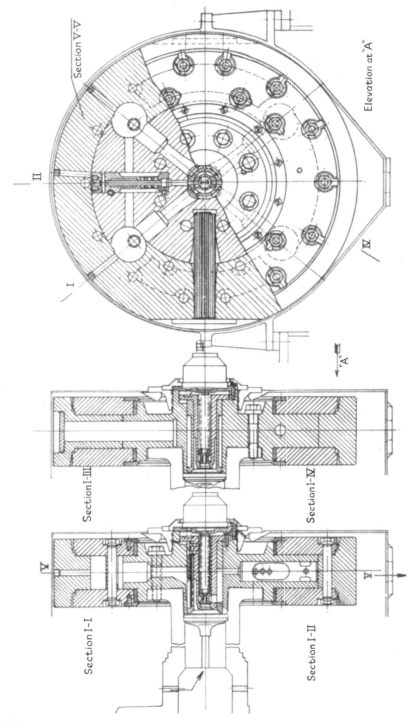

Section V-V

Elevation at "A"

II

I

Section I-III

Section I-IV

"A"

IV

V

Y

Section I-I

Section I-II

Fig. 4a. The Sandner pumping-chamber type damper.

Fig. 4b.　The SANDNER gear type damper.
(By courtesy of Vickers-Armstrongs (Engineers), Ltd.)

BIBLIOGRAPHY

Den Hartog, J. P. and Ormondroyd, J. *Trans. Amer. Soc. Mech. Engrs*, vol. 52 (1930), p. APM-133.
Jendrassik, G. *Z. Ver. dtsch. Ing.* 16 Sept. 1933, p. 1009.
Ker Wilson, W. *Practical Solution of Torsional Vibration Problems*, vol. 2, p. 467.
Klotter, K. *Ingen.-Arch.* vol. 9 (1938), p. 137.
Sandner, E. and Barraja-Frauenfelder, J. *S. A. E. Jl*, vol. 26, semi-annual meeting paper, 1931, p. 352.
Timoshenko, S. *Vibration Problems in Engineering*, 2nd ed., pp. 274–6.
Tuplin, W. A. *Torsional Vibration* (1934), pp. 206–17.

3·42   The Sandner damper (gearwheel type)

The main difference between the gearwheel type of Sandner damper (Fig. 5) and the pumping-chamber type is that in the former an internal gear is cut in the secondary member or rim, which meshes with a number of pinions arranged by a system of ports to act as gear pumps.

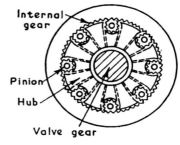

When the inertia torque of the rim reaches the critical value required to rotate the gearwheels so as to overcome the pressure to which the centrally located relief valves are adjusted, the rim is able to slip over a certain distance. Owing to the rotation of the pinions, oil is passed from one gear chamber to the next.

Fig. 5.

The valve gear is double acting, employing two relief valves so as to enable the motion of the secondary member (i.e. the rim) to be controlled in either direction. One relief valve is set at a slightly higher pressure than the other, in order to cause the secondary member to slip, relatively to the hub, at about one revolution per minute, thus avoiding uneven wear of the teeth. This gearwheel type design mounted on an eight-cylinder single-acting engine of 23×23 in. bore $\times$ stroke gave an amplitude reduction of about $6 \cdot 5 : 1$ for the $3\frac{1}{2}$, 4th, $4\frac{1}{2}$ and $5\frac{1}{2}$ orders.

The calculations for this damper are the same as for the pumping-chamber damper, except that the pressure is

$$p = T_{R\,\mathrm{max.}}/(8ra),$$

since there are eight gear pumps operating in the same direction, and a suitable value has to be determined for the effective area a of the gear pump recesses.

3·43   The Atlas damper (pumping–chamber type)

This damper (Fig. 6) consists of a hub, rigidly attached to the free end of the crankshaft, and a rim (or seismic member). It is so designed that, towards the end of the relative movement in either direction, the hub cuts off the main radial passages connecting to the dashpot recesses. Very fine radial clearances (of the order of 0·001 in.) are used between the hub and the rim.

The oil is supplied from the crankshaft, through annular and radial passages of the hub, to the pumping chambers. In contrast to the Sandner damper (see section 3·41), the Atlas damper has no locating blades

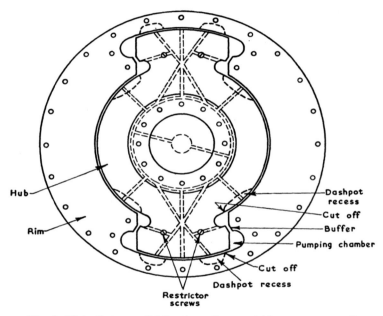

Fig. 6. Main features of Atlas-type damper (side cover removed).

and no pressure control valve in the hub. Consequently, slip does not occur at a critical pressure but begins as soon as any torque exists. The slip continues until sufficient relative movement occurs for the cut-offs to decrease the flow areas to their minimum values. The pressure forces on the rim then become sufficient to impose on it an acceleration which makes it almost follow the hub motion when the hub is in the region of its maximum amplitude. During the return stroke, as the hub acceleration decreases, relative flow can take place again with slip in the opposite direction.

Energy is dissipated during the slipping motion in the passages and in the recesses generally, and energy dissipation is greatest when the

[566]

cut-off has occurred and a strong dashpot action is obtained. The mechanical buffers should not normally come into contact. (See also Thege damper with liquid buffers, Brit. Pat. Spec. no. 447,188.) To enable adjustment of the flow resistance, some dampers have been provided with passages controlled by restrictor screws, as shown in Fig. 6. These adjustable restrictor screws are accessible through corresponding holes in the outer casing.

For an approximate assessment of damper dimensions, use is made of the following relation for rim torque:

$$T_R = 2\Delta par \simeq 0.5 J_R \omega^2 \theta_H \quad \text{[Lb.in.]},$$

where the pressure Δp representing flow resistance is taken as equal to the oil supply pressure p (e.g. 30 Lb./in.^{2}); also a = cross-sectional area of one pumping chamber, r = mean radius to a from the centre of the hub, J_R = inertia of damper rim, ω = natural phase velocity of engine with damper, and θ_H = vibration at hub or at cylinder no. 1 of the engine.

3·44   The Lanchester damper (semi-dry friction type)

Dampers of this type (Fig. 7) consist of two flywheels (or seismic masses), freely rotatable on shaft bearings, and a hub with friction plates which is keyed to the free end of the crankshaft.

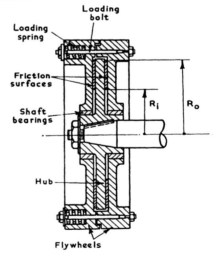

The flywheels are pressed against the friction surfaces of the hub by means of adjustable loading springs. As long as the inertia torque of the flywheels is smaller than the frictional torque, the flywheels vibrate as one mass with the hub.

When the vibration amplitude θ_H of the hub and the natural phase velocity ω of the engine system are such that the inertia torque of the flywheels is equal to the frictional torque, relative slip occurs, and energy is dissipated by friction.

As the hub motion is sinusoidal ($\theta_H = \theta_H \sin \omega t$), the inertia torque reaches the critical value of the frictional torque twice per cycle. The slipping and stopping motion

Fig. 7. Lanchester semi-dry friction damper.

thus also occurs twice per cycle, and during this period the damper also acts as a detuner, that is, owing to the cyclic variation of the effective inertia of the damper, the natural frequency of the engine system is continuously altered, or 'detuned', on either side of a mean value.

Basic design procedure. The calculations are based on the same theory as for the Sandner damper (see section 3·41, pp. 560–3), with the following minor modifications:

(1) The total inertia of the two flywheels J_{2F} is determined (instead of the rim inertia J_R) by the formula given on p. 560.

(2) The formula for the maximum torque $T_{2F\,\text{max.}}$ is the same as for $T_{R\,\text{max.}}$, on p. 561.

(3) The frictional torque T_f in the present case is obtained as

$$T_{2F\,\text{max.}} = T_f = \tfrac{4}{3}\mu P \times [R_o^3 - R_i^3]/(R_o^2 - R_i^2) \quad \text{[Lb.in.]},$$

where μ = friction coefficient, P = total spring load [Lb.], R_o and R_i = outer and inner radius [in.], respectively, of the friction surfaces; the factor $\tfrac{4}{3}$ being used because there are two friction surfaces.

Note. $P = N_s K_s \delta$, where N_s = number of loading springs, K_s = linear stiffness [Lb./in.] of one spring, and δ = deflexion [in.] of spring as assembled.

For hot, oily surfaces, $\mu \simeq 0·15$; for continuously lubricated metal surfaces $\mu \simeq 0·07$; for dry Ferodo friction linings $\mu \simeq 0·33$.

(4) The various expressions for U_{in}, U_G, U_H and U_D given on p. 561 remain valid; and the formula based on energy equilibrium now gives J_{2F} (the inertia of the two flywheels) instead of J_R.

General remarks

Friction dampers can be designed either as dry-friction dampers or as dampers with lubricated friction surfaces. Where dry-friction dampers are used, they are best mounted on the crankshaft extension outside the crankcase, to avoid having dust, due to wear, in the lubricating oil.

The two flywheels are usually designed to have equal moments of inertia ($J_{F1} = J_{F2} = \tfrac{1}{2}J_{2F}$) so as to exert equal inertia torques on the hub. In this way, one avoids differential loadings which might lead to fracture of the loading bolts between the flywheel masses (see Fig. 7).

Owing to surface wear, periodic resetting of the springs is recommended, in order to maintain the same spring deflexion δ and spring pressure.

The expression for frictional torque is obtained by evaluating the integral

$$T_{f1} = \mu \frac{P}{A} \times \int_{R_i}^{R_o} 2\pi R \times R \times dR = \frac{2\pi}{3} \mu \frac{P}{A} (R_o^3 - R_i^3),$$

and replacing A by $\pi(R_o^2 - R_i^2)$ = area of one friction surface. As in the design of Fig. 7 there are two such areas of contact, $T_f = 2T_{f1}$.

For a multiple-plate type damper (see Fig. 9), the frictional torque is

$$T_f = (N_P - 1) \times T_{f1},$$

where $N_P =$ total number of plates in the driving and driven members of the damper, each flywheel being also counted as one plate.

In the dampers of Figs. 10 and 11, the increased flexibility of the driving arrangement increases the relative amplitude between flywheel and driving member and thus renders the damper more effective.

EXAMPLES OF LANCHESTER-TYPE DAMPERS

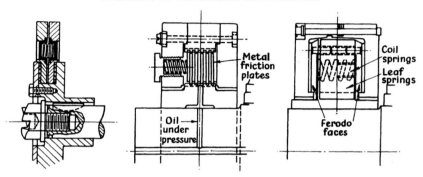

Fig. 8. Damper for high-speed engine.†

Fig. 9. Multi-plate damper.†

Fig. 10. Dry-friction damper with flywheels driven by leaf springs.†

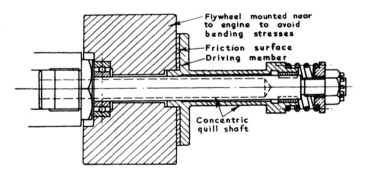

Fig. 11. Quill-shaft friction damper (Jackson‡).

† Figs. 8, 9 and 10 are reproduced from J. Smith, *Crankshaft Design and Manufacture*, (p. 44). (Published by Whitehall Technical Press Ltd, London, 1946.)

‡ Fig. 11 is from P. Jackson, *The Vibrations of Oil Engines*, Diesel Engine Users Association (D.E.U.A.), Paper no. S 115, 26 April 1933. Further bibliographical references: Den Hartog, J. P. and Ormondroyd, J., *loc. cit.* p. 565; Lanchester, F. W., British Patent no. 136,335.

3·5 PENDULUM DETUNERS

General remarks

The term 'pendulum' may be applied to any system consisting of a mass (or moment of inertia) capable of movement about a fixed point and of which the movement at any instant is opposed by a restoring force (or torque) proportional to displacement from the neutral position. The mass may be solid or liquid, and the restoring action may be due to spring force, the force of gravity, centrifugal force, etc.

Examples of pendulum arrangements are shown in Figs. 1, 2 and 3.

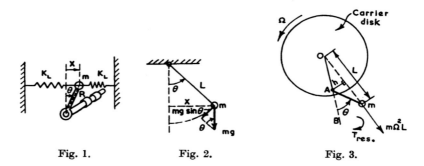

Fig. 1.　　　　　　Fig. 2.　　　　　　Fig. 3.

Notation:

$$m = \text{pendulum mass [Lb.sec.}^2\text{/in.]}$$
$$R = \text{radius [in.]}$$
$$J = mR^2 = \text{moment of inertia [Lb.in.sec.}^2\text{]}$$
$$L = \text{length [in.]}$$
$$x = \text{linear displacement [in.]}$$
$$\omega = \text{phase velocity of vibration [rad./sec.]}$$
$$\Omega = \text{phase velocity of shaft rotation [rad./sec.]}$$
$$K_L = \text{linear stiffness [Lb.in.] of one spring}$$

The spring pendulum (Fig. 1) has an inertia torque

$$T_i = -J\omega^2\theta$$

(where $J = mR^2$, $\theta = x/R$) and a restoring torque

$$T_{\text{res.}} = (2K_L x) \times R$$
$$= 2R^2 K_L \theta.$$

[570]

The gravity pendulum (Fig. 2) has an inertia torque

$$T_i = -J\omega^2\theta$$

(where $J = mL^2$ and $\theta = x/L$), and although it has no mechanical spring stiffness it has a restoring torque

$$T_{\text{res.}} = Lmg\sin\theta \simeq Lmg\theta$$

provided by gravity.

The centrifugal pendulum (Fig. 3) comprises a mass m freely attached by a light rod to a rotating disk. If the carrier disk vibrates during its rotation, the pendulum mass m swings through an angle θ. But centrifugal force $m\Omega^2L$ tries to move m radially outwards, back to its fully extended position (on line $\overline{OAB}$ of Fig. 3). Therefore a restoring torque is produced:

$$T_{\text{res.}} = hm\Omega^2L.$$

In the following, the features of two main types of pendulums used in engine applications will be considered, viz. the peripheral-spring pendulum, pp. 572–4, and the centrifugal pendulum, pp. 574–586.

It may be noted that both types are primarily designed to reduce vibrations of the engine system through 'detuning' action (i.e. by altering the frequency of the engine system) and not by dissipation of vibratory energy.

NOTATION

Symbol	Brief definition	Typical units	Symbol	Brief definition	Typical units
A	area	in.^{2}, cm.^{2}	P	force	Lb., Kg.
d	diameter	in., cm.	r	radius	in., cm.
D	diameter	in., cm.	R	radius	in., cm.
$\mathscr{D}$	denominator	—	t	time	sec., min.
F	vibration frequency	vib./min.	T	torque	Lb.in., Kg.cm.
g	acceleration due to gravity	in./sec.^{2}, cm./sec.^{2}	$|T_n|$	nth order component of tangential pressure	Lb./in.^{2}, Kg./cm.^{2}
h	distance	in., cm.	W	weight	Lb., Kg.
j	$\sqrt{-1}$	—	x	linear displacement	in., cm.
J	moment of inertia	Lb.in.sec.^{2}, Kg.cm.sec.^{2}	Δ	Holzer-table amplitude	rad.
K	stiffness	Lb.in./rad., Kg.cm./rad.	θ	vibration amplitude	deg., rad.
K_L	linear stiffness	Lb./in., Kg./cm.	$\overrightarrow{\Sigma\Delta}$	phase-vector sum	—
L	length	in., cm.	ϕ	phase angle	deg., rad.
m	mass	Lb.sec.^{2}/in., Kg.sec.^{2}/cm.	ψ	phase angle	deg., rad.
			ω	phase velocity of vibration	rad./sec.
n	critical speed order number	vib./rev.	Ω	angular velocity of shaft rotation	rad./sec.
N	shaft speed	rev./min.			

Subscripts are used in the text to define more precisely. Their meaning is fully explained in their context in each instance.

[571]

3·51   The transverse-spring pendulum detuner †

This detuner (see Fig. 4) consists of a carrier disk rigidly attached to the free end of the engine crankshaft and a seismic mass or pendulum centred by two pairs of preloaded springs.

The system operates in the same way as a tuning disk (see section 3·1). When the ratio K_P/J_P of the equivalent torsional stiffness of the springs to the moment of inertia of the pendulum is equal to ω_E^2 (where ω_E = natural phase velocity of the engine), the pendulum will suppress the engine resonance at that frequency, and produce two new frequencies above and below the original frequency. If the value of K_P is sufficiently high, the two new peaks may occur at speeds outside the running range.

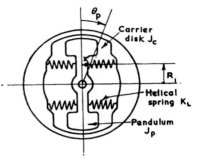

Fig. 4. Detuner with pendulum J_P and preloaded transverse springs of equivalent torsional stiffness $K_P = 4R_1^2 K_L$.

As explained in section 1·241, stiffness of peripheral-spring couplings (p. 104, remark (b)), the total equivalent torsional stiffness of the four springs is

$$K_P = 4R_1^2 K_L \quad \text{[Lb.in./rad.]}, \tag{1}$$

since all the springs are in parallel. In the above relation, R_1 = distance [in.] of springs from centre line of pendulum, and K_L = linear stiffness or 'rate' of one spring [Lb./in.]. (The restoring torque produced by a displacement $x = R_1 \theta_P$ (see Fig. 4) is $T = 4 \times (K_L x) \times R_1 = 4K_L \theta_P R_1^2$, hence $T/\theta_P = K_P = 4R_1^2 K_L$; θ_P being the relative angular motion.)

The resonant phase velocity of the detuner is

$$\omega_P = \sqrt{(K_P/J_P)} \quad \text{[rad./sec.]}, \tag{2}$$

and the system is designed so that ω_P will be equal to the natural phase velocity ω_E of the engine system without the detuner.

Derivation of eq. (2). The absolute angular motion of the pendulum J_P from its mean position is $\theta_{\text{abs.}} = \theta_P + \theta_C$, where θ_P = amplitude of pendulum relative to carrier disk, and θ_C = vibration amplitude of carrier disk. The spring torque on the pendulum is $K\theta_P$. Therefore the equation of motion of the pendulum is

$$J_P \ddot{\theta}_{\text{abs.}} + K\theta_P = 0 \quad \text{or} \quad J_P \ddot{\theta}_P + J_P \ddot{\theta}_C + K\theta_P = 0. \tag{3}$$

Substituting $\theta_C = \theta_{OC} \sin \omega_E t$ and $\theta_P = \theta_{OP} \sin \omega_E t$, where θ_{OC} and θ_{OP} = peak amplitudes of sinusoidal vibration, one obtains

$$\theta_{OP} \sin \omega_E t \times (-\omega_P^2 J_P + K_P) = +\omega_P^2 J_P \theta_{OC} \sin \omega_E t,$$

† Alternative designation: Frahm dynamic vibration absorber.

hence
$$\theta_{OP} = \frac{J_P \omega_E^2 \theta_{OC}}{-\omega_E^2 J_P + K_P} = \frac{\theta_{OC}}{\dfrac{K_P/J_P}{\omega_E^2} - 1},$$

which shows that the pendulum will resonate (with large values of θ_{OP}) when $\omega_E = \sqrt{(K_P/J_P)} = \omega_P$.

Design procedure

The design procedure may be outlined as follows:

(1) Choose a suitable value for the pendulum inertia J_P in relation to the equivalent engine inertia $J_E = \Sigma(J_{\text{cyl.}} \Delta_{\text{cyl.}}^2)$ (where $\Sigma(J_{\text{cyl.}} \Delta_{\text{cyl.}}^2) = $ sum of cylinder inertias multiplied by their corresponding relative amplitudes squared, from a Holzer table for the engine without the detuner), for instance
$$J_P = 0.1 J_E \quad \text{to} \quad 0.3 J_E.$$

(2) Taking account of the space available at the free end of the engine, decide on suitable values for the pendulum dimensions. Also determine the dimensions and moment of inertia J_C of the carrier disk (and front cover-plate).

(3) Evaluate a Holzer table for the engine system with the carrier-disk inertia J_C and the shafting of stiffness K_C between J_C and cylinder centre no. 1. This gives a corrected value $\omega_{E'}$ for the natural phase velocity of the engine system with the carrier disk (but without the pendulum inertia).

(4) Determine the equivalent torsional stiffness of the four springs as
$$K_P = \omega_{E'}^2 J_C \quad \text{[Lb.in./rad.].†}$$

In the pendulum design (already determined under 2) situate the springs at a suitable radius R_1 from the pendulum centre and obtain the linear stiffness of each spring as
$$K_L = K_P/(4R_1^2) \quad \text{[Lb./in.].}$$

(5) The engine system with the detuner will then be in an 'anti-resonant' condition at the phase velocity $\omega_{E'}$. Determine the new (one- and two-node) natural frequencies of the entire system by means of Holzer tables, with the pendulum inertia J_P as mass no. 1, followed by the pendulum stiffness K_P, and the carrier disk inertia J_C as mass no. 2, etc.

(6) Consider the critical speeds corresponding to these new frequencies in regard to their positions in the speed range of the engine. If they are still too close to the original critical, use a larger value for J_P (and hence for K_P) and repeat the calculations under (2)–(5) with this new value. If they are spread out more than necessary, use a smaller value for J_P.

† *Metric units.* All the relations in this section are valid when metric units are used; with for instance 'Kg.' instead of 'Lb.', and 'cm.' instead of 'in.'.

These remarks are based on the considerations set out in section 3·1 for 'tuning disks without damping'.

Spring loads. It may be assumed that, although they occur at different frequencies, the peak values of the new resonant amplitudes of the engine are unaffected by the detuner. Let $\pm\theta_{1\,\mathrm{cyl.}}$ = measured (or calculated) value of vibration amplitude [rad.] at cylinder no. 1, for the critical order considered (engine without a detuner).

Then, from the Holzer-table amplitudes evaluated under (5), the angular vibration amplitude of the pendulum springs will be

$$\theta_{sh} = \Delta_{sh} \times \theta_{1\,\mathrm{cyl.}}/\Delta_{1\,\mathrm{cyl.}} = (\Delta_P - \Delta_C) \times \theta_{1\,\mathrm{cyl.}}/\Delta_{1\,\mathrm{cyl.}}.$$

This corresponds to a linear deflexion $\pm x = R_1\theta_{sh}$ [in.] and to a dynamic load
$$P_L = \pm K_L x \quad [\text{Lb.}],$$

acting on each spring. The preload P_{static}, added to P_L, gives the maximum load on each spring. As shown in Fig. 3, p. 508, the maximum spring deflexion occurs at engine resonance, and not at pendulum resonance. Moreover, the spring surging frequency should be calculated and situated well away from the vibration frequencies of the engine system. In this connexion, one may make use of the fact that $K_P = R_1^2 K_L$ and alter R_1 for a different value of K_L if necessary.

3·52   Centrifugal pendulum detuners †

General remarks

The simple centrifugal pendulum detuner consists of (1) a carrier (e.g. a disk) rigidly attached to the engine crankshaft and (2) a pendulum mass fixed at one end of a link which is pivoted at a suitable point on the carrier (Fig. 5).

(1) When the carrier rotates at a uniform phase velocity Ω, the pendulum mass m_1 is subjected to a centrifugal force

$$P_{\mathrm{centrif.}} = m_1 \Omega^2 (R + r).$$

If, owing to an excitation torque $T_{(n)}$, the carrier J_2 vibrates through an angle θ_2 (Fig. 5), the pendulum will swing through a further angle θ_1. The centrifugal force is then no longer in the direction of the link $\overline{AM}$ and will therefore produce a restoring torque $T_{\mathrm{res.}}$ acting on the pendulum mass:
$$T_{\mathrm{res.}} = h P_{\mathrm{centrif.}} = h m_1 (R + r)\,\Omega^2,$$

where $h = \overline{AB}$ in Fig. 5. The pendulum thus has a 'dynamic stiffness' $K_P = T_{\mathrm{res.}}/\theta_1$ and can perform sinusoidal vibrations about its suspension

† Alternative designations: Pendulum dampers, pendulum absorbers, tuned centrifugal pendulums, undamped dynamic vibration absorbers, rotating pendulum vibration absorbers.

point A. This 'stiffness', however, depends on the shaft phase velocity squared (Ω^2). Consequently, the pendulum will have a resonant phase velocity ω_P which is not constant but varies proportionally to $\sqrt{K_P}$ (the pendulum mass being constant) and hence proportionally to the shaft speed of rotation N (where $N = \Omega \times 60/2\pi$ rev./min.).

(2) The carrier disk is subjected to a reaction torque

$$T_{\text{react.}} = \overline{OB}{}^* \times P^*_{\text{centrif.}},$$

where $P^*_{\text{centrif.}}$ is the component of centrifugal force acting via the link $\overline{MA}$ on the pivot point A (see Fig. 5). For small angles, $\overline{OB}{}^* = R\theta_1$ and $P^*_{\text{centrif.}} = P_{\text{centrif.}}$, so that

$$T_{\text{react.}} = R\theta_1 \times m_1 \Omega^2 (R+r).$$

It will be shown in the following that if this torque, which varies sinusoidally with θ_1, is equal and opposite to the excitation torque $T_{(n)}$

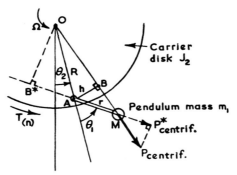

Fig. 5. $\overline{OA} = R$, $\overline{AM} = r$ and for small angles: $\overline{OM} \cong R+r$.

acting on the carrier, the latter will cease vibrating ($\theta_2 = 0$), while the pendulum will perform vibrations at its resonant phase velocity ω_P.

(3) If the distance between the pendulum mass and its suspension point is suitably chosen, the pendulum can be made to resonate at one particular engine speed $N_{\text{crit.}} = \Omega_{\text{crit.}} \times 60/2\pi$, at which the engine without the detuner was vibrating at its natural frequency. The engine system with the pendulum will then be in an 'anti-resonant' condition and the critical amplitude of vibration is suppressed.

(4) In contrast to other vibration-reducing devices, however, the centrifugal pendulum can only remove one critical order, since the pendulum is in resonance with the engine frequency only at one running speed. To eliminate more than one critical, it is necessary to use several pendulums, each of which is 'tuned' to one of the corresponding orders. Such detuners may be designed for use on crankwebs (where they also serve as balance weights) or arranged on a carrier disk at the free end of the engine crankshaft.

'Tuning' of a centrifugal pendulum

Let $n = \omega_E/\Omega_C =$ critical speed order number [vib./rev.],

$\omega_E =$ natural phase velocity of engine without detuner [rad./sec.],

$\Omega_C = N_C \times 2\pi/60$, where $N_C =$ critical engine speed [rev./min.],

$R =$ distance between shaft centre and suspension point of pendulum ($\overline{OA}$ in Fig. 5),

$r =$ distance between suspension point and mass centre M of pendulum.

It will be shown in the following that, for the simple pendulum, the vibration of the nth order engine critical will be suppressed if the pendulum is designed in such a way that

$$r = R/n^2 = R \times \Omega_C^2/\omega_E^2. \tag{1}$$

It may be noted that eq. (1) is strictly valid only for a simple pendulum with a point-like mass. For a 'physical pendulum' with a mass m_1 which has a moment of inertia J_M about its centre of gravity M, eq. (1) gives only the 'reduced' or 'equivalent' length, $r_{\text{equiv.}}$. The relation† between $r_{\text{equiv.}}$ and r ($= \overline{AM}$ in Fig. 5) is

$$r_{\text{equiv.}} = r[1 + (J_M/m_1 r^2)]$$

and in this case eq. (1) becomes

$$r_{\text{equiv.}} = R/n^2.$$

To fix the ideas it is convenient to consider the following example: It is proposed to suppress the $2\frac{1}{2}$ order critical by means of a centrifugal pendulum mounted on a crankweb (Fig. 6), the pendulum mass being supported at a distance $R = 5$ in. from the journal centre.

The 'length' of the pendulum is then

$$R/n^2 = 5/(2\cdot 5)^2 = 0\cdot 8 \text{ in.}$$

To suppress the 5th order, it would be necessary to design a pendulum with an effective length $r = 0\cdot 2$ in. This explains why particular attention is given to the design of the pendulum mass, which may, for instance, be given the form of a balance weight with a U-shaped cross-section, mounted on a roller (or rollers) passing through the crankweb thickness. Details of various designs used in practice are given in the illustrations facing p. 582.

Derivation of the tuning formula

Considering Fig. 7, let $L = \overline{OM}$, $h = \overline{AB}$, $R = \overline{OA}$ and $r = \overline{AM}$. For small angles θ_1 and ϕ, the restoring torque is

$$T_{\text{res.}} = m_1 \Omega^2 L h = m_1 \Omega^2 (R+r) h.$$

† For the derivation of this relation, see for instance, J. P. Den Hartog, *Mechanics*, 2nd ed., p. 235 (McGraw-Hill Book Co. Inc. 1948).

[576]

Now $h = R \sin \phi = r \sin (\theta_1 - \phi)$, so that for small amplitudes

$$h = R\phi = r\theta_1 - r\phi \quad \text{or} \quad \phi = \theta_1 r/(R+r)$$

and

$$h = \theta_1 Rr(R+r).$$

Therefore, $T_{\text{res.}} = m_1 \Omega^2 Rr\theta_1$ and the dynamic 'stiffness' is

$$K_P = T_{\text{res.}}/\theta_1 = m_1 \Omega^2 Rr.$$

The pendulum has a resonant phase velocity ω_P which is obtained from

$$\omega_P^2 = K_P/m_1 r^2 = \Omega^2 R/r. \tag{2}$$

The pendulum will resonate at the natural phase velocity ω_E of the engine without the detuner if $\omega_P = \omega_E$. Introducing the critical speed order number $n = \omega_E/\Omega$, one obtains from eq. (2):

$$\omega_P^2 = \frac{\omega_E^2}{n^2} \times \frac{R}{r},$$

so that $\omega_P = \omega_E$ if $R/(n^2 r) = 1$, or $r = R/n^2$.

Note. In the above derivation and throughout this section, it is assumed that the centrifugal forces are so large that the effects of gravity and Coriolis accelerations (due to radial velocity components) can be neglected.

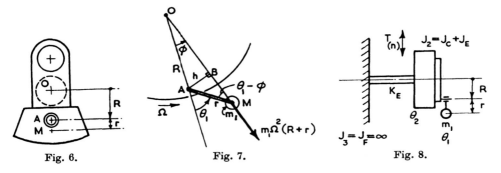

Fig. 6. Fig. 7. Fig. 8.

Two-mass system formulae for the simple pendulum detuner

The equivalent two-mass system for an engine with a simple detuner is shown in Fig. 8.

The system without its pendulum mass m_1 has a natural phase velocity

$$\omega_E = \sqrt{\{K_E/(J_C + J_E)\}} \quad \text{[rad./sec.]},$$

where $J_C =$ inertia of carrier disk [Lb.in.sec.^{2}] at free end of crankshaft,

$J_E = \Sigma(J_{\text{cyl.}} \Delta_{\text{cyl.}}^2) =$ equivalent engine inertia,

= sum of cylinder inertias multiplied by their corresponding relative amplitudes squared (from a Holzer table with J_C as mass no. 1), and

$K_E = \omega_E^2(J_C + J_E) =$ equivalent stiffness [Lb.in./rad.] of crankshaft.

The flywheel inertia J_F is regarded as being infinitely large. The system is excited by a sinusoidally varying torque with a magnitude

$$T_{(n)} = |T_n| A R_0 \overrightarrow{\Sigma \Delta}_n \quad \text{[Lb.in.]},$$

where $|T_n| = n$th order tangential-pressure component due to gas torque [Lb./in.^{2}], $A =$ piston area [in.^{2}], $R_0 =$ crank radius [in.], and $\overrightarrow{\Sigma \Delta}_n =$ phase-vector sum for the nth order, from the Holzer table for J_E.

Disregarding engine damping, at a critical speed N_C the system without m_1 will vibrate at its natural phase velocity ω_E with a resonant amplitude ($\theta_2 \to \infty$). With mass m_1 in operation, and the pendulum 'tuned' so that $n^2 = R/r$ when $\Omega_C = \omega_E/n = 2\pi N_C/60$, the results will be as follows:

(1) At the critical speed N_C (of order $n = \omega_E/\Omega_C$, with 'resonance tuning' of pendulum):

Vibration amplitude at engine mass:

$$\theta_2 = 0;$$

Vibration amplitude of pendulum:

$$\theta_1 = -T_{(n)}/[m_1 \Omega_C^2 (R+r) R] \quad \text{[rad.]}.$$

(2) At any other critical speed N_C^* (of order $n^* = \omega_E/\Omega^*$):
Vibration amplitude at engine mass:

$$\theta_2 = T_{(n*)}/\mathscr{D}_{(n*)};$$

Vibration amplitude of pendulum:

$$\theta_1 = \theta_2 \times (R+r) \Big/ \left\{ \left[\left(\frac{n}{n^*} \right)^2 - 1 \right] r \right\},$$

where $\quad \mathscr{D}_{(n*)} = -m_1(R+r)^2 \times \left[1 + \frac{1}{(n/n^*)^2 - 1} \right] \omega_E^2.$

(3) At any other non-resonant speeds N:
Vibration amplitude at engine mass:

$$\theta_2 = T_{(n)}/\mathscr{D}_{(n)};$$

Vibration amplitude of pendulum:

$$\theta_1 = \theta_2 \times (R+r) \Big/ \left\{ \left[\frac{\Omega^2 R}{\omega^2 r} - 1 \right] r \right\},$$

where $\quad \mathscr{D}_{(n)} = K_E - J_2 \omega^2 - \omega^2 m_1(R+r)^2 \left[1 + \frac{1}{\dfrac{\Omega^2 R}{\omega^2 r} - 1} \right].$

In these expressions, $T_{(n)} =$ excitation torque of nearest (nth order) critical in the neighbourhood of the engine speed $N = \Omega \times 60/2\pi$, and $\omega = n \times \Omega$. (In cases (1) and (2), $\omega = \omega_E$, hence $K_E - \omega^2 J_2 = 0$.)

[578]

Evaluation of pendulum mass

As previously stated on p. 576, at the critical speed $N_C = \Omega_C \times 60/2\pi$ corresponding to 'resonance tuning', the reaction torque

$$T_{reaction} = m_1 \Omega_C^2 (R+r) R\theta_1$$

of the pendulum on its carrier is equal to the excitation torque

$$T_{(n)} = |T_n| AR_0 \overrightarrow{\Sigma \Delta}_n.$$

Therefore the weight of the pendulum mass can be evaluated as

$$W_1 = m_1 g = \frac{|T_n| AR_0 \overrightarrow{\Sigma \Delta} \times g}{\Omega_C^2 (R+r) R\theta_1} \quad \text{[Lb.],}\dagger$$

where $g = 386\cdot4$ in./sec.^{2} and $\theta_1 =$ maximum allowable amplitude of the pendulum mass [rad.], based on design considerations. In practice, θ_1 may be a fairly large value, and the above expression is adequate for pendulum amplitudes as high as $\theta_1 = \pm 30°$.

'Effective inertia' of pendulum

The expression

$$J_{P\,eff.} = m_1(R+r)^2 \times \left[1 + \frac{1}{\left(\dfrac{\Omega n}{\omega_E}\right)^2 - 1} \right] \quad \text{[Lb.in.sec.}^2]$$

is frequently referred to as the 'effective inertia' of the pendulum, in view of the relation $J_{P\,eff.}\,\omega_E^2\,\theta_2 = T_{(n)}$ (see formulae under (2) above). At the critical speed N_C at which the pendulum resonance occurs ('resonance tuning') one obtains $J_{P\,eff.} = \infty$, hence $\mathscr{D}_{(n)} = \infty$ and $\theta_2 = 0$. At other speeds, $J_{P\,eff.}$ has a finite value, and $\theta_2 \neq 0$.

Note. The corresponding formulae for various types of centrifugal pendulums employed in practice are given on the sheet of illustrations facing p. 582.

Derivation of two-mass system equations

(a) *Equation of motion of pendulum mass m_1.* When the carrier disk vibrates through an angle θ_2 (Fig. 9), the disk periphery rotates from A' to A. The corresponding displacement of the pendulum is from M' to M'' and then to M, that is,

$$(R+r)\theta_2 + r\theta_1.$$

Hence the accelerating force of m_1 at right angles to OM is

$$-m_1[(R+r)\ddot{\theta}_2 + r\ddot{\theta}_1],$$

and if the tension in the arm is P, the above expression is equal to $P \sin \psi$.

---

† *Metric units.* The above expressions are equally valid in metric units, with W in Kg., $g = 981$ cm./sec.^{2}, and linear dimensions in centimetres.

Along $\overline{OM}$ we have

$$P \cos \psi = m_1 \Omega^2 \times \overline{OM} \cong m_1 \Omega^2 (R+r),$$

where $\Omega =$ phase velocity of shaft rotation. For small angles, $\sin \psi \cong \psi$, $\cos \psi \cong 1$; from the triangle OAM, one has

$$\theta_1 = \phi + \psi, \quad h = R\phi = r\psi = r(\theta_1 - \phi),$$

hence
$$\phi(R+r) = \theta_1 r \quad \text{and} \quad h = Rr\theta_1/(R+r).$$

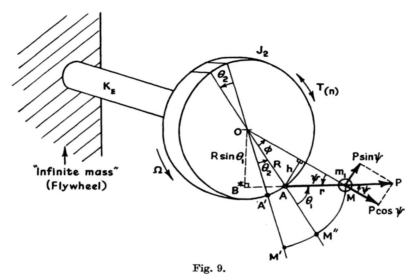

Fig. 9.

Therefore $\psi = h/r = \theta_1 R/(R+r)$. Eliminating P, we obtain

$$-m_1[(R+r)\ddot{\theta}_2 + r\ddot{\theta}_1] = \frac{\theta_1 R}{R+r} \times m_1(R+r)\Omega^2,$$

so that finally
$$(R+r)\ddot{\theta}_2 + r\ddot{\theta}_1 + \Omega^2 R\theta_1 = 0. \tag{1}$$

(b) *Equation of motion of carrier disk.* Let the disk be accelerated by an excitation torque $T_{(n)}$. The accelerating torque of the carrier disk is $J_2\ddot{\theta}_2$ and the motion of the carrier is given by

$$J_2\ddot{\theta}_2 = T_{(n)} + PR \sin\theta_1 - K_E\theta_2,$$

where $K_E\theta_2$ is the resisting torque due to shaft stiffness. As, from (a) above, $P = -m_1[(R+r)\ddot{\theta}_2 + r\ddot{\theta}_1]/\sin\psi$, one may write

$$PR \sin\theta_1 = -m_1 R[(R+r)\ddot{\theta}_2 + r\ddot{\theta}_1]\sin\theta_1/\sin\psi$$

$$\cong -m_1 R[(R+r)\ddot{\theta}_2 + r\ddot{\theta}_1](R+r)/R,$$

[580]

so that one has
$$J_2 \ddot{\theta}_2 + m_1(R+r)^2\, \ddot{\theta}_2 + m_1(R+r)\, r\ddot{\theta}_1 + K_E \theta_2 = T_{(n)}. \qquad (2)$$

If the excitation torque is sinusoidal, $T_{(n)} = T_{(n)\,0}\sin\omega t$ (where $\omega =$ phase velocity of the excitation), the usual substitutions can be employed, viz.
$$T_{(n)} = T_{(n)\,0}\, e^{j\omega t}, \quad \theta_1 = \theta_{01}\, e^{j\omega t} \quad \text{and} \quad \theta_2 = \theta_{02}\, e^{j\omega t},$$

and eqs. (1) and (2) may be rewritten as
$$[-\omega^2 r^2 + \Omega^2 Rr]\,\theta_{01} - \omega^2(R+r)\, r\theta_{02} = 0,$$
$$-m_1\omega^2(R+r)\, r\theta_{01} + [K_E - \omega^2 J_2 - \omega^2 m_1(R+r)^2]\,\theta_{02} = T_{(n)\,0}.$$

Solving by means of determinants and omitting the subscript 0 for simplicity, we have the system determinant
$$\mathrm{Det}_s = \begin{vmatrix} -\omega^2 r^2 + \Omega^2 Rr & -\omega^2(R+r)\, r \\ -m_1\omega^2(R+r)\, r & K_E - \omega^2 J_2 - m_1\omega^2(R+r)^2 \end{vmatrix}$$
$$= r(\Omega^2 R - \omega^2 r) \times [K_E - \omega^2 J_2 - \omega^2 m_1(R+r)^2] - m_1\omega^4 r^2(R+r)^2,$$

and the determinants for θ_1 and θ_2, viz.
$$\mathrm{Det}_1 = \begin{vmatrix} 0 & -\omega^2(R+r)\, r \\ T_{(n)} & K_E - \omega^2 J_2 - m_1\omega^2(R+r)^2 \end{vmatrix} = + T_{(n)}\omega^2(R+r)\, r,$$
$$\mathrm{Det}_2 = \begin{vmatrix} -m_1\omega^2 r^2 + m_1\Omega^2 Rr & 0 \\ -m_1\omega^2(R+r)\, r & T_{(n)} \end{vmatrix} = T_{(n)}r(\Omega^2 R - \omega^2 r),$$

which gives the vibration amplitudes $\theta_1 = \mathrm{Det}_1/\mathrm{Det}_s$ and $\theta_2 = \mathrm{Det}_2/\mathrm{Det}_s$ for the various conditions considered on p. 578. In the case of 'resonance tuning', $K_E - \omega^2 J_2 = 0$ and $\Omega^2 R - \omega^2 r = 0$. Thus $\theta_2 = 0$ and
$$\theta_1 = T_{(n)}/[-m_1\omega^2 r(R+r)] = -T_{(n)}/[m_1\Omega^2 R(R+r)].$$

But $m_1\Omega^2 R(R+r)\,\theta_1 = T_{\text{react.}}$ is the reaction torque exerted by the pendulum on the carrier disk (see p. 575). This proves the statement that, under tuning conditions, the main mass is stationary and the excitation is opposed at every instant by an equal reaction torque from the auxiliary system, with a dynamic magnifier of unity (since the excitation torque is not magnified by resonance). Similar 'anti-resonant' conditions were found for the 'tuning disk without damping', in section 3·1.

Additional remarks. The equations of motion were obtained directly from the geometry of the system. They could also have been derived by means of Lagrange's method† which requires only the determination of the kinetic and potential energies of the system.

† See, for instance, Jeans, J. H., *An Elementary Treatise on Theoretical Mechanics*, ch. 12 (Ginn and Company, 1907); Timoshenko, S., *Vibration Problems in Engineering*, 2nd ed., ch. 6 (D. Van Nostrand Company, 1944); Kármán, T. v. and Biot, M. A., *Mathematical Methods in Engineering*, chs. 3, 4, 5 (McGraw-Hill Book Company, 1940).

The use of Lagrange's equations for the investigation of pendulum detuners is shown in detail in a comprehensive analysis by Zdanowich and Wilson. Further extensive treatment of centrifugal pendulum detuners can be found in published papers by Den Hartog, Porter, Crossley (for wide-angle oscillations) and other investigators listed in the bibliography at the end of this section.

Holzer-table calculations

The use of Holzer tables for systems provided with centrifugal pendulum detuners will be described by means of the following example.

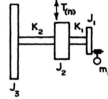

Fig. 10.

EXAMPLE. It is required to eliminate the 7th order (major) critical of a seven-cylinder engine by means of a detuner with four bifilar pendulums fitted on a carrier disk (as shown on the sheet of illustrations facing this page) attached to the free end of the crankshaft. The engine can be represented by an equivalent system (Fig. 10) consisting of the following inertias and stiffnesses:

Inertias	Stiffnesses
Main flywheel: $J_3 = 8000$ Lb.in.sec.^{2}	Crankshaft: $K_2 = 27 \times 10^6$ Lb.in./rad.
Engine cylinders: $J_2 = 480$ Lb.in.sec.^{2} (from a frequency tabulation giving $J_2 = \Sigma(J_{cyl.} \Delta_{cyl.}^2)$ for the engine without the detuner).	(from $\omega_0^2 J_2 = K_2$, where $\omega_0 =$ natural phase velocity of engine without detuner).
	Stiffness of shaft between cylinder no. 1 and carrier disk of detuner: $K_1 = 81 \times 10^6$ Lb.in./rad.

Excitation torque (7th order): $T_{(n)} = T_{(7)} = 12{,}000$ Lb.in.

(1) As a trial, it is assumed that a carrier disk of 24 in. diameter can be employed. From the graph of section 1·1, p. 30, it is seen that a steel plate of this diameter and a thickness of, say, 1 in., has a moment of inertia of about 22 Lb.in.sec.^{2}. Allowing for external shield plates rigidly attached to the hub of the carrier disk, let the total carrier inertia be $J_1 = 40$ Lb.in.sec.^{2}.

(2) The next step is to determine the natural frequency of the system with the carrier disk, by means of a Holzer table beginning at the driven-machine end (in this example, at the main flywheel):

Table 'A'. $\omega^2 = 0.0552 \times 10^6$ rad.^{2}/sec.^{2}

Mass no.	J [Lb.in. sec.^{2}]	$\dfrac{J\omega^2}{10^6}$ [Lb.in./ rad.]	Δ [rad.]	$\dfrac{J\omega^2\Delta}{10^6}$ [Lb.in.]	$\dfrac{\Sigma(J\omega^2\Delta)}{10^6}$ [Lb.in.]	$\dfrac{K}{10^6}$ [Lb.in./ rad.]	Δ_{shaft} [rad.]
3	8000	441·60	1·0000	441·60	441·60	27·0	16·3556
2	480	26·496	−15·3556	−406·86	+34·74	81·0	0·4289
1	40	2·208	−15·7845	−34·85	−0·11	—	—

(3) These results can now be used to evaluate the system under forced-vibration conditions, which are obtained for one critical order when the pendulums are operating, for the same value of ω^2. (For other examples of Holzer tables with

[582]

excitation torques, for non-resonant speeds, see p. 534.) It will be convenient to write $T' = T_{(n)}/10^6 = 0.012$ and to enter only the symbol T' in the corresponding column of the table. The relative amplitude $\Delta = 1$ radian on the first line is replaced by an unknown θ, which will be subsequently determined. The tabulation is as follows:

Table 'B'. *Forced-frequency tabulation to determine torque required of detuner pendulums*

$F = 2243.8$ vib./min., $\omega = 234.95$ rad./sec., $\omega^2 = 0.052 \times 10^6$, $T' = T_{(n)}/10^6 = 0.012$

Mass no.	J	$\dfrac{J\omega^2}{10^6}$	θ	$\dfrac{J\omega^2\theta}{10^6}$	$\dfrac{T_{(n)}}{10^6}$	$\dfrac{\Sigma[J\omega^2\theta + T_{(n)}]}{10^6}$	$\dfrac{K}{10^6}$	θ_{shaft}
3	8000	441.60	1.0000θ	441.60θ	—	441.60θ	27.0	16.3556θ
2	480	26.496	-15.3556θ	-406.86θ	T'	$+34.74\theta + T'$	81.0	0.4289θ $+0.0123T'$
1	40	2.208	-15.7845θ $-0.0123T'$	-34.85θ $-0.0272T'$	—	-0.11θ $+0.9728T'$	—	—

Use can now be made of the fact that the vibration amplitude at J_1 is zero, in order to obtain $\theta_3 = \theta$. Furthermore, the total torque $\Sigma[J\omega^2\theta + T_{(n)}]$ at mass J_1 will be the reaction torque required of the pendulums, to counteract the excitation of the system.

Therefore, from the last line of the table, the relation $-15.7854\theta - 0.0123T' = 0$ gives the vibration amplitude of J_3 as

$$\theta = 0.0123T'/(-15.7845) = -0.0123 \times 0.012/15.7845 = -9.351 \times 10^{-6}\,\text{rad.}$$

and the reaction torque of the pendulums is

$$T_{\text{react.}} = (-0.11\theta + 0.9728T') \times 10^6 = (+0.11 \times 9.351 \times 10^{-6} + 0.9728 \times 0.012) \times 10^6$$
$$= 11{,}675\,\text{Lb.in.}$$

Thus $\qquad \theta_3 = -0.000536°, \quad \theta_2 = -15.3556\theta_3 = +0.0008° \quad$ and $\quad \theta_1 = 0°.$

(4) It is now necessary to assume a value for the distance from the centre of the carrier disk to the point of suspension of each pendulum. Let this be $R = 10$ in. Then to suppress the 7th order critical, the pendulum length required is

$$r = R/n^2 = 10/7^2 = 0.205\,\text{in.}$$

The expression for the reaction torque is $T_{\text{react.}} = \theta_1 \times m_1 \Omega_C^2 (R + r) R$ (see p. 575). Assuming a permissible pendulum amplitude of $\pm \theta_1 = 25° = 0.43633$ radians, one obtains (with $\Omega_C = \omega/n = 234.95/7 = 33.56$ rad./sec.) the total weight $W_1 = m_1 g$ of the pendulums as

$$W_1 = T_{\text{react.}} \times g/[\Omega_C^2 (R + r) R\theta_1]$$
$$= \frac{1.1675 \times 10^4 \times 3.864 \times 10^2}{(33.56)^2 \times (10 + 0.205) \times 10 \times 0.43633} = 89.975 \simeq 90\,\text{Lb.}$$

The specific weight of steel is $w = 0.283$ Lb./in.^{3} Consequently, the volume of steel available for the four pendulums is

$$W_1/w = V = 90/0.283 = 318.0\,\text{in.}^3.$$

This should be adequate for four bifilar-type pendulum masses with U-shaped cross-sections having a wall thickness of about 1 in. and arranged at 90° to each other on the carrier-disk periphery. The detuner thus designed will neutralize the 7th order critical, at the speed $N_C = \Omega_C \times 60/2\pi = 320.5$ rev./min. It will also eliminate the flank amplitudes of the 7th order at all speeds, since it is constantly 'tuned' to the 7th order ($\sqrt{(R/r)} = 7$).

[583]

Additional remarks

(*a*) The Holzer table for the non-resonant condition considered can include all engine inertias $J_{cyl.}$ and crankthrow stiffnesses $K_{cyl.}$ with separate lines for each inertia. In this case, the gas torque column contains the term $T' = T_{(n)\,cyl.}/10^6$ on each line corresponding to the cylinder inertias. An example of a detailed table of this type is given in section 2·3, pp. 321–4.

(*b*) The positions of criticals other than that to which the pendulum is tuned are shifted somewhat owing to the effect of the pendulum inertia. At speeds below the tuning speed ($\Omega < \Omega_C$) the 'effective pendulum inertia' $J_{P\,eff.}$ is negative ($\Omega n/\omega_E$ is less than unity in the expression for $J_{P\,eff.}$ on p. 579, with $n^2 = R/r = $ a constant value) and the positions of criticals in this part of the speed range are shifted upwards; at speeds above the tuning speed ($\Omega > \Omega_C$) the value of $J_{P\,eff.}$ is positive and the positions of criticals in this part of the speed range are lowered.

(*c*) Where these other criticals are of sufficient importance to require further investigation, their peak-amplitude positions can be determined by ordinary Holzer tables in which, however, the carrier inertia $J_{carr.}$ is replaced by $J_{carr.} + J_{P\,eff.}$ for the particular speed related to each frequency tabulation. Expressions for $J_{P\,eff.}$ at any speed are given for various types of detuners on the sheet of illustrations facing p. 582. Owing to $J_{P\,eff.}$, the engine is a different vibrating system for each order of excitation.

(*d*) The effect referred to under (*b*) can be used for 'inertia tuning' of the engine system. Consider, for instance, a four-cylinder engine which has an 8th order critical in the upper part of its running range, as well as a fairly large flank of the 6th. 'Resonance tuning' to the 8th order will suppress this order but also bring in more of the flank of the 6th. On the other hand, 'resonance tuning' to the 6th order will suppress the flank of the 6th and may also, owing to the associated effect of 'inertia tuning', enable the peak of the 8th to be shifted upwards, outside the running range.

(*e*) The reaction torque of a pendulum placed at an intermediate mass can be conveniently determined as follows, for resonance tuning conditions: Evaluate an *ordinary* Holzer table using the natural frequency of the system without a pendulum, and beginning at the driven-machine end. Denote the Holzer-table amplitudes at each mass by Δ_n, Δ_{n-1}, ..., $\Delta_{carr.}$ (carrier mass), ..., Δ_1. Then the reaction torque to be provided by the pendulum will be

$$T_{react.} = T_{carr.} + T_1 \frac{\Delta_1}{\Delta_{carr.}} + T_2 \frac{\Delta_2}{\Delta_{carr.}} + \dots,$$

that is, it is equal to the sum of all the gas torques T_1, T_2, ... multiplied by the Holzer-table amplitudes of their corresponding masses Δ_1, Δ_2, ... and divided by the Holzer-table amplitude at the carrier $\Delta_{carr.}$. (This

[584]

equation can be obtained from energy considerations, by writing $\Delta_{\text{carr.}} \times T_{\text{react.}} = \text{work of pendulum} = T_{\text{carr.}}.\Delta_{\text{carr.}} + T_1\Delta_1 + T_2\Delta_2 + \ldots = \text{work}$ of gas torques.)

Application to the example of p. 582. The Holzer table now considered is Table 'A'; furthermore, $\Delta_{\text{carr.}} = \Delta_1, T_{\text{carr.}} = 0, T_2 = T' \times 10^6$ (no other gas torques in this system). Then

$$T_{\text{react.}} = T' \times 10^6 \times \frac{\Delta_2}{\Delta_1} = 1\cdot 2 \times 10^6 \times \frac{-15\cdot3356}{-15\cdot7845} = +11,764\,\text{Lb.in.},$$

which is, within the limits of accuracy of the calculation, the same result as that obtained by evaluation of the torque on the last line of Table 'B'. Moreover, the amplitude at the carrier mass (J_2) will again be $\theta_2 = 0$.

(*f*) In order to suppress different criticals, it is possible to use several pendulums of different lengths attached to different crankwebs. The inertias of the masses other than that considered for 'resonance tuning' to a particular order are then the cylinder inertias $J_{\text{cyl.}}$ plus their effective pendulum inertias $J_{P\,\text{eff.}}$ for the critical order investigated. (Formulae for finite values of $J_{P\,\text{eff.}}$ are given on the sheet of illustrations facing p. 582.) See also (*c*) above. If several pendulums $P_1, P_2, \ldots$, of different lengths (different n-values) are attached to a single carrier, its inertia is $J_{\text{carr.}} + J_{P\,\text{eff.(1)}} + J_{P\,\text{eff.(2)}} + \ldots$.

(*g*) In the design of centrifugal pendulums, it is also useful to consider the effects of the reaction forces. Moreover, any viscous-damping actions occurring in pendulum constructions should be reduced as far as possible, since their effect is to alter the phase relation between the pendulum motion and the excitation torque, and this may reduce the effectiveness of the pendulum. Also the surface stresses of the rolling bodies in contact require consideration. (See Ker Wilson, in the Bibliography given below.)

(*h*) It is of interest to note that roller and bifilar type pendulums have been designed to reduce bending vibrations as well as torsional vibrations of crankshafts. (See Wiegand and Olson, in Bibliography.) Also, centrifugal pendulums have been used at the main-flywheel position to suppress gearbox and transmission shaft vibration.

BIBLIOGRAPHY ON CENTRIFUGAL PENDULUM DETUNERS

Crossley, F. R. E. *A.S.M.E. Paper*, no. 52-F-11.
Den Hartog, J. P. *Stephen Timoshenko, 60th Anniversary Volume* (Macmillan Co. 1938), pp. 23–6.
Kammerer, H. *Jb. dtsch. Luftfahrt*. vol. 2 (1940), pp. 80–91; (1942), pp. 42–51.
Ker Wilson, W. *Practical Solution of Torsional Vibration Problems*, vol. 2 (Chapman and Hall, 1941), pp. 512–96.
Kimmel, A. and Lutzweiler, J. *Ingen.-Arch.* vol. 12 (1941), pp. 100–8.
Kleiner, A. *Sulzer Tech. Rev.* no. 1 (1938).

Kraemer, O. *M.T.Z.* vol. 1 (1939), p. 3; *Z. Ver. dtsch. Ing.* vol. 82 (1938), pp. 1297–1300; vol. 83 (1939), p. 901.

Lambrich, R. *Luftfahrtforschung*, vol. 18 (1941), pp. 262–74.

Porter, F. P. *Evaluation of Effects of Torsional Vibration*, vol. 1 (1945), S.A.E. War Engineering Board, pp. 269–377.

Salomon, B. *C.R. Acad. Sci., Paris*, vol. 203 (1936), pp. 1315–7; and vol. 206 (1938), p. 1614.

Schick, W. *Ingen.-Arch.* vol. 10 (1939), pp. 303–12.

Stieglitz, A. Dissertation, T. H. Dresden (1937); *D.V.L. Jahrbuch*, vol. 2 (1938).

Taylor, E. S. *S. A. E. Jl.* March 1936.

Wiegand, F. J. and Olson, E. H. *S. A. E. Quart. Trans.* Jan. 1950, pp. 8–14.

Zdanowich, R. W. and Wilson, T. S. *Proc. Inst. Mech. Engrs, Lond.*, vol. 143, no. 3, June 1940, pp. 182–210.

3·6 FURTHER TYPES OF DAMPERS AND DETUNERS

There are many others types of torsional vibration-reducing devices besides those described in the previous sections of this handbook and some brief descriptions are given below.

Rubber and dry-friction dampers

The seismic mass is connected by bonded-rubber material to a back plate which is rigidly attached to the crankshaft. The arrangement is designed in such a way that the periphery of the seismic mass rubs on the corresponding friction surfaces of the back plate.

Springs and dry-friction dampers

The seismic mass is elastically connected by means of peripheral or radial springs to a back plate on the crankshaft, and also has peripheral friction surfaces which rub against corresponding surfaces on the back plate.

Mercury-pendulum detuners

Circular-arc capsules† or conically shaped containers‡ or chambers are partly filled with mercury and fitted in corresponding recesses in crankwebs, etc., in order to obtain a detuning effect similar to that of the roller type pendulum.

Sound-deadening rings of cast iron

Inserted with a shrink fit in the inside periphery of gearwheel rims, act as dry-friction dampers and reduce the ringing noise of gears appreciably.§

'Clutch' type dry-friction dampers

Centrifugally operated contacting elements with friction surfaces rub against an outer annulus; the arrangement, fitted at the free end of a crankshaft, operates as a dry-friction damper. The spring loads on the contacting elements are adjusted so as to allow slip to occur only when the inertia torque due to vibratory motion is operative.

Ball-type pendulum detuners

These are variants of the roller type, with a ball oscillating in a curved track (see the sheet of illustrations facing p. 582). *Link-type pendulum*

---

† Brit. Pat. Spec. no. 331,075, Duesenberg, F.S.
‡ Brit. Pat. Spec. no. 632,313, The de Havilland Aircraft Co., Ltd.
§ Den Hartog, J. P., *Mechanical Vibrations*, 2nd ed. (The McGraw-Hill Book Co., Inc. 1940), p. 129 and pp. 259–61.

detuners, with two or more links, are variants of the bifilar or multifilar type (see Fig. 1). Detuner pendulums have also been designed to suppress axial and flexural, as well as torsional vibrations.

Radial-spring pendulum detuner

This has been suggested[†]. It consists of a vibratory mass attached to a helical spring vibrating in a radial recess in a carrier disk. The reaction torque acting on the carrier is produced by the side thrust of the pendulum mass under the action of Coriolis force.

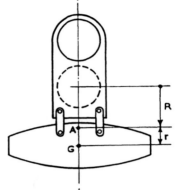

Fig. 1. Two link type pendulum detuner.

Fluid and electromagnetic couplings

These also reduce vibration amplitudes, see section 1·247. Formulae for damping coefficients for couplings of these two types are given in section 2·421. The theory of the vibration damping obtainable with fluid couplings is treated in detail by Den Hartog.[‡]

Non-linear spring couplings

Modified so as to have a large flywheel-type of seismic mass, are frequently used as detuners. For this they are fitted at the free end of the crankshaft. Examples:

Bibby detuners

(See section 1·245.)

Hülsenfeder detuners

(Fig. 2.)

Packs of slotted cylindrical-shaped springs[§] are fitted in recesses in the seismic mass and acted upon by the inner mass which is rigidly attached to the crankshaft. The stiffness of the spring packs increases with increasing deformation and provides the detuning effect. In addition, owing to oil lubrication, etc., there is a viscous-damping action with a relative damping coefficient (see section 3·2) γ of the order of 0·1. Peripheral deflexions of up to $\pm$ 6 mm. have been used in some designs.

Blade-spring detuners

One design (Fig. 3) comprises sets of blade springs acted upon by rods projecting from the carrier disk rigidly attached to the crankshaft, the seismic member being suitably supported on a crankshaft extension.

† Pringle, O. A., *A.S.M.E. Paper,* 52-SA-34.
‡ Den Hartog, J. P. *Mechanical vibrations,* 2nd ed. (The McGraw-Hill Book Co., Inc. (1940), p. 129 and pp. 259–261.
§ Pielstick, G., *Mitt. ForschAnst. Gutehoffn. Nürnberg,* vol. 4 (1936), pp. 123–8; and *M.A.N. Dieselmotoren-Nachrichten,* July 1937.

In another design (Fig. 4), packs of steel plate springs are fitted in grooves with curved sides, which give a non-linear (i.e. a variable stiffness) characteristic.

Junkers vane-type hydraulic damper†

This comprises a central disk, keyed to the free end of the crankshaft, and enclosed in a seismic mass designed as two halves of a casing supported on a crankshaft extension. The central disk is provided with radial vanes on either side near its periphery and the casing members also have corresponding sets of vanes which form a number of small chambers in recesses of the casing wall thickness, into which the vanes of the central disk project. The damper is continuously supplied with oil from the engine lubrication system and vibratory energy is dissipated by eddy formation and hydraulic resistance‡ in the flow of oil through narrow passages from one chamber to the other.

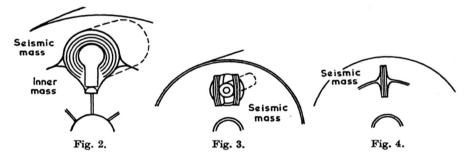

Fig. 2.   Fig. 3.   Fig. 4.

The relative amplitude of the casing members is limited by stops on the flat surfaces of the inner mass and the casing members are also connected to the inner mass by transversely arranged helical springs, which give a slight 'tuning' effect.

† For drawings of this damper, see Ker Wilson, W., *Practical Solution of Torsional Vibration Problems*, vol. 2, 2nd ed. (Chapman and Hall, Ltd, 1941), pp. 500–1.
‡ See: Neugebauer, F., *Tech. Mech. Thermodynamik.* vol. 1 (1930), pp. 137–47 and 184–97.

PART IV

INSTRUMENTATION

Note. The instruments mentioned in the text are confined to those types of which members of the B.I.C.E.R.A. Torsional Vibration Panel and the B.I.C.E.R.A. Laboratory have had experience in practical work. However, it is recognized that there are a number of other reliable types of mechanical and electrical equipment suitable for torsional vibration measurements.

Measurement accuracy. The accuracy of vibration measurements depends on a variety of factors and particularly on the following: the indication limit† of the equipment, the frequency and amplitude response of the measuring equipment, the experience and precautions of the user, and the extent to which test conditions may be disturbed or modified by external influences (spurious vibrations due to other motions than that being investigated, speed fluctuations, etc.).

In the following, these various aspects of vibration measurement will also be considered in reviewing various methods and equipments generally employed in torsional vibration investigations.

† I.e. the lowest amplitude of vibration that can be reliably indicated. For instance, this may be a total swing of 0·025° for a vibration pick-up; or a value corresponding to a stress of ± 200 Lb./in.² for a wire-resistance strain gauge on a steel shaft; or finally a value corresponding to a trace thickness (say 0·2 mm.) on recording paper. In all these cases, the accuracy will be good for large amplitudes, and will deteriorate gradually (to say 50 %) when measuring small amplitudes near the lower limit of reliable indication.

4·1 EQUIPMENT FOR THE MEASUREMENT OF VIBRATION AMPLITUDES

NOTATION

Symbol	Brief definition	Typical units	Symbol	Brief definition	Typical units
a	angle	rad.	r	radius, radius-vector	in., cm.
b	angle	rad.	R	radius	in., cm.
B	angle	rad.	T	torque	Lb.in., Kg.cm.
c	damping coefficient	Lb.in.sec./rad., Kg.cm.sec./rad.	v	velocity	in./sec., cm./sec.
c	angle	rad.	x	distance	in., cm.
D	diameter	in., cm.	y	distance	in., cm.
F	natural frequency	vib./min.	z	$=(1-\cos\phi)/(1+\cos\phi)$ in Hooke's joint formulae	—
G	modulus of rigidity	Lb./in.², Kg./cm.²	α	angle	deg., rad.
h	height	in., cm.	β	angle	deg., rad.
J	mass moment of inertia	Lb.in.sec.², Kg.cm.sec.²	γ	damping ratio	—
K	stiffness	Lb.in./rad., Kg.cm./rad.	δ	displacement	in., cm.
			Δ	Holzer-table amplitude	rad.
L	length	in., cm.	$\overrightarrow{\Delta}$	relative-amplitude vector	—
M	magnification factor	—	η	filter throughput	—
n	critical speed order number	—	θ	vibration amplitude	deg., rad.
			ϕ	angle	deg., rad.
N	speed	rev./min.	ψ	phase angle	deg., rad.
q	shear stress	Lb./in.², Kg./cm.²	ω	phase velocity	rad./sec.

Subscripts are used in the text to define more precisely, e.g. D_s = screw shaft diameter. Their meaning is fully explained in the context in each instance.

Metric units. The formulae given in this section can be used for computations in metric units (for instance those indicated in the above table). The numerical example of p. 598 is also valid as an illustrative example of evaluations if one replaces Lb. by Kg., and in. by cm., without altering the numerical values.

4·11 Mechanical torsiographs

4·111 *Geiger Torsiograph*

This mechanical recording instrument comprises a seismic mass or constant-speed flywheel (Fig. 1) freely supported on a shaft section to which it is elastically connected by means of a volute spring (in the low- or medium-speed models) or radial springs (in the high-speed model). This shaft section is coupled to the engine or driven-machine shafting, either by means of a belt-and-pulleys system, or (in the high-speed model) by a direct-drive arrangement at the free end of the engine shafting. The belt may be either of cotton (prestretched before use) or of steel 0·002 in. thick.

The seismic mass is enclosed in a casing which is rigidly attached to the vibrating shaft. Any torsional vibration transmitted to the latter will therefore result in relative motion between the casing and the seismic mass. In the absence of vibration, the three components (shaft, casing and seismic mass) rotate at the same uniform velocity.

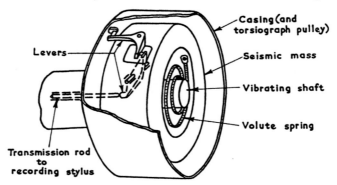

Fig. 1. Main features of a Geiger torsiograph for low and medium frequencies.

A stud situated on the inside periphery of the casing, in contact with a bell-crank arrangement pivoted on the seismic-mass periphery, transmits the vibratory motion to a second crank lever in a recess of the seismic mass. This second crank lever is pivoted in a direction at right angles to the shaft axis, and transmits vibratory motion axially, via a rigid rod passing through the hollow centre portion of the shaft, to a recording arrangement situated on the stationary body of the instrument.

The recordings, obtained either on ordinary paper by means of an ink-filled scribing pen or on special wax-covered paper with a stylus point, are 'unfiltered' records of the vibration. That is, they show the complete waveform variations, which may be due to one or more criticals, cyclic speed variations attributable to the engine or to malalignment of a pulley, etc., at the shaft position investigated. A timing system, with a second (electromagnetically-driven) stylus recording a 50 cycle/sec. vibration, is incorporated in the instrument. A third stylus, also electrically driven, and actuated by a contact-breaker on the engine shafting, is provided for the marking of engine revolutions.

Interpretation of records. Records of vibration at resonance, i.e. taken at a critical speed, are usually approximately sinusoidal and the maximum double-amplitude or total swing is easily determined (height $h = AA'$ in Fig. 2).

Recordings of flank amplitudes, that is, of vibrations in the neigh-

[594]

bourhood of a critical speed, are of a more complex shape (Fig. 3). In such cases, the total swing may be taken as the distance between the largest consecutive plus and minus peaks of the fairly smooth portions, determined by drawing 'envelopes', i.e. curves of the outer boundaries of the waveform.

In Fig. 3, the total swing due to torsional vibration is obtained at the position AA' of the recording, for the order investigated; it is not the value BB' which is due to other superimposed vibrations of lower frequency. (This was verified by harmonic analysis of such traces with a Stanley harmonic analyser at the B.I.C.E.R.A. Laboratory.) It is frequent practice, however, to use BB' (instead of AA') for conservative evaluations.

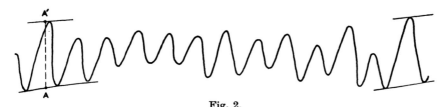

Fig. 2.

The shapes of very complex waveforms may be altered by phase lags (due to backlash, etc.) in the lever system of the instrument, so that recordings at non-resonant speeds cannot always be interpreted with the same accuracy as those for resonant, i.e. critical, speeds.

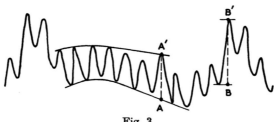

Fig. 3.

Evaluation of trace. From the measured height of the trace, the total swing is obtained by means of the relation

$$2\theta = \frac{h}{\frac{1}{2}D_P M_S M_D} \quad [\text{rad.}], \tag{1}$$

where h = total height of trace [mm.] ($h = AA'$ in Figs. 2 and 3), D_P = diameter of pulley on test shaft [mm.], M_S = static (pen and lever) magnification factor of instrument ($M_S = 3, 6, 9, 12$ or 24),† and M_D = dynamic

† These M_S-values are valid only if the normal torsiograph pulley of diameter D_T (148 or 100 mm.) is used. If a larger pulley of diameter $D_{T'}$ is fitted over the normal torsiograph pulley, the static magnification becomes $M_{S'} = M_S \times D_T/D_{T'}$, which should be used, instead of M_S, in eq. (1).

magnification factor of instrument. ($M_D \cong 1$ in the normal range of frequencies of the torsiograph, but for further accuracy it should be obtained from a response curve, such as that of Fig. 4.) The vibration amplitude is $\theta = \pm h/(D_P M_S M_D)$ [rad.]. If $D_P = D_T$ (= diameter of normal torsiograph pulley: 148 mm. for low-speed model, and 100 mm. for the high-speed model) then

$$\theta = \pm 0{\cdot}387h/M_S M_D \quad \text{[deg.]} \quad \text{for the low-speed model,}$$

and $\quad \theta = \pm 0{\cdot}573h/M_S M_D \quad \text{[deg.]} \quad \text{for the high-speed model.}$

Direct drive is equivalent to having $D_P = D_T$.

For further accuracy, the torsiograph may be calibrated before (or after) the test, using the same pulley system or direct-drive arrangement as for the test. If the calibrator produces a vibration amplitude $\pm \theta_{\text{cal.}}$ in the range of frequencies expected from the test, corresponding to a calibrated trace of height $h_{\text{cal.}}$, then the evaluation of the recordings can be made directly, by means of the relation:

$$\theta = \pm \frac{h}{h_{\text{cal.}}} \times \theta_{\text{cal.}} \quad \text{[degrees].}$$

Calibration. Mechanical torsiographs, owing to the fairly large inertia of their seismic mass, should be coupled only to calibrators capable of sustaining the corresponding inertia torques. Hooke's joint type calibrators, and parallel-misalignment type calibrators (described in section 4·12) are suitable in this regard, whereas eccentric or sine-cam calibrators used for smaller-size electrical pick-ups may not be able to sustain the loads, i.e. the high cam pressures. The vibration-amplitude and timing calibration-curves may vary from time to time and should be checked occasionally.

The frequency limits and amplitude response can be determined on a calibrator, for each particular instrument, on the basis of a calibration curve of the type shown in Fig. 4.

In this figure, the 'dynamic magnification'† of the instrument,

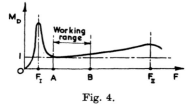

$$M_D = \frac{h_{\text{cal.}}}{D_P M_S \theta_{\text{cal.}(1)}},$$

for a constant value of $\theta_{\text{cal.}(1)} =$ vibration amplitude of the calibrator [rad.], is plotted against excitation frequencies F.

Fig. 4.

The vibration frequencies F_I and F_{II} correspond to resonances of the torsiograph system. F_I is the resonance due to the seismic mass and its spring arrangement, while F_{II} is the resonance caused by belt flap or

† Alternative designations. Frequency response, dynamic response.

vibrations of the lever system. The working range is usually taken between the frequency limits $A = 2F_{\mathrm{I}}$ and $B = 0.5F_{\mathrm{II}}$.

It is seen that the M_D-curve is nearly horizontal over the range AB, i.e. the instrument has a nearly 'flat (frequency) response' in this range of frequencies.

To verify that there are no additional effects on amplitudes, due to belt slip or other causes, it is useful to carry out further calibrations at other constant values of amplitude $\theta_{\mathrm{cal.\,(2)}}$, $\theta_{\mathrm{cal.\,(3)}}$, etc., also using various pen-lever ratios (i.e. different M_s-values). These will indicate the range of frequencies over which the 'amplitude response' of the instrument is linear.

Depending on the model (i.e. its flywheel size and spring stiffness), the lower frequency limit is between 150 and 30 vib./min., and the upper frequency limit is between 4500 and 10,000 vib./min., for vibration amplitude indication. The total height h of the trace is limited (by stops on the instrument) to 32 mm. for $M_S = 3$, and 16 mm. for higher M_S-values. If there is appreciable cyclic speed variation or any other periodic irregularity of the shaft rotation, this may cause low-frequency 'surging' of the trace and will make it necessary to use small M_S-values. At high engine speeds, the torsiograph can be coupled to the engine system by a belt-and-pulley arrangement giving a step-down ratio, although this also reduces the height of the trace. The recordings can be magnified photographically, if necessary, for evaluation.

4·112   The 'Cambridge' torsiograph

In this instrument, the driven pulley is coupled by soft springs to the small-size seismic mass or constant-speed flywheel. The relative angular movements between the pulley and the seismic mass are transmitted by a system of levers and converted into linear displacements by means of a rod acting on the recording stylus. The recordings are obtained on a moving celluloid strip driven by clockwork, and a second (electrically actuated) stylus provides 0·1 sec. time markings. Different stylus levers give (static) magnification ratios M_S of 5, 3 and unity. A third stylus is provided for the marking of engine revolutions. The torsiograph can be either belt-driven or directly coupled to the test shaft.

The records can be enlarged and measured by means of a viewer and photographic enlarger which is provided with the equipment.

For the interpretation of records, calibration, etc., the various considerations discussed in relation to the Geiger torsiograph are also applicable (with other numerical values) to the Cambridge instrument.

4·113   Basic components of mechanical torsiograph equipment

To obtain a clear idea of the functional arrangement of the systems, it is of interest to note that the mechanical torsiographs described above basically consist of the following components:

(1) a pick-up arrangement (seismic mass and transmission levers);

(2) an amplifying system (bell-crank levers of stylus and scribing pen); and

(3) an indicating system (stylus and wax paper or celluloid recording strip).

In addition, they may have:

(4) a time marking system, a shaft-revolution marking system;

(5) a clockwork (recorder drive) and an electric power supply.

The 'chain' or sequence of the assembly can be represented by a block diagram as follows:

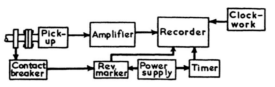

Fig. 5. Block diagram of mechanical torsiograph.

The similarity with corresponding block diagrams for electrical torsiograph equipment is evident (see section 4·12). In mechanical equipment, most of the components are incorporated in the body of the torsiograph. In electrical equipment, the pick-up unit is generally separated from the amplifying and recording unit. The 'output motion' (or signal) produced by mechanical pick-ups of the types considered above is proportional to displacement. These mechanical pick-ups can therefore be termed 'displacement pick-ups'.

4·114 *Measurements taken at intermediate points of the shafting*

Seismic types of torsiographs, mechanical or electrical, can be used in belt-drive arrangements to take vibration-amplitude measurements at intermediate points on the shafting, for instance on the line shafting of marine installations or on the transmission shafting of pump drives.

In such cases, it is necessary to interpret the recordings correctly by taking account of the phase angles due to driven-machine damping. An appropriate method for this purpose was evolved by Archer† and its use will be described by means of the following illustrative example.

NUMERICAL EXAMPLE. A marine propulsion system with a six-cylinder four-stroke cycle single-acting engine has a 3rd order one-node critical at 49 rev./min. Geiger torsiograph records were taken at two positions F and A on the line shafting. It is required to determine the total twist of the shafting between the engine and the propeller as well as the corresponding stresses in the line shaft and screw shaft.

† Archer, S., 'Contribution to further accuracy in the calculation and measurement of torsional vibration stresses in marine propeller shafting', *Proc. Instn Mech. Engrs, Lond.*, vol. 164, no. 3 (1951), pp. 351–6. [The above example is derived from this paper.]

‡ J_1 is obtained from a Holzer tabulation as the sum of the engine inertias multiplied by their corresponding Holzer amplitudes squared.

Data. Engine inertia $J_1 = 385,000$ Lb.in.sec.^{2}‡ Propeller inertia $J_2 = 100,000$ Lb.in.sec.^{2} (with entrained water). Natural frequency $F_1 = 147$ vib./min. (measured value). Shaft stiffness $K = \omega_1^2/[(1/J_1) + (1/J_2)] = 18 \cdot 8 \times 10^6$ Lb.in./rad., where $\omega_I = 2\pi F_I/60 = 15 \cdot 4$ rad./sec. Propeller damping coefficient $c_2 = 205,700$ Lb.in.sec./rad. (from example of p. 378, section 2·422). Excitation torque of 3rd order: $T_3 = 1 \cdot 76 \times 10^6$ Lb.in. Diameter of line shafting $D_e = 312$ mm. Diameter of screw shaft $D_S = 349$ mm.

For these tests, the torsiographs used were modified by fitting over their normal pulley (diameter $D_T = 148$ mm.) a larger aluminium pulley of diameter $D_{T'} = 220$ mm. Recordings were taken with a (static) pen magnification $M_S = 3$ and the dynamic magnification of the instruments was $M_D = 1$. The double-amplitudes of the recorded traces were $h_F = 18 \cdot 5$ mm. at the forward position and $h_A = 36 \cdot 0$ mm. at the aft position.

(1) *Evaluation of records.* The recorded trace heights can be evaluated by means of eq. (1), p. 595, which gives

$$\pm \theta = \frac{h}{D_P M_D M_S \times (D_T/D_{T'})} = \frac{h}{312 \times 1 \times 3 \times (148/220)} = 1 \cdot 59 \times 10^{-3} \times h \quad [\text{rad.}].$$

The ratio $D_T/D_{T'}$ takes account of the fact that the driven pulley, fitted on the torsiograph, was of a diameter $D_{T'}$ different from the normal diameter D_T. Consequently, the measured amplitudes at the forward and aft positions are

$$\theta_F = \pm 1 \cdot 59 \times 10^{-3} \times 18 \cdot 5 = 0 \cdot 0294 \text{ rad.}$$

and

$$\theta_A = \pm 1 \cdot 59 \times 10^{-3} \times 36 \cdot 0 = 0 \cdot 0572 \text{ rad.}$$

(2) *Interpretation by usual method not taking account of damping.* The total torque at either (end-) mass of the simplified two-mass system is $\Sigma(J\omega^2\Delta) = 0$ or $J_1 \Delta_1 = J_2 \Delta_2 = 0$ and this gives $J_1/J_2 = -\Delta_2/\Delta_1 =$ ratio of vibration amplitudes of the two masses. Therefore, with $\Delta_1 = 1$ unit,

$$\Delta_2 = -J_1/J_2 = -3 \cdot 85 \text{ units.}$$

The equivalent length L_e between J_1 and J_2 is obtained from $L_e = \dfrac{\pi}{32} G_e D_e^4/K$ (where $G_e =$ modulus of rigidity of shaft D_e) and can be taken as a base length for the deflexions diagram without damping (Fig. 6a). The 'swinging form' is obtained by plotting $\Delta_1 = 1 \cdot 0$ and $\Delta_2 = -3 \cdot 85$ in this diagram and drawing a straight line from the end-point of Δ_1 to that of Δ_2. The total shaft twist without damping is $\Delta_{sh} = \Delta_1 - \Delta_2 = 4 \cdot 85$ units.

The positions of the measuring points are determined from the L_e-values of the shaft portions $\overline{FD}$ and $\overline{AB}$. Their 'Holzer-amplitude' values can be read off the diagram: $\Delta_F = 0 \cdot 49$ and $\Delta_A = -1 \cdot 49$ units.

Therefore, for the system considered without taking account of damping, the total shaft twist can be obtained as follows:

(a) $\pm \theta_{sh} = \Delta_{sh} \theta_F/\Delta_F = 4 \cdot 85 \times 0 \cdot 0294/0 \cdot 49 = 0 \cdot 297$ rad.;

(b) $\pm \theta_{sh} = \Delta_{sh} \theta_A/\Delta_A = 4 \cdot 85 \times 0 \cdot 0572/1 \cdot 49 = 0 \cdot 186$ rad.

The considerable discrepancy between these two results shows that the usual method of evaluation is inadequate where there is appreciable driven-machine damping.

(3) *Interpretation by method taking account of damping.* The evaluation by Archer's method is as follows:

First the phase angle ψ between $\overrightarrow{\theta_1}$ and $\overrightarrow{\theta_2}$ is calculated from

$$\tan(180°-\psi)=\frac{\omega_1 c_2}{K}\times\frac{J_1}{J_2}=\frac{15\cdot4\times2\cdot057\times10^5}{18\cdot8\times10^6}\times3\cdot85=0\cdot6483;$$

hence $\psi=147°\,03'$ and $|\sec\psi|=1\cdot1925$. [Derivation in section 2·45, p. 437.] The ratio of the amplitudes $|\theta_2|$ and $|\theta_1|$, taking account of damping, is obtained from

$$\left|\frac{\theta_2}{\theta_1}\right|=\frac{(J_1/J_2)}{|\sec\psi|}=\frac{3\cdot85}{1\cdot1925}=3\cdot226.$$

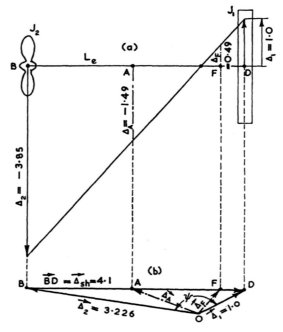

Fig. 6. (a)=deflexion diagram without damping; (b)=vector diagram with damping. $\overrightarrow{\Delta_A}=1\cdot295$; $\overrightarrow{\Delta_F}=0\cdot665$.

It is now possible to construct the vector diagram† of Fig. 6b by taking $\overrightarrow{\Delta_1}=1\cdot0$ unit, determining $\psi=147°\,03'$ and drawing $\overrightarrow{\Delta_2}=3\cdot226$ units with this angle relatively to $\overrightarrow{\Delta_1}$. Joining the tips B and D of the vectors, we obtain the total shaft twist as

$$\overrightarrow{BD}=\overrightarrow{\Delta_{sh}}=\overrightarrow{\Delta_1}-\overrightarrow{\Delta_2}=4\cdot1\text{ units.}$$

(This value can be measured on the diagram.) Using $\overrightarrow{BD}$ as the base length L_e,

† In practice, it will be found easier to construct the lower part of Fig. 6 first at an arbitrary scale and then to relate it to the other diagram (without damping) shown above it.

[600]

we mark on it the torsiograph positions F and A, and by joining these points to O we obtain

$$\overrightarrow{OF} = \overrightarrow{\Delta_F} = 0.665 \quad \text{and} \quad \overrightarrow{OA} = \overrightarrow{\Delta_A} = 1.295.$$

Considering first the records taken at the forward position F, we have

$$\pm \theta_{sh} = \pm \left| \frac{\overrightarrow{\Delta_{sh}}}{\overrightarrow{\Delta_F}} \right| \times \theta_F = \frac{4.1}{0.665} \times 0.0294 = 0.181 \,\text{rad.}$$

From the records at the aft position A

$$\pm \theta_{sh} = \pm \left| \frac{\overrightarrow{\Delta_{sh}}}{\overrightarrow{\Delta_A}} \right| \times \theta_A = \frac{4.1}{1.295} \times 0.0572 = 0.1813 \,\text{rad.}$$

That is, both records give the same total shaft twist.

The vibratory shaft torque is

$$T_{sh} = \pm K\theta_{sh} = 18.8 \times 10^6 \times 0.181 = \pm 3.4 \times 10^6 \,\text{Lb.in.},$$

and the vibratory stresses in the shafting can be evaluated:

$$\text{line shaft:} \quad q_{D_s} = T_{sh} \bigg/ \left(\frac{\pi}{16} D_s^3 \right) = \pm 9320 \,\text{Lb./in.}^2;$$

$$\text{screw shaft:} \quad q_{D_s} = T_{sh} \bigg/ \left(\frac{\pi}{16} D_s^3 \right) = \pm 6730 \,\text{Lb./in.}^2.$$

The evaluation without damping would lead to higher values for stresses, with disparity in the results due to the two different values for θ_{sh} derived by the method without damping.

FURTHER INFORMATION ON MECHANICAL TORSIOGRAPHS

M.A.N. Nachrichten, no. 27, January 1953, pp. 2–6. [In German.] (High-speed Geiger torsiograph with new transmission-lever arrangement.)

Ker Wilson, W. *Practical Solution of Torsional Vibration Problems*, vol. 2, 2nd ed. (1941). (Details of mechanical and electrical torsiographs.)

Evaluation of Effects of Torsional Vibration, S.A.E. War Engineering Board (1945), *Cooperative Testing of Torsiographs and Calibrators*, S.A.E. Technical Board (1947). Both published by the Society of Automotive Engineers, Inc., New York 18, N.Y. (Details of mechanical and electrical torsiograph equipment.)

Shannon, J. F. The effect of belt length on Geiger torsiograph records. *J. R. Tech. Coll. Glasg.* vol. 4, part 2 (January 1938).

4·12 Electrical torsiograph equipment

Before going into the features of electrical torsiograph instrumentation, it may be of interest to give a short list of books and papers which can be used as introductions to the types of electronic equipment employed in vibration engineering:

Stansfield, R. The measurement of torsional vibrations. *Proc. Instn Mech. Engrs, Lond.*, vol. 148, no. 5 (1942), pp. 175–93.

Briggs, G. A. and Garner, H. H. *Amplifiers—The Why and How of Good Amplification.* Publ. by Wharfedale Wireless Works, Bradford, Yorks. [Non-mathematical introduction to valves, amplifiers, feedbacks, causes of hum, instability and distortion, etc.]

King, A. J. The analysis of vibration problems. *J. Instn Elect. Engrs*, part 2, vol. 93 (1946), pp. 435–62. [Integrating and differentiating circuits.]

Van Santen, G. W. *Mechanical Vibration*. [Chapters 17, 18 and 19: Pick-ups and measuring techniques.]

Jones, Jr., H. J. *Prod. Engng*, March 1955, pp. 180–5. [Simple diagrams illustrating the working principles of pick-ups operating by modulation of amplitude, frequency, phase, etc.]

Dam, D. *Bull. schweiz. elektrotech. Ver.* 10 Jan. 1953, pp. 4–11. (In German.) [Diagrams of velocity and displacement type pick-ups.]

Wittke, H. Electrical Integration Methods. *Frequenz*, Feb. 1955, pp. 49–57. (In German.) [Wiegand-Hansen, Miller and Förster integrators.]

4·121 *Standard Sunbury torsiograph equipment*

This equipment consists of the following basic components:

(1) a velocity-type† torsional vibration pick-up, usually fitted at the free end of the crankshaft,

(2) a frequency selector,‡

(3) an integrator and amplifier,

(4) an oscilloscope,

(5) a power supply unit with a motor-type converter for H.T. supply, the entire equipment being operated from 12-volt storage batteries, and

(6) a contact-breaker unit fitted on the engine camshaft or crankshaft and operating the time-sweep of the oscilloscope.

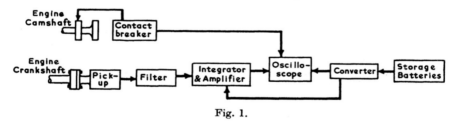

Fig. 1.

A block diagram of the assembly§ is given in Fig. 1.

The vibration pick-up is shown schematically in Fig. 2.

The axially arranged horseshoe-type field magnet is screwed into a brass casing serving as a seismic mass and freely supported on the shaft of the pick-up. The armature coils are mounted on stalloy laminations fitted into the shaft and their output leads are connected to slip-rings.

---

† This pick-up produces an output voltage proportional to the rate of cutting lines of force by the armature windings, i.e. proportional to the product of vibration amplitude θ and frequency $\omega/2\pi$, in other words to the vibration velocity $\dot{\theta} = \omega\theta$. In order to indicate vibration amplitude, an 'integrating circuit' is needed, which divides the pick-up output by ω $\left(\text{so as to have} - \int \theta_0 \omega \sin \omega t\, dt = \dfrac{1}{\omega}\,(\theta_0 \omega)\cos \omega t = \theta_0 \cos \omega t\right)$.

‡ Alternative designations. filter, band-pass filter, wave-analyser.

§ The details in the following refer to the Sunbury Indicator type 74301-B. Their new Indicator, 74702 Type F, has a push-pull input, a vibrator-type power supply, and an integrating circuit which is linear and without phase errors (integration of complex waveforms) down to about 2·5 c/s.

The brushgear is mounted on ball bearings and should be tied (to prevent its rotation) by means of an elastic band to some adjacent part of the engine. The plastic-material adaptor between the pick-up flange and the crankshaft flange is replaceable when its spigot and recess surfaces are worn by frequent assembling and removing.

The seismic mass has no elastic connection to the pick-up shaft, the coupling between these two components being entirely due to semi-viscous friction at the plain bearings. The inside recesses of the casing are packed with grease. The relative angular movement of the seismic mass is limited to a few degrees by stops (with flat packs of cushioning springs on the stops of the pick-up hub portion). The magnetic flux tends to centralise the casing around the armature midway between the stops.

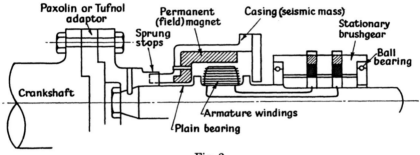

Fig. 2.

In an alternative arrangement, the pick-up can be driven by a belt-and-pulley system, but the direct drive is preferred, particularly for the indication of fairly high frequency vibrations.

The pick-up output leads are connected to the frequency-selector or filter unit, which is tuned by the operator to each vibration frequency investigated. This filter may be switched out when it is desired to obtain 'unfiltered' diagrams, in which case its input terminals are connected directly to its output terminals. The filter unit is a 'passive network', i.e. without valves or voltage sources in its circuit, and its 'throughput' (output voltage divided by input voltage) depends on the frequency of the vibration.† The 'filter calibration curves' supplied by the makers enable this attenuation to be taken into account in the evaluations.

From the filter, the signal voltage is applied to the input of the integrator and amplifier, and thence to the oscilloscope. The input impedance of the amplifier unit is fairly high (of the order of 100,000 ohms) and therefore prevents any loading of the low-impedance pick-up.

An engine-driven time-sweep is employed so as to obtain a stationary waveform on the oscilloscope screen; in this way the trace is periodically

† If there is any intermittent fault, however, in leads or connexions ahead of the filter, the filter will produce spurious sine-wave outputs at various frequency settings.

displayed at time intervals which vary in accordance with the engine speed. In order to obtain a complete trace of all the torsional vibrations occurring during one engine cycle, the contact breaker is mounted on the camshaft (or pump shaft) for four-stroke cycle engines, or on the crankshaft for two-stroke cycle engines. When the engine is running on a critical speed, one then observes that the number of peaks shown on the screen† is equal to the order number of the harmonic of the working cycle. Consequently, the

> critical speed order number = number of peaks,
> for two-stroke cycle engines,
> $= \frac{1}{2} \times$ number of peaks,
> for four-stroke cycle engines.

A sketch of the contact-breaker arrangement is shown in Fig. 3. Both contacts of the contact-breaker should be well insulated (and must not be earthed), since they carry an H.T. voltage.

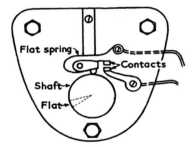

Fig. 3. Contact-breaker used with Sunbury equipment.

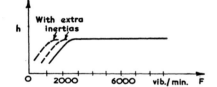

Fig. 4. Height h of oscilloscope trace at various frequencies, when the pick-up is vibrating at constant amplitude.

Contact is made only when the projection on one of the contact members goes over the small flat portion of the circular collar on the camshaft and this allows the triggering voltage of the time sweep to act. The gap should be carefully adjusted to avoid arcing. The flat on the shaft should be as small as possible so as to avoid losing more than, say, 20° of the 720° diagram.

Examples of torsional vibration oscillograms obtained with Sunbury torsiograph equipment are shown on pp. 289–90 of section 2·1 of this Handbook.

Working range. The equipment (i.e. pick-up, amplifier integrator and oscilloscope, without the filter) has a flat response down to a frequency of about 2600 vib./min. (Fig. 4). By fitting an additional inertia over the seismic mass (casing) of the pick-up, the low-frequency cut-off can be

† Care should be taken that the trace of the complete cycle can be identified on the screen.

shifted to 1500 or even close to 1000 vib./min. The inertia mass can be designed in the form of a split ring of brass, the two halves being joined together with socket screws.

It is also possible to operate on the sloping portions of the curves, provided that the calibrator is also run at the test frequency. The seismic mass of the pick-up should be allowed to swing freely in this calibration, that is, the pick-up should be vibrated from its flanged end and not at the seismic mass. In this frequency range below 2000 vib./min., the Sunbury filter gives little throughput, but other suitable types of filters (with an input impedance of 100,000 ohms or higher) can be used satisfactorily with the other components of the Sunbury equipment. Similar considerations apply in regard to the integrating circuit, and the Sunbury pick-up is now used with a variety of integrators, amplifiers and wave analysers.

Note. To avoid low-frequency surge or 'float' of the trace on the screen, a high-pass filter excluding vibration frequencies below, say, 20 cycles/sec. may be employed. In such cases, this filter should definitely also be included in the equipment when calibrations are made.

Calibration and evaluation of trace. The pick-up (and its associated equipment) is calibrated with one of the various types of calibrators described at the end of this section (pp. 618–29). The total height $h_{cal.}$ [cm.] of the trace on the oscilloscope screen corresponds to a vibration amplitude (half total swing) $\pm \theta_{cal.}$ [degrees]. Therefore, if in the engine test one obtains a trace of total height h, it will correspond to a vibration amplitude

$$\theta = \pm \frac{h}{h_{cal.}} \times \theta_{cal.} \quad \text{[degrees]}.$$

This applies strictly to 'unfiltered' results. It also applies to filtered readings if the calibration is made with the same filter settings (i.e. for exactly the same frequencies) as those of the test.†

If the calibration is made without the filter (which is the normal case) and 'filtered' traces of engine vibration are to be evaluated, it is necessary to use the 'throughput curves' of the filter (see Fig. 5) for the range-switch setting considered.

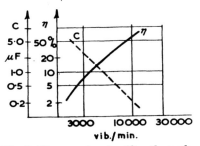

Fig. 5. Diagram showing filter throughput η and corresponding filter setting C (in microfarads) plotted against frequency F (vib./min.).

Let $h_1 =$ total height of trace on oscilloscope screen at an engine frequency F_1, obtained with a filter setting C_1, corresponding to a filter throughput η_1 [%].

† This subject is given further consideration under the heading 'Wave analysers', p. 611.

Then the value of the vibration amplitude of the pick-up on the engine is obtained as

$$\pm\,\theta_1 = \pm\,\frac{\theta_{\text{cal.}}}{h_{\text{cal.}}}\times\frac{100}{\eta_1}\times h_1 \quad \text{[degrees].}$$

An example of results of measurements is given in Fig. 6. The 'unfiltered' amplitudes indicate the overall value of torsional vibration and are not generally used to determine the stress due to a particular order, although they can serve as a basis for an assessment with a large safety factor.

The 'filtered' amplitudes for various orders are also shown. Their peak values are used as a basis for stress evaluations (see section 2·3).

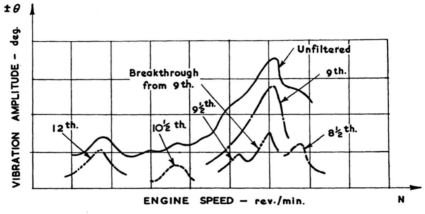

Fig. 6.

Technique of measurements. To obtain the resonance curve of, say, the 9th order critical in Fig. 6, the procedure is as follows:

(1) Examine the whole speed range without a filter, as a preliminary, to avoid missing any frequencies which are not expected and to have a plot of the unfiltered vibration conditions.

(2) Determine the peak amplitude of the resonance curve and the corresponding critical speed, by repeated measurements at various speeds around the peak and filter adjustments until the maximum amplitude value is found. The vibration frequency F_C should then be an integer or a half-order multiple n of the critical running speed N_C.

(3) For the flanks of this critical, beginning at an engine speed N lower than N_C, set the filter to a frequency $F = n \times N$ and take the corresponding amplitude reading. (This F is the forced-vibration frequency of the nth order.) Then increase the engine speed to N', set the filter to $F' = n \times N'$, and take a measurement. In this way, determine the trace height of the resonance curve from, say, $0 \cdot 9 N_C$ to $1 \cdot 1 N_C$.

The engine speed should be kept as constant as possible during each measurement,† and all measurements are taken at a constant b.m.e.p. (for instance, at full load) with a further set of measurements at another b.m.e.p. value if necessary. The constant-load curves, however, cannot be obtained in certain types of installations, and in such cases one determines the curves in accordance with the characteristics of the system (e.g. propeller law). Mark the test points in the results and do not smooth the plotted curves. The flat portions or secondary peaks in the resonance curve are frequently due to the filter selectivity characteristics (breakthroughs from other critical orders), which vary with each type of filter used.

Recommended procedure for measuring the amplitude of torsional vibrations

On the basis of the experience now available, the following procedure is recommended by the B.I.C.E.R.A. Torsional Vibration Panel for measurements of torsional vibration amplitudes by means of electronic equipment. The recommendations apply strictly to Standard Sunbury equipment. With minor modifications they are also applicable to other types of electronic equipment used for torsional vibration investigations.

Procedure

(1) Warm the pick-up on the engine for 15–30 min.

(2) Switch on amplifier and oscilloscope a quarter of an hour before use. Note the L.T. and H.T. voltages (or other voltages used as reference values). Ensure that the equipment and screened leads are effectively earthed.

(3) Remove the warm pick-up from the engine and calibrate it immediately, preferably at a fairly large amplitude, say $\frac{1}{2}$° total swing.

(4) Take the readings required with a number of check readings. A calibrated tachometer should be used for speed indication.

(5) Recalibrate the pick-up. If the calibration figures do not agree with those previously obtained, the test should be repeated. As a further check, note the H.T. and L.T. voltages (or other reference voltages of the equipment), which should be unaltered.

Precautions

Gain switch of amplifier. It is desirable to calibrate at the settings of the gain switch which are used for the test measurements. If different gains are used for calibration and testing, it is essential to verify the linearity of the gain settings and the repeatability of the results when the switch has been disturbed.

Battery voltage. The L.T. supply must be at least 11·5 volts when the battery is under load (i.e. as indicated by the voltmeter on the amplifier).

† For severe criticals the engine should be taken out of the control of the governor.

The capacity of the accumulator should preferably be not less than 75 ampere-hours. The H.T. supply should be between 220 and 240 volts.

External influences. As far as possible, avoid having large iron masses or sources of heat, such as exhaust pipes, in the proximity of the pick-up. If the engine has large masses which extend alongside the pick-up, such as some auxiliary drives, a stationary cylindrical mild steel shield surrounding the pick-up should be used to prevent interference with the magnetic field of the pick-up.

Magnetization of bearings. The ball bearings may become magnetized. A check should be made with the pick-up on a calibrator running at normal rotational speed. The brush gear assembly is turned slowly by hand through 360°. Any variation in the output signal will probably be due to magnetization of the ball races. After removal, the ball bearings can be explored with the aid of a compass, and then demagnetized if necessary.

Switches, contacts and terminals. All mechanical contacts should have clean surfaces. Contacts can be cleaned with acetone, carbon tetrachloride, Servisol, etc., where required. Soldered terminals are an advantage, except on the actual pick-up connexions which are susceptible to fractures due to vibration. Spade terminals or plug-in type terminals are preferred for pick-up connexions.

Frequency selector unit. With the Sunbury filter, it is recommended to use the lowest range setting for a given frequency, to avoid high-filter correction factors (that is, low η-values, see Fig. 5, p. 605). Occasional verification of the filter throughput curves is essential, using a variable frequency oscillator and measuring the output on the oscilloscope screen and without the filter.

Malalignment of pick-up. Laboratory tests show that an eccentricity of 0·010 in. or an angular malalignment of $\frac{1}{4}$° can give an error of indication of 3–4 % in the measurement of a vibration of 0·5° total swing. The flange of the engine shaft should be checked with a dial gauge on the mating face and in the spigot bore. The spigots on both sides of the plastic-material (e.g. 'Tufnol') disk supplied with the pick-up should be a good fit. Finally, the cylindrical surface of the seismic mass of the pick-up should be checked for true running with a dial gauge.

Calibrating equipment. The calibrator should give a fair amount of swing, say $2\theta = 0.5$°. A motor speed giving a vibration frequency of at least 3000 vib./min. is recommended for testing engines with natural frequencies above 3000 vib./min. For lower frequencies, the calibrator should be run at the expected frequency to minimize errors due to the integrating circuit. For very low frequencies, it is advisable to increase the seismic mass of the pick-up by adding an inertia ring (see p. 604).

Diagram stability. In some cases, the whole diagram is observed to swing up and down the oscilloscope screen. This may be due to faulty leads and connexions, or to a drop in the grid bias voltage. If it is due to external causes, such as excessive engine frame vibration, a high-pass filter admitting only the range of frequencies under study will give an improvement. However, this filter may affect the throughput curves of the frequency selector unit and these should be verified when such a filter is used.

Sensitivity of pick-ups. The sensitivity of various pick-ups can vary considerably but, in general, the response of a given pick-up should remain fairly constant if the unit is in good condition. Low pick-up output or spurious signals may be caused by dirty or defective brush gear. It has been found that a decrease in sensitivity and partial seizure of a pick-up can be due to the presence of grit in the bearings of the seismic mass.†

4·122   *Sperry-M.I.T. torsiograph equipment*

This equipment comprises the following units:

(1) a velocity-type torsional vibration pick-up, usually fitted at the free end of the crankshaft,

(2) an integrating amplifier,

(3) a Duddell (mirror-galvanometer type) oscillograph, with viewing screen and photographic film recording equipment, and

(4) a power supply unit, connected to the mains.

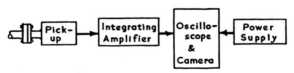

Fig. 7.

A block diagram of the assembly is given in Fig. 7. The equipment normally comprises two Duddell oscillographs, with separate channels for recordings from two different pick-up positions. In alternative arrangements, the amplifier output can be fed into a pen recorder, a valve voltmeter, or a wave analyser and cathode-ray oscilloscope.

The main features of the pick-up are shown schematically in Fig. 8. The brass casing carries two horseshoe-type permanent magnets and forms the seismic mass. It is supported on ball bearings and connected to a flange on the pick-up shaft by means of a soft helical coupling spring, which gives the pick-up a natural frequency of 6–10 cycles/sec. Frictional damping is provided by means of a spring-loaded friction plug rubbing

† Generally, on gain switch position 2, unfiltered, using the 'high engine speed' setting of the Sunbury amplifier, the pick-up response should be about 5 cm. per degree total swing or even higher, for a calibration frequency of the order of 2500–3000 vib./min.

against the flange surface. The slip rings are arranged concentrically on the flat surface of a bakelite disk.

The lower limit of the working range of the pick-up is around 900 vib./min. and the upper limit (with the Duddell oscillograph) is at about 60,000 vib./min.

As regards calibration, sensitivity, and measurement procedure, the considerations set forth in the previous pages are also valid, with minor modifications, for this equipment.

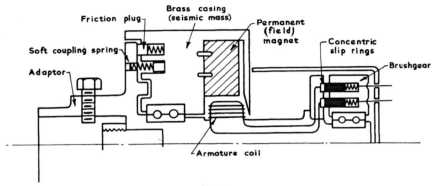

Fig. 8.

4·123   A.E.L. Mark III pick-up†

This velocity-type torsional vibration pick-up (see Fig. 9) comprises an axially arranged annular magnet which is screwed into a casing serving as a seismic mass. The seismic mass is supported on ball bearings and is elastically coupled to a flange rigidly connected to the pick-up shaft, by means of two helical coupling springs (not shown in the figure) situated in peripheral recesses between the flange and the seismic mass.

The system formed by the coupling springs and the seismic mass has a natural frequency of 6 cycles/sec. and is damped by means of two spring-loaded 'Tufnol' friction plugs so as to obtain 'dead-beat' damping (i.e. a damping ratio γ of 0·6).‡ This damping arrangement is not affected by temperature. Both the coupling springs and the friction-plug springs are easily accessible for adjustment. The vibration amplitude is limited to $\pm 10°$ by means of stops on the pick-up shaft.

The leads from the armature coils pass through the shaft and are connected to concentric slip-rings at the end of the shaft. The stationary

† By permission of The Superintendent, Admiralty Engineering Laboratory.
‡ The damping ratio for elastically-coupled systems is defined as $\gamma = c_{PU}/(2J_{SM}\omega_{PU})$, where c_{PU} = damping coefficient of pick-up, J_{SM} = inertia of seismic mass, and ω_{PU} = natural phase velocity of pick-up. With $\gamma = 0·6$ (which is determined by experimental adjustment of the spring-loaded friction plugs), the response is aperiodic, and practically flat over the resonance range of the pick-up, i.e. around 6 cycles/sec., besides being linear through the higher range of frequencies (above 12 cycles/sec.) in which the pick-up is used for measurements.

[610]

brushgear carrier is completely enclosed and also carries a small electro-magnetic pick-up which may be used to obtain revolution markings.

The pick-up sensitivity is 60 millivolts at 100 cycles/sec. for a vibration amplitude $\theta = \pm 0.5°$. The response is linearly proportional to vibration amplitude at constant frequency, and increases linearly with frequency at constant amplitude (i.e. the output signal is of constant amplitude with a 'perfect' integrating circuit). The pick-up can be used to indicate frequencies down to 12 cycles/sec.

For belt-drive applications, the flange member in contact with the friction plug is modified so as to form an outer housing suitable for use as a pulley and extending over the seismic mass.

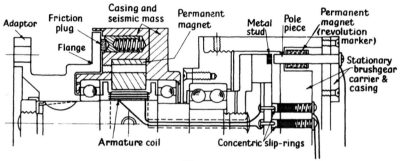

Fig. 9.

The pick-up output is fed into an integrating amplifier of suitably high impedance and then analysed with a wave analyser, or displayed on an oscilloscope screen, the arrangement being as shown in the block diagram of Fig. 10.

4·124   *Wave analysers*†

In many cases it is convenient to use a wave analyser in torsional vibration investigations. These filters, some of which were primarily designed for noise measurements, usually include compensating valve

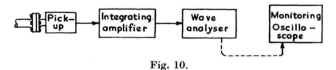

Fig. 10.

stages so as to obtain, in principle, a constant throughput (e.g. $\eta = 100\%$) at all frequencies. A suitable arrangement for a wave analyser used in conjunction with a velocity pick-up is shown in the block diagram of Fig. 10.

† Alternative designations: frequency selectors, waveform analysers, adjustable band-pass filters, vibration analysers.

Wave analysers of the following types are used for torsional vibration investigations by members of the B.I.C.E.R.A. Torsional Vibration Panel and the B.I.C.E.R.A. Laboratory:

Dawe analyser

This is a battery-operated unit, with a lower frequency limit of 2·5 cycles/sec. With velocity pick-ups, it requires an integrating amplifier (having an input impedance of about 100,000 ohms), which in many cases has been made by the users.

The Dawe analyser gives 'filtered' readings only. For calibration or for engine tests, the analyser should be tuned in each case to the calibration frequency or to the engine frequency investigated (using for the engine test the measurement technique described on p. 606).

Calibration. The pick-up, integrator and analyser should be calibrated together. The calibrator giving a vibration amplitude of, say, $\theta_{cal.} = \pm 0·5°$, the analyser is tuned to the calibration frequency and the analyser sensitivity control is adjusted so as to obtain a meter reading of 100 %. Then, if in the engine test the meter indicates a value of, say, 60 %, the corresponding vibration amplitude will be simply

$$\theta_1 = \pm \frac{0·5 \times 60}{100} = \pm 0·3°.$$

General Radio analyser

This battery-operated unit is employed in the same way as the Dawe, which has similar basic features. In both these instruments, the 'sensitivity control' is a continuously variable potentiometer and the usual practice is to leave this sensitivity control unaltered, at the adjustment used for the torsional calibration.

If the engine vibration is expected to have an amplitude exceeding that obtainable from the calibrator, for instance, an amplitude $\theta_1 = \pm 0·7°$, the calibration can be made with the analyser sensitivity control adjusted so as to obtain a meter reading of, say, 50 %, for $\theta_{cal.} = \pm 0·5°$. Then the test at this amplitude will give a reading of 70 %, so that one obtains

$$\theta_1 = \pm 0·5 \times \frac{70}{50} = \pm 0·7°$$

as required. Alternatively, the integrating amplifier may be designed to have several stage gains (which can be checked in the calibration test), in which case its gain switch can be changed to a low-gain position when required.

Note. These analysers both have a rectified output, which is not intended for display on an oscilloscope screen. However, the unrectified output can be obtained by taking tappings ahead of the rectifying circuit of the meter, and shown on an oscilloscope.

[612]

Muirhead-Pametrada analyser and accessory equipment

The equipment for use with velocity-type torsional vibration pick-ups consists of the following units:

(1) preamplifier and integrator (battery operated),

(2) wave analyser (mains operated), and

(3) power supply (mains fed).

The lower frequency limit is 19 cycles/sec., but it can be extended down to 2 cycles/sec. by using a mains-operated low-frequency modulator and integrator, instead of the normal preamplifier and integrator. The amplitude values are indicated on a meter. The analyser output is not rectified and can be displayed directly on an oscilloscope screen.

The analyser is provided with various selectivity controls, which enable one to obtain either unfiltered readings or filtered readings with various degrees of band-width selectivity (i.e. allowing wide-band or narrow-band tuning). The sensitivity control is by means of stud-type reistances ('attenuators'). The analyser also has a number of trimming resistances which can be preset, by the user or by the makers, so as to obtain a minimum variation in amplitude response over the usual range of mechanical vibration frequencies (e.g. up to 3000 cycles/sec.).

Characteristics of analysers and integrators

Tests of complete torsiograph equipment on a calibrator before and after an engine test are the most direct methods for verifying that the equipment is operating normally. However, in order to obtain optimum results, as in the case of the pick-ups, it is useful to consider the characteristics of integrators and analysers, since it is frequently necessary to introduce correction factors for these in evaluating test results.

Electronic equipment normally indicates voltages, and instrumental errors are conveniently expressed in decibels, with the definition: Throughput error $\eta_{dB.(volts)} = 20\log_{10}(V_{out}/V_{in})$, where V_{out} = output voltage and V_{in} = input voltage.† Errors less than ± 2 dB. (volts), or about $\pm 22\%$, are acceptable in electronic applications but require the use of correction factors in mechanical stress evaluations. An idea of the normal limitations of measuring equipment may be obtained by considering the curves of Figs. 11, 12 and 13.

Fig. 11 gives an example of the response curve of an integrating circuit to constant-amplitude signals at different frequencies. The input was from an oscillator and the output was measured on a valve voltmeter. The theoretical curve on a logarithmic base is a straight line.

Examples of throughput curves of an analyser are shown in Fig. 12. The oscillator and the analyser were tuned to the same frequency for each test point. The deviations of the particular instrument tested are of the order of $\pm 10\%$ in each range and can be taken into account by

† The decibel is basically related to power (e.g. watts), but the log values of voltage or current ratios are also termed 'decibels', though the factor is no longer 10 but 20.

means of an η-factor in the evaluations, as in the case of the passive filter (see Fig. 5, p. 605).

Fig. 13 gives examples of selectivity characteristics for a Sunbury filter, Dawe analysers and General Radio analysers (area between the

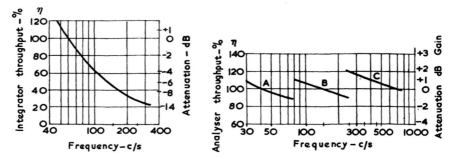

Figs. 11 and 12. Examples of throughput curves of an integrator and an analyser. Constant-amplitude input voltage at various frequencies.

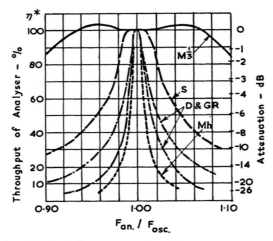

Fig. 13. Examples of selectivity curves of various types of analysers and filters. (D & GR, Dawe and General Radio; S, Sunbury; $M\frac{1}{3}$, Muirhead 'one-third octave'; Mh, Muirhead 'in-tune high'). $F_{an.}/F_{osc.}$ = tune frequency of analyser divided by oscillator frequency. Typical curves obtained with a constant-input signal of 100 c/s from a tuning-fork controlled oscillator.

two chain-dotted lines), and a Muirhead-Pametrada analyser (two curves only, for maximum and minimum selectivity).

These curves were obtained by tuning the analyser under test to various frequencies while feeding a 100 cycles/sec. frequency from an oscillator into the analyser. The corresponding curves obtained with an oscillator operating at 500 cycles/sec. are practically identical. Thus, the curves of Fig. 13, plotted against various values of off-tune ratios $F_{an.}/F_{osc.}$, are applicable in principle to all frequencies covered by the corresponding instruments. This is substantially the case for Sunbury filters, and strictly so for the other wave analysers considered.

Evaluation of 'filtered' test results taking account of filter selectivity

The resonance curve for each critical investigated can be determined by the 'Technique of measurements', described on p. 606. The engine frequency may be found to vary slightly for different criticals and also for different loads, but this does not alter the procedure since in all cases one should begin by determining the peak of the critical, and hence its corresponding natural-frequency value.

From the figures given above, it is seen that, in the general case, a filter may require the introduction of two correction factors;† the corresponding effects may be outlined as follows:

(1) When a filter is tuned to a particular frequency, its output voltage may, or may not, be equal to the unfiltered input voltage. This 'tuned-filter throughput' is taken into account by means of an η-factor, which may be smaller or greater than 100 % (see Figs. 5 and 11). The corresponding evaluation is described on pp. 605–6.

(2) In addition, the filter band-width‡ may be such that it allows a 'breakthrough' of signals which are at a frequency differing from that to which the filter is tuned.

This off-tune effect due to the filter selectivity characteristic results in increased apparent peak values for minor orders, if the filter accepts large flank-amplitude signals from an adjacent major order of torsional vibration. It may also raise the apparent flank values of a major order, owing to breakthrough of signals from the peaks of adjacent minors (see Fig. 14a). The effect requires compensation when passive filters are used.

Where necessary, it is possible to compensate this effect to some extent by introducing an additional 'breakthrough factor' η^* (or percentage throughput off resonance) in the evaluations, as shown in the following example of a numerical calculation:

Corrections for $9\frac{1}{2}$ order for breakthrough from 9th

$F_{anal.}/F_{eng.} = 9\frac{1}{2}/9 = 1\cdot055$, breakthrough factor: $\eta^* = 0\cdot42$
[From 'S'-curve of Fig. 13, for $F_{an.}/F_{osc.} = 1\cdot055$]

Engine speed rev./min.	$2\theta_{9th}$	$2\theta_{9th} \times \eta^* = \Delta\theta$	Apparent value $2\theta_{A(9\frac{1}{2})}$	Corrected value $2\theta_{c(9\frac{1}{2})} = 2\theta_A - \Delta\theta$
1100	0·016	0·067	0·120	0·053
1110	0·175	0·075	0·140	0·066
1120	0·185	0·078	0·160	0·082
1130	0·200	0·084	0·175	0·091
1140	0·220	0·093	0·160	0·067
1150	0·250	0·105	0·150	0·045

† In general, the frequency indication is correct (with less than 0·5 % error) for most filters, analysers, etc., investigated. In one case only, an error of 1–2 % in frequency was found, over a certain range.
‡ The 'band width' of a resonance curve (Fig. 13) is usually defined as the frequency-range between the two frequency values at which $\eta^* = 50$ % (with 6 dB. attenuation).

The corrected results are shown in Fig. 14 b.

Note. To emphasize the effect of breakthrough, the curve of the $9\frac{1}{2}$ order in the figures has been continued up to 1190 rev./min., i.e. some 50 rev./min. higher than normally required.

Furthermore, in Fig. 14 a one sees that from, say, 1100 to 1140 rev./min. the measured values for both the 9th and the $9\frac{1}{2}$ order are approximately the same, so presumably each curve needs correction for the effect of the other. The method given above will exaggerate the amount of

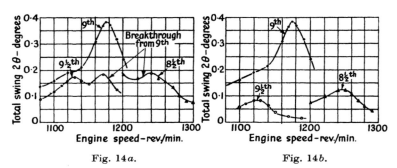

Fig. 14 a.   Fig. 14 b.

correction, since it subtracts from the $9\frac{1}{2}$ order a proportion of the incorrectly measured 9th order, and not of the true values of this order. Correct values can be obtained as follows:

Let x and y be the unknown true values of the 9th and $9\frac{1}{2}$ orders, respectively, at a certain speed. Let a and b be the known breakthrough factors. The measured values are then: $x + ay$ and $y + bx$. One thus has two simultaneous equations from which x and y can readily be found. The procedure is, of course, tedious but it may sometimes be desirable to use it at particular points of interest. Where the two measured amplitudes are very different in size, the simpler method of the above numerical example will be fairly accurate.

The calculations depend on the assumption that the two orders come fairly closely into phase at some point in the cycle, so that the figures combine arithmetically.

4·125 *Southern Instruments inductive pick-up and F.M. equipment*

The Southern Instruments pick-up is designed to operate with 2 Mc./s. frequency-modulating equipment. The pick-up comprises a seismic mass with radial arms, one of which holds a tangentially arranged magnetic core inside a small coil forming part of the outer casing. The cylindrical outer casing is rigidly connected to a flanged end portion which is bolted to the crankshaft under test. The restoring force which returns the seismic mass to its neutral position is provided by a cylindrical permanent magnet which forms the seismic mass; the magnetic path is radial into the outer casing and thence circumferential around this casing. Eddy-current damping (with an optimum value of about $\gamma = 0.6$) is also achieved through a path concentric to that of the restoring force. The natural frequency of the pick-up is about 6–9 cycles/sec. (360–540 vib./min.).

The output signal is obtained without brushgear by means of an inductive slipringless system (with rotating and stationary coils) which is fully enclosed in the free-end section of the pick-up. The pick-up can be calibrated either dynamically or statically. In the latter case a micrometer screw type jig is employed, and is inserted in the body of the pick-up to obtain a specified peripheral displacement.

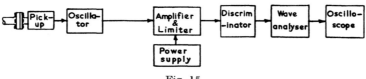

Fig. 15.

The inductance variations of the pick-up due to torsional vibration modulate the frequency of a valve oscillator. The frequency-modulated (F.M.) signal is then fed to an amplifier and limiter, which clips unnecessary amplitude fluctuations, and thence to a discriminator. The demodulated output voltage from the latter is at the frequency of the vibration and its amplitude is proportional to the vibration amplitude. This output can be either measured directly on an analyser or displayed on an oscilloscope screen. A block diagram of the assembly (shown with a wave analyser included) is given in Fig. 15.

4·126 *S.L.M. capacitive pick-up and F.M. torsiograph equipment*

The S.L.M. pick-up is of the capacitance type. The seismic mass, inside the casing, is designed as a slotted drum, the longitudinal metal bars forming one capacitor electrode. The other capacitor electrode is formed by corresponding radial projections in the casing. The seismic mass is carried on two bearings and connected to the casing by means of two soft springs giving the system a low natural frequency (about 2 cycles/sec.).

No slip-rings are used. One of the capacitor electrodes is earthed, while the electrical connexion to the other is made by means of an external spring-loaded needle pressing against a hollow metal contact at the end of the pick-up shaft. The spring-loaded needle is mounted on a separate stationary holder. The driving end of the pick-up is provided with a large-diameter internal thread for coupling to the engine crankshaft.

The capacitance changes of the S.L.M. pick-up, due to relative vibratory motion, modulate the carrier frequency voltage supplied from a valve oscillator. The block diagram of the F.M. equipment is similar to that of Fig. 15.

The pick-up can be calibrated statically or dynamically.

Other accessories and equipment

Tachometers. Various methods have been devised for tachometer calibration. Usually, the tachometer reading is compared with an average value of speed, obtained by counting the number of shaft revolutions occurring in a certain period of time.

Further accuracy can be obtained by noting the reading of a tachometer fitted to the shaft of an electric motor. The corresponding motor speed is determined by means of a stroboscopic disk (provided with a number of radial lines) fitted on the shaft and illuminated by flashes from a neon-tube stroboscope controlled by a tuning fork. If the disk has several groups of lines, calibration can be performed at a number of different speeds.

In another method the motor speed is recorded photographically, e.g. by means of magnetic pick-up indication of the shaft revolutions and time markings from stroboscope flashes controlled by a tuning-fork oscillator.

Errors of $\pm 1\%$ in speed indication result in corresponding errors in determining the natural frequency of a system, so that it is important to have at least one carefully calibrated tachometer as a reference instrument.

Reed vibrometers. Many cases of engine vibration are first thought to be due entirely to torsional vibration of the engine crankshaft, whereas subsequent inspection may show that the cause is mainly resonance of an engine frame, test bed, or accessory mountings. For such tests it is useful to have some simple instrument, such as a reed vibrometer, for linear-vibration indication.

Stroboscopes. A neon-flash tube stroboscope operated by an oscillator is useful also for a variety of applications. If the oscillator pulses are an integer multiple of the mean speed of rotation, the various phases of a shaft speed fluctuation can be identified. For this purpose, one may also place a thin metal strip, with black and white markings, around the shaft section considered, in order to assess the phase amplitudes. With further equipment (photocells, etc.) the arrangement can be developed into a 'phase-modulated system' and the vibration amplitudes can then be shown on an oscilloscope.

4·127 *Hooke's joint-type calibrator*†

A drawing of a Hooke's joint-type calibrator is shown in Fig. 1. The shaft carrying the torsional vibration pick-up is supported in needle or

† Alternative designations: universal joint-type calibrator, Hooke's coupling calibrator.

roller bearings mounted on a plate. This plate is arranged to swing on a pivot located perpendicularly below the centre of the Hooke's joint. The shaft carrying the pick-up can thus be set at various angles to the shaft of the variable-speed motor, in order to obtain any desired value of vibration amplitude indicated on a graduated protractor scale, the divisions of which are calculated and marked on the base plate. To reduce cyclical variations in the speed of the motor shaft, a substantial flywheel may be fitted on the driving shaft.

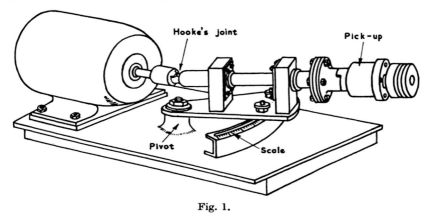

Fig. 1.

In alternative arrangements, the motor may drive an auxiliary shaft via a belt-and-pulley system or a gear box, so as to increase the range of driven-shaft speeds and vibration frequencies. The flywheel is fitted on this drive shaft ahead of the joint. In this arrangement (Fig. 2), it is also possible to keep the driven parts (with the Hooke's joint) stationary and to obtain the desired angular setting by mounting the motor, pulleys and flywheel on an adjustable slide pivoted at one end and with a slot for angular adjustment at the other end.

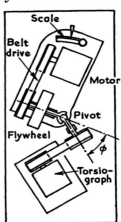

Fig. 2.

Where the Hooke's joint is to be used for the calibration of mechanical torsiographs with appreciable inertias, it should be provided with driving and driven shafts of adequate diameter to avoid shaft twists due to the reaction torque of the torsiograph inertias, particularly at high frequencies.

Calibration constants for Hooke's joint

ϕ = angle between driving shaft and driven shaft of Hooke's joint calibrator.

$\pm\,\theta_{(2)} =$ vibration amplitude produced by calibrator when set at an angle ϕ.

ϕ	$\pm\,\theta_{(2)}$	$\pm\,\theta_{(4)}$	$\pm\,\theta_{(6)}$
0°	0	0	0°
3°	0·039°	0·0000°	—
5°	0·109°	0·0001°	—
7½°	0·246°	0·0005°	—
10°	0·438°	0·0017°	—
13°	0·744°	0·0048°	—
15°	0·995°	0·0086°	—
18°	1·44°	0·0180°	0·0003°
20°	1·78°	0·0277°	0·0006°

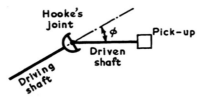

Fig. 3.

The output motion of the Hooke's joint consists of a sinusoidal vibration $\theta_{(2)}$ with a frequency $F = 2N$ [vib./min.] (where $N =$ shaft speed [rev./min.]) superimposed on the steady rotation.

In addition, the output contains small contributions of harmonics at higher frequencies, namely, $\theta_{(4)}$ and $\theta_{(6)}$ at $4N$ and $6N$ [vib./min.], respectively.

These values are based on calibration formulae which are derived at the end of this section. The $\theta_{(2)}$-values can be checked by rotating the driving shaft manually through 45°, starting from a particular position, or alternatively by taking a series of α and β readings and plotting them. One should then find $\theta_{(2)} = \alpha - \beta$ with a value for $\theta_{(2)}$ in accordance with that indicated in the above table for the angular setting ϕ employed.

To verify that the calibrator vibration is sinusoidal, a harmonic analysis can be made of the recorded trace obtained with a mechanical torsiograph, or of the photographed oscilloscope trace obtained with an electrical pick-up. In the case of an electrical pick-up, the output can also be analysed electronically by means of a wave analyser. The higher-frequency harmonics should be negligibly small. It may be noted, however, that the pick-up invariably also has a harmonic output due to its own deficiencies.

In making one's own calibrator one may have difficulty in finding a suitable Hooke's joint for the high degree of accuracy required. From a batch of joints one may select one which is entirely free from backlash. It should first be tested on the calibrator with the two shafts in line ($\phi = 0$) and should then give zero output. An output at the zero position may be produced by a number of small errors in the geometry of the apparatus. After a few hours' running, slight wear may occur in the Hooke's joint, causing backlash amounting to $\pm 0.05°$, which may be regarded as excessive if the apparatus is required to produce amplitudes of $\pm 0.25°$ with only a small percentage error. This wear is not produced by the alternating load necessary to oscillate the pick-up and shaft, but by greater loads arising from the small geometric discrepancies. To obtain satisfactory results a machine of this type requires an exceptionally high degree of precision in manufacture.

In practice, distortions of the sine-wave form[†] may occur to a noticeable extent in two extreme cases, namely, at very low frequencies and vibration amplitudes (irregularities in running smoothness of bearings, etc.) and at high frequencies[‡] and amplitudes, i.e. high loading conditions (irregularities of the Hooke's joint, gear box, etc.). Clearances or looseness at the pivot, the bracket-and-clamping arrangement, and other linkage points, should be reduced to the minimum possible values. Comparative checks of 'known amplitudes' on various calibrators, using the same indicating equipment, are useful.

Derivation of the Hooke's joint calibration formula

The Hooke's joint (see Fig. 4) basically consists of a cross with pivoted forks connected to the driving shaft S_1 and the driven shaft S_2. The two shaft axes are situated in the same plane and have a relative angle ϕ.

The relation required is one which will express the angular motion of S_2 in terms of the angular motion of S_1.

For this purpose, consider first Fig. 5, which shows two planes 'A' and 'B' intersecting along an axis OO' and forming an angle ϕ. Assume that on plane 'A', a displacement occurs from point O to point R_A. It is proposed to determine the projection of this displacement on plane 'B'.

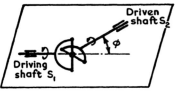

Fig. 4.

Any vertical displacement Oy_A on plane 'A' corresponds to an equal vertical displacement Oy_B on plane 'B', i.e.

$$Oy_B = Oy_A.$$

Any horizontal displacement Ox_A on plane 'A' is projected as a shorter horizontal displacement $Ox_B = Ox_A \cos \phi$ on plane 'B' (see plan view in Fig. 5).

---

† See also Johnson, D. C. and Bishop, R. E. D., 'A note on the excitation of vibrating systems by gearing errors', *J. Roy. Aero. Soc.* June 1955, pp. 434–5.

‡ Above, say, 5000–6000 vib./min. It may also be noted that the running speed should preferably be measured at the driven-shaft end (on an extension of the pick-up) in order to have a true figure for speed, if the machine is belt-driven.

Therefore, an angle α of a vector OR_A on plane 'A' corresponds to an angle β of a vector OR_B on plane 'B', and the relation between these two angles is given by

$$\frac{\tan\beta}{\tan\alpha} = \frac{Ox_B/Oy_B}{Ox_A/Oy_A} = \frac{Ox_B}{Ox_A} = \cos\phi,$$

so that
$$\tan\beta = \tan\alpha\,\cos\phi.$$

Now reverting to the Hooke's joint, consider Fig. 6 which represents a plane 'A' with the circular path of the cross-arms of the driving shaft S_1. The dotted path if the projection† of this path on the plane 'B'.

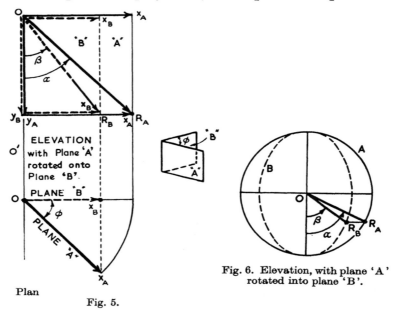

ELEVATION with Plane 'A' rotated onto Plane 'B'.

PLANE 'B'

Plan

Fig. 5.

Fig. 6. Elevation, with plane 'A' rotated into plane 'B'.

When a cross-arm of S_1 rotates through an angle α in plane 'A', the corresponding rotation in plane 'B' will be through an angle β, which is produced only by the projection of α imparted to the cross-arms of S_2, in accordance with the relation $\tan\beta = \tan\alpha\,\cos\phi$.‡

† The projection is an ellipse, since $r_A^2 = x_A^2 + y_A^2$ (circle in plane 'A') becomes $r_B^2 = x_B^2 + y_B^2 = x_A^2\cos^2\phi + y_A^2$, so that with $\cos\phi = 1/a$ one obtains $(x_A/a)^2 + y_A^2 = r_B^2$ (ellipse).

‡ A more direct derivation is obtainable by considering the spherical triangle in Fig. 7. Applying the cosine-theorem:

$$\cos b = \cos c \cos a + \sin c \sin a \cos B$$

(with $c = \beta$, $a = \pi/2 - \alpha$, $b = \pi/2$ and $B = \pi - \phi$) one finds

$$0 = \cos\beta\sin\alpha + \sin\beta\cos\alpha\cos(\pi - \phi)$$

which gives the above formula.

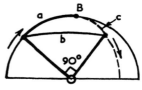

Fig. 7.

[622]

A visual interpretation of the type of motion produced is obtained by observing that the elliptical variation of the motion of the driven arms is similar to that produced by a two-peak sinusoidal cam (see Fig. 8).

This can be verified by taking, say, $\cos \phi = 0 \cdot 5$ and writing a simple table of relative values, as follows:

α	$\tan \alpha$	$\tan \beta = \cos \phi \tan \alpha$	$\theta = \alpha - \beta$
0°	0	0	0
45°	1	0·5	Positive
90°	∞	∞	0
135°	−1	−0·5	Negative
180°	0	0	0
225°	1	0·5	Positive
270°	∞	∞	0
315°	−1	−0·5	Negative
360°	0	0	0

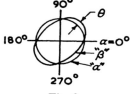

Fig. 8.

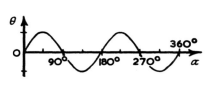

Fig. 9.

The vibration θ of the driven shaft (i.e. its variation from uniform rotary motion) thus occurs with two complete cycles per shaft revolution, and is very nearly sinusoidal (Fig. 9). As a torsional-vibration pick-up does not produce any output for uniform rotation, but only indicates deviations from the steady condition, when fitted to a Hooke's joint calibrator the pick-up will only indicate θ (the relative deviation) with a frequency equal to twice the speed of rotation of the pick-up shaft.

The next step is to express $\theta = \alpha - \beta$ as a function of ϕ, by means of the relation $\tan \beta = \cos \phi \tan \alpha$. This may be achieved as follows:

$$\tan \theta = \tan (\alpha - \beta) = \frac{\tan \alpha - \tan \beta}{1 + \tan \alpha \tan \beta} = \frac{\tan \beta \left(\dfrac{1}{\cos \phi} - 1 \right)}{1 + (\tan^2 \beta / \cos \phi)} = \frac{\tan \beta (1 - \cos \phi)}{\cos \phi + \tan^2 \beta}.$$

With
$$z = (1 - \cos \phi)/(1 + \cos \phi), \quad \cos \phi = (1 - z)/(1 + z)$$

and
$$1 - \cos \phi = 2z/(1 + z),$$

this becomes

$$\tan \theta = \frac{\dfrac{2z}{1 + z} \tan \beta}{\dfrac{1 - z}{1 + z} + \tan^2 \beta} = \frac{2z \sin \beta \cos \beta}{(1 - z) \cos^2 \beta + (1 + z) \sin^2 \beta} = \frac{z \sin 2\beta}{1 - z \cos 2\beta}.$$

Let $z \sin 2\beta /(1 - z \cos 2\beta) = r$ and $j = \sqrt{-1}$. Then $\tan \theta = \dfrac{e^{j\theta} - e^{-j\theta}}{j(e^{j\theta} + e^{-j\theta})} = r.$

[623]

Solving for $e^{j\theta}$:

$$e^{j\theta} = e^{-j\theta} = jr\,e^{j\theta} + jr\,e^{-j\theta}, \quad e^{j\theta}(1-jr) = e^{-j\theta}(1+jr),$$

hence
$$e^{j2\theta} = \frac{1+jr}{1-jr} = \frac{1 - z\cos 2\beta + jz\sin 2\beta}{1 - z\cos 2\beta - jz\sin 2\beta} = \frac{1 - ze^{-j2\beta}}{1 - ze^{j2\beta}},$$

since $e^{\pm j2\beta} = \cos 2\beta \pm j\sin 2\beta$. Taking logarithms, one obtains

$$j2\theta = \log_e(1 - ze^{-j2\beta}) - \log_e(1 - ze^{j2\beta}) = \log_e(1-x) - \log_e(1-y),$$

with $x = ze^{-j2\beta}$ and $y = ze^{j2\beta}$. As $\log_e(1-x) = -x - \dfrac{x^2}{2} - \dfrac{x^3}{3} - \dfrac{x^4}{4} - \dots$ for $|x| < 1$, we may write

$$j2\theta = -x - \frac{x^2}{2} - \frac{x^3}{3} - \frac{x^4}{4} - \dots + y + \frac{y^2}{2} + \frac{y^3}{3} + \frac{y^4}{4} + \dots$$

or
$$\theta = \frac{y-x}{2j} + \frac{y^2 - x^2}{4j} + \frac{y^3 - x^3}{6j} + \frac{y^4 - x^4}{8j} + \dots$$

$$= z\left(\frac{e^{j2\beta} - e^{-j2\beta}}{2j}\right) + z^2\left(\frac{e^{j4\beta} - e^{-j4\beta}}{4j}\right) + z^3\left(\frac{e^{j6\beta} - e^{-j6\beta}}{6j}\right) + \dots$$

Therefore $\quad \theta = \alpha - \beta = z\sin 2\beta + \dfrac{z^2}{2}\sin 4\beta + \dfrac{z^3}{3}\sin 6\beta + \dfrac{z^4}{4}\sin 8\beta + \dots,$

and the amplitude of the 2nd-order vibration is

$$\theta_{(2)} = \pm z = \pm(1 - \cos\phi)/(1 + \cos\phi).$$

These approximations can be used as a basis for the table of calibration values given on p. 620.

4·128 *Parallel-displacement type calibrator*

This machine, developed at the B.I.C.E.R.A. Laboratory, was constructed with a view to obtaining vibratory and rotary motion of the pick-up by a device simpler to manufacture than a Hooke's joint.

A drawing of the calibrator is shown in Fig. 10. A variable-speed electric motor drives through a flexible coupling a shaft mounted in ball bearings and carrying a flywheel. To the free end of this shaft, which during a measurement runs at a constant angular velocity, is attached an arm carrying a projecting pin and sliding block at its free end. For balance and for alternate use in the event of wear of the block, the other side of the driving arm is made in the same way.

The sliding block slides in a fork in a balanced arm attached to one end of a second shaft which carries at its other end the torsional-vibration pick-up unit to be calibrated. This shaft runs in ball bearings mounted on a plate which may be moved in a direction at right angles to the shaft axis, on the guides attached to the base plate of the machine, by turning a screw provided with a knurled head. The position of the plate is indicated by a clock micrometer reading to tenths of a thousandth of an inch, with its spindle bearing on the end of the plate remote from the adjusting screw. The two shafts are at exactly the same height above the baseplate.

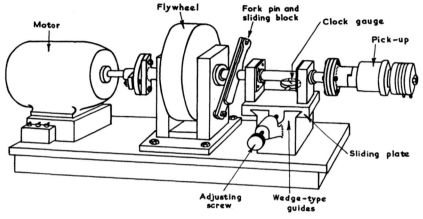

Fig. 10.

A drawing of the fork with the rectangular phosphor-bronze sliding block (which gives improved running and facilitates replacements) is shown in Fig. 11.

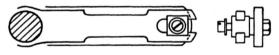

Fig. 11.

The axes of the shafts are set initially in line so that the shaft carrying the pick-up runs at constant angular velocity and there is no output from the pick-up. When the adjusting screw is turned, the axes of the shafts remain parallel but are displaced laterally. A cyclical variation of the angular velocity of the pick-up shaft then occurs, by virtue of the fact that the projecting pin of the constant-velocity shaft acts at a cyclically varying distance from the axis of the pick-up shaft.

Referring to Fig. 12a, in which the circle represents the locus of the pin on the constant-speed driving arm, one may analyse the motion as follows:

Let R be the fixed length of the constant-speed arm, measured from the axis of rotation to the axis of the projecting pin, L the varying length of the forked arm and δ the lateral displacement of the two shafts. Let

α be the angle the constant-speed arm makes with the horizontal, and β the corresponding angle of the forked arm.

The instantaneous amplitude of vibration θ is equal to the difference between the angles β and α, and is thus

$$\theta = \beta - \alpha = \tan^{-1}\left[\frac{\sin\alpha}{\cos\alpha - (\delta/R)}\right] - \alpha.$$

Fig. 12b shows the two arms in several positions. With the pin at position I, both arms are horizontal, the amplitude is zero and the forked arm is moving at its greatest angular velocity. As the arms rotate in the direction shown, the amplitude increases because β is increasing more rapidly

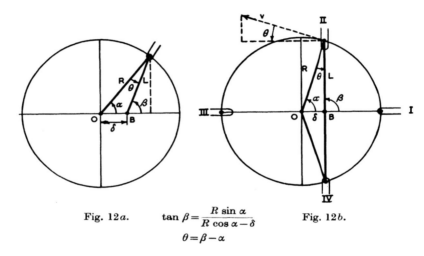

Fig. 12a.
$$\tan\beta = \frac{R\sin\alpha}{R\cos\alpha - \delta}$$
$$\theta = \beta - \alpha$$
Fig. 12b.

than α, and this continues until the angular velocity of the forked arm has become equal to that of the constant-velocity arm. This condition is reached at position II where the forked arm is vertical, since its angular velocity is equal to the horizontal component of the pin velocity v divided by the length of the arm, i.e. $v\cos\theta/R\cos\theta = v/R$, and this is the angular velocity of the constant-speed arm.

The condition for maximum amplitude is thus $\cos\alpha = \delta/R$, a result which may be obtained alternatively by equating $d\theta/d\alpha$ to zero, and the maximum amplitude is

$$\pm\theta_{(\text{max.})} = \sin^{-1}\frac{\delta}{R} \quad [\text{rad.}].$$

Beyond position II the forked arm travels slower than the constant-velocity arm and, at position III, the amplitude is again zero. Maximum amplitude with negative sign is reached at position IV. The total swing is $2\theta = 2\sin^{-1}(\delta/R)$.

[626]

In the machines used at the B.I.C.E.R.A. Laboratory, the length R of the constant-velocity arm is 3·75 in., so that for a total swing $2\theta = 1°$ the required lateral displacement is $\delta = 0·03271$ in. The micrometer measuring to 0·0001 in. gives satisfactory discrimination.

The frequency of vibration is equal to the running speed. Calibrators of this type give consistent readings and have performed satisfactorily at speeds up to 3000 rev./min.

In the construction of the machine, care should be taken to set the two shafts accurately in line in the initial position. Any discrepancy in the height of the two shafts above the base plate is immediately apparent, since this makes it impossible to obtain zero output from a pick-up. In use, the plate-adjusting screw is turned till zero output is obtained as indicated on the oscilloscope, and the dial of the micrometer is set to zero.† The plate is then moved by means of the screw till the calculated displacement for the required swing is shown on the dial. The height of the oscilloscope trace should be the same for equal movements in either direction away from the zero position.

The working surfaces of the fork are stellited and ground and the steel pin is case-hardened and ground. The fit should be such that the arms of the fork grip the sliding block very slightly. Wear in the pin and block arrangement can just be detected after a total of about 10 hours running and the driving arm is then rotated through 180° to bring the unused pin and block into operation. When it eventually becomes necessary, the fitting of a replacement pin and block is a very simple matter. The stellited faces of the fork generally show no signs of wear.

4·129   *Two-peak sine-cam calibrator*

Details of a machine of this type, constructed at the B.I.C.E.R.A. Laboratory, are shown in Fig. 13.

The flange of the pick-up is bolted to the flanged end of a shaft carried in ball bearings. This shaft has a radial oscillating arm of low moment of inertia, which is in contact with a two-peak sine cam.

The cam shaft is mounted in ball bearings and is driven through a Vee-belt and pulleys system. Contact between the oscillating arm and the cam is maintained by means of a restoring spring. The entire system is built up on a rigid base plate.

The cam is of a truly sinusoidal shape, i.e. its contour is obtained by plotting two cycles of a sine curve on a base circle. The frequency is thus twice the speed of rotation measured at one end of the cam shaft. The amplitude of vibration obtained is

$$\pm\, \theta = \delta/L \quad [\text{rad.}],$$

† This preliminary resetting to zero should be performed for every calibration test of a pick-up.

where $\delta =$ half total lift of cam [inches] and $L =$ effective length of the oscillating arm, that is, the distance from the cam contact point to the centre line of the pick-up shaft. For this type of calibrator, one can have a set of interchangeable cam shafts, if necessary, with lifts corresponding to vibration amplitudes θ of 0·25, 0·5 and 1°.

The restoring spring should be sufficiently strong to maintain good contact between the cam and the oscillating arm, with a minimum of

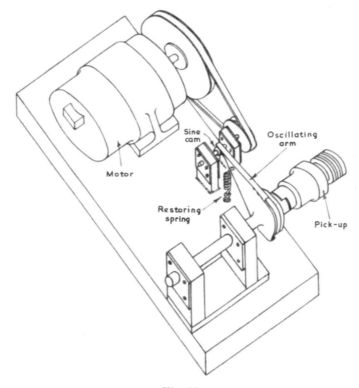

Fig. 13.

wear. It should also be chosen so as to avoid any spring surge, or resonance of the arm-and-spring system, within the range of frequencies of the calibrator.

For a given design, the upper frequency limit that can be used depends on the amplitude required. This in turn is limited by the return spring force and hence by the acceptable rate of wear at the cam and follower. The cam profile should be extremely carefully machined.

Machines of this type give consistent results with a minimum of trouble and can be operated up to frequencies of 6000–8000 vib./min. The lift 2δ of the cam can be verified with a dial gauge by rotating the driving pulley manually. The proper behaviour of the machine under running conditions can be checked by viewing with a neon-flash type stroboscope. The movement of the oscillating arm may be indicated with a 'proximity' type pick-up, or determined as regards dynamic amplitude by means of screw-type electrical contacts, in series with a battery and headphones.

4·130 *Other types of calibrators*

The oscillating-arm type calibrator described above may also be provided with a four-peak sine cam, but this is more difficult to manufacture than the two-peak sine cam.

A wide variety of other calibrators can be designed with kinematic systems giving approximately sinusoidal motion (eccentric, Scotch yoke and other crank type arrangements), operated manually or by means of an electric motor. Whatever the system used, the output from the pick-up on the calibrator should be analysed (by means of a wave analyser) to determine the harmonic distortion which may be present in the vibration, this method being more accurate than that of observing the waveform on an oscilloscope screen. For all types of calibrators, it may be said that the reliability of the results obtained depends on the care given to the equipment in manufacture and in subsequent use. It is therefore advisable to verify the condition of a calibrator occasionally, noting the observations in an instruments log book, which may also be used for notes on pick-ups, mechanical torsiographs, and electronic equipment.

In some arrangements, the oscillating arm of the calibrator is clamped directly on to the seismic mass of the pick-up, and the pick-up shaft is held stationary. The output obtained in this way would correspond exactly to that obtained with the pick-up in use on an engine only if the seismic mass were infinite.

To determine the effect of lateral vibration of the pick-up, tests were made at the B.I.C.E.R.A. Laboratory in which the base plate of a parallel-displacement type calibrator was shaken with a vibration exciter of the out-of-balance weight type. It was found that this shaking had no appreciable effect on the pick-up output until the up-and-down shaking amplitudes reached values of 0·060 in. total swing, which do not occur in practice.

The rotation of the unit may also be of importance as regards frictional conditions in the pick-up. Moreover, a rotating calibration test is more likely to bring to light any slight defect in the operation of the brush-gear, which may pass unnoticed in engine tests (and in stationary calibration). However, stationary-shaft type calibrators, properly used, give adequate results.

4·2 EQUIPMENT FOR THE MEASUREMENT OF VIBRATORY STRAINS

NOTATION

Symbol	Brief definition	Typical units	Symbol	Brief definition	Typical units
C	capacitance	farad	q	shear stress	Lb./in.^{2}, Kg./cm.^{2}
e	mechanical strain	—			
G	modulus of elasticity	Lb./in.^{2}, Kg./cm.^{2}	R	electrical resistance	ohm
			V	voltage	volt
k	gauge sensitivity factor	—	γ	shear strain	—

4·21 Electrical strain-gauging equipment

Strain gauges enable a direct determination of torsional strains to be obtained at the measurement position. Usually, the gauges are placed on plain cylindrical portions of the rotating shafting, so that peak stresses due to stress raisers in their vicinity still have to be evaluated by means of stress concentration factors (see section 2·32).

Moreover, the stresses and vibration amplitudes at any other position in the engine system still have to be evaluated by the methods described in sections 2·3111 (p. 319) or 4·114 (p. 598). The use of strain gauges is also limited to positions where sufficient space is available and becomes difficult for flanged-end shaft portions between the engine crankcase and the main flywheel. However, strain gauges indicate true strain values at the measurement position and this feature is sufficient to encourage their use wherever possible in torsional vibration investigations.

This section gives an outline of the equipment and techniques for strain-gauge measurements of torsional vibration. The precautions which have given good results are also indicated.

General features. Resistance strain gauges are usually made of thin Eureka or nichrome wire. The wire is wound in a helix and then flattened, or arranged in a zigzag pattern, so as to form a grid (Fig. 1) between two paper covers, which are glued together by means of a suitable adhesive. An alternative type is a 'foil gauge' obtained by etching a grid on a copper-nickel or gold-silver layer which has been deposited on a thin flexible plastic base.

When a strain gauge is cemented on a shaft, the strain at the shaft surface is transmitted through the cement and paper or plastic cover to

[630]

the wires of the strain gauge. When a tensile or compressive strain is applied, the cross-sectional area of the wire (and the distance between the molecules in its structure) is altered correspondingly and the electrical resistance of the wire is increased or decreased, respectively. If the gauge is connected across a constant voltage source, the relative change in resistance of the gauge, dR/R, will cause a corresponding variation of the current, dI/I, flowing through it. This variation, which within the usual limits is linearly proportional to the mechanical strain, can then be used for an electrical indication of the strain at the shaft position under test.

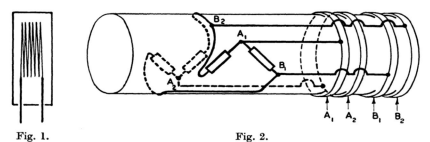

Fig. 1.   Fig. 2.

For the determination of torsional strain on the surface of the shafting, the usual arrangement is a Wheatstone bridge with four strain gauges on the shaft at 45° to the shaft axis, in order to obtain shear strain only. In order to neutralize bending strains, the strain gauges may be arranged at diametrically opposite positions on the shaft, as shown in Fig. 2.

The leads from the bridge are connected to slip rings A_1, A_2 and B_1, B_2 (of silver, silver-plated copper, or stainless steel) which are in contact with the stationary brushgear.† Silver slip rings are expensive but necessary with low-resistance gauges. Stainless steel tends to be noisy with all types of gauges. Monel rings are satisfactory with bridge terminal resistances down to 1000 ohms and possibly lower.

The shaft surface should first be chemically clean; it is then given a fine matt finish and finally degreased with acetone. The gauges are fixed with a solvent type adhesive, such as 'Durofix', a cold-setting 'Araldite' or a thermo-setting adhesive for wire gauges; 'Araldite' is also employed for foil gauges. The insulated leads are then anchored on the shaft, usually by binding.

Alternatively, a layer of adhesive (e.g. 'Durofix') is put on the shaft and allowed to dry; the gauge is then dipped in acetone to soften its adhesive coating and pressed on the shaft. Another convenient method

† Mercury contacts (i.e. steel disks rotating partly immersed in mercury in suitably closed containers), several brushes per slip ring, copper-carbon or silver-carbon brushes, etc., may also be used to improve results if necessary, particularly at high shaft surface speeds (above 100 ft./sec.).

for complicated patterns is as follows: The gauges are first stuck with weak glue in their correct 45° positions on squared (or graph) paper of a length equal to the circumference of the shaft considered. The paper with the gauges is then turned over and pressed into the cementing adhesive on the shaft. After drying, the paper is torn off.

For solvent type adhesives, one should allow at least 24 hours for setting before a test (the setting time may be shortened by moderate radiant heating). The gauges are then visually inspected for satisfactory adhesion (appearance at the edges). The amount of adhesive used should be sufficient to ensure adequate bonding strength but the layer should be as thin as possible.

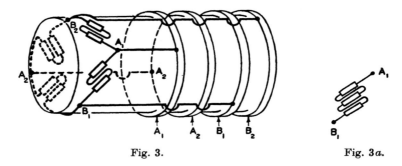

Fig. 3.　　　　　　　　　　　　　　　　　Fig. 3a.

Where external protection is needed, i.e. for permanent applications, the gauges can be encased in a coating, for instance of cold-setting 'Araldite'. External coatings provide mechanical protection and improved thermal stability.

The insulation resistance between the windings of the gauges and the shaft on which they are applied is tested and should be of the order of several megohms.

To increase the overall sensitivity of the bridge over the same shaft length, the number of strain gauges in each arm may be increased, with a 'multiple-chevron' arrangement, as shown in Fig. 3. This arrangement is also self-compensating for bending strains, provided that the gauges are placed in groups at 45° to the shaft axis and 90° to each other, so as to form an absolutely symmetrical pattern. The complete 'chain' around the shaft consists of four groups of n gauges ($n = 1, 2, 3$, etc.) and four connecting points to the leads between the groups of gauges. Fig. 3 shows groups of $n = 2$; Fig. 3a shows a group with $n = 3$.

Note. All external connexions should be made with insulated screened cables and the screening should be connected to a good earth, with a single earth-point for all the equipment. Any 50-cycle ripples, due to bad earthing or feedbacks, should be eliminated as far as possible before carrying out the test.

Gauge sensitivity factor. The change in resistance of a gauge for a given strain along its axis depends on its construction and is indicated by a 'gauge sensitivity factor' k, defined as

$$k = \frac{e}{dR/R},$$

where e = mechanical strain, and dR/R = relative change in electrical resistance. The value of k (usually between 1·5 and 3·0) is indicated by the makers and can be verified by a calibration test.

This k-value is for direct strains. For strain gauges placed at 45° to a shaft axis so as to indicate shear strains, the gauge factor is

$$k_s = \tfrac{1}{2}k,$$

because the strain gauge only indicates the linear strain e, which is one-half of the actual shear strain γ ($\gamma \cos^2 45° = \gamma/2 = e$). The shear stress can then be evaluated as $q = G\gamma$, where G is the modulus of rigidity [Lb./in.²] of the shaft under test.

4·22 Equipment used with strain gauges

In the following, two main types of equipment will be considered for torsional strain measurements on rotating shafting, viz. (a) amplitude-modulated equipment, and (b) a simple direct-current circuit with an R.C.-amplifier.†

4·221 *Amplitude-modulated system*

A block diagram of the amplitude-modulated system is shown in Fig. 4.

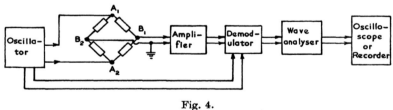

Fig. 4.

The carrier frequency, usually between 1 and 10 kc./sec.,‡ is supplied by a monitored or constant-voltage oscillator to two opposite arms of the strain gauge bridge. The bridge unbalance variations due to the torsional strains modulate the amplitude of the carrier-frequency

† Various other combined systems can also be employed, for instance a d.c.-excited bridge with a direct-coupled amplifier. See, for instance: Miller, S. E., 'Sensitive d.c. amplifier with a.c. operation', *Electronics*, Nov. 1941, pp. 27–31, 105 et seq.
‡ Giving a frequency response of the signal up to 60 c./sec. at 1 kc./sec., and 600 c./sec. at 10 kc./sec.

voltage and the output signal is fed to an amplifier (which has a blocking condenser in its input stage) and thence to a phase-sensitive demodulator, where it is compared with the unmodulated voltage from the oscillator. The demodulated output can then be either applied directly across the input of an oscilloscope for an 'unfiltered trace', or analysed by means of a wave analyser. It is advisable to take a permanent record of the demodulated waveform wherever possible.†

As the strain variations are indicated by corresponding resistance variations of the strain gauges, the latter are in effect 'amplitude-type pick-ups' and their output requires no integrating circuit.

Amplitude-modulated equipment has the advantage that it can be used for any mechanical frequencies down to zero cycles/second and enables static calibration to be made. Thus, a known static torque can be applied to the crankshaft under test and the corresponding 'static' output measurement gives a calibration reading for the entire equipment.

The phase response of the demodulated output depends on frequency, so that the 'unfiltered diagram' may be somewhat affected if the vibration frequency is at the upper end of the permitted range. With a wave analyser, however, each of the component amplitudes and frequencies is correctly indicated, independent of the phase effect.

Before each test, it is necessary to verify that the bridge is properly balanced, i.e. that it gives a zero output signal in the absence of strain. This is usually achieved by adjusting the balance resistors and capacitors of a balancing unit, which is provided in the amplifier unit, in order to cancel out the distributed resistances and capacitances of the wiring.

A minimum of balancing is needed if the strain gauges are suitably 'matched'. A good practice is to select from a batch of gauges a number which are suitably matched, i.e. which all have practically identical resistances. The resistances of the strain gauges may be determined by means of an ordinary d.c. Wheatstone bridge coupled to a galvanometer. All such measurements should be made at a constant temperature, with suitable precautions to avoid air currents so as to obtain stable readings.

The bridge circuits of Figs. 2, 3 and 4, as well as the potentiometer circuit of Fig. 5, are substantially insensitive to temperature effects, particularly if the shafts are relatively small and the gauges are close together (chevron arrangement of Fig. 3), and if the carrier or polarizing current is also fairly low (say not greater than 10 mA.). In applications where space is limited, slip rings have been applied over strain gauges, after the latter have been satisfactorily attached and wired. In cases involving high rubbing speeds, and hence high brush pressures and heating of the rings, 'temperature drift' may occur in a.c. bridges,

† The practice of 'filtering' waveforms can lead to dangerous under-estimation of stresses. In any event, strain gauges can only respond to strains and not to torsional swing, so that the peak-to-peak strain is the only criterion of the fatigue condition of the shaft under test.

owing to heat flow from the rings to the gauges. Final balancing adjustments, therefore, should be made when the shaft is rotating (at a speed free from criticals) and has reached thermal stability. Recalibration at the operating temperature is necessary if the test temperature differs appreciably from the original calibration temperature.

4·222 *Simple d.c. bridge with capacitive coupling*

A diagram of the circuit with a block diagram of the equipment required for measurements is shown in Fig. 5. The numerical values for the blocking condenser C and the load resistance R are only representative values. The load resistance R is given a fairly high value in order to maintain linearity. The condenser should be of high quality and a stable voltage source should be employed for the excitation of the strain gauges.

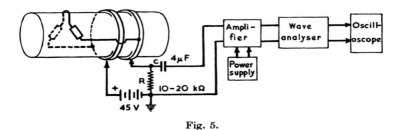

Fig. 5.

The two strain gauges at diametrically opposite positions on the shaft ensure that the bending strains are compensated. Owing to the blocking condenser, the system is insensitive to temperature effects. The amplifier is an ordinary R.C.-amplifier with a fairly long time constant.

The system has the advantage of requiring only simple equipment. Its main limitation, however, is its frequency response, which is suitable for dynamic measurements down to about 5 cycles/sec. but does not allow static calibration of the strain gauges on the test shaft.

A dynamic calibration of gauges and equipment is therefore necessary. This is achieved, for instance, by fixing one of the strain gauges of the same batch on a vibrating beam type calibrator, the deflexion (and hence the strain) of which is known.† If the strain of the calibrator gauge $e_{cal.}$ corresponds to a meter reading $V_{cal.}$ of the measuring equipment, the shear strain on the test-shaft surface will be obtained as

$$\gamma = e_{cal.} \times \frac{V}{V_{cal.}} \quad [\text{rad.}],$$

† For further accuracy, the dynamic response of the cam-excited cantilever beam used for the calibration of the strain gauge may then be checked against the static response of the same gauge and calibrator in a static deflexion test, using a d.c. Wheatstone bridge with the test gauge in one of the arms of the bridge. This method gives a true gauge factor k, and is described at the end of this section.

where V is the meter reading corresponding to γ. This equation is valid if there are two gauges on the test shaft as shown in Fig. 5, while only one gauge is used on the calibrator. The shaft surface stress is then obtained as $q = G\gamma$, where G is the rigidity modulus of the shaft material.

Note on use of Wheatstone bridge for the static calibration of a strain gauge

$G =$ galvanometer, in Fig. 6.

The bridge resistances are chosen so that

$$R_2 = R_3 = R_4 \cong 1{\cdot}01R_S,$$

where $R_S =$ resistance of the strain gauge in the unstrained condition.

Then for bridge balance, the variable resistance R_x in series with the strain gauge is adjusted to a value

$$R_x \cong 0{\cdot}01R_S.$$

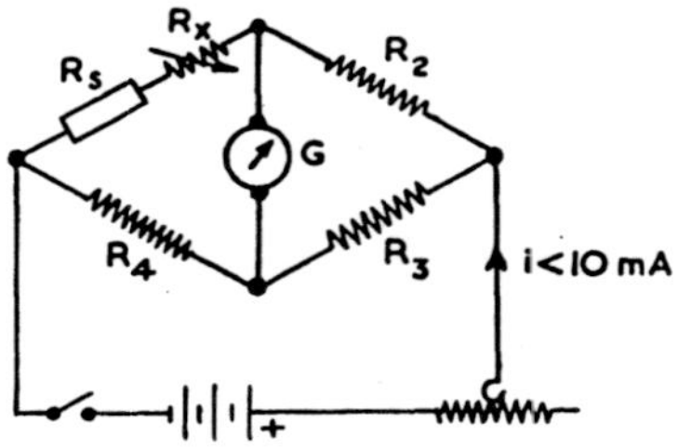

Fig. 6.

When a mechanical strain e is applied to the strain gauge, the value of R_x has to be varied by an amount $\pm dR_x$ to maintain the balance of the bridge. Hence the gauge sensitivity factor is obtained as

$$k = \frac{dR_x/R_2}{e}.$$

In the static calibration with a d.c. excited bridge, care should be taken to avoid thermo-electric effects at the strain gauge connexions, since any differences in temperature at two such junctions create electromotive forces. The gauge connexions can be protected from temperature influences due to air circulation by surrounding them with a suitable substance, such as cotton wool, plasticine, etc.

BIBLIOGRAPHY

There are various excellent publications on the subject of strain gauge measurements, and the following can be referred to for further details:

Dobie, W. C. and Isaac, P. C. G. *Electric Resistance Strain Gauges* (The English Universities Press, Ltd, 1948). [110 pages, with bibliography.]

Hetényi, M. *Handbook of Experimental Stress Analysis* (John Wiley and Sons, Inc., and Chapman and Hall, Ltd, 1950). [Chapter 5, pp. 160–237, with bibliography.]

Koch, J. J., Boiten, R. G., Biermasz, A. L., Roszbach, G. P. and Van Santen, G. W. *Strain Gauges: Theory and Application* (Philips Technical Library, 1952), [93 pages].

Perry, C. C. and Lissner, H. R. *The Strain Gauge Primer* (McGraw-Hill, 1955).

Yarnell, J. *Resistance Strain Gauges* (published by Electronic Engineering, 1951).

INDEX

Back-to-back tests of generator shaft, 142
Balance weight
  effect on vibration amplitudes, 349, 357
  moment of inertia of, 8
Balance-weight pendulum, 575
Balancing
  of crankshaft, 308
  of strain-gauge bridge, 634, 635
Ball-type pendulum, 582
  detuner, 587
BANASHEK, K., 88
Band-pass filter, 602, 611
Band-width of a filter, 615
Banks of cylinders, 305
BARRAJA-FRAUENFELDER, J., 564
'Barred' speed range, 472, 480
Basic
  rectangular crankweb, 61
  system, 186, 212
BATHO-BREDT formula
  for equivalent length of thin-walled
    cylindrical sections, 135
  for welded shaft sections, 336
BAUD, R. V., 89
BAUTZ, W., 327, 329, 331
B.E.A.M.A., see BRITISH ELECTRICAL AND
    ALLIED MANUFACTURERS' ASSO-
    CIATION
Bearing
  constraint, 368
  damping, 368
  load, 308
  restraint, effect on crankthrow stiffness,
    73
Beat-frequency amplitudes, 488
BECK, E., 130
BEHRENS, H., 222
BEHRMANN, W., 69, 222
Belt
  cotton, 593
  drive
    damping due to, 368
    HOLZER tables for systems with,
      181
    relative-amplitude curves of system
      with, 185
  equivalent stiffness of, 91
  flapping frequency of, 94
  slip, 193
  steel, 593
  stiffness of, 81
Bending
  fatigue strength of crankweb, 344
  of crankweb, 73
  strain, strain-gauge circuit compensated
    for, 631
  stresses, 462

vibration
  detuner, 588
  frequencies due to, 142
BENZ, W., 89
BERNHARD, J. M., 89
Bevel gear, 88
  deflexion of, 84
BIBBY, 353
BIBBY
  coupling and detuner, 125
  coupling, detuning effect of, 369
  detuner, 588
  and coupling: formula for estimating
    torsional stiffness of, 127
B.I.C.E.R.A., see BRITISH INTERNAL COM-
    BUSTION ENGINE RESEARCH ASSO-
    CIATION
BIERMASZ, A. L., 636
Bifilar pendulum, 582
  method for determining moment of
    inertia, 16
BIOT, M. A., 173, 243, 581
BISHOP, R. E. D., 238, 399, 621
Blade-spring detuner, 588
Block diagram
  of amplitude-modulated system for
    strain gauging, 633
  of capacitively-coupled strain-gauging
    equipment, 635
  of electrical torsiograph equipment,
    609
  of frequency-modulating equipment,
    617
  of mechanical torsiograph, 598
Blower
  drive, geared, 192
  gear-driven, 225
BOEGEHOLD, A. L., 88
BOITEN, R. G., 636
BOLLENRATH, 329
Bolt, coupling flange, 340
Bonded-rubber damper, 508
Bored
  crankweb, 64
  shaft, 38, 40
    stiffness of, 38
    with tapered bore, 40
Boss, moment of inertia of, 14
BOUMARD, B., 230
BOYD, G. M., 337
BRADBURY, C. H., 8, 155, 197, 496
BRADBURY, G. V., 89
BRADBURY's method for estimating na-
    tural frequencies, 155
Brake
  damping due to hydraulic, 369
  eddy-current, 33
    damping due to, 368

Coupling (*cont.*)
peripheral-spring, 572
stiffness of, 103
pin-type, stiffness of, 104
restoring torque, 110
rubber
damping due to, 370
sandwich type, 115
selection of, 115
stiffness of, 95
test, 455
test rig, 107
tooth type, 130
with rubber blocks in compression, 119
Couplings and clutches of various types, 122
Cox, L., 344
Crank angle
effect of, 368
spacing, 291
Crankpin
equivalent length of, 57
fillets, 58
Crankshaft
balancing, 308
of auxiliary engines, permissible stresses, BUREAU VERITAS, 463
position of concentrated masses on, 52
stiffness
B.I.C.E.R.A. method for evaluating, 55
B.I.C.E.R.A. provisional formula, 54
comparative results of evaluations by B.I.C.E.R.A. method and other methods, 70
stress
engines with rubber dampers, 518
permissible: BUREAU VERITAS, 464; DET NORSKE VERITAS, 479; GERMANISCHER LLOYD, 467; LLOYD'S REGISTER OF SHIPPING, 469, 474
with shrink-fit pins and journals, 338
Crankthrow, equivalent shaft length of, 57, 71
basic theoretical formula, 72
B.I.C.E.R.A. method of evaluation, 55–71
B.I.C.E.R.A. provisional formula, 54
CARTER's formula, 73
CONSTANT's formula, 73
GEIGER's formula, 74
HELDT's formula, 74
HOLZER-FÖPPL formula, 75
JACKSON's formula, 76
KER WILSON's formulae, 76
NORMAN-STINSON formula, 77
SEELMANN's formulae, 78
SHANNON's formula, 79

SOUTHWELL's formulae, 80
TIMOSHENKO's formulae, 80
TUPLIN's formula, 81
ZIMANENKO's formula, 81
stiffness
B.I.C.E.R.A. investigation of, 67
see also Crankshaft stiffness
Crank-type calibrator, 629
Crankweb
basic rectangular, 61
bending, 73
bending fatigue strength of, 344
biconcave, 57
bored, 64
chamfer of, 57
circular, 57
hollow-bored, 58
inclined centre: formula for equivalent length of, 76
moment of inertia of, 4
oval, 57
shear stress in, 343
stiffness, effect of
crankpin fillets, 58, 65
journal and crankpin of unequal diameter, 63
oil passage, 69
side and back chamfer, 66
web width, 62
test rig, 67
with crankpin and journal of unequal diameter, 58
with non-uniform thickness between shaft centres, 57
with radiused corners, 69
with radiused top faces, 69
Criteria of performance for
rubber damper, 518
viscous-shear damper, 530
Critical
major, 311
minor, 311
order numbers of Vee engines, 307
speed order number, 264, 292, 294
speed shifted outside the running range, 178
CROFT
clutch, 126
coupling, 125
CROSSLEY, F. R. E., 582, 585
Cruciform cross-section
equivalent length of shafts with, 136, 141
point of rigidity of shafts with, 132
stress in shafts of, 339
Cubic equation for branched systems, 189
CURRIE, C. C., 543

41 BTV

[641]

Decrement, logarithmic, 458
Deflexion
    diagrams, 169
    of gear teeth, 84
    of shafts rotating at different speeds, 82
    static, 313
DE HAVILLAND AIRCRAFT CO. LTD., THE,
        587
DEN HARTOG, J. P., 94, 111, 265, 308, 333,
        340, 354, 400, 519, 525, 562, 564,
        569, 576, 582, 585, 587, 588
DENMAN, R. P. G., 266
Dermatine rubber, 117
Determinants, notes on, 153, 231, 436, 437,
        454
DET NORSKE VERITAS, 478
Detuner, 497
    ball-type pendulum, 587
    BIBBY type, 125
    blade-spring type, 588
    centrifugal-clutch type, 587
    centrifugal pendulum, 574
    engine system with, 174
    HÜLSENFEDER type, 588
    link-type, 587
    mercury-pendulum type, 587
    non-linear spring type, 588
    pendulum, transverse-spring type, 572
    stress in engine systems with—(DET
        NORSKE VERITAS), 482
    type coupling, 119
Detuning
    action, 571
    due to
        coupling, 111
        slipping action, 129
    effect of
        BIBBY coupling, 369
        slipping torque type damper, 559
Diagram
    of deflexions with damping, 600
    stability, 608
DICKSEE, C. B., 254
DIETZ, W., 329
Direct-injection engine, 289
Direct, non-integrating, circuit, 489
Disk
    area ratio of a propeller, 376
    brake, 251
    clutch, torque-carrying capacity of, 341
    tuning, see Tuning disk
Displacement
    admittance, 238
    impedance, 237
    type pick-up, 598
Dissimilar alternators, combination of, 495
Dissipated energy, 312
Dissipative systems, 84

Distributed-mass
    method, 199
        basic theory of, 204
        tabulations, 202
    shafts with, 52
DOBIE, W. C., 636
Dog clutches, 130
DOLAN, T. J., 89, 329, 345
DOREY, S. F., 89, 355
Double-spring-rate couplings, 128
DOUGLAS, L. M., 89
DOXFORD opposed-piston engines, tan-
        gential-pressure components of, 284
DRAMINSKY, P., 357, 368, 371, 373, 400
Driven-machine
    damping, 319, 434
        in 2-mass system, 438
    'isolated', 501
    natural frequency of an engine coupled
        to various types of, 172
    point of rigidity of, 53
    power absorbed by, 372
    shafting, equivalent length of, 131
Drum camera, 23
Dry-friction damper, 559
Dual analogies, 234
DUDELL mirror-galvanometer type oscillo-
        graph, 609
DUESENBERG, F. S., 587
DUNCAN, W. J., 238, 455
DUNLOP couplings, 122
Duplex pendulum, 582
Durability
    of crankshaft, 344
    of gears, 88
    see also Fatigue strength
Dynamic
    characteristics of couplings, 108
    magnifier, 313
        and damping factors, 364
        curves for 1-mass systems, 431
        experimental determination of, 430
        for electrical resonance, 494
        of engine with viscous-shear damper,
            523, 530
        of 1-mass system, 429
        of 2-mass systems, 437, 438; experi-
            mental determination of, 440
        stiffness, 576; of a pendulum, 574;
            of non-linear couplings, 111
    vibration absorber, 499
Dynamometer
    eddy-current, 33
    hydraulic, 33
    see also Brake

EAGLE, A., 265
Earthing of electronic equipment, 632

Eccentrically-bored straight shaft, 38
Eccentricity of pick-up, 608
Eccentric type calibrator, 629
ECHHARDT, 344
Eddy-current
  brakes, 33
  damping of pick-up, 616
Effective-inertia method
  determination of natural frequencies by, 212
  diagrams, 208
  for systems with rubber damper, 509
  for gear-branched systems, 221
Effective inertia of pendulum, 579
Effective length of belt, 91
Effective stress, 318
Efficiency
  mechanical, 250
  of a filter, see Filter throughput
Effort, tangential, 247
Eight-mass system, 400
Elastically-mounted gearing, 88
Elasticity modulus
  of belts and chains, 94
Electrical
  cyclic speed deviation, 493
  frequency, 493
  hunting frequency, 495
  resonance, 494, 495
  strain-gauging equipment, 630
  torsiograph equipment, 601
ELECTRICAL RESEARCH ASSOCIATION
    (E.R.A.), 148
ELECTRIC CONSTRUCTION CO. generators, 32
Electric slip coupling, 131
Electromagnetic coupling, 95, 130, 131, 588
Electromechanical
  analogies, 231
  system, equations for, 242
Electronic torsiograph equipment, 601
Elliptical cross-section
  shafts, equivalent length of, 135
  stresses in shaft of, 325
Energy
  dissipated by viscous-shear dampers, 554
  dissipation, 312
    of rubber, 508
    of viscous-shear damper, 558
    per vibration cycle, 458
  input, 312
Engine
  damping, 434, 562
    inherent, 505
    in 2-mass system, 438
  hysteresis damping, 375
  impulse, 493

torque, 313
  with front-end pulley
    mass damping, 447
    shaft damping, 441
  with geared blower drive, 192
  with main flywheel, 448
  and driven machine, 449
ENGLISH ELECTRIC generators, 32
Envelopes of wave forms, 595
Epicyclic gears, 88
Equilibrium amplitude, 313
  see also Static deflexion
Equivalence function, reduced-inertia method, 208
Equivalent
  disk method of calculating crankweb inertia, 7
  electrical and mechanical systems, 231
  electrical stiffness, 495
  length
    of belt, 91
    of chain, 94
    of crankthrow, 57; B.I.C.E.R.A. method for evaluating, 55; B.I.C.E.R.A. provisional formula, 54; GEIGER's formula, 74; HELDT's formula, 74; HOLZER-FÖPPL formula, 75; JACKSON's formula, 76; KER WILSON's formulae, 76; NORMAN-STINSON formula, 77; SEELMANN's formulae, 78; SHANNON's formula, 79; SOUTHWELL's formulae, 80; TIMOSHENKO's formulae, 80; TUPLIN's formula, 81; ZIMANENKO's formula, 81
    of elliptical-section shaft, 135
    of gear teeth, 85
    of journal, 57
    of junction, 44, 51
    of keyed coupling, 95
    of rectangular-section shaft, 134
    of shaft, 36; of cruciform cross-section, 136, 141; of irregular cross-section by GRIFFITH and TAYLOR method, 137; with convex periphery, SAINT-VENANT's formula, 136; with distributed masses, 52; with fillets, 45; with flat side, 43; with gradual change in cross-section, 41; with keyway, 43
    of solid shaft with linear taper, 40
    of splined shaft, 43
    of square-section shaft, 134; with fillets, 46, 47
    of stepped cylindrical shaft, 44
    of stepped shaft, 36
    of straight cylindrical shaft, solid and hollow, 38

Flexural vibration detuner, 588
Flicker, 178
  limits of, 493
  of electrical generator sets, 483
Flickering of lights, 493, 495
Float of trace on oscilloscope screen, 605
Fluid
  coupling, 130, 588
    damping due to, 370
    engine with, treated by graphical
      methods, 395
  reservoirs in viscous-shear dampers, 544
Flywheel
  auxiliary, engine with, 441
  experimental determination of moment
    of inertia of, 15
  inertia, subdivision of, 178
  moment of inertia, 12
    estimated by means of chart, 31
    required for a specified value of cyclic
      speed variation, 485
  point of rigidity of, 52
  stresses in, 340
  tuning, see Tuning disk
Foam rubber, 119
Foil gauge, 630
FÖPPL, O., 332
FÖPPL-HOLZER formula for equivalent
  crankthrow stiffness, 75
Force, tangential, 247
Forced
  frequency tables, 231
    with damping, for engines with
      dampers, 532, 534
    with gas torque and damping, 413
  frequency tabulations, 323
    for centrifugal pendulum, 582, 583
  motion, 227, 228
  vibration
    characteristics of couplings, 108
    on flank of a critical, 461
Forging, stress increases due to unequal,
  465
Formulae
  based on analytical methods for calcu-
    lations of systems with damping,
    428
  for equivalent length of crankthrow, 72
  for gears, 88
FORSYTH, G. H., 89
FÖTTINGER couplings, 131, 588
Foundation damping, 373
Four-mass system, natural frequencies of,
  153
FOURIER
  analysis of tangential-pressure curve,
    262
  series of tangential-pressure curve, 277

FRAHM dynamic vibration absorber, 572
FRANK, B., 223
FRAZER, R. A., 455
FREBERG, C. R., 238
Free
  damped vibration, 455
  vibration characteristics of couplings,
    108
Frequencies due to cyclic speed variation,
  227
Frequency
  alteration of 1-node or 2-node, 178
  antiresonant, 500
  calculations, 150
    for branched systems, 188
  determination
    by effective-inertia method, 212
    by reduced-inertia method, 206
    further methods for, 232
  electrical, 493
    hunting, 495
  equation
    for three-branch system, 191
    obtained by determinants, 155
  excitation
    by gearwheel backlash, 224
    due to propellers, 230
  modulating equipment, 616
  of an engine coupled to various types of
    driven machines, 172
  of belts, flapping, 94
  of gear-branched systems by inertia-
    torque diagrams, 223
  response of calibrator, 597
  selector, 602, 611
  tabulations with damping, 400
  2-node, 165
  3-node, 167
Fretting, 374
Friction
  clutches, 129
  coefficients for disk clutches, 341
  damper, LANCHESTER, 567
  non-viscous, 84
Frictional
  mean effective pressure, 250
  torque
    at no-load, 23
    determined by running-down test,
      255
FROCHT, M. M., 329
Front-end pulley, engine with, 441, 447
Fuel cam, 226
FULLAGAR opposed-piston engine, tangential-pressure components of, 285

GAGNE, Jr., A. F., 341
GARNER, H. H., 601

HUMPHREY-DAVIES, M. W., 241
Hunting frequency, electrical, 495
HÜTTE, 265
Hydraulic
  brakes, 33, 149, 250, 369
  couplings, 95, 130, 193
Hyperbolic contour, stiffness of rubber
  annulus with, 100
Hysteresis
  damping, 562
    of shafting, 375
  loops of couplings, 105, 107, 108
    of rubber, 509
    of steel, 358

IDE, H., 89
Identically-built engines, vibration amplitudes of, 372
Impact
  loading of couplings, 119
  torsion experiments, 344
Impedance
  locus, 238
  mechanical, 237
  method, 454
Increase in amplitude with running time,
  372
Indicated
  mean effective pressure (i.m.e.p.), de-
    termination of, 250, 251
  work per revolution, 485
Indication limit of measuring equipment,
  591
Indicator diagrams, 251
Inductive type pick-ups, 616
Inertia
  force components, 485
    due to reciprocating parts, 267,
      274
  measuring apparatus, 17
  region, one-mass system vibration, 387
  torque
    and damping torque expressed as
      complex numbers, 415
    at node estimated by LEWIS formula,
      160
    curve determined by tabulation me-
      thod, 276
    diagrams for natural frequency of
      gear-branch systems, 223
    per radian, 162
    tables for determination of, 259
    tangential-pressure components due
      to, 274
  see also Moment of inertia
Infinite damping, curves of system with,
  511
INGLIS, Sir C., 400

Inherent engine damping, 372, 375, 505
  see also Engine damping
Injection
  nozzle springs, pressure, adjustment of,
    290
  timing
    effect on tangential pressure com-
      ponents, 289
    effect on vibration amplitudes, 374
In-line engine, electrical analogue, 239
Input
  energy, 312
  quantity, 242
  torque, 247, 291
Instability and distortion in amplifier
  stages, 601
Instrumentation, 591
Integrating circuits, electrical, 602
Intermediate points along the shafting, ex-
  tension of HOLZER-table method
  for, 174
Intermediate shaft, permissible stresses in
  BUREAU VERITAS, 465
  DET NORSKE VERITAS, 480
  GERMANISCHER LLOYD, 467
  LLOYD'S REGISTER OF SHIPPING, 470,
    475
Intermittent faults, signals produced by,
  603
Internal
  friction of rubber, 120
  phase angle, 407
Interpenetration, see Junction
Interpretation of records, 594
  from systems with heavy damping,
    600
ISAAC, P. C. G., 636
'Isolated' driven machine, 501

$j = \sqrt{-1}$, remarks on, 431
JACKSON, P., 76, 495, 496, 569
JACKSON's formula for equivalent length
  of crankthrow, 76
JACOBSEN, L. S., 328, 343
JAPANESE MARINE CORPORATION, THE,
  477
JEANS, J. H., 581
JENDRASSIK, 564
JENNINGS, J., 89
JOHNSON, D. C., 621
JONES, Jr., H. J., 602
Journal
  bearing displacement, effect on gear-
    wheel, 88
  centre
    cycloidal motion of, 368
    transverse movement of, 308
  equivalent length of, 57

[649]

Junction
effect
of stepped cylindrical shafts, 45
on square-section shafts, 46, 47
flexibility, 44
test methods for determining, 51
JUNKERS hydraulic damper, 589

KALICHMAN, S. L., 222
KAMMERER, H., 585
KARAS, F., 90
KARELITZ, G. B., 332
KÁRMÁN, see VON KÁRMÁN
KAUERMANN clutch, 126
KAUPPI, T. A., 543
KELLEY, O. K., 131
KEMLER, E. N., 238
KERSEY, A. T. J., 308
KER WILSON, W., 76, 199, 208, 223, 265, 276, 308, 344, 347, 358, 496, 564, 585, 589, 601
KER WILSON's formulae
for equivalent length of crankthrows of normal and opposed-piston engines, 76
for inclined centre crankweb, 76
for stresses, 351, 358
Keyed
assemblies, damping obtained with, 374
coupling on straight shaft, 95
coupling on tapered shaft, 97
shafts, point of rigidity of, 132
Keys and keyways, stresses in, 332
Keyways
in gears, 88
shafts with, equivalent length of, 43
KIENZLE, W., 338
KIMMEL, A., 69, 308, 585
Kinematic viscosity of silicone fluids, 539
KING, A. J., 602
KLAMT, J., 131
KLEINER, A., 208, 243, 496, 585
KLOTTER, K., 69, 234
KNIBBE, K., 90
KOCH, J. J., 636
KOFFMANN, J. L., 131
KÖNIG, H., 265
KORN, A. H., 332
KRAEMER, O., 586
KURZEMANN, W., 400

LAGRANGE's equations, 582
LAIBLE, T., 496
LAMBRICH, R., 586
LANCASHIRE DYNAMO generators, 32
LANCHESTER, F. W., 569

LANCHESTER damper
semi-dry friction type, 559, 567
viscous fluid type, 520
LAPLACE's expansion, determinants, 454
LAUGHARNE THORNTON, G., 308
LAURENCE SCOTT generators, 32
Leakage, precautions against, in viscous-shear dampers, 542
LEHNERT, H., 90
LEHR, E., 329, 344
LEHR-GREINACHER strain gauge, 328
LEVEN, M. M., 332
LEWIS, F. M., 158, 358, 373, 525, 562
LEWIS formula for estimating inertia torque at node, 160
LEWIS formulae for stresses in gear-tooth fillets, 342
LEWIS method for estimating natural frequencies, 158
LIN, S. N., 230
Linear
deflexion of a rubber bush, 105
network, analogue, 239
spring stiffness, 573
Link-type detuner, 587
LIPSON, C., 344
LISSNER, H. R., 636
LLOYD's REGISTER OF SHIPPING, 468
Load
capacity and load distribution, gears, 88
reversals, 484
variation, electrical, 495
Locus of impedance vector, 238
LOFGREN, K. E., 90
Logarithmic decrement, 458
Loose coupling, 501
Loosely-coupled systems, 447
Loosening of keyed-on or screwed-on connexions, 374
Loss
angle of rubber, 120, 508
of filter, 605
Low-flank amplitude positions, 435, 452
Lumped systems, 181
LUNDBERG, S., 359
LUTZWEILER, J., 585

MACGREGOR, C. W., 90
Magnetic
powder couplings, 131
restoring force, 616
Magnetization of ball races of pick-up, 608
Magnification ratio
dynamic, 595
static, 595
Magnifier, see Dynamic magnifier
Major
and minor criticals, 311

Major (*cont.*)
order
HOLZER table with damping for, 400
vibration amplitudes for, determined
by HOLZER tables with gas torques
and damping, 424
Malalignment
of coupling, parallel and angular, 116
of pick-up, 608
of tooth-type coupling, 130
M.A.N., *see* MASCHINENFABRIK AUGSBURG-
NÜRNBERG
MANLEY, R. G., 223, 265
MARCH, H. W., 137
Marine
propeller
chart for estimating moment of
inertia of, 27
chart for estimating weight of, 28
formulae for estimating moment of
inertia of, 29
moment of inertia of, 24
shafting, evaluation of records from,
598
propulsion system, 192
example of, 174
with distributed masses, 202
MARQUARD, E., 400
MARTIN, L. D., 90
MARTYRER, E., 130
MASCHINENFABRIK AUGSBURG-NÜRNBERG
double-acting engine, tangential-
pressure components of, 286
Mass
damping and shaft damping, 446
density
equivalent for distributed mass moments of inertia, 200
distributed, 52
elastic line, 169
reduction method, 232
of frequency determination, 206
Matching of pick-up and filter, 605
Mathematical centrifugal pendulum, 582
MATHIESSEN, O., 110
Matrices, 454
Matrix methods, 231
MATTHAES, K., 344
Maximum damping of viscous-shear dam-
pers, 555
MCARD, G., 90
MCCAIN, G. L., 334
MCCANN, G. D., 231
MCGREGOR, R. R., 543
MCLACHLAN, N. W., 111
Mean
effective pressure, 250
engine torque, 250

frictional pressure, 250
indicated pressure, 250
piston pressure, 249
tangential
crankpin pressure, 249
effort, 249
transmitted torque, stress due to, 347
Measurement
accuracy, 348, 591
of cyclic speed variation, 483
of vibration amplitudes and stresses,
315
taken at intermediate point of the
shafting, 598
torsiograph, 606
Mechanical
admittance, 238, 455
efficiency, 250
impedance, 237, 242
torsiographs, 593, 597
vibration exciter, 142
'Medium' coupling, 501
MELDAHL, A., 90
Mercury
contacts for strain gauges, 631
pendulum detuners, 587
MERRITT, H. E., 90
Metaduct coupling, 124
METALASTIK coupling, 123
Metal-disk type coupling, 124
Metastream metal-disk type coupling, 124
METZ, 370
MEYER, J., 69
MEYER, W., 332
MICHELSON-STRATTON harmonic analyser,
265
MILLER, S. E., 633
MILLS, K. N., 90
Minimum-amplitude
conditions, rubber dampers, 510
parameters
for rubber dampers, 512
for viscous-shear dampers, 523
point of, 408
Minor
criticals, 311
orders, HOLZER tables with damping for,
400
Mirror
steel, 50
supporting rings for static and dynamic
tests, 49, 50
Misalignment, *see* Malalignment
MITCHELL, R. W. S., 360
Mixing of silicone fluids, 544
Mobility method, 238, 454
MODERN WHEEL DRIVE couplings, 128
Modes of vibration, 173

Propeller (*cont.*)
  pitch, effect of, on propeller damping, 376
  power, 375
  shaft, permissible stresses
    BUREAU VERITAS, 466
    DET NORSKE VERITAS, 479, 481
    GERMANISCHER LLOYD, 467
    LLOYD'S REGISTER OF SHIPPING, 471,
      474, 476
  slip, 379
  spare, 473, 478
  torque, 375
Pulley, 91
  engine with front-end, 348, 441
  point of rigidity of long-hubbed, 134
Pulsatance, *see* Phase velocity
Pulsating drives, 118
Pump
  damping due to, 374
  drives, 193
Pumping-chamber damper, SANDNER, 562
PUTNAM, H. E., 496
pV diagrams, 251

Q-factor of rubber, 120
Quill-shaft friction damper, 569

RABBENO, G., 230
Radial link couplings, 123
Radial-spring pendulum detuner, 588
Radian frequency, *see* Phase velocity
Radius of gyration of generator armatures
    and rotors, charts for, 31
Radiused
  corners, crankwebs with, 69
  top faces, crankwebs with, 69
RANZI powder coupling, 128
RASMUSSEN, A. C., 90
Rate of shear, *see* Shear rate
RAUH, K., 88
Receptance
  mechanical, 238
  method, 454
Recessed fillet, stresses in, 343
Reciprocal theorem, 454
Reciprocating
  mass, moment of inertia of, 9
  part, inertia force due to, 267, 274
Recording
  at constant b.m.e.p., 607
  of cylinder-pressure diagrams, 251
  of deceleration process, 254
  of engine speed and total number of
    revolutions, 255
  on celluloid strip, 597
Records
  evaluation of, 598
  interpretation of, 594

Recovery of rubber, 122
Rectangular cross-section
  shaft, equivalent length of, 134
  stress in shaft of, 325
Reduced-inertia method of frequency de-
    termination, 206, 232
Reduced values of torques, 291
Reduction
  of certain orders of vibration in Vee
    engines, 305
  ratio, 179
Reed vibrometer, 618
REGISTRO ITALIANO NAVALE, 482
Relative amplitude, 292
  curves of systems with belt drive or
    chain drive, 185
  diagrams, 169
  (HOLZER-table amplitudes), 161
Relative motion of a pendulum, 572
Relative peak amplitude, 518
Relative torque, 310
Relaxation method, 231
Reluctance-type magnetic coupling, 131
Residual torque, 163
  diagram, 164
Resistance
  analogue, 240
  torque of driven machine, 225
Resonance
  curve
    shape of, 435, 451
    with finite damping, 363
    with zero damping, 362
  electrical, 494, 496
  region, 1-mass system vibration, 387
  tuning, 578, 579
Resonant speed, stress occurring at,
    317
Restoring torque, 570
  of coupling, 108
Resultant
  components of tangential-pressure
    curve, 277
  cyclic speed variation, 488
  excitation torque, 304
    of a multi-cylinder engine, 291, 293
  torque curve, 167
REUF, 344
Rib, moment of inertia of, 14
Ribbed cross-section, equivalent length of
    shaft with, 136
Rigidity
  modulus of rubber, 508
  point of, 52
'Rigid' or 'close' coupling, 501
Rings, sound-deadening, 587
Ring-type pendulum, 582
Ripple, 50-cycle, in output signal, 632

ROARK, R. J., 344
ROBERTSON, D., 400
Roller-type pendulum, 582
Rolling
    method for determining moment of
        inertia, 18
    motion, see Cyclic speed variation
Roots
    of polynomials, 230
ROOTS-type blowers, 226
Rope brake, 250
ROSSBACH, H. F., 333
ROSZBACH, G. P., 636
Rotating
    pendulum, 574
    see also Centrifugal pendulum detuner
    vectors, 431
        representation by means of, 291
Rotation, direction of, 83
Rotor
    charts for estimating radius of gyration
        of, 31
    moment of inertia of, 14
Rounded edges, crankweb with, 69
ROWETT, F. E., 358, 562
Rubber, 116
    and dry-friction dampers, 587
    annulus in concentric sleeves, stiffness
        of, 98
    block coupling, 122
    bush
        coupling, 123
        tests, 104
    coupling, 115
        damping due to, 370
    damper, 502
        criteria of performance, 518
        material, twist in, 512
        minimum-amplitude parameters for,
            512
    damping ratio, 508
    gum stock, 120
    heat dissipation, 509
    hysteresis, 509
    loss angle, 508
    materials, properties of, 120
    rigidity modulus, 508
    sandwich coupling, stiffness of, 101
    stresses in, 509
    temperature rise, 509
    tread, 120
RUBBER BONDERS coupling, 122
RUNGE, C., 265
Running-down test
    for determination of frictional torque,
        254, 255
    for determining moment of inertia,
        22

Running time, increase in amplitude with,
    372
RYBNER, 371

S.A.E., see SOCIETY OF AUTOMOTIVE
    ENGINEERS
SAINT-VENANT, B. DE, 134, 324
SAINT-VENANT'S
    formula for equivalent length of shafts
        having an entirely convex peri-
        phery, 136
    theory of torsion, 47
SALMON, E. H., 338
SALOMON, B., 586
SALOMON pendulum, 582
SANDEN, K. V., 69
SANDNER, E., 564
SANDNER
    damper test rig, 546
    gearwheel type damper, 559, 565
    pumping-chamber damper, 559, 564
Sandwich coupling, rubber, stiffness of,
    101
SANKEY, J. O., 344
SARAZIN pendulum, 582
SAWIN, N. N., 90
SCHARBACH, K., 90
SCHEIL, E., 329
SCHICK, W., 586
SCHJOLIN, H. O., 131
SCHLESINGER, G., 90
SCHMITTER, W. P., 90
SCHWEIZERISCHE LOKOMOTIV-UND MAS-
    CHINENFABRIK capacitance pick-up,
    617
Scotch yoke type calibrator, 629
SCOTT, A. W., 338
Screwed-on assemblies, damping obtained
    with, 374
Screw gears, 88
Screwshaft, see Propeller shaft
Sectional modulus, 317
SEEGER, G., 329
SEELMANN, 78
SEELMANN'S formulae for equivalent
    length of crankthrow, 78
Seismic member of a damper, 504
Seizures of dampers, 547
Selection of couplings, 115
Selectivity
    correction for filters, 615, 616
    curves of analysers and filters, 607, 614
Semi-graphical methods, 399
Sensitivity
    factor, see Gauge sensitivity factor
    of pick-up, 609
Separation point, 212, 214
Serrated shaft, 43, 333

Stiffness (*cont.*)
with fillets, 45
with flat side, 43
with keyway, 43
with linear taper, 40
of splined shaft, 43
of square-section shaft with fillets, 46, 47
of stepped shaft, 36
of circular cross-section, 44
of straight shaft, 37
with eccentric bore, 38
with tapered bore, 40
of tapered shaft
with straight bore, 40
with tapered bore, 40
per unit length, 176
region, 1-mass system vibration, 386
see also Equivalent length
STINSON, K. W., 77
STOKER, J. J., 111
STONE, M., 496
STONE-WALLWORK
coupling, 122
powder coupling, 128
Straight
cylindrical shaft stiffness of, 37
geared system, 179
shaft with eccentric bore, 38
systems with belt drive or chain drive, 181
Strain gauge
bridge with d.c. coupling, 635
calibration of, 634, 635
measurements, 315
multiple chevron arrangement, 632
temperature compensation of, 634
Strain-gauging equipment, 630
STRAUB, J. C., 88
Stress
amplitudes predicted, *see* Stress prediction
at node, 318
expressions for, 320
per degree amplitude at mass No. 1, 317
at resonant speeds, 317
bending, 462
concentration factors, 324
determined by HOLZER table with gas torque and damping, 400, 408
distribution diagrams, 320
due to mean transmitted torque, 347
effective, 318
evaluation from vibration measurements, 314
from vibration amplitudes measured at intermediate positions along shafting, 319

in circular shaft
with one flat side, 335
with two diametrically opposite flat sides, 335
increases due to unequal forging, 462
in fillets, 328
in flywheels and coupling flanges, 340
in geared systems, 180
in gears, 88
in gear-tooth fillets, 342
in keys and keyways, 332
in normal fillets, 343
in recessed fillets, 343
in rubber dampers, 509
in shaft
of cruciform cross-section, 339
with circumferential groove, 330
with deep circular grooves, 332
with narrow rectangular slot, 333
with transverse hole, 329
in shrink-fit assemblies, 337
in splined shaft, 333
in U-section grooves, 332
in Vee engines, 348
in welded shaft sections, 336
nominal, 318
prediction, 346
ARCHER's formula, 353
BIBBY's formula, 353
B.I.C.E.R.A. provisional formula, 347
DEN HARTOG's formula, 354
DOREY's formula, 355
DRAMINSKY's formula, 357
KER WILSON's formulae, 351, 358
LEWIS-ROWETT formula, 358
LUNDBERG's formula, 359
MITCHELL's formula, 360
SHANNON's formulae, 352, 361
ZDANOWICH's formula, 361
raisers, 324
estimated by GRIFFITH and TAYLOR method, 339
reduction at a certain position, 178
resultant, 318
rules and recommendations regarding permissible vibration, 459
total shear, 347
Strip-spring couplings, *see* BIBBY couplings
Stroboscope, 618
STROMAG coupling, 122
STRUNZ, L., 78, 223
STUART MITCHELL, R. W., 360
Subscripts, *see* Notation
Subsidiary system, 186, 212
Suction cam, 226
SUHR, F. W., 131

Twist
and stiffness, relation between, 36
in rubber damper material, 506
Two-cylinder engine with main flywheel,
448
Two-mass system, 400, 432
dynamic magnifier curves, 438
natural frequency of, 151
vector diagram for, 437
with damping, 440
Two-node
frequency of system with belt drive or
chain drive, 181
vibration frequency, 165, 362
TYMSTRA, S. R., 91

Ultimate tensile strength and fatigue
strength, approximate relation between, 344
Unbalance, see Balancing
Undercut fillets, 59, 65
stresses in, 343
Unfiltered diagrams of vibration amplitudes, 289, 318, 603, 605
Universal-joint type calibrator, 618
U-section grooves, stresses in shafts with,
332
Untuned
damped vibration absorber, 520
viscous-shear damper, 520

VAN SANTEN, G. W., 602, 636
Variation of vibration amplitudes with
running speed, 433
Vector diagram
for 2-mass system, 437; see also Graphical methods
of vibration amplitudes and torques,
401
methods for frequency determination,
232
rotating, 431
Vee
angle, effect on phase-vector sum, 306
belts, 91
engines
moment of inertia per line, 4, 352
phase-vector sums for, 304
stresses in, 348
with articulated connecting rods, 307
Velocity
admittance, 238
impedance, 237
type pick-up, 602, 609, 610
Vibration
amplitude
and phase-vector sums, relation between, 310

at non-resonant speeds, 362
at resonance, 347
determination by graphical methods,
385
diagram, 318
effect due to injection timing, 373
estimation by empirical formula,
112
formulae: for 1-mass system, 429; for
2-mass systems, 432
from HOLZER tables with gas torque
and damping, 400, 424
increase with running time, 373
independent of damping, 532
measurements, 315
predicted value of, 313
prediction, 346; ARCHER'S formula,
353; BIBBY'S formula 353;
B.I.C.E.R.A. provisional formula,
348; DEN HARTOG'S formula, 354;
DOREY'S formula, 355; DRAMIN-
SKY'S formula, 357; KER WILSON'S
formulae, 351, 358; LEWIS-ROWETT
formula, 358; LUNDBERG'S formula,
359; MITCHELL'S formula, 360;
SHANNON'S formulae, 352, 361;
ZDANOWICH'S formula, 361
2-node frequency, calculations, 362
vector diagram of, 401
analysers, 611
axial and flexural, 588
condition independent of damping, 511
diagram
of engine with nozzle of one cylinder
cut out, 290
unfiltered, 289
frequency
calculations, 150
2-node, 165
3-node, 167
see also Frequency
in gears, 88
measurements evaluated for stresses,
314
modes of, 173
reducing-devices, see Dampers; Detuners
spectrum, 318
torques, LLOYD'S REGISTER OF SHIP-
PING, 475
transient, 455
Vibrator, mechanical, 142
Vibrometer, 235
reed type, 618
Viscometers, 543
Viscosity
effective, of silicone fluids, 539
nominal, silicone fluids, 539, 551

Viscous
damping, 458
coefficient, viscous-shear dampers, 551
fluid damper, see Viscous-shear damper
shear damper, 520, 548–50
calculation taking account of driven machine, 534
clearances, 538, 544
common-point frequency and amplitude by HOLZER table with damping, 532, 534
criteria of performance, 530
design features, 541
design procedure, 537
dynamic magnifiers of engines with, 523, 530
effect of casing inertia, 526
effect of using various values for damping coefficient, 557
energy dissipation, 554
energy equilibrium, 555
flank-amplitude evaluation, 534
fluid reservoirs, 545
formulae for peak and flank amplitudes, 525
loads acting on damper casing, 543
location of inner mass, 545
maximum damping, 555
minimum-amplitude parameters, 523
operating experience, 541
optimum damping, 525, 528, 555
optimum phase velocity, 526
peak amplitude at common-point, by HOLZER table with damping, 534
peak-amplitude calculation, 530
performance of, 546
phase-reducing effect, 555
power dissipated by, 538, 554
proportions for dealing with high or low values of shear rate, 547
relative amplitude of inner mass, 554
side surface correction factor, 539
size of, 530
steady rotation of inner mass, 547
system with, 236
transient conditions, 547
variation of shear rate on lateral surfaces, 552
vibratory torque in crankshaft, 528
with elastically-coupled seismic mass, 504
torque transmitted by fluid, 520
Voltage variations, 493
VON KÁRMÁN, T., 173, 243, 581
VREEDENBURGH, C. G. J., 337

WALKER, H., 91
WANG, C. T., 325, 341
WARING, J. R. S., 122
WATTS, F. G., 91
WATTS, J. L., 131
Wave analyser, see Waveform analyser
Waveform
analyser, 602, 605, 611
complex, 595
envelopes, 595
WD^2 of a flywheel, 176
Wear
of gears, 88
of rubber bushes, 118
Web, see Crankweb
WEBER, C., 91
Webbed cross-section, equivalent length of shaft with, 136
WEIGAND, A., 43, 328, 335, 343
Weighing method, for connecting rods, 9
Weight of marine propellers, chart for estimation of, 28
WELBOURN, D. B., 399
Welded
arms, point of rigidity of shafts with, 133
generator shafts, 149
shaft sections, stresses in, 336
WENNERBERG, J., 367, 496
WEYMOUTH, H. P., 131
WHEATSTONE bridges, a.c. and d.c. types, 631, 634, 635, 636
WHITACKER, J., 29
WIEGAND, F. J., 585, 586
WIGGLESWORTH couplings, 123
WILLAN's fuel consumption diagram used to determine f.m.e.p., 253
WILSON, T. S., 38, 328, 582, 586
WILSON, W. KER, see KER WILSON, W.
Winch drive, 193
WITHERS, J. G., 266
WITTKE, H., 602
WOODFORD, D. E., 120, 122
WORK, C. E., 230, 345
Work
done by torque harmonics, 263
input of viscous-shear dampers, 555
per crankshaft revolution, 249
per vibration cycle, 263
Working
cycles per crankshaft revolution, 249
range of mechanical calibrator, 597
strokes per cycle, 493
Worm
gears, 88
wheels, deflexion of, 85

[663]

9 780521 203524